Astrophysics and Space Science Library

Volume 416

EDITORIAL BOARD

Chairman

W. B. BURTON, *National Radio Astronomy Observatory, Charlottesville, Virginia, U.S.A. (bburton@nrao.edu); University of Leiden, The Netherlands* (burton@strw.leidenuniv.nl)

F. BERTOLA, *University of Padua, Italy*
C. J. CESARSKY, *European Southern Observatory, Garching bei München, Germany*
P. EHRENFREUND, *Leiden University, The Netherlands*
O. ENGVOLD, *University of Oslo, Norway*
A. HECK, *Strasbourg Astronomical Observatory, France*
E. P. J. VAN DEN HEUVEL, *University of Amsterdam, The Netherlands*
V. M. KASPI, *McGill University, Montreal, Canada*
J. M. E. KUIJPERS, *University of Nijmegen, The Netherlands*
H. VAN DER LAAN, *University of Utrecht, The Netherlands*
P. G. MURDIN, *Institute of Astronomy, Cambridge, UK*
B. V. SOMOV, *Astronomical Institute, Moscow State University, Russia*
R. A. SUNYAEV, *Space Research Institute, Moscow, Russia*

More information about this series at http://www.springer.com/series/5664

Jacobus Cornelius Kapteyn around 1910 (Kapteyn Astronomical Institute)

Pieter C. van der Kruit

Jacobus Cornelius Kapteyn

Born Investigator of the Heavens

 Springer

Pieter C. van der Kruit
Jacobus C. Kapteyn Professor of Astronomy
Kapteyn Astronomical Institute
University of Groningen
Groningen, The Netherlands

ISSN 0067-0057   ISSN 2214-7985 (electronic)
ISBN 978-3-319-10875-9 ISBN 978-3-319-10876-6 (eBook)
DOI 10.1007/978-3-319-10876-6
Springer Cham Heidelberg New York Dordrecht London

Library of Congress Control Number: 2014950149

© Springer International Publishing Switzerland 2015

This work is subject to copyright. All rights are reserved by the Publisher, whether the whole or part of the material is concerned, specifically the rights of translation, reprinting, reuse of illustrations, recitation, broadcasting, reproduction on microfilms or in any other physical way, and transmission or information storage and retrieval, electronic adaptation, computer software, or by similar or dissimilar methodology now known or hereafter developed. Exempted from this legal reservation are brief excerpts in connection with reviews or scholarly analysis or material supplied specifically for the purpose of being entered and executed on a computer system, for exclusive use by the purchaser of the work. Duplication of this publication or parts thereof is permitted only under the provisions of the Copyright Law of the Publisher's location, in its current version, and permission for use must always be obtained from Springer. Permissions for use may be obtained through RightsLink at the Copyright Clearance Center. Violations are liable to prosecution under the respective Copyright Law.

The use of general descriptive names, registered names, trademarks, service marks, etc. in this publication does not imply, even in the absence of a specific statement, that such names are exempt from the relevant protective laws and regulations and therefore free for general use.

While the advice and information in this book are believed to be true and accurate at the date of publication, neither the authors nor the editors nor the publisher can accept any legal responsibility for any errors or omissions that may be made. The publisher makes no warranty, express or implied, with respect to the material contained herein.

Cover illustration: Jacobus C. Kapteyn in an undated illustration from the biography by Henriette Hertzsprung-Kapteyn.

Printed on acid-free paper

Springer is part of Springer Science+Business Media (www.springer.com)

To Adriaan Blaauw
To Corry

Preface

> *In science, moreover, the work of the individual is so bound up with that of his scientific predecessors and contemporaries that it appears almost as an impersonal product of his generation.*
> Albert Einstein (1879–1955)[1]

In the September 1925 issue of the British journal *The Observatory*, Willem de Sitter, director of Leiden Observatory, wrote a short 'letter to the editor' in the section 'correspondence' [2], in which he announced a plan to write a biography of Jacobus Cornelius Kapteyn. The text of the note reads as follows:

A Biography of Kapteyn

GENTLEMEN, –
 May I, through your columns, make an appeal to friends of Kapteyn, and to astronomers generally, for their help in carrying out a plan which, I like to think, will have the sympathy of all interested in our science?
 My friend J. Huizinga, Professor of History in this University and myself intend as joint authors to write a biography of Prof. J.C. Kapteyn. Kapteyn has undoubtedly occupied a position of great prominence and influence in the astronomical world in the last quarter of a century, as well by his scientific attainments as by his personal qualities of character and mind. It will be our endeavor to set forth the development of his remarkable personality from both the scientific and the general human point of view, and especially to show the fundamental unity of this two sides of his personality.
 We shall be greatly indebted to any friends of Kapteyn, or other astronomers, who may be willing to assist by communicating to us any detail they may happen to know, or any important point of view or material they

may have at their disposal, regarding Kapteyn's many national and international connections with astronomers and institutions all over the world, or his personal relations with his many friends and acquaintances, or any information, even of anecdotal character, which may help us to make the picture of the man as complete as possible. Letters written by Kapteyn will, of course, be most valuable.

Any letters or documents entrusted to us will be very carefully preserved, and will be returned after use. Communications should be sent to the undersigned at the University Observatory, Leiden, Holland.

Leiden.   Yours faithfully,
1925, August 24.   W. DE SITTER

Willem de Sitter (1872–1932) was the first PhD student of Kapteyn and was appointed professor of astronomy at Leiden University in 1908. Johan Huizinga (1872–1945) was a prominent Dutch historian, appointed professor of history in Leiden in 1915; he was the son of Dirk Huizinga (1840–1903), professor of physiology in Groningen and great friend of Jacobus C. Kapteyn. They were ideally positioned to write a comprehensive and authoritative biography of Kapteyn, but as it turned out the biography, for reasons unknown, was never written.

Sources

Three years after de Sitter's appeal, in 1928, Kapteyn's daughter Henriette Hertzsprung-Kapteyn published a biography in Dutch of her father: *J.C. Kapteyn; Zijn leven en werken* [3]. This work, by an admiring daughter, contains many details on Kapteyn as a person, although undoubtedly colored by the love and admiration for her famous father. Very little astronomical details are described in the book, as could not really be expected. Henriette Mariette Augustine Albertine Kapteyn (1881–1956) had studied law and English at the universities of Groningen and Amsterdam. Some more background on her and the biography can be found in Appendix B. I refer to this volume as the *HHK biography*.

This biography has been translated by the American historian Erich Robert Paul, first as a journal contribution [4], later as a book *The life and works of J.C. Kapteyn by Henriette Hertzsprung-Kapteyn: An annotated translation with preface and introduction by E. Robert Paul* [5]. It contains a list of publications about Kapteyn and remarks about astronomy in his day. E. Robert Paul was definitely an authority on the subject of Kapteyn and the astronomy of his time and he published a few authoritative papers on the subject: *The death of a research programme – Kapteyn and the Dutch astronomical community* [6]; *Kapteyn and statistical astronomy* [7] and *Kapteyn and the early twentieth-century Universe* [8]. He also wrote a scholarly book *The Milky Way Galaxy and statistical cosmology, 1890–1924* [9]. This is an excellent source for information and I refer to this in the following as *Paul's Statistical Astronomy*.

Preface

In 1999 a 'Legacy' symposium has been held in Groningen, entirely devoted to Kapteyn with the title *The Legacy of J.C. Kapteyn: Kapteyn and the development of modern astronomy*. The proceedings [10] were published by myself and K. van Berkel (editors). This volume contains a number of very important articles about Kapteyn; it will be referred to below as the *Legacy*. The quality of Paul's translation of Henriette Hertzsprung-Kapteyn's biography has been seriously criticized in that volume (see its appendix B for more details). I have recently produced a new, improved translation that I have posted on a special Web-page dedicated to Kapteyn, that I am maintaining [11].

De Sitter and Huizinga seem to have moved most of Kapteyn's archive of papers and correspondence to Leiden in preparation of the biography they were planning to write, but it was never returned to Groningen; so most of Kapteyn's letters are no longer present here. In contrast to this, the archives at the University of Groningen do hold the letters of Sir David Gill to Kapteyn. These were apparently not included in the material that de Sitter and Huizinga took to Leiden, or had been returned to Groningen later. This important set of letters from Gill fortunately has survived and I have used it extensively. What happened to the rest? It is assumed that at a later time de Sitter's son Aernout was planning to write the biography and had been transporting Kapteyn's correspondence – maybe together with other papers – in a big crate to Indonesia, where he had become director of Lembang Observatory. This crate was then supposedly lost in the bombing of Rotterdam in May 1940, while it was awaiting shipment. Some more discussion of this can be found in Petra van der Heijden's contribution to the *Legacy*. The story remains speculative; in particular, there is no good reason why the letters of Gill, and only those of Gill, were left in (or returned to) Groningen.

The unavailability of much of Kapteyn's correspondence might have played a role in the fact that no further attempts at a thorough biography have been undertaken. As part of the activities around the 'Legacy symposium' of 1999, Petra van der Heijden has made an extensive inventory and collected selected copies of letters from Kapteyn and drafts of letters to him from archives around the world, to a large extent correcting this lack of correspondence. Her chapter in the *Legacy* proceedings details this exercise.

The remains of the Kapteyn archives reside at Groningen University. It has been described in some detail in Adriaan Blaauw's chapter in the *Legacy*, where he presents the 'Kapteyn Room' in the Kapteyn Astronomical Institute. This room contains furniture and printed works from Kapteyn's (and later van Rhijn's) office. Actually this room has since been relocated in the Institute, while a fair fraction of the books, notes, letters and other publications has now actually been moved to a repository of the University of Groningen Archives, which like the Kapteyn Astronomical Institute is located on the Zernike Campus, to ensure more protection and safe storage under controlled climate conditions. Among this is also the set of 'Kladboeken' (see Adriaan Blaauw's chapter in the *Legacy*), which are bound volumes in which Kapteyn and his staff made all sort of notes (compiling tables, drafting letters or publications, record measurements or results of calculations, etc.). Often they were used from both ends. As Blaauw notes, the staff discarded the oldest ones (numbers

1 to 100) before he assumed the directorship of the Kapteyn Laboratory (as it was then called) in 1957 and when it moved to its present location on the Zernike Campus of Groningen University (in 1983) only a selection was kept that was judged to be sufficiently important for historical purposes. As a result, what is available is all of Kapteyn's notes in the Kladboeken between 1907 and 1912 and about half of them up to 1921. In this book I will use the notation that Blaauw adopted in the *Legacy*, so that 'KB-no.104, inverted, p.6 (1907)' means Kladboek number 104, page 6 starting from the back and the notes involved are from the year 1907. Some examples of pages have been reproduced in the *Legacy* in the chapter *Meeting Kapteyn in the Kapteyn Room* by Adriaan Blaauw. Although fascinating to browse, I have found the amount of useful information in the Kladboeken limited; there is little that is not known otherwise.

Recently a collection of 'Love letters' that Kapteyn wrote to Catharina Elisabeth (Elise) Kalshoven has been published under the title *Lieve Lize: De minnebrieven van de Groningse astronoom J.C. Kapteyn aan Elise Kalshoven, 1878–1879* [12] by Klaas van Berkel and Annelies Noordhof-Hoorn, referred to below as the *Love Letters*. These were written in 1878–1879, when Ms Kalshoven was still Kapteyn's fiancée and future wife, while Kapteyn himself already had moved to Groningen as a young professor.

Other important sources of information, especially for the later parts of Kapteyn's life, are two very well-written history books. The first is the biography of Henry Norris Russell (1877–1957) by David DeVorkin, *Henry Norris Russell: Dean of American astronomers* [13], hereafter referred to as *DeVorkin's Russell biography*. The other is the history of the Mount Wilson Observatory, written by Allan Rex Sandage (1926–2010), *The Mount Wilson Observatory: Breaking the code of cosmic evolution* [14]. I will refer to this below as the *Sandage's Mt. Wilson History*.

The present book describes Kapteyn's scientific work in detail and for this it relies heavily on Kapteyn's publications, many of which are available as reprints in the Kapteyn Room. Some of these are copies for his wife with regularly compliments written on them, but there are also leather-bound volumes in which Kapteyn collected his publications (see Appendix A). In addition to this and the archive of letters, which have been a major source of information, much on Kapteyn as a person can be gleaned from his daughter's biography and from the set of letters he wrote to his future wife. Further personal information is contained in an article in Dutch by journalist and amateur astronomer Cornelis Easton regarding his personal memories of Kapteyn. Cornelis Easton (1864–1929), a Dutchman in spite of his last name, was a prominent Dutch journalist and newspaper editor. He had no formal training in astronomy but still did original research in that field and in climatology. In recognition of his contributions, he received an honorary doctorate in 1903 from the University of Groningen with Kapteyn as his 'promotor' (supervisor). This interesting source on the Kapteyn as a person has been added to this book in the form of excerpts in English translation as Appendix C. The full text in English (and much other material on Kapteyn) is available on a special Website that I am maintaining at www.astro.rug.nl/JCKapteyn/. And of course the studies in the *Legacy* are of great value and extremely useful.

Preface

Aim of This Book

I am an astronomer, not a historian; actually my professional career is closely related to Kapteyn. I have worked at the Kapteyn Astronomical Laboratory of the University of Groningen since 1975 and have led the process of formalizing its status as a recognized research institute within the University of Groningen under the name Kapteyn Astronomical Institute. I have been its scientific director for two separate periods, together adding up to more than 10 years. In 2003, I was appointed distinguished Jacobus C. Kapteyn professor of astronomy [15] by the University of Groningen upon nomination of the Faculty of Mathematics and Natural Sciences. Finally, my research has been much along the lines of Kapteyn's (see my 'valedictory lecture' *It all started with Kapteyn* [16]).

So, the aim of this book is to discuss the astronomical work of Kapteyn, and put it in two contexts. The first is the astronomy of his day; after all the development of our knowledge about the stars and universe is a background against which we can judge his contributions. In this process I will try to separate as much as possible his contributions from those of his predecessors and his contemporaries, keeping in mind the difficulty pointed out by Albert Einstein in the quote at the beginning of this preface. The second context is his personality and personal life. We have some information about that in reminiscences about him in obituaries and other articles that address his personality. Much can be learned from his daughter's (Henriette Hertzsprung-Kapteyn) biography, as long as we concentrate on parts that are not too much influenced by the writing about him by an admiring and loving daughter. I have therefore chosen for a structure of the book in which for sections concerning his personal life I make extensive use of the text in the *HHK biography* in my new English translation. I could have paraphrased Henriette Hertzsprung-Kapteyn, but I rather chose to use her own words and at the same time provide an improved English translation of those words. She must have consulted her mother many times, since they lived in the same street in Hilversum while she was writing it (see Appendix B), the Oude Amersfoortsche Weg, respectively on numbers 66 and 29, which according to *Google Maps* is only 230 meters and 3 minutes walking apart. The information contained in the biography is almost certainly reliable, although colored. I have then left out parts that are affected too much by the time in which it was written or by the relationship between father and daughter. Parts from the *HHK biography* that have been quoted have been typeset in a sans serif font to distinguish them from my own texts; I have added sometimes explanatory remarks in the texts in the regular font and then enclosed these in square brackets. I have done the same thing also whenever I quote from other sources.

Nicknames, References, Notes and Currencies

In what follows, whenever I refer to 'Kapteyn' I will mean J.C. Kapteyn, the subject of this book, and when to 'Mrs Kapteyn' I will mean his wife C.E. Kapteyn-Kalshoven. Other Kapteyns will be identified by using their first name(s).

Fig. 0.1 This book is dedicated to Professor Adriaan Blaauw, former director of the Kapteyn Astronomical Laboratory, and to my wife Corry. This picture was taken during the reception after the symposium 'Challenges', celebrating my 65th birthday (Photograph Elmer Spaargaren [19])

An important point concerns the use of nicknames, etc. Many authors refer to Kapteyn's oldest daughter as 'Dody', the second as 'Hetty' and his son as 'Rob', which were indeed the way they were addressed by family members and friends. They signed letters to those who knew them well in this way. I will refrain from using these nicknames, as I feel this should be reserved to intimates, which I am not. Reading the *Love Letters* feels like enough of an infringement on other people's privacy.

The audience I have in mind while writing this book encompasses historians and astronomers, but I particularly kept those in mind that are interested in Kapteyn without having a formal training or professional expertise in any of these two disciplines. I will therefore explain fundamental astronomical and astrophysical matters, and when these are too technical, use separate boxes. I also assume that readers are not familiar with the Netherlands and Dutch society, traditions and structure (I chose the English language and foreign publisher to make the book accessible to an international audience). Mathematical equations can be very enlightening for those familiar with their use and I will not shy away from using them. However, I put them also in separate boxes, such that those not so apt in working with mathematics and equations can continue the main text.

I use two kinds of notes in this book. Remarks or information, that may be of immediate use to the reader, but which does not seem suitable for inclusion in the main text, will be given in footnotes to the text. I have kept these to an absolute

Preface   xiii

minimum and the use of footnotes is only by exception. All information concerning details of how to locate publications and other sources referred to – and exclusively that – is collected at the end of the book in Appendix D (page 661), items being numbered like references, enclosed in square brackets. Publications by Kapteyn and his collaborators have been listed in Appendix A; when in the text I refer to e.g. Kapteyn (1899), then this reference can be looked up in that appendix.

Then there is the question of the conversion of currencies. In this book I convert quoted amounts of money to current values using the 'Historic Dutch Currency Calculator' on the Dutch Ancestry Magazine site *How much did you say?* [17] when it concerns salaries or other income. I prefer this site for this, because it expresses the sums of money in terms of the 'income of an unskilled worker'. I will use that without quoting an amount of €s. In other cases, costs of equipment, general budgets, etc., I use the converter of the 'International Institute of Social History' of the Royal Netherlands Academy of Arts and Science [18], which aims at comparing 'cost of living' or 'purchasing power' and this I will express in current €s.

Title and Dedication

The (sub)title of this book derives from the statement made by Simon Newcomb, in his book *The Stars: A study of the Universe* [20], published in 1901:

'This work [the Cape Photographic Durchmusterung] of Kapteyn offers a remarkable example of the spirit which animates the born investigator of the heavens. Although the work was officially that of the British Government, the years of toil devoted to it were, as the writer understands, expended without other compensation than the consciousness of making a noble contribution to knowledge, and the appreciation of his fellow astronomers of this and future generations.'

This book is dedicated to Professor Adriaan Blaauw (1914–2010), third director of the Kapteyn Astronomical Laboratory between 1957 and 1970, who brought it back to the national and international forefront, and who encouraged me over a long period to write this biography after my formal retirement and – last but not least – to my wife Corry for her love and support and for her understanding that active professional life should not end at 65.

Assen, The Netherlands/ Baltimore, USA/   Pieter C. van der Kruit
Beijing, China / Canberra, Australia/
Groningen, The Netherlands /
La Laguna, Spain / Pasadena, USA
Spring 2010 – Summer 2014
(northern hemisphere seasons)

Fig. 0.2 Photograph from Kapteyn Astronomical Institute

Kapteyn lived at the address 'Ossenmarkt 6, Groningen' from 1910 to 1920. In 1999, three of his great-great-great-grandchildren unveiled this memorial on the occasion of a special 'Legacy Symposium' in Kapteyn's honor during the celebration of the 385th anniversary of the University of Groningen. The plaquette is placed on the facade of that house. In English translation it reads: Here lived and worked [incorrect years] Professor Jacobus Cornelius Kapteyn, inspiring Groningen astronomer. The quote translates into English as:

*When you don't have what you love,
you have to love what you have.*

It has been attributed to French memoirist Roger de Rabutin, Comte de Bussy (or Roger Bussy-Rabutin) (1618–1693); see page 496.

Acknowledgments

I am grateful to many individuals for help, advice and assistance and institutions for support. In no special order, except that it is more or less chronological, I like to mention the following persons and institutions.
Prof. Klaas van Berkel, professor of History of Science at the University of Groningen for much advice, help and insistence I write this book. The staff of the Library of the University of Groningen, especially those at the branch of the Faculty of Mathematics and Natural Sciences, and in particular Harri Preuter for scanning Kapteyn's and his brother Willem's theses. Gerard Huizinga and Alf Tent of the archive repository of the University of Groningen. Kor Holstein, expert on architecture in Kapteyn's times. Wija Friso for help to locate information on Kapteyn in various archives. Prof. Leen Dorsman, professor for university history at Utrecht University, and Dr. Paul Lambers, curator of natural history, University Museum Utrecht and his retired predecessor Dr. J.C. Deiman. Dr. David Baneke for help with archives in Leiden. Dick Veldhuizen of the municipal archives of Barneveld. The staff of the 'Groninger Archieven', of the section Special Collections of the Leiden University Library and of the National Archives in Den Haag for cheerfully assisting me in my archival research. Adam Perkins, curator at the Royal Greenwich Observatory archives, Cambridge University Library, Cambridge, UK. Frank Bowles and the staff in the Manuscript Room have been very helpful when I consulted the Gill-Kapteyn correspondence kept there. Paul Haley of 'The Share Initiative' of Herefordshire (tsieuropean.co.uk) in Hereford, UK, who sent me complete transcripts of all letters of Gill to Kapteyn from 1906 onwards (as kept in the Royal Geographical Society archives), and who together with his wife Ann and daughter Clarissa have been very hospitable, receiving my wife and myself in their home for a delightful day of discussing Gill, Kapteyn and the history of astronomy. Dr. Juha Hannikainen and Jouni Nikula of the Keskusarkisto (Central Archives) of the Helsingin Yliopiston (University of Helsinki), Finland for providing scans of letters between Kapteyn and Anders Donner. O.J. Nienhuis, archivist at the municipality of Tynaarlo, for locating the summer house of the Kapteyns in Vries. Dr. Stephan P.C. Peters, my last astronomical PhD student, but also oracle and technical assistant in

computer matters, for installing Linux (Ubuntu) on my laptop and flatfielding some photographs I had made in archives. Dan Lewis and Molly Gipson of The Huntington Library, Art Collections, and Botanical Gardens, San Marino, California for help with accessing Hale, Seares and Adams correspondence with Kapteyn. Charlotte (Shelley) Erwin and Loma Karklins of the Archives at the California Institute of Technology, Pasadena, California for support with the examination of the Hale papers. John Grula, Librarian of the Carnegie Observatories for help with the use of the library. Larry Webster of the Center for High Angular Resolution Astronomy for showing me and my wife around the telescopes and the Kapteyn Cottage on Mount Wilson. The staff of the Pasadena Public Library and of the Research Library and Archives of the Pasadena Historical Society at the Pasadena Museum of History for help locating the December 19, 1908 issue of the Pasadena Star. Susanne Elisabeth Nørskov, AU Library, Fysik & Steno, Institut for Fysik og Astronomi, Aarhus Universitet, Denmark for scanning all over 100 letters between Hertzsprung and Kapteyn free of charge and providing a photograph of a very young Hertzsprung. Rolf ter Sluis and Jan Waling Huisman of the University Museum Groningen for being so helpful in many respects. Professors Jan Borgman, Harm Habing and Tjeerd van Albada, emeritus colleagues, for giving descriptions of what they remembered from the insides of the Astronomical Laboratory at the Broerstraat in van Rhijn's time. Prof. Joan Booth and Prof. Manfred Horstmanshoff, emeritus professors of respectively Latin and ancient history at Leiden University, translated the Latin text on the postcard found in the album for Kapteyn's 40th anniversary as professor. Dr. Dirk van Delft, director of the Boerhaave Museum in Leiden and Paul Steenhorst, head of the conservation department, who kindly showed me the Universal Instrument of Leiden Observatory that Kapteyn used. Jan Willem Noordenbos, great-grandson of Kapteyn, helped me with information, photographs and other matters concerning Kapteyn's relatives and descendants. Prof. Eckhard Limpert of Zürich, first author of a paper on skew frequency distributions and Kapteyn has patiently explained his work and why he and his co-authors dedicated his paper to him. Annie and Johan Weelink, current owners of the second home that the Kapteyns owned in Vries for providing information and photographs of the house. Dr. Inge de Wilde, author of a PhD thesis on women at Groningen University in Kapteyn's day, for help to find the photographs of Kapteyn's oldest daughter Jacoba. F.A.M. (Frans) van den Hoven of the University Museum Utrecht for searching archives for these photographs. Dr. Inge de Wilde for supplying the photograph in Fig. 8.1, when it could not be located by Mr van den Hoven. Tony Misch, director of the Lick Observatory Historical Collections for providing scans of the original plates containing Nova Persei 1901. And I thank my contacts at Springer Publishers: Harry Blom for his continued interest over a number of years to publish a biography of Kapteyn, Prof. Butler Burton for suggestions and his enthusiasm to publish this book in the Astrophysics and Space Science Library, of which he is editor-in-chief, and finally my editor, Jennifer Satten, for her efficient help, assistance and pleasant interaction.

A few persons have read drafts of (sometimes parts of but in a number of cases the complete) manuscript of this book. I am extremely grateful for the large amount

Acknowledgments

Fig. 0.3 For archival or library research a number of passes need to be acquired. Here are the most colorful examples that I received during my research for this book

of time they invested and the advice they provided. They are Klaas van Berkel, professor of history after the Middle Ages in Groningen, Wessel Krul, professor of modern history of art and culture, also in Groningen, David Baneke, assistant professor of the history of modern natural sciences in their cultural and social context (with special attention to astronomy) at the Free University Amsterdam, Jan Willem Pel, emeritus professor of observational astronomy in Groningen, Scott Trager, professor of astronomy and astrophysics in Groningen, François Schweizer, staff member of the Carnegie Observatories, Pasadena, California, Albert Jan Scheffer, emeritus associate professor of molecular biology and virology in Groningen, Ton Schoot Uijtenkamp, emeritus professor of chemistry and environmental science in Groningen, Jan Willem Noordenbos, retired biology teacher and great-grandson of Kapteyn, Gerard van den Hooff, dear friend and emeritus teacher of English and currently professional translator, who did a very thorough and time-consuming job checking my use of the English language, and last but not least my wife Corry. It goes without saying that the blame for any mistakes is due to me and responsibility for any inaccuracies is definitely mine.

The preparations for this book started in February 2010 with a conversation on how to proceed with Klaas van Berkel, after infrequent and friendly, but persistent encouragement to write the biography by Adriaan Blaauw and sometimes Klaas. I first made an inventory of what to do during a conference on 'Galaxies and their Masks' in Sossusvlei, Namibia in April of that year, especially during the five days afterwards, that I found myself stuck in Johannesburg when the Eyjafjallajökull volcano in Iceland disrupted almost all air traffic to and from Europe.

Most of the work for this book was performed at my home institution, the Kapteyn Astronomical Institute of the University of Groningen, where I have profited from the support of its staff, especially the computer and administrative staff and in particular Hennie Zondervan, Jackie Zwegers and Wim Zwitser. In my two successors as director, Prof. Thijs van der Hulst and Prof. Reynier Peletier, I gratefully acknowledge all support, help, interest, advice, suggestions, encouragement and interaction in the excellent scientific environment of a renowned institution.

I also thank various institutes for hospitality when I made preparations for and worked on this book away from my home affiliation. These were in alphabetic order of the city names: the Space Telescope Science Institute, Baltimore, Maryland, USA; the Kavli Institute for Astronomy and Astrophysics at Peking University, Beijing, China; the Mount Stromlo Observatory, Research School for Astronomy and Astrophysics of the Australian National University, Canberra, Australia; the Instituto de Astrofisica de Canarias at La Laguna, Tenerife, Spain and the Carnegie Observatories of the Carnegie Institution for Science, Pasadena, California, USA. I am grateful to the directors of these institutions, Matt Mountain, Doug Lin, Harvey Butcher, Paco Sanchez and Wendy Freedman for their hospitality and my local hosts Ron Allen, Richard de Grijs, Ken Freeman, Johan Knapen and Barry Madore for their help, interest and friendship (and Wendy and Barry for presenting me with a private copy of *Sandage's Mt. Wilson History*). I have often worked on this book while traveling, in numerous hotel rooms across the world when I was attending committee meetings, symposia and celebrations, such as in Ameland, Beijing, Brussels, Buenos Aires, Johannesburg, La Palma (Canary Islands), Leiden, München, Napoli, Noordwijkerhout, Santiago de Chile, San Pedro de Atacama (Chile), Sossusvlei (Namibia), etc. These visits were often, but not always, unrelated to the writing of this book. Both my astronomical research and that related to this Kapteyn biography have been funded from an annual grant from the Faculty of Mathematics and Natural Sciences of the University of Groningen between 2003 and my formal retirement in 2009 in connection with my distinguished named Jacobus C. Kapteyn professorship and a grant from the Netherlands Organization for Scientific Research (NWO) as compensation for my membership of its Area Board for Exact Sciences between 2007 and 2012. I thank Louis Vertegaal, Nico Kos and Christiane Klöditz for their help. Further financial support has been provided by the Kapteyn Astronomical Institute, the Leiden Kerkhoven-Bosscha Fund and the Royal Natural Sciences Society at Groningen.

The Internet, the search engine Google, the NASA Astrophysics Data System and the document preparation and typesetting system LaTeX have been indispensable. I gratefully acknowledge the Dutch educational system of my generation, so that as a teenager in secondary high school (the HBS; see page 62) I had to extensively study French, English and German, which together with my Dutch background provided the essential ability to read the material necessary for writing this book, which included documents in all of these four languages. It would have been very difficult had I not had that background.

Many of the unpublished sources used for this book are available at www.astro.rug.nl/JCKapteyn.

Contents

Preface		vii
Acknowledgments		xv
1	**Growing Up in Barneveld**	1
	1.1 Barneveld	3
	1.2 Henriette Hertzsprung-Kapteyn's Biography	5
	1.3 Parents	5
	1.4 Boarding School	8
	1.5 Religion	15
	1.6 Brothers and Sisters	16
	1.7 Jacobus Cornelius Kapteyn	18
	1.8 Star Map	20
	1.9 Kapteyn as a Boy	22
	1.10 Preparing for University	25
2	**Studies in Utrecht**	27
	2.1 Universities in the Netherlands	27
	2.2 Kapteyn and His Academic Lineage	28
	2.3 Kapteyn Entering Utrecht University	30
	2.4 Kapteyn as a Student	33
	2.5 Kapteyn's Supervisor and Astronomy in Utrecht	34
	2.6 The Thesis of Willem Kapteyn	39
	2.7 Kapteyn's PhD Thesis	41
	2.8 Propositions	44
	2.9 Catharina Elisabeth Kalshoven	49
3	**Astronomer in Leiden**	53
	3.1 Military Service	53
	3.2 Kapteyn's Height	56
	3.3 Looking for Employment	57

	3.4	Leiden Observatory	59
	3.5	Positional Astronomy	63
	3.6	Kaiser and van de Sande Bakhuyzen	67
	3.7	'Observator' at Leiden Observatory	69
	3.8	Kapteyn's Work at the Observatory	74
	3.9	Solar Parallax	79
	3.10	Personal Life in Leiden	81
4	**Professor in Groningen**	85
	4.1	The City of Groningen	85
	4.2	The 1876 Law on Higher Education	87
	4.3	Astronomy in Groningen	89
	4.4	A Professor in Astronomy	92
	4.5	Taking Up the Professorship	97
	4.6	First Attempt at an Own Observatory	99
	4.7	Marriage Proposal	102
	4.8	Finding a Place to Live	105
	4.9	The Marriage	106
5	**From Kepler to Parallax**	111
	5.1	Kepler's Equation	112
	5.2	Copernicus	118
	5.3	Higher Order Sines	119
	5.4	Tree Rings and Climate	122
	5.5	Absolute Declinations	126
	5.6	Parallaxes	138
	5.7	Family Life	147
6	**Cape Photographic Durchmusterung**	153
	6.1	Astronomical Photography	155
	6.2	Kapteyn and the Cape Photographic Durchmusterung	159
	6.3	Kapteyn's Funding of the Work on the CPD	166
	6.4	Execution of the CPD	169
	6.5	The Parallactic Method	173
	6.6	Production of the CPD	181
	6.7	The 1887 Carte du Ciel Congress	184
	6.8	Execution of the Carte du Ciel	192
	6.9	Gill, Christie and the CPD	196
	6.10	Concluding Remarks	203
7	**An Astronomical Laboratory**	205
	7.1	Stellar Structure and Evolution	206
	7.2	The Milky Way and Other Galaxies	209
	7.3	The Herschels and the Sidereal Problem	211
	7.4	Aftermath of the CPD	216
	7.5	Rector Magnificus	218

Contents   xxi

	7.6	One Last Attempt at an Observatory	219
	7.7	An Astronomical Laboratory	226
	7.8	Refraction and Aberration	234
	7.9	Kapteyn's Star	237
	7.10	Gill Visits Groningen	246
	7.11	Parallax Again	247
	7.12	A Durchmusterung for Parallax	254
	7.13	Enter de Sitter	256
	7.14	Statistical Parallaxes	260
8	**Colors and Motions**		263
	8.1	Vries	265
	8.2	Visits by Gill, Donner and Newcomb	269
	8.3	Natural Sciences Society	276
	8.4	Colors of Stars	284
	8.5	Spectral Types of Stars	288
	8.6	Parallactic Motion	290
	8.7	Secular Parallax	292
	8.8	Praise by Newcomb	296
	8.9	A Resume for Gill	297
	8.10	Personal Matters	302
	8.11	German to English	309
9	**Star Streams**		311
	9.1	Nova Persei 1901	311
	9.2	Cosmic Velocities	315
	9.3	The Solar Apex Revisited	320
	9.4	Reactions to the Apex Determination	323
	9.5	Mean Parallaxes and Luminosities	324
	9.6	Absolute Magnitude	326
	9.7	Stellar Densities as a Function of Distance	328
	9.8	The Louisiana Purchase Exposition	333
	9.9	International Congress of Arts and Sciences	335
	9.10	Aftermath of the Announcement of Star Streams	342
	9.11	Impressions of America	343
	9.12	Kapteyn Homes in Groningen	346
10	**Selected Areas**		351
	10.1	R.A.S. Gold Medal	351
	10.2	Visit to the Cape	353
	10.3	Acceptance of the Star Streams Concept	356
	10.4	Plan of Selected Areas	359
	10.5	Finding Support for the Plan	362
	10.6	Enter George Ellery Hale	364
	10.7	Early Progress of the Plan	368

	10.8	The Harvard-Groningen Durchmusterung 369
	10.9	Further Progress of the Plan 375
	10.10	The Bergedorfer Spectral Durchmusterungs 376
	10.11	The Unofficial End of the Plan 379

11	**Extinction** 383
	11.1 Weersma and the Distance to the Hyades 384
	11.2 Comparison to the Results of Boss 388
	11.3 Star Counts, Extinction and the Star Ratio 390
	11.4 Luminosity and Density Laws 392
	11.5 Star Density or Extinction? 396
	11.6 Interstellar Extinction and Dust 400
	11.7 Cornelis Easton and Spiral Structure 404
	11.8 Kapteyn and Nebulae 408
	11.9 On the Absorption of Light in Space 409
	11.10 The Thesis Work of van Rhijn 413
	11.11 Shapley's Absence of Extinction 415
	11.12 The Laboratory's Permanent Location 418
	11.13 Some Routines 422
	11.14 Philosophy, Music and Poetry 424

12	**Students** 429
	12.1 Kapteyn as an Educator 429
	12.2 Courses Kapteyn Taught 432
	12.3 Thesis Projects in Groningen: Weersma 434
	12.4 Yntema and the Background Starlight Brightness 436
	12.5 Directorship of Leiden Observatory 444
	12.6 De Sitter's Appointment to Leiden 447

13	**Mount Wilson** 453
	13.1 George Ellery Hale 453
	13.2 Mount Wilson and the Selected Areas 456
	13.3 Hale Invites Kapteyn to Mount Wilson 460
	13.4 Kapteyn's First Visit to Mount Wilson 463
	13.5 The Kapteyn Cottage 469
	13.6 The Luminosity Curve Revisited 478
	13.7 The Lecture Before the Royal Institution of Great Britain 480
	13.8 The 1910 Solar Union Meeting 483
	13.9 The Mount Wilson Catalog 487
	13.10 The 'Pipeline' 490
	13.11 The Structure of the Universe 493
	13.12 Pickering's Harvard Northern Durchmusterung 495
	13.13 End of the Mount Wilson Visits 497

14	**Tides, Statistics and the Art of Discovery**	501
	14.1 Origin of the Tides	501
	14.2 Skew Frequency Distributions	505
	14.3 Work with van Uven	512
	14.4 The Correlation Coefficient	513
	14.5 Tercentennial of the University of Groningen	517
	14.6 World War I	519
	14.7 Induction Versus Deduction	521
	14.8 Continuation as a Research Associate	526
15	**First Attempt**	529
	15.1 Radial Velocities	529
	15.2 Kapteyn's Developing View of the Sidereal System	534
	15.3 In the Mean Time in Groningen	538
	15.4 Helium Stars	542
	15.5 Friction with Adams	545
	15.6 Hale's Reaction and Seares	549
	15.7 The Adams: Strömberg Result	552
	15.8 Preparations for the First Attempt	555
	15.9 The Distribution of Stars in Space	558
	15.10 The Crowning Achievement: Galactic Dynamics	562
	15.11 The Kapteyn Universe	567
16	**Finale**	569
	16.1 Orden pour le Mérite	569
	16.2 Germany and International Organizations	571
	16.3 Forty Years as a Professor	576
	16.4 Reorganization of Leiden Observatory	581
	16.5 Seventieth Birthday	586
	16.6 The Kapteyn Universe and Shapley's Globular Clusters	589
	16.7 Kapteyn's Last Years in Groningen	593
	16.8 Return to Mount Wilson?	598
	16.9 Retirement and Illness	601
	16.10 International Astronomical Union	603
	16.11 The End	607
Publications by and about J.C. Kapteyn, His Honors and Academic Genealogy		611
	A.1 Publications by J.C. Kapteyn	611
	A.2 Publications About Kapteyn	624
	A.3 Kapteyn and His School: PhD Theses up to 1946	626
	A.4 Kapteyn's Honors	629
	A.5 Institutions, Objects etc. Named After Kapteyn	631
	A.6 Academic Genealogy of Kapteyn	635

Henriette Hertzsprung-Kapteyn's Biography of J.C. Kapteyn 639
    B.1   Henriette Hertzsprung–Kapteyn 639
    B.2   Ejnar Hertzsprung and First Marriage 641
    B.3   Joost Hudig and Second Marriage 645
    B.4   The Biography and Hale 645
    B.5   Paul's Translation 647

Cornelis Easton: Personal Memories of J.C. Kapteyn 651

Notes and References 661
       References 663

Index 685

Chapter 1
Growing Up in Barneveld

> *The best inheritance a person can give to his children is a few minutes of his time each day.*
> O.A. Battista (1917–1995).[1]

> *Your children need your presence more than they need your presents.*
> Jesse Jackson (1941–present).[2]

In August of the year 1922, prominent British astronomer and astrophysicist Sir Arthur Stanley Eddington (1882–1944) published an obituary of Jacobus Cornelius Kapteyn [1]. The opening sentences were:

'Holland has given many scientific leaders to the world; it is doubtful whether any other nation in proportion to its size can show such a record. J.C. Kapteyn was among the most distinguished of its sons – a truly great astronomer, whose death on 1922 June 18 has removed from our midst a pioneer and leader in the problems of the sidereal system. which now claim so large a part of astronomical effort.'

And only days after Kapteyn's demise and funeral, famous Dutch historian Johan Huizinga (1872—1945), professor in Leiden and son of a very close friend of Kapteyn, wrote in the Dutch cultural and literary magazine *De Gids* [2]:

'Will later generations be surprised to find out that there has not been a widespread mourning this June month over the death of one of the greatest Dutchmen? Whoever listens to astronomers describing the importance of Kapteyn's researches, discoveries and theories, would expect nothing

---

[1] Orlando Aloysius Battista was a Canadian author and chemist.
[2] Jesse Louis Jackson, Sr. (b. Burns), American baptist minister and civil rights activist.

Fig. 1.1 Map of the Netherlands around 1900. The capital letters indicate the location of towns and cities where Kapteyn spent major parts of his life: B = Barneveld, where he was born and grew up; U = Utrecht, where he went to university; L = Leiden, where he first worked as an astronomer; G = Groningen, where he was a professor at the university for more than 40 years; V = Vries, where the Kapteyns owned a second house; H = Hilversum, where he and his wife bought a house to live after retirement and A = Amsterdam, where he was nursed during his fatal illness and died. Other cities and villages that occur in this book are in lowercase: Alkmaar (al), Amersfoort (am), Apeldoorn (ap), Arnhem (ar), Assen (as), Bodegraven (bo), Breda (br), Delft (de), Den Haag (hg), Den Helder (he), Dordrecht (do), Eindhoven (ei), Enschede (en), Franeker (fr), Haarlem (hl), Harderwijk (hw), Gouda (go), Leeuwarden (le), Maastricht (ma), Rotterdam (ro), Vlissingen (vl), Wageningen (wa) and Zwolle (zw) (Adapted from 'Kaart Nederland Jan clip art' [3])

else but that his name will survive as of one of those very few that everybody knows. But the fame of an astronomer leaves little impression on the present-day audience, even though magazines and newspapers have over the recent years more than once reported stories about this person.'

Huizinga expressed the opinion that '… the Life of J.C. Kapteyn, which posterity would want to have available, could be one of the most beautiful books ever written. But will it ever come about, the complete picture of this honorable and happy life?'. As documented in the Preface to this book, he had planned to write a biography of Kapteyn with astronomer Willem de Sitter, also professor in Leiden and first PhD student of Kapteyn. They never completed this task and it is unlikely that the present book comes even close to what these two men had in mind. But if we want to give it a try anyway, we need to start at the beginning, halfway through the nineteenth century in the rural village of Barneveld not far from the geometrical center of the Netherlands (see Fig. 1.1).

1.1 Barneveld

The village of Barneveld is located on the western side of the Veluwe, the central part of the Netherlands that was, and still is, relatively sparsely populated. The Veluwe has forests and woodland, large areas covered with heath, and here and there hills, sand drifts and lakes. The municipality of Barneveld as a whole currently has a population of about 50,000 and the village proper about 30,000 [4]. In the middle of the nineteenth century the Netherlands had a population of about one-fifth of the present 17 million, Barneveld would have had a few thousand inhabitants.

Barneveld's location is about halfway what now is often designated with the term Bible Belt (Bijbelgordel or Refoband in Dutch) which is roughly a diagonal, running from the province of Zeeland in the south-west of the country through areas south of Utrecht to the northern parts of the province of Overijssel close to the city of Zwolle, with an additional small pocket north of the city of Groningen. The Bible Belt is currently inhabited by a relatively large percentage of Protestants (particularly members of the orthodox Reformed Protestant Church). It can easily be recognized on maps that show the percentage of voters for the most orthodox Protestant political party SGP (Staatkundig Gereformeerde Partij). They are characterized by a reluctance to have their children vaccinated against measles and rubella, an unwillingness to accept birth control, homosexuality, euthanasia and forms of assisted reproductive technology and to volunteer as donors of organs after death.

There is a wealth of literature on the Dutch Bible Belt, but the works that I have found the most informative are only available in Dutch, unfortunately [5]. The Bible Belt more or less forms the boundary between the parts of the Netherlands where the Christian population is predominantly Roman Catholic (to the south) or predominantly Dutch Reformed (to the north). Today it forms the location of the highest concentrations of members of orthodox reformed churches, usually indicated with the adjective 'bevindelijk'. This relates to the word 'experience'; in this tradition, the life-changing pietistic experience of a person's own guilt and sin and his/her personal salvation through Jesus Christ, assumes a central role. The Bible Belt as such did not yet exist in the present form in the middle of the nineteenth century, but there was a wider and more extended 'Belt of Protestants' (Protestantenband) running up to the provinces of Groningen and Friesland (Frisia) in the north. The Bible Belt formed around the end of the nineteenth century as the southern part of this

Fig. 1.2 The old church of Barneveld, the Netherlands, and the statue of Jan van Schaffelaar, who jumped from this tower. The statue dates from a later time (1903) than Kapteyn's youth (Photographs by the author)

Belt of Protestants [6]. Barneveld in the middle of the nineteenth century was part of all this and strict as well as orthodox reformed and protestant Christians would dominate its population.

Barneveld lives up to its location in the Bible Belt; in 2010 eight out of the 31 municipal councilors were members of the Staatkundig Gereformeerde Partij, and another ten belonged to two other Christian parties.

Apart from being known as a strong protestant community, Barneveld has two claims to fame. It is famous as a center of poultry farming and commerce; actually there is a very special breed of chicken, known as the 'Barnevelder'. This bird was bred around 1850 as a cross between Dutch chickens and imported Asian breeds. The Barnevelder was in demand and exported because of its ability to produce about 200 large, brown eggs per year. The other fact for which Barneveld is well-known (featuring in every elementary school history book) is connected to the name of Jan van Schaffelaar and the so-called 'Hook and Cod Wars', which comprised a number of battles between 1350 and 1490 over who was entitled to be the 'Count of Holland'. These 'Hoekse en Kabeljauwse Twisten' were fought between two fractions, one calling themselves the 'Cods', presumably a re-appropriation related to the arms of Bavaria, which supported Margaret, wife of the emperor of Bavaria, and the 'Hooks'. This name probably refers to the hooked instrument used in catching cod. When Philip the Good of Burgundy appointed his son Bishop of Utrecht, this was opposed by the Hooks. In 1482, the Cod Jan van Schaffelaar and his men hid in the tower of Barneveld, which was subsequently besieged by the Hooks. Rather than surrender, he bought free withdrawal for his men by jumping from the tower.

In the middle of the nineteenth century Barneveld had also made a name as a place that harbored excellent (and numerous) boarding schools. At that time the number of inhabitants was a mere three thousand [7]. A prominent boarding school for boys was named 'Benno'. It had been established in 1840 by Gerrit Jacobus Kapteyn and his wife Elisabeth Cornelia Koomans. In this village, and to these parents, Jacobus Cornelius Kapteyn was born on the 19th of January, 1851.

1.2 Henriette Hertzsprung-Kapteyn's Biography

What we know about Kapteyn's youth is almost entirely based on the biography by his daughter Henriette Hertzsprung-Kapteyn (see Appendix B for a detailed account of the background). In this book I will quote extensively from her biography in a new translation that I have prepared; I chose to use direct quotes rather than try to paraphrase the author. The *HHK biography* not only contains recollections of things Kapteyn may have told his daughter himself, but also Henriette's mother Catharina Elisabeth (Elise) must have supplied much information after Kapteyn died. It is important to note in this context that when Henriette Hertzsprung-Kapteyn wrote the *HHK biography* (it was published in 1928), she lived very close to her mother; she and her husband, the astronomer Ejnar Hertzsprung, had split up (in late 1925 or maybe even somewhat earlier) and she had moved with their daughter Rigel to Hilversum. This is known from the address at the top of the letters she wrote to George Ellery Hale between 1925 and 1930 about the biography (see Appendix B). It may reasonably be assumed that Kapteyn's widow and daughter discussed the progressing biography frequently and in detail. Kapteyn had bought this house, planning to spend his years in retirement in Hilversum. His fatal illness prevented him from actually moving there. Hilversum is a moderately sized town about 20 km north of Utrecht and 30 km south-east of Amsterdam. During the nineteenth and early twentieth centuries it prospered as a result of textile and tapestry industry, while it increasingly became a place for rich commuters working in Amsterdam to build expensive homes.

1.3 Parents

For purposes of reference, I list some details about Kapteyn, his wife, parents and in-laws in the table on page 6. The *HHK biography* starts as follows.

Table 1.1 The Kapteyns and their parents

Jacobus Cornelius Kapteyn	Jan. 19, 1851 Barneveld	June 18, 1922 Amsterdam
Married July 17, 1879, Utrecht		
Catharina Elisabeth Kalshoven	June 19, 1855 Amsterdam	March 2, 1945 Amsterdam
Parents		
Gerrit Jacobus Kapteyn	Jan. 4, 1812 Bodegraven	July 26, 1879 Barneveld
Married March 12, 1837, Bodegraven		
Elisabeth Cornelia Koomans	Nov. 5, 1814 Rotterdam	Nov. 24, 1896 Barneveld
Jacobus Wilhelmus Kalshoven	June 18, 1813 Nieuwer-Amstel	Febr. 14, 1869 Abcoude-Proostdij
Married (I) June 21, 1838, Amsterdam		
Catharina Elisabeth Brandt	June 30, 1814	Nov. 9, 1849 Amsterdam
Married (II) June 1, 1853, Zwolle		
Henriëtte Mariëtte Augustine Albertine Frieseman	April 7, 1822 Harderwijk	April 13, 1895 Dieren

'The Kapteyn family had always been educators. Throughout the centuries and generation upon generation they acted as teachers – education was in their blood and if they did not teach at a school or some other teaching institute, they taught at home.

The earliest known teacher in the family was Paulus Captijn, who was born in the village of Berkenwoude in 1712, which is situated in the Krimpener Waard. The archives of the 'Schepenen' of the village of Heukelom for 1756 mention him as Paulus Captijn, 'schoolmaster here in this town'.'

The Krimpenerwaard is the rural area just south of Gouda and east of Rotterdam. It obtains its name from the town located in the western corner, Krimpen aan den IJssel. 'Krimp' used to mean a bend in a river. A 'waard' is an inter-fluvial area between rivers, in this case the Lek and the Hollandse IJssel. A 'schepen' is a municipal senior magistrate and ruler and Heukelom is a small municipality in the province of North-Brabant in the south of the Netherlands.

'But before him many generations were probably educators. Archives do mention names, but seldom occupations. In those days it was customary for sons to follow in their fathers' – or indeed relatives' – footsteps profession-wise, and this is the way it went with the Kapteyns. They became quite well-known as educators; in an Amsterdam newspaper of July 12, 1868 for example, a Kapteyn was put forward in response to a request for suggestions for a teacher in English, with the recommendation: 'one who carries the name of a family, in which pedagogical excellence is a tradition'.

Such an excellent reputation as an educator had likewise been obtained by G.J. Kapteyn, who owned a boarding school in Barneveld. Gerrit Jacobus

1.3 Parents

Fig. 1.3 Picture of Gerrit Jacobus Kapteyn (1812–1879), father of Jacobus Cornelius Kapteyn (Fotocollectie gemeentearchief Barneveld [9])

Kapteyn [see Fig. 1.3] was born in Bodegraven in 1812, where his father was the head of the municipal elementary school. As a child he suffered from 'English disease', which resulted in an unusually large skull'.

Bodegraven is a municipality in the province of South-Holland, not far from Gouda. Wikipedia notes that it lies roughly equally distant (about 30 kms) from the four major cities in the Netherlands: Amsterdam, Rotterdam, The Hague, and Utrecht. The 'English disease' is rachitis or rickets and results from a deficiency of vitamin D or calcium; it occurs mainly among young children. It affects the bones, often by softening them, leading to fractures and deformations.

'Later he used to tell the story that a peddler who noticed him on his rounds in Bodegraven to sell his healing products, offered his mother a 'medicine' to help the boy get rid of the disease. This, he related, worked miraculously, especially given the fact that the disease usually disappeared spontaneously. He completely recovered, as is clear from his further life when he was an example of physical and mental strength.

At a young age he became an assistant in his father's school and passed an unusual number of exams to become a fully-qualified teacher. In those days this meant having to obtain diplomas for various levels of expertise and qualification, from fourth level up to first level. Since high-school and secondary education was not organized according to the modern structure of the educational system, this was the highest attainable level. This first level was awarded only in exceptional cases; only two persons passed, Kapteyn being one of them.

In the same period he traveled to Leiden on a regular basis to attend courses in the classical languages, Greek and Latin. He had to travel by night using a horse-drawn boat in order to arrive in time, since the lectures started at 9 o'clock in the morning. This boat transported farmers and others from the surrounding villages, who went to Leiden's market place, where they had to arrive very early to build up their stalls and arrange their products and be ready in time for the market to open. As a result, Kapteyn arrived much too early for the lectures and in order to make good use of his valuable time he took organ lessons. He had a good talent for music and he already played the organ in his local church in Bodegraven. In this way he helped the widow of the former organist, as he donated his earnings to her. His organ teacher in Leiden, who felt this was much too early, refused to come out of bed for the lessons, and shouted his instructions and comments from his bed. After having attended the lectures he traveled back to Bodegraven, taught his classes and studied in the evenings.

When he was appointed assistant-teacher in the boarding school of Mr van Wijk at Kampen, he had to give up attending the lectures and had to study the subject independently. His daily work was intense, but still he worked every evening for his exam. Sometimes, when he was very tired and found it hard to stay awake, he put his feet in cold water. Although his life was busy and demanding, he still found time to read the complete works of Shakespeare, especially during the hours that he was watching over pupils doing their written examinations, which shows his ability to study efficiently and his discipline to organize his activities. He had indomitable energy and perseverance, and he passed his Candidaats exam in the humanities in Leiden with excellent marks.'

Kampen is an old town in the province of Overijssel, at the northern end of the continuous part of the Bible Belt. It is situated on the mouth of the river Ijssel, where it flows into the Zuiderzee; Kampen is an old Hanseatic town. Candidaats is the equivalent of the current Bachelors degree.

1.4 Boarding School

'When he was 25 years of age he married Elisabeth Cornelia Koomans, a 23 year-old farmer's daughter from Bodegraven. They settled in Voorschoten, where he opened a boarding school for boys. [Voorschoten is a village between Den Haag and Leiden] This turned out to be too heavy a task for the young couple. The boys attending the school were from The Hague,

1.4 Boarding School

spoiled, used to luxury and difficult to control. After a few difficult years they gave up. Kapteyn applied for a position at a school in Barneveld, where he was held out the prospect of a salary of 600 Guilders and the use of a house from the municipality free of charge. [This corresponds to about two annual salaries of an unskilled worker at the present time.] He was selected for the position and appointed, and they moved to Barneveld with their two little sons and never regretted this change.'

There is a short biography of G.J. Kapteyn in the *Biografisch Woordenboek Gelderland* [8], but much of this biography has actually also been taken from the *HHK biography*. According to this, the number of applicants for the position in Barneveld was 27 and Gerrit Kapteyn seemed to have been preferred because he had the full qualification of first level and had been an established teacher; in any case he did not have to take any tests and compete with other applicants. This also played a role when it came to a further, additional allowance he had requested and which he was awarded by the municipal board, in recognition of his very rare qualifications. He also was exempted from teaching girls; the town council was considering the founding of a separate public boarding school for girls.

'The school was founded in order to offer boys from Barneveld and environs the opportunity to obtain more formal education than was possible at the elementary school. The school made quite a name for itself in subsequent years and the number of applications to attend became so high that boys had to be lodged elsewhere in the village. After some time Kapteyn became unsatisfied with this situation and he decided to resign from his appointment and build his own boarding school intended for a limited number of students. He bought a substantial piece of land and had a stately, large house erected on it that he christened 'Benno', which means 'strong child' (see Fig. 1.4).'

In 1851, according to the short biography of Gerrit Kapteyn, a Latin school was being associated with the boarding school, which was a privilege previously restricted to major cities. This was very welcome in the village of Barneveld, since it made it possible to prepare children for universities.

The archives of the municipality of Barneveld contain a brochure of the Benno boarding school from 1852, *Educational Institute* (Opvoedings-instituut). It lists the subjects that were taught in ten(!) categories. These are: I. Languages (Dutch, English, French, German, Latin, Greek), II. Arithmetics, mathematics and corresponding disciplines, III. History, IV. Geography, V. Physics and natural history, VI. Style and literature, VII. Ethics and religion, VIII. Writing, reading and recitation, IX. Song and drawing, and X. Dancing, fencing, marching and gymnastics. Under II., the subjects listed in a detailed description include 'differential, integral and variation calculus and higher geometry. Also physics, engineering and astronomy, from a mathematical point of view. For those destined for the Navy, also navigation will be offered'.

The fame of the boarding school Benno grew and eventually it recruited not only pupils from all over this country – some of whom actually became famous in later life – but it also attracted famous teachers and assistants. Among them (in either

Fig. 1.4 'Benno', the boarding school of G.J. Kapteyn, father of J.C. Kapteyn, in Barneveld (This picture is from Kapteyn's biography by his daughter Henriette Hertzsprung-Kapteyn)

of these groups) were the poet Petrus Augustus de Genestet (1851–1922), physiologist Adriaan Heynsius (1831–1885), theologian Johannus Hermanus Gunning (1829–1905) and Johan Hendrik Meijer (1831–1892), who was one of the founding fathers of a teaching system for blind children. The fee was considerable though. The 1852 brochure advertises the full package of accommodation, food, care and lessons at 600 Guilders a year (this is roughly equivalent to 2.2 yearly wages of an unskilled worker, and a purchasing power of 14,000 € today), apart from the entrance fee of 30 Guilders. For many lessons, such as music and singing, fencing, carpentry or dancing, extra amounts between a few and fifteen Guilders per month were charged.

'Students came from all over the country and that was no wonder. After all, he [Gerrit Kapteyn] was a natural teacher and educator (a heritage from generations of educators), a man of justice and love for his work and with such energy that he was always fully committed. His versatility was incredible. It was not superficial, but of a quality of broad and deep knowledge that rarely goes together.

He spoke and had a thorough knowledge of the three modern languages [English, French and German] and for students who were preparing for the entrance exam to the Gymnasium, he also taught Latin and Greek. Furthermore he lectured in geography and history, economics and accounting. He was widely known as a mathematician. His handwriting was like printed

Fig. 1.5 Picture of Barneveld from the castle 'de Schaffelaar'. On the left the tower of the Old Church, on the right 'Benno' (Fotocollectie gemeentearchief Barneveld [10])

work. He never used drafts and made no mistakes. He had taught this to himself by never presenting to others anything that had corrections in it. Whenever he had made a mistake he tore up the piece of paper and started anew.

At times when Barneveld happened to be without a minister, he taught the catechism to his students and his children and did that in a exemplary way. On Sunday evenings he also gave a sort of catechism that he called 'lectuur' [literally: reading material]. And after that, if it was not too late, he told stories. He was very gifted at that. His stories were so gripping and stirring that every listener experienced the story as if it was happening to them and was held spellbound. He mostly told mythological stories, ones involving robbers and ghosts and full of the most thrilling adventures. Whoever heard him tell the story of Cartouche, the 'eight-sided peasant' or the ghost who wanted to be shaved in order to be freed of a curse, never forgot them.'

Louis Dominique Cartouche (real name Bourguignon) was a well known French bandit around 1720, who robbed transports on various roads, but especially that between Versailles and Paris. The eight-sided peasant was Cornelisjansz van Swieten, a well-known thief in Amsterdam around 1680, who stole many precious objects from the homes of the rich. He was called 'achtkanten' as a superlative of square because of his tough and rude manners.

Fig. 1.6 Picture of Barneveld taken around 1888/89 from the Grote Kerk. In the background we see the castle 'de Schaffelaar' and on the left, among the trees, 'Benno' with the observatory (Fotocollectie gemeentearchief Barneveld [11])

'The library on the top floor of the house was exceptionally extensive and rich. Everything that was fashionable in literature in those days was present: [Jean-Jacques] Rousseau, [Pierre] Corneille, [Jean] Racine, Molière, [Johann Wolfgang] Goethe, [Johann Christoph Friedrich von] Schiller, [Gotthold Ephraim] Lessing, etc.

Conversations took place in French, which was very valuable for the boys, who learned in this way to express themselves in the foreign language that was used most often. Mr Kapteyn himself knew the French language very well, but it happened on one occasion that he was asked a particular word which he did not know, so he subsequently memorized the entire French dictionary. Boys who knew this tried to trap him, but they always failed. He simply knew all the words.

His aim was to shape his students 'with the blessing of the Lord into virtuous and able men'. He was constantly aware that he personally had to set an example, so he always behaved correctly, giving much attention to good manners. He was a stately figure with a stern expression, a straight posture with a skullcap over his gray hair. The typical schoolmaster, in the best sense of the word.

His students as well as his own children had great respect for him. He was one of those who possessed a natural authority based on his character and example. He did not punish; his authority was acknowledged without

1.4 Boarding School

questioning by all and was taken for granted. If something in the boys met his disapproval he made that known by a certain coolness in his morning handshake.

He had commissioned the layout of a large garden, but he still found that it was not sufficiently large for the boys, so he bought a piece of forest just outside Barneveld, where he let workers erect a tent with benches and construct a bowling alley. On Sundays and free half-days they had their breakfasts early to go there. Some girls brought bread and a local lady served tea and milk. The boys constructed huts, climbed trees with rope-ladders, played the most enjoyable games and in general delighted in the free atmosphere.

His students were well aware that his life was devoted to the well-being of the school, for which he reserved all the dedication and funds necessary. For the rest of their lives they remembered this man who had given them so much with fondness and gratitude.

But Mrs Kapteyn had no less a part in the success of the school. When a child, she had learned on her parents' farm what keeping a large household involved. This was very useful, as it was a heavy task that was demanded of her. She gave birth to no fewer than 15 children, but all the same she directed the household, which at times was made up of 70 members, with a determined and able hand. She was not particularly beautiful and had those long eyelids that my father inherited from her and that gave her complexion a strangely absentminded touch. She possessed a natural, calm dignity and had a strong sense of humor.

In the house as well as in the school there was always a high level of order and punctuality. This was difficult to accomplish on her own and she was helped enormously by two of her daughters, four maids who worked with her for many years, a nanny for the small children and two servants, who took care of the garden and stocked the kitchen with an abundance of vegetables and fruits.

Every day 13 large loaves of bread were baked. A baker from the village came in every evening to prepare them. The house possessed a real baker's oven. The bread tasted so well that the boys were of the opinion that there was no place where bread tasted better. Later they wrote about it and sometimes even had loaves sent to them. The bakers themselves liked it as well and used to say 'It seems so heavy, but it is so light'. Mrs Kapteyn was able to realize an unbelievable variety in the meals that were provided, and was understandably famous for it. The mothers of the boys, who mentioned the variety and extraordinary quality of the meals, frequently asked her how she managed to do it. At home, food never tasted as well as it did in school, the boys complained, and they had recipes sent to their homes, but it never turned out the same.

There were also two deaf-mute seamstresses, who had a full job of mending the clothes. The children thought that all seamstresses were deaf-mute. One of them came home once and was full of amazement [from visiting another household?], saying: 'Mother, mother, they had a seamstress who can talk!'.

For mother and daughters there was more than enough needlework to be done for this large family and in the evening they mended clothing while Mr Kapteyn was reading out stories.

During the evenings, the boys were in the classroom with the assistants, preparing for the lessons of the next day. This way the sons saw very little of their parents and family life. There was no distinction between sons and students, as a result of some misguided sense of justice. The more sensitive ones, such as my father, suffered from this and from the lack of warmth and togetherness, for which there was no time in this busy family.

They worked hard and the days and years passed. In addition to their father and his five assistants, the older sons also started teaching at the school, while they studied for their exams at the university.

All year round, breakfast was at 7 o'clock, during which there was a reading from the Bible, as the parents were very religious. Drawing lessons started at half past 7 and the rest of the lessons at 8 o'clock. The assistants were required to be well-prepared, following the example of Mr Kapteyn himself. The sons, who taught Latin, Greek and mathematics, also had to prepare well, since their father checked up on them regularly and often came in to listen to the lessons.'

The village of Barneveld and the boarding school 'Benno' has another close and interesting link with Kapteyn and astronomy. Egbert Adriaan Kreiken (1896–1964) was one of Kapteyn's students, and obtained his doctorate at Groningen University in February 1923, a little more than six months after Kapteyn's death. Kreiken defended his PhD thesis *On the colour of the faint stars in the Milky Way and the distance of the Scutum Group*, with Kapteyn's successor Pieter van Rhijn as his thesis supervisor. Kreiken had also been born in Barneveld, and his father Willem Rudolph Kreiken (1852–1898), and mother Ada Christina Mathilde Ilcken (1858–1941), had bought the boarding school 'Benno' from Kapteyn's mother Elisabeth Kapteyn-Koomans in 1885, after she had become a widow, and run it until Kreiken died in 1898. Egbert Kreiken was appointed 'privaatdocent' (private lecturer, a basically unpaid position, usually for a term of five years, to teach in a discipline that is not formerly covered) at the University of Amsterdam, but later moved to the Lembang Observatory in the Dutch East Indies and afterwards to Liberia and Turkey. For more information on Kreiken, see page 557.

The building that housed 'Benno' no longer exists. It has given way to a car park on the edge of the town center of Barneveld. But the street that runs around it is fittingly named 'Kapteynstraat' in recognition of Gerrit Jacobus Kapteyn (see Fig. A.12).

1.5 Religion

The website of the municipality Barneveld provides a list of personnel who work, or used to work at the boarding schools [12]. It also names Kapteyn's father Gerrit Jacobus as a member of the 'Gereformeerde Kerk' (reformed church). A larger number of the persons in the list are registered as members of the 'Nederlands Hervormde Kerk' (Dutch Reformed church), which is regarded *at present* as less orthodox. The history of splits that have occurred in the reformed churches in the Netherlands over the years is complicated and long. The secession of 1834, which began in the northern village of Ulrum in the province of Groningen under the Rev. Hendrik de Cock is a prominent one, where the 'Christelijk Gereformeerde Kerken' (the Christian Reformed Churches) split from the 'Nederlands Hervormde Kerk'. It is generally regarded as having had a major influence on the development of the reformed religion in the Netherlands and as the one that gave rise to a number of further splits (as well as a few mergers). The list of personnel of the Barneveld boarding schools predominantly contains members of either of these two church organizations, although a few of them have a different denomination altogether or are not listed as church members at all.

Gerrit Jacobus Kapteyn was not only known for his reputation as the teacher who owned and managed the best-known boarding school in Barneveld; he also was a prominent member of his church and of social organizations. In addition, he had a good relationship with the mayor of Barneveld at the time. For some years (1859–1862) he chaired a society for poetry and literature, 'Erica', and other societies such as the local chapter of the 'Society for the Common Good'. 'Het Nut', as this society is often referred to, had been founded in 1784 as a 'Society of Arts and Sciences with the motto: 'To the Common Benefit' (Genoodschap van Konsten en Wetenschappen, onder de zinspreuk: Tot Nut van 't Algemeen) and it is still active today. Originally it focused on the improvement of life of the working classes by founding elementary schools, printing schoolbooks and training teachers, but also operated libraries, organized general education courses in many areas, etc. Many activities have since been taken over by successive national governments, but the organization is still active in elementary education by operating special schools and organizing activities to educate the general public. It is not associated with any religious or political views or institutions. It currently has about 100 local chapters and a total of more than 10,000 members [13].

In short, Kapteyn grew up in an orthodox protestant environment, at home as well as in the village he lived in. We will see later that he turned away from the church at a young age and that he originally resolved to have a civil wedding only and not to marry in church; this led to a severe clash with his parents, especially his father, who refused to provide the parental permission that was still legally required in those days (see page 106). Yet it is not likely that Kapteyn Senior and his wife belonged to the 'bevindelijke', extremely orthodox Protestant, church. Wessel Krul, professor of history in Groningen, provides three arguments to support this: First, to be teachers, they would have had to conform his ideas and teaching to the official church ('Staatskerk'), which is the Hervormde Kerk. Second, such a board-

Table 1.2 Jacobus Cornelius Kapteyn and his brothers and sisters. The first two were born in Voorschoten and the others in Barneveld

	Born	Deceased	
Paulus Huibert	Jan. 4, 1838	July 16, 1864	Barneveld
Huibert Paulus	July 10, 1839	Sept. 23, 1914	Utrecht
Adriaan Pieter Marinus	Nov. 21, 1841	Dec. 19, 1910	Utrecht
Maria Adriana	April 3, 1843	Dec. 31, 1931	Hilversum
Machtelina Elisabeth ('Bet')	Aug. 21, 1844	March 13, 1905	Utrecht
Nicolaas Pieter ('Piet')	Oct. 28, 1845	May 18, 1916	Hilversum
Albertina Maria ('Bertamie')	Febr. 28, 1847	Jan. 20, 1927	Hilversum
Albertus Philippus	March 31, 1848	Jan. 12, 1927	Den Haag
Willem	Aug.16, 1849	Dec. 6, 1927	Utrecht
Jacobus Cornelius (Ko)	Jan. 19, 1851	June 18, 1922	Amsterdam
Frederik Willem Hendrik (Frits)	Jan. 13, 1853	Dec. 5, 1920	Amsterdam
Johannes Catharinus (Johan)	Jan. 11, 1854	June 26, 1895	Abcoude
Cornelia Louise Alexandra (Cor)	Aug. 11, 1855	Nov.14, 1909	Kaiserswerth
Marius	May 13, 1857	July 25, 1859	Barneveld
Agatha	June 21, 1862	May 1, 1921	Hilversum

ing school relied on rich parents, who would not have chosen a school for their sons with an extremely orthodox director. And thirdly, the library contained books by more 'liberal' writers such as Lessing and Roussau. Orthodox Calvinists would not own such books. The incident related to Kapteyn's marriage does show that the environment followed a strict doctrine and the parents Kapteyn were as well and would not allow deviations from church rules.

1.6 Brothers and Sisters

So the household in which Kapteyn grew up, was large (see Table 1.2). In Kapteyn genealogies on the Web [14] there are three recurrent errors. Maria Adriana was born in 1843 (not 1841) and consequently she is number 4, Johannes Catharinus on June 11 (not January 11) and Marius on March 13 (not May 13). I have checked these dates in official records using FamiliSearch [15].

'Gerrit J. and Elisabeth C. Kapteyn had 15 children; the youngest was born after their silver wedding anniversary. One son died at a very young age of hypothermia [This is the 14th child Marius (1857–1859), see Table 1.2.], but the others all grew up prosperously. [...]

During their childhood, there was no particularly strong band between the brothers and sisters. They regarded each other as students in the same school, following the example of their parents. It is characteristic of this situation that in later life they remembered rather little of each other as children. They lived their own lives, and since they all were more rational than emotional individuals, they did not suffer too much from this. However, those who were more sensitive by nature, often felt lonely and

1.6 Brothers and Sisters 17

Fig. 1.7 Reunion of the Kapteyn family. The date is not specified. We see that father G.J. Kapteyn, in the front (fourth from the left), is still alive and that Catharina Elisabeth (Elise) Kalshoven is not in the picture. This puts it at around 1878. Jacobus Kapteyn is in the back row just to the right of the center (Fotocollectie gemeentearchief Barneveld [16])

misunderstood. This was certainly the case for my father and it would be a shadow from the past that he carried with him all of his life.

They were all successful in their careers. The boys were given the opportunity to study at university. The third son, Adriaan, wanted to be a sailor, to which his father strongly objected. He nevertheless persisted, not being intimidated by the threat of his father that for him that would be the end of his chances of a academic study. He came back after one year as a sailor, sobered and disillusioned by the experience, and decided he wanted to be an engineer. The only route open to him was to become an ordinary worker in a plant. He actually rose through the ranks of the organization and ended up as chief engineer with the national railroad company.

The fifth son, Albert, was judged by his father to be unsuitable for a university education. So he became an apprentice with a local blacksmith in Barneveld, and then worked at the Rhijn railway factory in Utrecht as a voluntary worker in exchange for his training. A friend, who had seen him at work, persuaded his father to give him a chance to study in Liège (Belgium). Kapteyn Senior made the condition, however, that he would pass his entrance exam within three months. Giving his all, Albert managed to complete his studies, which was a tremendous achievement for one who had had no education beyond elementary school. But he passed with the highest marks of his group, studied for three years at the same technical university and followed this up with an extraordinary career. Five of the boys studied at the university of Utrecht, and none of them ever failed to pass an exam. There were not sufficient funds for the two youngest boys to study at a university and they were engaged at a merchant's office in Amsterdam when they were 15 and 16. The sisters likewise passed many exams and excelled thanks to their intelligence, perseverance and responsibility.

They all succeeded splendidly in whatever they undertook. The doctor, the engineer, the scientist, the teacher, the nurse, the businessman, they were all among the best in their trade and were given many responsibilities; the name 'Kapteyn' had a very good reputation and they all had a share in this. They were proud of each other, and after the death of their father they made it a habit to have a reunion every year with their mother, putting her in the center of attention. Even after her death they continued that tradition. So, that was the environment in which my father grew up.'

1.7 Jacobus Cornelius Kapteyn

Kapteyn's birth certificate is shown in Fig. 1.8. The father's last name is spelled 'Kapteijn'; I address this rather less interesting issue in Appendix B (page 641). Throughout this book I use 'Kapteyn' because that is invariably what Jacobus Cornelius himself did.

'He was born in 1851, as the ninth child [This in incorrect; Kapteyn was the tenth child, see Table. 1.2.] and was called Jacobus Cornelius, after the sister of his father [Jacoba Cornelia Kapteyn (1818–??).]. His Aunt 'Ko' (a shortened name which was also used for the boy but which he himself hated), was a very gifted woman. She had a good head on her shoulders and a big heart. She lived frugally, using her savings – she ran a girls' boarding school – to send her second brother to medical school. '

Jacobus Cornelius was a delicate child and the only one that caused his parents concern in terms of health, although he was never seriously ill. He was of a slender build and pale, and had a high-pitched voice. In bearing and complexion he made an absent-minded and introverted impression, which was enhanced by his long upper eyelids.

His fellow students do not remember much of him, except that he was very smart and went through elementary school in an incredibly short period, so that in no time there was little left for him to learn. And they remember that he was a slender, pale young man with a thin and long neck, absent-minded look and poorly kept appearance. He paid little attention to the way he dressed; there were other things that interested him much more. He would lie awake for hours making calculations or considering problems. He was always absorbed in profound thoughts, and could be so caught up in them that he forgot everything around him, noticing nothing except whatever was occupying his mind. One day, when he was on a walking trip with his schoolmates, he was so completely absorbed by his thoughts that he walked into the canal that ran around 'de Schaffelaer', only coming to his senses when he lost solid ground below his feet, and

1.7 Jacobus Cornelius Kapteyn

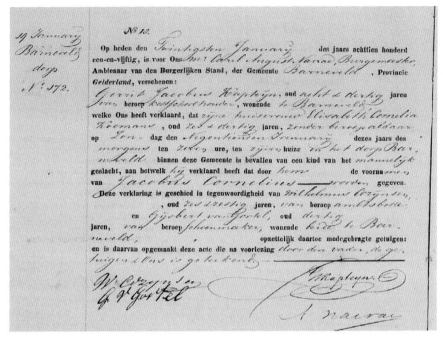

Fig. 1.8 The birth certificate of Jacobus Cornelius Kapteyn from the archives of the municipality of Barneveld. The birth, which took place at the parental home, is dated Sunday January 19, 1851, at 7 o'clock in the morning. Witnesses were Wilhelmus Corijnsen, aged 66, a municipal messenger [''Ambtsbode'; a person employed by the town administration, whose primary task was to deliver messages and documents.], and Gijsbert van Gortel, aged 30, a cleaner by profession. Note that here the last name is spelled Kapteijn (From FamilySearch [17])

was ridiculed without restraint by the other boys. Some other time he was on an excursion with his school to the Nellestein monument, not too far from Doorn. He was focusing so intently on the details of the monument that he forgot that he was on top of a small hill when he made a few steps backward to have a better view of it. He fell on his back, fortunately without any serious consequences.'

The 'Schaffelaar' is a small castle just outside Barneveld. It is named after Jan van Schaffelaar, the cavalry officer who, in 1482, when besieged with his men in the tower of the church of Barneveld, threw himself from the tower to give his men free withdrawal. The castle was built in the 16th century, so after van Schaffelaer's time. In the village of Leersum (about 25 km south-east of Utrecht) there is a tower, the 'Nellestein', built in 1818 on a small hill ('Donderberg'), which affords a view over the land of Broekhuizen. Underneath this tower the Nellesteins have a family grave.

'He was an animal lover, recognized birds by their song, and actually kept birds and rabbits. Obtaining sufficient amounts of food for the animals was a real problem, since the many children did get not much in the way of pocket money. He then got the idea to sell 'krakelingen', on which he made a very small profit. Still, this did not deter him.'

Krakelingen are a kind of Dutch pastry cookie, coated with sugar, rolled out into a rope and twisted into the form of a figure eight.

'For two cents a week he delivered the newspaper for the local minister every day, and he received a small amount of money from his mother for collecting the eggs in the chicken coop. In this way he managed to get the necessary funds. Whatever he undertook, he was always successful, since he never gave up, was not discouraged by disappointments or failures and overcame all difficulties. He was always successful at breeding canaries and everyone trying to breed such birds knows that this is very demanding in terms of care and dedication.

Once he had caught a squirrel, which he kept locked in a small room in the attic, where he looked after it lovingly. He brought the animal the best and most desirable food a squirrel could wish for, but then saw that the animal was in obvious distress and proceeded to sit quietly and sadly in a corner. He became very sad himself, but then suddenly he realized that the animal needed water to drink. Others had told him that squirrels, just like rabbits, have no need to drink water. He quickly fetched a small bowl of water, which the squirrel drank eagerly. As soon as it had finished drinking it stretched in satisfaction and moved to the top of the curtains in no time. That happy outcome he would always remember.

Some other time he had caught an owl. He had heard that these animals are blind during the day, and he wanted to find out if that was true. So he stretched wires criss-cross through the attic room and released the owl. But it evaded all the strings, flew under or over them without touching a single one. Blindness was just a fable, he concluded. In this manner he learned invaluable lessons, the importance of experimenting for yourself instead of unquestioningly believing in the authority of others.

1.8 Star Map

When he was 14 years old, his sister Albertine Maria (Bertamie), his senior by four years, returned from England and brought a star map that they studied together. [HHK's text mentions a map, but the next sentence seems to suggest it was in fact a globe.] The celestial globe was put in the garden and stars were localized. With enormous care and great accuracy he produced a star map, on which he indicated stars of the first magnitude [the brightest] with star figures cut from gold-colored paper, second magnitude stars from silver-colored paper and painting in the positions of third magnitude star with white paint. It was a beautiful piece of work [see Fig. 1.9] which has fortunately survived, and shows us how, even at the age of 14, he already possessed the same love and accuracy of work that he later demonstrated in

1.8 Star Map

Fig. 1.9 Picture of the star map that Kapteyn made when he was 14 and which now hangs in the Kapteyn Room in the Kapteyn Astronomical Institute (Kapteyn Astronomical Institute)

his astronomical research. When his father noted how seriously he studied the stars, he bought a large telescope for him, which was placed in the attic room, and with which he very diligently observed the stars.'

These days the star map hangs on a wall in the Kapteyn Room in the Kapteyn Astronomical Institute at the University of Groningen (see Fig. 11.19). The 'magnitude', according to which the symbols used for the stars were chosen, is an astronomical designation; it is a widely used measure for the brightness of a star. It will be used extensively in this book; I give an explanation and its definition for those not familiar with it later (see Box 2.4). For a general idea, the brightest stars have magnitude zero, the faintest the eye can see under optimal condition, magnitude six. The map has indeed been made with great care and attention to detail, expressing the young boy's obvious fascination with the nocturnal sky. What about this telescope that Henriette Hertzsprung-Kapteyn refers to? The short biography of Gerrit J. Kapteyn, (*op. cit.*), mentions that the boarding school 'Benno' actually accommodated a small observatory, where astronomy was taught. Similarly, in a series of articles in the local Barneveld newspaper on boarding schools in the village, [18], the observatory is also mentioned. The caption in the Barneveld archives to Fig. 1.6 likewise mentions the observatory, by which a structure built on top of 'Benno' or a part of the roof may have been meant. Unfortunately, nothing further on the subject can be found in these archives. Various old pictures show an extension at the rear of the building (see also Fig. 1.4), but it is unknown what kind of telescope, if any, this structure housed and when it was installed there.

It is not clear whether Kapteyn, when a boy, ever considered becoming an astronomer. As has been further documented later on, many writings about Kapteyn and the early phases of his career take the view that his entering the field of astronomy was very much a matter of chance (or luck). Kapteyn studied mathematics and wrote a PhD thesis on an applied mathematics/physics-related subject. That he developed a keen interest in the heavens at an early stage is obvious from the care and detail exercised in the making of the star map, and he may very well have had a special passion for the starry sky. Besides, all facilities necessary to learn about astronomy and to do some actual star-gazing were available to him at home during his youth.

1.9 Kapteyn as a Boy

'It was difficult to argue with him [Jacobus Kapteyn]. There was a little shop in Barneveld – everyone in those days knew the shop of Jan and Maatje den Oudsten – where people came together to discuss politics and local affairs. He was able to win arguments effortlessly from older people, who certainly knew what they were talking about. This superiority was noted by others, although he was hardly aware of it himself. [...]

His older brother Huibert, who was a university student and only came home during holidays, used to play chess with his father in the evening. Ko, who was about 10 years old at the time, watched this and Hubert said: 'Do you also want to play a game with me, little boy?'. The father Kapteyn smiled knowingly, as he knew what this little boy's capabilities were, and the game started. In no time the student lost. That irritated him and he proposed they play a rematch. Huibert lost again and he became angry, saying: 'If I lose another time, I will never play you again.' He lost a third time, and he refused to ever play again against this younger, talented brother.

Some other time the boy went to his father with an arithmetical problem that he was unable to solve. Since his father was busy, a friend of his oldest brother said: 'Come here and I will help you.' But when he saw the problem he realized it was well beyond his abilities.

The brothers found it difficult to accept that they did not measure up to their younger brother, neither in discussions nor in games, and they started to ignore him. At school he had no real friends either. Nobody really cared about him and he became quiet and shy, keeping to himself most of the time and becoming more and more introverted and withdrawn. But he did not enjoy being by himself, as he actually was a sociable person, not just rational but longing for warmth and love more than anything else.

1.9 Kapteyn as a Boy

His parents too had little time for him. Only when he was ill would his mother be loving and caring. She would sit by his bedside and call him 'my little servant' and he would bask in this beneficence. But as soon as he got better that was over and the motherly love was obscured again by the daily demanding tasks. This feeling of loneliness in a large family was a deep sorrow that forever left its scars for the rest of his life. His family probably never noticed this, since they were so obviously different in this respect. [...] This lack of love and the unfulfilled longing for it during his youth, is the key to understanding many aspects of his later life. It explains his joy at any display of love towards him, and his quick response to anyone who approached him with warmth. He was not very critical and analytical in his dealings with other people, he accepted and responded, and gave much from the warmth of his heart. Later, in America, he met many people much like him. He found them less critical and complicated, less prone to taking a reserved attitude unlike the Dutch manner of waiting and seeing whether others could be trusted. [...]'

According to many who have written about him, this lack of parental attention in his youth was experienced by Kapteyn as a major factor in the development of his character during his youth and a serious handicap for the rest of his life.

'As a young man he preferred female company, probably since male contemporaries could not always stand up to him and left him alone. He was painfully aware of his shyness and his slovenly appearance. He really never cared about his clothing and appearance, but in company of young females this bothered him and he saw with silent admiration how easily his brothers and the boys from the school behaved in those circumstances. One day there was a gymnastics event in the main hall of the school, to which many boys and girls had been invited. The boys performed all kinds of courageous stunts and all girls cheered; the boys talked easily and gracefully to the young beauties and made an impression of enviable grace and ease on the timid boy. He was standing in a corner and felt ignored and left alone. Then he got an idea; he wanted to show he was a man as well and took a heavy pole upon his shoulders that was actually much too heavy for him. With incredible effort he carried it around the hall. Only later did he understand why this act of bravery did not bring him the success he had anticipated, but at the time it was a disappointment that returned him to his lonely state of feeling rejected.

He was a brave and resolute child. One day when he had hurt himself in the leg with a small ax, he did not tell his mother, but went to the pharmacy to obtain an adhesive plaster. Of course she found the hole in his trousers and the blood stains on his clothes, and confronted him with that. This attitude of not wanting to complain at times of physical suffering remained typical for him throughout his life.

When he was 13 years of age he had been playing in the church tower and had tied down the rope of the bell. He was woken in the middle of the night by the sound of rain beating on his window, and he remembered with a shock that he had tied the rope, which he imagined might now be shrinking. He had all sorts of fearsome visions, such as that the bell might be jerked loose and would suffer enormous damage through his fault. He got up in the middle of the night, went through the sleeping village to untie the rope. He never had the courage to tell his parents about this.

In those days he also thought very seriously about what reforms would be required to make society a better place. He did not know exactly how, but in his childlike inexperience he decided to campaign against top hats, which he regarded as silly and unworthy. When he told his very practical mother about his world-reforming plans, she said: 'And you think you can reform the world?' He replied in all earnest: 'Yes mother, even with my limited opportunities, if only I could move one straw to open new possibilities, I would be satisfied.' His mother looked at him surprised and questioningly and for a fleeting moment may have been impressed by the determination and emerging abilities of her son, whom, she realized, she hardly knew. He actually wore a top hat later himself and he did not reform society, but he did give his will-power to those he knew in order to help bringing about his conciliatory and idealistic ideas. He moved his straw out of the way and the world became a better place through his life and works.

This quiet, but devout boy had an artless belief in the presence of God as a savior in times of trouble, but when he grew older he lost this belief, when he realized it would make him as rigid and dogmatic as his parents. [...] He became dissatisfied with the religion in which he was brought up.

Every young generation battles the older ones. In those days it was particularly focused on the battle against orthodox articles of faith, such as compulsory attendance of church services, reading the Bible, praying, [...] And he thought and battled, trying to find the only way to reach the most important thing: the supreme freedom of the soul to define its own rules and laws and to choose the manner in which to fulfill them. And with his own freedom he developed a respect for the freedom of others, a right that had to be unassailable and unquestioned.

Hence all prejudices to do with rank or class had to be abolished. This was not a conviction that had come forth out of advancing insight; it was entirely natural and instinctive, since his spirit went beyond all pettiness. Once, when he could not be found in the house and they had been looking for all over the village, he was found in the tabernacle of the poor Jewish boy Sammie, sincerely taking part in the ceremonies of the *Sukkot*, the Jewish Feast of Tabernacles. For him man was sovereign and his worth transcended qualifications in terms of rank, class or religion. Even at a young age these thoughts rooted themselves in his thinking. He considered everything to be natural; the strange distortions introduced by civilized society did not affect him. He remained himself throughout his life. His success and the important positions he was to hold later, invoked no change in him. He greeted his old maid as politely as he greeted the highest dignitaries, and his clerks remembered with adoration how he often raised his hat for them. The true democratic spirit lived in his soul, and this humble attitude stayed with him all his life.'

We have no further independent information on how Kapteyn started to question the religious conviction of his parents and the environment in which he grew up. The only further piece of information comes from his letters to his future wife, Catharina Elisabeth (Elise) Kalshoven, when he had already moved to Groningen. I will go into this more deeply further on in the book when I look at his early time as a university professor in Groningen in more detail. As already mentioned, there had been a very deep and emotional battle between him and his parents, his father in particular. This resulted in their refusal to give the required permission for Kapteyn's marriage, when they learned that he planned to have a civil marriage only and would not marry in church. Henriette Hertzsprung-Kapteyn does not mention this and provides no further details of Kapteyn's religious beliefs and philosophy of life.

1.10 Preparing for University

'When he was 16 he passed the entrance exam of the University of Utrecht, but his father considered him too young to become a university student. Consequently he stayed in Barneveld for another year. Having finished school he now had much spare time, so he often just attended lessons when he felt like it, went on walks with other schoolboys, studied a little just for fun, but so thoroughly and focused that by the time he took up his studies at university, he had already mastered much of the material for the first year. He was very talented in all subjects, especially mathematics, for which he developed his own methods. There was a very difficult book on mathematics by Sturm: *Cours d'analyse de l'École Polytechnique*, which all his brothers had studied.'

This book on analysis had been written by Charles-François Sturm (1803–1855), who was a professor of analysis and mechanics at the École Polytechnique and later at the Faculté des Sciences, both in Paris.

'He only examined the synopsis and not the book itself, finding the solutions to the problems in his own manner. When he had to take his examination, Prof. Grinwis, the professor of mathematics, rejected his solutions since the methods deviated completely from the usual ones. However, in the end he had to admit that the results were correct. In the year he spent at home Kapteyn made extensive use of his father's comprehensive library. If no one knew where Ko was, they could be sure that he was there, absorbed in some book. All his life he used to quote parts he would never forget from works he had read on his sixteenth, by Goethe, Rousseau, Lessing, etc.

He had a special love for Lessing's *Nathan der Weise*, which brought tears to his eyes even on his deathbed, *Der König von Sion* by Hamerling, whose stately hexameters made it so easy to recite; how familiar to us was

Fig. 1.10 Photograph of 'Benno' taken in the nineteen-twenties, when it was used as a guesthouse/hotel (From fotocollectie gemeentearchief Barneveld [19])

'der mächtige Knipperdolling!' the book with the most delicious boasting ever produced. [...]', and somewhat later *Tartarin de Tarascon.*'

Nathan the Wise, a play by the German writer Gotthold Ephraim Lessing (1729–1781), published in 1779, is a strong plea for religious tolerance and promotes free thought and freemasonry. Its performance was forbidden by the church during Lessing's lifetime. Robert Hamerling (1830–1889) was an Austrian poet. The set of poems referred to is a narrative of the Anabaptist movement of 1534. Bernhard Knipperdolling (1495–1536) was the German leader of the Münster Anabaptists. Anabaptists rejected conventional Christian practices such as wearing wedding rings, taking oaths, and participating in civil government. They required that candidates be able to make their own confessions of faith and so refused baptism to infants. As a result, Anabaptists were heavily persecuted during the 16th century and into the 17th by both Roman Catholics and Protestants alike. *Tartarin of Tarascon* is a novel written in 1872 by the Frenchman Alphonse Daudet (1840–1897).

The picture that emerges of Kapteyn as a child and young man is that of a quiet person, who suffered from the lack of attention and affection on the part of his parents, due to the size of their household which included boys from the boarding school they were running. He seems to have been an exceptionally bright and clever student, who developed a broad interest in many areas of science and in astronomy in particular. At a young age turned away from religion.

Chapter 2
Studies in Utrecht

> *Mathematics is the door and key to the sciences.*
> Roger Bacon (ca.1214–1294).[1]
>
> *In theory, there is no difference between theory and practice.*
> *But, in practice, there is.*
> Jan L. A. van de Snepscheut (1953–1994).[2]

2.1 Universities in the Netherlands

At the time when Kapteyn was preparing to start his studies at a university, there were only three such institutions in the Netherlands. The oldest, the University of Leiden, had been founded in 1575, the one in Groningen in 1614 and the university of Utrecht in 1636. In Amsterdam an *Athenaeum Illustre* had been established in 1632, but this had not been given the status of university. The 'illustrious school' had no authority to bestow degrees upon students, although it was recognized as an institution for higher education. Amsterdam has only had a university since 1877, when the Athenaeum was finally given that status.

There had been two more universities in the Netherlands, the one of Franeker in Friesland, which had been founded in 1585, and the university of Harderwijk, not far from Barneveld, founded in 1648. During the French occupation (1795–1813) all universities, except those in Leiden and Groningen, were closed, in 1811, by decree

---

[1] Roger Bacon, British Franciscan friar and philosopher.

[2] Johannes Lambertus Adriana van de Snepscheut, computer scientist, originally at the University of Groningen, later California Institute of Technology (Cal Tech). He died in a fire in his Pasadena home under circumstances that remain unclear. Some people attribute the quote to Yogi (Lawrence Peter) Berra (b. 1925).

of the Napoleonic government. Eventually those at Franeker and Harderwijk continued as athenea, but did not prosper and were closed in 1818 and 1843 respectively (see G. Jensma & H. de Vries, *Veranderingen in het hoger onderwijs in Nederland tussen 1815 en 1940* [1].). The university of Franeker had originally thrived, but during the eighteenth century had suffered from strongly decreasing student numbers. The university of Harderwijk had never been much more than a cheap place to graduate, its reputation as a learned institution being very limited. Indeed, most students only went there for a short period to take advantage of the low tuition fees, obtained their doctor's degree and left. The most prominent among these were the physician and botanist Herman Boerhaave (1668–1738), who graduated in 1693 and Carl Nilsson Linnæus (1708–1778), founder of the scheme of binomial nomenclature for living species. He graduated in 1735, reputedly since it was not possible to obtain such a degree in Sweden at the time. Linnæus only stayed in Harderwijk for a week to get his thesis printed.

After the University of Utrecht had been closed by the Napoleonic occupation, it was actually downgraded to a secondary school (école secundaire). After the defeat of Napoleon, when the Netherlands re-emerged as an independent state, the University of Utrecht, unlike those of Franeker and Harderwijk, was reinstated as a university. This was part of the resolution of August 2, 1815, in which the Dutch government created a new system of the 'higher education in the northern provinces' (the current Netherlands). At the 'Congress of Vienna' in 1815, the United Kingdom of the Netherlands was created, encompassing the area covered today by the Netherlands and Belgium. The latter country gained independence again in 1830. The fact that it would have been very expensive to support the two universities of Groningen and Franeker, located so near to one another in the sparsely populated North of the country, must have played a role in the fact that the University of Franeker was not re-instituted. In the so-called *Organiek Besluit* – 'organizational resolution' – (see e.g., G. Jensma & H. de Vries; M. Groen, *op. cit.* or H.A.M. Snelders, *De schei- en natuurkunde aan de Utrechtse universeit in de negentiende eeuw* [2]), the universities at Leiden, Groningen and Utrecht were instituted with five Faculties, including a separate one for mathematics and physics. In this Faculty, four professors taught the subjects of mathematics, physics, astronomy, chemistry, biology and agriculture. It was separate from the faculties of philosophy and literature, medical sciences, theology and law.

2.2 Kapteyn and His Academic Lineage

In Appendix A, I have traced the academic lineage of Kapteyn. It goes all the way back to persons associated with Johannes Kepler, which is very gratifying as Kapteyn's genealogy applies to a large number of Dutch astronomers, including myself, and I am a great admirer of Kepler. In this Academic Genealogy the first person in the Netherlands was Johann Samuel König (1712–1257), who had studied under Johann (1667–1748) and Daniel Bernoulli (1700–1782) in Basel. König had been a professor in Franeker. His student Antonius Brugmans (1732–1789) had been a professor in Franeker and in Groningen and was the supervisor of Johannes Theodorus Rossijn (1744–1817). Rossijn had written a thesis *De tonitru et fulmine ex nova electricitatis theoria deducendis* (On thunder and lightning according to the new theory of electricity), which he defended at Franeker in 1762. Subsequently he was appointed professor in philosophy, mathematics and astronomy in Utrecht in 1775, after a period as professor in Harderwijk. Under Rossijn, mathematics and physics obtained an excellent reputation in Utrecht.

Gerard (also Gerrit) Moll (1795–1838) received his doctoral degree *honoris causa* in Utrecht in 1815 under Rossijn. Moll had also studied extensively in Amsterdam under physicist Jean Henri (also Jan Hendrik) van Swinden (1746–1823), and after having obtained a 'Candidaats' (Bachelor degree) in Leiden, he went to study in Paris under the mathematician and astronomer Jean-Baptiste Joseph Delambre (1749–1822), among others. Moll was appointed professor of mathematics, astronomy and physics in Utrecht in 1815. When he received a tempting offer to come to Leiden in 1826, he decided to stay in Utrecht, however. The city of Utrecht expressed its gratitude by awarding him the sum of 10,000 Guilders (equivalent to about 230,000€ today), to buy instruments for his scientific research. He was primarily famous for his work on the speed of sound and electromagnetic experiments, but also well-known as a practitioner of applied sciences such as mechanical and civil engineering, architecture, etc.

This tradition of mathematics, astronomy and physics, established by Moll, was continued by his student Richard van Rees (1797–1875), who obtained his PhD under Moll on a thesis entitled *De celeritate soni per fluida elastica propagati* (On the speed of sound in an elastic fluid) in 1819. Van Rees was a professor of mathematics and experimental philosophy in Utrecht from 1831 to 1867. He was a very active professor and he had 22 students, including the famous physicist and meteorologist Christophorus Henricus Didericus Buys Ballot (1818–1890) and the mathematician Cornelis Hubertus Carolus Grinwis (1831–1899) (see Fig. 2.1). Buys Ballot had obtained his doctorate in 1844 on a thesis *De synaphia et prosaphia* (On cohesion and adhesion) and Grinwis defended a thesis *De distributione fluidi electrici in superficie conductoris* (On the distribution of electricity over the surface of a conductor) in 1858. Buys Ballot was appointed professor of mathematics in 1847 and professor of physics in 1867; Grinwis became a professor of mathematics and physics in 1867. Kapteyn was to become a student under Buys Ballot as well as Grinwis.

Physics had been practiced extensively in Utrecht for quite some time. This was much facilitated by the existence of the 'Natuurkundig Gezelschap' (the Physics Association) in Utrecht [3]. It has been founded in 1777 as an initiative of Rossijn. The original full name was 'Gezelschap ter Beoefening en Bevordering van de proef-ondervindelijke Natuurkunde': Association for the practice and promotion of experimental physics. Originally, its primary aim was to collect funds to buy instruments for scientific research. The Association is still active today as one of the oldest of such societies in the Netherlands. It was an important reason why physics in Utrecht was a prominent field of study when Kapteyn entered its university.

Fig. 2.1 Christophorus Henricus Didericus Buys Ballot (1818–1890) and Cornelis Hubertus Carolus Grinwis (1831–1899) (Universiteitsmuseum Utrecht [14])

2.3 Kapteyn Entering Utrecht University

Kapteyn's choice for Utrecht was probably the obvious one; his two older brothers Nicolaas Pieter (1845–1916) and Willem (1849–1927) had studied there as well. Utrecht was the university closest to Barneveld, and the mathematics and natural sciences faculty as indeed the University of Utrecht as a whole, had an excellent name. In Dutch the faculty was designated 'wis- en natuurkunde', which currently would be translated 'mathematics and physics'. However, the faculty also comprised chemistry, biology and pharmacy. 'Natuurkunde' here refers to studies of Nature in a broad scene, so the appropriate translation of 'natuurkunde' here is 'natural sciences'.

Apart from Buys Ballot and Grinwis, the university had among its professors a group of prominent scientists (the 'School of Utrecht'). They included the physiologist Franciscus Cornelis Donders (1818–1889), an authority on eye-diseases, biologist Pieter Harting (1812–1885), who devised improved microscopes and is seen as the earliest supporter of Charles Darwin's evolution theory in the Netherlands, and chemist Gerardus Johannes Mulder (1803–1880), an expert on the chemical composition of proteins. In his study on the development of natural science in the Netherlands, *In het voetspoor van Stevin; Geschiedenis van de natuurwetenschap in Nederland 1580–1940* [4], Klaas van Berkel writes: 'With Mulder, Buys Ballot, Donders and Harting, Utrecht had become the most important university of the country halfway through the nineteenth century, as far as the natural sciences were concerned.' Before I turn to Kapteyn's studies, I reproduce part of the *HHK biography* first, which deals with Kapteyn's choice to study mathematics and physics in Utrecht.

2.3 Kapteyn Entering Utrecht University

'His parents wanted him to study theology. None of the other sons had wanted to do that and they hoped that this quiet, devote son would fulfill their deepest wish and become a preacher. They had no idea of the boy's inner thoughts, or indeed of the changes that had taken place in him. They had no idea that he had in fact chosen to follow his own beliefs long before and would not be able to follow the path that they had wished him to choose. He knew that, like his brothers, he was more suited for a study in exact sciences. However, he did not have the courage to tell his very stern father, and together with another boy from the school who had the same problem, he agreed that they would present their wishes to their respective fathers at roughly the same moment. Taking heart from the fact that someone else was faced with the same difficult task, he found the courage. It turned out to be not difficult at all, since the old man Kapteyn was a wise man, who, although being thoroughly disappointed in his greatest wish, realized very well that he should not force his children to do a thing they did not want to do. And he agreed that this son would likewise study mathematics and physics. As a result he was enrolled at the University of Utrecht in 1868, where two of his brothers were students already.

The brothers did not see too much of each other and lived their lives as independently as they used to at home. Like all previous Kapteyns he was given the nickname 'Dux' [Latin for leader/captain and origin of the later 'Duke'], which was a name they were very much attached to. He studied under Professor Buys Ballot, the renowned meteorologist, and the mathematician Prof. Grinwis. He had little difficulty with his studies, since, as has already mentioned, he had done a lot of work at home in the previous year. So he spent little time studying and enjoyed life. Being the only student enrolled in the Faculty of philosophy (mathematics and physics), he joined a social club (society) of students in law, where he had great fun and enjoyed a free life, which brought him much happiness after his life of duty and obedience. This sudden, big change did little harm; it was common in those days that students did not take everything too seriously and Kapteyn went along with that. As soon as he noticed that the leisurely life was getting too much of a grip on him, he would find the strength to stop.'

There are two comments to make in relation to what is mentioned in these paragraphs. The first has to do with the interaction with his brothers. When Henriette Hertzsprung-Kapteyn tells us that the contacts with his brothers were very infrequent, this is not because they lived in different quarters. At least not where it concerned Willem. Van Berkel and Noordhof-Hoorn in the *Love Letters* note that in Utrecht Kapteyn moved into a room in the same house as his brother Willem, i.e., with the Huisman family at the address Predikherenkerkhof. If they were leading separate and different lives, this would be no more than continuing the way they grew up. A fact that may serve to illustrate this is that according to the *Love Letters*, Kapteyn, unlike his brother, became a member of the Utrechts Studenten Corps, the oldest student corporation (fraternity) in Utrecht, founded in 1816.

The second point concerns the remark that Kapteyn was the only student enrolled in his faculty. Indeed, the number of students in those years was small. Kapteyn was enrolled in the Faculty of mathematics and natural sciences, but it is incorrect to assume that he was the only student there. Maybe Kapteyn was the only *first year* student in that faculty; however, the total number of students that academic year (1868/69) was actually 63 and it is not likely that Kapteyn was the only one who had enrolled in his Faculty that year [5]. With regard to student numbers I note that these 63 compare to a total of 147 students for the three Dutch universities together in the mathematics and natural sciences faculties, and a total of 501 students for Utrecht university as a whole. By the time Kapteyn obtained his PhD degree during the academic year 1874/75, the first number had decreased to 52, while the national total in mathematics and natural sciences had grown to 175 and Utrecht as a whole to 527. The total number of students at the three universities of Leiden, Groningen and Utrecht grew from 1319 to 1684 [6]. The student numbers were small, but it is highly unlikely that Kapteyn was the only student in his cohort in his faculty.

'A great joy in those days was his friendship with 'Willy' Andree Wiltens. This young man was immediately charmed by the character and greatness of the Kapteyn boy and showed this with great warmheartedness. The heart of the lonely boy was warmed with joy and he basked in the ensuing love and friendship, which was new to him, but which he had longed for all his life. Wiltens was a son of a civil servant from the East Indies, whose three sons all studied in Utrecht and were members of prominent clubs. [...] He was a noble and good person and he had a rare characteristic, namely that he could admire and love with all the warmth of his heart. [...] Wiltens had a very special place in Kapteyn's life as the one who for the first time showed him what was most important in life, namely love.'

The combination Andrée Wiltens is the full double-barreled last name, sometimes written with the acute accent. The father, Henry Maximiliaan Andrée Wiltens (1823–1889), was a civil servant in the East Indies, in Padang on the island of Sumatra. With his second wife Euphemia Clementina Towsend (1822–1857) he had four sons. The study of the fourth and youngest, Jacob Willem Gerard Hendrik (1858–1923), would hardly have overlapped with Kapteyns, if he went to study in Utrecht in the first place, so the three must have been the older ones. The person referred to by Henriette Hertzsprung-Kapteyn is almost certainly the oldest, Henry William Andrée Wiltens (1851–1917), who had the same age as Kapteyn. In the *Love Letters* Kapteyn refers to him as 'Willie Wiltens'. The other two were Albert John (1853–1915) and Maximiliaan Leonard (1856–1935). 'Willie' Wiltens studied law and became a lawyer or attorney in his birthplace Padang, where he married Johanna Josephine van Hulsteijn (1855–1929) in 1880. He died in the Hague, so he must have returned to the Netherlands at some time. Kapteyn kept in close touch with the Andrée Wiltens family; Willie's brother-in-law was in fact a witness at Kapteyn's wedding (see page 109).

2.4 Kapteyn as a Student

'Kapteyn did not study too hard, as there was so much socializing and joy to have with friends as student life in the Netherlands was offering then, and his inherently joyful, social character enjoyed all this happily. His sense of humor and sociability, the happy side of his character, now started to come forward. He could be very sharp, which hurt others where his criticisms were wrapped in strict logic. He made enemies this way, which bothered him much, since he could not bear animosity and preferred to be on good terms with everybody. [...]

His first exam [Candidate (Bachelor), I presume] was only average. In those days exams were public and the audience had come in large numbers, so that the room was almost full. His father had come as well and was sitting close to him. He did not perform brilliantly and people started to become restless. The examiner kept going on about the same subject, although it was clear there was not much to gain from that. Kapteyn became irritated by that lack of logic and he said loud and clear: 'Professor, I don't know anything about this subject!'. It was absolutely quiet for a short moment and he saw his father turn pale. The professor, however, changed the subject to one Kapteyn knew a great deal about and in this manner he was enabled to save the day and obtain his degree.

After the exam the father and his sons went to the room where Kapteyn lived. The atmosphere was happy and lively and the largest surprise happened to the sons when they saw their very formal father taking an easy chair, propping his feet on the window sill and tipping the ashes from his cigar through the window. Was that really their formal, always correct father? 'Yes, boys, the upbringing has been completed and I no longer have to set an example', he said, laughing when he saw their surprise. In the evening the father went along with the friend to the student pub to celebrate the event. He was the center of attention and those present said unanimously: 'The old Dux will have to tell a story.' Everyone knew his talent, and the old Dux started telling stories so that everyone lost all sense of time.

The necessity to read out texts aloud in public proved a major difficulty for the young Kapteyn. He had a rather high-pitched, somewhat shrill voice, his intonation was not perfect and his reciting restless. One day he recited *Der Taucher* by Schiller, in a meeting of a students' reciting society, when someone in the audience remarked: 'I thought it impossible for anybody to recite so poorly.' He did not take this strong criticism as an insult, but as an inspiration to practice. A few years later, when a young professor in Groningen, he had to deliver his inaugural lecture, and he did all he could to be well prepared. He recited his lecture up to thirty times in front of two of his sisters, who were patiently sitting in the two farthest corners of the room as his audience and made comments. [...]'

Fig. 2.2 Title page of the PhD theses *On the theory of vibrating slabs and their relation to experiments* by Willem Kapteyn, submitted to the University of Utrecht to defend at 3 o'clock in the afternoon (15:00) on Friday, June 14, 1872 and *Study of vibrating flat membranes* by Jacobus Cornelius Kapteyn, submitted to the University of Utrecht to defend at 3 o'clock in the afternoon (15:00) on Thursday June 24, 1875 (Both theses were printed in Barneveld by P. Andreæ Menger)

2.5 Kapteyn's Supervisor and Astronomy in Utrecht

In the 1870s, the decade in which Kapteyn and two of his brothers obtained a doctor's degree, a number of persons in this country who were later to become famous scientists, wrote their PhD theses in natural sciences. Chief among these were the contemporaries of Kapteyn, and later Nobel Prize winners Johannes Diderik van der Waals (1837–1923) of Leiden University (*On the continuity of the gaseous and the liquid state*; 1873), Jacobus Henricus van 't Hoff (1852–1911) of Utrecht University (*Contribution to the understanding of cyanoacetic acid and Malonic acid*; 1874), Hendrik Antoon Lorentz (1853–1928), also of Leiden (*On the theory of reflection and refraction of light*; 1875), and Heike Kamerlingh Onnes (1853–1926) of the University of Groningen (*New proofs of the rotation of the Earth*; 1879).

In Utrecht, Kapteyn and two of his older brothers all obtained their doctorates with Grinwis as their supervisor. Willem Kapteyn (1849–1927) and Nicolaas Pieter Kapteyn (1845–1916) obtained their doctorates on the same date, June 14, 1872. The titles of their theses were *Over de theorie van trillende platen en haar verband met experimenten* (On the theory of vibrating slabs and the relation to experiments) and *Over de rekening met symbolen en de toepassing daarvan op de integratie van differentiaal-vergelijkingen* ((On the calculus with symbols and the application thereof on the integration of differential equations). Nicolaas Pieter re-

2.5 Kapteyn's Supervisor and Astronomy in Utrecht

ceived the judicium *magna cum laude* and Willem *cum laude*. Kapteyn followed three years later (June 24, 1875), also *magna cum laude*, with the thesis *Onderzoek der trillende platte vliezen* (Study of vibrating flat membranes).

Theses are being accompanied by propositions ('stellingen'). Such propositions (which have survived in Dutch universities to this day, except – as it turns out – in the University of Utrecht!) are a set of statements of a scientific nature, the first few of which may be related to the subject of the thesis itself and the rest is usually of a more general nature. The candidate should be prepared to defend these at the promotion ceremony. In the days of the Kapteyn brothers it was also possible to obtain a doctor's degree solely on presenting and defending a set of propositions. We find these identified as such in the *Album Promotorum Utrecht* [7], which lists all PhD theses, and shows that this happened almost exclusively in the Faculty of law and not at all in the natural sciences.

In the year 1872, a total of thirty doctor's degrees were awarded by the University of Utrecht. Seven were on propositions only (in the Faculty of law), eleven *cum laude* and another eleven *magna cum laude* (and one without a judicium). Of these thirty, six were in the Faculty of mathematics and natural sciences, three *cum laude* and three *magna cum laude*. In 1875 the university awarded 51 doctor's degrees. In this year 18 were on propositions only and one *honoris causa*, 16 *cum laude* and 16 *magna cum laude* (and, once again, one without a judicium). Of these, six were in the Faculty of mathematics and natural sciences again, albeit only one *cum laude* and five *magna cum laude*. It appears that the judicium *magna cum laude* was an honorable but not exceptional distinction.

Grinwis had 18 PhD students, according to the *Mathematics Genealogy Project* [8]. When we look at the titles we see that only five theses were in the field of pure mathematics (the one by Nicolaas Pieter Kapteyn being one of them), but most of them were in what we would now call applied mathematics. The ones by Willem and Jacobus Kapteyn are closely related. As we will see, the further lives of these two brothers show more parallels; they both became university professors, for example. Willem and Jacobus would later collaborate on some research projects in mathematics and publish a few joint papers. They even wrote one on an astronomical subject (*On the distribution of cosmic velocities*; Kapteyn & Kapteyn (1900), see appendix A.). The question arises why Willem was awarded the 'judicium' *cum laude* and Kapteyn *magna cum laude*. I will therefore discuss and compare the two theses.

But before looking into them in more detail, I will first address the question as to what could possibly have been the reason that Kapteyn did not produce a thesis in astronomy. Astronomy has a long tradition and history in Utrecht, as documented in detail by astronomers C. (Cees) de Jager, H.G. (Henk) van Bueren and M. (Max) Kuperus: *Bolwerk van de sterren* [10]. This book was written on the occasion of the celebration of 350 years of astronomy in Utrecht, in 1993. It is the basis of most of the contents of the next few paragraphs.

The first observatory was installed on a tower (the 'Smeetoren') on the walls of Utrecht in 1642, not long after the university had been founded. This tower was used by the guild of blacksmiths (the name Smee-tower comes from the Dutch word

Fig. 2.3 Sonnenborgh not long after the Observatory and the Meteorological Institute were established here. The two towers were built to enable observing (Koninklijk Nederlands Meteorologisch Instituut [9])

'smeden' or forging); it had been built in 1145 and the sharp steeple was removed to leave an eight-sided platform fitted with a roof to enable astronomical observing. It was torn down in 1855; the observatory and the associated meteorological institute had been moved to a new location in 1854, on top of the old Sonnenborgh bastion or bulwark, which dates from 1552 and was built as part of the city fortification walls to provide cannon emplacements. Buys Ballot was very much involved in this, establishing the first meteorological institute there in 1854, together with the astronomical observatory (see Fig. 2.3). It has been a museum since the Utrecht astronomy department moved to the Uithof campus outside and to the east of Utrecht in the 1970s. Sonnenborgh is the oldest cupola (dome) observatory in the Netherlands. Eventually there were four large telescopes to study the Universe. Until 2002, the Merz Telescope from 1863 was one of the larger (night) telescope of its kind in the world [11]. Sadly, Utrecht University decided in 2011 to close its astronomy department, abruptly ending a tradition of 370 years of excellence and recognized first-class research (see the symposium *370 Years of Astronomy in Utrecht* [13]).

Adolf Stephanus Rueb (1806–1854) had been appointed lecturer in astronomy in 1843, relieving Buys Ballot from his astronomy teaching. Rueb was also in charge of the astronomical observations at the observatory. His appointment was an upgrade of the position of observator, which had been instituted under Moll. This means that in 1843 astronomy was actively practiced in Utrecht as a discipline for the first time. Rueb had supervised the construction of the building of the observatory on the Sonnenborgh, but died in 1854 before it was completed. Again Buys Ballot had to take over the teaching of astronomy until in 1856 Jean Abraham Chrétien Oudemans (1827–1906) was appointed Rueb's successor, but now as an extraordinary profes-

2.5 Kapteyn's Supervisor and Astronomy in Utrecht

Fig. 2.4 Jean Abraham Chrétien Oudemans (1827–1906) and Martinus Hoek (1834–1873) (Universiteismuseum Utrecht [12])

sor. Oudemans had been a student of Frederik Kaiser (1808–1878) in Leiden, who regarded Oudemans as his best student. However, Oudemans accepted in 1857 a position as senior engineer for the Geographical Service of the Dutch East Indies, and after a period in which Buys Ballot was once again forced to take over the teaching of astronomy, Martinus Hoek (1834–1873) was appointed in 1859. Figure 2.4 shows Oudemans and Hoek.

The first doctor's degree that was awarded in Utrecht on an astronomical subject or under a supervisor who was professor of astronomy dates from 1880, according to *Bolwerk van de sterren*. The title of the thesis was *Planeet 182 (Elsbeth)*, the candidate J. Robbers and the supervisor Oudemans. Minor planet 182 was discovered on February 7, and is now designated 'Elsa' 1878 [15] (see Box 2.1 for objects in the Solar System known in 1870). Oudemans must have taken up this PhD thesis project after he returned from the Indies. Buys Ballot had PhD theses written under his supervision when Kapteyn was a student in Utrecht; the nearest in time to Kapteyn's were those by Abraham Johan Verweij, 1874, *De waarnemingen der bevolkingsstatistiek* (Observations of population statistics) and Eiso Henricus Groenman, 1877, *Iets over den invloed van de temperatuur op het magnetisme van ijzer en staal* (On the influence of temperature on the magnetic properties of iron and steel). The only astronomical one under Buys Ballot had been much earlier, in 1871, by Adrianus Jacobus Sandberg *De orbita Undinae* (On the orbit of Undina). Undina is a minor planet, no 92, discovered on July 7, 1867. The list of astronomical theses in Utrecht in *Bolwerk van de sterren* has overlooked this one.

So we see that Hoek was the professor of astronomy during most of the time that Kapteyn was a student of Utrecht University. But Hoek was in poor health and

> **Box 2.1 The Solar System in 1870**
> As far as people around 1870, when Kapteyn entered university, were aware of, the **Solar System** looked as follows:
> Apart from the Sun in the center, all eight **planets** were known; in addition to Mercury, Venus, Earth, Mars, Jupiter and Saturn, which could be seen with the naked eye, another two that could only be seen with telescopes, had been discovered: Uranus (accent on first syllable) in 1781, by Sir William Herschel (1738–1822), and Neptune in 1846, by Johann Gottfried Galle (1812–1910), after its position had been predicted by Urbain le Verrier (1811–1877) on the basis of perturbations of Uranus' orbit.
> Apart from the Moon, the only known **moons** or **natural satellites** of other planets in 1870 were four for Jupiter, eight for Saturn (in addition, its ring had been known since Christiaan Huygens (1629–1685) recognized it as such in 1655), two for Uranus and three for Neptune.
> **Comets** had been known to be objects on highly eccentric orbits ever since Edmond Halley (1656–1742) realized that the comets of 1531, 1607 and 1682 had very similar orbits and actually involved the same object, which subsequently became known as Halley's comet. His prediction in 1705 that it would reappear in 1758/59 was confirmed.
> **Minor Planets** or **asteroids** reside in the large gap between the orbits of Mars and Jupiter. The first one, Ceres, was accidentally discovered on the first day of the eighteenth century, January 1, 1801, by Giuseppe Piazzi (1746–1826). By 1870, about 200 were known.

did little observing, concentrating on experiments and theory, not only in astronomy but also in physics. He is widely known for his interferometric work, published in 1868 (for a description of his experiment see *The Hoek Experiment (1868)* by Doug Marett [16]), in which he looked for the effect of a dragging by the Earth of the ether that was supposed to propagate light; he found no such effect, but his results were not sufficiently convincing to take it as proof of the non-existence of ether. As is well-known, this issue was settled in the end by the famous experiment of Albert Abraham Michelson (1852–1931) and Edward Williams Morley (1838–1923) in 1887. Hoek also advanced the concept of groups of comets (comets that have very similar orbits resulting from the breaking up a single 'mother-comet') and on this basis worked out a theory for the origin of comets. Hoek's last publication listed in the Astrophysics Data System (ADS) is a paper on this subject of 1868, *On the phenomena which a very extended swarm of meteors coming from space presents after its entry into the Solar System* [17], and it is very likely that he did little astronomical research after that; in fact he never had a PhD thesis written under his supervision. Hoek died in 1873 at the very young age of 39 years and Oudemans, who was wrapping up his work in Java, was asked to return to Utrecht. He did not arrive until 1875, around the time Kapteyn was *completing* his PhD thesis. Buys Ballot had once more taken over the responsibility for astronomy in Utrecht. If Kapteyn had been interested in doing a PhD thesis on an astronomical subject in addition to the astronomical courses he had followed, it would have been very difficult to do so. Hoek had died in 1873, Buys Ballot had little interest in astronomical research and

Oudemans had not returned from his work in the East Indies yet. A mathematical thesis under Grinwis seemed the only realistic option.

We know from the *Utrechtse Studenten Almanak* [18] that astronomical subjects were taught extensively in Kapteyn's (undergraduate) student days (I consulted the Almanak for the years 1868 through 1874). There is very little variation from year to year. Buys Ballot gave lectures on 'experimental physics', 'analytical engineering', 'meteorology', 'analytical geometry and higher algebra' and sometimes selected topics in physics, while he also gave practical courses in experimentation. Grinwis taught 'mathematical physics', 'fundamentals of mathematics', 'integral and differential calculus' and 'spherical geometry'. During the majority of Kapteyn's years in Utrecht, Hoek offered courses in 'general astronomy' (described as 'popular astronomy', but not in the current meaning of the word popular), 'theoretical astronomy' and 'practical astronomy', while he usually offered (daily!) 'exercise in observational astronomy' as well. During most of Kapteyn's years, van Rees offered 'to assist students with their studies' even after he had retired. After Hoek's death in 1873, Buys Ballot again took over astronomy and in 1974 he gave three series of astronomical lectures, viz. 'popular astronomy', 'theoretical astronomy' and 'practical astronomy', in addition to four courses on physics-related subjects.

So, astronomy was definitely not neglected in the curriculum. Kapteyn had a wide choice and ample opportunity to be taught astronomy. There is no record of which students followed which courses, but it is very likely that Kapteyn made use of this broad range of subjects to choose from and consequently received extensive training in astronomical subjects as well as in mathematics, physics and possibly meteorology. His interest in astronomy, which was obvious even before he entered university – as is evident from the facts that as a teenager he made a map of the stellar sky and that his father made him a present of a telescope when he was a boy – must have developed to the extent that he formulated a relatively large number of propositions on astronomical matters in his thesis (as we will see below).

2.6 The Thesis of Willem Kapteyn

I will discuss the thesis by Willem Kapteyn first. His subject was the theory of vibrating slabs (see Fig. 2.2) and concerned the observation by the German physicist Ernst Florens Friedrich Chladni (1756–1827) that when a flat plate is covered with a powder and then made to vibrate by a violin bow, for example, the powder forms regular, symmetric patterns. Various famous scientists, such as Jacob (II) Bernoulli (1759–1789), Joseph-Louis Lagrange (1736–1813), Siméon Denis Poisson (1781–1840), Augustin Louis Cauchy (1789–1857) and Gustav Robert Kirchhoff (1824–1887), had worked on a theory for this. A prize competition on the subject of vibrating slabs, organized by the French Academy of Paris in 1810, had been won by the French female mathematician Marie-Sophie Germain (1776–1831). Germain, being a woman, met with great opposition on the part of her parents when she wanted to enter the field of mathematics (later on, she was likewise frustrated in her plans by the Paris École Polytechnique) and she never came to hold a formal position. She learned mathematics from her father's library and corresponded with Lagrange.

Willem's thesis started with a presentation of the basic theory behind the physics of vibrating slabs. He considered the mechanical work expended by the internal elastic forces when each infinitely small element is displaced by an infinitely small amount and equating that to the work exercised by the external forces. This had to do with behavior characteristic of the (elastic) properties of the material involved. The resulting equations were then treated first for a situation where the slab is infinitely thin and subsequently for the general case, but reduced to the mid-plane of the slab. This was valid as long as one limits oneself to transverse vibrations. Since the slabs were finite in extent, the results depended on the shape of the slab through a different set of equations that were applicable to the elements at the outer edges.

Willem Kapteyn compared his results to those of earlier studies. He concluded that the theory of Germain started from an incorrect assumption (viz., that the slab is in equilibrium), but in spite of that she 'accidentally' arrived at the correct equation of motion. In his book *Vibrations of Shells and Plates* [19], Werner Soedel agreed: 'Germain (1821) gave an almost correct form of the plate equation. The bending stiffness and density constants were not defined. Neither were the boundary conditions stated correctly. These errors were the reason that her name is not associated today with the equation, despite the brilliance or her approach'. (The book by Soedel has a excellent introductory chapter on the history of vibration analysis.) Willem Kapteyn then criticized a theory proposed by Jacob (II) Bernoulli as a generalization of work by his uncle Daniel Bernoulli (1700–1782), who had put forward a theory of transverse vibrations of flexible thin beams in 1735. Jacob had taken the important step to extend this by treating a slab as a collection of connected beams (incidentally, Willem Kapteyn invariably spelled the name Bernoulli incorrectly as Bernouilli). Without detailed discussion Willem Kapteyn discarded this theory, claiming it disagreed with observations. Soedel simply stated that Bernoulli's equation was incorrect. Willem Kapteyn mentioned the theories of Poisson and Cauchy omitting to discuss them in detail and demonstrated that these gave correct results for the case of a circular slab, for which the theory was the same as that developed by Kirchhoff. The conclusions of Willem Kapteyn corresponded to the historical section in Soedel's extensive and definitive book on the subject of vibrations.

Willem then went on to review experiments (performed by others) to which solutions of the equations given in his first chapter should be compared. Prominent are the observations by Chladni on rectangular and elliptical slabs (including square and circular ones as special cases) and by the British physicist Charles Wheatstone (1802–1875). The results of the experiments were classified according to patterns of the (lines of) nodes. Nodes are the positions on the slab where vibration is absent. An example of a node is the stationary point halfway along a string that vibrates in the first overtone. Nodes are observable by 'strewing sand on vibrating surfaces, commonly called acoustic figures' (taken from the title of Wheatstone's publication). Chladni used metal plates, those employed by Wheatstone were wooden.

On the final eleven pages Willem presents solutions to the (differential) equations for the case of circular slabs. These solutions had been given earlier by Poisson and in a different form by Kirchhoff. After comparing them to the experiments, he concludes that the analogy is 'good enough to assume that the actual differences can be ascribed to unavoidable measuring errors.'

The work in this thesis displays a good understanding of the issues, but does not really provide a great deal of new and original material; it is a re-discussion, occasionally presenting newly formulated results that have been obtained earlier. In addition, the comparison between theory and observation is very limited.

I conclude by considering the set of propositions accompanying Willem Kapteyn's thesis. He presents fourteen such propositions, the first four being directly related to the content of the thesis itself. Nos 5 through 9 are statements of a mathematical nature. The final five are of a diverse nature. No 10 is astronomical; it reads: *The explanation for the colors of stars as given by Christian Andreas Doppler is improbable*. Doppler (1803–1853) had discovered the effect named after him that light is shifted to shorter wavelengths when the object is approaching the observer and vice versa. He surmised that all stars are white and that their colors derive from different motions with respect to us. This was also criticized by Buys Ballot, among others, who noted that for a significant shift in color the radial velocity of the star would need to be a significant fraction of the speed of light and that this was highly improbable. Evidently, Willem Kapteyn agrees. His further propositions concerned matters such as calorimeters, explanations for the Ice Ages, the origin of life and the division of humankind into species rather than races.

Neither the propositions nor the thesis itself present much original work, as far as I can tell. However, it should be remembered that this was by no means uncommon in these days.

2.7 Kapteyn's PhD Thesis

The thesis of Jacobus Kapteyn is much along the lines of Willem's. Kapteyn begins by pointing out that studying membranes rather than slabs or plates is more straightforward, the reason being that experiments with slabs are less conclusive due to the variations in thickness, lack of homogeneity of the material and difficulty to support them. Membranes are supported along the edges and less vulnerable to inhomogeneities. Because of their relative simplicity, membrane vibrations are useful to study for the development of a theory of elasticity and in particular for a better understanding of hearing and sound.

The theory Kapteyn elaborates on is to a large extent a re-discussion of that of Poisson, who concentrated on the general rectangular case and the special case of a circular membrane. But Kapteyn had also available to him the work of Gabriel Léon Jean Baptiste Lamé (1795–1870), in particular on triangular membranes, Georg Friedrich Bernard Riemann (1826–1866) and Émile Léonard Mathieu (1835–1890), the latter for treating elliptical membranes.

Kapteyn's exposition is easier to follow than Willem's. This is no surprise, as the membrane analysis is a two-dimensional problem, whereas that of Willem Kapteyn's slabs is not (although it is not fully three-dimensional, since the slab is assumed to be of constant thickness and Willem mostly restricted himself to transverse vibrations in the central plane). In order to understand the import and purpose of the thesis, I discuss his approach in some detail. Kapteyn noted a simple equation of motion for a membrane, in which the co-ordinates of an infinite element are x,y, which experiences an infinitely small displacement w in the perpendicular direction. The equation is no more than Isaac Newton's law in the vertical direction, which says that the force equals the mass times the acceleration. For a more detailed treatment see Box 2.2 (and also chapters 2 and 6 of M.H. Sadd: *Wave Motion and Vibration of Continuous Media* [20]). Kapteyn showed how upon integrating this (differential) equation, one obtained an expression containing a superposition of waves of various frequencies. There are different ways of the membrane to vibrate. In Box 2.3 the mathematical details are given for a rectangular membrane. Figure 2.5 shows modern illustrations of some of the simplest modes in which a square membrane can vibrate. Note that (except in the first fundamental mode) there are straight lines which are stationary; these are the lines of nodes and they depend on the shape of the membrane. They can be used (as they are by Kapteyn) to classify the various modes. Generally speaking, these lines of nodes need not be straight.

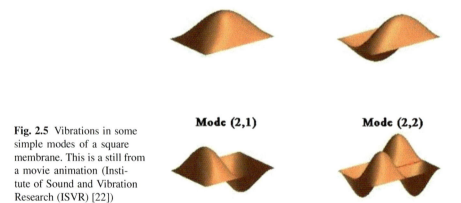

Fig. 2.5 Vibrations in some simple modes of a square membrane. This is a still from a movie animation (Institute of Sound and Vibration Research (ISVR) [22])

Kapteyn treated the square membrane in detail. In Fig. 2.6 I have collected some of Kapteyn's illustrations of nodes. The interesting conclusion is the occurrence of nodal *points*. This seems new and appears to be an original contribution to the theory. In addition, he supplied a detailed treatment of the theory of unilateral triangular and circular membranes, and a less in-depth treatment of elliptical membranes.

Box 2.2 The equation of motion in Kapteyn's vibrating membrane study
Take an infinitely small area of size ∂x and ∂y. Then we have

$$F dy \frac{\partial^2 w}{\partial x^2} dx + F dx \frac{\partial^2 w}{\partial y^2} dy = \rho dx dy \frac{\partial^2 w}{\partial t^2}.$$

Left is the force and right is mass times acceleration, both in the (vertical) *w-direction*. Start with the right-hand side. Here ρ is the mass surface density (mass per unit area) and this is multiplied by the area (so we get the mass of the element), and $\partial^2 w/\partial t^2$ is the *w*-acceleration. If *w* is the position, the first derivative $\partial w/\partial t$ is the change in *w* per unit time *t* and thus the velocity; the second derivative $\partial^2 w/\partial t^2$ is the acceleration (the change per unit time of the velocity).
Now look at the left, where we have two very similar terms. It is the total force *in the vertical direction*, one term comes from the *x*-direction, one from *y*. The first one shows the force arising from the stretching of the membrane in the *x*-direction, where *F* is the tension per unit length in the membrane (assumed the same everywhere in both *x* and *y* directions). The force in the *x*-direction on the area equals $F dy$ and is multiplied by a factor that projects it onto the *w*-direction (it equals the differences between the sinuses of the projection angles onto the *w*-direction at either end of the area). This then gives the component of the force in the *w*-direction arising from the stretching of the membrane in the *x*-direction. The second term is the same in the *y*-direction. This equation Kapteyn rewrites as

$$\frac{\partial^2 w}{\partial t^2} = c^2 \left(\frac{\partial^2 w}{\partial x^2} + \frac{\partial^2 w}{\partial y^2} \right) \text{ with } c^2 = \frac{F}{\rho}.$$

The fundamental equation was first derived in this form by Poisson. It must be complemented with the conditions that the displacements *w* are zero at the edges only and the initial conditions that all *w* and dw/dt are zero at the start (time $t = 0$).

Kapteyn finally compared his theoretical results extensively to experiments in the relevant literature, in particular a comprehensive set of experiments by Justin Bourget and Félix Bernard ('Professeurs à la Faculté des Sciences de Clermont-Ferrand'), *Sur les vibrations de membranes carrées* [21], published in 1860. On the whole, these observations compared reasonably well to the theory, but no more than that, and he attributed this to problems with imperfections in the experiments, particularly inhomogeneities in the thickness and tension across the membranes used. It was the basis of his first proposition: *I. The agreement between theory and observation concerning vibrating membranes is not completely satisfactory.* Concerning the nodal *points*, Kapteyn noted that Bourget and Bernard observed them although, as far as I can see, only with respect to tones $N(p, 1)$–, but provided an incorrect interpretation. In Kapteyn's own words: 'Bourget and Bernard, who

> **Box 2.3 Kapteyn's solutions for the rectangular membrane**
> Assume a rectangle with sides a and b.
>
> $$w = \sum_{i=0}^{\infty}\sum_{j=0}^{\infty}(A_{ij}\cos\gamma t + B_{ij}\sin\gamma t)\sin i\pi\frac{x}{a}\sin j\pi\frac{y}{b}.$$
>
> The constants A_{ij} and B_{ij} are found by applying the boundary and initial conditions. Each term ij then represents a vibration with frequency
>
> $$N(i,j) = \gamma/2\pi = \frac{c}{2}\sqrt{\frac{i^2}{a^2}+\frac{j^2}{b^2}}.$$
>
> Restricting to *square* membranes ($a = b$) gives the following results. The tones allowed have frequencies in the ratio $\sqrt{2}:\sqrt{5}:\sqrt{8}:\sqrt{10}:\ldots$, which equals $1:1.58:2:2.24:\ldots$. The system of lines of nodes for tone $N(p,q)$ is determined by the equations
>
> $$A_{p,q}\sin\frac{p\pi x}{a}\sin\frac{q\pi y}{a}+A_{q,p}\sin\frac{w\pi x}{a}\sin\frac{p\pi y}{a}=0,$$
>
> $$B_{p,q}\sin\frac{p\pi x}{a}\sin\frac{q\pi y}{a}+B_{q,p}\sin\frac{w\pi x}{a}\sin\frac{p\pi y}{a}=0.$$

restricted themselves to Lamé's theory, have, just like him, overlooked the theoretical necessity of nodal points.' Kapteyn's second proposition expressed this: *II. the theory of the membrane can be pronounced complete only with the theoretical derivation, given in this thesis, of nodal points of the square membrane.* Here we have something in his thesis that appears original. He also remarked that such nodal points could only be seen in experiments when the tension across the membrane was extremely uniform. In modern textbooks, such as those by Soedel or by Sadd (but see also He & Fu [23]), little attention is paid to these nodal points. These may be formal solutions, whose usefulness or applicability in practice is limited.

2.8 Propositions

I look at Kapteyn's propositions in some detail. He had 18 of them. As we have seen, the first two were on his thesis subject. Nos 3 through 9 were concerned with mathematics or mathematical physics. But his propositions 10 through 16 were on astronomy, and the last two had a biological scope.

2.8 Propositions

Fig. 2.6 Lines of nodes and nodal points for a square vibrating membrane from Kapteyn's thesis. In the top-left the wavelength is twice the sides of the square, so that the only nodes are the edges. This mode is designated as $N(1,1)$, according to a classification scheme of Lamé. The next five (the next two in the top row and the three in the second row) are the cases $N(2,1)$. The third and fourth rows have $N(2,2)$, three cases for $N(3,1)$, $N(3,2)$ and $N(4,3)$

No VII merits a closer look, not so much for reasons of the content but rather because of the person he was referring to. It reads: *VII. The assertion by Multatuli that the profit of the casino is not a result of the advantage of the presence of a zero which, according to him 'can for my part be completely abolished', is incorrect.* Eduard Douwes Dekker (1820–1887), who used the pen name Multatuli (literally: 'I have suffered much'), was a 19th century Dutch novelist and essayist and had been a major source of inspiration to Kapteyn (see page 110). He was quickly drawn to the non-conformist views upon society and religion promoted by Multatuli, in particular those protesting the exploitation of the indigenous population of the Dutch East Indies (the current Indonesia) by the Dutch administrators, in his most famous book *Max Havelaar*. Reading Multatuli (and also Jean-Jacques Rousseau) must have played a major role in Kapteyn's rejection of the dogmatic religion of his parents. We know from the HHK biography that Kapteyn had read almost all of Multatuli's works and may have stumbled on this remark, which was probably meant by Multatuli as a protest against a suspected unfairness of the operators of casinos.

Five of the seven astronomical propositions concerned matters such as the theory and orbits of comets, apparent size of the Sun and the Moon on the sky and sunspots (which he thinks are cyclones). It demonstrated an active interest in astronomy and astronomical publications, for it seems unlikely that he would have chosen propositions without it. Clearly, these subjects must have been close to his heart and it

> **Box 2.4 Brightness and magnitudes in astronomy**
> For historical reasons, astronomers express the **apparent brightness** of a star in **magnitudes**. This concept has come down to us from antiquity, which is why the scale was not rigorously defined at first. Eventually the scale became accurately established keeping the historic values roughly conserved as a logarithmic one, such that 5 magnitudes corresponds to exactly a factor 100 (and thus one magnitude to a factor $10^{0.4} = 2.512$). This definition was established by the British astronomer Norman Robert Pogson (1829–1891), who proposed this scale in 1856.
> To the naked eye the brightest star is Sirius at a magnitude of −1.5; the faintest that can be seen are about magnitude 6. This represents a factor of 1000 or so in brightness. To give an idea, the Pole-star (Polaris) is of magnitude 2.0. The majority of the visible stars in the well-known constellations Ursa Major (Big Dipper), Cassiopeia or the belt of Orion (accent on the second syllable 'ri') have magnitudes between roughly 1.7 and 2.4 or differ by only a factor two or so. But their distances range from 65 to 1300 lightyears, or by a factor 20.
> In order to correct the apparent magnitude of a star for its separation from us the **absolute magnitude** is defined by calculating the magnitude of a star if its distance were 10 parsecs or about 32.6 lightyears (see Box 5.6 for the definition of the parsec). This definition of absolute magnitude was introduced by Kapteyn.

seems he was actively pursuing and extending his knowledge and understanding of this discipline. In this context, there are two propositions that I have examined in more detail, viz., nos 10 and 15.

X. The best photometer is that of Zöllner. For a good understanding of this, we need to go into the practice of astronomical photometry in some more detail. Johann Karl Friedrich Zöllner (1834–1882) developed a photometer to accurately determine the brightness of stars in 1858. Originally the assignment of stellar magnitudes (see Box 2.4) was done by a visual estimate when comparing two stars in the telescope. However, in the early nineteenth century this started to change. For example, Sir John Frederick William Herschel (1792–1871), son of Sir William Herschel, used a device that he called an 'astrometer', in which he compared the particular star to a reduced image of the Moon, reflected by a prism and focused into the field of view with a lens. Carl August von Steinheil (1801–1870), of Munich, had developed a device in which out-of-focus images of stars were compared to estimate brightness ratios. His device was a telescope with an objective lens (at the front of the telescope) cut into two parts which could be moved independently. He brought two star images into the same field of view and then, with both stellar images out of focus, he adjusted one of the half-objectives until both images had the same surface brightness.

Zöllner's photometer (see Fig. 2.7) made use of a suggestion by François Jean Dominique Arago (1796–1853) to use polarization of light , which he had discovered together with Augustin-Jean Fresnel (1788–1827). This is the property of light, when regarded as a wave, that the wave can oscillate in different planes. Sometimes

2.8 Propositions

Fig. 2.7 In the Zöllner photometer, the light of a flame (in the tube on the right) is focused into the field of the telescope through a diaphragm. Between the flame and telescope a rotating polarizer was used to decrease the brightness of the resulting artificial star. The structure on the left acts as a counterweight (Efron Encyclopedic Dictionary [24])

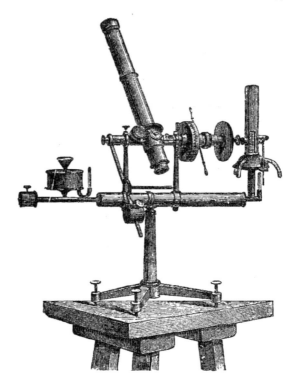

the discovery is credited to Christiaan Huygens (1629–1685), who discovered the double refraction that occurs when light passes a crystal such as Iceland spar. This effect, in which the different polarizations of light have different refractive indices depending on how they are aligned with the crystal, indeed results from polarization, but Huygens proposed a different explanation. By using a transparent plate that transmits light of only a particular polarization, one selects a limited range of planes of vibration. Using another such plate (or prism), the light can be fully extinguished if the two polarizations are perpendicular or completely transmitted when the two are parallel. In intermediate cases a certain percentage of the light is transmitted. Zöllner, in his photometer used a controllable flame (in a Bunsen burner so that it is very stable) and diaphragms to produce an artificial star image that was projected into the field of view. Using a fixed and a rotating polarizing element, the apparent brightness of the artificial star could then be adapted until it was the same as that of the 'real' star. The amount of rotation of the polarizer can be calibrated using stars of known brightness and this way the magnitude of stars can be determined quite accurately. A detailed and interesting discussion of the instrument, including the construction of a replica, can be found in two papers by Klaus B. Staubermann: *The trouble with the instrument: Zöllner's photometer* and *Making stars: Projection culture in nineteenth-century German astronomy* [25].

The Zöllner photometer revolutionized stellar photometry and remained in use for many decades. In 1922, J. van der Bilt in *Note on the photometric scales of Pickering and Parkhurst* [26] reports on the use of the Utrecht Zöllner photometer (at some point in time Utrecht apparently owned such a photometer) mounted on a Leiden telescope to compare the inconsistent magnitude scales as given by Yerkes and Harvard observatories. These photometers were complemented and gradually replaced by photographic techniques and modern photo-electronic devices. An excellent review of *the development of astronomical photometry* was written by Harold F. Weaver in 1946 [27].

This proposition demonstrates Kapteyn's early interest in astronomical instrumentation and the measurement of stellar properties. Zöllner photometers were extremely popular at the time when Kapteyn was working on his thesis and they were used on a large scale; Utrecht Observatory almost certainly possessed one. In spite of the kind help of the staff of the Utrecht University Museum, it has not been possible to find an inventory that proves that Utrecht actually possessed a photometer as early as the 1870s. It cannot be denied that the proposition is hardly original, since there were no other accurate photometers to compete with Zöllner's. In that sense Kapteyn's proposition was the equivalent of forcing an open door. However, knowledge of the existence and purpose of such instruments was limited to a very specialist circle and it is unclear how Kapteyn would have known about them, let alone be acquainted with their use, had he not had a lively interest in astronomy and been aware of the instruments at the observatory, their operation and pitfalls.

The other proposition to consider in more detail is no 15. It reads: *XV. The average proper motions of stars of various magnitudes is not inversely proportional to their distance.* This is remarkable, as the matter of distances of stars as inferred by their proper motion would later become a major focus of his scientific efforts. It already shows that he foresees that deriving distances of stars in a statistical manner requires careful treatment; at least he must have looked into the possibility and maybe have tried it seriously on actual data in order to formulate this proposition. In addition, there was some literature on the subject that Kapteyn just may have seen which had been published in the major astronomical journal 'Monthly Notices of the Royal Astronomical Society', and which was present in the Utrecht University Library (the catalog shows ownership of a copy back to the first volume in 1827). Relevant papers that Kapteyn may have used to formulate this proposition include: George B. Airy: *On the movement of the Solar System in space*, Edwin Dunkin: *On the movement of the Solar System in space, deduced from the proper motions of 1167 stars* (1859) and Richard A. Proctor: *Note on the Sun's motion in space and on the relative distances of the fixed stars of various magnitudes* (1869) [28]. The last one in particular is relevant. Richard Anthony Proctor (1837–1888), a British writer and self-employed astronomer, concluded from a discussion of the fundamental data on proper motions of 1167 stars, that large proper motion is a better indication of a small distance ('argument for proximity') than apparent brightness. This study and this proposition were very much along the lines of Kapteyn's much later research, and his remarks in this proposition were a prelude to his extensive attempts to find a statistical method to derive stellar distances using their proper motions (see page 292). Once again, this proposition implies more than just a superficial knowledge of and interest in astronomy.

Kapteyn's thesis demonstrates some originality in both the work itself and the propositions, which may be the reason for his *magna cum laude*. He must have had an eye for experimental data, but above all it suggests that even at this early point in his career he was thinking much about astronomical matters. It is interesting to note that the Kapteyn Room in the Kapteyn Astronomical Institute (see Preface) does not contain a copy of the thesis, although Kapteyn collected reprints of all his publications and even had them bound in leather volumes. Maybe he simply did not feel this piece of work was special and worth keeping.

2.9 Catharina Elisabeth Kalshoven

While a student in Utrecht, Kapteyn became acquainted with his future wife, Catharina Elisabeth (Elise) Kalshoven. He was probably introduced into the Kalshoven family through Elise's half-sister Jacqueline (see Table 2.1.). The latter married Simon Brouwer (1833–1920) in 1873, who was a notary and, although Kapteyn's senior by almost twenty years, may still have been active in the student society of which Kapteyn was also a member (see page 31). Brouwer lived in Maarssen, in the same village close to Utrecht where the Kalshoven family lived. There is reason to believe that this is the way Kapteyn and Catharina Elisabeth Kalshoven met, for in one of the *Love Letters* (of November 22, 1878) Kapteyn expresses his concern with regard to Jacqueline being ill, noting that he 'owes much, probably the largest share in his life's happiness' to her. I again quote from the *HHK biography*, since this narrative must almost certainly be based on what Henriette Hertzsprung-Kapteyn was told by her mother.

'During his last year in Utrecht he made the acquaintance of the Kalshoven family, who turned out to be completely different from the Kapteyns. They could be described as antipodes. No heavy responsibilities were laid on the shoulders of the two young daughters, they did not study or have a job, the atmosphere was cheerful and cozy, life was enjoyed in a simple and relaxed manner. Gaiety, music and courtship, all those enchanting things in a young person filled the atmosphere. Mrs Kalshoven was a woman of fine, civilized manners and culture. As a young girl she had married the wood merchant Kalshoven, who was a widower with five children. Together they had another three, two girls and one boy. It had been a heavy task for the lively young girl, who came from a quiet, untroubled life in Harderwijk, where her father was the 'conrector' of the local Gymnasium, and now she had moved to a large old-fashioned house on the Singel in Amsterdam. Her husband's business was not flourishing and the family decided to move to Abcoude, where life was expected to be easier and less expensive.'

Table 2.1 Kapteyn's brothers and sisters-in-law. They were all born in Amsterdam. See Table 1.1 for their parents

	Jacobus Wilhelmus Kalshoven and Catharina Elisabeth Brandt	
Willem Johan Christiaan	Sept. 13, 1839	Dec. 11, 1881 – Batavia
Elisabeth Wilhelmina	Oct. 17, 1840	June 4, 1860 – Amsterdam
Johan Christiaan	June 27, 1842	Sept. 10, 1842 – Amsterdam
Johannes Christiaan	Febr. 7, 1844	March 9, 1920 – London
Jan Anton	Jan. 31, 1845	April 17, 1928 – Amsterdam
Jacqueline Wilhelmine	June 4, 1846	April 2, 1926 – Den Haag
	Jacobus Wilhelmus Kalshoven and Henriëtte Mariëtte Augustine Albertine Frieseman	
NN	March 5, 1854	(stillborn)
Catharina Elisabeth (Elise)	June 19, 1855	March 2, 1945 – Amsterdam
Marie Gabrielle (Marie)	June 7, 1857	Nov. 17, 1940 – Borger
Jacobus Wilhelmus (Jacques)	Dec. 30, 1859	Febr. 7, 1941 – Den Haag

A 'conrector' is the deputy of the rector or principal of a gymnasium (or grammar school). Harderwijk, which is also where the university was located mentioned in the beginning of this chapter, is a small town a little north of Barneveld, situated on what was then called the Zuiderzee. Most inhabitants made a living out of fishing and the related industry. Singel in Amsterdam is a canal that originally served as a protective moat that ran around the city in the Middle Ages, situated inside the system of canals that was dug in later centuries as Amsterdam expanded. Abcoude is a small municipality to the south-east not far from Amsterdam, in the province of Utrecht.

'Mrs Kalshoven was a strong woman, who accepted the circumstances and adjusted to them, and [she and her family] enjoyed happy years after the difficult times in Amsterdam. Her stepchildren grew up and left home, and after the death of her husband, Mrs Kalshoven and her own children moved to Utrecht [really Maarssen near Utrecht], where she had to live a simple and frugal life. As she became older, she had more and more difficulty walking and became restricted to her chair. She underwent these tribulations with resignation, but there were also days when she found it difficult to accept her handicap and hated having to rely on help and assistance. She would then complain and moan, but when she had visitors she became gay and happy, so that the two healthy daughters would look at each other and conclude it was not so serious after all. Being healthy youths they could not imagine what it meant to be an invalid.

The mother was surrounded by a warm love, especially from her elder daughter Elise, who was the quieter and more serious of the two. She did her domestic duties without complaining, and the greatest joy was to sit at the piano and to play whatever came to her mind. She had a great talent for music and she further developed it by herself, since there was no money for music lessons in this simple family. Marie, the younger sister, had a beautiful singing voice and the two girls could spend hours playing and singing. Marie would think of a song and start singing, Elisa followed

2.9 Catharina Elisabeth Kalshoven

Fig. 2.8 Kapteyn as a young man, probably sometime during the 1870s (Kapteyn Astronomical Institute [20])

on the piano, which went without much effort, since she was a very gifted musician. However, she failed to develop these talents by having lessons and much practice . And maybe that was not really necessary, as they were not persons given to serious study and work. They rather lived as the birds in a forest, singing and cheering when they wanted to, when the sun was shining and the sky was blue; and music brought them happiness throughout their lives. The mother had an impulsive character too and lived from one day to the next. She could not resist a beautiful summer's day and would then call out: 'Children, the Sun is shining, it is so lovely outside. We will have a carriage prepared and we will go on a tour.' The thrift and worries were forgotten for a moment and they went outside cheerfully.

Kapteyn enjoyed visiting the family and he became a regular guest. In this home he found the carefree way of living that he had never experienced at his home. There it had only been duty and work, here they did not immediately switch on the lamps to hover over books, but sat in the twilight as long as possible – with the voices of young girls singing and making fun in front of the open window, while all sorts of jokes were invented.

It was a lot of merriment, but Kapteyn did not fail to notice the more serious side of the character of the older girl. The small dark girl with the thick braids and the striking brown eyes possessed a dignified confidence, a caring soul and an absolute lack of self-awareness, so that she did not realize how beautiful she was. Since the girls had never gone to school and all they knew had been taught by a governess, they had never lost their authenticity, which gave them a certain noticeable charm.

One evening, when Kapteyn rang the doorbell of the Kalshoven house, the door was opened by Elise, who radiated happiness and joy. 'What happened?', he said, moved by her happiness. 'Oh, I am so happy, my brother [Jacques] has returned from the Indies.' Then he realized he wanted no other person to be his wife than this woman, who was the picture of happiness and, unlike many others, did not require to be made happy first. He had thought much about life, and made clear for himself what it was that constituted value and what not. His unique, genuine common sense had afforded him discernment and insight at an early age, when to others life was still a chaos of impulses and temptations. And he kept the picture of this young radiant girl in his heart, abiding his time.'

Chapter 3
Astronomer in Leiden

> *No known roof is as beautiful as the skies above.*
> Mícháel Ó Muircheartaigh (1930–present).[1]

> *... if it happened to clear off after a cloudy evening,*
> *I frequently arose from my bed at any hour of the night or morning*
> *and walked [...] to the observatory to make some observation ...*
> Simon Newcomb (1835–1909)[2]

3.1 Military Service

After obtaining his doctor's degree, Kapteyn had to decide what to do next. He did not have to perform his military service. In Fig. 3.1 we see a copy of an decision stating that he was exempted from military service, as he had offered a substitute in his place.

In the Netherlands, military service (or conscription) had been instituted by the French occupation in 1810. Dutch men served in the French army; they were selected by drawing lots. After the Kingdom of the Netherlands had been proclaimed in 1813, conscription did not cease to exist, however. The selection of those required to serve in the army continued to take place on the basis of drawing lots.

---

[1] Mícháel Ó Muircheartaigh, Irish sports commentator, particularly of Gaelic games.

[2] Simon Newcomb, American astronomer and contemporary of Kapteyn. Contrast this with the statement: *Sunday Nov. 9, 1828. A most beautiful day and splendid evening. I had intended to observe a number of small stars, but on walking home after dining in Hall I found myself totally unable to observe, and did nothing.* Discovered in Astronomer Royal Airy's manuscript journal in Cambridge, according to the department 'Here & There' in 'The Observatory' [1].

Fig. 3.1 Certificate showing how Kapteyn performed his obligation for military service, filed in the archives of Utrecht together with his marriage certificate. See text for details (From FamilySearch [2])

In the eighties of the nineteenth century about one in three young men had to report for actual military service. The drawing system remained in use until 1938.

Until 1898 it was possible to put forward a *'remplaçant'* or a replacement, someone who had been released by the draw, but was being paid to take the place of a person who had been selected. The system of drawing lots and replacements has been described extensively, but unfortunately only in Dutch [3]. Among the upper classes it was common practice to have such men take one's place in the army. Apparently Kapteyn did this as well. Remplaçants were often young men who had ended up free from military service, sometimes because they had drawn a 'lucky' number in the same draw, a so-called 'number-exchanger', or otherwise a volunteer unrelated to that particular draw, called a 'substitute'. This was an even more advantageous situation, because it automatically exempted the brother of the substituted person from military service. Having a remplaçant required substantial funds, especially in the case of a substitute, and was obviously only reserved for the higher classes; the costs were between 600 and 800 Guilders, or the equivalent a purchasing power of 14,000 to 19,000 € and to about two years of labor by an unskilled worker. The procedure involved a formal registration with a notary and the system

3.1 Military Service

gave rapidly rise to a complete industry of intermediaries and procurement agencies. The replacement often received no more than 400 Guilders after all expenses had been deducted. During the years that the replacement system was in force, about one in five of the young men who were 'selected' for military service, actually made use of it. In addition, the system was increasingly abused, for example by falsifying records so that men shorter than the minimum height for military service (1.65 meters) could be used. The system was discontinued in 1898, because of changing social awareness and growing discontent with the power of the intermediaries. The military also supported this abolition, arguing that the Prussian army – considered to be the strongest in Europe – had no system of replacements, unlike most European countries, and therefore did not end up with recruits entirely unsuitable and useless as combat soldiers.

In the nineteenth century, at any rate around Kapteyn's time, military service lasted 5 years nominally, one of which had to be spent in actual military service. After that the men were required to return twice for a period of a month or so [4]. Those who had accomplished their military service, together formed a substantial reservoir of men who could at any time be mobilized to form an army in case of a threat of war.

The document in Fig. 3.1 says among other things that the Royal Commissioner for the province of Gelderland (in which Barneveld was located) certifies that Kapteyn had been drafted into the army in 1871, that in the draw he had ended up with number 5 (a low number, which meant that chances were very high that conscription would automatically follow), but that he had put forward a replacement, who had been drafted in the 8th infantry regiment on May 12, 1871, and who had voluntarily signed a contract with the *koloniaal werfdepôt* on April 23, 1875. This 'koloniaal werfdepôt' was the department in charge of recruiting soldiers to serve in the 'Royal Netherlands East Indies Army'. The document in Fig. 3.2, which reports Kapteyn's draw for the draft, also records (not shown here) that the name of the remplaçant was Johannes Anthonie van Winden, a peddler or market vendor by profession.

The document is dated May 27, 1875, while the designation of a replacement had started in 1871. May 1875 was the month before Kapteyn defended his thesis. This may have been purely accidental, but perhaps he needed to produce the document in order to be allowed to defend his thesis, or to obtain the position he hoped to find (he was employed by the University of Leiden in October 1875; see below). It served as a final statement that Kapteyn had fulfilled the requirements of the conscription and was free to do whatever he wanted. This document was attached as an appendix to Kapteyn's marriage certificate (he got married in July 16, 1879). The fact is that at that time young men had to prove that they had fulfilled the requirements for military service if they wanted marry. Kapteyn had obviously acquired the document long before the date of his marriage.

Fig. 3.2 The entry for Kapteyn in the Register for the draw of the 1871 draft for young men from Barneveld (From Lotingsregister voor de Nationale Militie, 1871, [5])

3.2 Kapteyn's Height

Before continuing, a few words on Kapteyn's height. Interesting, perhaps, particularly in the light of relevant military service-related records. The question comes up because Cornelis Easton in his publication *Persoonlijke herinneringen aan J.C. Kapteyn* (see Appendix C) describes Kapteyn as very tall. His first meeting with Kapteyn was in 1894, when Kapteyn had come to Dordrecht as an external examiner of highschool exams. Easton was editor-in-chief of the local newspaper *de Dordrechtsche Courant* at the time. Easton writes: 'How tall he is! He seemed to be looking over the heads of almost all his peers and that impression was strengthened by his upright head on a long neck. Already then his eyelids where long and hanging low, which could give the impression that he looked down upon the world, not just from high up, but also somewhat disparagingly.'

However, photographs of Kapteyn at meetings or conferences do not at all give the impression that he was particularly tall compared to his contemporaries (see e.g. Figs. 6.18 or 13.18). Furthermore Henriette Hertzsprung-Kapteyn in the *HHK biography* has two paragraphs that are relevant and supply rather different information. The first is:

'The afternoons were for the contacts with the rest of the world: his lectures, discussions, exams, his walks, including his usual stroll on Monday afternoon with his trusted friends [Gerard] Heymans and [Ursul Philip] Boissevain. The three of them could be seen every Monday walking along the Harenschen weg. Heymans in the middle, the tall imposing figure in a pelarine coat with a distant view in his eyes and his mind rising above humanity. Boissevain, who was small, but always moving and full of lively interest. Kapteyn on the other side, slim and entertaining, enjoying the open air and the interesting things that each of them told about their work, while at the same time noting the songs of the birds and recognizing them.'

This seems to imply that Heymans may have been exceptionally tall and Kapteyn more or less of average height. The second reference in the *HHK biography* is:

'Mother looked at them with joy, as she was standing in front of the house when they left arm in arm for the Laboratory (the two small rooms had this grandiose name in the Kapteyn family). Gill talking loudly and gesticulating, stared at by the Groningen citizens, Kapteyn, short and modest, quietly happy next to him.'

This does not seem to support Easton's description.

So, what can we say about Kapteyn's height? We can use the bricks in the wall behind him and his daughter Henriette in the photograph of Fig. B.1. Kor Holstein of *Holstein Restauratie Architectuur*, a business specializing in the renovation of old buildings, informs me as follows: 'The details of the pilaster, the parapet of the large window and the window frame all point clearly at a late nineteenth century construction. The masonry confirms this. If the photograph is from the province of Groningen, ten layers of brick would correspond to 480 to 520 mm (brick plus masonry joint). Each brick is between 45 and 49 mm. Bricks have become 'higher' in the course of time, early nineteenth century brick being 435 mm and mid-twentieth century brick 50 mm. Currently they usually measure 50 mm.'

Kapteyn measures 38 layers of bricks. With bricks of 45 mm each Kapteyn is 182 cm tall, with 43.5 mm this is 176 cm. A few centimeters should be deducted if allowance is made for the heels of his shoes. According to the Netherlands Central Bureau for Statistics the height of an average man towards the end of the nineteenth century was about 165 cm. Henriette's height, allowing for her high heels, would be 170 and 165 cm respectively, compared to an average of 155 cm at the time.

There is one known request by Kapteyn for a passport (which states a person's height nowadays), but all it mentions is that he needs it to travel to Germany in 1915 (as it turns out, to visit the same daughter as in the picture and son-in-law, the Hertzsprungs, in Potsdam) and gives no further details. Men in the Netherlands were measured as part of the medical that determined whether they were suitable for military service. Luckily, whereas all records for the draft registrations and draws in the Province of Gelderland were lost, the ones for Barneveld, in the archives of that municipality, have survived. The entry for Kapteyn is shown in Fig. 3.2, and states that he was 1 meter and 715 millimeters tall. The corresponding records for the (1871) cohort of men from Barneveld (a total of 56 persons) state the height of 53 of them. They range from 152 cm to 185 cm and the median is 167.2 cm. Of the 53 men, 11 are taller than Kapteyn.

Kapteyn (as well as his daughter) seems to have been slightly taller than average, but there is no reason to think that he was exceptionally tall. The background of Easton's remark is not clear; maybe he was somewhat short himself or misled by the fact that Kapteyn was rather thin.

3.3 Looking for Employment

Relieved of the obligation to spend time in the army, Kapteyn was free to look around for employment. Before discussing this period in some more detail, I first quote the relevant paragraphs from the *HHK biography*.

'So then his wonderful years as a university student drew to an end. He had worked hard, passed his doctoral exam [This is the equivalent of the current Masters] *cum laude* and on the 24th of January 1875 he defended his PhD thesis *A study of vibrating flat membranes* and became a doctor of mathematics and natural philosophy. [Obviously, Henriette Hertzsprung-Kapteyn makes a mistake here; the defense took place on *June* 24, 1875.] Strangely enough it only now dawned upon him that he would have to decide what profession he wanted to choose. His father had been looking around for him and knew where he could get a job as a teacher. But he would have nothing of that. A schoolteacher, no; he felt this to be neither his calling nor his talent.

His father was extremely put out over this refusal; after all, the Kapteyns were all born to be teachers and the time had come for him to realize that he now had to find his way in life by himself. But the young man persisted; he wanted to work in a scientific environment and thought he would be able to find a way to accomplish that. He had heard that a new astronomical observatory was going to be founded in China, so he contacted the people in charge, who invited him to come to Peking to become the director. He accepted enthusiastically, there was some correspondence and an exchange of ideas, but then it all came to naught, as the observatory was never built. He then heard that a post as observator would become available at Leiden Observatory. He applied and was offered the position. In this way he returned to the passion of his youth, and full of determination he accepted. He remained at the Observatory in Leiden for two years and all this time he worked with youthful vigour, fortitude and determination that was so typical of him. 'I never worked harder in my whole life', he said later. He never took a rest, he observed at night whenever the weather allowed, usually came to work in the morning before all the others and lived so intensely in this grand science that in those two years all sorts of plans grew in his head that never released their grip on him and on which he worked the rest of his life.'

Other accounts tend to describe Kapteyn's appointment at Leiden Observatory and his entry into the world of astronomy as accidental more than anything else. For example, van Maanen (*J.C. Kapteyn, 1851–1922* [6]) writes: 'From 1869 to 1875 Kapteyn was a student at the University of Utrecht, where his principal teachers were Buys Ballot and Grinwis, so that it is no wonder that his doctoral thesis was in physics: *Onderzoek der Trillende Platte Vliezen*. Just at this time, however, the position of observator at the Leiden Observatory was vacant, and Kapteyn applied for and obtained the position. By this accidental circumstance astronomy secured the privilege of counting Kapteyn as one of its workers and before long one of its foremost leaders.'

Pannekoek (*J.C. Kapteyn en zijn astronomisch werk* [7]) describes the situation along similar lines: 'It was by pure accident that Kapteyn ended up in astronomy. After completing his studies he was looking for a scientific job; he wanted to work in some field as a researcher of natural science, without knowing at first which

3.4 Leiden Observatory

Fig. 3.3 Leiden Observatory in 1864 (Archives Leiden Observatory)

discipline. He applied for example for a position as meteorologist in Batavia, which post however, had just been filled. He also started negotiations about a job in China; but then, when [...] the position of 'observator' became vacant at Leiden Observatory he accepted an appointment in 1875.' Batavia was the capital of the Dutch colonies in the East Indies on the island of Java. It is the current Jakarta, capital of Indonesia.

It is by no means unlikely that Kapteyn looked for employment outside the field of astronomy. Indeed, the number of research jobs in Utrecht and Leiden (the only two universities with an observatory in the Netherlands) were extremely limited and even if he had had a primary wish to become an astronomer, he would have to be prepared to leave the country and work abroad. His early interest in astronomy, as evidenced for example by his 'star map' (see Fig. 1.9), the abundance of propositions on astronomical matters with this thesis and by his apparent willingness to move to China in order to work in an astronomical environment, suggest that he had a strong preference for working as an astronomer.

3.4 Leiden Observatory

Before discussing Kapteyn's activities in Leiden, we first need to look into the question what the place where he worked looked like and therefore I will treat in some detail the history, facilities and ongoing research at Leiden Observatory. It was founded in 1632, when a quadrant, originally constructed by Willibrord Snellius (1580–1626), was bought by Jacobus Golius (1596–1667) and installed on top of the central building of Leiden University. A quadrant is an apparatus that can measure angles accurately; it had been used by Snellius to perform a new measurement of the radius of the Earth. The way this works is as follows: From two positions on Earth that are on the same meridian (i.e. if they are exactly north-south with respect to each other) we first need to know their distance. If we then also measure the altitude

Fig. 3.4 Leiden Observatory as it appeared on the frontispiece of the Annals of Leiden Observatory, volumes 1 through 8 (1868–1902) (Archives Leiden Observatory)

of the pole above the horizon (see page 65 for a description how in principle that is done) in both places, we calculate the difference between the two. This angle is a fraction of the full 360° and the distance between the two places is the same fraction of the Earth's circumference. This method, first applied by Eratosthenes of Cyrene (276BC–195BC), is referred to as triangulation. To this end, Snellius performed measurements of the angle between two church spires from one another as seem from a third tower. In this way he measured the shape of the triangle between these three towers and if one side of the triangle is known, the others can be calculated. In this way he eventually found the distance between the cities of Alkmaar north of Amsterdam to Bergen op Zoom somewhat north-east of Antwerp. His measurement of 107.3 km is actually too small by about (only) 4%.

It took more than two centuries before the famous 'Sterrewacht' was built under the leadership of Frederik Kaiser (1808–1872); it was officially opened in 1861 (see Fig. 3.3). Kaiser (see Fig. 3.5) is seen as the modernizer of astronomy in the Netherlands, which was still relatively insignificant when he was appointed. For descriptions of the history of Leiden Observatory, see for example H. Kleibrink & J.H. Oort, *Honderd jaar Leidse Sterrewacht*; G. van Herk, H. Kleibrink & W. Bijleveld: *De Leidse Sterrewacht: Vier eeuwen wacht bij dag en bij nacht* or F.P. Israel: *De Leidse Sterrewacht: Glorieus als vanouds* respectively [8].

Kaiser was born in Amsterdam in an immigrant family from Germany. His father died when he was eight (for an excellent introduction to Kaiser, see Hans Hooij-

3.4 Leiden Observatory

Fig. 3.5 Frederik Kaiser (1808–1872), director of Leiden Observatory and initiator of its new building (Archives Leiden Observatory)

maijers: *Een passie voor precisie: Frederik Kaiser en het instrumentarium van de Leidse Sterrewacht* [9]) and his upbringing was taken over by his uncle Jan Frederik Keyser (1766–1823), who also lived in Amsterdam but had changed his name into the Dutch equivalent. This uncle gave him a thorough education, including much astronomy which was one of his hobbies and passions. Professor Gerard Moll of Utrecht, who also owed some of his education to Keyser, pushed Kaiser forward for the position of 'observator' at Leiden Observatory. In spite of his lack of formal education, Kaiser was duly appointed in 1826. An observator belongs to the scientific staff, working as an observer at the observatory under the direction and responsibility of the director and professor. A formal translation of observator into English would be 'observer'; however, I prefer to use the term observator when it refers to a formal appointment at an astronomical observatory. The observator had no duties in the educational program, his primary task being to perform observations and reduce the data into publishable form. Kaiser went to university and obtained the 'Candidaats' degree (comparable to a Bachelor's degree) in 1831. He made quite a name for himself when Halley's comet was due to return in 1835. He revised calculations made earlier by renowned scientists and predicted the time of passage of the perihelion to an astounding accuracy of 1.5 hours, where many celebrated astronomers had been wrong by much larger time intervals, sometimes even 9 days. [The perihelion is the point in the orbit of a solar system object where it is closest to the Sun.] In the same year Leiden University promoted him to Doctor *honoris causa*.

When the director of the observatory, Pieter Uylenbroek (1797–1844), resigned in 1837 to become full-time director of the physics department, Kaiser was appointed in his place. In 1840 followed an appointment as extraordinary professor of astronomy and in 1845 as full professor. But it was not until 1853 that a new observator was appointed, Jean Oudemans. However, Oudemans left again in 1856 to take up a professorship in Utrecht and his successor, Martinus Hoek did the same thing in 1859. His place was taken by Nicolaas Mattheus Kam (1836–1896), who remained active as an astronomer, albeit in his spare time after he took up a position as teacher at the 'Hoogere Burgerschool' (HBS) in Schiedam in 1869. This was the new type of secondary education introduced in 1863 as an alternative to the gymnasium. At the HBS no Greek and Latin were taught, its aim being to educate young men for leading positions in industry and commerce. Eventually it offered the possibility to enter university and the major focus on mathematics and science at the HBS possibly accounts for the unusually large number of Dutch Nobel laureates in the early twentieth century (see *De tweede Gouden Eeuw: Nederland en de Nobelprijzen voor natuurwetenschappen 1870–1940* or *Origins of the Second Golden Age of Dutch Science after 1860* by Bastiaan Willink [10]). This HBS in Schiedam is the same one, incidentally, where I myself received my secondary education.

Kaiser felt very restricted in his work by the location of the observatory on an observing platform of the central Academy Building. The position was far from ideal when it came to the stability of his instruments, and he appealed regularly to the Curators, the governing body of the university, and to the Government, to provide funds for the construction of a dedicated observatory building on solid ground. Curator can also be translated as 'trustee'; however, I prefer a translation as close as possible to the original Dutch term. Kaiser often referred to the building of the Pulkovo Observatory near St Petersburg as a model for what he envisaged for Leiden. At the end of the day his efforts paid off and funds became available from a whole range of sources. But the major share had to come from the Government and in 1857 the minister (of Internal Affairs, who was responsible for universities) finally reserved sufficient funds in his budget to enable Kaiser to realize his dreams. This minister was Gerrit Simons (1802–1868), himself a student of Gerard Moll in Utrecht. However, this budget was voted down in Parliament for other reasons and further frustrating delays followed. But then the new minister, Anthony Gerhard Alexander Ridder [Knight] van Rappard (1799–1869), who succeeded Simons in 1856 when the latter's health deteriorated, happened to be a strong supporter of Kaiser, and the costs for the construction of a Leiden Observatory were finally securely included in the government's budget.

Then things moved faster and the observatory was in place by 1861 (see Figs. 3.3 and 3.4). Kaiser's main activity in astronomy concerned positional astronomy or astrometry. This indeed was central in the research of the cosmos at the time, astronomers being very much occupied with measurements of positions of stars, planets, comets and increasingly of minor planets (see Box 2.1) to find their orbits. Minor planets, whose numbers were increasing, were at the risk of being lost again soon after their discovery, if their orbits were not determined relatively quickly. This was a real threat until in the 1890s the Astronomisches Rechen-Institut in Berlin under Friedrich Tietjes (1834–1895) and Julius Bauschinger (1860–1934) took the responsibility of keeping track of asteroids and started the publication of circulars with positions, etc.

3.5 Positional Astronomy

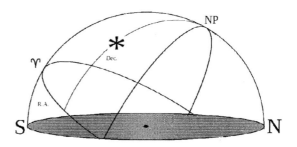

Fig. 3.6 The coordinate systems in positional astronomy

The new observatory's main instrument was the meridian circle, but there was a variety of other telescopes and instruments. Before continuing with an inventory of what Leiden Observatory looked like when Kapteyn joined it, it may contribute to a good understanding of what follows to make a short excursion into the fundamentals of observational positional astronomy and its context in the middle of the nineteenth century.

3.5 Positional Astronomy

Positional astronomy is concerned with measuring accurate positions of stars and other objects on the sky. In Fig. 3.6 the sky is depicted as a sphere with the observer in the center. The horizon is the circle where the horizontal plane (the Earth's surface) cuts through the celestial globe and the meridian is the half-circle on it going from the south on the horizon through the zenith and the celestial pole (NP) to the north point back on the horizon (the situation depicted here applies to the northern hemisphere). The Pole is at an angle above the northern horizon that is equal to the geographical latitude of the observer. The line 90°a way from it is the equator, it crosses the meridian at an angle of 90° minus the latitude. In astronomy, positions are expressed in terms of *Right Ascension* (R.A.) and *Declination* (Dec.) as the fundamental coordinates. These are defined much like those we use for positions on Earth, where geographical longitude now is Right Ascension and geographical latitude is replaced by Declination. For a more complete presentation of astronomical coordinate systems I refer to Box 3.1.

There are two fundamental types of measurements of astronomical positions. The basic one is the *absolute* measurement of such positions to construct so-called *Fundamental Catalogs*, the positions having been measured directly and absolutely with respect to the zero points of the coordinate system (see Box 3.3). The other is of course *relative*, meaning that a star's position is measured with respect to that of one or more stars of known (absolute) positions. In principle, the measurement of absolute positions is straightforward and deceptively simple, viz., by observing the star cross the local meridian. Let us start with the declination. Imagine that the celestial sphere has rotated in the figure until the star is at the meridian. As can

Box 3.1 Astronomical coordinate systems
Celestial positions of stars are designated in coordinate systems on a sphere, much like geographical longitude and latitude (see Fig. 3.6). The sky is viewed as a spherical dome (also called the celestial sphere) and one could then use the angles *azimuth*, measured along the horizon, usually from the north, and *altitude*, the angle above the horizon, to define an object's position. However, these angles change with time and with position on Earth.
The fundamental coordinates for an object are based on the rotation axis on the Earth, which gives celestial poles (seen in the zenith when standing at the corresponding Earth pole) and the equator. The coordinates are *Right Ascension (R.A. or α)*, measured along the equator, and *Declination (Dec. or δ)* as the angular distance from the celestial equator. The zero point of R.A. is the *vernal equinox*, which is the position on the sky where the Sun crosses the equator at the time of the northern hemisphere start of spring (roughly 21 March). Figure 3.6 depicts the situation when the equinox is on the meridian. R.A. is measured anti-clockwise and is expressed in hours (the full circle of 360° corresponding to 24 hours), minutes and seconds.
The Sun moves along the celestial sphere, following a great circle that is called the *ecliptic*. This makes an angle of about 23° with the celestial equator. The equinoxes are the points of intersection of the equator and the ecliptic. Using the ecliptic as the great circle, we have *ecliptic longitude* (usually λ) with also the vernal equinox as zero point and *latitude* (β). The angle between ecliptic and equator (the *obliquity*) varies over time between about 22°.5 and 24°.5 because of *nutation* (see below). Since the planets in the solar system move in planes very similar to that of the orbit of the Earth (i.e. the plane of the ecliptic), these coordinates are especially useful for Solar System studies.
For studies of the Galaxy, the coordinate system is based on the Milky Way. The corresponding coordinates are *Galactic longitude* and *latitude* (l and b). The Galactic equator has been redefined since Kapteyn's day, but the difference is small. The longitude zero point is the direction of the Galaxy's center as defined by the IAU (see Box 16.1); in Kapteyn's day the zero point was the intersection of the Galactic and celestial equator (33° different). The Galactic equator makes an angle of 62°.9 with the celestial one.
The position of the equinox changes with time due to the *precession* of the Earth's rotation axis like the wobbling motion of the axis of a spinning gyroscope a flat surface om Earth. This leaves the obliquity unchanged, but the equinoxes move along the equator in about 26,000 years; a shift (the *precessional constant*) of about 50 arcsec per year. Precession was discovered around 130 BC by Hipparchus of Nicaea. The obliquity varies by small amounts as a result of perturbations by the Sun and planets and this is called *nutation*.

be appreciated from the figure, the star crosses the meridian at an angle above the horizon (the altitude) that equals the sum of the altitude of the equator above the southern horizon and the declination of the star.

In the meridian the angle between the celestial equator and the southern horizon equals 90° minus the geographical latitude. So once that latitude is known, one just needs to measure the altitude (angle above the horizon) of a star at its highest crossing of the meridian. Subtracting (90°-latitude) then gives the declination [for a meridian crossing below the pole one would have to add (90°-latitude)]. This could

3.5 Positional Astronomy

Fig. 3.7 The meridian circle at Leiden Observatory as it can be found in the *Album Amicorum* presented at the retirement of H.G. van de Sande Bakhuyzen as professor of astronomy and director of Leiden Observatory in 1908 (Archives Leiden Observatory [11])

actually be the 'Pole Star' *Polaris* itself, which is a bright star that is by accident located very closely to the celestial north pole (at present about three quarters of a degree away from it). Observe the altitude of that such a circumpolar star as it passes the meridian 'above' and 'below' the pole and take the mean. For this you have to wait twelve hours, so this can only be done in winter, when the nights are long.

The angle between the pole and the southern horizon is 180° minus the geographical latitude. So, determine the altitude (angle above the horizon) when the star crosses the meridian and subtract it, and the answer is the declination. Obviously, it is sufficient for this to have a telescope that always looks at the meridian and therefore only needs to move around one axis that is perpendicular to the plane of the meridian. Such a telescope is called a meridian circle.

The measurement of right ascension is somewhat more complicated. Figure 3.6 shows the situation when the vernal equinox is on the meridian. Due to the rotation of the Earth it will move through a full circle as the Earth makes a full rotation with respect to the stars. This is different from our solar day, however, as the latter is based on return of the *Sun* on the meridian and the Sun after all moves with respect to the stars. Astronomers define this time interval when the equinox (and also every star) returns to the meridian as the sidereal day. This lasts 23 hours and 56 minutes and 4.091 seconds on our (solar) clocks. This 'sidereal time' is set to zero when the vernal equinox is on the meridian and a full day of 24 *sidereal* hours then lasts until it has returned there. At any moment the sidereal time is different for two places on Earth with different geographical longitudes. In their domes, astronomers used to have special clocks that showed the sidereal time; nowadays sidereal time is displayed electronically. Now, from the geometry in Fig. 3.6 it can be deduced that the right ascension of a star is equal to the sidereal time as it crosses the meridian. So calibrate a sidereal clock and read off the exact time the star passes the meridian on that clock. This manner of measuring the right ascension is the reason it is expressed in hours, minutes and seconds rather than degrees etc. Again all we need as far as a telescope is concerned, is one that only moves in the plane of the meridian, a 'meridian circle'. Figure 3.7 shows the meridian circle that Kaiser obtained for Leiden Observatory.

The observer's procedure of measuring the time involved counting the ticks of a (sidereal) clock. This is not very accurate, and around the time Kaiser built his observatory this was replaced with electrical devices that recorded the ticks and the signal from the observer of the star's meridian passage, where the observer pressed a key, much like those used for Morse code telegraphs. This not only gave an improved intrinsic accuracy, but also removed an important 'personal' bias and added to intrinsic accuracy. To improve accuracy the telescopes were often fitted with a set of parallel cross-wires at equal spacings, so that the median passage could also be measured as the mean of the times the star passed two wires at equal distances from the one in the center of the field. Usually there were five cross-wires, together giving three independent measurements of the time of meridian passage.

In practice the measurement of fundamental coordinates (especially declination) is much more complicated due to a number of effects. One is the obvious problem that the weight of the telescope causes flexure in its tube, so that the altitude at meridian crossing is incorrect. There are two more effects, which have been described in some more detail in Box 3.2. Atmospheric refraction is the phenomenon that the path of the light from the star is bent when it passes through the atmosphere. The other has to do with the fact that the Earth moves in space and is called aberration . The latter is not difficult to correct for, but although refraction in the atmosphere was understood in the days of Kapteyn, the actual amount was not known with sufficient precision.

The second type of study of positional astronomy is that of *relative* positions, which can be used not only for stars, but also for planets, comets or minor planets for example, where positions are measured as angular distances on the sky with respect to stars of known coordinates. Obviously, this is only possible in practice when the object can be seen in the field of view of a telescope together with a star, or preferably more than one star. This requires a telescope that can be pointed at all possible positions on the sky, and such telescopes have so-called *equatorial mounts*; these are telescopes that can move along two perpendicular axes, one of which is pointed towards the celestial pole. In this manner the daily motion of the sky as a reflection of the Earth's rotation can be followed by moving the telescopes around this axis only. The other axis is used to set the telescope at the declination of the object.

To measure angular distances between objects (e.g. a planet and a background star with a known position) in the field of view, the telescope then needs measuring apparatus in the focal plane. Traditionally this could be a set of movable

Box 3.2 Refraction and aberration
Measurements of star positions are affected by refraction in the atmosphere and by aberration of star light.
Refraction of starlight in the Earth's atmosphere arises from the fact that the refractive index of air is slightly higher than 1 (the index for vacuum), that the air density changes with height, and that the atmosphere follows the curvature of the Earth. Together this gives rise to a curved path of the light through the atmosphere, which results in the apparent position being slightly higher in the sky. The effect changes with altitude (height above the horizon), being zero at the zenith and maximum on the horizon. At optical wavelengths it amounts to about a minute of arc at 45° altitude. This effect had been known for quite some time, but when Kapteyn entered astronomy its magnitude was not known with sufficient accuracy.
Aberration of light results from the fact that the speed of light is finite (300,000 km/sec), while the Earth goes around the Sun with approximately 30 km/sec. Usually the effect is likened to the case of raindrops falling on a window of a moving train; even if the drops fall vertically they come down at a slanted angle on the window. The effect results in a shift of the star's position because the telescope has moved in the transverse direction during the finite time it takes for the light to travel the length of the telescope tube. The effect is maximal in the direction perpendicular to the Earth's motion and of the order of a ten-thousandth (30/30,000) of a radian or about 20 arcsec. It was discovered by James Bradley in 1725.

cross-wires to position one on the planet and the other on a background star, in such a way that micrometer readings tell their distance in the focal plane; from this the angular separation on the sky can be calculated. The orientation of the set of cross-wires can then be used to find the direction of the arc between the planet and star. In Kaiser's days, there were telescopes with built-in devices which produced two separate images, which by adjusting the optics in the micrometer in the focal plane could be moved into coincidence; the angular separation could then be calculated from reading the settings of the system. Usually this was achieved by splitting the objective lens in two, so that they could be moved with respect to each other. This was called an *heliometer*, since the first design was to measure the variation of the diameter of the Sun during the year. An increasingly important application – in addition to measuring the positions on the sky of planets or minor planets in this manner – was that of mapping the orbits of stars around each other in binary systems.

3.6 Kaiser and van de Sande Bakhuyzen

The construction of the Leiden observatory building was completed in 1860. The total costs had originally been estimated at well over 100,000 Guilders (this is equivalent to a purchasing power of about 2.2 million € today), as Kaiser writes in his communication *Nachrichten über die neue Sternwarte in Leiden* to the German journal 'Astronomische Nachrichten', where he announced the completion of his observatory to the astronomical world [12]. He does not state the final costs. His major instrument was the meridian circle (see Fig. 3.7), which had arrived in 1861; it had been constructed by the German firm of *Pistor & Martins* from Berlin. The diameter of the aperture lens was 15 cm and the focal length about 2 meters, which means it was a state-of-the-art instrument when it was delivered. It served as the main instrument of Leiden Observatory until it was finally decommissioned and taken apart in the fifties of the twentieth century; it has been in the Boerhaave Museum in Leiden since. There were two telescopes in the domes of the Observatory, which could be used for measuring relative positions of celestial object, the largest being a new 18-cm telescope (from the well-known firm of *Merz & Sons* in Munich) in addition to the older 15-cm telescope from the same company that had been acquired in 1838. Essential for positional astronomy is the availability of excellent clocks. The main clock was one that gave sidereal time; it had been constructed by the famous Andreas Hohwü (1803–1885), a Danish clock-maker who had settled in Amsterdam. Furthermore there was a large suite of smaller instruments for astronomy and geodesy.

In 1868 Kaiser obtained funds to start the publication of the series *Annalen van de Sterrewacht te Leiden*, in which many of the observations were published. The first volume opened with a description by Kaiser of the history of Leiden Observatory, a summary of its status and an inventory of the instruments in 1868 and a description of the meridian circle [13].

After the new observatory had been completed, Kaiser requested the appointment of a second observator. He had a young student, Hendricus Gerardus van de Sande Bakhuyzen (1838–1923) in mind, but the request was denied. Van de Sande Bakhuyzen had already done some voluntary work for Kaiser between 1861 and 1862, but had left and become a teacher at a secondary school. The position was allowed towards the end of 1863 and Kaiser chose for Andreas van Hennekeler, who eventually (in 1868) left as well to go into teaching. Kaiser complained much about the problem of finding suitable candidates for his observator positions; students were much more interested to finish their studies as quickly as possible so as to be able to work in secondary education. In the end he had to resort to German astronomers and in 1868 he appointed Karl Wilhelm Valentiner (1845–1931).

Valentiner took over as temporary director when Kaiser died in 1872 until van de Sande Bakhuyzen was appointed as Kaiser's successor a few months later. Under Kaiser, Leiden Observatory established itself as a major center for positional astronomy, not in the least as a result of Kaiser's extraordinary ability to determine extremely accurate positions of celestial objects. This prominent international reputation was expressed in the fact that the *Astronomische Gesellschaft*, the leading astronomical society of continental Europe, decided to hold its bi-annual meeting in Leiden in 1875, the first time it convened outside of Germany and Austria.. This brought some 30 leading astronomers, mainly from Germany and Austria, to Leiden [15].

Van de Sande Bakhuyzen appointed his younger brother Ernst Frederik (1848–1918) as observator. By that time the latter had not yet completed his PhD thesis, which he finally did in 1879 on the subject *Bepaling van de helling der ecliptica, uit waarnemingen verricht aan de sterrenwacht te Leiden* (Determination of the obliquity of the ecliptic from observations obtained at Leiden Observatory). But prior to this – in 1875 – Valentiner left to take over the directorship of the Mannheim Observatory and the resulting vacant position was the one that was filled by Kapteyn. For the sake of completeness I mention that the Curators had allocated funds to appoint an 'assistant' on a permanent basis to make all the necessary calculations for the reduction of observations and other matters to alleviate the workload of the observators. In March 1876, Dr. A. Eecen was appointed in this position. Adrianus Eecen (1846–1918) was a mathematician, who wrote a thesis *Over de torsie van een elliptischen cylinder* (On the torsion of an elliptical cylinder). He left in December 1877 to become a teacher at the gymnasium in the city of Assen (my current residency, some 25 km south of Groningen), and was replaced by Thomas Joannes Stieltjes (1856–1894). According to van de Sande Bakhuyzen in his Annual Report, Stieltjes was 'a very promising mathematician'. He was to become a famous mathematician afterwards, his name surviving as one of the eponyms of the 'Riemann-Stieltjes integral' in mathematics.

3.7 'Observator' at Leiden Observatory

The archives of Leiden Observatory are kept in the University Library (see the article *Leiden Observatory Archives* by David Baneke [16]). In them we find some information on the circumstances of Kapteyn's appointment, especially the correspondence between van de Sande Bakhuyzen and the Curators. The letters to the Curators are drafts, full of erasures and changes, in van de Sande Bakhuyzen's atrocious handwriting. On June 22, 1875 he informed the Curators of the intended departure of Valentiner and wrote that he hoped to put forward proposals for a suitable replacement soon. The Curators replied on July 28 that the Minister had taken note of the situation and would grant Valentiner an honorable discharge as soon as he had received van de Sande Bakhuyzen's proposal for his replacement. This was made in a letter of August 10, 1875, from van de Sande Bakhuyzen to the Curators, from which I quote: 'In response to the letter of Your Noble Great Honorables of July 28, 1875 (No. 329) I have the honour to inform YNob. Gr. Hon. that Dr. J.C. Kapteyn has been working at the Observatory for some time at my request, so what he has accomplished here places me in a position to form an opinion on his suitability for the duties of an observator at the Observatory. As a result of the experience I have had with him, I believe I can decide that, although he is not yet capable in all respects to perform the tasks that can be demanded from an observator, this will be the case in not too long a period, so that from the various candidates who are under consideration for this position, he should be recommended as the best.'

Fig. 3.8 H.G. van de Sande Bakhuyzen (From Wikimedia Commons, the free media repository [14])

Apparently, Kapteyn had been on probation for some time, although it is hard to see that this can have been for more than a few months. And it appears that there were other candidates as well, who lost out to Kapteyn. The case was not shut and closed, however. Van de Sande Bakhuyzen received a letter from an official of the Ministry of Internal Affairs, which I found in the filed correspondence of van de Sande Bakhuyzen with the Ministry. It is dated September 21, 1875, and the official (probably named Vollenhoven, although the signature is hard to read) reports that the Minister had suggested that 'maybe Dr. Gleuns should be considered as well for the position'. The letter refers to the fact that Gleuns had been involved in calculations for van de Sande Bakhuyzen, but was currently unemployed and looking for a job. Indeed, in the Annual Report for 1875 [17], it is reported first that one Dr. J. de Jong had been employed for performing calculations on special funds from the Ministry and that Dr. W. Gleuns had done the same work for a few months. The person referred to is Willem Gleuns (1841–1906), born in Groningen. The official notes that the Curators had only suggested one name, that of Kapteyn, without further explanation, 'which for a nomination like this is in fact very little'. Indeed, the letter from the Curators requests no more than the honorable discharge of Dr. Valentiner, the appointment of E.F. van de Sande Bakhuyzen as first observator and Kapteyn's as second observator, without further motivation.

The letter from the Minister gives the impression that somebody had been lobbying quietly for Gleuns. Van de Sande Bakhuyzen responds the next day. He is clearly unhappy with the prospect of having Gleuns as an observator, so he responds im-

3.7 'Observator' at Leiden Observatory

mediately and in no uncertain terms. The letter in the Leiden archives is a draft with many deletions and in a virtually illegible handwriting. Fortunately I have been able, albeit with some difficulty, to locate the final in the archives of the Ministry in the National Archives in Den Haag (see Fig. 3.9). Van de Sande Bakhuyzen opens the letter as follows: 'I hasten to give you the further information with respect to the two gentlemen who, according to your letter, are being considered for the position of observator. In my firm belief there cannot be a single moment of doubt when it comes to making this choice.' He states that Dr. Gleuns had not done any astronomy during his mathematics studies in Groningen, which is no surprise considering the lack of instrumentation there and the little amount of astronomy that was being offered at the time of Prof. Mees's predecessor. We will see later (page 92) that the predecessor of Mees, Prof. Enschedé, was not greatly interested in astronomy, although he was responsible for the teaching of it. Gleuns had inquired if he could do some calculating work at Leiden and had been accepted 'since there was much to do and both observators happened to be abroad.' However, van de Sande Bakhuyzen notes, this work could be performed even without any knowledge of astronomy, since it was done according to some fixed recipe and 'it is only necessary to know how to use logarithms'. Gleuns did finish a thesis in Leiden, but 'there was no time for work in mathematics and he had opted for a study on some historical compilation in astronomy'. Books and journals were available for this at the Observatory. It 'has merit, but apart from this the dissertation adds very little. Any critical remark that could betray an indication of astronomical knowledge is totally missing.'

'I certainly do not say this to accuse Dr. Gleuns in any manner – he simply has never been trying to do anything astronomical and is just as little suitable for the position of observator as any other doctor in philosophy, [meaning mathematics and natural sciences] or teacher at a highschool ['Burgerschool' or HBS; see page 62], who, of course, may have excellent qualities in other areas. I have heard that Mr Gleuns has very good qualities as a teacher. I would not be able to work with him as observator.'

I quote the paragraphs devoted to Kapteyn in full:

'Vis-a-vis Dr. Gleuns we have Dr. Kapteyn. During his studies in Utrecht he has, as much as opportunities allowed, devoted himself to astronomy and has observed with instruments at the observatory so that he has obtained considerable familiarity with the use of astronomical equipment. Last year he took up the plan to continue his studies at the Observatory in Leiden, but financial considerations prevented this. Mr Kapteyn stayed on in Utrecht and obtained his degree there in June of this year with an excellent thesis. [Here the official at the Ministry had underlined the 'excellent' and added: 'so not on astronomy?'] When the appointment of Dr. Valentiner created a vacancy, I immediately thought of Mr Kapteyn as one of the candidates for the position of observator and the very positive words I received from his professors Buys Ballot and Grinwis about him made me decide to invite Mr Kapteyn to work at the Observatory in order to decide to what extent he was a suitable candidate for the position of observator.

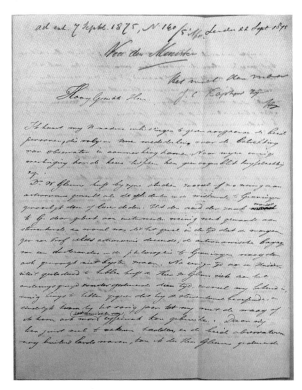

Fig. 3.9 Letter of van de Sande Bakhuyzen to the Minister to support his choice for Kapteyn as observator in Leiden (From 'Nationaal Archief' [18])

Mr Kapteyn accepted this offer and worked here for about 3 months, during which time he has completely confirmed the favourable expectations I had of him from the beginning. Although he is not yet familiar with all the observing techniques he will have to perform, he has shown to be second to no more experienced astronomer.
In theoretical astronomy Dr. Kapteyn has been working in various areas, while his mathematical skills will enable him to fill any gaps in his knowledge in a very short time.
As a result of all this it is with no hesitation and with great confidence that I recommend Dr. Kapteyn for the position of observator.
As to his background I can add that K.'s father used to own a very well-known boarding school in Barneveld and that two of his older brothers are also doctors of philosophy and teachers at highschools ['Hoogere Burgerscholen' or HBS], one I believe in Maastricht, and these are also excellent persons ['flinke lui']. The future observator is still young, I believe about 24 years.'

3.7 'Observator' at Leiden Observatory

The letter brought about what van de Sande Bakhuyzen was aiming for. The Ministry official pencilled on the front page1 'To the Minister. It will have to be J.C. Kapteyn then'. The resigned tone, added to the fact that the Minister himself was to be informed and should read the letter, confirms that there had been some pressure to appoint Gleuns.

This information, which clearly implies that Kapteyn very actively pursued astronomy in Utrecht, and was even prepared to move to Leiden to do a PhD there on an astronomical subject is extremely relevant. It makes it very likely that it simply was impossible for him to do an astronomical thesis in Utrecht and that instead he did his thesis work on an applied mathematics subject with Grinwis, who had already supervised two older brothers. It also shows beyond doubt that Kapteyn did not enter astronomy accidentally and had been preparing himself for an astronomical career as much as was achievable in Utrecht, as I already inferred from his large number of astronomical thesis propositions. Note that the letter was written only three months after Kapteyn defended his thesis, and since he had been on probation in Leiden, this must have started before or almost immediately after this defense at the latest. It could even be that Kapteyn came up with the proposition concerning the Zöllner photometer because he had seen (or maybe even used) one in Leiden. Leiden Observatory at that time had such an instrument; in an article by van de Sande Bakhuyzen, listing the instruments available at Leiden Observatory in 1868 (*Verzeichniss der Instrumente der Sternwarte in Leiden, beim Anfange des Jahres 1868* [19]), there is under the heading 'apparatus for special studies that can be used at various instruments' an item 'astrophotometer and calorimeter after the principle of Zöllner, built in 1867 by Ausfeld in Gotha'. Hermann Ausfeld (1816–1899) of the Gotha Observatory (in Germany often referred to as the Seeberg-Sternwarte) in Thüringen in central Germany, is known to have built more such (and other) instruments.

The reply of the Curators to van de Sande Bakhuyzen on September 29, 1875: 'We have the honour to inform Your Honourable that by decree of the Minister of Internal Affairs of the 28th of this month (No 30, fifth section), Dr Valentiner has been granted at his request an honourable discharge as temporary observator at the Observatory per October 1 of this year, and furthermore that Mr E.F. van de Sande Bakhuyzen has been appointed temporary first observator with an annual salary of 1800 guilders, and Dr. J.C. Kapteyn temporary second observator with an annual salary of 1600 guilders'. For the year 1875, 1700 guilders was equivalent to 4.6 yearly wages of an unskilled worker. For 1600 guilders this is 4.3 years.

In 1877 the salary seems to have gone up. In a letter to the Curators of April 10, 1877, van de Sande Bakhuyzen starts by explaining the operation of the Observatory and the need for observators to work at night and be available when the weather clears in the middle of the night, etc. He then continues: 'to all three of these appointments [so including the one of A. Eecen as computational assistant] is attached the use of a residence at the Observatory exclusive of heating and lighting. In relation to this I have the honour to propose to you that first observator E.F. van de Sande Bakhuyzen keep his salary of 1800 guilders for the moment, under the condition that it will be raised to 2000 guilders when he completes

> **Box 3.3 Fundamental Catalogs**
> These are catalogs of positions and proper motions of stars measured in an absolute manner; i.e., they have been measured as absolute coordinates, not related to those of other stars. An excellent introduction and review has been presented by W. Fricke: *Fundamental Catalogs. Past, present and future* (1985) [21]. The oldest such catalog is that of Hipparchus of Nicaea (ca 190–ca 120 BC), who measured fundamental positions of 850 to 1,000 stars. He was the discoverer of the *precession of the equinoxes* (see Box 3.1). These star positions were later collected and extended by Ptolemy (Claudius Ptolemaeus, ca 90–ca 168) in his famous work the *Almagest*, which was the most influential book and standard for astronomy until its basis of the geocentric model of the Universe) was removed by Nicolaus Copernicus (1473–1543), Johannes Kepler (1571–1630) and Galileo Galilei (1564–1642).
> Modern astrometry dates back to James Bradley (1693–1762), who performed 60,000 observations of 3,000 stars from Greenwich between 1750 and 1765. Bradley was Astronomer Royal and discoverer of the aberration of light and the nutation of the Earth's axis (see Box 3.1). He never reduced his observations into a catalog of fundamental positions, but this was done by Friedrich Wilhelm Bessel (1784–1846), who published the results in a catalog *Fundamenta Astronomiae* in 1818. Later, Canadian-born astronomer Simon Newcomb (1835–1909), director of the U.S. Naval Observatory in Washington D.C., improved the determination of the position of the equinox and published a Fundamental Catalog in 1898.
> In 1888, German astronomer Georg Friedrich Julius Arthur von Auwers (1838–1915) produced a fundamental catalog from observations made at Greenwich and Berlin (Babelsberg) of Bradley's stars between 1836 and 1880. This resulted in a careful revision of Bessel's catalog, published in three volumes of which Volume III of 1888 contained the proper motions of 3268 'Auwers-Bradley' stars.
> Fundamental catalogs served as reference frames for **Durchmusterungs** with relative positions. In 1862, Friedrich Wilhelm August Argelander (1799–1875) completed the *Bonner Durchmusterung*, a catalog of stars north of declination −2°. In 1881 this was extended to declination −23°, also from Bonn, in the *Südliche Durchmusterung* and in 1914 from Argentina to −90° in the *Córdoba Durchmusterung*. These Durchmusterungs were all observed visually and contained about a million stars to roughly magnitude 10.

his PhD thesis, that the salary of second observator Dr. J.S. Kapteyn's salary be raised to 1700 guilders, and that the assistant for the computations Dr A. Eecen keep his salary of 1200 guilders, all three of them with the continued use of their residence.'

3.8 Kapteyn's Work at the Observatory

So, on his arrival in Leiden, Kapteyn found an observatory that was renowned for its accurate astrometric program and was equipped with excellent, first class instrumentation. Together with eleven other observatories, Leiden Observatory took

3.8 Kapteyn's Work at the Observatory

Table 3.1 Astronomical publications in the NASA Astrophysics Data System ADS in 1877

	#	%
Instrumentation	26	4
Observing techniques/photography	29	4
Mathematical techniques	16	2
Meteors	48	7
Moon (incl. occultations and eclipses)	53	8
Sun	23	3
Planetary orbits, solar parallax	109	16
Minor planets	31	5
Comets	98	15
Physical studies of planets	41	6
Positional astronomy (incl. double stars)	92	14
Stellar photometry (incl. variable stars)	35	5
Stellar spectra	7	1
Nebulae	10	1
Geology, meteorology, physics, biology	52	8
	670	100
Announcements, commentaries/columns, errata, obituaries	340	
Annual reports of observatories, etc.	23	
Total	1033	

part in what at that time was one of the most important projects, the *Astronomische Gesellschaft Katalog* or AGK. This project had been initiated by Friedrich Wilhelm August Argelander (1799–1875), who was also one of the founding fathers of the Astronomische Gesellschaft. Argelander had founded Bonn Observatory in 1844 and between 1849 and 1863 he had measured positions of all stars in the northern hemisphere (north of declination $-2°$), complete down to apparent magnitude nine and with an incomplete sample of stars of magnitude ten. The resulting star catalog, known as the *Bonner Durchmusterung*, contained 324,198 stars (see A.H. Batten, *Argelander and the Bonner Durchmusterung* [22].). The AGK was the follow-up with improved positions using meridian circles at various observatories, which was started in 1861 and encompassing about 200,000 stars, complete down to magnitude 9.

Before turning to Kapteyn's work in Leiden, I first look at the astronomical research of the period. To this end I compiled an inventory of the astronomical literature as published in 1877, using the NASA Astrophysics Data System (ADS). In Table 3.1, I collect the counts of publications (both 'refereed' journals as well as 'non-refereed' publications such as Observatory Annals and Annual Reports). The category announcements include publications such as the weekly 'Our Astronomical Column' in *Nature*. I count 670 scientific publications. Of these, more than half concern objects in the solar system. In addition to work aimed to refine the orbit of the Moon from occultations of stars (exact timing and localizing of the moment and position, where stars disappear behind the Moon or – harder still – reappear, can be used to constrain the orbit of the Moon) and to determine orbits of newly discovered minor planets, the effort went largely into the orbits of the planets, but in particular the determination of the 'Solar parallax' (the distance to the Sun).

From measurements of positions of planets or minor planets as a function of time on the sky one determines the orbit of that object. However, only positions on the sky are used here, that is to say *directions* but not distances. So we can find from such data the object's orbit only *relative* to that of the Earth. In other words, we derive a map of the solar system from such research, in principle with orbits of all objects (planets and minor planets, but also comets), but we do not know the scale of the picture. This needs to be determined from independent measurement of the distance between the Earth and any object at any time. This fixes the scale of the Solar System and consequently the size of the Earth's orbit, in other words the (mean) distance to the Sun in meters. This property is also called the Solar parallax. Now this measurement of a distance can be obtained most accurately by using objects that come relatively close to the Earth, and by making measurements of its position on the sky at the same time from two (or more) observatories on Earth. For this the distance between them should be as large as possible (and known accurately). If the circumference of the Earth is known, we then have the length of the base of the triangle formed by these two observatories and the object, and for this triangle we can measure all angles and hence calculate the distance to the object and the scale of the Solar System. This is performed with for example Mars when it is in opposition;, Mars then is on the sky opposite the Sun and therefore as close as possible to us. It can be done even better, using minor planets that happen to have orbits that can take them relatively near to us. The solar parallax can be determined very accurately also, using accurate (timing) measurement from more than one position on Earth, when Venus is moving in front of the solar image. This is rather rare, but occurred in 1874 and 1882, and the Netherlands were involved significantly in this (see below, page 79).

The category 'positional astronomy' refers to research related to measuring accurate positions of fixed stars. This is either work involved in the construction of Fundamental Catalogs (see Box 3.3), but also the mapping of stars in binary systems so that their relative orbits can be determined. This is of fundamental value in the measurement of the masses of stars, which is a necessary property for theories of stellar evolution. Until the end of the nineteenth century very little research was done in the area of properties of stars. The changes in case of variable stars were duly recorded and analyzed to see if these variations were periodic, but no physical explanations could be found. Likewise there were experiments with stellar spectra, but this also was preliminary to any attempt at understanding stellar structure and physics.

The bulk of the observational effort at Leiden was in positional astronomy, for the construction of fundamental catalogs as well as the determination of orbits of stars in binary systems, but also to provide absolute positions of stars that could be used as references for measuring relative positions (in particular of the planet Mars) in efforts to determine the Solar parallax. This was the work in which Kapteyn was going to take part when he was appointed at Leiden Observatory.

Kapteyn was in an excellent position to gain experience with astronomical observing. The Annual Reports that van de Sande Bakhuyzen published (*Verslag van de staat der Sterrenwacht te Leiden en van de aldaar volbrachte werkzaamheden,*

3.8 Kapteyn's Work at the Observatory

in het tijdvak van den eerste Juli 1874 tot de laatste dagen der maand Juni 1875 and similarly for the periods 1875–1876, 1876–1877 and 1877–1879 [23]), are very useful for defining the work Kapteyn did in his Leiden period. In the report for 1874–1875, before Kapteyn joined the Observatory, van de Sande Bakhuyzen reported in detail on the structural work on the building that was necessary for optimal observing. He described the state of the walls of the Observatory, which were very sensitive to humidity, so that 'thick layers of salpeter' (various salts such as nitrates, chlorides and sulfates) would form, releasing small particles which would damage the instruments. Although temporary solutions had been found, the problems persisted to the extent that the layers of the offending salpeter had to be removed and the walls treated by applying special materials, such as plastering with Portland cement. For that purpose the telescopes had to be temporarily removed and stored. This thorough process started in late 1874, but the room where the meridian circle was located was designated to be treated later. The process was completed in the course of 1875.

The roof of the room where the meridian circle was located was also in need of improvement, since it was found that it was extremely difficult to equalize the temperatures inside and outside during observing. This also required the temporary removal of the meridian circle, which enabled certain changes in the instrument that were found essential after the experience of 15 years of observing. The downside was that much time had to be devoted to accurate measuring of the characteristics of the instrument before it was taken apart, in order to be able to fully and accurately reduce the data that had been taken earlier with it. In the Annual Report for 1875–76, van de Sande Bakhuyzen describes these activities, which were performed by the two observators (E.F. van de Sande Bakhuyzen and Kapteyn). This involved detailed and accurate measurements of the lines on all the relevant dials, distortions of the bearings, the threads of the screws, flection of the telescope tube, etc. This took all of the first half of 1876. After that the telescope was dismounted and sent to the A. Repsold and Sons factory in Hamburg.

Effectively the meridian circle was unavailable for observing at the time when Kapteyn took up his duties in Leiden. And during the first half of 1876 he was fully occupied measuring up the instrument before it was taken apart. But he did have extensive possibilities to become intimately acquainted with telescopes and observational techniques and practices. From the Annual Report by van de Sande Bakhuyzen for 1875–76, we know that Kapteyn used the 'Universal Instrument' of Repsold to determine the latitude (of Leiden Observatory) from 85 sets of measurement of altitudes. A universal instrument (see Fig. 5.7 for the one Kapteyn most likely used) is primarily designed for use in geodesy to measure azimuth and altitude for determining positions on Earth or measurements of time. It can also be used for astronomical purposes (which is why it was sometimes called an 'astronomical theodolite'). It has a 'broken' telescope, which means that the eye-piece was often not at the end of the telescope tube, but that light is reflected into the horizontal axis, where the eye-piece is located. In the specimen in Fig. 5.7, owned by the Boerhaave Museum in Leiden, that eyepiece is missing, as are the two that were used for reading off positions on the dials on the circular structures. In the inventory by Kaiser in the first volume of the *Annalen der Sternwarte in Leiden* in 1868, it is listed under number C1 [25]. In Leiden it was used primarily for educational purposes. Indeed, the archives of the Observatory contain two hand-written documents by Oudemans with instructions on how to use the Universal Instrument for the measurement of azimuths and of time, almost certainly for the benefit of students studying observational astronomy with the help of the instrument. In his Annual Report, van de Sande Bakhuyzen adds the sentence 'the excellence of this small instrument has been demonstrated [from Kapteyn's measurement] again very clearly'. In the same report, van de Sande Bakhuyzen relates how the two observators and Dr. Eecen had spent much time on the observations of planets and comparison stars from 1874 and 1875 and of the fundamental stars observed between 1864 and 1868.

During the second half of 1876 and the first half of 1877, while the meridian circle was absent, various other telescopes were used for observations. The van de Sande Bakhuyzen brothers as well as Kapteyn, for example, were recorded in the Annual Report to have fully observed an occultation of the Pleiades by the Moon (see Fig. 7.11; see also Fig. 11.1 for a picture of the constellation Taurus and the Pleiades) and there had been a search for a possible planet closer to the Sun than Mercury (provisionally named Vulcanus). But during that period much time was devoted once again to reductions of older observations. Kapteyn conducted an investigation into the temperature sensitivity of the screws that hold the microscopes to read the dials. This turned out to be a significant effect that was subsequently corrected for. The work on the 'zone-observations' (in the zone allocated to Leiden for the AGK) eventually led to two volumes in the series 'Annals of Leiden Observatory', which was not published until 1890. In view of Kapteyn's contributions, I have included them in the list of his publications in Appendix A, where they appear as van de Sande Bakhuyzen, Kapteyn *et al.* (1890a), respectively van de Sande Bakhuyzen, Kapteyn *et al.* (1890b). The first concerns the measurements of positions in the zone between declinations 30° and 35°; this was part of the Leiden contribution to the Astronomische Gesellschaft Katalog (AGK), and the latter volume contains observations of fundamental star positions over other parts of the sky.

When the meridian circle returned, there was again much work to be done before it could be put back into full operation, and Kapteyn was undoubtedly heavily involved in this. Many aspects needed careful consideration. Van de Sande Bakhuyzen notes in the Annual Report 1876/87 that he introduced new shields of the piers of the telescope against the body-heat of the observers. This had been done on earlier occasions by using paper shields, but now he arranged for the piers to be covered with a thick layer of felt, which in its turn was covered with thin copper plates to avoid dust from the felt from going freely around. This is illustrative for the extreme care required to perform positional astronomy.

3.9 Solar Parallax

Kapteyn was involved in a number of other projects. In van de Sande Bakhuyzen & Kapteyn (1877a) and van de Sande Bakhuyzen & Kapteyn (1877b) both observators report observations of the positions of Comet b, 1877 (Winneke's). Since the meridian circle was not present at that time, these were performed with one of the other (the 6- and 7-inch) refractors of the Observatory. This is also apparent from the re-publication of these measurements by van de Sande Bakhuyzen in 1883 (*Cometenbeobachtungen am 6- und 7-zöll. Refractor der Leidener Sternwarte* [26]). After the meridian circle was back in operation, Leiden Observatory embarked on a few projects that had been decided upon earlier by van de Sande Bakhuyzen, in consultation with his two observators, but had to await the work on the 'zone observations' to be completed. Prominent among these were the following: First a program to perform accurate position determinations of 84 stars near the celestial pole; these are 'circumpolar', which means that they are so close to the pole that they never set and are therefore available as comparison objects throughout the year, as seen from many observatories in the northern hemisphere. Further, they intended to determine positions of stars in various fields where minor planets were found and for which positions of these minor planets had been measured with respect to faint background stars. Also, such fields could be used to calibrate heliometers (see page 67), as with the observations published in van de Sande Bakhuyzen & Kapteyn (1879a). Finally they observed stars that were used to determine the position of the island of Réunion to which an expedition had been organized to observe the Venus transit before the Sun in 1878. Réunion is an island in the Indian Ocean, about 800 km east of Madagascar.

These Venus transits are very important to determine the scale and distances in the Solar System or the determination of the solar parallax that I described above (see page 75). These transits come in pairs of about eight years apart, but with long gaps of time (either 105 or 120 years) in between. One group occurred in 1874 and 1882 and Leiden had participated in two expeditions related to the December 8, 1874 transit. Observator Valentiner had joined a German expedition to Zhifu, a small island in the Shandong Province in north-eastern China, not far from Yantai. The other, E.F. van de Sande Bakhuyzen, had joined a Dutch expedition under the leadership of Oudemans to Réunion. Unfortunately, the weather did not co-operate and clouds allowed only a minimum of observations. In a telegram Oudemans transmits: 'Clouds. Photographs only two partially successful. .....'. Furthermore, as van de Sande Bakhuyzen notes in his (bi-)Annual Report for 1877/79, comparisons of the results for the value of the solar parallax from various expeditions were less satisfactory than expected (and hoped for), and other means to determine the distance to the Sun had to be devised.

One other method, in which Leiden (and Kapteyn) participated, was to observe Mars during close opposition, which was the other method to derive the solar parallax. Mars has a somewhat more elliptic orbit than the Earth, and when Mars is in opposition (i.e. directly opposite the Sun on the sky as seen from Earth), and at the same time near its perihelion (or closest point to the Sun in its orbit), the dis-

Fig. 3.10 E.F. van de Sande Bakhuyzen during the expedition to Réunion to observe the Venus transit of 1874 (Detail from a group photograph; Archives Utrecht Observatory [27])

tance from us is as small as possible. Whereas Venus can come as close as about 40 million kilometers from the Earth during a transit, Mars is some 55 million kilometers from the Earth at minimum, due to its rather elliptic orbit. If Mars' orbit were circular, this distance at closest approach would have been 79 million kilometers. The year 1877 offered a new opportunity. During the previous one in 1862, John Robie Eastman (1836–1913) of the U.S. Naval Observatory had organized a similar campaign. For this project, Kapteyn and E.F. van de Sande Bakhuyzen obtained the declinations of Mars and some comparison stars and these, among other data, were used by Eastman to derive the solar parallax. This was published as *Beobachtungen von Gill's Mars-Sternen* in van de Sande Bakhuyzen & Kapteyn (1879b) and later further discussed by H.G. van de Sande Bakhuyzen (*Die Rectascensionen von Gill's Mars-Sternen und die Änderung des persCölichen Fehlers bei der Beobachtung von Sternen verschiedener Helligkeit* [28]). The final publication by Eastman (*The solar parallax from the meridian observations of Mars at the opposition in 1877* [29]) quotes a value for the solar parallax in the range of 8''.611 to 8''.971. The current value is 8''.794143. The British astronomer Arthur Matthew Weld Downing (1850–

1917) of the Royal Greenwich Observatory also requested the data and he combined it with observations from Melbourne to derive a solar parallax of $8''.960 \pm 0''.051$ (*A determination of the Sun's Mean Equatorial Horizontal Parallax from meridian Declination observations of Mars and neighboring stars made at the Observatories of Leiden and Melbourne near the time of Opposition 1877* [30]).

The problem with such observations is that the measurements are carried out by different observers from different places with different instruments. British Astronomer Royal Sir George Biddell Airy (1801–1892) had proposed a different approach, in which an observer performed measurements early in the evening, when Mars was low above the eastern horizon, to repeat this later the same night in the early morning, when Mars would be in the west. This would mean observing Mars with the same equipment and the same person about one Earth diameter apart. Of course Mars would have moved during this time interval with respect to the stars, but that could be corrected for since the orbit of Mars was known to sufficient accuracy. David Gill (1843–1914), director in Capetown (who was to play a very important role in Kapteyn's career) had proposed to perform such observations again from the island Ascension in the South Atlantic Ocean, but needed accurate positions of comparison stars. These were provided by Leiden Observatory and eventually Gill reported a value for the solar parallax of $8''.76$ to $8''.78$ (*On the solar parallax derived from observations of Mars at Ascension* [31]).

Kapteyn's years in Leiden must have been very intense and he must have worked very hard, but there can be no doubt that he learned much and became very experienced in the field of observational positional astronomy. Leiden, although having quite a name for the accuracy of the determination of (stellar) positions it produced, was not really involved in astronomical interpretation of the data they obtained. This was probably more a matter of restricted resources and manpower than lack of interest.

3.10 Personal Life in Leiden

I would like to conclude this chapter with a few paragraphs from the *HHK biography* by Henriette Hertzsprung-Kapteyn, leaving out the astronomical parts of her text.

'In Leiden he laid the foundations for his scientific career. But the social aspects were not neglected. He made some very good friends, who remained friends throughout his life. He had his dinners with other young men, who all were hard workers on the thresholds of their careers, as enthusiastic as he was himself, a true round table of courage and power. Among them were W. Burck, the later renowned botanist, Hubrecht, who would become a professor of zoology in Utrecht, and Hoek, the later leader of North Sea marine research in Den Helder, all men of weight and importance.'

William Burck (1848–1910) spent much of his career in Java in the Dutch Indies, at the School of Agriculture at Buitenzorg, and became a scientific adviser for the government. Ambrosius Arnold Willem Hubrecht(1853–1915) was a professor of

Fig. 3.11 The official Government Gazette, *Nederlandse Staats-courant* in which official decrees, laws, etc. were published. This is the issue of Saturday December 15, 1877, in which the appointments of the Kapteyn brothers as professors were announced. The lower part zooms in on the actual announcements, dated the day before, December 14; they are in the lower-left corner in the upper panel (From: Koninklijke Bibliotheek, Den Haag)

zoology and anatomy in Utrecht, well-known for his studies of embryology. Paulus Peronius Cato Hoek (1845–1914) became well-known as a fisheries expert and carcinologist, and highly noted for his work as secretary of the International Council for the Exploration of the Sea. He became director of both the Fishery Research Institute and the Zoological Station at Den Helder, which later became the NIOZ, a national marine sciences research institute on the West-Frisian island of Texel, just north of Den Helder.

'One day, in late 1877 – Kapteyn was 26 years old – when they were sitting at dinner again, Hubrecht jumped up and to everyone's surprise ordered a bottle of champagne. When the glasses were filled he said: 'I propose a toast to the brother of one of us, who has been appointed professor of mathematics in Utrecht, Willem Kapteyn'. That came as a

3.10 Personal Life in Leiden

complete surprise to Ko, who had not yet heard of this. He rejoiced for his older brother, since in their eyes a professorship was the most desirable position that one could be appointed to. When the excitement settled down, Hubrecht ordered another bottle of champagne and when once more the glasses were filled, he said: 'And I want to propose another toast, but now to someone who is among us and who also has been appointed professor. To Ko Kapteyn, who has been appointed professor of astronomy in Groningen.' That was an even bigger surprise, not in the least to Kapteyn. Indeed, Hubrecht, who was the son of the Secretary General of the Ministry of Internal Affairs, sometimes knew things before others did. Both the brothers were appointed by two subsequent Royal Decrees, 14 December 1877, Nos 35 and 36 of the *Staatscourant* [the Staatscourant is the official newspaper published by the government of the Netherlands, in which new laws and other official announcements are published]. So it became a double celebration.

After dinner they all hit the streets of Leiden, singing and cheering. An old lady watched the noisy young men with disapproval. Hoek went up to her and said: 'Would you believe that one of them is a professor?'. Laughingly, he pulled the young Kapteyn, who certainly had not been the quietest, to the front. 'That is difficult to believe!' said the woman, clasping her hands together. 'Is that what a professor looks like?' This did not agree with what she had imagined. She kept staring at them as they went on, while their voices died away in the dark evening.

That same evening, Kapteyn Senior was lecturing for the Barneveld chapter of a reforming society called *het Nut* [see page 15 for a description of this *Maatschappij tot Nut van 't Algemeen*, Society for the Common Good]. In the middle of his lecture a telegram arrived, saying that his son Willem had been appointed professor in Utrecht. He mentioned this full of joy to his audience and, after he had been congratulated, proceeded with his lecture. But not long after that a second telegram arrived, telling him about the appointment of his son Ko. Great excitement followed, while the father beamed with joy and pride. The chairman took the floor and said: 'Mr Kapteyn, I propose that you end your lecture and tell us about your life.' Mr Kapteyn did just that and, excellent story teller as he was, he started an improvised account of his life and that of his family.'

Fig. 3.12 J.C. Kapteyn at age 35 (Illustration and caption reproduced from the *HHK biography*)

Chapter 4
Professor in Groningen

> *'It is no prudence, but a waste of the Kingdom's money to appoint a professor without supplying him with the indispensables for his teaching.'*
> Jonkheer Bernard H.C.K. van der Wijck (1836–1925).[1]

> *'O, telescope, instrument of much knowledge, more precious than any sceptre, is not he who holds thee in his hand made king and lord of the works of God?'*
> Johannes Kepler (1571–1630).

4.1 The City of Groningen

According to written documents, the city of Groningen has been in existence since 1040, but probably was a Saxon village for many centuries before that. It is situated at the north-western tip of the Hondsrug (lit. 'Dog's Back'). It is the most prominent in a set of parallel ridges of sand, running from south-east to north-west, through the current provinces of Drenthe and Groningen and are thought to be glacial ridges formed by the last Ice Age. North of the city the land is low, and when the sea level rose after the Ice Age had ended and a new warm period commenced, it was easily submerged during heavy storms. This prompted the inhabitants of the low-lying areas to build artificial hills that were called 'terp' (in the province of Groningen) or 'wierd' (in Friesland).

In the Middle Ages, particularly the thirteenth century, the city of Groningen was an important center for commerce and trade, being part of the Westphalian or Rhineland quarter of the Hanseatic League. It was conveniently situated with a view to trade, having access to the North Sea and being on the routes to Hamburg and the

---

[1] Bernhard Hendrik Cornelis Karel van der Wijck was a Rector Magnificus of the University of Groningen.

Fig. 4.1 The central Academy Building of the University of Groningen at the time when Kapteyn was appointed there. It had served as such since 1850, but was completely destroyed by fire in 1906 (University of Groningen [1])

Baltic, with which most of the trade was done, and Holland. Groningen also came to dominate the regions surrounding it (the so-called 'Ommelanden'), which gave rise to much resentment, especially among the local leaders and nobility.

In the Eighty Years' War, the war of independence between the Netherlands and Spain (1568–1648), the city of Groningen initially chose the side of the Spanish, as it was felt that in the long run this would best serve its interests . The 'Ommelanden' joined the Union of Utrecht of 1579, in which the northern Netherlands joined forces to fight the Spanish domination. The City of Groningen changed its mind and likewise joined the Union of Utrecht, in 1594. This Act is known as the 'Reduction of Groningen'. Groningen and the Ommelanden together formed the seventh province of the Republic of the Seven United Netherlands in 1594. It was governed by the 'Staten', a sort of regional assembly, in which the city and the Ommelanden had equal representation, so that the overarching 'States General' in Holland had to cut the knot if the two fractions could not come to an agreement.

Not long after the Reduction of Groningen, in 1614, the University of Groningen was established by the States of Groningen and Ommelanden. This was one of the few cases where the two fractions were in agreement. It started with four faculties, viz., of (Protestant) theology, law, medicine and philosophy. It was successful from the very beginning, attracting some 100 students every year, although it must be said that about half was foreign. The first Rector Magnificus, Ubbo Emmius (1547–1625), came from a neighboring region in what is now Germany. Groningen went through a significant decline in the seventeenth and eighteenth centuries, the university losing much of its reputation. It is true that professor Petrus Camper (1722–1789) was a world-renowned and versatile scientist, working as a physician, anatomist, zoologist and paleontologist. But apart from this towering peak the landscape of Groningen university was flat, unassuming and mediocre, at least in the natural sciences.

I have described in Chap. 2, how Dutch universities fared during the French Napoleontic occupation – Groningen and Leiden were the only two not closed by the French administration – and how the 'Organiek Besluit' or organizing resolution of 1815 re-instituted the University of Utrecht, but did not promote the Athenaeum Illustre of Amsterdam to a full university. This Organiek Besluit was meant as a temporary measure, awaiting a final decision on the structure of higher education in the Netherlands. However, little progress was made for decades, and it was not until 1876 that it was replaced by the *Law on Higher Education*. This law introduced a profound change in the organization of universities and was the direct cause of Kapteyn's appointment in Groningen. It has been discussed in detail in the literature [2], although unfortunately with one exception inaccessible to the non-Dutch speaking world.

4.2 The 1876 Law on Higher Education

In the important revision of the Constitution of the Netherlands in 1848, a new law on education had been envisaged. However, it lasted until 1857 before a law on the organization of elementary education was adopted and until 1863 before secondary education was reorganized; the latter included the important introduction of the HBS (see page 62). But a new law on higher education, which was essential to accommodate changing practice at the universities, was delayed time and again. Professors in natural sciences were increasingly orienting their teaching towards practical work in laboratories, etc., not only to provide students with a university education, but also to train them as practitioners of scientific research. However, politicians were loath to make more room for the development of scientific research.

Another problem was the cost of higher education. The number of students at the three universities of Leiden, Groningen and Utrecht during the 1850s and 1860s remained relatively constant at around 1,300 nation-wide. Conservative politicians of the liberal party felt that there was no market for more university-trained individuals. There were even doubts whether sufficient qualified professors could found in the Netherlands to justify an expansion of the university. Between 1815 and 1876 the number of professors had slowly increased from about 70 to about 90. Groningen university had by far the smallest number of students, fluctuating between about 150 and 200 (numbers are from Jensma & de Vries, *op. cit.*), and there was certainly a growing body of opinion that the smallest and most remote university (as seen from the seat of government) should be closed. Moreover, if two universities were sufficient, it would be very unlikely that the government would approve of the wish of the Governing Board of the city of Amsterdam to give their Athenaeum Illustre full university status, even though the city was financing it from its own pocket. Also Groningen's enrollment had been growing to a total of 400 or so, comparable to that of Leiden and two-thirds of Utrecht.

In the end a law on higher education was passed in 1876 and, contrary to the expectation that only the universities of Leiden and Utrecht would survive, the University of Groningen was continued and the city of Amsterdam was allowed its (Municipal) University. So in 1877 there were not two or three, but even four full-fledged universities in the Netherlands. In addition, the law provided for another very important change, which involved the expansion of opportunities for scientific research. Academic studies were to be seen as training and preparation for independent practice of scientific scholarship and performing societal service in positions for which a scientific background was required. The emphasis was put on science rather than a broad general education, and this in turn led to increasing specialization in the curriculum. As a result, students received a more thorough but inevitably more specialized training.

The number of subjects taught increased greatly and, rather unlike what was deemed necessary in the early 1870s, the new law provided for a sharp increase in budget for the universities, starting with double the amount in 1877 and 1878. In part this was due to an increase in the number of professors, but their salaries went up as well. Before the 1867 law, professors supplemented their university earnings with fees that students had to pay; these were in the order of 15 Guilders for a course of two hours a week and 30 Guilders for four or more. For 1875 this corresponds to a purchasing power of 350 and 700€ today respectively. The number of professors went up to almost 200 nationwide, also twice the previous number. Not only were they required to offer more courses, as research was now an increasingly important part of their duties. But gradually publications had come to be a major criterion in the selection for appointments.

The law did not specify the curricula to be offered; this was done in the 'Academisch Statuut' or Academic Statute of April 1877. The Kapteyn Room at the Kapteyn Astronomical Institute has a copy of this document with remarks and other thoughts and clarifications by Kapteyn himself. It appears to contain notes from a university-wide discussion (probably in the Senate, the body that was composed of all professors), presumably on how to introduce the new structure. The Faculty of mathematics and natural sciences now offered six separate doctorates, viz., in mathematics and astronomy, mathematics and physics (the first years had the same courses for these two disciplines), chemistry, geology and mineralogy, botany and zoology, and pharmacy. With regard to mathematics and astronomy Kapteyn notes that 'de Boer feels that the designation differential calculus is incomplete and wants integral calculus added'. This either refers to Prof. Petrus de Boer (1841–1890), who was professor of botany from 1871 to 1890, or to Prof. Floris de Boer (1846–1908), professor of algebra between 1884 and 1908, but then the discussion took place much later than 1877. The doctorate in mathematics and astronomy involved courses in geology; Prof. Friedrich Julius Peter van Calker (1841–1913), professor of crystallography, mineralogy, geology, paleontology and physical geography between 1877 and 1911, objects to the name geology, since it should include

Fig. 4.2 Nicolaus Mulerius was the first professor to teach astronomy in Groningen between 1614 and 1630. He holds two small models of celestial spheres (University of Groningen [3])

crystallography and mineralogy and that is not covered under geology. As has been the case throughout history, professors had definite and strong opinions.

The law came into effect in October 1877. The number of students in Groningen of mathematics and physics had increased from only 17 in 1870/71 to 37 in 1876/77. After the introduction of the 1876 law, this number initially decreased to 20 in 1881/82, but then went up to 41 in 1886/87. The University of Groningen as a whole showed a similar pattern, viz., 142, 196, 252 and 326 respectively. On the national level too, the number of students fell at first after 1876 (from 1782 in 1876/77 to 1596 in 1881/82 and 2120 in 1886/87) (from Jensma & de Vries, *op. cit.*)

4.3 Astronomy in Groningen

Astronomy had been taught in Groningen from the University's earliest years. For a more complete overview than can be provided here see the article by J. Schuller tot Peursun-Meijer in *Sterrenkijken bekeken* (*op. cit.*). The first to do so was Nicolaus Mulerius (1564–1630), starting at the time of the founding of the university (see Fig. 4.2). He had been born as Nicolaas des Mullers in Brugge (Bruges) in Flanders and studied in Leiden, where he obtained a doctor's degree in medicine. After having been chief physician of the city of Harlingen (near Franeker in Friesland) and of the Ommelanden he became principal of the gymnasium in Leeuwarden, the

Fig. 4.3 Jacob Baart de la Faille, a professor of mathematics, physics and astronomy at the University of Groningen between 1790 and 1823 (University of Groningen [4])

current capital of Friesland. In Groningen he had become acquainted with Ubbo Emmius, who would become the first Rector Magnificus of the University of Groningen. Mulerius had a keen interest in astronomy, and produced astronomical tables for Ubbo Emmius, helping him with calender issues. So, when the university was founded, Ubbo Emmius brought him back to Groningen. The strongest evidence that Mulerius taught astronomy is the publication of his book *Institutionum Astronomicarum* in 1611, which in all respects appears to be a textbook accompanying his lectures.

Religion-wise, Groningen university was very orthodox in those days. Mulerius as well as Ubbo Emmius and professor of theology Hermann Ravensperger (1586–1625) opposed the views of the liberal Leiden professor Jacobus Arminius (±1555–1609). The view of Calvinism was that humans were predestined for Heaven or Hell and had no free will to choose to live either in virtue or in sin. The followers of Arminius (sometimes called the 'Rekkelijken' or Flexibles) opposed this and a few related points of view, which were collected in the 'Five Articles of Remonstrance' in 1610, which caused them to be designated Remonstrants. Their views were particularly disputed by Franciscus Gomarus (1563–1641), whose followers were called 'Preciezen' (Strict Ones) or Counter-Remonstrants. Gomarus already taught in Leiden when Arminius arrived there. Mulerius had tried to prevent the appointment of a successor to Arminius who held the same views. Later, after Gomarus had left Leiden, he was appointed in Groningen in 1618; Mulerius was Rector Magnificus at the time and must have played an important role in this appointment.

4.3 Astronomy in Groningen

Fig. 4.4 Rudolph Adriaan Mees (1868–1886), professor of physics, meteorology and higher mathematics (University of Groningen [5])

Mulerius also opposed the Copernican view of a heliocentric Universe, stating in his *Institutionum Astronomicarum*, that this could not be true as a moving Earth did not fit in with the biblical account. Yet, Mulerius also published an annotated third edition of Copernicus' famous book *De Revolutionibus*, which exposes the heliocentric view of the Universe. This may seem contradictory, but it would seem to me he may have taken the point of view expressed in the preface by Andreas Osiander (1498–1552) to *De Revolutionibus*, to the effect that this heliocentric model could be seen as a convenient trick to simplify the calculation of planetary positions and not necessarily a description of the true situation, which could still be geocentric.

After Mulerius, astronomy was on the curriculum on a permanent basis, but it seems that the professors were primarily interested in other subjects, and/or the number of students was extremely small. Natural science in general was not very prominent, except in periods such as the decade between 1694 and 1705, when Johann Bernoulli (1667–1748) was a professor in Groningen, who began to perform experiments during his courses. Matters took a turn with the appointment of Jacob Baart de la Faille (1757–1823) as professor of mathematics, physics and astronomy (see Fig. 4.3). He should not be confused with his son, also named Jacob Baart de la Faille (1795–1867), who was a professor of medicine in Groningen. He was the first to express the wish to build an observatory in Groningen. He and his successor Seerp Brouwer (1793–1856) were unable to raise the necessary funds, but during their days astronomy seems to have been attracting attention, if not students. Baart de la Faille is recorded to have given practical lessons in astronomy, which at the time must have been by using a telescope or some other instrument that he owned privately, as there was no such instrumentation for teaching available. In 1820 a plan was presented to the minister for a new building, to extend the very limited space the University had available for teaching, laboratories and collections. It included options to build an observatory, but a shortage of funds forced the minister to reject the plan. When the need to construct a new central Academy Building presented itself in 1844 (the old one was in a very bad state of repair), the design featured a telescope dome on the roof. The professor responsible for astronomy at that time, Jan Willem Ermerins (1798–1869), rejected the design, however, since it would be too expensive to build a dome that would have all the necessary qualifications. The building (see Fig. 4.1) was eventually opened in 1850 without an observatory, but with some facilities for physics experiments.

When the 1876 law was passed, there was still no observatory and no professor who had been hired specifically to teach astronomy. The person responsible for the teaching of astronomy was Rudolf Adriaan Mees (1844–1886) (see Fig. 4.4). However, he was also a professor of physics and mathematics, having succeeded Willem Adriaan Enschedé (1811–1899) in 1868. Enschede was not particularly interested in astronomy, but all the more so in the University library, for which he was responsible.

4.4 A Professor in Astronomy

Soon after the introduction of the law of 1876, the Faculty of mathematics and natural sciences requested an increase in the number of professors. The following is based on the archives of the Faculty and the Board of Curators in the University of Groningen; but also see the article by J. Schuller tot Peursun-Meijer, *op. cit.*. The Faculty of mathematics and natural science stated on March 19, 1877, that the appointment of a professor of astronomy was considered 'urgent and most necessary', insisting strongly, on March 28, that in addition the appointment of a professor of mathematics and theoretical mechanics was urgent. This did not immediately meet with the approval of the Minister (of Internal Affairs at that time) and initially the chair in astronomy was denied. On September 25 the Faculty wrote to the Curators. They started out by noting that as a result of the new law of 1876 'new' subjects had to be taught, viz., mechanics, physics, astronomy and statistics, and meteorology. The professor of physics, Mees, was already responsible for teaching astronomy and would have to take on statistics and meteorology as well. The need for specific professors of physics and of astronomy was obvious. In addition, the need was restated for new professors of mathematics (within which more subjects had to be taught), pharmacy and biology. Indeed, on September 25, 1877, the Curators wrote to the Faculty, saying that they agreed that Mees' tasks were too extensive and they would try their best to convince the Minister that the astronomy chair was necessary. The chair of the Curators was held at the time by L. Graaf van Heiden Reinestein, the King's Commissioner in the province of Groningen, and the secretary was Jhr. Witius Henrik de Savornin Lohman (1842–1895), a lawyer, and later a judge in

4.4 A Professor in Astronomy

Groningen. Both had studied law and were politically of a conservative, protestant tradition. They took their duties seriously.

After this things moved fast. On October 31, 1877, and at the request of the Curators, the Faculty sent a list of 'Wishes of the Faculty of Mathematics and Natural Sciences', signed by the chair, Prof. Enschedé (who, as we saw above, had taught astronomy before for a while, but much against his interests), and acting secretary Hendrik Jan van Ankum (1845–1940), professor of zoology. This put the new professor of astronomy prominently in first place, but with additional teaching duties, including statistics. Second priority was pharmacy. The next day, the Curators informed the Faculty that the minister had asked them in a letter of October 29, to send a nomination for a new professor. They also asked which subjects had to be taught. Finally, they urged the Faculty to reply with the 'utmost haste' so that a decision could be taken 'in a few days'. On November 3 they sent a reply to the earlier missive of the Faculty, of October 31, concurring with the view that professor Mees's workload was indeed excessive.

On November 19, the Faculty sent a nomination, running to nearly four densely written pages, almost half of that dealing with considerations regarding the new professor's profile. The discussion centered on who was to be responsible for the teaching of theoretical mechanics. The most preferred option of the Faculty was to include it in the duties of the professor of astronomy and statistics. They argued that mechanics could just as well be taught by an astronomer as by a physicist, pointing out that it was part of the profile 'mathematics and astronomy' as well as 'mathematics and physics' in the 'Academisch Statuut'. He would, they noted, also have the time for such lectures, since there would be no training in practical astronomy, at least for the time being, as no instrumentation was available. In view of the expected retirement of a professor of mathematics within a few years' time, the Faculty expected that theoretical mechanics could be included in the profile of the new professor. After explaining these considerations, the document continued as follows (see also Figs. 4.5 and 4.6):

> 'Of those who can be considered suitable for the professorship, we feel that in the first place we have to bring to your attention Dr. J.C. Kapteyn, currently observator at Leiden Observatory. Dr. Kapteyn obtained his doctorate in Utrecht two years ago on a thesis entitled *Study of vibrating flat membranes*. His skills as a mathematician are highly praised by his professors in Utrecht [plural; so apparently this refers to both Grinwis and Buys Ballot], and in their view he would be very well qualified to teach statistics. After his doctorate he worked as an observator at Leiden Observatory for two years, under the responsibility of the Professor van de Sande Bakhuyzen. He was primarily occupied with practical astronomy and according to the testimony of the Director of the Observatory, has shown himself to be a good observer. Both as a practical astronomer and as a mathematician he can be recommended.

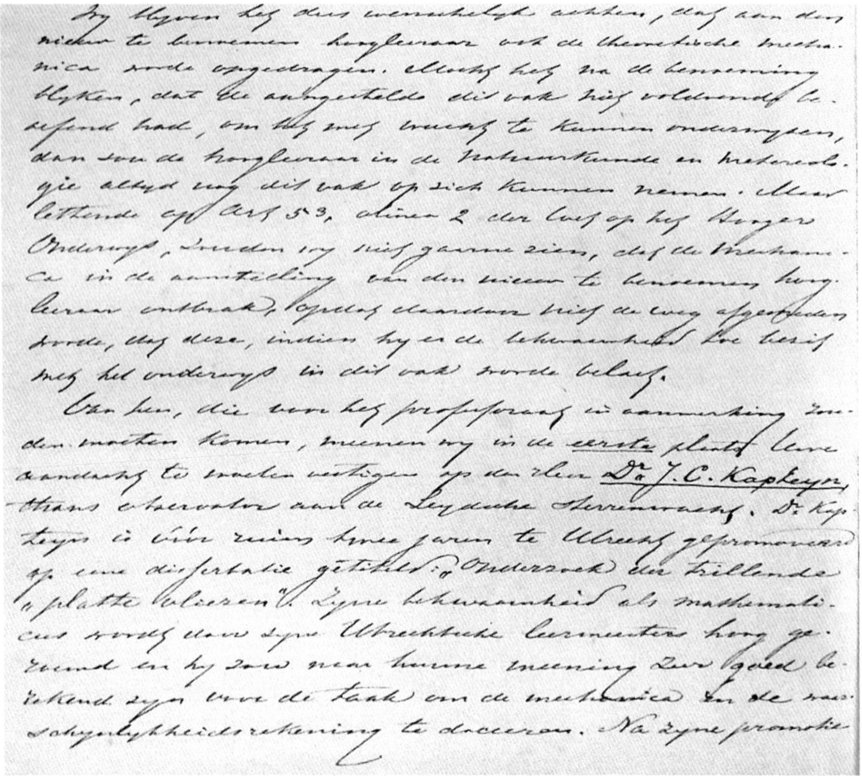

Fig. 4.5 Part of the letter with the nomination of Jacobus Cornelius Kapteyn for his appointment as professor, sent by the Faculty of Mathematics and Natural Sciences to the Curators of the University of Groningen (From the Archives of the University of Groningen in 'Groninger Archieven')

In the second place we bring to your attention Mr Ch.M. Schols, a teacher of geodesics and statistics at the Military Academy in Breda. [The Royal Military Academy in the city of Breda is a semi-academic institution that trains officers for the Dutch army (and at present the Air Force as well).]. Mr Schols is a civil engineer trained at the Polytechnical School in Delft [the current Technical University]. His teachers have a high opinion of him, because his exposition of statistics in the publications of the Royal Academy of Sciences in Amsterdam shows him to be an excellent mathematician, to whom the teaching of statistics and mechanics can be fully entrusted. His intellect, according to someone who knows both candidates, is higher than Kapteyn's, but with real astronomy he has never occupied himself. As a geologist he has studied some practical astronomy, so that the teaching of the fundamentals of astronomy could very well be entrusted to Mr Schols. If you should prefer the latter, we suggest that the teaching of statistics and mechanics be entrusted to him as his main discipline and in addition to this assign him the teaching of astronomy. When in a few years' time an-

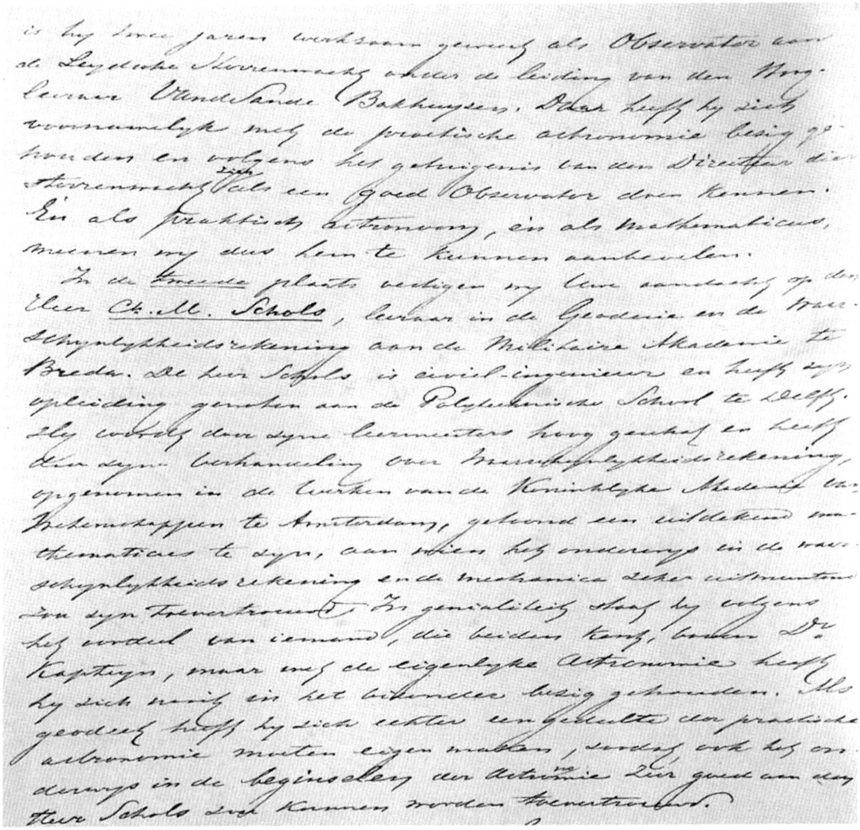

4.4 A Professor in Astronomy

Fig. 4.6 Part of the letter with the nomination of Jacobus Cornelius Kapteyn for his appointment as professor, sent by the Faculty of Mathematics and Natural Sciences to the Curators of the University of Groningen (continuing Fig. 4.5) (From the Archives of the University of Groningen in 'Groninger Archieven')

other professor will be appointed, as has been described above [the matter of the retirement of the mathematics professor], a separate professorship of astronomy could be instituted so that Mr Schols can exchange astronomy for parts of mathematics.

Since our Faculty is aware of your opinion that astronomy and statistics are subjects that should be kept in mind first and foremost when it comes to finding a candidate for the professorship, we conclude that, notwithstanding the excellent qualities of Mr Schols, Dr. Kapteyn should be suggested to you with first priority.

In addition to the two persons just presented as candidates for the professorship, we also feel that the three Germans presented next would be suitable: Dr. Valentiner, director of the Observatory at Mannheim; Dr. Becker, working now at the Observatory of Berlin and Dr. Seeliger, working now at the observatory of Bonn and as a Privatdozent at the local University. These three persons, the first two of whom also worked as observators at Leiden Observatory for some time, have been strongly recommended as being very well qualified. Especially the last two, Dr. Becker and Dr. Seeliger, have been praised for their astronomical and mathematical qualities. Since the appointment of a foreign astronomer of high international standing at a university as ours, where facilities for practical astronomy do not yet exist, might seem rather awkward, while on the other hand there are persons in our country who not unsuited to a professorship, we have concluded to draw your attention to these first two [meaning Kapteyn and Schols].

The Faculty of Mathematics and Natural Sciences
W.A. Enschedé,
Chair
M.J. Rina
Secretary.'

Who were the other candidates?

Charles Mathieu Schols (1849–1897) studied at the Polytechnical School at Delft, the predecessor of the current Technical University, and obtained a degree there as a civil engineer. He became a professor at the Royal Military Academy in Breda, where he was employed at the time when he was being considered for the position in Groningen. In 1878 he would be appointed professor of geodesy and (land) surveying in Delft. He was also interested in and published on the theory of probability, but eventually became quite famous for a textbook on surveying and leveling (*Landmeten en Waterpassen*), which was published in 1879.

Karl Wilhelm Valentiner (1845–1931), already referred to earlier in this book, was Kapteyn's predecessor as an observator in Leiden. In 1876 he was director of the Observatory of Mannheim. One of his major accomplishments was to have the observatory moved to Karlsruhe, but that was at a later time than when he was considered for the position in Groningen. In 1898 he assumed the responsibility for the astronomy section of the Königstuhl Observatory in Heidelberg.

Ernst Emil Hugo Becker (1843–1912), worked at the Berlin Observatory at the time. Eventually he became director of Strasbourg Observatory, from 1887 to 1907. Before he went to Berlin, he worked as an observator at Leiden Observatory for a while (in 1870), and in 1871 at Neufchatel in Belgium.

Hugo von Seeliger (1849–1924), whom we will meet again later in this book, worked as an observator at the Observatory of Bonn from 1873, where he left in 1877 to settle in Leipzig as a university teacher. After a short spell as director of Gotha Observatory in central Germany, he became director of München Observatory in 1882, which post he held until his death. Incidentally, people often leave out the 'von' in his name (as he sometimes did himself in his publications), but I will retain it for the sake of consistency. He would turn out to become the most important German astronomer in the field of statistical astronomy and the structure of the system of stars, of which Kapteyn and he would eventually be the major pioneers.

4.5 Taking Up the Professorship

The Curators did not waste much time over the nomination and went on to discuss it as early as November 21, 1877, during their regular meeting. The minutes do not specify how they arrived at their decision, but for the professorship of astronomy, statistics and theoretical mechanics they resolved to put forward Dr. J.C. Kapteyn from Leiden as the first candidate, and Dr E. Becker from Berlin as the second. The reason for not nominating Schols as the second candidate may have been strategic. The Faculty had supplied some arguments in favor of Schols, even though they preferred Kapteyn. Presumably they took into account the possibility that the minister might be persuaded to reverse the order (this was known to happen, even in those days), ignoring Groningen's preference. The chance that this would happen with a German in second place was considered much smaller. However, it may equally well have been testament to the view that this new professorship should be fully focused on astronomy.

Kapteyn was appointed professor by the Royal Decree of December 14, 1877. On page 82 I have quoted Henriette Hertzsprung-Kapteyn, relating how Kapteyn first heard about his appointment (and that of his brother Willem in Utrecht). In those days vacancies of professorships were not usually advertised formally, neither was it customary to send written applications and submit letters of reference. The Faculty consulted third parties who were well acquainted with the professional sector, asking them to draw up a list of possible candidates. In practice this was usually done by word of mouth, so that as a rule no written documents are available. In this case, van de Sande Bakhuyzen in Leiden and Oudemans in Utrecht had no doubt been asked to name high potentials, as well as his former supervisor Grinwis in Utrecht, when Kapteyn's name came up. It is unlikely that Kapteyn or any of the others were aware of this and it would have been highly unusual had they been let in on the procedure. So, it must have come as a complete surprise when Hubrecht ordered champagne, first to celebrate Willem's appointment in Utrecht and next that of Jacobus in Groningen.

4.5 Taking Up the Professorship

When Kapteyn moved to Groningen, he found a boarding place with a family named Raken (see *Love Letters*, page 42) in Peperstraat (number 116, which now is 11). This location has been indicated in Fig. 4.7 together with other homes and workplaces on a map of Groningen in 1868. Kapteyn's employment started with the oath of allegiance to the King of the Netherlands, Willem III (1817–1890). He took this oath on February 20, 1877, before the Curators of the University, who immediately sent a letter to the Senate that Kapteyn was ready to be inaugurated into that body. The minutes of the Senate (meeting at 13.30 hours) indeed duly record this: 'During the assembly the Senate receives the notice from the Honorable Gentlemen the Curators that Dr. J.C. Kapteyn has taken the oath of allegiance and can therefore be inaugurated into the Senate. The new professor, who has been ushered in by the Secretary, receives the good wishes of the Senate at the commencement

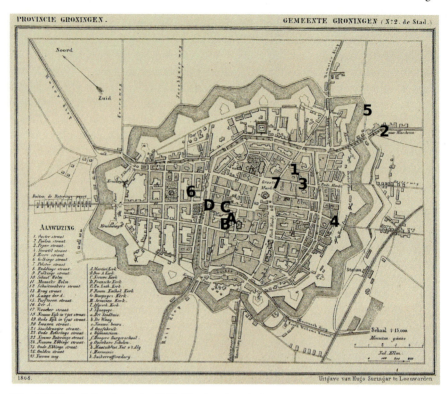

Fig. 4.7 Map of Groningen in 1868. The locations indicated are the living quarters (numbers) of Kapteyn and that of his working place/laboratory. 1 = Peperstraat 116 (1878–1879), 2 = Winschoterkade WZ, X 132 (1879–1885), 3 = Oosterstraat OZ, D 42 (1885- < 1891), 4 = Heerestraat S 6a, later 113a (< 1891–1906), 5 = Eemskanaal/Oosterhaven ZZ 16a(1906–1910), 6 = Ossenmarkt 6a (1910–1920), 7 = Grote Markt 35 (1920–1921), A = Academy Building, B = Broerstraat Physiological Laboratory (1886–1896) and Astronomical Laboratory (1911 onwards), C = Oude Boteringestraat (1896–1904), D = Oude Boteringestraat/Spilsluizen (1904–1911). Note that the map is oriented, i.e., east is at the top (University Library Groningen and Gemeente Atlas Jacob Kuyper (1865–1870) [6])

of his employment through the Rector Magnificus.' The Rector Magnificus was Prof. Frederik Willem Bernard van Bell and the secretary Jhr. Prof. Bernhard Hendrik Cornelis Karel van der Wijck. Van Bell (1822–1896) was a professor in the faculty of theology and van der Wijck (the prefix Jhr. stand for Jonkheer or Esquire) (1836–1925) was a professor of (the history of) philosophy.

Afterwards, Kapteyn gave his inaugural lecture. Unfortunately this lecture has not been preserved. In all likelihood it was never printed; in any case it is not among his collection of publications in the Kapteyn Room (see Appendix A), but then neither is a copy of his PhD thesis. The title of the inaugural lecture was *The annual parallax of the fixed Stars* (de jaarlijkse parallaxis der vaste sterren), as reported in the Yearbook of the University [7]. Of the obituaries written shortly after his death (see Appendix A.2), two do actually mention the title, viz., that by Adriaan van Maanen (1922), and the one by Sir James Jeans (1923), but both omitted the word 'annual'. In Henriette Hertzsprung-Kapteyn's *HHK biography* the title is not mentioned either. She does, however, describe how he practices reading out the text before his sisters (see page 33), as he felt extremely uncomfortable at the time about speaking in public. From the title we may infer that he must have been presenting the problems of the distances to stars, which were indeed among the most important issues related to the arrangement of stars in space. We have seen above (page 48) that one of his propositions at his PhD defense concerned this issue, in the sense that he was considering whether large brightness or proper motion was a better indicator of proximity. Apart from his thesis, Kapteyn published no other article until 1884 (in fact he became a professor without any astronomical publication of substance). So, we cannot be fully certain what Kapteyn tried to convey in the lecture. The assumption has been made that he wished to describe the line of research that he was to follow for the rest of his life, viz., with regard to the problem of the 'Structure of the Sidereal System', or the distribution of stars in space. On introducing Kapteyn's inaugural lecture, van Maanen (1922) (in *J.C. Kapteyn, 1851–1922, op. cit.*) says: 'The problem of the stellar distances was naturally of first importance to him, his ideal being to throw some light on the structure of the Universe. We do not know when this idea began to ripen in Kapteyn's mind, but it probably dates from the time that he decided to devote his life to astronomy.' Again, we have no further direct evidence, but he had been interested in the distances of stars ever since his days in Utrecht and presumably the subject of the structure of the Sidereal System was never far from his thoughts.

4.6 First Attempt at an Own Observatory

No more than three days later, on February 23, the Senate met again. Kapteyn is not listed among those absent, so we may assume he was present. The major business was the draft reply of the Curators to the Minister concerning the budget. The Minister had asked what investments (in terms of buildings, apparatus, etc.) were necessary to bring the facilities up to a reasonable standard. And what would the costs be, approximately? The draft reply, which was approved, first noted that a substantial increase in the annual funding was required and then listed a number of one-off projects requiring subsidies. Among them, under the Faculty of mathematics and natural sciences, there were ten priorities, the third of which was the 'construction of an astronomical observatory'. It was estimated at 14,000 Guilders or about 320,000€ purchasing power today, but is was noted that this was exclusive of any real estate. Higher up on the list were a physics laboratory (of about 85,000 Guilders or about 2 million €) and a mineralogical museum and laboratory of 5,000 Guilders (115,000€). Under a further heading of extraordinary subsidies required there were three categories, the first involving astronomy. There was a request of 21,300 Guilders (488,000€) for astronomical instruments, divided as follows (current € are purchasing power):

	Guilders	current €
heliometer	19,200	440,000
clock	1,000	23,000
chronometer	500	11,500
Universal Instrument	400	9,000
Prism binoculars with support	200	4,500

For zoology and mineralogy there were requests for 15,000 and 9,000 Guilders respectively, and it was noted that more would be required for instruments in the new physics building as soon as it was completed.

This constituted a good start, as there seemed to be widespread support for an astronomical observatory for the new professor. However, appearances may be deceptive and the observatory was never realized during Kapteyn's life. Henriette Hertzsprung-Kapteyn, in the *HHK biography*, described this as follows:

'On the 20th of February of the year 1878 he presented his inaugural lecture at the University of Groningen. However, things did not look promising for a young and enthusiastic professor in astronomy. As the professorship was newly created, he had no observatory waiting for him and not many prospects of getting one soon. And a professorship in astronomy without an observatory was unthinkable. But what was not here now, could come to be and with all his energy and unshakable optimism he started the task of obtaining an observatory. Many letters to the Curators stem from this period, plans were presented and proposals made, in the course of time becoming less elaborate and asking for lower funds. But it never went beyond these Curators, since he did not receive any support for requests to the government. Both other directors of observatories, Prof. van de Sande Bakhuyzen, director of Leiden Observatory and Prof. Oudemans of Utrecht gave negative advice. They did not like the idea of a third observatory in this small country. The response of the disappointed young astronomer was bitter: 'On the basis of your advice the installation of an acceptable instrument in Groningen will be rejected for quite some time. In all of the Netherlands, the astronomer in Groningen will be practically the only person who has no decent workplace for scientific research.'

There is no evidence that the requests were blocked by the Curators. The source of the bitter remarks is not known. There is no letter to this effect from Kapteyn in the papers of H.G. van de Sande Bakhuyzen as preserved in the archives of Leiden Observatory. In the *Love Letters* there are a few references to Kapteyn's attempts to obtain funds for an observatory. In his letter of October 19, 1878, he wrote:

'Tomorrow is the day that Bijlsma comes [According to the *Love Letters*, page 55, this is Dirk Klazes Bijlsma (1851–1879), a student of mathematics and physics. He died of an insidious and debilitating disease, probably tuberculosis, according to in the Annals of Groningen University for 1879/80. It has been suggested that Bijlsma might very likely have become Kapteyn's first PhD student, had he not died prematurely.], but it seems that the gentlemen students do not approve of my lectures. Yesterday Bijlsma already paid me a visit. He proposed to delay our scientific meeting until next week

4.6 First Attempt at an Own Observatory

since he had used all his time this week to study and digest what he had learned. Very praiseworthy. Until then I can enjoy myself as if it is vacation: since also the others, the first-year students, did not turn up this week. I don't know what the reason is, but it will probably not be a reason that is very pleasant for the gentlemen freshmen themselves. In the meantime, in order to occupy myself, I have taken Bijlsma to the roof of the Academy Building. It was reasonably clear, at least sufficiently so to distinguish the major constellations. [...] Soon to be repeated. [...]

I have not written you, I think, about the sad results of my attempts to obtain an observatory. On the budget as it has just been published there is not a single cent assigned to me. I don't know what to do. Go to His Excellency and tell him that I will not be able to have any student write a PhD thesis under my supervision if I don't get at least 6,000 Guilders for the basic essentials? That is plainly impossible. I have talked to the Honorable Gentlemen the Curators about my grievances; they were friendly enough in the way they expressed their impotence in this matter, but that's no good to me. At the moment the coffers are empty. I hope for the best. Maybe, with a lot of trouble, I can get 6,000 Guilders outside the regular budget.'

On May 6, 1879, Kapteyn noted that he had received notice that funds had been made available for his astronomical instruments for teaching. It turned out that they were intended for the maintenance of existing installations. But he had none! The Curators told him that is was for the instrument on the roof of the Academy Building. However, that was not an astronomical instrument, but a weather vane! He did receive 250 Guilders for the purchase of a telescope, though, which is about 5,700€ purchasing power in current value. In his *'Love letter'* of December 6, 1878, Kapteyn wrote that he had purchased the telescope out of his own pocket, but will receive the money back from the University later. So what was Kapteyn's income by comparison? Krul, in his chapter in the *Legacy* made the following comparison: 'To give an impression of the value of Dutch money at the time: a skilled worker earned 10 to 15 Guilders a week, a primary schoolteacher was paid about 1,000 Guilders a year, a grammar school teacher 2,000 Guilders; a young professor like Kapteyn had a salary of 4,000 Guilders; for a senior professor, the maximum was 6,000 Guilders. In a town like Groningen, at the turn of the century, anyone with an annual income over 4,000 Guilders would be considered well-to-do.' Indeed, in a letter of July 1, 1879, he notes that he has received a telegram that he will be reimbursed for the 250 Guilders 'tomorrow', which is 'like rain in desert'; after all he and Catharina Elisabeth were about to get married and were furnishing their home. On December 6, 1878, Kapteyn wrote: 'It is not a big instrument, but I expect much pleasure from it.' In January 1879 he was busy installing it, but no details of its use are available.

The Annals of the University of Groningen for 1878/79 record that Kapteyn received a sum of 250 Guilders for the purchase of a telescope and 200 Guilders for the purchase of other instruments. Rector Magnificus van der Wijck added: 'But he needs 6,000 Guilders, according to the most conservative estimates, to be only moderately equipped. It is clear that eventually he will receive that modest sum.

Why not allocate it now? It is no prudence, but waste of the Kingdom's money to appoint a professor without supplying him with the essentials for his teaching. The way things are now, practical astronomy has had to be removed from the Series Lectionum (the list of courses) for the simple reason that more is needed for that course than just a professor in astronomy.'

It is not known whether the advice given by van de Sande Bakhuyzen and Oudemans as quoted by Henriette Hertzsprung-Kapteyn, lay at the bottom of the refusals to fund the observatory in 1878, or that they refer to some later situation/occasion. In the archives and the Annals of the University of Groningen some more records of requests for the necessary funds can be found (see also *Love Letters*, p. 20–21). In 1880 Kapteyn submitted a revised request. He argued that this instrumentation was absolutely vital, rather to enable students to obtain a doctor's degree in astronomy, than that it was essential for performing astronomical research. His requests were to remain unanswered throughout his career, even a much later one for a photographic telescope, which went to Leiden.

4.7 Marriage Proposal

It appears that Kapteyn and Catharina Elisabeth (Elise) Kalshoven were not engaged when he moved to Groningen to start his career as an astronomy professor. We have much information on this period through the *Love Letters*. The first one was written only shortly after he had delivered his inaugural lecture, i.e., on March 15, 1878. In this letter he appealed to her to write to him by return. He stated that he did not much enjoy the oath, lecture and congratulations afterwards, being much relieved that it was over. He confessed that his first lecture was likewise a failure. He went to the student's dance, though, meeting Groningen's beauties ('and there are some really very nice ones') and described what he felt was a delightful local custom: 'Prior to the dance there is a supper where you pay for the lady who accompanies you and then you have a long dance. The next morning you meet the parents of the beauty and then you receive permission to take her out that evening to the Harmonie to commemorate the dance with further food and drink.' The 'Harmonie' was a clubhouse in Groningen; it was housed in a major building in the city center, and accommodated a well-known concert hall. Was he teasing her or trying to make her jealous? He concluded by asking her to not forget a lonely friend and sent his best wishes to a few people.

The next letter was from Utrecht, on August 13, 1878. It was very short, containing only a marriage proposal. I concur with van Berkel & Noordhorn-Hoorn, that the probable situation was that he spent the summer in Utrecht, where their relationship deepened. Kapteyn apologized for his strange behavior the evening before, which he could not get himself to explain. 'Lize, be my dear friend forever; will you share my joys and sorrows and be my wife? How often have I longed to ask you that.' He explained that on another occasion two souls (one of whom, he says, she knows) had become unhappy since they did not know each other well enough.

4.7 Marriage Proposal

Fig. 4.8 Jacobus Kapteyn and Catharina Elisabeth (Elise) Kalshoven in pictures taken before their marriage (From the *HHK biography*)

He seemed to refer to his earlier engagement with one 'George'. But, he said, this time he has thought it over much longer and been more cautious. 'I promised that to myself when I lost George. But I have kept that promise, I think. I am sure: nobody except you can make me happy. Would I be able to make you happy? You will understand, Lize, in what tenseness I am awaiting your reply. If you need more time to think, send me a few words anyway, if you please?'

Catharina Elisabeth (Elise) Kalshoven must have answered within a few days, since Kapteyn designated August 16 as the date of their engagement. Elise felt it was August 13, the date of Kapteyn's letter. The *Love Letters* were kept in a bundle on which was written 'Blessed memory; reread on 13 August 1933 [later added – 1934–1935 –], 55 years after the day of engagement.' They give a very revealing impression of Kapteyn as a private person. We get to know Kapteyn as the rather unconventional young professor, trying to find his way in the Groningen circles of academics, often uncertain about his qualities as a teacher and his social skills, all the time longing for love and appreciation. We have already learned of the unhappy aspects of his youth, when he was a 'child like all the others' (see W. Krul, *Kapteyn and Groningen; a portrait* in the *Legacy*) in a large household with parents who had no time for individual attention. But what also emerges is the wish to make himself useful, and the resolve to remain optimistic that one day he will be able to accomplish important things. But frequently he felt unimportant, insignificant and superfluous.

During the better part of his first year in Groningen, his sister Cornelia Louise Alexandra (Cor), four and a half year his junior, lived with him at the same address. Presumably she ran his household while he was still single. She must have been a great help and supporter, listening to his complaints, worries and uncertainties, encouraging and advising him when required and providing company when he felt lonely. She must also have advised him on the selection of a house the future couple would move into. But apart from this Cornelia also studied while in Groningen. She took lessons from Barend Sijmons (1853–1935), who was a private teacher of German and English language and literature (in 1881 Sijmons was appointed professor of comparative Indo-German linguistics).

In November 1878 Kapteyn expressed his fear to his future wife that the government would close down one or maybe even two universities, which would almost certainly involve the closure of Groningen University. In his letter to Elise of November 7, he intimated to her that he may soon be out of a job:

'If, for example, only Groningen is closed, then Leiden will get additional positions. So it could be that I will be transferred there to share astronomy with van de Sande Bakhuyzen. If that would happen I will have to do all the teaching and Bakhuyzen will be director of the observatory (I would rather stay here then). If they have no use for me, I will simply remain idle, which is not so bad. All in all, I feel that we should hope things will remain the way they are, although personally I have always felt that it would be expedient for the country and for the students to close 1 or even better 2 universities.'

He regularly admitted that he felt his lecturing was not of a high quality and, weeks before the event, he wrote about his anxiety and nervousness in relation to a lecture he had promised to give to the 'Natural Sciences Society'. This is the current Royal Natural Sciences Society (see page 276).. He eventually gave the lecture on November 28, 1878. The archives of the society show that he talked about the 'relation between various groups and clusters of stars, binaries and multiple star systems, the Sidereal System and of nebulae' (see also footnote on page 71 in the *Love Letters*). A few days later (in his letter of November 30, 1878) he wrote that he was reasonably happy with it, having decided not to read the text but improvise on the basis of prepared notes.

Elise's background was more Anglo-Saxon-oriented than Kapteyn's. He even described her as 'half-English'. At his request she occasionally wrote to him in English, but he never wrote back in that language. On February 23, 1879 he noted: 'I have been foolish enough to start a letter in English. What nonsense. What would others think of me: English home decoration, English carpets and curtains, English sisters [his sister Albertina Maria (Bertamie) had an English address in the letters], half-English wife and now English letters on top of everything!'

4.8 Finding a Place to Live

Two main issues play a major role in the *Love Letters*, namely the preparations for their marriage and the difficulties with his parents on the issue of religion on the one hand, and the preparations for their future home on the other. Let me start with the latter subject.

The first reference was in a letter of October 27, 1878, in which he listed an inventory of what is needed. 'It seems to me unavoidable that first what we need for myself is: 1 study, 1 workshop, 1 room for physics experiments, 1 room for sports and physical exercise, 1 bathroom, and then of course for further use 1 living room, 1 lounge, 1 bedroom, 2 guestrooms, and in addition a kitchen, a loft, a cellar and further essentials. That already makes 10 rooms in total. But if you would sit with me regularly in my study, and if we do our physical exercise in the workshop, which would also be the room for physics experiments (watch out for the glass-works), and buy some room dividers and make no friends who love lounges, we can do with 5 or 6.' As a sideline, he also referred to a constellation M.G. that he wanted to see in the sky, but had failed to do until then. In the *Love Letters* this constellation remains unidentified. I believe he actually meant M.Q. for 'Muur Quadrant' (Quadrans Muralis), an extinct constellation between Draco and Bootes. It should indeed be visible from the Netherlands at that time of year.

He managed to find a suitable house that was under construction on Winschoterkade, which was in an area outside the traditional city limits and would seem to be somewhat unusual for a young professor to go and live (see Fig. 4.7). The address in the archives of the City of Groningen is WZ(= West Zijde, West Side) X132. The location made for a relatively low rent, however. The part where this house was situated (it has been demolished in the meantime) is now called 'Winschoterdiep'. It is along a canal of that name running eastwards. It had two floors, which could be rented separately, but he proposed to rent them both. The rent amounted to 500 Guilders (11,500€ purchasing power today) (per year). Willink (*De Tweede Gouden Eeuw, (op. cit.*) quotes an average (annual) income for professors after the 1876 law of 4,900 Guilders (equivalent to 10.6 yearly wages of an unskilled worker). There were also preparations for Elise to visit Groningen in December 1878 (Kapteyn was to sleep elsewhere!) and see the house. Kapteyn had had some changes made while the house was being built.

On January 24, 1879, he sent a letter with the lay-out of the house (reproduced in Fig. 4.9). In an enclosure Kapteyn noted that he has measured the rooms himself as 'the carpenters do their measuring too inaccurately to my taste'. In further letters there is talk about what colors to paint the house ('floors: doesn't matter, the rest everything in light gray, mantelpieces completely black') and there is an exchange of ideas with regard to wallpaper and furniture, etc., and reports of purchases. Also he complained about the poor job that the painters are doing. At last the house was ready, as he reported on June 14, 1879 (albeit not yet fit to live in), and he asked Elise to come to Groningen to select wallpaper and carpets, among other things. A few days later he wrote to congratulate her on her birthday (June 19 was her 24th), and more letters around that time show much activity to get the wedding organized.

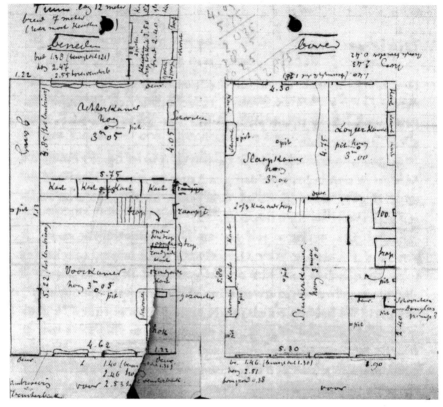

Fig. 4.9 Lay-out of the house Kapteyn rented on the Winschoterkade. This is a copy of the drawing he sent to his fiancée Catharina Elisabeth (Elise) Kalshoven as an enclosure to a letter on January 24, 1879 (Illustration reproduced from the *Love Letters* [8])

In the same period, physics professor Mees became seriously ill and Kapteyn had to stand in for him on a number of occasions, such as 'playing a major role' at the thesis defense of 'one Onnes'. Heike Kamerling Onnes obtained his doctorate on July 10, 1879 – only 6 days before Kapteyn's wedding – on a thesis on *New proofs of the rotation of the Earth*. Mees was the supervisor, but that role was taken over by professor Enschedé. Kapteyn's sister Cornelia moved to different quarters; it is not clear when she left Groningen, but it must have been well after the marriage took place. Kapteyn and his wife were forced to live at Kapteyn's quarters for a short period of time, since the home on Winschoterkade was not finished yet.

4.9 The Marriage

We have seen that in his youth Kapteyn must have started to seriously doubt the religion of his parents. As of 1811, legal marriages were performed by an official

4.9 The Marriage

of the municipality where they took place and registered by this organization in the Registry of Births, Deaths and Marriages. A church wedding before a priest or pastor was optional, although considered compulsory by church officials for everyone in their 'flock'. For persons under the age of thirty, a civil marriage could only take place with the parents' permission and approval. This is still the case at present, except that the maximum age that this permission is required is now eighteen. Kapteyn and his fiancee had decided not to marry in church, which met with the severe disapproval of Kapteyn's parents and resulted in their refusal to give the required permission. It had been an important subject of discussion, when Kapteyn's parents came to visit him in Groningen in October 1878. In his letters to Elise Kapteyn described the visit as very pleasant, but awkward when the issue of religion came up. Father Gerrit Kapteyn actually wrote to his son on this subject in January 1879, saying that he and his wife would refuse to be present at the wedding, since he believed that their presence could be seen as an approval of their son's principles. Kapteyn asked his father (*'Love Letter'* [22] of January 17, 1878, had a copy attached to it of a letter to his father of the previous day.) to take the point of view that if the parents made some formal statement to the rest of the family that their presence should definitely not be seen as such, there could be no reason not to come to the wedding. He implored him to accept his proposal to simply make it known to everyone that he did not agree with his son's principles, rather than refuse to be present and end their relationship. His mother wrote that 'we do not refuse our permission, but we will not in person be present at the wedding'. However, a little later father Kapteyn retracted that statement. Kapteyn also considered asking his brother Adriaan Pieter Marinus (Piet) for help and advice. Whether he did or not is not known.

Some of Kapteyn's brothers and sisters sided with the parents. His sister Cornelia Louise Alexandra had spent the greater part of his first year in Groningen with him, as we have seen. Most of the time their relationship was fine, but it seems that whenever the subject of religion came up, Cornelia expressed her vehement disapproval of Kapteyn's point of view. The subject had apparently been avoided in most of their conversations. The relationship between Kapteyn and his sister must at times have been uncomfortable at the very least.

As appears from a letter, Kapteyn's sister Albertina Maria (Bertamie) also chose the side of the parents. Kapteyn actually wrote to Elise in excited terms on February 6, 1879, saying that she should urgently see her half-sister Jacqueline Wilhelmine, as Bertamie had written her begging her to break off her friendship with Kapteyn! Further, Elise apparently visited the parents in Barneveld in January 1879, but that did not make them change their minds either, as might have been expected.

In the end Kapteyn and Catharina Elisabeth Kalshoven married in church. In the last *Love Letters*, Kapteyn twice mentions the name 'Trénité', who was a priest of the Eglise Wallone, the Walloon Reformed Church. The 'Walloon' refers to the fact that members of this church came from the Southern Netherlands or France and French was the language commonly used in the church. The priest is Jean Gédéon Lambertus Nolst Trenité (1830–1889). Van Berkel & Noordhof-Hoorn in the *'Love Letters'* spells his name with another acute accent on the first letter 'e'. This seems incorrect (see [11]). In a footnote to his letter of July 3, 1879 (two weeks before

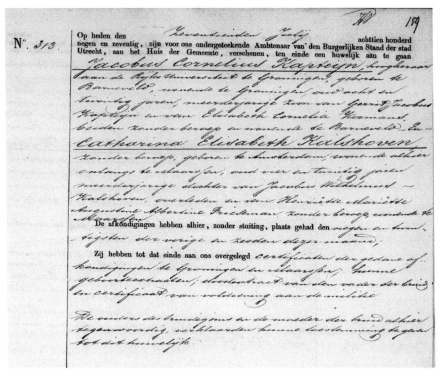

Fig. 4.10 Marriage certificate of the wedding of Jacobus Kapteyn and Catharina Elisabeth (Elise) Kalshoven on July 16, 1879 (upper part) (From FamilySearch [9])

their marriage), Kapteyn wrote 'Have you been to Trénité? What time in the church? Ask Simon or Jacques [Simon Brouwer and Jacques Kalshoven] to make a reservation as quickly as possible for the wedding chamber at the town hall half an hour or an hour before the ceremony in the church. Then write me the time, so I can inform W. Burck.' Kapteyn maintained a close friendship with Burck since his Leiden days (see page 81).

Jacobus Kapteyn and Elise Kalshoven married on July 18, 1879 in Utrecht. The marriage certificate (reproduced in Figs. 4.10 and 4.11) shows the signatures of Kapteyn's parents. The handwritten text just above the signatures identifies the witnesses. It reads: '... and in the presence of Johan Christaan Kalshoven, shipping agent, age thirty five years, living in Rotterdam, brother of the bride, Simon Brouwer, notary, age forty six, living in Maarssen, brother-in-law of the bride, Henrik van den Broeke, teller, age forty, living in this city and Willem Burck, teacher at the Hoogere Burgerschool [HBS], age thirty two, living in Apeldoorn.' The names are familiar, except that of van den Broeke.

Who was he? This has not been easy to answer. Henrik van den Broeke, as it turns out, was the husband of Elisabeth Henriette Andrée Wiltens (1846–1940), the older half-sister of Kapteyn's close friend Henry William (Willie) Andrée Wiltens

4.9 The Marriage

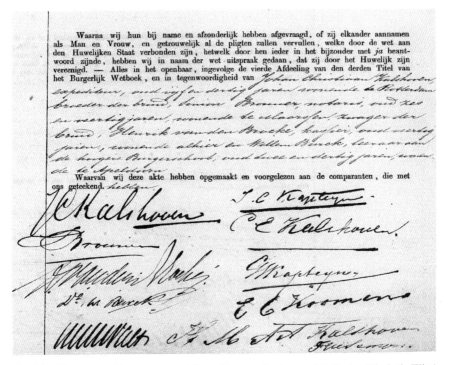

Fig. 4.11 Marriage certificate of the wedding of Jacobus Kapteyn and Catharina Elisabeth (Elise) Kalshoven on July 16, 1879 (lower part) (From FamilySearch [10])

(see page 32). Father Andrée Wiltens had first been married to Sara Wille (1822–1857), also while he lived in the East Indies and whom he apparently divorced. In 1850 he married his second wife, before the death of his first wife. Henrik van den Broeke (1839–1903) married Elisabeth in 1870. His first name is sometimes spelled as Hendrik, but his marriage certificate in Utrecht clearly states his name as Henrik. He and his wife later moved to South Africa, which is where they both died. It would be fairly obvious for Kapteyn to ask his old friend Willie Wiltens to be his best man. He featured in the *Love Letters* a number of times, but at that time he was back in Padang, where he had bought a law firm and he probably could not afford the voyage to the Netherlands. Kapteyn, in one of his letters, mentions him often in relation to his (=Wiltens') fiancee, identified as 'Trui', for whom Wiltens was saving money so that she could move to Padang. However, at the time of Kapteyn's marriage their relationship was in heavy weather. We know that in 1880 Wiltens married Johanna Josephine van Hulsteijn (1855–1929), born in Semarang, also in the Dutch East Indies. This would seem to be a different person; the first name 'Trui' usually is short for Geertruida or so and not for Johanna. Van Berkel and Noordhof-Hoorn identify 'Willie' Wiltens as *Willem Hendrik Andreas Wiltens*, who enrolled in the University of Utrecht in 1869. He may have rearranged his first names while still in the Netherlands.

Why the Kapteyns eventually decided to marry in church is not clear from the *Love Letters* (no more love letters were written after they had got married). However, it seems surprising in the light of their strong views on the issue. It may have been the result of efforts on the part of some intermediary between the couple and Kapteyn's parents. It is also likely that the health of father Kapteyn played a role; he died ten days after the marriage and the young couple might have felt that with his impending death it was prudent to put their objections and concerns aside.

In her *HHK biography*, Henriette Hertzsprung-Kapteyn devoted a few paragraphs to the early relationship between her parents. She and her mother must have been in touch on a regular basis when she was writing the book, and it is only natural that she based the text below on information supplied by Catharina Elisabeth Kapteyn-Kalshoven.

'In the summer of 1878 he proposed to Elise Kalshoven and she accepted. Both knew it felt good, and they looked upon the future with much confidence. They were engaged for a year, during which time Kapteyn traveled frequently to Utrecht. They were also quite serious, in spite of all the joy, discussing the great dreams Kapteyn had and which Elise listened to attentively. A new world of scholarship, and science in particular, opened up to her, a world of an intellectually higher level than the one she had grown up in.

The enormous revolution that Darwin's [(1809–1882)] *Origin of Species* had brought about left a strong mark on their times. It was the era of rationalism, and a young generation had chosen against worn-out forms and dogmas. They felt like gods in the kingdom of intellect, and old conventional traditions were thrown aside as dead branches on the perpetually green tree of life. In the Netherlands, the voice of Multatuli thundered and swept along all who were young and ablaze and not yet immune to it in the safe care of old familiar traditions. Kapteyn revered Multatuli and agreed enthusiastically with his 'Freedom, everybody's greatest right!'.

He introduced his future wife to this world of new ideas. The first book he suggested to her was Drapers' *History of the Conflict between Religion and Science*. These problems were too difficult for her, but all the same they had a strong influence on her thinking and filled her with admiration. She felt uplifted in an atmosphere that felt to her earnest and pure soul as the most beautiful way to live, searching for truth and freedom from narrow-minded traditional ways that mankind had had to endure for so long. Together they bought the complete *Household Edition* of [Charles] Dickens's [(1812–1870)] work, which became a source of happiness throughout their lives. The bust of Dickens later decorated their living room and his spirit lived in the family.'

Chapter 5
From Kepler to Parallax

> *You fathom the stars and measure their lane*
> *no high nightly vigil reveals any course*
> *your paper discloses the paths that never wane,*
> *you weigh sphere and sphere: their distance, mass and force.*
> Albert Verweij (1865–1937)[1]

> *To be idle is a short road to death and to be diligent is a way of life;*
> *foolish people are idle, wise people are diligent.*
> Gautama Siddhartha (ca.563–ca.483BC).[2]

So, by the end of 1879, Kapteyn had settled in Groningen as a married man and a university professor. What next? He had few students, no facilities to teach practical astronomy (at least not to the extent he felt he should) and no facilities to perform the astronomical research of the kind he did in Leiden. But this by no means deterred him from actively involving himself in astronomy. As we will see, at this stage of his career he also turned his hand to research in other subjects, viz., mathematics and theoretical astronomy, but also tree rings and climate. At some later point in his life he published on subjects such as statistics, and all along he kept a keen interest in disciplines such as geology and ornithology. At the same time he maintained an active research interest in observational astronomy, even though he had no observatory at his disposal. As early as 1880, this resulted in the development of a new method to derive the position of the celestial pole, which he tested during a vacation period in 1882 when he was allowed to use the Leiden facilities to obtain the necessary observations. Also in these first years in Groningen, Kapteyn experimented with a method of measuring right ascension differences between stars with meridian passage timing to determine parallaxes (distances) of large numbers

---

[1] From the poem 'To an astronomer'. Dutch poet Albert Verweij was a member of the so-called Generation of the Eighties (c.1880–1894). Translation by Gerard van den Hooff.

[2] Gautama Siddhartha, founder of Buddhism.

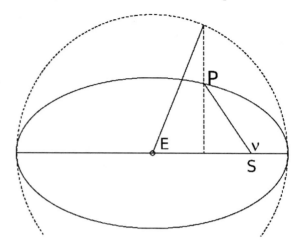

Fig. 5.1 Diagram explaining the definition of eccentric anomaly E and the true anomaly v. The ellipse is the orbit of planet P and the Sun S is in one of its foci

of stars. But before turning to his early work in Groningen, I quote a paragraph from Wessel Krul's chapter in the *Legacy*.

'Why did Kapteyn stay in Groningen at all? During his first years at Groningen, he had very little to do. The number of students in physics and mathematics was small, and it was impossible to specialize in astronomy. Lacking the necessary instruments, he could not offer a proper training in observation, and worse, he could only continue the research he had been doing at Leiden for short periods during the vacations. In the circumstances, a choice between two options seemed inevitable: either to take things easy and settle down to contented mediocrity, or to look for the nearest opportunity to find a job elsewhere.'

In fact, Kapteyn actually did neither of the two. He looked for interesting problems (in astronomy and elsewhere) that he could address without immediate access to a telescope; and not without success.

5.1 Kepler's Equation

The theoretical astronomical problem that caught Kapteyn's interest is that of Kepler's equation. This equation has to do with the orbits of planets and other bodies in the Solar System and looks deceptively simple, probably even to those less familiar with mathematics, viz. (the meaning of these symbols has been explained in Box 5.1)

$$M = E - e \sin E.$$

The description of the motion of a planet (or comet or asteroid) around the Sun is based on the laws that Johannes Kepler (1571–1630) wrote down for the planetary orbits (see Box 5.1). The first law says that the orbits are ellipses with the Sun in

Box 5.1 The two-body problem

The **Kepler Laws** of planetary orbits are the following.
1. Planets move in elliptical orbits with the Sun in one of the foci.
2. The planet sweep out equal areas in equal times. This means that the angular velocity as seen from the Sun is inversely proportional to the distance of the planet, so moving faster when nearer the Sun and vice versa.
3. The *Harmonic Law* says that the square of the period in the orbit is proportional to the cube of the semi-major axis.

Isaac Newton (1643–1727) showed that these three laws all followed directly from his theory of the gravity when used to solve the **two-body problem**. If r is the distance of the planet from the Sun at any particular time, a the semi-major axis, P the period in the orbit, V the velocity in the orbit, e the eccentricity, $M_\odot$ the mass of the Sun and G Newton's gravitational constant, in mathematical terms the Kepler Laws read:

$$\text{(I.)} \quad r = \frac{a(1-e^2)}{1+e\cos v} \; ; \quad \text{(II.)} \quad V^2 = GM_\odot \left(\frac{2}{r} - \frac{1}{a}\right) ; \quad \text{(III.)} \quad \frac{a^3}{P^2} = \frac{GM_\odot}{4\pi}$$

The first equation can be simplified using the eccentric anomaly E (see Fig. 5.1) to become:

$$r = a(1 - e\cos E) \; ; \quad \tan\left(\frac{E}{2}\right) = \sqrt{\frac{1-e}{1+e}} \tan\left(\frac{v}{2}\right).$$

Kepler's equation is a tool to derive where a planet or other body is in its orbit at any given time. Define a virtual, uniformly increasing *mean* anomaly M, moving with angular velocity $n = 2\pi/P$. Kepler's equation then is

$$M = n(t - T_\circ) = \frac{2\pi}{P}(t - T_\circ) = E - e\sin E,$$

where t is the time corresponding to the value of M and $T_\circ$ the time when the planet went through perihelion.

---

one of the foci. In Fig. 5.1 the ellipse is the orbit of the planet and the point labeled with the 'Sun-symbol' $\odot$ is the position of the Sun. The angle v is called the *true anomaly*, and is the angle as seen from the Sun between the line towards the planet and that of its perihelion, where the planet is in its orbit closest to the Sun. The flattening of the ellipse is indicated with the symbol e and is called the eccentricity. The ratio of the axes of the ellipse is $\sqrt{1-e^2}$, so that for a circle $e = 0$ and for a line $e = 1$. The *true* anomaly v does not increase uniformly with time, as the planet moves faster when closer to the Sun. This is Kepler's second law, which we now know as expressing conservation of angular momentum.

If one wants to calculate where a planet, asteroid or comet at a particular time, say t, would be in its orbit, then one needs to find the corresponding v. The trick

Box 5.2 Series expansion of Kepler's equation
A simple expansion is the following, based on the Lagrangian inversion theorem [due to Joseph-Louis Lagrange (1736–1813)]:

$$E = M + \sum_{n=1}^{\infty} a_n e^n,$$

$$a_n = \frac{1}{2^{n-1} n!} \sum_{k=0}^{n/2} (-1)^k \binom{n}{k} (n-2k)^{n-1} \sin[(n-2k)M].$$

The binomial coefficients are

$$\binom{n}{k} = \frac{n!}{k!(n-k)!} = \frac{n(n-1)(n-2)\ldots(n-k+1)}{k!}.$$

So you proceed by evaluating the terms $a_n e^n$ in the first equation, using the second one to calculate the a_n, starting with $n = 1$ you continue this process until the terms become so small that they can be neglected. Needless to say this is very cumbersome and time-consuming in actual practice, especially when e is not small. In fact, in this particular example, when $e > 0.663\ldots$ (the Laplace limit), the series diverges and becomes completely useless. Obviously, to a clever mathematician it constituted a challenge to develop more practical procedures, and this is exactly what Kapteyn set out to do.

devised by Kepler is to use the eccentric anomaly E, which is found with the construction as indicated in Fig. 5.1 using the enclosing circle. This is a useful change of variables, as E is related to v by a simple relation, which can be easily evaluated. But then the problem is to find E at a given time and that is where Kepler's equation comes in. Kepler defined this (in different terms and notation) in his book *Epitome Astronomiae Copernicanae*, which he published in several installments between 1617 and 1621. This work is a textbook to introduce students to planetary astronomy according to Kepler (in spite of the title), presenting – among other things – his laws on the motion of the planets and explaining how to calculate the positions of the planets from the properties of their orbits. Volume V presents what we now know as Kepler's equation, although there it is introduced as a geometrical construction and a recipe in Latin of how to proceed.

For any object, as long as we know its period around the Sun and one moment of passage through its perihelion we can call upon Kepler's equation to find what the corresponding E is and therefore v. In the late nineteenth century this had to be done not only for planetary positions, but also to determine the orbits of (newly discovered) asteroids and comets, and subsequently to keep track of these objects.

The problem is that Kepler's equation is a so-called transcendental equation when using it to find E for a known M. If you know E, it is easy to calculate M: just fill in the value of E and you find M. But the other way around is not so easy. The reason

5.1 Kepler's Equation

is that E cannot be expressed as an algebraic function of M. In other words, no formula exists in which you can fill in M and calculate E. Other methods were then required, involving numerical analyses and series expansions, and these were very cumbersome and time consuming. Therefore methods to speed up this calculation were urgently needed. Various methods had been devised to solve Kepler's equation (see e.g., F.R. Moulton, *An Introduction to Celestial Mechanics*, 1914 [2]). Theodor von Oppolzer (1841–1886) wrote the standard text on the determination of the orbits of planets and comets (*Lehrbuch zur Bahnbestimmung der Kometen und Planeten, 1870/1880* [3]) and that was basically the state of affairs at the time when Kapteyn engaged this problem.

Before describing Kepler's approach, let me first explain in general terms what procedures were devised. In order to present the problem Kapteyn wanted to tackle in understandable terms, I will use as little algebra as possible. If you are not primarily interested in the technical details, you can skip to the ¶ symbol on page 117.

There were two general classes of procedures to solve Kepler's equation (for known eccentricity e), the first being by *iteration*. So, for a particular time you want to calculate the position of an object in its orbit. If you know the object's orbital period and time of perihelion passage, you can easily calculate M. What you need is E. An example of a very simple iteration procedure is the following: Rewrite Kepler's equation by bringing E to the left and M to the right:

$$E = M + e \sin E.$$

Then take as a first guess that on the right-hand side $E = M$. Fill that in in the equation and you get a new – and, as it turns out – improved estimate for the value for E, viz. $E' = M + e \sin M$. Fill that in again in the equation and you get $E'' = M + e \sin E'$, which is an even better estimate for E and then repeat this process until the subsequent values you find converge. That then is the solution to Kepler's equation; you have found E and from that it is easy to calculate v and the position of the object in its orbit at the time t.

However, the procedure to solve Kepler's equation this way is slow and time-consuming, especially when e is not small, which is often the case for asteroids and in particular for newly discovered comets. Now, there existed much-improved versions of iteration procedures, but even the best were still unacceptably time-consuming, considering the other work that astronomers were faced with.

The other approach is through *power expansion*, where one writes E as a power series in e, based on mathematical properties of Kepler's equation. This means that one designs a formula for calculating E for known M using an infinite set of terms, but such that these terms become smaller and smaller so that after a small number of terms the remaining ones can be safely neglected and ignored. A relatively simple example of this, just to show what is involved, is given in Box 5.2.

The convergence of such algorithms can be illustrated by the following table for the case that the mean anomaly $M = 30°$ and the eccentricity $e = 0.1$.

step	0	1	2	3	4	exact
iteration	30°00'00"	32°51'54"	33°06'32"	33°07'48"	33°07'52"	33°07'52"
power series	30°00'00"	32°51'54"	33°06'47"	33°07'49"	33°07'52"	33°07'52"

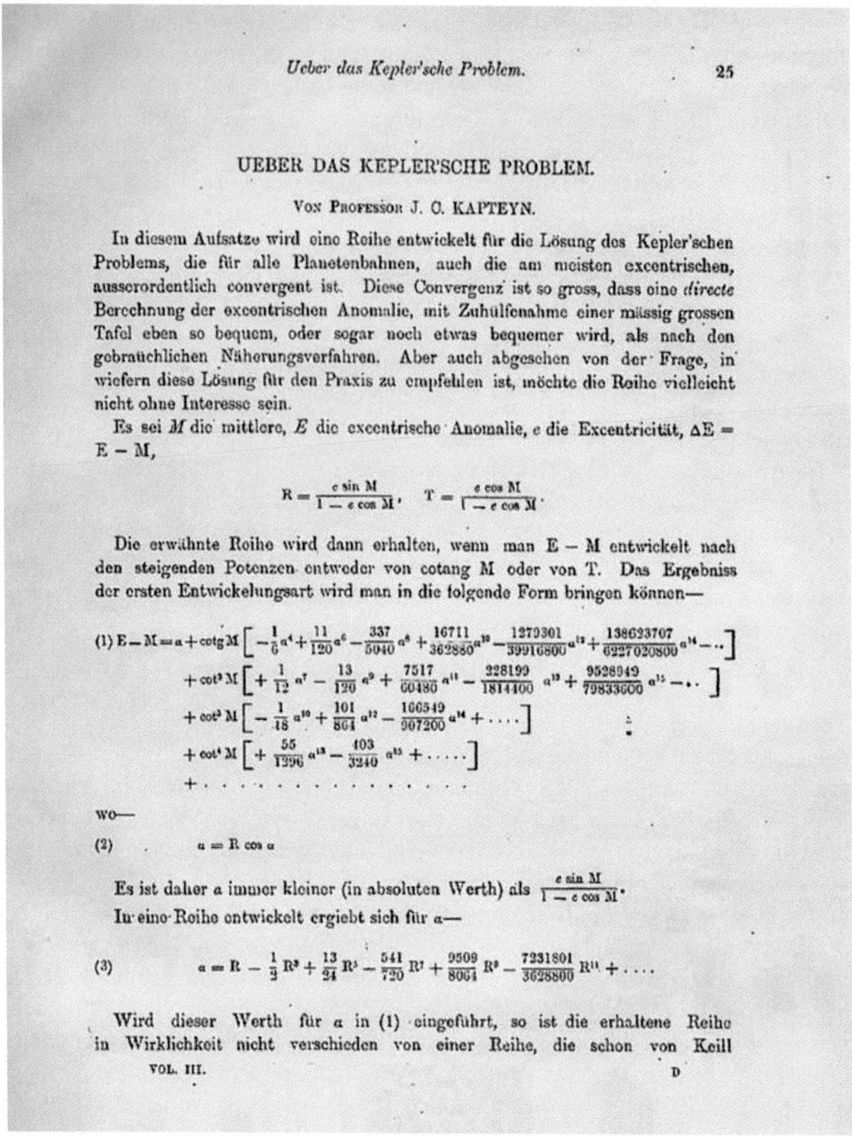

Ueber das Kepler'sche Problem. 25

UEBER DAS KEPLER'SCHE PROBLEM.

Von Professor J. C. KAPTEYN.

In diesem Aufsatze wird eine Reihe entwickelt für die Lösung des Kepler'schen Problems, die für alle Planetenbahnen, auch die am meisten excentrischen, ausserordentlich convergent ist. Diese Convergenz ist so gross, dass eine *directe* Berechnung der excentrischen Anomalie, mit Zuhülfenahme einer mässig grossen Tafel eben so bequem, oder sogar noch etwas bequemer wird, als nach den gebräuchlichen Näherungsverfahren. Aber auch abgesehen von der Frage, in wiefern diese Lösung für den Praxis zu empfehlen ist, möchte die Reihe vielleicht nicht ohne Interesse sein.

Es sei M die mittlere, E die excentrische Anomalie, e die Excentricität, $\Delta E = E - M$,

$$R = \frac{e \sin M}{1 - e \cos M}, \quad T = \frac{e \cos M}{1 - e \cos M}.$$

Die erwähnte Reihe wird dann erhalten, wenn man $E - M$ entwickelt nach den steigenden Potenzen entweder von cotang M oder von T. Das Ergebniss der ersten Entwickelungsart wird man in die folgende Form bringen können—

(1) $E - M = a + \cotg M \left[-\frac{1}{6}a^4 + \frac{11}{120}a^6 - \frac{337}{5040}a^8 + \frac{16711}{362880}a^{10} - \frac{1279301}{39916800}a^{12} + \frac{138623707}{6227020800}a^{14} - \ldots \right]$

$\quad + \cot^3 M \left[+\frac{1}{12}a^7 - \frac{13}{120}a^9 + \frac{7517}{60480}a^{11} - \frac{228199}{1814400}a^{13} + \frac{9528949}{79833600}a^{15} - \ldots \right]$

$\quad + \cot^5 M \left[-\frac{1}{18}a^{10} + \frac{101}{864}a^{12} - \frac{166549}{907200}a^{14} + \ldots \right]$

$\quad + \cot^7 M \left[+\frac{55}{1296}a^{13} - \frac{403}{3240}a^{15} + \ldots \right]$

$\quad + \ldots$

wo—

(2) $\qquad a = R \cos a$

Es ist daher a immer kleiner (in absoluten Werth) als $\frac{e \sin M}{1 - e \cos M}$.

In eine Reihe entwickelt ergiebt sich für a—

(3) $\qquad a = R - \frac{1}{3}R^3 + \frac{13}{24}R^5 - \frac{541}{720}R^7 + \frac{9509}{8064}R^9 - \frac{7231801}{3628800}R^{11} + \ldots$

Wird dieser Werth für a in (1) eingeführt, so ist die erhaltene Reihe in Wirklichkeit nicht verschieden von einer Reihe, die schon von Keill

VOL. III. D

Fig. 5.2 First page of Kapteyn's paper on the problem of Kepler's equation in the journal *Copernicus*

As it happens, the first terms in the two approximations happen to be the same, but then the results become different. This does not look too bad, although we have to remember that at that time all calculations were done by hand (or using tables or a slide rule). Furthermore, for this kind of work the required accuracy was comparable

5.1 Kepler's Equation

to that of measurements of positions and that is about 0″.1. For larger eccentricities the convergence is slower. For $e = 0.3$:

step	0	1	2	3	4	exact
iteration	30°00′00″	38°35′40″	40°43′21″	41°12′50″	41°19′31″	41°21′27″
power series	30°00′00″	38°35′40″	40°49′38″	41°18′41″	41°25′44″	41°21′27″

We see that for this larger eccentricity the power series expansion method converges much more slowly. These simple examples show, that there was a great need to make the calculations easier and faster.

Kapteyn investigated the possibility that there might be more convenient ways of expressing Kepler's equation as a series. Indeed the first page of the paper (see Fig. 5.2) shows that much algebra was involved. The equation in the middle of that page in fact is the series he proposes to use, which does not look less impressive than that presented in Box 5.2. However, he shows that it can be written as

$$E - M = \alpha + \beta \cot M + \gamma \cot^2 M + \delta \cot^3 M + \ldots,$$

(where cot is the trigonometric cotangent function) and the coefficients α, β, etc. can be calculated using a variable R, which is a combination of e and M (viz. $R = (e \sin M)/(1 - e \cos M)$). In principle these α's, β's, etc. can be tabulated once and for all for various values of e. Kapteyn then shows that for the planets no more terms are necessary than that of the fourth order (one beyond what is written out in the equation) to get an accuracy in the solution for E of 0″.01 or less and this constitutes a significant improvement. He works out three examples in detail based on the asteroids Juno ($e = 0.2559$) and Aethra ($e = 0.389$) and the periodic comet Faye (he refers to it as 'Faye-Möller'), which is a comet discovered in 1843 with a period of 7.55 years and an eccentricity of 0.5666. Even for the last one he needed no more than the sixth-order term.

¶ After presenting the method, Kapteyn compared it to the one described and recommended in the classical textbook by von Oppolzer. He did so by working out a typical example, and concluded that his proposed formula constituted a worthwhile improvement. However, close reading of the text tells me that this is a conclusion that he really had some difficulty justifying and the improvement can probably be best described by 'marginal'. Indeed, Kapteyn's paper never attracted much attention and the method has not found wide application. Or, to be more precise, no application whatsoever. In the Astrophysics Data System I have found 20 papers on the subject of the solution to Kepler's equation for the period 1885–1900 (which illustrates the urgency of the problem); only one contains a reference to Kapteyn's paper, but that is no more than a compilation of all the – approximately – 140 publications on the subject since Kepler's own. Neither does the massive *Handwörterbuch der Astronomie in vier Bänden* by Karl Valentiner, mention it. This work (of five volumes actually, as Volume III consists of two parts) is a comprehensive compendium, written by many contributors and with Valentiner as the final editor, which covers all of astronomy in 2770 pages of text and 240 pages of tables, index, definitions, etc., published between 1897 and 1902 [4]. This is a clear indication that the work did not attract special attention, but what is even more telling is the following. In 1993, a book was published by Peter Colwell entitled *Solving Kepler's equation over three centuries* [5], but in spite of being a comprehensive study and historical review of the subject, Kapteyn's work is not mentioned at all! Remarkably, Kapteyn's brother Willem, on the other hand, *is* mentioned in the book, but I will come to that below.

5.2 Copernicus

For this paper on Kepler's equation, Kapteyn (1883), together with another one, published in 1884, which will be discussed below, we have to search in an unusual place, namely in the journal *Copernicus*. This journal was granted only a short life and is not included in the NASA Astrophysics Data System (ADS). Many modern astronomers have not even heard of it. It originated from Dunsink Observatory (or Trinity College Observatory) of Dublin University [6], that was built between 1783 and 1785 using a bequest to Trinity College from Provost Andrews [7]. The journal *Copernicus* was set up by two astronomers at Dunsink, John Louis Emil Dreyer (1852–1926), born as Johan Ludvig Emil Dreyer in Copenhagen, who had worked in Ireland since 1874, and Ralph Copeland (1837–1905), a British astronomer. The story of the journal *Copernicus* has been described as follows by Wolfgang Steinicke in his book *Observing and cataloguing nebulae and star clusters: From Herschel to Dreyer's New General Catalogue* [8]:

'In August 1878, aged scarcely 26, Dreyer moved to Dunsink, the site of Trinity College Observatory. [...] Ralph Copeland – who had been assistant at Dunsink from 1874 to 1876, before changing to Lord Linsay's Dun Echt Observatory in Scotland – might have recommended him for the job. [...] Dryer and Copeland edited the Dublin magazine *Copernicus* bearing the subtitle 'International Journal of Astronomy'. From January to July 1881 it was first titled 'Urania'. Since a publication with the same name already existed (though with astrological content), a new name was chosen. Lord Linsay, the owner of Dun Echt Observatory, financially supported the project, but obviously the enthusiastic publishers had underestimated the international competition in the shape of the *Astronomical Journal, Bulletin Astronomique* and *Astronomische Nachrichten* [Respectively published in the USA, France and Germany.] [...]. Thus, after a notable initial success, *Copernicus* was abandoned in 1884, after only three volumes.'

Kapteyn must have sympathized with the initiative to establish this journal, as he published two papers in it. On an earlier occasion, Kapteyn had published in the renowned *Astronomische Nachtrichten* in conjunction with van de Sande Bakhuyzen, while – apart from the two other major journals mentioned above – he could have opted for the British *Monthly Notices of the Royal Astronomical Society*. The currently largest astronomical periodical, *The Astrophysical Journal* (USA), had not yet been founded; this did not take place until 1895, by George Ellery Hale (1868–1938) and James Edward Keeler (1857–1900). Kapteyn's preference for *Copernicus* over the more natural choice of *Astronomische Nachrichten* may be based on a characteristic tendency to like and support initiatives that ignored beaten tracks, but even so it is remarkable that he chose to publish his papers in German, in a journal that was, after all, published in an English-speaking country ('Copernicus' published papers in English, German and French). Maybe Kapteyn felt that he was still out of his depth in the English language.

5.3 Higher Order Sines

In the same years Kapteyn also carried out some purely mathematical work in co-operation with his brother Willem (see Fig. 5.3), who was, as I have mentioned before, a professor of mathematics in Utrecht. It concerned what they called in their publications 'sine functions of higher orders'. For the convenience of mathematics affectionados, the basics are summarized in Box 5.3. It concerns series expansions that look like those of the classical sine and cosine functions, but with orders left out in regular patterns. It would have been gratifying to find that this would somehow or other relate to Kapteyn's work on Kepler's equation, but the connection is not obvious at all.

Kapteyn and his brother Willem published two long papers about this, Kapteyn & Kapteyn (1884) of 98 and Kapteyn & Kapteyn (1886) of 62 pages full of mathematical equations. Both papers have Jacobus as the first author, but it seems to me that this does not imply he also did the major share of the work. It is probably simply the alphabetic order. This conclusion is supported by the choice of language. In the early stages of his career Kapteyn published almost exclusively in German; the first paper with his brother, however, was in French. It was published in the proceedings of the Royal Netherlands Academy of Sciences, which accepted papers in various languages, including French. The second paper was in German, but that was published with the German Kaiserliche Akademie. It is not unlikely that they felt compelled to submit a manuscript in German to this body. The only other cases where Kapteyn published in French were in the proceedings of the Carte du Ciel Permanent Committee (see later). But there *every* paper was in French and the organizers of the Carte du Ciel project at the Paris Observatory must have insisted on French, and would certainly have had papers translated if they had been submitted in another language. But Willem Kapteyn published many papers in French, as a search on *Google Scholar* confirms, even on occasions when, apparently, French was not strictly necessary. Furthermore, studying infinite series was a subject very close to Willem Kapteyn's heart and on which he wrote many papers. I therefore believe that the conclusion that Willem actually wrote the papers, and probably did the lion's share of the work contained in them, is justified.

Box 5.3 Higher order sine functions
It is well-known that the trigonometric sine and cosine functions can be expressed as an infinite series, as follows

$$\sin x = \sum_{n=0}^{\infty} \frac{(-1)^n x^{2n+1}}{(2n+1)!} = x - \frac{x^3}{3!} + \frac{x^5}{5!} - \frac{x^7}{7!} + \cdots$$

$$\cos x = \sum_{n=0}^{\infty} \frac{(-1)^n x^{2n}}{(2n)!} = 1 - \frac{x^2}{2!} + \frac{x^4}{4!} - \frac{x^6}{6!} + \cdots$$

In these the powers of the subsequent terms differ by two. In their paper of 1886, the Kapteyn brothers studied the **general series**:

$$\varphi_\mu(x) = \sum_{n=0}^{\infty} \frac{x^{kn+\mu}}{(kn+\mu)!} = \frac{x^\mu}{\mu!} + \frac{x^{\mu+k}}{(\mu+k)!} + \cdots$$

$$\psi_\mu(x) = \sum_{n=0}^{\infty} \frac{(-1)^n x^{kn+\mu}}{(kn+\mu)!} = \frac{x^\mu}{\mu!} - \frac{x^{\mu+k}}{(\mu+k)!} + \cdots$$

The special case of the **fourth order sine functions** that the brothers Kapteyn studied in their paper of 1884 have $k = 4$:

$$\mathfrak{A}(x) = \sum_{n=0}^{\infty} \frac{(-1)^n x^{4n+1}}{(4n+1)!} = x - \frac{x^5}{5!} + \frac{x^9}{9!} - \frac{x^{13}}{13!} + \cdots$$

$$\mathfrak{B}(x) = \sum_{n=0}^{\infty} \frac{(-1)^n x^{4n+2}}{(4n+2)!} = \frac{x^2}{2!} - \frac{x^6}{6!} + \frac{x^{10}}{10!} - \frac{x^{14}}{14!} + \cdots$$

$$\mathfrak{C}(x) = \sum_{n=0}^{\infty} \frac{(-1)^n x^{4n+3}}{(4n+3)!} = \frac{x^3}{3!} - \frac{x^7}{7!} + \frac{x^{11}}{11!} - \frac{x^{15}}{15!} + \cdots$$

$$\mathfrak{D}(x) = \sum_{n=0}^{\infty} \frac{(-1)^n x^{4n+4}}{(4n+4)!} = \frac{x^4}{4!} - \frac{x^8}{8!} + \frac{x^{12}}{12!} - \frac{x^{16}}{16!} + \cdots$$

With a view to what follows, it is not necessary to understand in detail what the Kapteyn brothers did in these papers. Suffice it to say that they determined the mathematical properties of these series in detail. The paper of 1884 concerns the special case of the fourth-order sine functions and that of 1886 the general case of higher order functions. It may very well be that Kapteyn consulted his brother on what series expansions might be useful to use in the solution of Kepler's equation, and that they ended up considering these higher order sine-functions as a possible new approach. But it is also evident that these did not prove very useful in this respect.

5.3 Higher Order Sines

Fig. 5.3 Willem Kapteyn (1849–1927), brother to Jacobus Kapteyn, was a professor of mathematics at the University of Utrecht, as he appears in the *Album Amicorum* presented at the retirement of H.G. van de Sande Bakhuyzen as professor of astronomy and director of Leiden Observatory in 1908 (Archives Leiden Observatory; see caption Fig. 3.7)

Willem Kapteyn remained interested in infinite series, and he actually developed a set of series was named after him. These are too complicated to discuss here in any detail; they involve so-called Bessel-functions, which are a particular kind of advanced mathematical function (see for more details e.g. D. Dominici, *Some observations on a Kapteyn series* (2005) [9]). The irony of the matter is that these Kapteyn series turned out to be very useful for solving Kepler's equation, as Willem Kapteyn did note himself (W. Kapteyn, *Recherches sur les Fonctions de Fourier-Bessel*, and *Over Bessel'sche Functiën* [10]). So, in the end Jacobus Kapteyn was never mentioned in the book *Solving Kepler's equation over three centuries* by Peter Colwell, whereas Willem Kapteyn did! It is interesting that Colwell cited Kapteyn with the initials M.W. – as is indeed clearly printed on the title page of the (French) article –, being unaware that the M. was short for Monsieur. This has actually percolated into a number of further citations of the paper.

5.4 Tree Rings and Climate

So, Kapteyn's alternative to observational astronomy and find a better way to solve Kepler's equation were not really successful. And working with his brother on mathematical issues may have been a reasonably good investment of time, but in the end it probably did not give him the satisfaction he was looking for. One may well wonder what else he did in this period? Well, during his first years in Groningen, Kapteyn did extensive research into the connection between tree rings and weather. Why he turned to this is not recorded, but it may safely be assumed that his interests in this subject had much to do with the fact that he used to be a student of Buys Ballot in his days at the University of Utrecht. This work was not published until much later. In 1908, while visiting Mount Wilson Observatory, where he spent part of his time for a number of years, he gave a public lecture in Pasadena, California, on this piece of old research and actually published it in the local newspaper, the 'Pasadena Star' of December 19, 1908. The newspaper is now called the 'Pasadena Star News' and its old issues are kept on microfilm in the Pasadena Public Library. Unfortunately, these go back no further than 1911. However, the Research Library and Archives of Pasadena Historical Society at the Pasadena Museum of History has paper copies well back into the nineteenth century. Figure 5.4 shows the relevant pages from the newspaper.

The header above the article is 'Dutch scientist says growth rings of trees form weather record'. Undertitles are 'Dr Kapteyn describes investigations of the subject' and 'Believes study of cross sections of trees will allow prognostication of weather for years to come'. The article starts: 'Although the subject has been hinted at by other investigators, Professor J.C. Kapteyn, director of the astronomical observatory at the University of Groningen, Holland, is probably the only scientist who has made a careful study of trees as records of the weather'. And in the box: 'Prof. Kapteyn decides that weather is the same every twelve years and tells how he made the discovery'. After a few more words the full text of the lecture follows.

Kapteyn later decided to publish more formerly, and a reprint is available in the Kapteyn Room. However, it has no indication of where the article appeared (probably as a special publication), nor when. In it Kapteyn referred to his public lecture in Pasadena in 1908 and on some later occasion he referred to a publication in 1912, so it should be no earlier than that year. He mentions that he decided to publish when a similar paper appeared in 1909, so it should also not be too long after that. He further mentions that the work had taken place at a different place, over 30 years ago, and that this was in 1880 and 1881. So, 1912 seems a good guess. Cornelis Easton, in an article with recollections of Kapteyn published in the popular astronomy magazine *Hemel & Dampkring* in 1928, quotes it as having been published in 1916, but that seems too long after 1909 (Easton's article is partly reproduced in Appendix C, but the paragraph on this matter is not included there). Easton may easily have been confused with the publication that year of another non-astronomical paper, notably on skew frequency curves in biology and statistics.

5.4 Tree Rings and Climate

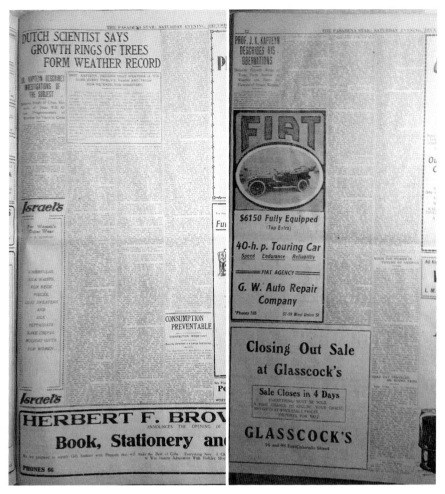

Fig. 5.4 Parts of pages 11 and 12 of the 'Pasadena Star' of December 19, 1908, reporting on Kapteyn's public lecture (Pasadena Historical Society [11])

I adopt 1912 as the most likely year for the tree rings paper. Since it was essentially all work he did in 1880 and 1881, I will discuss it here.

At the time Kapteyn was doing this work, weather prediction was in its infancy. Making any prediction of the weather, even one day ahead, was a hazardous undertaking. It was probably safest to say that the weather tomorrow would be the same as today, since any other prediction was very likely to be far from correct. In the reprint we have, Kapteyn first devotes a few pages to justifying his decision to publish, and gives some insight in why he conducted this research in the first place. He states that the goal of the investigation is to see whether or not there are systematic patterns in the weather over long timescales and maybe even regular periodicities. He compares the weather regularities he is looking for to the occurrence of solar and lunar

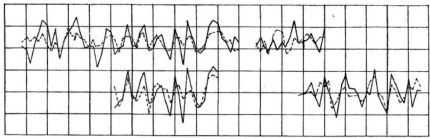

Fig. 5.5 Figure from Kapteyn's paper on growth rings in trees. The data apply to the neighborhood of Treves (Trier) in Germany. The horizontal scale is time and runs from 1770 to 1880. The data on the rings is in the full-drawn lines and is in fact the deviation for a particular year with respect to the average of the 7 previous and the 7 subsequent years; the dashed lines represent the number of wet days per year (top) and amount of rainfall (bottom) (Kapteyn Astronomical Institute)

eclipses in astronomy. After all, after examining records over long periods of time and from various places on Earth, ancient astronomers found that eclipses repeat themselves in the same pattern with a period of 6585 days and about 8 hours (a little over 18 years), which is known as the Saros period. The approximate 8 hours ensure that the pattern of eclipses repeats on a particular spot on Earth (more precisely, a spot with the same geographical longitude) after three Saros periods.

The reason for the periodicity is well understood. An eclipse occurs when at the same time the Moon is in one of its nodes (that is to say when in its orbit, it crosses the plane of the orbit of the Earth around the Sun, so that the Sun, Moon and Earth are in the same plane) *and* in the same phase (Full Moon for a lunar eclipse and New Moon for a solar eclipse). The Moon returns to a particular node every 27.21 days (the so-called draconic month) and a lunar phase repeats every 29.53 days (the synodic month). Now 223 synodic months are almost exactly equal to 242 draconic months and that is the Saros period. Also the Moon returns to the same point in its elliptic orbit every 27.55 days (the anomalistic month) and the Saros period just indicated is again very close to 239 anomalistic months. So after this Saros period the Sun, Moon and Earth return to almost exactly the same relative positions, and eclipses repeat with almost identical circumstances. It is not exact, so the Saros cycli do not last forever; old ones go away and new ones appear. This was determined in antiquity in an empirical way, by examining records of eclipses over long periods of time. Kapteyn explained that his motivation was to see to what extent there are similar periodicities in weather patterns. This may sound naive today, but at the time so little was understood about weather that this would appear to be a valid question. Kapteyn infers that many meteorological stations (recording temperatures and rainfall) are being erected and that long series of weather records will become available, but as that will take decades at least, he admits to being too impatient and therefore decides to look into the possibility that nature has kept such records in tree rings.

5.4 Tree Rings and Climate

Kapteyn (1912c) is in fact the article in the Pasadena Star, preceded by a few introductory remarks. For this research he collected disks from recently cut oak trees from various areas in Germany and the Netherlands. The material he discussed is only part of what he had available and came from regions in Germany, predominantly along the rivers Main and Moselle and – not far from there – near the Rhine. In total he reported on data obtained from disks of 52 trees, measuring in all of them the thickness of the annual growth rings. By averaging and adding results extracted from trees from the same areas, he is able to produce continuous records for seven areas ranging, except for one, from 1640 to almost 1880.

In her *HHK biography*, Henriette Hertzsprung-Kapteyn refers to these studies as well, albeit in a single paragraph.

> 'He was also busy with meteorological studies and he traveled to Worms and to Paris in order to do research on the growth of trees in relation to the climatic circumstances. He sent a request to the [Dutch] government to ask the French government for slices of trees, that were two hundred years old or more, from the surroundings of Paris. That was where the weather station was located that had the longest historical records of meteorological data, in particular of the amount of rainfall. He occupied himself for some time with research into growth rings of trees, but he never came to formulate a theory. Only much later did he publish the results.'

The study reported in the lecture mentioned no samples from Paris, so apparently these never materialized. Kapteyn measured the thickness of the tree rings for all samples he had and his first finding was that the profiles he got from these separate places in Germany correlated well with each other. He interpreted that to mean that the cause of different thicknesses of tree rings must be meteorological. He did find some records that classify summers as hot or cool, and winters as cold and mild, but saw no correlation with his patterns. So, temperature was not the determining factor. For Treves, which today is the city of Trier on the Moselle near the border of Luxembourg, has records of rainfall (see Fig. 5.5). To his satisfaction he noted a large degree of 'parallelism' between the measurements of ring thickness and amount of rainfall. Further he arrived at the important conclusion that in all the profiles there was a large degree of regularity with cycles of a well-defined period of 12.4 years. He did not speculate on what may have given rise to this, preferring to leave it to 'some competent men to point out the true causes of these regularities'. In the notes preceding the lecture he remarks that the recurring patterns are not correlated to the 11-year solar cycle of the number of sunspots.

Kapteyn concluded his Pasadena lecture by pointing out that the very old trees of California offer an excellent opportunity for this kind of work. He actually showed a 'representation of rings of a Sequoia of 1244 years old', which he obtained from Prof. McAdie. This undoubtedly is the meteorologist Alexander George McAdie (1863–1943), head of the US Weather Bureau in San Francisco at the time. To his 'no small surprise' he noted that peaks occur that are all multiples of 12.4 years apart; in the relevant period they even correspond exactly to the years of the maxima in his European data.

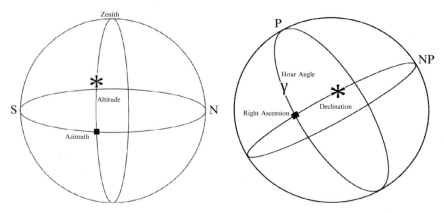

Fig. 5.6 The coordinates on the sky for a star in azimuth and altitude (left) and in right ascension (or hour-angle) and declination (right). The symbol ♈ indicates the position of the vernal equinox

To the best of my knowledge no such well-defined cycles are currently accepted as established fact.

5.5 Absolute Declinations

Kapteyn did embark also on two pieces on astronomical research. First he considered the matter of the measurement of declinations of stars. In the end this turned out to have profound consequences for his career, since it led to an association with David Gill, His Majesty's astronomer at Royal Observatory at Cape of Good Hope. The paper in which he reported this work was also published in the journal *Copernicus*, Kapteyn (1884). It was written again in German and the title was (in translation) *On a method to determine the altitude of the pole as much as possible free from systematic errors*. So, what was this paper about?

For a good understanding of the background and the importance of the matter, I first need to add some remarks about astronomical coordinate systems and the measurements of stellar positions on the sky.

Recall from Box 3.1 and Fig. 3.6 that the local measurements of stellar positions were azimuth and altitude with respect to horizon and zenith on the celestial sphere and those in catalogues right ascension and declination, relative to equator and poles. In Fig. 5.6 on the left I illustrate the first two. Azimuth Az and altitude h relate to the horizon and the zenith and are a *local* set of coordinates. Project the position of the star onto the horizon using a great circle through the zenith. This gives the small filled square in the figure. The azimuth is measured as an angle along the horizon, usually from the south point. The altitude is the angle between the square and the star perpendicular to the horizon. Because of the diurnal motion of the Earth, reflected in the revolution of the celestial sphere and the rising and

5.5 Absolute Declinations

setting of objects not too close to the pole, the azimuth and altitude of any star or other celestial object change constantly and therefore are only suitable for local use. Now Kapteyn did not use the altitude, but the zenith distance, which is simply 90° minus the altitude.

Box 5.4 Determining one's geographical latitude
The **Horrebow-Talcott method** to determine a geographical latitude, which is the same as the determination of the pole's altitude in the sky works as follows.
Take two stars that have declinations such that they pass the meridian at about the same time and at very closely the same altitude (or zenith distance), one south of the zenith and the other north of it. We need stars for which the declinations have been determined elsewhere.
Let these be δ_{south} and δ_{north}, and measure their zenith distance (which is 90° - altitude) z_{south} and z_{north}. Then the geographical latitude ϕ is

$$\phi = \tfrac{1}{2} H(\delta_{south} + \delta_{north}) + \tfrac{1}{2}(z_{south} - z_{north}).$$

Since the flexure and refraction are the same because of the similar altitudes and times, they cancel when the two zenith distances z's are subtracted. However, the problem is that these systematic errors change with time, because they are sensitively dependent upon temperature, density (atmospheric pressure) and maybe even humidity. Therefore this only works when observations are obtained very closely spaced in time. This method requires *absolute* declinations of the stars used.

At any place on Earth the 'visible' pole (north pole in the northern hemisphere) is at an altitude equal to the geographical latitude (ϕ) of the observer. On the right in Fig. 5.6 the (north) pole is illustrated for an observer at geographical latitude 30°. For reasons of clarity the horizon is left out. At 90° from the pole we have the equator, also shown in the figure. Remember that similar to the local coordinates we define right ascension and declination, the latter as the angle the star is from the equator. Right ascension is measured along the equator (similarly as azimuth along the horizon), the zero point being the vernal equinox γ. But there is also an angle that is called the *hour angle*, which is the angle between the point 'P' where the equator crosses the local meridian and the projection of the star onto the equator.

For any celestial object the declination remains constant, but the hour angle changes with time. When the vernal equinox is on the meridian (γ and 'P' coincide), by definition the hour angle is equal to the right ascension. During the sidereal day (a full turn of the stellar celestial sphere) at any time the hour angle is the difference between the right ascension of a star and the (local) sidereal time, since by definition sidereal time is zero when the vernal equinox is on the meridian.

The zenith angle of the equator on the meridian (the zenith angle of 'P') is equal to the geographical latitude ϕ of the observer and the altitude is then $90° - \phi$. The zenith distance when a star crosses the meridian is the declination (δ) plus that altitude (or $(90° - \phi + \delta)$. *In this business accurate knowledge of the observer's geographical latitude (ϕ) or altitude of the pole is essential.* Kapteyn started his paper by recalling that there are major discrepancies between declination determinations at different observatories. The declination is usually measured with meridian circles by reading of the altitude or zenith angle of the telescope when the star crosses its field of view at the meridian (and the right ascension by noting the sidereal time when this occurs). The problem with declinations is that this altitude is affected by two important systematic effects, viz. the flexure in the telescope tube and its mounting and atmospheric refraction. The latter (see Box 3.2) is the effect that the path of the light from a star is bent when it travels through the atmosphere and enters areas of increasing density and changing composition, humidity and temperature, just as light is bent as it enters water or glass. At the zenith the effect is zero, but it increases with larger zenith distances and also when the color of the light is bluer. Now, this was all known when Kapteyn wrote his paper, but there was no reliable calibration of the magnitude of the effect as a function of temperature, humidity and atmospheric pressure. Both telescope flexure and atmospheric refraction make a star appear higher in the sky than where it would be without these effects. Kapteyn wanted to eliminate these two systematic errors.

The assumption, discussed in some detail in Kapteyn's paper, is that both these effects only operate in latitude but not in azimuth, which to an excellent approximation is indeed true. There is of course also the effect of aberration, due to the finite speed of light, which depends on the direction in which the Earth moves through space (see Box 3.2). But this was well known and understood and correcting for it was standard procedure.

How did astronomers then determine the pole's altitude at the time Kapteyn was conducting his research? Remember that the straightforward method is to take a circumpolar star and measure its zenith angle when it crosses the meridian above and below the pole and take the mean. But that presents the problem that in both measurements the altitudes are different and consequently the systematic affects as well. In addition, to calculate the mean you have to add these zenith angles and even if they were the same, they would not cancel. What was used extensively is the so-called Horrebow-Talcott method, first proposed by the Danish astronomer Peder Horrebow (1679–1764) and rediscovered and extended by the American civil engineer Andrew Talcott (1797–1883), who used it for geodesy. This uses two stars that cross the meridian at about the same time at opposite sides of the zenith at about the same altitude or zenith angle. The altitude of the pole then follows from the difference of these two zenith angles. Since telescope flexure and refraction are essentially identical (approximate simultaneity assures equal atmospheric conditions such as temperature, pressure and humidity, etc.) this is eliminated by the subtraction. The disadvantage, however, is that the absolute declinations of these two stars are required. The method is explained in more detail in Box 5.4. As in the Horrebow-Talcott method, Kapteyn devised a procedure in which systematic errors are also eliminated by subtracting observations that are likely to be affected equally. Moreover, he makes use of measurements of azimuth and timing, both of which are not affected by flexure and refraction.

5.5 Absolute Declinations

Fig. 5.7 Universal instrument, produced by A.&G. Repsold, in Hamburg, acquired by Leiden Observatory in 1853. It measures 50 cm by 78 cm. This is most likely the instrument used for Kapteyn's work on absolute declinations (Boerhaave Museum, Leiden [24])

At the outset Kapteyn mentioned that the method he presented was first designed by him in early 1880. But at the time he complained about his lack of a telescope or even a moderate modest universal instrument, so that there was no way for him to test it. He expressed his indebtedness to van de Sande Bakhuyzen, the director of Leiden Observatory, for showing the 'extreme willingness' to allow him the use of the universal instrument in the period July 21 to August 18, 1882. I am not sure whether this was indeed an act of extreme generosity, but in any event Kapteyn was exceedingly grateful. This must have been during the summer vacations; the instrument was otherwise used for instruction and Kapteyn could not afford to spend time in Leiden in view of his teaching and other commitments in Groningen.

I will now explain in some detail how the method that Kapteyn proposed works, but should you be less interested in technical complications, please refer to the ¶-symbol on page 131. In what follows I use the numerical example Kapteyn gives in his paper, based on a geographical latitude (and therefore altitude of the pole) of 40° on the northern hemisphere. The first step is the question of how you can determine absolute declinations once you do know the altitude of the pole. This is the opposite of the Horrebow-Talcott procedure.

(1). For stars not too far from the pole, the declination can be inferred from the maximum difference in azimuth that these stars attain from that of the pole. This only works for stars that are close enough to the pole, so that the altitude at maximum azimuth difference is not too different from that of the pole itself. Kapteyn thought that the limit to this was about 10°, so that absolute declinations could be measured this way between declination 80° and 90°.

(2). Take stars that culminate very close to the zenith and time their passage through the *prime vertical*, i.e., the great circle from east on the horizon through the zenith to the west back on the horizon (similar to the meridian, but perpendicular to it). Kapteyn did not specify what measurements needed to be taken, but it probably was the time interval (in sidereal time) between the passage through the meridian and the prime vertical. The azimuth in the prime vertical is then 90° (or 270°) and since we measure the hour angle H, we can, for known ϕ, determine δ. Since we have only measured time, the outcome is independent of the systematic errors. The time interval should be kept minimal, which means this can only be done for stars that pass very close to the zenith. Kapteyn took a range of 2°; then the hour angle at the crossing of the prime vertical is already 1.5 hours. This method yields absolute declinations between 58° and 60°.

The next three measurements should be performed using the Horrebow-Talcott method described on page 127, but then to solve for declination δ with known ϕ.

(3). Determine declinations for stars that have about the same altitude on the meridian south of the zenith as the stars in (1.) have north of the meridian. Since one can observe upper and lower culmination (above and below the pole) this gives declinations between 40° and 60°.

(4). Do the same for stars corresponding to those in (2). This results in declinations between 60° and 62°.

(5). Use the stars from (2) and (4) to find declinations for stars in the range −2° and 2°.

This covers much of the sky, leaving only three bands (of 18° width in the example) over which one would need to interpolate.

With all this Kapteyn had reduced the problem of determining absolute declinations to that of finding the absolute altitude of the pole. So he returns to that problem. The principle of his method is straightforward. It starts out with a star, designated Ω that is circumpolar. Then he needs two stars (S_1 and S_2) that have the same declination within a few minutes of arc, chosen such that within a few minutes S_1 passes through the meridian at the same time that Ω has its upper culmination (i.e. goes through the meridian between the pole and the zenith), and S_2 that almost simultaneously crosses the meridian while Ω has its lower culmination between pole and horizon. Then a series of measurements need to be done that I have tabulated in Box 5.5. All these measurements are either of timing or of azimuth, and are therefore free from the systematic errors in altitude measurements due to flexure and refraction. The problem of course is to find a star first that has almost exactly the same right ascension as the circumpolar star *and* another star with almost the same declination, while at the same time having a right ascension that differs by almost exactly 12 hours. Kapteyn noted that the method only worked in theory, when ϕ exceeds 45° and preferably by a significant amount.

5.5 Absolute Declinations

The usual consequence of such a practical impossibility in astronomy (and comparable sciences) is to look for an adapted procedure in which a larger number of stars is employed. Kapteyn worked out a procedure in which n circumpolar stars Σ_k passed the meridian reasonably regularly over a 24 hour period (both in upper and lower culmination, so we have $2n$ culminations), and $2n$ stars S_k that crossed the meridian south of the zenith, but which line up closely with the circumpolar stars in culmination times. Then observe at the time of each crossing of the meridian the difference in zenith distance between one circumpolar star Σ_k and a corresponding 'southern' star S_i. At a later time (since near simultaneity is then no longer critical) measure the azimuths' difference and the difference in time when any pair of stars S_i and S_{i+1} has the same zenith distance, where every star has been measured once to the east and once to the west of the meridian. Altogether this gives $6n$ measurements and on the basis of this Kapteyn showed that one can solve for the altitude of the pole. All $6n$ measurements are azimuths, timings or differences of zenith angle, and are all independent of systematic errors. Of course, one cannot observe stars continuously for 24 hours, as would be done ideally, so one performs the observations during the length of a night and during various observing sessions throughout the year. ¶ Kapteyn tested his scheme with the Universal Instrument of Leiden Observatory (see Fig. 5.7 for the one Kapteyn most likely used), but he experienced unusually poor weather in the summer of 1882. He mentioned that he could make use of a series of observations he had made previously (in 1875) with that same instrument and he did some additional observations. He never showed any outcome but he concluded that this was enough to show that the observing scheme could be executed and that the method was feasible.

The paper, Kapteyn (1884), appeared in the journal Copernicus in February of 1884. He apparently wanted to bring it to the attention of Sir David Gill, the director of the Royal Observatory at Cape of Good Hope. Gill had been born in Aberdeen as the son of a clock-maker. After studying at the university in Aberdeen, where he was taught among others by James Clerk Maxwell, he joined his father's business but became dissatisfied with the profession. He worked for a while at the Dun Echt private observatory of James Ludovic Lindsay, 26th Earl of Crawford and 9th Earl of Balcarres (1847–1913). Lindsay was an astronomer and president of the Royal Astronomical Society. While working there in charge of the equipment, Gill took part in expeditions to Mauritius to observe the Venus transit of 1874 and the closest approach of Mars to the Earth to Ascension Island in 1877. With these observations he hoped to determine the average distance of the Earth to the Sun more accurately. He became well-known for his excellent work and subsequently was appointed Her Majesty's Astronomer at the Cape of Good Hope in 1879. He remained there until his retirement in 1906. Undoubtedly, Kapteyn had heard of Gill through his publications and judged that he would be interested in his method to determine absolute declinations.

The letters from Kapteyn to Gill are kept in the archives of the Royal Greenwich Observatory (not in Cape Town) at the Cambridge University Library. Actually, Gill kept 'press copies' of all his letters. This technique had been invented by James Watt (1736–1819) and involved pressing a semi-transparent, damp piece of paper against

> **Box 5.5 Kapteyn's procedure for absolute measurement of the pole**
> **a.** Determine the moment when $S1$ and S_2 have the same zenith distance (say with S_1 east and S_2 west of the meridian) and measure their azimuths. What is used is:
>
> $$h = \text{Azimuth of } S_2 - \text{Azimuth of } S_1,$$
>
> $$\lambda = \text{Sidereal time of } S_2 - \text{Sidereal time of } S_1.$$
>
> **b.** Do the same when S_1 is west and S_2 east of the meridian:
>
> $$h' = \text{Azimuth of } S_1 - \text{Azimuth of } S_2,$$
>
> $$\lambda' = \text{Sidereal time of } S_1 - \text{Sidereal time of } S_2.$$
>
> **c.** When S_1 crosses the meridian and Σ in upper culmination:
>
> $$\beta = \text{Zenith distance of } \Sigma - \text{Zenith distance of } S_1,$$
>
> **d.** and when S_2 crosses the meridian and Σ in lower culmination:
>
> $$\beta' = \text{Zenith distance of } \Sigma - \text{Zenith distance of } S_2.$$
>
> The important thing to note is that these are all measurements of azimuth and time, except the last two, but these are differences in two zenith distances that are almost identical, eliminating the systematic errors. So, also β and β' are free from these errors. Kapteyn then found that the latitude or polar height ϕ can be found from solving
>
> $$\sin\phi \sin\tfrac{1}{4}(\lambda+\lambda') - \cos\phi \cot[2\phi + \tfrac{1}{2}(\beta+\beta')] + \cos\tfrac{1}{4}(\lambda+\lambda')\cot\tfrac{1}{4}(h+h') = 0.$$

a letter, so that the ink of the letter is partly transferred onto the copy. This works well for letters not too long after they have been written. Many of the press copies of Gill's letters in the RGO Archives are in poor shape or downright useless, partly because of fading and in many cases because the Gill papers were damaged by sea water during the transfer from Cape Town to Britain. Kapteyn's letters to Gill also show some damage, but by and large they are in reasonably good shape. Of course originals of Gill's letters to Kapteyn are available in the archives of the Kapteyn Astronomical Institute.[3]

The first time Kapteyn wrote to Gill was on April 30, 1884. Figure 5.8 shows the first part. The association of Kapteyn and Gill, which started at this point, was to become of overriding importance in Kapteyns career, so I reproduce the letter in full.

---

[3] I have scanned all 200 or more of them and sent them electronically to the Cambridge archives. I also provide them on my J.C. Kapteyn Website.

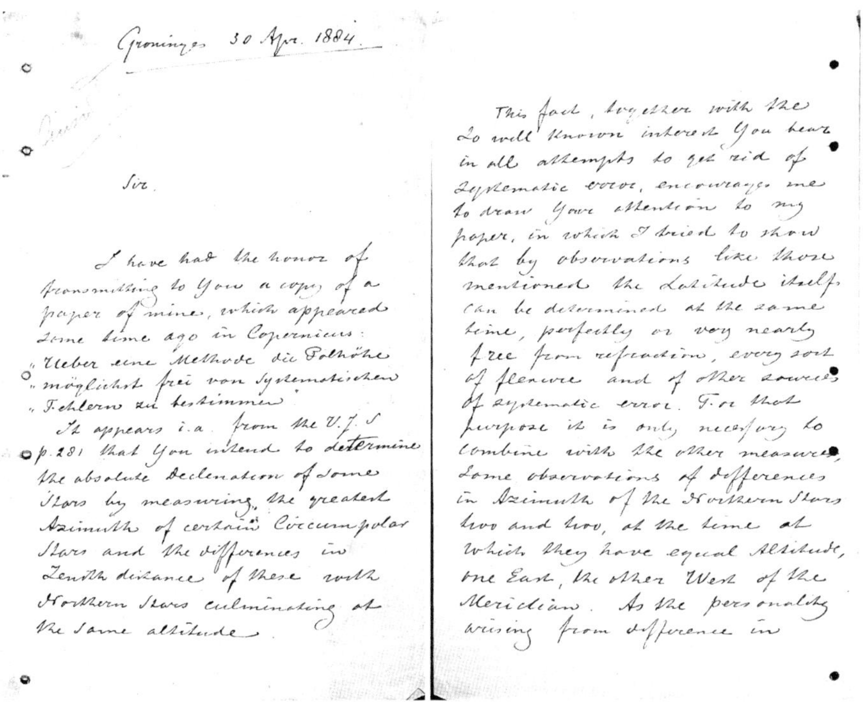

5.5 Absolute Declinations

Fig. 5.8 First part of Kapteyn's first letter to Gill dated April 30, 1884, in which he wrote he had sent his *Copernicus* paper on polar altitude and absolute declinations to the Cape (Royal Greenwich Observatory Archives [12])

'Sir,
I have had the honor of transmitting to You a copy of a paper of mine, which appeared some time ago in Copernicus: 'Über eine Methode die Polhöhe möglichst frei von systematischen Fehlern zu bestimmen'.

It appears i.a. from the V.J.S. p.281 [this is the Vierteljahrsschrift der Astronomischen Gesellschaft.], that You intend to determine the absolute Declination of some stars by measuring the greatest Azimuths of certain Circumpolar Stars and the difference in Zenith distance of these with Northern Stars culminating at the same altitude.
This fact together with the so well known interest You have in all attempts to get rid of systematic error, encourages me to draw Your attention to my paper, in which I tried to show that by observations like those mentioned the Latitude itself can be determined at the same time, perfectly or very nearly free from refraction, every sort of flexure and of other sources of systematic error. For that purpose it is only necessary to combine with the other measures some observations of differences in Azimuth of the Northern Stars two and two, at the times at which they have equal Altitude,

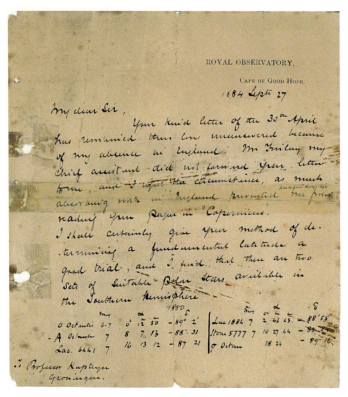

Fig. 5.9 First page of Gill's first letter to Kapteyn on September 27, 1884, where he acknowledged receipt of Kapteyn's letter and his *Copernicus* paper on polar altitude and absolute declinations (Kapteyn Astronomical Institute)

one East, the other West of the Meridian. As the personality arising from difference in magnitude, which at first sight would seem to vitiate the conclusions drawn from these observations, vanishes absolutely in the result for the latitude (vid.p. 160) it seems possible to secure for these observations a freedom from systematic errors, as great if not greater as is attainable in the observing of the greatest Azimuths and differences of Zenith distance. The Editors of Copernicus have thought fit (probably from want of room) to omit the observations and almost everything connected with them. In some instances these omissions have somewhat prejudiced the clearness of the explanations. In those case I have taken the liberty to add some explanatory words in margins in the copy I have sent to the Cape.
Very respectfully,
    Your obedient servant,

                                                                                        J.C. Kapteyn'

5.5 Absolute Declinations

Note that Kapteyn capitalized the first letter in 'You' and 'Your', very much as in the polite form of address in Dutch. In the course of time, as they became close friends, this 'Y' started to shrink and became a small letter 'y'. But this process took many years. Gill took a long time in replying; on September 27 (see Fig. 5.9) of the same year he wrote that this was because he was on leave in England and his chief assistant, who was William Henry Finlay (1849–1924), did not forward the letter to him. 'I shall certainly give your method of determining a fundamental latitude a good trial' and proceeded to indicate pairs of suitable stars near the south pole. He described the various instruments he had at his disposal, finishing by saying:

> 'I thus propose to devote the next two years to this work – that is to the establishment of a fundamental system of Declinations, or as far as I can do in that time, and to reduce the work to a system that can be carried on by my assistants, because early in 1887 I expect to have my new Heliometer of 7 inches aperture by Repsold, and shall be devoting myself to the parallax of the fixed stars.'

Kapteyn wrote back on December 10, 1884. He opened as follows: 'Many thanks for Your and Dr. Elkins' admirable paper on Stellar Parallaxes [this is the paper *Heliometer-Determinations of Stellar Parallax in the Southern Hemisphere* by Gill and W.L. Elkin (so he should have written 'Elkin's'), submitted in January 1884 [13]; I rejoice most heartily in the plan proposed by you in the concluding remarks, the execution of which will give at last a firm basis to all researches on the arrangement and constitution of the Stellar Universe.'

The paper of Gill & Elkin has an interesting history that I will briefly deal with as an aside, as it described how heliometers (see page 67) can be used to measure parallaxes, a major concern of Kapteyn. In the introduction to the paper Gill describes how he met William Lewis Elkin (1869–1940) while visiting European observatories prior to leaving for the Cape. Elkin was preparing an 'Inaugural Dissertation' in Strasbourg on the parallax and orbit of α Centauri, a subject that would interest Kapteyn enormously. At a little over 4 lightyears, α Centauri is the nearest star to the Sun, but is also a multiple star system, consisting of a binary (A and B) and a third component that is currently closer to the Sun (called Proxima Centauri). The binary system (α Centauri A and B) and Proxima are separated at the moment by about 2 degrees on the sky as seen from Earth and Proxima is a little more than 0.1 lightyears closer to us than the α Centauri AB system. Now, Gill had just purchased a heliometer and Elkin was coming to the Cape to use it. After he had arrived by the end of 1880, a month was spent on preparations. But then (Gill described) 'I was suddenly recalled to England on urgent private matters. [...] Dr. Elkin occupied my house in my absence, and remained as my guest, and as a member of the family circle, till the completion of our programme. He sailed from the Cape on May 16, 1883. His work from first to last has been a labour of love.' This project was aimed at determining parallaxes of nine stars, among which α Centauri and Sirius. Elkin proceeded to have a prominent career in astronomy and eventually became director of Yale Observatory. The paper ended with an exposition by Gill that heliometer measurements could give parallaxes to an accuracy of $0''\!.2$ and therefore a larger program to observe a large number of stars appeared feasible.

Kapteyn's letter continued: 'Many thanks too for Your kind letter of Sept. 27. I congratulate myself very much on Your plan of giving my method of determining a fundamental Latitude a good trial.' He described observing procedures in detail, also clearing up alleged misunderstandings of Gill, and at the end produced a list of suitable stars for Gill to use. The final paragraph reads: 'I am heartily glad that the establishment of a fundamental system of Declinations is undertaken by Yourself, by new methods, to avoid refraction and flexure altogether. May You equally succeed in this important work, as You have done in so many others. Yours very sincerely, J.C. Kapteyn.' To this Gill replied on January 18, 1885 in a very long (18 pages) letter. He laid out his observing program describing many details and asked for Kapteyn's opinion. Gill disagreed with some of Kapteyn's remarks and discussed these, so that the tone of the correspondence is much like that of two scientist carefully setting up an investigation with consideration of all details. This is the start of a successful collaboration between these two men.

There is an amusing twist to the fact that Kapteyn had chosen to write his Copernicus paper in German. At the end of the (his first) letter of September 1884, Gill wrote: 'Thank you very much for writing me – your paper is a model of clearness.' But the January 1885 letter says: 'I am a poor linguist. I have great difficulty in acquiring language and though I read French easily, German is very very difficult for me. I have therefore had a translation of your paper made and now it is all clear to me.' This letter has an postscript worth quoting: 'Since the above was written I have met with a serious loss – no less than that of your list of Latitude stars. It must have slipped from my table and have fallen into my wastepaper basket and so has been destroyed. As you probably have the original besides you I beg you to send me another copy without delay.'

Gill and Kapteyn exchanged another four letters during 1885 on the procedure and execution of the project. On October 5, 1885 Gill wrote: 'It was a very great pleasure to receive your letter of Sept. 30. I begin to think you had not received my last letter on the list of stars, and so it is also the more pleasant to find that you are still in fullest active sympathy with what we may call 'our work'.' He also reported that the project would proceed much slower than expected. 'Meanwhile I find from actual trial that the Azimuth work is a more considerable matter than I at first anticipated, and that it will take quite a year [does he mean 'quite a few years?'] to determine the latitude fundamentally, including the fundamental determinations of 8 fundamental Azimuth stars and 11 Polar Stars [...].' Obviously the collaboration was very much in place. Also Gill referring to it as 'our work' shows his dedication to the undertaking.

The annual reports of the Royal Observatory at Cape of Good Hope do indeed mention quite a few observations for this project [14]. For 1885 we read:
Number of observations of meridian marks   129
Number of azimuths of circumpolar stars at elongation   210
Number of azimuths [observations] for latitude by Kapteyn's method   63
The results appear to be very satisfactory. A fundamental latitude is being determined by Prof. Kapteyn's method ('Copernicus,' vol. iii. p. 147). The declinations of circumpolar stars are determined by observing azimuths of great-

5.5 Absolute Declinations

est elongation. Declinations of northern stars will be determined by connecting them by Talcott's method with those of southern stars determined as above. In this way it is hoped that a new and trustworthy system of fundamental declinations, entirely independent of refraction, will be established.

For 1886:

With the zenith telescope 849 pairs of stars have been observed by Talcott's (Horrebow's) method, partly in connection with Kapteyn's method of latitude determination, partly for control on the law of flexure of the transit-circle, and for the connection of the northern and southern systems of declination.

Azimuths of N. stars for latitude by Kapteyn's method   32 pairs

For 1887:

77 azimuth pairs of N. stars for latitude (Kapteyn's method)

With the zenith telescope 365 pairs of stars have been observed in connection with the latitude (Kapteyn's method), and for control on the law of flexure of the transit-circle.

For 1888:

Azimuths of pairs of N. stars for latitude (Kapteyn's method)   12

For 1889:

With the zenith telescope 863 pairs of stars have been observed by Talcott's (Horrebow's) method, partly in connection with Kapteyn's method …

And for 1890:

With the zenith telescope 638 pairs of stars have been observed by Talcott's (Horrebow's) method, partly in connection with Kapteyn's method …

Obviously a great deal of effort was put in by the Cape observers to use Kapteyn's method. In 1894, Gill published a paper, *Note on the latitude of Cape of Good Hope, Royal Observatory* [15], in which he used observations over the period 1886 to 1891. Curiously, Kapteyn was not mentioned. Nonetheless, there is no doubt that there was significant collaboration between Kapteyn and Gill as far back as 1884.

How did this method proposed by Kapteyn fare elsewhere? According to the Kapteyn obituary (longer Dutch version) by Antonie Pannekoek (1923b), the only other application of the method was in the PhD thesis of Leopold Courvoisier, (1873–1955): *Untersuchungen über die absolute Polhöhe von Strassburg i.E* (im Elsass) [16] in 1901. The method did indeed get mentioned extensively in Valentiner's *Handwörterbuch der Astronomie*, but not in Simon Newcomb's fundamental 1906 textbook on spherical astronomy *A Compendium of Spherical Astronomy, with its Application to the Determination and Reduction of Positions of the Fixed Stars* [17]. However, eventually it led somewhat indirectly to the major effort of the Kenya expeditions that Leiden Observatory organized in the 1930s and 1940s. These have been described and discussed in much detail by Jet Katgert-Merkelijn in *The Kenya Expeditions of Leiden Observatory* (1991) [18]. Coffee plantation owner C. Sanders in Portuguese Congo (presently the province Cabinda in Angola), an amateur astronomer, devised a method to obtain absolute declinations at a site near the Earth's equator. The basic idea behind that is that at a position exactly on the equator, the poles are on the horizon so that the azimuth of a star at rising and setting is directly (and exclusively) dependent upon the declination. Sanders was in close contact with

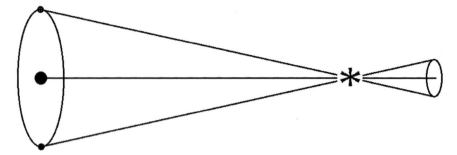

Fig. 5.10 The annual motion of the Earth around the Sun (left) is reflected on the sky by a small ellipse that is traced out by the star in the course of a year

E.F. van de Sande Bakhuyzen in Leiden, who brought his ideas to the attention of W. de Sitter. Sanders published a short note on the subject in 1917, *Determination of latitude and declinations* [19]. This was taken up by de Sitter and Jan Hendrik Oort, who published a paper on the method in 1925, *Provisional scheme for the determination of fundamental declinations from azimuth observations* [20]. They related the Sanders approach to the 1884 paper by Kapteyn, presenting the first as an elaboration of the latter. For further details I refer to the Katgert-Merkelijn paper. Clearly Kapteyn's work was widely read and applied.

5.6 Parallaxes

Obviously knowing distances of stars is the first, vital step towards understanding their distribution in space. The fundamental measurement of the distance of a star is by mapping its annual motion on the sky as a reflection of the orbital movement of the Earth around the Sun (see Box 5.6 and Fig. 5.10). With respect to faint and distant background stars the annual motion is an ellipse, whose major axis depends on the distance and its flattening on the astronomical latitude. If the latitude is 90° (the star is in the ecliptic pole) the ellipse reduces to a circle; when it is 0° (star in the ecliptic) it degenerates into a line. The major axis is always parallel to the ecliptic. By definition the distance of a star is 1 parsec (=3.26 lightyears) when the parallax is 1 arcsec.

Now, since the ellipse lines up with the ecliptic, the parallax is always larger in right ascension than it is in declination. So, a parallax can be measured in principle by comparing a star's right ascensions half a year apart. In practice a third measurement is required one full year after the first one to measure the proper motion as well (see also Box 5.6) and correct for this.

Right ascension is measured by timing the passage through the meridian with a meridian circle. In principle, parallax is also measurable by timing then, and more precisely if the difference in meridian passage of the star and some faint compari-

> **Box 5.6 Distances and motions of stars**
> In the course of a year a star describes an ellipse on the sky as a reflection of the motion of the Earth around the Sun (see Fig. 5.10). The semi-major axis of this ellipse depends on the star's distance and is designated as its *parallax*. The first parallaxes were measured in the 1830s by F.W. Bessel (61 Cygni from Königsberg Observatory, Germany), Friedrich Georg Wilhelm von Struve (1793–1864) (of Vega from Dorpat Observatory, currently Tartu in Estonia) and Thomas James Henderson (1798–1844), Astronomer Royal for Scotland (α Centauri from Cape Observatory).
> The distance of a star is expressed in *parsec*, which is the distance at which the parallax is one parsec. Herbert Hall Turner (1861–1930), a British astronomer is quoted as the one who introduced the parsec. A parsec is 3.1×10^{16} meters. Alternatively the *light-year* is used as the distance traveled by light through vacuum in a year. This is 9.5×10^{15} meters. One parsec is equal to 3.26 light-years.
> In the course of time stars change their positions with respect to each other due to their motions in space. This is called a star's *proper motion*. It was first noted by Edmond Halley in 1718.
> In addition to the proper motion due to the random velocity of a star through space, there is also a systematic pattern of streaming across the sky as a result of the Sun's motion with respect to the whole of the stars in its neighborhood (the *Local Standard of Rest*.). This pattern consists of a systematic tendency of stars to be streaming away on average from the point on the sky of the direction of the Sun's motion through space and this is called the *Solar Apex*. It lies in the constellation 'Hercules' (see Fig. 8.14) and the velocity of the Sun with regard to the Local Standard of Rest is of order 20 km/s. This streaming pattern can be used to derive statistical distances for well-defined groups of stars, the so-called *secular parallax* (see further on page 290 for more on the Solar Apex and page 292 for the concept of secular parallax).

son stars (more distant so with small expected parallaxes) is recorded. What is the magnitude of the effect? Take a star at a distance such that its annual parallax is 0.1 arcsec and its declination 50°. Depending on its actual right ascension, the parallax in the direction of right ascension is equal to or slightly less than 0.1 arcsec. The major axis (so twice the parallax) of the ellipse on the sky is 0.2 arcsec, which corresponds at the assumed declination to 0.02 seconds of time. This means that in order to measure the parallax the timing has to be done to an accuracy of better than one hundredth of a second! It had been tried before by von Auwers on the star 34 Groombridge between 1863 and 1866 (*Bestimmung der Parallaxe des Stern 34 Groombridge* [21]), but he concluded that the method was too difficult and inferior to parallax measurements with a heliometer (see page 67).

In practice the measurement of meridian passage is performed using a set (often five) of parallel, vertical cross wires in the focal plane of the meridian circle with the same spacing. The times of passing of the star behind all of these cross wires is recorded. Using more than one cross wire improves the accuracy significantly. There are differences in the way different observers perform in the timing,

Fig. 5.11 The recording apparatus (Registrir-Apparat) of Mayer & Wolf, described by van Littrow (see text), of which an example was in use at the Leiden meridian circle, at the time when Kapteyn performed his observations of parallaxes of stars (From the proceedings of the Vienna Imperial Academy; see text for full reference)

so in order to compare data between persons corrections are required for the so-called observer's 'personal equation', which are determined empirically. Obviously, the most accurate measurement is a relative one by the same person at the same telescope preferably a short time interval apart. The actual data recording is done with a so-called 'Registrir-Apparat' (apparatus for recording), which is a kind of strip recorder; the one used by Kapteyn on the Leiden meridian circle has been described by van de Sande Bakhuyzen as part of an inventory of instrumentation at the Observatory [22]. It is interesting to have a look at this description (translated from the German):

'The recording apparatus of Mayer & Wolf has been described by Prof. von Littrow [this is Austrian astronomer Karl Ludwig von Littrow (1811–1877)] and illustrated in the proceedings of the Vienna Academy [23] (see Fig. 5.11). With this small, handy and easily transportable instrument a paper tape is moved along by an electric motor. The pin for the observations has a joint enabling it to still produces points when the current is interrupted somewhat longer than just a moment, by making it possible for the pin to follow the paper for a short while. The pins for the recording of seconds is a wheel with 60 pins, of which the first is triple, every tenth double and the others single.'

Kapteyn's work on parallaxes of stars dates back to 1884, although it took until 1891, before a paper was published (in German), *Determination of parallaxes*

5.6 Parallaxes

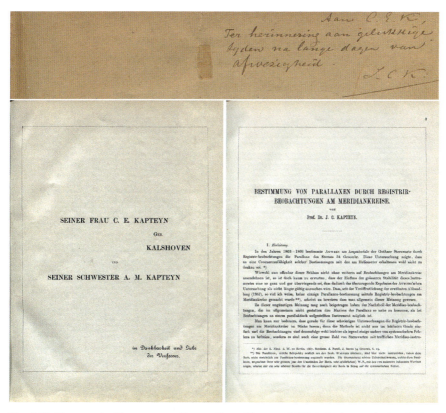

Fig. 5.12 The paper by Kapteyn on stellar parallaxes Kapteyn (1891a). At the top the handwritten compliments to his wife (see text). Below that the printed dedication and on the right the first page (From the bound copy of the paper in the Kapteyn Room)

with '*registrir*' *observations using a meridian circle* Kapteyn (1891a). This was in volume 7 of the *Annals of Leiden Observatory*, which is dated 1897, and therefore the paper was published under that year in the ADS. However, the introduction to the 'Annals' clearly states that this particular paper was published in 1891. As a matter of fact, Kapteyn published a preliminary paper earlier on (in 1889), which was in fact no more than an announcement and did not contain much more than the results and very little of the methods and discussion of the accuracies, Kapteyn (1889a). He started the paper by discussing of the feasibility of measuring parallaxes from timing observations of meridian passages not only based on general considerations, but also on his experience in Leiden in determining positions for stars in Perseus in van de Sande Bakhuyzen & Kapteyn (1879a). He concluded that even though the *individual* measurements of differences in right ascension have uncertainties of more than $0''.1$, with repeated measurements an accuracy of $0''.1$ could be attained.

There is another complication with this approach, that harms the final accuracy. The parallax ellipse on the sky has its major axis parallel to the ecliptic and the

Fig. 5.13 The meridian circle of Leiden Observatory that Kapteyn used for his parallax measurements, as it resides now in the Museum Boerhaave (Museum Boerhaave Leiden [24])

geometry (see Fig. 5.10) is such that the maximum deviation from the mean position occurs when the Sun is 90° (or 270°) away from the star in ecliptic longitude. But then meridian passage of the star occurs very close to sunrise or sunset.

The idea of measuring parallaxes by timing had also been subject of discussion with Gill in their letters. Kapteyn ended a long letter on the polar altitude program (paying much attention to the mathematics of reduction of the data) of March 2, 1885 with a short paragraph to announce the matter. 'I had hope[d] to find time to submit to Your judgment a plan of my own for a determination of stellar parallaxes by means of RA differences, but as it is high time to close the letter I ask Your leave to do so on a subsequent occasion. I hope too that You will write to me if You think the method explained to be practicable or not.' He returned to the matter in his letter of September 20, 1885. 'Since several years I have had the impression that the well known unfavourable results of Auwers when testing the method, were not wholly 'sans appel', it being acknowledged by Auwers himself that his case was (for different reasons) a very unfavourable one. [...] I was confirmed in my idea after the observations of the Heliometer Stars in Perseus.' [This is the publication van de Sande Bakhuyzen & Kapteyn (1879a)]. 'These stars were observed in Leiden by <u>one</u> person (almost all the observations were made by myself). [...] by comparing the results of different nights, the Prob. Error of

5.6 Parallaxes

a <u>difference</u> of RA of two stars was found to be only 0.″21 (arc of the gr.[eat] c[ircle]). [...] Might it not be possible by more careful observations to reduce the Prob. Err. still further?'

Kapteyn mentioned to Gill that van de Sande Bakhuyzen had given him access to the Leiden meridian circle for parts of nights (either morning or evening hours) during his academic vacations. He described in great detail how he performed his observations and asked Gill for advice on the procedures. A good example of the amount of detail they went into, is Kapteyn's question whether or not Gill sees any danger 'in taking the winter observations Clamp West, when the summer observations have been taken Clamp East?' This refers to a clamp of the telescope axis to an approximate declination. The telescope is regularly inverted (lifted from the mount and put back in a reversed orientation). The clamp is on one side of the pivot and, depending on the orientation, it is near the eastern or the western pier. This reversal was also used for aligning the telescope wires in the focal plane to the meridian by observing a distant object on the horizon in both orientations. It was common practice to observe any star more than once and to eliminate errors due to imperfections in the pivot in both positions. This remained an issue until recently, see e.g., B. Loibl (1978), *A new method of determining pivot errors of meridian circles* [25]. One would think that in Kapteyn's case where he measured *differences* in time for passage of stars, the orientation of the meridian circle would not matter too much.

Yet, on October 15, 1885, Gill writes: 'Now about your parallax measurements. I have very little experience in chronographic observing. Our chronograph is a very troublesome instrument and has not been used since 1874, and I really must try to get a new chronograph.' Gill made a point about accuracy, questioning the dependence of the personal equation upon the brightness of the stars and the amount of twilight during observations and he stresses the need to select stars of about equal magnitude. Also he worried about temperature effects (both 'optical and mechanical') since observations have to be taken half a year apart, so one in summer and one in winter. After this he added (the details themselves that he mentioned are less important here than the fact that he felt he has to go into such great detail):

'I should strongly recommend you to observe your winter & summer observations in the same position of the instrument, even if you have to reverse it half an hour before you begin work. The symmetry of an instrument is optical as well as mechanical, and there might be slight differences in the symmetry of the images which may come into the result, besides minute pivot errors which may affect differences of RA even over so small a difference of Decl. as $1°$ or $1\frac{1}{2}°$. I think it is more of a superstition than a reality that the strains in a transit circle are changed by reversing. If you plan a transit instrument with pivot I on bearing W and pivot II on bearing E I have never been able to see that there can be any change in the strains when pivot I is placed on bearing E and pivot II on bearing W, so long as the zenith distances are equal & opposite. The moment however that you change the zenith distance of the pointing of the instrument you change all the strains, yet nobody dreams of letting a transit circle remain some days at a particular ZD before venturing to observe with it. In my opinion this supposed change of strains by reversal is a suggestion of the Devil to lazy astronomers to excuse them from the toil and trouble of reversing their instruments!!, at least I have never been able to otherwise account for this dogma.'

This not only illustrates the level of detail that was necessary for Kapteyn to consider, but also the extent to which Gill and Kapteyn were exchanging opinions and advice as early as 1885.

Kapteyn started the definitive publication, Kapteyn (1891a), by presenting calculations showing what procedure needs to be followed depending on the celestial position of a star, so that the condition is met that the Sun should be at least 5° below the horizon. Also he noted his experience of azimuth observations with the universal instrument when he was still working in Leiden. Kapteyn described in the article how he showed his conclusions to van de Sande Bakhuyzen by the end of 1884, who immediately agreed to make the meridian circle available to test the scheme provided it did not interfere with the most pressing work at the Observatory. 'For this', Kapteyn wrote, 'and for so much else that I owe my honorable teacher, I express here my sincere thanks.' He proceeded to make observations during his vacations in four periods: March to April 1885, November to January 1885/6, November to January 1886/7 and March to May 1887. The Kapteyn Institute has a copy of the 1891 paper that was especially printed and bound and in which he had written in handwriting 'To C.E.K, In memory of the happy times after long days of absence. J.C.K.'. It also contains a printed page in which he dedicated the paper to 'his wife C.E. Kapteyn (born Kalshoven) and his sister A.M. Kapteyn, in gratitude and love from the author' (see Fig. 5.12). The sister is Albertina Maria (Bertamie), who is also the sister who brought him a star map (see page 20) when he was a teenager.

The program concerned 15 stars for which he determined parallaxes to an accuracy around $0\rlap{.}''03$ (and probable errors of individual measurements of differences in Right Ascension between $0\rlap{.}''16$ and $0\rlap{.}''11$). These stars have been selected as probably nearby based on their large proper motions. In Table 5.1 I compare the final values Kapteyn derives to modern observations from the astrometric ESA satellite Hipparcos. The HD numbers are those in the Henry Draper Catalogue, from Harvard College Observatory, published between 1918 and 1924. I have identified the stars after applying the appropriate precession correction to Kapteyn's coordinates to the year 2000 and searching around that position in the Simbad data base [26]. The modern values for the parallaxes have been also taken from Simbad. The first star is a binary consisting of two stars of roughly equal magnitude and here Kapteyn must have been confused (Kapteyn indicates he has taken the primary, but the magnitude difference is very small, only about 0.1 magnitudes when these stars are not flaring). Except for this case, the comparison shows that by and large Kapteyn's observations were in general extremely good, especially considering the primitive instrumentation he had at his disposal and the difficulties in doing such measurements. He must have been an extraordinarily accurate and careful observer!

5.6 Parallaxes

Table 5.1 Modern parallax measurements $p_{\rm modern}$ of the stars studied in Kapteyn (1891a), compared to final values $p_{\rm JCK}$ in Table 65 of that publication
('BB VII' stands for a list of 'Bonn Beobachtungen, series VII')

Star	$p_{\rm JCK}$	HD	$p_{\rm modern}$	Remarks
BB VII 81 (pr.)	74 ± 27	79210	172.06 ± 6.31	Flare star; binary
		79211	156.45 ± 8.58	Flare star; binary
θ Ursa. Maj.	52 ± 26	82328	74.19 ± 0.16	Spectroscopic binary
BB VII 85	64 ± 22	84031	54.89 ± 0.92	Variable star
20 Leon. Min.	62 ± 29	86728	66.46 ± 0.32	High proper-motion star
BB VII 89	176 ± 24	88230	205.21 ± 0.34	Flare star
BB VII 94	101 ± 26	90508	43.65 ± 0.43	High proper-motion star
BB VII 95	38 ± 27	91347	26.48 ± 0.59	High proper-motion star
Lal. 20670	-6 ± 28	92855	26.84 ± 0.50	Star in double system
BB VII 104	428 ± 30	95735	392.64 ± 0.67	Flare star
BB VII 105	168 ± 27	–	206.27 ± 1.00	High proper-motion star
BB VII 110	30 ± 27	101177	43.01 ± 0.73	Spectroscopic binary
BB VII 111	16 ± 32	102158	20.29 ± 0.70	Star in double system
BB VII 112	139 ± 26	103095	109.99 ± 0.41	High proper-motion star
BB VII 114	-28 ± 42	104556	17.5 ± 0.51	High proper-motion star
BB VII 119	56 ± 34	105631	40.77 ± 0.66	High proper-motion star

The concluding paragraph of the 1891 paper is worth quoting (translated from the German):

'With the advances of practical astronomy in recent times it is already now apparent, that a *systematic* measurement of the parallaxes of all stars, probably even up to stars of magnitude 8.5 or 9, with a probable error of for instance 0″.05 for each individual value no longer is a task beyond the capabilities of astronomers. By such determinations it will now be possible not only to answer in a comprehensive manner the 'great cosmical questions' [Kapteyn writes this in English], that have been posed by Gill and Elkin in their paper cited above. [This paper is *Heliometer-determinations of stellar parallax in the 'southern hemisphere' (1885)* [27]]:

a. What is the median value of the parallaxes of stars of the 1st, 2nd, 3rd, …. magnitude relative to the parallaxes of fainter stars?
b. What is the relation of the parallax of a star to the magnitude and direction of its proper motion or can it be proven that such a relation does not exist? But also the following no less interesting one:
c. Which are the fixed stars closest to the Solar System?'

Here we have of course a Kapteyn writing in around 1890, but it may be safely assumed that these thoughts, although possibly less matured, were already with him well before those days.

The paper on the determination of parallaxes using relative timing of meridian passages certainly definitely attracted attention. Only shortly after it was published in 1891, it was already introduced to the English speaking world through two literature discussions. The first, Plummer & Kapteyn (1891), was published in *The*

Observatory by William Edward Plummer (1849–1928) from Oxford, later Liverpool. The article started with: 'We have to congratulate Prof. Kapteyn on the successful inauguration of a new method of determining relative stellar parallax, by the employment of a transit instrument for the observation of differences of Right Ascension between selected stars and neighboring stars of comparison.' Towards the end: 'Prof. Kapteyn has given a scheme by which the comparatively rapid determination of many parallaxes of stars may be obtained. We beg to express a very sincere hope that the process may have a fair trial by some astronomer who has greater leisure than has Prof. Kapteyn, or, better still, we trust that that astronomer may soon find himself provided with suitable instruments to enable him to carry out an inquiry to which he has already applied himself to such good purposes and under somewhat adverse circumstances.' However, even this support of his attempts to obtain his own observatory failed to create the effect he was craving for. The other discussion, Boss & Kapteyn (1891), was published by Lewis Boss (1846–1912), who also wrote a favorable review. Boss was the director of the Dudley Observatory in Schenectady, New York.

In 1896 a paper appeared in *Science* by Albert Stowell Flint (1853–1923) on *Some values of stellar parallax by the method of meridian transits* [28]. He stated: 'The method employed is that of the differences of meridian transits, and it is believed this is its first application since it was introduced in its present detail by Prof. J.C. Kapteyn of Leiden Observatory in 1885–87.' Obviously the publication in the Leiden Annals led Flint to believe that Kapteyn was at Leiden. It was followed up in 1902 by a further detailed account in the Publications of the Washburn Observatory at Madison, Wisconsin, where Flint worked: *Meridian observations for stellar parallax* (1902) [29]. I quote again: 'The present work was suggested some years ago by Professor Comstock, director of this observatory, and Professor Asaph Hall, U.S.N., then consulting director, after the appearance of the excellent results obtained by Professor J.C. Kapteyn who was the first to make this particular employment of the meridian instrument.' The persons mentioned, George Cary Comstock (1855–1934) and Asaph Hall (1829–1907), were his superiors at Washburn.

During this period up to the mid 1880s, one other astronomical exercise resulted in a publication in the professional literature, although it was not explicitly under Kapteyn's name. It concerned visual observations of comets and the work referred to was done with H.J.H. Groneman, teacher at the secondary school of the type HBS (see page 62) in Groningen. Hendrik Jan Herman Groneman (1840–1908) taught mathematics, mechanics, cosmography and technical drawing and was an amateur astronomer. He should not be confused with his brother Florentius Goswin Groneman (1838–1929), the director of the same HBS, whom we will see later in connection with Kapteyn's association with the Natuurkundig Genootschap (see page 276). Before and after this, Groneman published extensively on atmospheric effects, aurora and meteors, but in 1881 he published a paper *Beobachtungen der Krömmung der Schweifaxe von Comet 1881 III*, while he was 'angestellt in Gröningen' (employed in Groningen) [30]. The subject was the curvature of the comet's tail. The visual observations were for a substantial part done with 'my friend Prof. J.C. Kapteyn'; they were reported, but not discussed in any detail.

There is a general belief that during his first years in Groningen Kapteyn led an unproductive life, as far as astronomical research was concerned, until he stumbled upon a paper by Gill in late 1885. It prompted him to offer his services in the production of the Cape Photographic Durchmusterung and completely changed his career. There is no doubt that this is precisely what Henriette Hertzsprung-Kapteyn was referring to in her *HHK biography*. 'Kapteyn suffered under the impossibility to bring his scientific plans to reality', she said. And then: 'But all this did not satisfy him. It was only child's play and he wanted to do something much greater and he knew he was able to do that. Then, suddenly the solution came and it gave an entirely new direction to his life.' I hope that the discussion in this chapter has thrown a different light on this subject.

After settling down with his newly-wed wife in the summer of 1879, he spent much time studying tree rings and putting in a moderate amount of effort (at least, but maybe more) on mathematical work with his brother Willem. His failure to acquire funds for his own observatory must have been a source of considerable frustration. But before the end of 1885 he also initiated three extensive investigations that all led to novel approaches. The method he designed for the solution of Kepler's equation was novel, although not entirely successful, and went largely unnoticed. But two other projects did attract attention and involved important, new techniques to improve determinations of star positions and distances. And these were vital to the problem that may have been taking shape in his mind as the fundamental one he wished to attack, i.e., the distribution of stars in space and the construction of the Sidereal System. His new concepts attracted attention and were adopted and implemented by others, giving credit to Kapteyn for their design and proof of concept. Moreover, the first of these on the absolute determination of an observatory's geographic latitude, led to Kapteyn's association with David Gill. His first years in Groningen were far from unproductive.

5.7 Family Life

The private life of Kapteyn and his wife during this period is best described by daughter Henriette in her *HHK biography*. I have selected some more passages. The Kapteyn children and their later marriages have been collected from various sources in Table 5.2.

'The earnest, somewhat stiff Groningers, who were always behind in fashion and strongly pronounced the *'n'* at the end of words, were not regarded as first class citizens. Young Mrs Kapteyn did not feel at home there at all. It seemed as if she had emigrated to a different country. The Groningers spoke and dressed differently as in Utrecht, they were stiff and rigid and put on airs. The last thing one could say about the Kapteyns was that they put on airs. The first thing they did was to rent

Fig. 5.14 Kapteyn at his desk. This picture was used as frontispiece in the *HHK biography*. He seems to be in his thirties here

against all unwritten rules, as it was Winschoterkade in a neighborhood that was occupied by sailors and common folks. It had never occurred to any professor to choose such an ordinary neighborhood. But they had little money and a lot of common sense and preferred this large house in humble quarters over a small one in a higher-class part of town, and lived very happily there. […]

Elise Kalshoven had been a quiet, calm girl, but Elise Kapteyn developed into a spontaneous, lively woman. Once, after having stayed with them for some time, Marie, the younger sister saw her and was very surprised. When she came home she told her mother: 'You would never recognize Lise, she is a completely different person.' However, this was not entirely fair. At home, her originality and lively spirit had been eclipsed by the others, but in her own environment these qualities were allowed to freely come to the fore. It represented a new style that the Kapteyns were displaying and many in Groningen noticed this young, happy couple with a smile, and possibly with some envy.

After a year their first child was born, a daughter. Together they had acquainted themselves with the principles of child care. This was exceptional in those days and they were ahead of their peers with their ways

5.7 Family Life

Table 5.2 Jacobus and Elise Kapteyn's children and their partners

Jacoba Cornelia ('Dody')	Groningen, May 31, 1880	Amsterdam, Oct. 4, 1960
	Married July 14,1906 – Vries to	
Willem Cornelis Noordenbos	Hallum, June 10, 1875	Amsterdam, Aug. 18, 1954
Henriette Mariette Augustina Albertine ('Hetty')	Groningen, Nov. 16, 1881	Utrecht, Oct. 15, 1956
	Married on May 16, 1913 – Groningen to	
Ejnar Hertzsprung	Fredriksborg, Oct. 8, 1873	Roskilde, Oct. 21, 1967
	Divorced January 19, 1937.	
	Married on April 17, 1937 – London to	
Joost Hudig	Rotterdam, March 18, 1880	??, June 4, 1967
Gerrit Jacobus ('Rob')	Groningen, Dec. 14, 1883	Davos Platz, Dec. 25, 1937
	Married January 30, 1918 – Groningen to	
Wilhelmina Henriette van Gorkom	Johannesburg, Jan. 7, 1895	Utrecht, Dec. 21, 1953
NN (stillborn)	Groningen, Aug. 28, 1895	

of thinking, so that often they would correct a know-all midwife, who was used to being the uncontested ruler according to antique custom. This was no minor thing, since in those days midwives ruled with an iron hand and did not leave any room for other opinions. Kapteyn also disagreed with the doctor. He would only take advice from an established authority if his common sense agreed. When the child, only a few months old, developed a serious intestinal infection and the doctor prescribed a mix of proteins, he had the courage and the determination to disregard this prescription. His common sense told him that the weak stomach of a almost dying child would not be able to cope with proteins. He used a diet of sugar-water and that at an absolute minimum. The father and mother watched with fearful tenseness. What responsibility had they taken upon themselves! But the child recovered and with that their trust in their own common sense.

They weighed their child regularly, which was not common in those days and they read books on feeding and child raising. Allebé was extensively studied and for the higher education also Rousseau and dreams were dreamed as would any other young pair of parents do at the cradle of their first child.'

The book 'Allebé' is *De ontwikkeling van het kind naar ligchaam en geest; eene handleiding voor moeders bij de eerste opvoeding* (The development of the child in body and soul: companion for mothers raising their first child), first published in 1845 and written by the Amsterdam physician Gerardus Arnoldus Nicolaus Allebé (1810–1892). It was the standard text for young parents for many decades (see Fig. 5.15). Jean-Jacques Rousseau (1712–1778) was the well-known French philosopher, writer and musician.

'Mrs Kapteyn was the first among her friends to push the pram herself rather than leaving that to a governess as the silly etiquette then dictated.

Fig. 5.15 The title page of the very influential book on the development of children by the Amsterdam physician Gerardus Arnoldus Nicolaus Allebé (1810–1892), first published in 1845. The cover on the left is the 1887 edition (From Canon Sociaal Werk [31])

And when they went out together, the young professor pushed the cart despite the scorn of the street-boys who were less tame than they are now, which meant a lot. But the criticisms of the street youth affected them just as little as the admiration of the colleagues.

To the contrary, Mrs Kapteyn found a welcome challenge in going against the tide of the conventions as she knew that her independence from the judgment of the rest of the world would make her happier. [...]

Another theory of Kapteyn was that a child would not have a natural affection for dolls, but that this was inflicted by the parents. He did not want to present his child with a doll. Poor Dody [Dody was the nickname of Jacoba Cornelia] had to go through life without a doll. Once he noticed that the child had an ugly Japanese doll, one that could be moved up and down on a stick, and which she was slowly rocking in her arms much like a mother would do with her child. He did not say anything, but when he came back in the afternoon from town he had a small wrapped-up package as a present for Dody, in which the delighted child found a beautiful doll. His theory was wrong and he was able to let it go immediately when life itself in its beauty showed him otherwise. [...]

5.7 Family Life

In the house on Winschoterkade, where they lived for six years, two more children were born, another girl and a boy. [Henriette Mariette Augustine Albertine herself in 1881, and Gerrit Jacobus Kapteyn in 1883. It is not mentioned in this biography that the Kapteyns also had a male stillborn child on August 29, 1895.] Kapteyn was an educator by nature, just like his forefathers had been. He taught them things in a playful manner, during meals, during walks; he told them about everything that was interesting and had the spark of life. He was interested in everything that concerned them and never sent them off with an easy answer. He treated every child as a human being that was growing up, took them very seriously, respected them and gave them the necessary attention.

The law of causality was the path he followed in his educational effort. He seldom inflicted punishments; the child should learn to understand the consequences of its actions and accept them. *Life* itself punished, not *he*, in this way he prepared them for the life that was ahead of them. His earnestness and dedication, his patience and love, his common sense and practical outlook made his children trust him like a solid rock. He embodied a never failing or never-disappointing trust for them, but he was also a cheerful companion and a loving friend.

Our mother supported him in every sense and smoothed the more sharp and cutting edges of his character with her warmth and spontaneity and unselfish love. She was always busy and caring, to make ends meet, which sometimes was not easy in the early years. She was merry, but also illogical and inexact in a funny way, and could well endure a joke or a little teasing at her expense. Her joyful originality enlightened his intellectual and critical mind, although they sometimes had difficulty to grasp what the other meant. Their backgrounds, characters and upbringings were too different for that. But they grew into a better mutual understanding, since both of them had a lot to offer and they found out how to learn from each other to the benefit of both of them.'

According to the archives of the city of Groningen the Kapteyns moved from the Winschoterkade to a home on Oosterstraat OZ(=East Side) 42. From *the HHK biography* I infer that this took place in 1885. Henriette Hertzsprung-Kapteyn mentions: 'In the house on the Winschoterkade, where they lived for six years, two more children were born, a girl and a boy.' Since the Kapteyns married in 1879 and the son Gerrit Jacobus Kapteyn was born December, 1883, this suggests it must have been in 1885. It is in agreement with the archives, which in the limited records available locate them on Winschoterkade in 1880 and on Oosterstraat in 1886. That house in the Oosterstraat still exists; it is a municipal monument (has a special protection status) because its oldest parts date back to the 14th century. The current lay-out and front facade date from the last quarter of the nineteenth century [32]. The ground floor currently houses a shop for design furniture; the upper part contains two floors where Kapteyn and his family lived. Curiously enough, Henriette Hertzsprung-Kapteyn does not mention this home (the family moved there when she was about five years of age). On the other hand, they did not live there for very

Fig. 5.16 Jacobus Cornelius Kapteyn around 1910 (Kapteyn Astronomical Institute)

long. After listing the Kapteyns in the Oosterstraat in 1886, the archives of the city of Groningen has as the next address the Heerestraat in 1891. They lived in the Oosterstraat probably for no more than four or five years.

Chapter 6
Cape Photographic Durchmusterung

> *His path was marked by the stars in the southern hemisphere.*
> Paul Simon (b. 1941).[1]

> *The Great Bear remained the Great Bear – and unrecognizable as such –*
> *for thousands of years; and people complained about it all the time,*
> [...] *Congress changed it to the Big Dipper, and now everybody is satisfied,* [...].
> [...]. *I would change it* [the Southern Cross] *to the Southern Kite.*
> [...] *for up there in the general emptiness is the proper home of a kite,* [...]
> Mark Twain (1835–1910).[2]

Astronomy, especially stellar astronomy, benefited very much from the technological advances of the nineteenth century. In the context of Kapteyn's propositions accompanying his thesis, we have already seen how photometers, although using flames and undoubtedly awkward and cumbersome to use, provided more precise magnitudes of stars, greatly improving the older visual estimates. Machines such as the 'registrir apparat' as described in the previous chapter, driven by electric motors, improved the timing of meridian transits to a great extent. Yet, the major revolution in that period was without doubt the photographic plate. It actually dominated astronomical observing for about a century, and it was not until ca 1980 that it was replaced by charge-coupled devices (CCDs) and other photon counting techniques. For astronomers of the late-eighteenth century the advent of photography must have felt similar to that of electronic, digital recording a century later, when astronomers like myself, who grew up with photographic plates, were liberated from the limitations of photography. In spite of techniques of hyper-sensitizing, developed later in the twentieth century, photographic emulsions always suffered from low quantum efficiencies (only a few per cent of the photons were effectively used to darken the

---
[1] From the song *African skies* of the album 'Graceland' by Paul Simon.
[2] Mark Twain, born Samuel Langhorne Clemens, American author. Quote from *Following the Equator*. See Fig. 10.1 for a picture of the constellation Southern Cross.

emulsion). Furthermore, the dynamic range or useful range in brightness over which the emulsion is usable, was limited to a factor to the order of just a hundred or so between under- and overexposure. And most importantly, there was the so-called 'low intensity reciprocity failure', the effect that – at low levels – the increase in the required exposure times is faster than the decrease in intensity. Likewise, irregularities impeded the work on faint objects and extended structures. Those problems were equally acute when Kapteyn was a young astronomer, and when his generation was pushing limits just as much as astronomers are now. An elementary introduction to the subject is the classic book *Astronomical Photography; from the daguerreotype to the electron camera* by the French astronomer Gérard de Vaucouleurs (1961) [1]. A deeper and comprehensive introduction is *The Theory of the Photographic Process* by Thomas H. James & Charles E.K. Mees (1977) [2].

The problem with night-sky astronomy is the faintness of the objects. Early photography involved the daguerreotype, named after Louis Jacques Mandé Daguerre (1787–1851), where a thin silver coating on a copper plate was sensitized with vapors containing iodines, and after exposure developed in mercury vapor and fixed with a sodium thiosulphate bath. Pictures of the Moon and the Solar corona during an eclipse feature as the prime examples of this technique applied to astronomy. The collodion process increased the sensitivity substantially. Initially this was a wet process – one spoke of wet plates or wet emulsions–, whereby a coating containing grains of silver halides (iodide or bromide) was produced. This was not very practical for astronomy, as the coatings had to be used wet. Nevertheless it was utilized on a wide scale and the images that received the most attention were those of the Venus transit of the Sun of 1874.

The photographing of the starry sky itself was not possible until dry plates were invented, and became common when they became mass-produced after George Eastman (1854–1932) had found a manufacturing process for this and founded the Eastman-Kodak Company. Dry plates would be instrumental for astronomical photography as long as it thrived. In no time, pictures appeared of planets like Jupiter and Saturn, as well as the picture made in 1880 by medical doctor, amateur astronomer and pioneer of astrophotography Henry Draper (1837–1882) of the Orion Nebula [3] (for a recent picture of the Orion Nebula see Fig. 11.8).

For our story the most important development in the early 1880s was the comet of 1882, and the photographs that were made of it. The development I am turning to has been described in an excellent way by C.A. Murray in *David Gill and Celestial Photography* (1988) [4]. This comet was photographed by many pioneers and amateurs of astrophotography. David Gill (Fig. 6.1) realized that good pictures needed accurate guiding of the telescope, keeping the comet in the same place on the photographic plates, and he proceeded to obtain such pictures.

6.1 Astronomical Photography

Fig. 6.1 Sir David Gill (Picture is taken from *the HHK biography*)

6.1 Astronomical Photography

In Chap. 3 I have outlined how astrometric catalogues had been constructed during the first part of the nineteenth century. I will provide a summary below, but also make reference to an excellent general article on the subject, *The History of Astrometry* by M. Perryman [5] for more details. Friedrich Argelander had founded Bonn Observatory and between 1849 and 1863 he had measured positions and (visually estimated) magnitudes of all the – more than – 300,000 stars in the northern hemisphere (north of declination $-2°$) brighter than apparent magnitude 9.0 and in addition an incomplete sample of another 150,000 stars, approximately, down to magnitude 10.5. The result was the *Bonner Durchmusterung*. This was subsequently extended to southern declinations −23° in the *Südliche Durchmusterung*. Other than this extension, the southern hemisphere was missing. A further southern extension was contemplated, and was eventually executed from Argentina to −62° in the *Córdoba Durchmusterung*, but that was not completed until 1908. In the meantime, the *Astronomische Gesellschaft Katalog* or AGK was started in the north by a group of observatories as a follow-up with improved positions using meridian circles. Gill came to realize that photographic means might make it possible to extend the Durchmusterung efficiently and accurately to the celestial south pole. This resulted in the Cape Photographic Durchmusterung or CPD.

Gill himself has described these developments (including earlier ones on the use of celestial photography and subsequent ones in the area of star mapping) in a lecture he gave in 1887 before the Royal Society, London on *Applications of Photography in Astronomy* [6]. I quote from the Introduction to the first part of the Cape Photographic Durchmusterung, Gill & Kapteyn (1896), published in the Annals of the Cape Observatory. Gill wrote:

Origin and History of the work.

'On the early morning of the 8th September 1882 (civil time), Mr W.H. Finlay, First Assistant at this Observatory, when on the way to his house after observing an occultation of the star 5 Cancri, saw a bright comet-like object in the constellation Hydra, which proved to be the afterwards celebrated Comet of that year. It appears that the Comet was seen by various less responsible observers several days before its discovery by Mr Finlay, but the fact remains that the accurate observations of this object which he secured, by returning to the Observatory on the morning in question, are the first of any scientific value that exist. [...]

So early as October 4 several photographers in South Africa had obtained impressions of the comet with their ordinary apparatus; [...]. Their photographs have no scientific value as representations of the Comet, since they were taken without means for following the diurnal motion during exposure. [...] I had, at that time, no suitable lens and no experience in the development of modern dry plates. I accordingly called upon Mr Allis, a photographer in the neighbouring village of Mowbray, of whose skill as a photographer I had previous experience. No sooner had I explained to him the object of my visit and the conditions necessary for success, than he at once volunteered all requisite aid, and entered into the work with heart and soul. The most suitable lens in his possession was a doublet by Ross (the work, I believe, of the late Mr Dallmeyer) of $2\frac{1}{2}$ inches aperture and 11 inches focal length. This lens was mounted on an ordinary camera, and the latter was attached to a stout board which was then clamped to the counterpoise of the Declination axis of the 6-inch Grubb-Equatorial. This counterpoise could be rotated with respect to the Declination axis and then be clamped; it was thus easy to adjust the optical axis of the photographic lens in any required position with respect to the axis of the 6-inch telescope, so that, whilst the latter axis was directed to the nucleus

Fig. 6.2 The great comet of 1882, photographed by David Gill (Picture taken from the website of the South African Astronomical Observatory [8])

of the Comet, the general image of the Comet occupied the centre of the field of the photographic lens. [...]

Apart from their scientific interest as representations of the Comet itself, these photographs appeared to have a still wider interest from the fact that, notwithstanding the small optical power of the instrument with which they were obtained, they showed so many stars, and these so well defined over so large an area, as to suggest the practicability of employing similar, but more powerful means for the construction of star-maps, on any required scale and to any required order of magnitude.

A short paper expressing these views, and accompanied by paper copies of the six photographs, was forwarded to Admiral Mouchez, and by him communicated to the Paris Academy of Sciences on the 26th December 1882 (Comptes Rendus, Vol. xcv., pp. 1342–43 [7]). In his accompanying remarks, Admiral Mouchez endorses the view that these photographs point to the possibility of producing excellent star-charts by means of photography, and he afterwards told me that these pictures led him to encourage Messrs Henry to devote their attention to the construction of

Fig. 6.3 The great comet of 1882, photographed by David Gill (Picture taken from the website of the South African Astronomical Observatory; see Fig. 6.2)

Astrophotographic lenses, and the application of photography to astronomical work generally. The brilliant results which Messrs. Henry soon attained are still fresh in the minds of astronomers and mark an epoch in the history of astronomy in the 19th century.'

Admiral Ernest Amédée Barthélemy Mouchez (1821–1895) was a naval officer who observed the transit of Venus from St. Paul Island in 1874, while working for the Bureau des Longitudes. In 1878 he became the director of the Paris Observatory (l'Observatoire de Paris). The printed version of the paper referred to in *Comptes Rendus* does not contain the photographs. The comment Gill is referring to reads (my translation from the French): 'It was both the well-known skill of Mr Gill and the purity of the skies of Cape of Good Hope that made it possible to obtain such a beautiful result, which leaves no doubt that it will soon be possible to produce excellent celestial maps using photography.'

The 'Henry brothers' are Paul-Pierre (1848–1905) and Mathieu-Prosper Henry (1849–1903); they are usually referred to as Paul and Prosper Henry. They were astronomers and very accomplished telescope builders. We will meet Mouchez and the Henry's below in relation to the so-called Carte du Ciel project. I continue with Gill's introduction to the CPD.

'Meanwhile, in November 1882 I wrote to Mr J.H. Dallmeyer requesting him to send me a lens of moderate dimensions for further experiment, as a preliminary step to ordering for definitive work.

Mr Dallmeyer, in reply, kindly forwarded a 'Rapid Rectilinear Lens' of 4 inches aperture and 33 inches focal length, which reached the Cape in April 1883, and preliminary experiments were made with it in course of the year. Early in 1884 I went on leave of absence to England, and after some conversation with Mr Dallmeyer, he undertook to make a special doublet which he hoped would prove to be better adapted for Astrophotographic work than the ordinary 'Rapid Rectilinear Lens' but, as he could not at once obtain the glass which he required for the new lens, he very kindly lent me a 'Rapid Rectilinear Lens' of 6 inches aperture and 54 inches focus which he had in stock.'

6.2 Kapteyn and the Cape Photographic Durchmusterung

Gill managed to obtain the funding for the project itself, but was obviously at a loss on how to reduce the observations into a finished product, which would be the CPD. This is were Kapteyn came in. Let me first continue by quoting from Gill's Introduction. Where £ are mentioned in the following, one may apply a factor of the order of 50 to convert to current £, as inferred from the Currency Converter of the (UK) National Archives [9].

'In September 1884 I applied for a grant of £300 from the Government Grant Fund of the Royal Society, partly for the purpose of making attempts to photograph the Solar Corona [...]

In January 1885, the Government Grant Committee placed the sum of £300 at my disposal, and Mr C. Ray Woods, whose services as my photographic assistant had been provisionally secured beforehand, reached the Cape on the 18th February of the same year. After a number of preliminary experiments had been made, systematic work was begun on the 10th of April and continued without interruption till its completion.[...]

Meanwhile the following correspondence had taken place between Professor Kapteyn and myself (see Fig. 6.4):

J.C. Kapteyn to David Gill
Leiden, December 16th, 1885.

'I must here to break off because this letter has to be dispatched an hour earlier than I expected. I will therefore write you another letter that will reach you a week later. In that letter I will make bold to make and explain to you a proposal that I hope you will not deem indelicate. It is, in the main, what follows: If you will confide to me one or two of the negatives

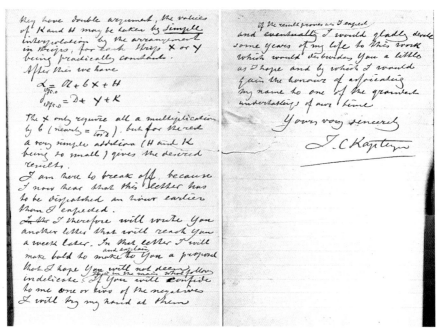

Fig. 6.4 Final two pages of Kapteyn's crucial letter to Gill of December 16, 1885 (From the RGO Archives at the Department of Manuscripts & University Archives, University Library, Cambridge, UK)

I will try my hand at them, and, if the result proves as I expect, I would gladly devote some years of my life to this work, which would disburden you a little, as I hope, and by which I would gain the honour of associating my name with one of the grandest undertakings of our time.'

J.C. Kapteyn to David Gill

Leiden, December 23rd, 1885.

'I have still to explain to you the proposal in my former letter, which I thought it better not to postpone, my resolution being taken. In doing this, you will excuse me in premising so much about my private circumstances as seems necessary for the purpose.

In the year 1878 I was appointed Professor of Astronomy and Theoretical Mechanics at the University of Groningen, having been before, during a couple of years, observer at the Leiden Observatory. Directly on my appointment I proposed to the Government to fit out a little Observatory where, besides instruments for teaching purposes, a Heliometer of 6 inches aperture would be mounted. [...]

Perhaps I shall succeed after some years in getting one or two instruments with which truly scientific research may be prosecuted; but at all

6.2 Kapteyn and the Cape Photographic Durchmusterung

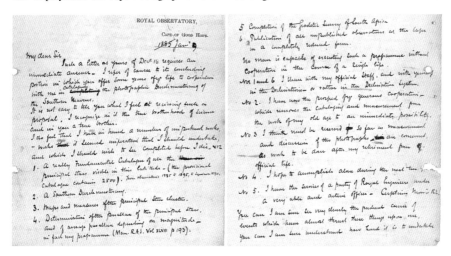

Fig. 6.5 Part of Gill's reply to Kapteyn's letter of December 16, 1885. The year of the date at the top is incorrect; someone has corrected the original in pencil (Kapteyn Astronomical Institute)

events a very long time will have to elapse before any such result may be looked for.

The first years of my Professorship once passed, my lectures left me considerable leisure which it has been always my desire to devote to astronomical observations. [....]

Now, after your success in Stellar photography, and especially after your letter in which you tell me 'I am obliged to crave help where I can get it,' it has occurred to me that by measuring and reducing your photographs I could contribute very effectually towards the success of an enormous and eminently useful undertaking. Since then I have constantly revolved the idea in my mind and I have come to the conclusion that if you will let me, and if I can secure the necessary help, then no one can be in better conditions to undertake this work than myself.

The former point being granted, it is my plan as to the latter—

1st. To request the Government to double the yearly subsidy of somewhat more than £40 that is granted for the acquisition of books and small astronomical instruments.

2nd. To request another subsidy of £80 for a series of consecutive years from the Society of Teyler, a very rich society, that is always very willing to bestow some money on scientific pursuits.

With this sum I think I can procure the constant help of three persons to do the most mechanical part of the work, the copying and the simple arithmetical processes, while I myself would execute all the measurements, the computation of the tables of reduction, the comparison of catalogues, &c., for which work perhaps now and then the help of a student could be secured. [...]

Having once got the necessary information and tried the necessary experiments, a rough estimate of the time to be expended on the work may be made, and I will then be able to make a definite proposal, stating the approximate time in which, and the approximate accuracy with which, I could undertake the whole business. Supposing that you are willing to leave the thing to other hands at all, I do not doubt but that we will soon agree as to these points. [...]

I have kept this letter here some days to talk the matter of the Photographic Durchmusterung over with Professor Bakhuyzen and his brother. I am bound to say that they were not very enthusiastic about the matter; of course they thought the results, once reached, of immense value, but the drudgery to be gone through before these results are once got into the form of a catalogue almost unbearable. However, I think my enthusiasm for the matter will be equal to (say) six or seven years of such work. [...]'

Part of the next letter is reproduced in Fig. 6.5. There are two minor transcription errors by Gill that I have corrected.

David Gill to J.C. Kapteyn.

Cape of Good Hope, January 9th, 1886.

'Such a letter as yours of December 16th requires an immediate answer; I refer, of course, to its concluding portion, in which you offer some years of your life to co-operation with me in cataloging the Photographic Durchmusterung of the Southern Heavens. [...]

Naturally, before you commit yourself to so serious a work, you desire to see a sample of the photographs on which so much labour is to be expended, accordingly I send you two photographs- representing the same area. [...]'

Here follows a long account of instrumental details.

David Gill to J.C. Kapteyn.

Cape of Good Hope, January 22nd, 1886.

'It will, I hope, be as satisfactory to you as it has been to me, that we have mutually, and almost simultaneously, confided to each other the objects of our work, our hopes, and our difficulties. I with too much on hand, you with too little – both interested in precisely the same kind of work, and both intent on having such work done. [...]

I think you will find that my letter of January 9th anticipates most of your questions, and that the new apparatus fulfills all the requirements which you have suggested the desirability of realising [Here some technical matters are discussed.] [...]

For my part I will undertake the re-observation or re-photographing of all doubtful points, all necessary re-examination of discordant magnitudes, all re-observation of stars with the Transit Circle, and so on. All such things are within the limits of my time and that of my staff. I think it would be of very great interest to continue the work to the Equator, for sake of comparison

6.2 Kapteyn and the Cape Photographic Durchmusterung

with Schönfeld, but not until the work has first been completed from -90°0 to -23°0. It will be time enough, however, to settle that when we have seen how the work goes on. You probably know that I have obtained the money for this work from the Government Grant Fund of the Royal Society. The President, Professor Stokes [Sir George Gabriel Stokes (1819–1903), known from the Navier-Stokes equations in fluid dynamics.], takes great interest in it, and I think he will be greatly pleased when he hears of your offer. Of course I shall not mention the matter till I hear definitely from you on the subject. I do not think it would be right to go on without the full consent of the Royal Society, but I do not think they will object to the plan if I do not ask for a larger grant than I receive at present. [...]'

Eduard Schönfeld (1828–1891) was assistant to Argelander in the construction of the Bonner Durchmusterung and succeeded him as director. He subsequently carried out the southern extension (Südliche Durchmusterung).

This sequence of events was extremely important for Kapteyn's career. There is, however, a serious misunderstanding in some accounts with regard to the question whether Kapteyn's offer to Gill was entirely unsolicited or not. E.R. Paul, for example, wrote in his paper *Kapteyn and statistical astronomy* [10]: 'Recognizing the importance of Gill's work, in 1886 Kapteyn offered without solicitation his aid to Gill for measuring many photographic plates and cataloging numerous stars.' This undoubtedly is the result of the account of the story as presented in Henriette Hertzsprung-Kapteyn's biography. She describes Kapteyn's decision as rather impulsive after he had read a paper by Gill, when Kapteyn was spending time in Leiden around Christmas 1885. The impression is also given that Gill and Kapteyn were complete strangers. Neither does Gill's account, as related above, go back to their correspondence before the end of 1885. Kapteyn was in fact in Leiden around Christmas 1885, working on his parallax observational project. But there is no such paper by Gill, published before December 1885, in which he describes his plans for the Cape Photographic Durchmusterung and which could be the one referred to in the *HHK biography*.

We have seen that Gill and Kapteyn had been in much contact in the preceding year, working together on a project to do an absolute measurement of the pole's altitude, and this has been noted before in Murray's account (referred to at the beginning of this chapter) and – on the basis of this – also in the chapter on 'Kapteyn and South Africa' by Michael Feast in the *Legacy*. Murray quotes from this early correspondence, which is in turn quoted by Feast. Both fail to appreciate the significance of the existent, extensive collaborations and correspondence between Kapteyn and Gill. In my view Kapteyn's offer was far from unsolicited and I will document this point below.

In 1884 the *Catalogue of 4,810 stars for the epoch 1850 : from observations made at the Royal Observatory, Cape of Good Hope, during the years 1849 to 1852* based on extensive observations with the meridian circle, had been published [11]. Gill wrote to Kapteyn on January 18, 1885: 'I send you by post a copy of the 1850 Cape Catalogue; before using it please employ the Table of Errata in the beginning. I hope to send you the Cape Meridian Results for 1879, 80 and 81

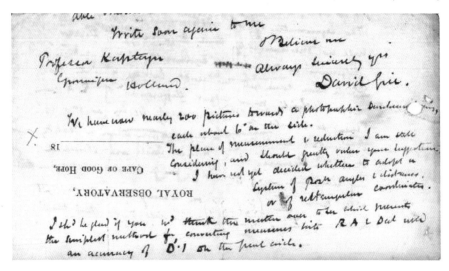

Fig. 6.6 Part of Gill's letter of October 15, 1885, in which he asks Kapteyn to comment on possible ways of measuring and reducing the photographic plates for the Cape Durchmusterung (Kapteyn Astronomical Institute)

very soon.' Gill's photographs of the comet of 1882, and his realization that this could be used for star cataloging, were a few years old and he was looking for a way of enlisting help with this project. Whether or not this was a prelude to a deliberate attempt to entice Kapteyn to do the work, is not sure, but it is likely and certainly aroused Kapteyn's interests. On February 5, 1885, the latter wrote in reply: 'Many thanks for Your present of the 1850 Cape Catalogue. I could only glance at some results of the Preface.' And then he went on to comment on accuracies in Right Ascension measurements, which are less relevant in this context. In his letter of April 3, 1885, Gill became more specific:

> 'Meanwhile I send you some of our first attempts in producing a photographic Durchmusterung of the southern Heavens. I hope to improve very greatly on this attempt. If you have any suggestions to give me on the best plans of measurement I shall be glad and grateful to have them. An accuracy of $\pm 0\overset{''}{.}2$ would be ample.'

In his letter of May 15, 1885 (already quoted above, when he comments on Kapteyn's parallax project), Gill wrote in a postscriptum (see Fig. 6.6):

> 'We have now nearly 200 pictures towards a photographic Durchmusterung, each about 6° on the side. The plans of measurement & reduction I am still considering, and should greatly value your suggestions. I have not yet decided whether to adopt a system of position angles and distances or rectangular coordinates. I should be glad if you could think this matter over as to which presents the simplest method for converting measures into RA & Decl. with an accuracy of $0\overset{''}{.}2$ on the great circle.'

6.2 Kapteyn and the Cape Photographic Durchmusterung

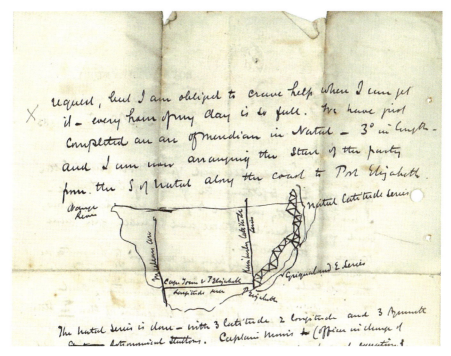

Fig. 6.7 Part of Gill's letter of November 2, 1885, in which he stresses his need 'to crave help where I can get it' (Kapteyn Astronomical Institute)

This is certainly not a formal request to Kapteyn to take a burden upon himself in the data reduction, but it comes close. Kapteyn's reaction to this in his letter of September 20, 1885 was a single paragraph:

'I have still to thank You for Your splendid Stellar photographs? [Kapteyn clearly writes a question mark, but it should probably be an exclamation mark.] Have You already devised a method for their measurement? A Photographic Durchmusterung! It seems too splendid to be believed in and yet how near in fulfillment [it] seems to be after your first attempts!'

On November 2, 1885 Gill reported on the progress with the altitude study:

'I have arranged and been at work for two months on the latitude observations. So far as the theodolite is concerned, i.e. in the Azimuths observations of the S stars and determinations of the Azimuths of elongations of the Σ stars. But the Talcott observations cannot be begun till Mr Finlay returns in the end of the year, and so I would ask you a great favour, to enter all your alterations, corrections and suggestions in red ink upon the lists I sent you, and return them to me. I will then send you a clean copy of the finally adopted list. It may seem to you absurd to make such a request, but I am obliged to crave help where I can get it – every hour of the day is so full.'

He adds that he is involved in a triangulation effort over South Africa, which he then describes in detail (see Fig. 6.7). After noting that 'it takes a great deal of pressure to get the necessary money', he writes: 'So do not think I am neglecting astronomy. I am only rather overbusy at present, and would ask you to assist me in the way I propose.'

It never hurts to stress the need for assistance! As Michael Feast has put it: 'This is clearly as near to a direct appeal to Kapteyn as Gill could allow himself and it elicited the hoped-for response.' The conclusion presents itself that the crucial point in Kapteyn's career was not his reading about Gill's plan to produce the CPD in some publication, but rather Kapteyn's initiative to send his polar altitude paper to Gill. After all, from that point onward an extensive collaboration and mutual understanding evolved, which subsequently developed into one of the most productive partnerships in astronomy. On January 9, 1886, Gill wrote: (see Fig. 6.5):

'It is not easy to tell you what I feel at receiving such a proposal. I recognize in it the true brotherhood of science and in you a true brother.'

Following this, he gives an exposition of all the work he is engaged on. In his letter of January 22, 1886 he expands on this:

'[…] it is a very great and honourable work that will forever claim the gratitude of astronomers, but it is greatly to the interest of science that it should be done without delay – or with as little delay as possible – and as you have the desire, the ability, and above all the necessary enthusiasm of the work which converts apparent drudgery into a glorious labour of love. I should be sinning against the clearest light not to accept such partnership, and such a noble offer as yours.'

The impression that Kapteyn's offer to help Gill measure the Durchmusterung plates came unsolicited is reinforced by Gill's choice of citations from the correspondence with Kapteyn. It may not be surprising that he did not want to spell out in detail how he had tried to lure Kapteyn into making that offer. Although Gill quoted from Kapteyn's letter of December 16, 1885 (Fig. 6.4), as if this remark was added as an afterthought to a letter on a different subject and which had to be completed in haste, that particular letter is in fact a very extensive and detailed answer by Kapteyn to Gill's request to think about how to go about measuring the plates (see page 174).

6.3 Kapteyn's Funding of the Work on the CPD

We have seen in the correspondence above how Gill found funds for the CPD from the Royal Society. How did Kapteyn go about funding his part? He wrote about it to Gill, and we may conveniently continue the quotation of Gill's Introduction to the CPD. However, the letter is quoted by Gill without the first sentences. These read: 'I have referred writing so very long, firstly because of the state of my eyes and

6.3 Kapteyn's Funding of the Work on the CPD

Fig. 6.8 The Royal Observatory at the Cape of Good Hope in a drawing of 1865 (Picture taken from the website of the Astronomical Society of Southern Africa [12])

secondly in expectation of an answer from the government regarding the money. The inflammation of my eye has lasted a long time, however I am now able...' This is the point where Gill started his quotation.

J.C. Kapteyn to David Gill.

Groningen, April 19th, 1886.

'[...] I am now able to report almost perfect recovery. (Professor Kapteyn had been ill). The answer from the Minister has only just now arrived. He writes that he thinks my proposal fit for acceptance, and that for the next seven years he will propose to raise my subsidy to £80 [note that 500 Guilders was the actual sum spread over six years; in current currency this is roughly 13,500€ in terms of purchasing power]. For this year, on the contrary, the extra subsidy of £40, which I requested, cannot be conceded. I have waited for this answer before addressing myself to the Teyler Society. So this money question is still in a somewhat unsatisfactory state [...].
I am very grateful for your offer to endeavour to procure the money from the Government Grant Fund of the Royal Society, or even from your own pocket, [...] still you will understand that for the honour of my University and country, I cannot accept this proposal before having tried to get from Dutch funds the means for executing the Dutch part of the intended work. If I fail in this, however, I certainly will reconsider your proposal [...]
In the meantime I wish to begin the work, even with the prospect that at first it will not advance at a very rapid rate. So if you will send to me the

pictures $-77°$ to $-90°$ I will immediately set to work. With the help of one assistant, I think I can get through that, perhaps, in half a year; this work will enable me to try various methods to definitively on the method to be adopted for the rest.

By the pressure of other work accumulated during my illness, I have not examined the photographs of the Orion region as thoroughly as I wished. Still I have examined them closely enough (I have measured and discussed some three hundred stars) to have a tolerably clear idea what they will yield. They will certainly give as many stars as the Durchmusterung, up to magnitude 9.4, and by a single examination of one of the plates, I have not found a single star wanting. Star-like specks of the magnitude 9.0 or 9.2 seem to be very rare. On the contrary, the number of specks equal to the fainter stars has somewhat exceeded my expectations. I think, as you seem to do, that it will be absolutely necessary to examine two plates at the same time.'

The remainder of this letter discusses purely technical matters.

The Teylers Foundation that Kapteyn refers to was founded with the heritage of the Dutch cloth merchant and banker Pieter Teyler van der Hulst (1702–1778) to support people in need and encourage worship, science and arts. According to the *HHK biography*, he did not receive any money from the Teyler Society, but rather from two other private foundations. Indeed, in the acknowledgments of the publication of the CPD Kapteyn writes: 'Besides the Government, which, on the recommendation of the curators of our University, has gradually augmented its annual subsidy, financial help was generously given by the Bataafsch Genootschap der Proefondervindelyke wijsbegeerte te Rotterdam and the Provinciaal Utrechtsch Genootschap van Kunsten en Wetenschappen.' These were societies in Rotterdam and Utrecht that supported research in general and apparently Kapteyn's applications there were successful. As can be seen from the list of memberships of Kapteyn in Appendix A, he eventually became a member of these societies, the names of which may be translated into English as Batavian Society for Experimental Philosophy, Rotterdam and the Society of Arts and Sciences of the Province of Utrecht.

Cape of Good Hope, May 20th, 1886.
'David Gill to J.C. Kapteyn.

'[...] I am pleased to hear that your Government is prepared, so far, to meet your wishes. The best thing is to proceed with the means available, and prove by 'something attempted, something done,' that the work is worthy of continued support. [...]

I have sent the duplicate pictures $-77°$ to $-90°$, of which the necessary particulars are enclosed. I shall send to Professor Stokes (Pres. R.S.) extracts from the correspondence that has passed between us on this subject, and request the approval of the Royal Society for what I have done, and what is proposed to be done.'

6.4 Execution of the CPD

The telescope used for the CPD is shown in Fig. 6.9. The mount is that of an older telescope at the Cape, which was no longer in use. The main structure is a wooden box of about 12 by 12 inches (30 cm) with a structure at the top that contained one of the Dallmeyer lenses; this lens had a diameter of 6 inches (about 15 cm) and a focal length of 54 inches (about 137 cm). It could be focused by shifting the small tube with respect to the wooden box; the focus could be read off with the dial at the side, designated with the letter 'S'. A guiding telescope was attached to the whole. This was a 9 cm telescope, approximately, that had originally belonged to another telescope that was no longer used. The lamp 'L' illuminated the focal plane of the telescope after reflection off a small elliptical mirror. In this plane the telescope did not employ the usual set of cross-wires to guide on a star in the field, but a 'flat piece of watch-spring perforated near one rounded extremity with a small hole subtending an angle of about 30'''. The star, which was seen through a red glass for convenience, was kept in the middle of the hole.

In the course of testing the methods, another Dallmeyer lens was required, also with a diameter of 6 inches but with a focal length of 5 feet 9 inches (about 175 cm). This was mounted in the telescope shown in Fig. 6.10. This lens, however, proved to be inferior to the first one and although Dallmeyer undertook some re-polishing, the decision was taken to revert to the camera of Fig. 6.9. That lens was also re-polished and all the final CPD plates were taken with that lens (either before or after the re-polishing).

The field of view of the telescope and the size of the photographic plates (about 14 cm at a side) was about 6° and here one millimeter corresponded to about 2.5 arc-minutes. The arrangement on the sky was such that there was an overlap of 1° between adjacent exposures. Each field was photographed at least twice, so that plate faults that could be mistaken for stars could be identified. The first exposure for the definitive CPD was taken on April 15, 1885; except for some areas that Kapteyn requested to be retaken, the collection of the photographic plates was finished in December of 1888. The declinations covered were from $-18°$ to the south pole at $-90°$. The first experiments had been undertaken in 1883. The total number of plates that were eventually accepted for the CPD and actually measured by Kapteyn was 613. But there is a list of about 1100 plates that were rejected and another 100 or so used for other purposes such as calibrations or selection of comparison stars or for measurement of parallaxes of some of the brighter stars. Reportedly, the reason for listing the rejected plates was to provide a complete record, so that researchers of variable stars, for example, would know that these did in fact exist and could be put at their disposal. A few different emulsions were used and for uniform results the exposures varied between 30 minutes and one hour, depending on circumstances.

I now turn to the developments in Groningen. We have seen that Kapteyn received some financial support from the Government. He was in urgent need of three things to get to work on the Durchmusterung. First, he needed an assistant, because the workload would be prohibitively large for a single person. The funds he had received from the government would allow him to hire an assistant. However, experienced technicians were very expensive to hire and his funds went only a little way towards covering those expenses. Henriette Hertzsprung-Kapteyn wrote: 'Since he could not afford to pay a high salary, he decided to look for an 'unskilled' worker. But where could such a person be found? Intelligent and able men working for little money were rare. But he had an idea; he went to the director of the 'Ambachtsschool' and asked him if he had a good pupil whom he could recommend for the work.' The Ambachtschool (ambacht is handicraft) was a secondary school for young men that prepared them for technical professions. 'The director knew of one and put forward the name of a 19-year old boy whom he felt had the necessary qualities. Young de Vries (see Fig. 6.11) came as a temporary apprentice and was directed to do the measurements with the machine, which he did excellently. He turned out to be very talented when it came to performing extremely accurate measurements and was an superb observer; he also carried out the calculations with such accuracy and diligence that he had made himself indispensable in no time.'

De Vries started work for Kapteyn in 1888, according to the University Yearbooks, and was a very important factor in completing the CPD and for the Astronomical Laboratory in general. His full name was Teunis Willem de Vries, born in 1862 in the small village of Anloo, where his father was a tailor. He was 19 in 1881, so there is an inconsistency in Henriette Hertzsprung-Kapteyn's account. To continue the *HHK biography*: 'Still Kapteyn was aware that de Vries with his capabilities deserved a better job with a higher salary than he could ever expect to offer him. So, in 1901 he recommended him to his colleague Nijland when a better paid job came available at his observatory.' Albertus Antonie Nijland (1868–1936) was a professor of astronomy in Utrecht and director of Utrecht Observatory. After having praised de Vries extensively, Kapteyn wrote: 'Heaven knows that I will see this man leave with much sorrow, but I don't feel at ease to be in the way of his career. Such an opportunity will probably not be coming again soon'. Fortunately for Kapteyn, the position was filled by someone else and after having sent a splendid testimonial to the government Kapteyn was able to offer him the position of 'amanuensis [a skilled assistant in a laboratory or educational institution; in this case it was the highest position that could be created for a skilled worker such as de Vries] in his laboratory in 1911. De Vries remained his senior associate and dedicated assistant for 36 years until Kapteyn's retirement. He worked another six years under Prof. van Rhijn, after which he enjoyed his well-deserved retirement.'

Pieter Johannes van Rhijn (1886–1960) was a student of Kapteyn, and succeeded him as professor of astronomy. Assuming that de Vries had been born in 1862, he would be entitled to retire at the age of 65, so in 1927. This is indeed 6 years after Kapteyn's retirement (professors retired at the age of 70). De Vries moved with his son to Eelde in 1926, and later to Leeuwarden (Friesland), where he died on January 1937 at age 74.

But Kapteyn also had no adequate spaces for the measuring machine and his assistant. Here he was helped out by his good friend Dirk Huizinga (1840–1903), professor of physiology (see Fig. 8.10), who offered him two workrooms in his own

6.4 Execution of the CPD 171

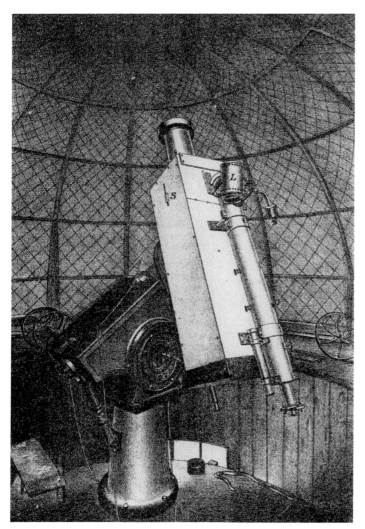

Fig. 6.9 The instrument at the Cape Observatory for taking the plates for the Cape Photographic Durchmusterung (Plate I from the Cape Photographic Durchmusterung at the Kapteyn Astronomical Institute)

physiological laboratory. This laboratory was situated next to and at the back of the University Academy Building. Ironically, it would later become the home of Kapteyn's own 'Astronomical Laboratory', when Huizinga's successor Hartog Jacob Hamburger (1859–1924) moved into a new and larger physiological laboratory in 1911 (see page 418 and Figs. 11.15 and 16.7). In Kapteyn (1914e), part of a the volume published on the occasion of the 300th anniversary of Groningen University, we find a floorplan of the building (Fig. 11.17), where these two rooms are indicated by numbers 2 and 6. In his statements of gratitude at the end of the Introduction to

Fig. 6.10 The instrument at the Cape Observatory for taking the plates for the Cape Photographic Durchmusterung (Plate II from the Cape Photographic Durchmusterung at the Kapteyn Astronomical Institute)

the CPD, Kapteyn included the following words: 'Last, and not least, I am happy to acknowledge my deepest obligations to my friend and colleague, Professor Huizinga, who, at the beginning of the work, when practical astronomy was absolutely houseless at Groningen, graciously, and at no small inconvenience to himself, offered me the use of two rooms in his laboratory.' This help had been vital to the completion of the work. Kapteyn therefore concluded: 'In these two rooms the whole of the measurements, and by far the greater part of the computations, have been made, so that, strange as it may sound, the Cape Photographic Durchmusterung, as far as the catalogue is concerned, may be said to have emanated from the Physiological Laboratory of Groningen.'

6.5 The Parallactic Method

Fig. 6.11 Mr T.W. (Teunis Willem) de Vries, technical and research assistant to Kapteyn (University Museum Groningen)

The fact that he had to rely on the generosity of colleagues and friends when it came to working space goes to show that Kapteyn received very little support from his University indeed, or from the Government, for that matter.

6.5 The Parallactic Method

Kapteyn also needed an instrument to measure the plates. It is here that we again see Kapteyn's inventiveness and originality, already displayed in the novel approaches to absolute measurement of the polar altitude and determinations of stellar parallaxes. The problem is the following: The straightforward way of measuring would be to locate each star on the plate and register its position in a rectilinear grid or Cartesian coordinate system, say x and y, relative to some reference point on the plate such as the center or one of the corners. One would then use stars of known positions to fix a right ascension, declination grid on the plate.

> **Box 6.1 Box Measurements of star positions on photographic plates**
> If α is right ascension and $\alpha_\circ$ that of the center of the plate, and δ and $\delta_\circ$ the same for declination, the procedure is usually to first define 'standard coordinates' ξ and η which are measured in units of the focal length of the telescope and relate to the sky coordinates through
>
> $$\xi = \frac{\sin(\alpha - \alpha_\circ)}{\sin\delta_\circ \tan\delta + \cos\delta_\circ \cos(\alpha - \alpha_\circ)}$$
> $$\eta = \frac{\tan\delta - \tan\delta_\circ \cos(\alpha - \alpha_\circ)}{\tan\delta_\circ \tan\delta + \cos(\alpha - \alpha_\circ)}$$
>
> Now, the plate scale should be assumed to be unknown, as it changes with temperature, and the center of the x and y coordinates is also unknown, while the x and y axes may not line up accurately with the sky coordinates. The standard coordinates then are fitted to the measured coordinates x and y using
>
> $$\xi - x = Ax + By + C$$
> $$\eta - y = Dx + Ey + F$$
>
> The six constants A through F are known as the *'plate constants'* and they can in principle be determined by using three stars of known sky coordinates. In practice this is not accurate enough and one has to use a larger set of such calibrating stars and use special techniques ('least-squares fits', see page 322 for a brief presentation) to get the best possible values for the plate constants. For any star with measured x and y, one can find the corresponding ξ and η and from this their right ascension and declination, α and δ, from inversion of the first set of two equations:
>
> $$\tan(\alpha - \alpha_\circ) = \frac{\eta}{\cos\delta_\circ - \eta\sin\delta_\circ}$$
> $$\tan\delta = \frac{(\eta\cos\delta_\circ + \sin\delta_\circ)\sin(\alpha - \alpha_\circ)}{\eta}$$

However, the conversion of the rectilinear grid to one of spherical, celestial coordinates is somewhat complicated and requires a lot of arithmetic. To get an impression of what is involved, I refer to Box 6.1. Kapteyn avoided all this by using what he called the principle of 'parallactic measurement', the idea of which – he wrote in his Introduction to the publication of part I of the CPD – 'occurred to me very soon after the first plates were placed in my hands by Dr. Gill'. He noted that viewing the plate from a distance that is the same as that in the telescope between the plate and the objective or primary lens, recovers *exactly* the relative positions of the stars on the sky. So, by using a piece of apparatus that works like a telescope, one can – so to speak – 'undo' the original image formation in the telescope when the plate was taken, by pointing it at a star on the plate and recording the orientation

6.5 The Parallactic Method

of the device. Then the measurement of the position of a star is the same as in the case of a telescope pointed at the sky; the right ascension and declination can then be read off directly from the orientation of the telescope with respect to its two axes. This constitutes an enormous saving of time and effort. The price to pay is a loss of accuracy of the measurement of positions because micrometer measurements are more accurate than determinations of these angles, but for the purpose of a catalog this is acceptable.

To quote Kapteyn from the Introduction to the CPD:

'The possibility of measuring spherical coordinates directly on the plane plates results from an evident property of these plates, viz., that by placing them at the proper distance (the focal distance of the photographic telescope) with the film farthest from the eye, it is always possible to cover the stars in the sky by their corresponding images. Therefore, if for the eye we substitute an instrument by which spherical coordinates are measured in the sky, it must be possible to measure these coordinates as well on the plates as in the sky itself.

The one fundamental condition – depending on the finite distance of the plates – which the instrument will have to fulfill, in addition to those of an ordinary equatorial, is, that the three axes: the polar axis, the declination axis, and the optical axis of the telescope, prolonged, must intersect in one and the same point. Moreover, the distance of the objective from this point must be small. For the rest, it will be advantageous to construct the instrument for latitude zero. Every plate can then be observed in the horizon and may even be mounted in an absolutely fixed frame if we give the whole instrument a vertical axis about which it can rotate.'

How Kapteyn's thinking actually developed towards the parallactic method can be gleaned from his correspondence with Gill. Remember (page 164) that in his letter of May 15, 1885, Gill had asked Kapteyn to think about how to measure the plates, either by two rectangular coordinates (like x and y) or in polar coordinates (angle seen from the center as hours are indicated by the hands of a clock, and distance from the center). Kapteyn's famous letter of December 18, 1885 (Fig. 6.4) was a very extensive reply to this. Gill had succeeded in making Kapteyn think about the problem long and thoroughly, during which time he was probably becoming increasingly enthusiastic about it, and the possibility to offer his assistance had slowly been taking shape in his mind. I quote from the letter of December 16, 1885:

'I have somewhat deferred answering your letters, because in one of them You kindly once again requested me to think over the question of the measurement and reduction of Your star-photographs. [...]

The vastness of the undertaking seems to me to require

1° – The greatest possible expedition in taking the measures without sacrificing the required accuracy.

2° – The utmost simplicity in the reduction to a fixed Epoch.

3° – Very little liability to mistakes. [...]

The second condition seems to me to exclude every other method than that of measuring rectangular coordinates.

> These considerations led me to the following plan, [...] The negative is mounted on a frame, which, by means of screws of correction, maybe somewhat raised or lowered as well as rotated about an axis vertical to the plate. In front of the plate, say at about one or two millimeters distance is a ruled glass plate to be mounted. [...] The reading off of the plates can now be very conveniently effected by means of a telescope placed at some distance (say 15 or 20 meters).'

It should be remembered that measuring up star positions on photographic plates was not a tested procedure, as this was the first time a photographic Durchmusterung was contemplated. The letter is nine pages long (Fig. 6.4 is the upper half of the final page, but written with a rather larger letter), mostly concerned with the mathematics of the reduction. Since he had to send off the letter before completely finishing it, the discussion of procedures continued in Kapteyn's letter December 23, 1885, which is another twelve pages long. It is important to realize that at that time the parallactic method had not yet occurred to him.

This comes to his mind a few months later as an adaption and improvement of what he had described in the letters just mentioned. On May 17, 1886, Kapteyn wrote:

> '[...] an idea has occurred to me that would shorten the work of reduction by more than half, while the work to be expended on the measuring themselves will not be <u>very</u> much greater. As I am anxious to have Your opinion and Your <u>suggestions</u> on this plan I will indicate the alterations which are being made in my equatorial stand:
>
> 1st – The telescope is replaced by a smaller telescope of $\pm 24^{cm}$ focal distance which is mounted so that its axis coincides with the hour axis prolonged. Of course this telescope must be what the Germans call 'ein gebrochener Fernrohr'.
>
> 2nd – To make the two axes <u>perfectly</u> parallel, the diaphragm with the reticle has corrections in every <u>direction</u>.
>
> 3^{d} – To make the two axes perfectly coincident the whole telescope can be slightly displaced parallel to itself.
>
> 4. – The hour axis is made horizontal. By this arrangement it will always be possible to mount the centre of the photograph in the same horizontal plane with the instrument.
>
> 5. – The whole of the instrument can be rotated round about a vertical axis. [...]'

A reticule is a set of fine wires in the focal plane. After some more details, Kapteyn concluded:

> 'If this instrument and the position of the plate is carefully adjusted, the measurements can evidently be made to give at once the <u>mean</u> Right ascensions and Declinations for any required Epoch, which will hardly require any corrections at al.'

The instrument is shown in Figs. 6.12 and 6.13. It is currently in the possession of the Museum of the University of Groningen. The observer looks through the

6.5 The Parallactic Method

Fig. 6.12 The parallactic instrument constructed by Kapteyn to measure the plates for the Cape Photographic Durchmusterung (Plate III from the Cape Photographic Durchmusterung at the Kapteyn Astronomical Institute)

ocular (eye-piece) at the extreme right (label 'J'). The telescope is 'broken', which means that a prism is used to deflect the light at right angles; the lens 'H', so to speak, 'looks' at the photographic plate. This is illustrated in Fig. 6.13. The distance between the plate from the center of the instrument is not to scale there; it should be equal to the focal length in the telescope (137 cm). The measuring has two axes around which it can be rotated. The vertical one, sticking out just to the right of the center of the picture in Fig. 6.12 and labeled 'B', serves as the declination axis and the position of the star in that coordinate can be read off with the (small) horizontal wheel 'D' at its bottom. The horizontal axis on top of the stand 'A' acts as a right ascension (or hour angle) axis and the big wheel 'C' on the left is used to find that position. There are two microscopes since there were two plate-holders, slightly displaced from each other in the direction of right ascension. This double plate-holder (see Fig. 6.13) helped to speed up observations. The apparatus was kept in the correct position with respect to the plate by the reading of a scale in the distance through the small telescope 'L' that stick out vertically in the middle.

It is evident that the plate and the instrument have to be positioned carefully, but after that is done the (relative) right ascensions and declinations can be measured directly. Using stars of known positions on the plates, celestial coordinates can be determined for all stars. Between plates adjacent on the sky sufficient stars in the overlap region were measured to ensure an accurate transfer of coordinates from one plate of the next.

Fig. 6.13 The parallactic instrument and the positioning of the plate holder, as used to measure the plates for the Cape Photographic Durchmusterung. Their distance is not the 137 cm required for the method of measurement (University Museum Groningen [13])

The way Kapteyn did the measurements is that he always mounted two plates in the plate-holder (see Fig. 6.14, which was especially designed for this). These were plates of the same area of sky and he used the deepest of the two in the back, so that this one was viewed through the emulsion of the first. The plates were separated by a tiny distance to the order of one millimeter. The plate-holder was designed in such a way that the two plates were shifted with respect to each other by a small amount (so that the images of the brightest stars were only just separated). In this manner it was not complicated to distinguish stars from plate faults and moving objects (asteroids). The measurements were then performed by three persons at all times; one observer looked through the instrument and positioned it on a star, estimated its diameter on the plate (to be used later to determine the brightness) and read the position on the declination axis. The second observer read off the position of the instrument on the right ascension axis using microscopes that made it possible to view the dials and a clerk recorded the results that were called out by the two observers on paper. This was a tedious business, which had to be performed in a darkened room. All plates were measured twice.

The operations at Kapteyn's Laboratory during the times of the measuring of the CPD plates have been described in a very interesting way in an article by Henry Sawerthal (1867–1919), in which he describes visits to a number of observatories in Europe (in Germany, Bohemia, Austria, Italy and France) in 1889. At the start of the note, he went extensively into his visit to Groningen [14]. Sawerthal had been

6.5 The Parallactic Method

Fig. 6.14 The plate holder in a different viewpoint, where it can be seen that two plates were mounted simultaneously (University Museum Groningen [13])

Gill's secretary and night assistant at the Cape, but later became Surveyor General of Rhodesia. I quote from this article to illustrate the primitive nature of the equipment Kapteyn had at his disposal, but also his ingenuity.

> 'Professor J.C. Kapteyn's Observatory consists of two small rooms placed at his disposal by his colleague, the Professor of chemistry, on the ground floor of Groningen University [should be the laboratory of the professor of physiology, of course]. One room is for the computers and the other is the Observatory. In these days of elaborate details in small and stupendous contrivances in large instruments, expectancy suffers almost a shock on entering the genial Professor's sanctum. Imagine a small equatorial, over a century old; a combination altazimuth and equatorial, used two hundred years ago to demonstrate lectures on astronomy, taken to pieces and supplemented by a few odds and ends of an ancient five-inch reflector; two paper scales and a wooden straight-edge divided to millimetres! The measurement of the plates is effected by observing them with a small telescope movable about two perpendicular axes, corresponding to R.A. and N.P.D. [right ascension and north polar distance] on the plates; which axes intersect at a point in the optical axis of the telescope, at a distance from the plate

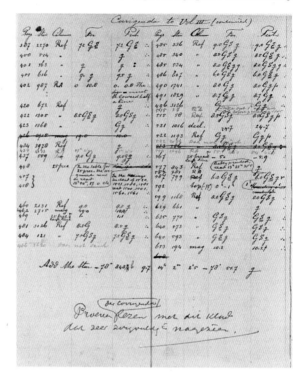

Fig. 6.15 Notes on the corrections for the Cape Photographic Durchmusterung. The bottom sentence says: 'Read proofs for corrections on this draft that has been checked very carefully' (University Museum Groningen)

equal to the focal length of the objective with which the photographs were taken. The angle between two positions of the observing telescope when directed to the images of two stars on the plate is therefore the same as that between the two stars in the sky. Professor Kapteyn has written a paper on the theory of this method in the 'Bulletin of the Congress' [this refers to the Carte du Ciel, which will be discussed in great detail below]; but a description of the actual apparatus may be of interest. For the two axes of the small telescope the corresponding parts of the old equatorial mentioned above are made use of; the polar axis is placed horizontal, so that its hour-circle is in a vertical plane and is easily read by two micrometers. This axis gears into a brass block, and is made to work very smoothly by cone bearings. The brass block is not a fixture, but can be rotated round a vertical axis, carrying the whole instrument with it, the use of which we shall see presently. To the end of the polar axis, remote from the hour-circle, is attached the declination axis, carrying at one end its circle and at the other end its slow motion. Two arms fixed to this axis carry a broken telescope of about $2\frac{1}{2}$ inches aperture. The optical axis of its objective passes through the point of intersection of the R.A. and declination axis, thereby fulfilling the condition imposed by the finite distance of the object to be observed.

The plates to be measured are placed in pairs, one behind the other, in light frames, and held in their places by springs; the front negative, for which the more clearly developed is selected, is separated from the back by a thin piece of brass, and can be adjusted by screws relative to it so as to present a series of close double stars. The back plate rests on a rigid horizontal plane. The vertical sides of the frames are produced as arms to carry a small telescope and thus read the millimetre scales, indicating the adjustment required to make the hour-circle for 1875 of the plate pass through the prolonged hour axis of the observing instrument. Then the whole observing instrument is turned about the vertical axis mentioned before through the N.P.D. of the centre of the plate, so that the hour axis prolonged now passes through the pole (1875.0) of the plate. This angle is read off by a small telescope attached to the brass block. For mounting the plate-frames the horizontal circle of the veteran combination instrument, the foot of the five-inch, and some old sextants have been utilized. It takes 10 to 15 minutes to adjust the apparatus before dashing into zone observations. Each set of plates is observed once by the assistant and once by the Professor, and in case of discrepancy once more by the Professor. The wooden and paper scales, on which some of the adjustments referred to depend, would not of course be used for very accurate measurements, but are sufficient in the present instance to give results more accurate than those of the Northern Durchmusterung, a remark which applies not only to positions, but to magnitudes [also].'

The plates were measured sequentially in declination zones. The first measurement was that of the plate centered on the south pole; this was done on October 28, 1886. The declination 85°-zone was completed on June 9, 1887. Kapteyn took part in these measurements, but he left it to assistants afterwards. The measuring was completed on June 11, 1892, but various shorter sessions of repeat measurements were found to be necessary and these took place in 1892, 1896, 1897 and 1898. Henriette Hertzsprung-Kapteyn quotes a letter from Kapteyn to Gill in June of 1892:

'Finished! – The job of measuring the plates done, done at last. The work has been to me a source of no end of good things, but still its being done at last is one among the best ... the number of observations we got, must be upwards of a million – and the truth is that I find my patience nearly exhausted.'

Gill's reply was a short note, that could just make it in the mail (see Fig. 6.16):

'This is only a jubilant shout – a hurrah – a God bless you, my boy, –and long may you continue to prosper!'

6.6 Production of the CPD

The careful reductions of the recordings to star positions and magnitudes took place over subsequent years. The Cape Photographic Durchmusterung was published in three parts in the *Annals of the Cape Royal Observatory* in Gill & Kapteyn (1896)

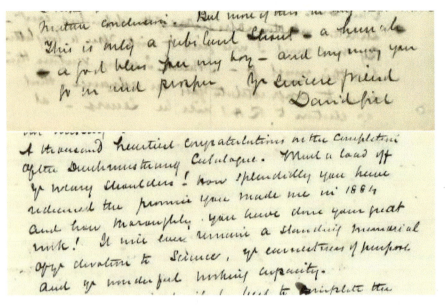

Fig. 6.16 Parts of the letters from David Gill, often quoted, in which he congratulates Kapteyn on the completion of the measurements for the CPD (June 6, 1892; top) and on the completion of the calculations for the catalogue (April 6, 1899; bottom) (Kapteyn Astronomical Institute)

the declination zones $-18°$ to $-37°$, in Gill & Kapteyn (1897) $-38°$ to $-52°$, while in Gill & Kapteyn (1900) the southern sky was completed with the declination zones $-53°$ to $-89°$. Meanwhile, reports on the progress had been forwarded to the Astronomische Gesellschaft, Kapteyn (1888c), Kapteyn (1890d), Kapteyn (1892g) respectively, and a final, extensive summary of the whole project in Gill & Kapteyn (1898). On April 6, 1899, Gill wrote (see Fig. 6.16):

'A thousand heartiest congratulations on the completion of the Durchm. Catalogue. What a load off your weary shoulders! How splendidly you have redeemed the promise you made me in 1884 and thoroughly you have done your great work! It will ever remain a standing memorial of your devotion to science, your earnestness of purpose and your wonderful working capacity'

The final publication and completion of the project occurred just before the start of the twentieth century (it started on January 1, 1901) and Kapteyn's 50th birthday on the 19th of January, 1901.

In the 'Introduction' to the final volume, Gill & Kapteyn (1900), Gill wrote:

'I now realize that my many other duties, and the difficulty of obtaining adequate assistance, would have compelled me to defer a great part of the work to the years of my retirement from official life. I feel assured that Kapteyn has not laboured in vain, and that astronomers will duly appreciate what he has done for science.'

6.6 Production of the CPD

Fig. 6.17 A box of CPD plates that is kept at the University of Groningen depot at Zernike campus (University Museum Groningen)

In this Introduction Kapteyn provided a long and detailed discussion of the overall quality of the catalogue. Final positional accuracies were determined to be to the order of 0.28 seconds of time in right ascension at the lower declinations (and correspondingly more at larger declinations, because the lines of equal right ascension become more closely spaced), corresponding to about 4 arcsec, and about 3 arcsec in declination. The magnitudes of the stars were determined from the diameters of the images on the plates and were finally judged to have an accuracy of about 0.06 magnitudes. Remember the magnitude scale is a logarithmic one and that therefore this value corresponds to a percentage of intensity, viz., 6%. There is an issue, however, with regard to the magnitudes, and this concerns the wavelength response of the photographic plate, which is different (bluer) than that of the eye. Consequently the relative brightness of stars may be different in the two systems or magnitude scales (which were called 'photographic' and 'visual' respectively) as a result of their different colors. The CPD has the photographic magnitude scale defined such that the number of stars of visual magnitude m is in the mean equal to the number of stars of photographic magnitude m. This means that the assumption is that the distribution of star colors is the same at all apparent magnitudes. However, systematic differences in photographic and visual magnitudes of stars were seen as a function of Galactic latitude. After careful discussion Kapteyn concludes that the CPD is complete to at least photographic magnitude 9.2. The total number of stars in it is 454,875.

The Introduction of the Cape Photographic Durchmusterung mentioned that the decision had been taken to store the plates in Kapteyn's Laboratory on a permanent basis. They are still in Groningen as a matter of fact, and are preserved at the depot of the Groningen University Museum at the Zernike Campus, not far from where the Kapteyn Astronomical Institute is currently located. They are stacked away in rows of small brown wooden boxes, each containing about 35 to 40 plates (remember they measure 14 × 14 cm). An example of such a box is shown in Fig. 6.17.

6.7 The 1887 Carte du Ciel Congress

At the outset, the CPD project ran into some fundamental and threatening difficulties, having to do with a major international project that was initiated at about the same time and which was closely related to the CPD. This major undertaking, in which Gill and Kapteyn both participated, almost stopped the CPD where it had only just started. It was the *Carte du Ciel*. The background has been described in various places. The early stages, for example, were described by David Gill in his already cited lecture [15] for the British Association on June 3, 1887, and more recently by T. Weimer in a review as part of a international symposium on mapping of the sky [16]. Another very informative book is, *The Great Star Map, being a brief general account of the international project known as the Astrographic Chart*, written in 1912 by the British astronomer Herbert Hall Turner (1861–1930) [17].

We have seen that Gill had been in contact with Admiral Ernest Mouchez, director of Paris Observatory as early as 1882, on the matter of using photographic plates in astronomy, pointing out the possible use of such photography for cataloging stars. In his lecture just referred to, Gill states: 'There was thus nothing really new either in my suggestion or in the *modus operandi*, [...], and Admiral Mouchez tells me that these Cape photographs and my suggestions first directed his attention and that of the brothers Henry to the application of photography to the work of star-charting, which had for many years been carried on at Paris by the older methods of astronomy.' Gill next described the work of the Henry brothers in improving photographic methods.

> 'With an exposure of forty-five minutes, pictures of stars were obtained to the 12th magnitude, in which the star disks were quite round and sharply defined. Fully appreciating the beauty of this result, and seeing its importance, Admiral Mouchez boldly faced many administrative difficulties, and accepted without delay the proposals of the brothers Henry to construct an object-glass of thirteen inches aperture and about 11 feet focal length, as well as the offer of M. Gautier to mount the same on a suitable stand. The new instrument was mounted in May 1885. Both from an optical as well as a mechanical point of view, the new instrument was admirably adapted

6.7 The 1887 Carte du Ciel Congress

for its intended work, and the results obtained by the brothers Henry, and rapidly published and circulated by Admiral Mouchez, at once astonished and delighted the astronomical world. [...]

The means of rapidly obtaining the data for an accurate survey of the heavens on a very large scale were now within the reach of astronomers, and the time for decisive action had arrived. The work, however, was too extensive to be undertaken at a single observatory, or even by a single country, and it was agreed on all hands that international co-operation was essential for its execution in a sufficiently short space of time. I need not enter into the details of preliminary consultation or correspondences, but at last a time was fixed, and invitations were issued by Admiral Mouchez, Director of the Paris Observatory, under the auspices of the Paris Academy of Sciences, for an International Congress of Astronomers to be held at Paris.'

There is a letter from Gill to Mouchez that he published as an 'open letter' in the *Bulletin Astronomique* [18]. The letter, dated March 1, 1886, reported that he had obtained photographs from the Henry brothers, of a part of the Milky Way, which he had shown at the Royal Society in London. These had made a very strong impression. Gill then suggested that these should be used to produce celestial charts of two classes (my translation from the French):

'1. A plan that could be completed within a limited and well-defined period, and which would not just include a photographic survey, but also measurements of right ascensions and declinations, resulting in the publication of a catalogue reduced to a common equinox.

2. A programme that would directly aim at the formation of a photographic atlas of the entire heavens with a plate scale that enables one to find the brightness of the stars photographically and determine their relative positions on the images with great precision.'

In a second open letter in that journal [18], Gill mentioned that he had written to Otto Wilhelm von Struve (1819–1905), director of Pulkovo Observatory near St Petersburg, about support for a congress of astronomers in early 1887, and that he had informed Admiral Mouchez of his strong support. Gill noted that the atlas should be published on paper, an idea that had been suggested by Kapteyn. Mouchez then added that he fully agreed to proceed with such an enterprise.

Indeed, Mouchez organized a conference in Paris, the Congrès Astrophotographique International tenu á l'Observatoire de Paris pour le Levé de la Carte du Ciel [20]. It opened on April 16, 1887. Astronomers from all over the world were invited. Most (20) came from France or the rest of Europe (Austria, Belgium, Denmark, Finland, Germany, Great Britain, Italy, the Netherlands, Portugal, Russia, Spain, Sweden, Switzerland), but there were also three astronomers from the United States, and one from Argentina, Brazil, South Africa and Australia respectively. In all 55 astronomers took part. Figure 6.18 shows the group photograph that Kapteyn kept in his archives. This picture is usually published without the set of signatures at the bottom; these and the list of participants make it possible to identify all persons. I show some of them who figure prominently in a few subsequent pictures that follow separately below. In Fig. 6.19 we see Kapteyn. The two men above Kapteyn are

186 6 Cape Photographic Durchmusterung

Fig. 6.18 Group photograph of the meeting of the Carte du Ciel in 1887. Each person has added his signature at the bottom (Kapteyn Astronomical Institute)

6.7 The 1887 Carte du Ciel Congress

Fig. 6.19 Kapteyn in the group photograph of the meeting of the Carte du Ciel in 1887. The persons above him are Paul (left) and Proper Henry (Full photograph in Fig. 6.18)

the Henry brothers, Paul on the left and Prosper on the right. Kapteyn seems ill at ease; but it should be noted that at the time he was only 36 years of age, which is significantly younger than most of the other participants, who all were very senior astronomers. One exception is Anders Severin Donner (1854–1938) of Helsingfors Observatory (Helsingfors is Swedish for Helsinki), who was actually younger than Kapteyn and with whom he developed a long lasting friendship at the Carte du Ciel meeting. Donner and David Gill are shown in Fig. 6.20). Figure 6.21 shows van de Sande Bakhuyzen and Oudemans, the two other participants from the Netherlands. Evidently, they were considered prominent enough to be seated in the front row. In the middle of that row we find the most powerful men; Fig. 6.22 has on the left William Henry Mahonie Christie (1845–1922), director of the Royal Greenwich Observatory and Astronomer Royal, next to him the host Admiral Ernest Mouchez, director of the Observatoire de Paris, then Otto Wilhelm von Struve (1819–1905), director of Pulkovo Observatory near St Petersburg and on the right Georg Friedrich Julius Arthur von Auwers (1838–1915), secretary to the Berlin Academy.

There is a curious issue with this group photograph. In his book *La Carte du Ciel* [21], Jérôme Lamy includes a copy of the same group photograph. However, Kapteyn is missing from this copy! Closer inspection shows that another six people are likewise missing. In both pictures all the others are seated or standing in the same arrangement and they all show more or less the same posture. However, their faces are often turned in a (slightly) different direction. Clearly, that picture is not the same one as the one reproduced here, but must have been taken only shortly

Fig. 6.20 Two astronomers who played a very important role in Kapteyn's career, here in the group photograph of the meeting of the Carte du Ciel in 1887. On the left David Gill and on the right Anders Donner (at the top) (Full photograph in Fig. 6.18)

before or after it. There is no particular clue as to why certain persons are absent in the picture. The article referred to above by Weimer has a detail from the group photo showing the Henry brothers, but presumably that is also taken from the copy without Kapteyn.

What can be learned from the record of the congress as to the reason for the two photographs? The logistics of the meeting are also well documented [22]. The meetings were held in the 'Bâttiment Perrault', the old building of Paris Observatory, completed in 1671 (but extended in later years). It defined the French meridian and has been designed by architect Claude Perrault (1613–1688) (who should not be confused with his brother Charles Perrault (1628–1703), the famous author of fairy tales). The meeting took place in the large gallery on the first floor. The state-owned furniture was supplied by the City of Paris and included tables (60) and chairs, as well as plants. Receptions also took place here, as well as a musical recital on April 19*, followed by a ball (would Kapteyn have danced?). There was also a dinner, on April 21. There is a bill from a fitter who took care of lighting, and from this it can be deduced that the attendants met all day on the 19th, followed by the evening program until one in the morning; then, in the afternoon of the 20th, a meeting a meeting took place on astrophotography itself and on the 21st there were sessions in the morning and the afternoon, followed by the banquet. The meetings took place between Saturday April 16 and Monday April 25 (see below).

The table seating of the banquet has also survived [23]. Curiously, Kapteyn is also missing on this list. The total seating plan included 70 places, the host and his wife facing each other in the middle of the long table. In addition to the participants of the conference there were a number of important guests who would not have taken

6.7 The 1887 Carte du Ciel Congress

Fig. 6.21 H.G. van de Sande Bakhuyzen in the group photograph of the meeting of the Carte du Ciel in 1887 (left) and J.A.C. Oudemans (right) (Full photograph in Fig. 6.18)

part in the scientific sessions; I count 21 of such persons (Mouchez's wife one of them). But again a few astronomers are missing. These add up to six, two of whom were members of the Academy of Science. The overlap with the persons missing on the group photograph is significant, viz., four. In addition to Kapteyn, these are Isaac Roberts (1829–1904), pioneer of photographing nebulae and owning a private observatory in Maghull near Liverpool (he is known to have had an observatory near his homes at Rock Ferry, Birkenhead, near Liverpool and Crowborough, Sussex), Eduard Schönfeld (1828–1891), director of Bonn Observatory, and François Perrier (1835–1888), geodesist and a member of the Academy of Science. Absent at the dinner were James Francis Tennant (1829–1915), a British astronomer and member of the Royal Astronomical Society, and Hervé Faye (1814–1902), a French astronomer specializing in comet and solar system studies at the Paris Observatory. Why Kapteyn is absent from the dinner and photograph remains unclear. Maybe the picture was taken before the dinner started and re-taken for some reason after Kapteyn, together with some colleagues who had perhaps declined to attend the dinner, had left. Three others may have been too late for the photograph, but in time for the dinner. These would be Luís Ferdinand Cruls (1848–1908), a Belgian astronomer and director of the Rio de Janeiro Observatory, Father Stephen Joseph Perry (1833–1889), director of Stonyhurst College Observatory in Lancashire, and the French hydrographer Jean Jacques Anatole Bouquet de La Grye (1827–1909).

There were also dignitaries involved. The Minister of Foreign Affairs opened the proceedings and the attendants were received by the President of the Republic. The total costs of the conference added up to a few thousand Francs, probably for

Fig. 6.22 The center of power at the Carte du Ciel meeting in Paris in 1887. From left to right: William H. Christie, Astronomer Royal and director of the Royal Greenwich Observatory, Admiral A. Ernest B. Mouchez, host and director of the Paris Observatory, Otto W. von Struve, director of Pulkovo Observatory near Saint Petersburg and G.F.J. Arthur von Auwers, secretary to the Berlin Academy (Full photograph in Fig. 6.18)

the greater part paid for by the Ministry. One thousand French Francs of 1887 were equivalent to about half to two-thirds of an annual salary of a Paris worker [24].

At the time the Academy of Sciences in Paris, together with the French Cabinet, was very much occupied with initiatives it had been taking for the organization of international conventions. The French had lost the battle for the zero meridian, which went to London (Greenwich) in 1884, but were very much involved in establishing the standards for the 'weights and measures', for which they had adopted the kilogram and the meter. In 1875 the *Convention du Mètre* had been signed in Paris, where an institute in Sévres near Paris had been established to co-ordinate international matters concerning metrology in general. During the ensuing first *Conférence Générale des Poids et Mesures* (CGPM) in 1889, an 'International Prototype Kilogram' and an 'International Prototype Meter' were to be selected. At the same time, France gave much attention – to no avail, as it turned out – to the promotion of the decimalization of time and circumference, a process that (like the meter and the kilogram) had been initiated at the time of the French revolution. Initiatives such as dividing the right angle into 100 rather than 90 degrees, the division of the year into 12 months with three weeks of 10 days each (plus 5 or 6 extra holidays), and the day into 10 hours of 100 minutes of 100 seconds, seemed rational but received no support at the end of the day. In this system a degree on the Earth's surface would have corresponded to a nicely round 100 kilometers (now it is 111.111111). The adoption of the Celsius or centigrade temperature scale, named after Swedish scientist Anders Celsius (1701–1744), in the Système Internationale was only partly successful. The Anglo-Saxon world in particular stuck to that of Fahrenheit, named

6.7 The 1887 Carte du Ciel Congress

after the German physicist Daniel Gabriel Fahrenheit (1686–1736). In any case, the fact that the French were so keen on hosting the Carte du Ciel conference and taking the lead in this project, fits in well with their broad interest and preference to coordinate and lead international initiatives and undertakings.

The Carte du Ciel conference took place at a time when such large international meetings and projects started to become part of the way science was conducted. Major progress depended on international and often intercontinental co-operation. Obviously, it is no surprise that differences in cultures came to light in this context. It is amusing to read the reminiscences of one participant of the meeting. Herbert H. Turner from Oxford (who later coined the term parsec) in his book, *The Great Star Map* of 1912 on the Carte du Ciel that I have already mentioned above, gave the following description:

'Astronomers from distant quarters of the globe, speaking different languages, none of them with much experience of photography or of its possibilities, but most of them with opinions more or less formed, met together to try and secure unanimity, not only in generalities but equally in small details. The discussions were, to say the least of it, animated. There are no universal rules for conducting public business, and astronomers of one country were not familiar with rules in use elsewhere. It interested Englishmen, for instance, who are accustomed to have resolutions moved by any one rather than the chairman, to learn that this was by no means a universal rule. On the contrary, the chairman of the first session considered it part of his duties to move all resolutions. After listening to a discussion, he took it to be his function to summarise the sense of the meeting in a resolution which he put from the chair and in favour of which he held up his own hand. Unfortunately for his success his was sometimes the only hand held up, and the discussion was necessarily resumed.'

The initiators of the Carte du Ciel conference, Mouchez in particular, strongly insisted on uniformity in the way the project was to be conducted at the various observatories, and that quickly gave rise to resistance, not only as a matter of principle but, with the taking of the plates for example, for very practical and justified reasons. Exposure times depended on location, atmospheric conditions, etc. Indeed, in the beginning the attitude had been that resolutions once passed were never to be changed. This proved untenable and eventually resulted in changes to some of the rules. Actually, some observatories simply ignored them and used their experience to arrive at results comparable to those of others. The leading role of France, insisted upon by the French political leadership of the Third Republic, was acknowledged but it was also recognized that democratic procedures had to be followed to organize an undertaking such as this. The publication of long articles, defending certain points of view, in the proceedings of the Permanent Committee (see below) played an important role in this process. It is beyond the scope if this biography, however, to pursue this issue further, but see e.g., *The role of the Conferences and the Bulletin in the modification of the practices of the Carte du Ciel project at the end of the nineteenth century* by Jérôme Lamy [25].

6.8 Execution of the Carte du Ciel

The minutes of the 1887 meeting were published in full (*op. cit.*) – albeit entirely in French – so we can monitor Kapteyn's involvement in some detail. Kapteyn participated in the general (afternoon) sessions of Saturday April 16 and Tuesday April 19, but not in those of Saturday April 23 and Monday April 25. The first meeting was chaired by the Minister of Foreign Affairs, Émile Flourens (1841–1920), the others by von Struve, the vice-chairs rotating between the Brit Christie, von Auwers from Germany and the Frenchman Faye. The secretaries were François Félix Tisserand (1845–1896) and van de Sande Bakhuyzen. In the minutes of the first meeting a remark by Kapteyn was recorded on the text of the resolution; there is no mention of a contribution on his part for the second session.

There was a meeting of a small group, the 'Technical Committee', in the morning of Monday April 18, which Kapteyn attended, but no input in the discussion on his part was recorded. Two meetings of the 'Section Astrophotographique' were held under the chairmanship of the director of the Observatoire de Meudon, Pierre Jules César Janssen (1824–1907), one in the morning of Wednesday April 20 and one in the afternoon of Thursday April 21. During the second meeting Kapteyn made a remark on the problem of elliptical shapes of stellar images on photographic plates and related this to the determination of magnitudes. The three sessions by the Section Astronomique were in the afternoon of Wednesday April 20, in the mornings of Thursday April 21 and Friday April 22, and were chaired by von Auwers. Kapteyn was present at all three. It may be assumed that Kapteyn must have left Paris and skipped the meeting of Saturday April 23.

During the final meeting of the plenary sessions the participants elected a 'Permanent Committee' to work out details, coordinate the work and prepare the next meeting. To begin with they selected six members – directors of participating observatories – and then decided to elect eleven more from those present. The full vote was listed; this seems to indicate that the procedure was that every person present could vote for eleven names. It was noted that 45 persons voted, but that one of them abstained. At the top of the list is Gill, who collected 44 votes, followed by Christie with 39 votes, etc. Obviously, Gill must have voted for himself. This incidence is also described in the *HHK biography*:

> 'Although he [Gill] was modest and humble, he was nevertheless in the possession of an appropriate sense of self-consciousness. That was evident at the Astronomical Congress in Paris in 1890. A vote had to be taken who would become the president of the *Carte du Ciel* undertaking. When the votes were counted, it turned out that all votes were for Gill. Obviously he had voted for himself. 'Yes', he said without restraint, 'I thought I was the most qualified person', which he was and proved to be without doubt. It is a sign of greatness to do a thing like that without giving rise to offense. Gill was one who could do and say anything without harming others.'

Henriette Hertzsprung-Kapteyn does not represent the situation quite correctly here. In the first place there was no such meeting in 1890. The Permanent Committee met in 1889 and 1891, the proceedings of which were reported briefly in the *Bulletin As-*

6.8 Execution of the Carte du Ciel

tronomique [26]. Secondly, the vote was not about a chairmanship. Gill was never chairman of the Permanent Committee, although he obviously played a very important role in the project. The committee met on Tuesday April 26, 1887, with von Struve in the chair and Charles Trépied (1845–1907), director of Algiers Observatory, as secretary. Gill was present, of course. This meeting selected a 'Bureau du Comité Permanent', made up of eleven persons with von Struve as chair and with three secretaries, one of whom was David Gill. This bureau met in the morning of Wednesday April 27. Its most important business was to define a work plan and designate persons to perform specific tasks. The Permanent Committee met again in 1889 and in 1891. In 1891 three persons were absent, including Gill and Christie, and they were replaced by three others, one of whom was Kapteyn. In 1891 the meeting was once again presided over by Mouchez, with Gill and van de Sande Bakhuyzen as vice-presidents, and with Kapteyn and Trépied as secretaries. Clearly Kapteyn had become a prominent participant in these international discussions.

At the 1889 meeting, Kapteyn was elected in two committees. One was together with van de Sande Bakhuyzen, Christie, Gill, and Lœwy (1833–1907), deputy director of the Paris Observatory to: '1° determine the positions of all the plate centers for the equinox of 1900,0.' '2° and choose the guide stars and assure their reduction'. The second committee also counted van de Sande Bakhuyzen, Christie and Gill among its members, but in addition there were Prosper Henry and Hermann Carl Vogel (1841–1907), director of the Potsdam Astrophysikalisches Observatorium. Their assignment was to make a study of possible measuring instruments and all related questions.

Kapteyn strongly promoted his parallactic method. He published an article in the weekly illustrated journal *'Engineering'* (London), Kapteyn (1893b), describing the method in detail, and even went out of his way to build a prototype for the measurement of Carte du Ciel plates, with the generous support of Mr Bischoffsheim, Kapteyn (1892b). He also published extensive studies in the Bulletin of the Permanent Committee on the parallactic method, Kapteyn (1888a) and Kapteyn (1888b), a possible work plan for its use in the Carte du Ciel, Kapteyn (1892a), and a theory on the associated random and systematic errors, Kapteyn (1892b) and Kapteyn (1892c). Gill supported this, (see Gill & Kapteyn, 1892), but others were more critical. In particular van de Sande Bakhuyzen published an extensive study [27]. in the Bulletin on the 'measurement of plates according to the method of rectangular coordinates'. He quoted Gill's explanation that there were three general methods: rectangular coordinates relative to plate center (or corner), polar coordinates with respect to the center (so distance and angle seen from the center), and Kapteyn's parallactic direct method, adding Gill's conclusion that the first was inferior to the second. Van de Sande Bakhuyzen disagreed, however, and stuck to the first method. Kapteyn (1889b) replied to this in great detail.

There must have been much discussion that went unrecorded, but the project at the 1896 meeting (where Kapteyn was not present) decided the following:

'2.(a). The Committee regarded it necessary to publish coordinates of the stars on the photographs as soon as possible *rectangular* [my emphasis].
(b). It was considered desirable that this publication should contain data

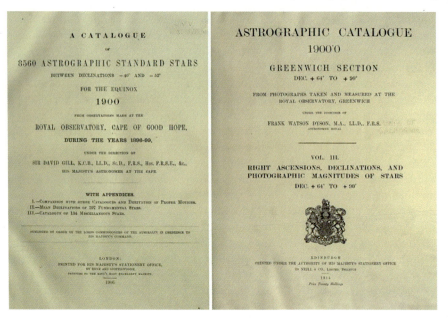

Fig. 6.23 The covers of two of the volumes of the Astrographic Catalogue

required for the conversion of the results to equatorial coordinates. (c). The Committee expressed the desirability that a Catalog with provisional right ascension and declinations be published by those observatories that had sufficient resources.
3. Each observatory would have the freedom to choose the positions of guide stars for the catalogs which seemed most convenient and suitable. For the calculation of the plate constants they should adopt a minimum of 10 guide stars. They should also publish the adopted positions of these guide stars.'

The parallactic method was not adopted, however.

The Carte du Ciel project as an international undertaking had its successes and failures. It succeeded in enlisting observatories all over the world to do coordinated sets of observations. However, the organizers failed to enlist American observatories in the project, although there were contacts with the director of Harvard College Observatory, Edward Charles Pickering (1846–1919). In the end twenty observatories participated. They had settled on a common telescope (with a 33 cm lens with 3.4 meters focal length), either from the Henry brothers or from the Grubb factory in Dublin, founded by Howard Grubb (1844–1931) and a resulting uniform plate scale of about 1 arcmin per mm. Plates covered a square area on the sky of just over 2° on a side. There was considerable overlap, the corner of each plate being the center of an adjacent one.

The project's aim was to deliver two products. The first was the so-called *Astrographic Catalogue*, which would provide the positions and magnitudes of all stars down to magnitude 11. For this, before the stellar properties were derived from the

6.8 Execution of the Carte du Ciel   195

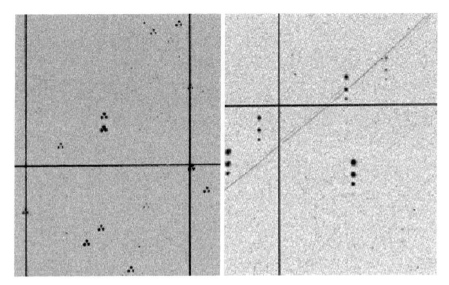

Fig. 6.24 Examples of star images on Carte du Ciel (left) and Astrographic Catalogue (right) plates from the Observatorio Nacional Argentino, Córdoba. (From Orellana *et al.*, 2010 [28]

photographic plates, the positions of a set of reference stars had to be determined first, using meridian circles; in some descriptions this is mentioned as a separate activity. After a substantial delay, this was finally completed in 1961, the roughly 22,000 plates having been taken between 1891 and 1950. Its result can be found in over one hundred printed volumes; an inventory has recently been provided by Donata Randazzo [29]. Figure 6.23 shows two covers of these publications.

The second part is the *Carte du Ciel* proper, which envisaged the provision of photographic images of the whole sky, distributed as photogravure plates (engraved images in copper plates). However, this turned out to be prohibitively expensive and this part of the project was never finished. Of course it was superseded by the *National Geographic Society – Palomar Observatory Sky Survey* of the 1950s and its later southern extensions, which were produced quickly and distributed on paper copies without much delay. This immediately made the Carte du Ciel proper unnecessary and outdated.

Figure 6.24 shows the method of observing. All plates had *réseau's* imprinted photographically on them, consisting of a square matrix of thin lines; these were to check against distortions in the photographic emulsion. In the end this turned out to be entirely unnecessary, so the réseau's could be used instead to measure star positions against. This was fortunate, since measuring positions over the full dimension of the plates would have required microscopes on long, accurate screws, which would have been very expensive and also subject to wear in the course of the project, with attendant inaccuracies. For the Astrographic Catalogue plates, there were always three different exposures, lasting up to 6 minutes and shifted fractionally in position to allow for stars of different brightness to be recorded in such a way as to make them accessible to accurate measurement. This also made it possible to distinguish stars from plate defects and asteroids. The deeper Carte du Ciel exposures were for three times twenty minutes, producing a small equilateral triangle, again to eliminate plate faults and asteroids (see Fig. 6.24).

The Astrographic Catalogue was never a success even though it was completed. Until very recently it was almost completely ignored and not used. Not only were the magnitudes very inaccurate and even lacking in consistency between zones, but a further important reason was that in addition to being much delayed, the publications that were available eventually contained only the rectangular coordinates on the plates in almost all cases, and although plate constants were also provided, deriving right ascensions and declinations from those was cumbersome to say the least. In more recent times, the catalogue has become a goldmine. After they were made accessible to computers, the positions were combined with the results of the European Space Agency ESA's astrometric satellite Hipparcos The Astrographic Catalogue now serves as a first epoch determination of the star positions and offers the possibility to have an of order a century time base for the measurement of accurate proper motions. Kapteyn no longer contributed to the Carte du Ciel effort, after a theoretical paper on corrections for refraction and aberration images taken with photographic telescopes, Kapteyn (1896a), probably because he felt he could not convince the project sufficiently to adopt any of his views and maybe also because the CPD might have been taking too much of his time anyway.

6.9 Gill, Christie and the CPD

The Carte du Ciel project, to a large extent conceived and brought about by David Gill, turned out to have a very serious effect on the execution of the Cape Photographic Durchmusterung. Please remember from the above that the Government Grant Fund of the Royal Society granted to Gill for the execution of the Cape Photographic Durchmusterung was renewed for the year 1886, but in November of the same year a resolution was passed postponing a decision as to its continuation until after the meeting of the Astrophotographic Congress in Paris in May 1887. This might well stop the entire project. At the basis of all this lies the very poor relationship of Gill with Astronomer Royal Christie, the director of the Royal Greenwich Observatory.

Michael Feast, in the *Legacy*, wrote:

'To understand this we have to go back to 1879 when the post of H.M. Astronomer at the Cape became vacant. There had been two candidates for the post: Gill, and Christie (see Fig. 6.25) who was then Chief Assistant at Greenwich and who had the support of the Astronomer Royal, Airy. [George Biddell Airy (1801–1892) was Astronomer Royal from 1835 to 1881.] Much to Gill's surprise he, rather than Christie, got the job. Later Christie became Astronomer Royal and created a number of difficulties for Gill. It is not now easy to decide how much of this was due to genuine scientific disagreement

6.9 Gill, Christie and the CPD

Fig. 6.25 William H.M. Christie, Astronomer Royal, as he appears in the *Album Amicorum* presented at the retirement of H.G. van de Sande Bakhuyzen as professor of astronomy and director of Leiden Observatory in 1908 (Archives Leiden Observatory; see caption Fig. 3.7)

and how much to obstructiveness on Christie's part. [...] Both the CPD and the Carte du Ciel sprang from Gill's enthusiasm to apply photography to astronomy on the grand scale. The Carte du Ciel was planned to go to magnitude 11.0, whereas the CPD was only to go as faint as 9.5 mag. Some saw this an unnecessary duplication of effort. Gill argued that whilst the CPD could be finished and made available for study in just a few years, the Carte which went fainter and covered the whole sky would take much longer. In this he was correct, the Carte du Ciel catalogue was completed only in 1964. In addition, others, it would appear, were not happy to see visual observations superseded by photography. When Gill's grant for the CPD project came up for renewal, Christie managed to get it stopped though apparently only after a stormy committee meeting (see the Murray (1988) paper [30]). A resolution at the Paris conference, itself, in favour of the CPD, was proposed by two of the leading astronomers of the time, [von] Auwers and Struve, but had to be withdrawn since Christie threatened to withdraw his official support for the Carte du Ciel if it was carried. Auwers then offered Gill financial support from the Berlin Academy, but Gill declined this because: '[...] I should probably receive a reprimand and be ordered to stop the work.' Rather than stop the project, Gill paid for its continuation out of his own pocket.'

Gill's letters to Kapteyn contain a wealth of information on this issue, since he related the events as they develop. I will quote the most interesting parts verbatim below. The Murray paper mentioned above contains a long quote (taken from a

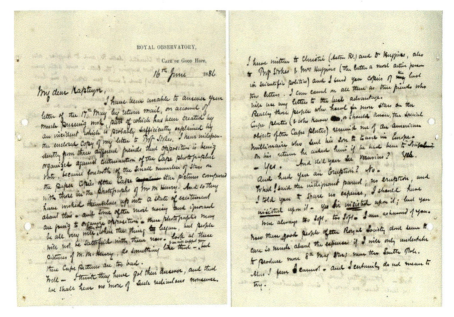

Fig. 6.26 First two pages of the letter by David Gill of June 16, 1886, in which he makes mention of the opposition against the CPD in the Royal Society for the first time (Kapteyn Astronomical Institute)

biographical text on Gill from 1916), supposedly from a letter of Gill to Kapteyn on June 5, 1887. However, no such letter is to be found in the Kapteyn archives, and I presume it concerns a letter addressed to someone else. The first mention of trouble is 11 days later in a letter by Gill to Kapteyn of June 16, 1886, and because of its revealing nature it seems appropriate to quote a larger part of this letter (see also Fig. 6.26.)

> 'I hear independently from three different friends that opposition is being organized against continuation of the Cape photographic vote, because forsooth [Gill spells this foresooth] of the small number of stars on the paper copies of the Cape star pictures compared with those in the photographs of M.M. Henry. And so they have worked themselves up into a state of excitement about this – and some of the most noisy and ignorant are going to organize opposition – these photographs may be all very well, they say, when the thing began, but people will not be satisfied with them now – look at these pictures of M.M. Henry, do something like that and we will support you– but these Cape pictures are too bad.
> Well – I think they have got their answers, and that we shall hear no more of such ridiculous nonsense. I have written to Christie (Astron. R.) and Dr Huggins, also to Prof. Stokes & Mrs Huggins (the latter a most active person in scientific politics) [Stokes was president of the Royal Society (see above) and Sir William Huggins (1824–1910) its secretary] […]. I can count on all these as true friends who will use my letters to the best advantage.

Really those people who howl for some stars on the Cape plates (& who know, or should know, the special objects of the Cape plates) remind me of an American millionaire when sent his son to travel in Europe. On his return he asked him if he had been to Naples? – Yes. And did you see Vesuvius? Yes. And had you an eruption? No. What! said the indignant parent, no eruption, and I told you to spare no expense. I should have <u>insisted</u> upon it – Yes Sir <u>insisted</u> upon it! but you were always too soft, too soft. I am ashamed of you.

Now these good people of the Royal Society don't seem to care too much about the expense if I will only undertake to produce more 8th magnitude stars near the South Pole. Alas I fear I cannot – and I certainly do not mean to try.'

So, Gill apparently put his full trust in Christie and the Royal Society. On September 12, 1886, Gill wrote to Kapteyn:

'In the first place I am happy to tell you that I have very satisfactory letters from various friends including Prof. Stokes & Dr Huggins, and I think the opponents to the Cape photographs are now sufficiently ashamed of themselves – and that we shall hear no more of the proposed opposition.'

But the matter did not rest here, unfortunately. There is no mention of it in Gill's letters of September 24 and December 1, which concern the progress of the CPD itself and Kapteyn's parallax program that I have already discussed. On February 19, 1887, a few weeks before the Paris meeting, Gill wrote from London on his time schedule, which was to include a visit to Kapteyn in Groningen (see the next chapter). He mentioned that he needed to go to Hamburg, where he had ordered a new heliometer from the Repsold factory. The device had to be packed and the money had to be paid before the end of the fiscal year. He also stated that he was

'very anxious to pay a private visit to Paris before the photographic congress, as I wish to be fully prepared on all points before the congress begins. How necessary this is for full support of our scheme I will explain fully to you when we meet. Not that our scheme will be stopped, because I have made up my mind to carry it out even if necessary at my private expense and my wife has made up her mind to give up her carriage. But there is a strong opposition by some men who should know better – they are incapable of appreciating the value of the Durchmusterung – and they carried a resolution to suspend my vote till my arrival in England. A committee is appointed and our final meeting is on Thursday.'

Kapteyn wrote about Gill's visit on March 2, 1887:

'I need not tell You how keenly I enjoyed Your visit and our trip to Hamburg. It has been a thorough treat to me from beginning to end and it will remain among the happiest recollections of my life' [...]

Certainly, if a year ago I had had an observatory granted, I would hardly have considered myself at liberty to undertake a work of such extent, which will absorb the whole of my energy for years to come. But in my present circumstances I consider it an extraordinary piece of good fortune to be still able to contribute so effectively to the progress of astronomy; and whatever anxiety I may ever have felt about the possibility of bringing this work to a happy end, has been absolutely removed by the ready manner in which You take upon Yourself anything, whatever troublesome that may contribute to the ultimate success of the work.

I think it only fair to assure You that if I have health, I will <u>never</u> give up this work before its completion, whatever may happen. If, for instance, it might come to pass that I were appointed director of an observatory, I will make sure first of the possibility of prosecuting and bringing to an end the work undertaken. If this proves impossible You may be sure that I will not accept the appointment.'

It had been a very important visit indeed. After having visited Kapteyn in Groningen and having been to Paris, Gill stated from Aberdeen on March 29, 1887:

'I think I am sufficient judge of character to find out what manner of man you are [crossed out and replaced by were] during the happy days which we spent together. Still it is very pleasant to be assured in plain English, and in the only terms which you employ, your fixed resolution to stick to the work you have undertaken through thick and thin – and that, having put your hand to the plough, no consideration will move you from the work you have begun – and no temptation cause you [to] turn back from it.'

Gill visited Leiden after the Paris congress and on May 10, 1887, he wrote to Kapteyn from there:

'I am happy to tell you however that the Durchmusterung is safe. [Von] Auwers is to write me an official letter to say that if the Royal Society cannot find the money for the prosecution of the Durchmusterung, it will be provided by the Humbol[d]t fund of the Berlin Academy.'

The Alexander von Humboldt Stiftung für Naturforschung und Reisen (Gill misspelled it) was established after the death of Friedrich Wilhelm Heinrich Alexander Freiherr von Humboldt (1769–1859), the Prussian explorer and geographer. Gill added:

'I had the great pleasure of seeing your sister here the other evening, and fell as much in love with her as a man who is very much married may do. I hear there is a large family of Kapteyns, and I hope it may be my good fortune to meet more of them if there is a strong family likeness to the two I know.'

Presumably this was Albertina Maria (Bertamie), Kapteyn's bachelor sister who seemed to be fond of traveling. On September 6, 1887, Gill wrote:

'The money matter stands <u>in status quo</u>, except that the R' Society have paid me now the grant up to the end of June of this year, and now the work is being done at my own cost – but my wife has gone thoroughly hand in hand with me in the matter. We have carried out a great many domestic economics, and with a little sacrifice of capital we can manage. I shall be truly thankful if in any way we can manage together to do this great and necessary work.

Of one thing you can be quite sure, that if I am spared in health and if I am left alone without official interference, the work shall be done – yes

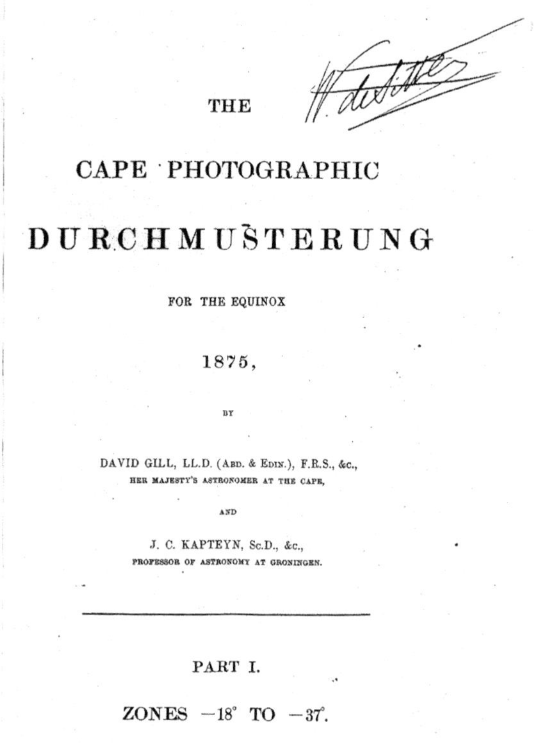

Fig. 6.27 The cover page of the Cape Photographic Durchmusterung in the former library of the Kapteyn Astronomical Institute in Groningen. This copy carries the signature of Willem de Sitter on it (Library Kapteyn Astronomical Institute)

THE

CAPE PHOTOGRAPHIC

DURCHMUSTERUNG

FOR THE EQUINOX

1875,

BY

DAVID GILL, LL.D. (Abd. & Edin.), F.R.S., &c.,
HER MAJESTY'S ASTRONOMER AT THE CAPE,

AND

J. C. KAPTEYN, Sc.D., &c.,
PROFESSOR OF ASTRONOMY AT GRONINGEN.

PART I.

ZONES −18° TO −37°.

even in spite of a very great deal of interference. Of another thing you may be quite sure, that Auwers will not change his mind, and that if you require help & he can procure it. He will most certainly do all that he has promised. There is no more reliable man living, no man more free from little mindedness or petty jealousy than Auwers.'

Further on in the letter, Gill referred to the point of view that Kapteyn had expressed, that he 'feels it honorable' to raise the funds for his involvement in his own country. Gill says about that:

'I see no narrow-minded prejudice in your wishing to raise the money for your work in your own country and by aid of your own enthusiasm. On the contrary: I admire your work and I am sure Auwers will do the same. I have told you that I will do all that I can do to help you. I have written Auwers telling him exactly what you think of the matter, and how fully I enter into your feelings –and that I am sure he will do the same – and if Auwers even fails you (which I do not for a moment believe he will), don't let that even stop you. We will somehow manage it together even if I have to give up tobacco, –which God forbids.

Yes, my good friend, you have all the hard work of the business, and your part is the all important part. The Catalogue is and must be too. In

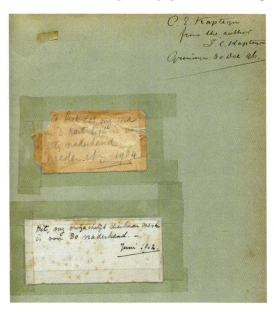

Fig. 6.28 The front page of the copy of the CPD-Part I that Kapteyn gave to his wife, dated 30 December 1896. There are two small notes pasted in on a later occasion. The top one says: 'This book that is very close to my heart is for Hetty afterwards, August 1904'. The bottom one reads: 'This work that is very precious to me is for Do afterwards. June 1912.' Kapteyn intended the book to go to his youngest daughter Henriette first and later to his oldest daughter Jacoba after his (and presumably his wife's) death (Kapteyn Astronomical Institute)

this affair the pictures are merely a means towards the Catalogue. In the Paris work the 11th Mag. Catalogue will be the really important part of the business. On this point Christie and I are at war. What do you think I find him doing but writing to the Admiralty ridiculing the idea of the Catalogue, and proposing to determine the places of the necessary 'etoiles de réfère' for the photo. charts pari passu ['hand-in-hand'] with the taking of the photographs. But how can he do so without the necessary sufficient number of stars determined on the meridian, and this he can only manage for the N Hemisphere when the Bonn Catalogue is published.

I have sent the Admiralty a copy of the Procès Verbaux and pointed out where the discussion on the point is contained and the division 44 to 6 against Christie's amendment. I have also pointed out that any isolated action such as Christie proposes is to be depreciated as out of harmony with the international character of the undertaking.'

I have included all of this to document the extent of the animosity between Gill and Christie. But in fact Gill went further than this. In the following letter by Gill references are made to two powerful persons: James Ludovic Lindsay, 26th Earl of Crawford and 9th Earl of Balcarres (1847–1913) was president of the Royal Astronomical Society at the time and a member of the Royal Society. The other is James Hamilton, 9th Lord Belhaven and Stenton (1822–1893).

'Further 'entre nous' I have, at Lord Crawford's (formerly Lord Lindsay) request sent a full account of Christie's proceedings at Paris to him, with copies of letters of Auwers and Struve, and he is to explain the matter to Lord Hamilton (First Lord of the Admiralty) and a motion is on foot to refer all points on which the Admiralty wants advice about the Cape

Observatory – <u>not</u> to Christie but to the Board of Visitors of the Greenwich Observatory.'

Gill recorded this in the Introduction of the first volume of the CPD:

'I was officially informed that the Government Grant Committee was not prepared to recommend that the vote of £300 should be granted for 1887. The sum of £150 for the first half of 1887 was subsequently provided from the donation of the Royal Society.

I had no doubt in my mind as to the desirability of this work, nay, its urgent need in the existing state of Astronomy, and, indeed that the results of the Paris Congress were to increase rather than diminish the urgency of the need.

I found that my most distinguished colleagues were unanimously of the same opinion, and generous offers of pecuniary and other aid were not wanting. [...] The work was carried on without external aid till 1889 September 31, [...] But the tax on my private resources, which resulted from the withdrawal of the Government Grant Fund, rendered me unable to procure the Repsold apparatus which we had designed. Consequently, the measurement of the plates to the single second of arc, which otherwise would have been carried out, had to be abandoned, [...]

The statement that the Cape plates were not recording sufficient numbers of stars to produce anything useful, was manifestly incorrect. When Kapteyn at some stage felt that his progress was too slow, due to the amount of work resulting from the unexpectedly large number of stars on the plates, Gill reassured him (April 24, 1888) that he was 'doing famously – and I think we shall have to laugh over the Royal Society who complained that there were not enough stars on the plates!'

Obviously both men suffered from this, Gill primarily in a financial sense and Christie in terms of reputation. Gill came back to London after his retirement from the Cape Observatory and served as president of the Royal Society (1909–1911). He was knighted in 1900 and received the Gold Medal of the Royal Astronomical Society in 1882 and in 1908, as well as the Catherine Wolfe Bruce Gold Medal of the Astronomical Society of the Pacific in 1900. Christie was knighted in 1904, but did not receive any of these (nor any other) prestigious medals.

6.10 Concluding Remarks

The work on the CPD was not finished with the publication of part III in 1899. In fact, three revisions were published in 1903, but Kapteyn did not take part in any of them. The first one, *Revision of the Cape Photographic Durchmusterung, Part I. Results of examination of questions which have arisen from a comparison of other star-catalogues with the Cape Photographic Durchmusterung*, was authored by Robert Thorburn Ayton Innes (1861–1933), who was an observer at the Cape, and Gill [31]. Innes started as an amateur astronomer, but received a honorary doctor's degree at Leiden in 1923. In 1903 he was appointed director of the 'Republic

Observatory', later the Transvaal Observatory, near Johannesburg. The second revision, Gill & Kapteyn (1903), for the greater part concerned variable stars which Kapteyn had identified in the course of the measuring of the plates. *Revision of the Cape Photographic Durchmusterung, Part III. Errata in southern star catalogues (south of -19 deg declination)* were published by Gill [32].

A final issue concerning Kapteyn's work on the CPD is the persistent story that Kapteyn used inmates of Groningen prison to help with the reductions. Adriaan Blaauw has always sustained its credibility, reminding others that this story was told to him by Pieter van Rhijn, who claims he had heard the story from Kapteyn himself (see Blaauw's article on Kapteyn in the *Dictionary of Scientific Biography* [33]). This matter was taken up by Wessel Krul in the *Legacy* and I quote from his contribution :

> 'In this way, it is sometimes said, even certain well-known political prisoners, like the radical socialist F. Domela Nieuwenhuis, (1846–1919) contributed to the CPD project. The story is highly improbable. The inmates of the Groningen penitentiary were for the greatest part almost illiterate or innumerate, and could not be trusted with even the most simple calculations. They were employed in routine menial tasks, like the folding of cardboard boxes or the weaving of straw mats [34]. Exceptions were rarely made. When Domela Nieuwenhuis, who had a degree in divinity, in 1887 served a term of almost eight months in prison in Utrecht, the prison authorities set him to work in the same way. His biographers do not mention any excursions into astronomy: e.g. Meyers [36] stresses the utter unwillingness of the prison authorities to have the inmates do any other than the usual mindless jobs. Presumably, the source of the legend is a casual remark by [British astronomer Arthur Stanley] Eddington [(1882–1944)]: 'If I remember rightly, some of the work was done by prison inmates' [35]. Perhaps Eddington's memory was letting him down, or he misunderstood something Kapteyn once told him. In a different recollection of Kapteyn he was more cautious, speaking only of 'a certain amount of irregular and chiefly unskilled help' [37] It is significant that neither Hertzsprung-Kapteyn, Gill or Pannekoek, who certainly would have known about it, mention anything of the sort. [...] The Groningen archives contain no indication of any exception from the usual labor made for specific prisoners, or of any payments made by Kapteyn or the University to the prison authorities. As long as no archival material has come to light to prove Kapteyn's reliance on prisoners, the story should better be treated with caution.'

On balance, I tend to believe that the position taken by Krul provides the most reasonable and satisfactory interpretation of the issue.

Chapter 7
An Astronomical Laboratory

> *It has always irked me as improper that there are still so many people*
> *for whom the sky is no more than a mass of random points of light.*
> *I do not see why we should recognize a house, a tree, or a flower here below*
> *and not, for example, the red Arcturus up there in the heavens*
> *as it hangs from its constellation Bootes, like a basket hanging from a balloon.*
> Maurits Cornelis Escher (1898–1972).[1]

> *I have looked further into space than ever a human being*
> *did before me. I have observed stars of which the light,*
> *it can be proved, must take two million years to reach the Earth.*
> Sir William Herschel.[2]

Before continuing with Kapteyn, I provide some astronomical background information on the grand theme that constituted Kapteyn's research efforts; it is sometimes called the 'Construction of the Heavens', but also the 'Structure of the Sidereal System'. In modern terms, that would be the study of the distribution of stars in space or the structure of the Milky Way Galaxy. Let me first outline what we now know about the formation, structure and evolution of stars, as well as the structure of the Milky Way Galaxy and external galaxies as far as it is necessary to follow Kapteyn's contributions. Most if this was completely unknown to Kapteyn and his contemporaries, working in the final decades of the nineteenth century. A major reason for this was the absence of large numbers of reliable parallaxes or distances. The astronomical background I am about to cover will of course be very familiar to astronomers reading this book or others who have a good understanding of astronomy, and I refer those to the ¶ symbol on page 211.

---

[1] M.C. Escher was a Dutch graphic artist, often inspired by mathematical concepts and regularities.
[2] Quoted by Agnes Mary Clerke (1842–1907) in *Dictionary of National Biography*, Vol. 26 [1].

7.1 Stellar Structure and Evolution

Stars have a very wide range of properties. For example, look at the light they emit. They can be intrinsically very bright or very faint; the brightest being some ten thousand times more luminous than the Sun and the faintest only one ten-thousandth. As will be true for all properties, the Sun is very average. The light from the stars has a general color, which can be relatively blue or relatively red. In this respect, too, the Sun is average. The color is a measure for the temperature on the stellar surfaces, and ranges from 30,000 Kelvin for the hottest to less than 3,000 K for the coolest. For the Sun the surface temperature is about 5,800 K.

This range of luminosities and temperatures is most conveniently displayed in the so-called Hertzsprung–Russell (or HR) diagram (see Fig. 7.1), named after Ejnar Hertzsprung (1873–1963) and Henry Norris Russell (1877–1957). It was independently found by these astronomers. Hertzsprung (Henriette Kapteyn's husband) worked at the Potsdam Astrophysikalisches Observatorium at the time.

In the Hertzsprung-Russell diagram, most stars are on the Main Sequence, running from top-left where the hot luminous stars reside, to the cool, faint stars in the bottom-right. Stars spend most of their lives on this sequence. While there, they 'burn hydrogen' in their extremely hot (of order ten million Kelvin) interiors; in these nuclear reactions, hydrogen is converted into helium and this releases energy, which – after slowly drifting up to the surface by repeated absorption and re-emission – is eventually radiated away into space. The temperature inferred from the color of the light corresponds with that at the surface of the star. Stars are made up of almost three quarters of hydrogen and about one quarter of helium, the other chemical elements accounting for almost nothing to a few per cent. But, interestingly, this is always in roughly the same relative proportion. In the Sun the chemical elements heavier than helium account for 2% of the mass.

There also is a large range in stellar masses and lifetimes; the bright, hot ones at the top-left of the Main Sequence turn out to be the most massive stars (dozens of solar masses) and the least luminous, red stars have masses of one tenth of that of the Sun. Stars have very different lifetimes on the Main Sequence, the massive, bright hot stars 'burning' their hydrogen into helium at a much faster rate than the light, faint and cool ones. The life ends when the central part, about ten percent of the star, has been consumed. The Sun again is average; its lifetime on the Main Sequence is about 10 billion (1×10^{10}) years. Currently it has an age of about 4.6 billion (4.6×10^9) years and is about halfway through its Main Sequence stage. The most massive stars live on the Main Sequence for one million years or less and the lightest ones one hundred billion (1×10^{12}) years, which is about ten times longer than the current age of the Universe.

Stars consist for the greater part of hydrogen and helium, and may have hardly any other chemical elements or only up to a few percent. They can have large or small velocities through space with respect to the mean. Why this is so, is of no direct consequence for our story, but a very brief explanation is provided below.

Stars form from collapsing gas clouds that heat up as a result of loss of gravitational energy when they contract, and which they cannot radiate away rapidly

7.1 Stellar Structure and Evolution

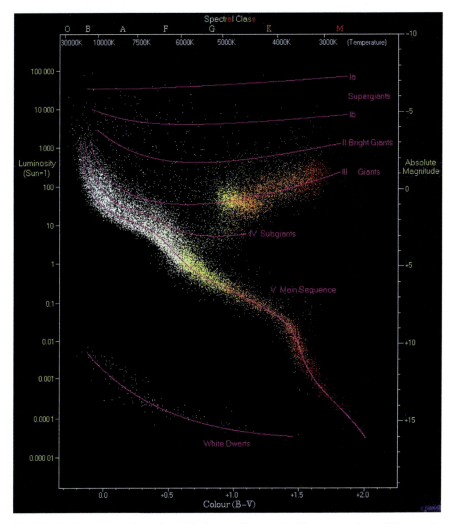

Fig. 7.1 The Hertzsprung–Russell- or HR-diagram (Source: The Hertzsprung Russell Diagram by Richard Powell [2])

enough. On the Main Sequence, stellar interiors have become sufficiently hot to 'burn' hydrogen into helium. What happens when the hydrogen is exhausted? Stars comparable to the Sun in terms of mass become red giants on a 'branch' running up to the top-right of the HR-diagram. After further contraction and heating they will burn helium into carbon and oxygen and by that time they are situated in the HR-diagram in the clump of stars half-way the giant branch. They stay there for a relatively short period of time, when they throw off their outer envelopes which then form a shell around the star. This phase is called 'planetary nebula', because of its appearance in primitive telescopes. Subsequently they slowly extinguish as very compact white dwarfs (bottom of the HR-diagram)..

Fig. 7.2 A panoramic view of the Milky Way, as seen from the southern hemisphere. This view of the full 360°, made up of many separate photographs, has been produced by the European Southern Observatory ESO. The two spots below the Milky Way right of the center are the Magellanic Clouds, companions to our Galaxy. The two just below the Milky Way on the left are the star cluster the Pleiades (extreme left) (see also Fig. 7.11) and the nearest spiral galaxy, the Andromeda Nebula (see also Fig. 8.16). The blob just below the Milky Way to the extreme right is the constellation Orion with the Orion Nebula (see also Fig. 11.8). Other bright objects are planets that happened to be in the field when the pictures were taken (European Southern Observatory)

Stars that are initially more massive than the Sun will become neutron stars rather than white dwarfs; in neutron stars, gravity is so large that matter becomes even more compact as the electrons are, so to speak, 'compressed' into the protons to form neutrons. But stars more massive than, say, 5 times the Sun will after exhausting the hydrogen go through a number of short phases (Supergiants) where they 'burn' their atomic nuclei into heavier and heavier chemical elements (up to iron). When these nuclear reactions stop and the production of energy in the center ceases, they will contract violently and the energy released in that process will cause them to explode as a supernova. During this phase they produce chemical elements including the heaviest existing in nature. Since they throw out most of their matter into space, they enrich the gas between the stars with the products of the nuclear reactions. This results in subsequent stars having more elements than just hydrogen and helium. This process is called (stellar) nucleosynthesis. What remains after a supernova is a neutron star when the original star is not too massive, but if it is very heavy, a black hole is formed that is so compact that even light cannot escape from it.

7.2 The Milky Way and Other Galaxies

The stars we see at night are all part of the Milky Way Galaxy. Its main structure is revealed as a faint band of light from which it derives its name. Since the Galaxy is spatially predominantly a flat structure and as we are in the middle of it, the Milky Way stretches as a great circle over the full sky. In Fig. 7.2 we see a composite photograph of the full 360°. The most prominent part in the middle can best be seen from the southern hemisphere; this corresponds to the direction towards the center of the Galaxy. In the northern hemisphere it can only rise above the horizon at a small angle (for the Netherlands slightly less than 10°).

By and large, stars can be divided into two Stellar Populations. The stars that we see in the sky almost all belong to the flat structure that we see as the Milky Way, and this is called the disk of the Galaxy or Population I. The halo stars are distributed over a more or less spherical volume, but much concentrated towards the center of the Galaxy. These stars are less prominent, except when they form large conglomerations of some one hundred thousand stars that are called globular clusters. The Populations II stars are roughly as old as the Galaxy itself (i.e., up to 12 billion years) and have less (and often much less) than 0.1% in terms of heavier elements. These form the oldest parts of the Galaxy.

The disk formed later from the gas that was left behind after the star formation in the halo stopped, and this gas settled in a flat volume about 10 billion years ago. Ages of disk stars that formed in this gas range from very small to about 10 billion years and they contain about a few per cent in chemical elements beyond helium.

The Universe is full of galaxies, some similar to our own, some very different. Figure 7.3 is intended to indicate what the Galaxy would look like if we could view it from the outside. Here I have chosen two other galaxies that resemble our own

Fig. 7.3 Two galaxies that resemble our own Milky Way galaxy. On the left we see the galaxy NGC6744, which is seen close to face-on; on the right NGC891, which has the disk edge-on (Left European Southern Observatory ESO, right National Optical Astronomy Observatories, [3])

fairly well. The picture on the left shows the view if we would be situated such that we would see the disk more or less from above (or below), and on the right if it would be seen edge-on. The face-on view shows a so-called spiral structure, which is closely related to where star formation is currently taking place. This dominant part is the disk. On photographs, like the one on the right of Fig. 7.3, this flat disk also dominates the light. The halo or Population II is not very prominent; its central parts can be seen as the bright concentration near the center.

So the picture is the following: The disk is Population I; its stars are very young to rather old (up to about 10 billion years) and form a flat disk, which rotates with a velocity of a few hundred km/sec, while the relative velocities of stars are of the order of tens of km/sec. The Sun is part of Population I. The halo stars of Population II are all very old and since they occupy a more or less spherical volume, this does not rotate rapidly as a whole, but the stars have velocities up to a few hundred km/sec with respect to each other and to the stars in the disk. Very early on, the universe consisted almost entirely of hydrogen and helium (with only a tiny trace of lithium and beryllium), so these first stars of Population II have very little or sometimes virtually no chemical elements heavier than helium.

The canonical picture of galaxy formation is that Population II formed first, from a collapsing gas cloud, supplemented by in-falling and merging smaller companion systems. Originally this gas consisted only of hydrogen and helium, which was formed about three minutes after the Big Bang. The gas was enriched by stellar nucleosynthesis in the first generations of stars of Population II, so that eventually it contained significant amounts of chemical elements heavier than helium. This gas dissipated its random motions by collisions of gas clouds, so that rotation became dominant and settled in a flat disk, where stars have formed up to the present time as Population I stars.

There is a wide range of morphologies among galaxies that are classified as the so-called Hubble morphological types, proposed in the 1920s by Edwin Powell Hubble (1889–1953), although to a significant extent based on earlier work by John Henry Reynolds (1874–1949). The various morphological types in this classification scheme can be interpreted as a range in the relative proportion of the two Stellar Populations, ranging from pure Population II in elliptical (or early type) galaxies to exclusively Population I in 'pure-disk' (sometimes called late type) spiral galaxies. The picture of galaxy formation and evolution is evolving rapidly at present as a result of the enormous improvement in modern observations and numerical simulations.

7.3 The Herschels and the Sidereal Problem

¶ The interpretation of the Milky Way as a flat structure composed mostly of stars, dates back a long time. In fact, Galileo Galilei (1564–1642) was the first to see that the Milky Way was actually a feeble glow from many stars that are too faint individually to be seen by the naked eye. It was only possible to see this when Galilei turned a telescope to the sky – perhaps he was not the first person to do this but certainly the first to report this publicly. The first detailed inquiry into the distribution of stars in space was by William Herschel (1738–1822), who conducted an extensive study into the 'Construction of the Heavens' from counts of stars. This is a prelude to Kapteyn's most important work, so we need to look at this in some detail. The authority on William Herschel, his sister Caroline and son John, is Michael Hoskin from Cambridge University, who wrote many books and articles on them. The most important are *William Herschel and the Construction of the Heavens* (1963), *Discoverers of the Universe: William and Caroline Herschel* (2011), and *The Construction of the Heavens: William Herschel's Cosmology* (2012) [4].

Sir (Frederick) William Herschel (1738–1822) was a musician who played the oboe in the Military Band of Hanover before he emigrated to Britain at the age of 19. He fled the country because of the threat of war with France and was accused of desertion from the Hanoverian Guard. In addition to building up a reputation as a musician (he played also harpsichord and organ) and composer, he became an accomplished telescope builder and astronomer, first as an amateur but later as his main occupation,. His efforts were rewarded by his accidental discovery of the planet Uranus in 1781. He designed and built very large reflecting telescopes, of which the '20-foot' telescope was used in the 1780s for systematic counts of stars. This telescope had a 20-foot focal length (6.10 meters), a 18.7-inch (47 cm) 'mirror' (actually made of speculum metal which was a two-thirds copper and one-third tin metal alloy). It was a smaller version of his later 40-foot telescope, of which a picture has been reproduced in Fig. 7.5. Herschel conducted this work with the help of his sister Caroline Lucretia Herschel (1750–1848), who recorded his shouted remarks from the telescope in a room with an open window in their house in Slough, about 35 km west of London.

Fig. 7.4 Pictures of William (left; 1785), Caroline (middle; 1929) and John (right; 1846) Herschel (From Wikimedia Commons [5])

William Herschel's contributions are very fundamental and far reaching. He used his increasingly larger and better telescopes to examine vast portions of the sky. His first work had to do with double stars. He examined many bright ones to see whether there were fainter stars nearby (which he assumed would be further away from us) that could be used to measure parallaxes or to return to after a number of years to see proper motions. Herschel was interested in measuring the motion of the Sun with respect to the system of nearby stars as a whole, and argued that this could be found from studying proper motions. Secondly, he cataloged nebulosities. This he did by setting his telescope at a particular elevation towards the south and let the sky drift by. He called out what he saw and Caroline Herschel recorded his descriptions and the (sidereal) time. On the basis of this, the right ascensions and declinations could be recovered and a catalog produced. This was an extensive piece of work that took him many years to finish. It concerned only the northern hemisphere, of course, but his son John Herschel (1792–1871) carried on his father's work. He left for South Africa in 1833, and here he set up the 40-foot telescope and 'swept' the southern skies. He returned to Britain in 1838 and made numerous important contributions to astronomy, but also to other fields such as biology and the development of photography. John Herschel published a *General Catalogue of Nebulae and Clusters* combining his father's work in the north with his own in the south. At first, William Herschel convinced himself that all these nebulosities were collections of faint stars that could not be distinguished individually, but later he started to have doubts about this.

Thirdly, and most importantly for our story, William Herschel made star counts in order to map the distribution of stars in the 'Sidereal System'. He published a crosscut of the Sidereal System in his paper *On the construction of the Heavens* in 1785 [7]. This crosscut was derived as follows: Using his 20-foot telescope, he counted stars in various positions on the sky along a great circle that was more or less perpendicular to the Milky Way. He called these 'star gauges'. He assumed

7.3 The Herschels and the Sidereal Problem

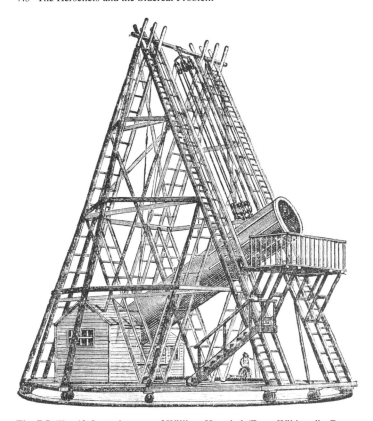

Fig. 7.5 The 40-foot telescope of William Herschel (From Wikimedia Commons [6])

all along that all stars are intrinsically equally luminous. Now he made two crucial further assumptions. The first is that stars are distributed uniformly in space (the density is the same everywhere) and the second that his telescope could fathom the Sidereal System out to its edges. Under these assumptions the number of stars seen in any particular direction depends only on the extent of the system. The longer the sight-line and the farther the boundary of the system, the more stars you see per area of the sky (either per square degree or per field of view of his telescope). This resulted in the famous diagram reproduced in Fig. 7.6.

It may be of interest to compare the counts that Herschel made two centuries ago to modern data; not only will it allow us to see how accurate they were, but in this way we can also find out to what faintness he could still observe. In a paper in 1986, *Surface photometry of edge-on spiral galaxies. V – The distribution of luminosity in the disk of the Galaxy derived from the Pioneer 10 background experiment* [8], I performed the following exercise. First I measured up Herschel's cross-cut to find the radii ('rays') of his 'star gauges'; Herschel expressed these in units of what he thought was the distance to Sirius (which is given in inches on the figure). From this it is possible – using Herschel's description – to convert them into the number of

Fig. 7.6 The famous crosscut through the 'Sidereal System' published by William Herschel on the basis of his counts of stars ('gauges') along a great circle on the sky (From Herschel: *On the Construction of the Heavens*, 1785; cited in the text)

stars counted in his field of view (which was 15 arcmin diameter) and hence stars per square degree. We also need to determine the coordinates of the great circle observed for the crosscut. Herschel gave the curious description that from Slough (which is at geographic latitude 55°), this circle was traced out by the horizon when the star τ Ceti culminated. It is somewhat complicated to derive from this that his great circle crossed the Galactic equator at galactic longitudes (see Box 3.1) $l \sim 45°$ and $l \sim 225°$ (for longitude zero at the Galactic Center) and that it missed the Galactic pole by only about 5°.

The comparison is made in Fig. 7.7. This figure is derived from a model of the Galaxy that can quite accurately represent the actual star counts on the sky. On the horizontal axis the figure shows the galactic latitude along the great circle, and the vertical axis is the logarithm of the number of stars per square degree brighter than a certain magnitude V (which is close to the visual magnitudes that our eyes would see). Remember (see Box 2.4) that five magnitudes correspond to a factor 100 and that the naked eye sees stars down to about magnitude 6. The lines are derived from the model (ignoring low latitudes since the counts will be affected by extinction from the interstellar dust), and the dots are Herschel's counts. The conclusion is that Herschel made his counts rather consistently to an apparent magnitude of $V \sim 15$, which is 4000 times fainter than the naked eye can see. The scatter in the figure is likely to reflect the fluctuations in actual star density on the sky, to a large extent, although Herschel did not admit areas where 'stars happened either to be uncommonly crowded or deficient in number'.

We can also compare this limiting magnitude to Herschel's own estimates, which he published in a paper in 1817 (*Astronomical Observations and Experiments Tending to Investigate the Local Arrangement of the Celestial Bodies in Space, and to Determine the Extent and Condition of the Milky Way* [9]). In the first place he quoted values for the 'gaging or space-penetrating powers' of his telescopes that expressed the increase in distance that stars could be seen compared to his unaided dark-adapted eye. The latter he determined from his method of *'equalisation of starlight'*. In this method he looked through two telescopes, one called the 'standard', through which a star was observed with the full aperture, and the other, the 'equalizing telescope', pointed at a brighter star but with a part of the aperture covered, so that the 'light was equal' because of the smaller aperture. In this way he

7.3 The Herschels and the Sidereal Problem

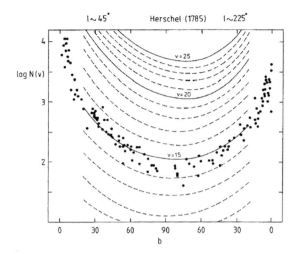

Fig. 7.7 The limiting magnitudes of stars corresponding to the counts of William Herschel from which he derived his crosscut (see Fig. 7.6) (From P. C. van der Kruit (1986); see text)

determined that with naked eye he could look out to stars at about 12 times the distance of Sirius. His experiments also gave him a value for his pupil-diameter of 0.2 inch and he also experimentally derived values for the absorption of light in his mirrors (33% per reflection). The observer, usually Herschel himself, of course, looked down into the telescope tube; the light was directed to a focus close to its edge, so he did not need an extra reflection to get the light out of the tube. Using the blocking of his primary mirror by the observer, he found a value of about 75 for the space-penetrating power of his 20-foot (18.7 inch aperture) telescope. This is a decrease in brightness by 5.6×10^3 or an increase of 9.4 magnitudes over the faintest stars his unaided eye would see. Further he noted that 'stars beyond its [his 20 feet telescope] reach much have been farther from us than the 900th order of distances' or fainter by a factor of 8.1×10^5 or 14.8 mag than 'stars of the size and lustre of Sirius, Arcturus, Capella, etc.'. These values indicate that he would also have quoted a limiting visual magnitude of 15 or so had he expressed his 'extent of telescopic vision' in this way.

In 1789, his 40-foot telescope was commissioned (see Fig. 7.5) and he immediately noted he had to abandon his earlier view of the Sidereal Universe. First he found that he saw more stars, so that previously he had not seen out to the edge of the stellar system, as he had assumed. Secondly, he revisited his double stars that he had observed a few decades earlier and found that they moved around each other. This not only showed that gravity is at work in these distant places, but also that intrinsically stars are not equally luminous, since the components of the double stars may be very unequal in brightness. Finally, he convinced himself that some of the nebulosities, notably the Orion Nebula (see Fig. 11.8 for a recent picture) were not collections of faint stars. Herschel considered the case that *statistically* stars of brighter magnitude were closer than fainter ones. Eventually he had to conclude that the telescopes that he used for his star gauges did not have 'the space penetrating power to fathom the profundity of the Milky Way'. His final description of the Sidereal System was a stratum of stars (and nebulae), which – at least in two of the three dimensions – is infinite in extent.

Another highlight in the quest to understand the distribution of stars in space prior to Kapteyn, is the work of Friedrich Georg Wilhelm von Struve (1793–1864), whom we have already met as one of the earliest astronomers to determine stellar parallaxes. He was the director of Pulkovo Observatory and the father of Otto von Struve, his successor at Pulkovo, whom we have met in connection with the Carte du Ciel. In 1847 Von Struve senior wrote an influential book, *Études d'Astronomie Stellaire: Sur la Voie Lactée et sur la distance des étoiles fixes* [10]. In this work he studied the distribution of stars as a function of distance from the Milky Way using Bessel's catalogs. He noted that the concentration to the Milky Way was essentially absent for stars visible to the naked eye, but with fainter and fainter magnitude the concentration became more and more pronounced. He combined Bessel's counts in a band of 30° from the Milky Way with those of Herschel and concluded that the Stellar System is composed of parallel layers of stars. In the central layer the density is highest and it falls off away from it. The Sun is situated near this central plane. Von Struve was also one of the first astronomers to speculate that there was some kind of interstellar absorption of starlight.

Although more research was performed during the nineteenth century, these most important studies served as useful descriptions of the state of affairs concerning the arrangement of stars in space when Kapteyn entered the field of sidereal astronomy.

7.4 Aftermath of the CPD

The work on the Cape Photographic Durchmusterung took a major part of Kapteyn's efforts over a period of a dozen years or so. His assistants did most of the measurements, but Kapteyn took part in and kept a close look at the reductions. He also conducted a series of detailed investigations into sources of errors, as documented in much detail in the introductions to the CPD volumes. The help of the assistants was a major concern to Kapteyn. In the Introduction to Part III, Gill & Kapteyn (1900), he discusses the reason for the extended timescale in producing the CPD. The first is the fact that the number of stars turned out to be about two times as high as had been anticipated at the start. But then he writes:

> 'The slender resources at my disposal did not allow the employment of more than two, sometimes, three clerks. Of these only *de Vries* has been with me from beginning to end, and it is largely to his untiring zeal and devotion that the ultimate success of the work is due. The others left me, generally after a very short time, for more remunerative posts. It must be evident that this frequent change of assistants, all of whom were at first absolutely inexperienced in work of this kind, was most detrimental to the rapidity, if not the quality, of the work.'

7.4 Aftermath of the CPD

Had he used prisoners, it seems that he would have mentioned it here. His complaint about the time needed to train inexperienced workers indicates that it is very unlikely that he would ever have have contemplated using such persons. Next he adds:

'No one can regret this long delay more than myself. My only consolation is that, however long the delay, the man who originally planned the whole work, but who feared that the measurement and cataloging must be relegated to his old age, now sees it finished, secure against the uncertainties of all human affairs, at a time of life in which his activity is unabated, so that, if the work proves to be a useful one, he may not only see it bearing fruit, but may himself be enabled to bring some of that fruit to maturity.'

Kapteyn was ready to do some serious astronomy. However, his health was affected by the work, and it also put a lot of stress on his family life; his wife and children were weighed down by the constant nature of his duties. To illustrate this I quote a passage from the *HHK biography*.

'Indeed it was a demanding effort; his family could testify to that. The otherwise homely father and comrade who always knew a new game or joke, gave exercises and problems to solve and told exciting stories, especially during meals, never had time for them now. The meals were eaten hastily, since he was always in a hurry to go back to work. It kept him occupied and busy. For a while he got up at four o'clock in the morning in order to be at the laboratory at five. He then worked until nine, rested till noon so that he could pick up with new strength. He liked this regime very much, as it placed him in a position to use his energy efficiently and to be enormously productive. [...]

But the tension was too high. He became easily irritable and would then be unreasonable to others in the house. It was no wonder that his temper would suffer under the heavy demands he put on his work-power, but the children did not understand the reasons behind this and felt unfairly treated and unhappy. Mrs Kapteyn talked to her husband about the fact that his children were complaining. She herself did not complain and never asked anything for herself, and precisely because of that the seriousness of her complaint on behalf of her children made a deep impression on him. He called the children to come to him and told them he had done them injustice. A happy atmosphere in the house was after all the most important thing in the world and he promised to try his utmost to behave better. [...]

Physically the effort was also too great; he started to complain of nervous problems with his stomach and pain in his eyes. Taking a moment's rest was out of the question. Sometimes he was lying on the floor on his stomach, working with his books spread around him as there was no other comfortable posture. He lost weight and slept poorly, work was haunting him and he often complained about that in his letters. To [Anders] Donner, the Finnish astronomer at Helsingfors, whom he had met during a congress in Paris and with whom he had become good friends, he wrote among other things: 'I am being consumed by the work of the Photographic Durchmusterung, and that makes me ignore and neglect all other matters...'.

The Durchmusterung consumed him completely and would not let go until, in 1899, he had the three impressive quarto [the book size quarto measures 30.5 × 24.15 cm] volumes lying in front of him.'

7.5 Rector Magnificus

It would be incorrect, however, to say that Kapteyn spent all his time on the Durchmusterung. Not only did he have his teaching duties at the university, he also was involved in other organizational matters. In the academic year 1890–91 he acted as rector magnificus of Groningen university, which must have taken a major share of his time. And also in 1890, as we will see below, he undertook a new, but what proved to be final and unsuccessful attempt to obtain his own observatory.

Traditionally the term of rector magnificus ended in September at the start of the next academic year, with two presentations to the Senate. In Kapteyn's case this took place on September 15, 1891. The first presentation involved a discussion on a scientific subject of the departing rector's choice, and the second an overview of the highlights of previous academic years. At the end of the latter the formal transfer of the office of rector magnificus to his successor took place. I start with the second presentation, Kapteyn (1891c), and leave the first one to later when I have introduced his final attempt to obtain an observatory.

Kapteyn started his presentation on the *Report on the fate* ('lotgevallen') *of the University* recalling that at the beginning of the previous academic year the King of the Netherlands, Willem III (Willem Alexander Paul Frederik Lodewijk), had passed away. Willem III (1817–1890) had been King of the Netherlands since 1849. The three sons with his first wife (and full cousin), Sophia Frederika Mathilde van Württemberg (1818–1877), had all died young without offspring; the middle one died from meningitis when he was 6, the other two at the ages of 38 and 32, both of typhoid. Willem III, who had many extramarital affairs and lived separated from his wife at the time she died, married Adelheid Emma Wilhelmina Theresia (Emma) von Waldeck und Pyrmont (1858–1932) in 1879. Their daughter Wilhelmina Helena Pauline Maria, the later Queen Wilhelmina, (1890–1962) was still under age in 1890 and Emma became Queen regent until Wilhelmina's 18th birthday in 1898. Of course, Kapteyn praised the deceased King extensively, recalling the peace that the Netherlands had enjoyed, the progress that had been made and the prosperity of the country. But he also had some very nice words for Queen Emma, who had actually visited Groningen in May of that year.

Kapteyn also drew attention to the display of the new banner of the volunteer students who had served as soldiers, fighting the Belgian revolutionaries in 1830/31, who demanded the independence of Belgium. The banner had been put on display in the Senate room a few months before, 'kept under surveillance by the Professors of the past', when the veterans who were still alive were honored with a festive celebration. Next he reviewed the illnesses, deaths, but also birthdays of important professors and curators and the appointments of new professors. He recalled important festivities abroad where the university had been represented, such as the recognition of the academy of Lausanne as a full university and the 70th birthday of the 'world famous physicist and physician Helmholtz' (Hermann Ludwig Ferdinand von Helmholtz; 1821–1894). He also listed the number of students per faculty, totaling 415 according to the faculties themselves and 473 according to the Student Almanac of that year.

Interesting is his summary of the University's equipment, apparatus and instrumentation. He rejoiced in the new physics laboratory that was about to become operative, but also noted difficult situations elsewhere. This included the room where students practiced anatomy with half of them having to stand with their back to the window, unable to see small organs in their specimens clearly. And the lack of proper lecture rooms for the increasing number of students in anatomy, mineralogy and chemistry. But then he said:

'Is the lack of rooms for teaching a weak spot for the medical and natural sciences at our university? This as you know is not *the* weak spot. [...] Those who are curious about the state of affairs concerning a hospital and an observatory only have to consult these reports over the last dozen or so years. Taking everything together, one would have to admit that regardless of the nice buildings for pharmacy, hygienics and physics that have been erected since the implementation of the new law, the equipment, the non-living army of science, remains the weak spot of the university. What is lacking is not an army of science, what is lacking is weapons. How can one expect that the men with no more than deficient weaponry or even without arms at all, can, in spite of their strength, determination and courage, compete with those soldiers elsewhere who are armed to the teeth?'

He next mentioned two cases, involving a pharmacist and a physician, who had left Groningen in exchange for places where better facilities were available.

7.6 One Last Attempt at an Observatory

Kapteyn's earlier plan of 1881 to have an observatory built for the astronomy professor in Groningen had been refused year after year. In the 'Programme of requirements for a small observatory in Groningen' of November 25, 1881, he had stated that the observatory should be located such that it had 'a free southern horizon and is situated such that observing is disturbed as little as possible by movements in the air or by vapours etc.'. The location he selected is identified on the map that Kapteyn submitted together with his plans (see Figs. 7.8 and 7.9). It is that of the present offices of the Dutch equivalent of the Internal Revenue Service on Kempkensberg; it offered an unblocked view towards the southern horizon at the time,

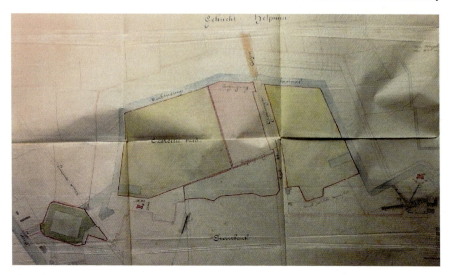

Fig. 7.8 Drawing of possible locations of Kapteyn's observatory. Note that north is to the bottom at a slight angle. The positions are two red outlines; the one on the left, labeled 'a' is his primary choice (University of Groningen archives)

overlooking the drill and parade grounds of the adjacent Army barracks. A secondary location has also been identified on the map, viz., the current grounds of the Groningen municipal open-air swimming pool 'Papiermolen', the former site of a paper mill.

The building he envisaged had a maximum length of 17.90 meters and a maximum width of 11.10 meters. It should be surrounded by a fence and the total area of real estate should be at least 500, but preferably 1000 m^2. The building had two floors, the top floor with living quarters for the 'custos' (custodian[3]) and the upper one having rooms for teaching, etc. In 1890 he revised the plan in two respects. In the first place, it was much more modest and the building had just one floor (but included living quarters for the custodian) and the telescope should be photographic, so the plan included a darkroom. Figure 7.10 shows the drawings accompanying the 1890 plan. These had been drawn up by Jacobus van Lokhorst (1844–1906), the Chief Government Architect ('rijksbouwmeester') of the Ministry of Internal Affairs, who was responsible for the design of all educational buildings in the country. The plan was neither adopted nor realized. The government did support the erection of a photographic telescope as envisaged by Kapteyn, but it went to Leiden (see page 225).

I now return to the proceedings at the end of the academic year, when Kapteyn ended his term as rector magnificus. The scientific lecture Kapteyn (1891b) on that occasion was entitled *De beteekenis der photographie voor de studie van de hoogere delen des hemels* [The importance of photography for the study of the higher parts

---

[3] Wessel Krul in the *Legacy* incorrectly infers that Kapteyn had designed living quarters for himself.

7.6 One Last Attempt at an Observatory

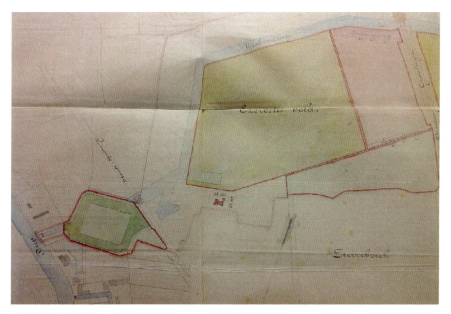

Fig. 7.9 Close-up of the drawing of the preferred location of Kapteyn's observatory. To the north (towards the bottom at a slight angle) is a stretch of woodland called 'Sterrebosch' ('Star Wood'), between the location and the city of Groningen, and to the south, where the observing mostly would take place, are the drill and parade grounds ('exercitie veld') and to the left the Army barracks (University of Groningen archives)

of the heavens]. Higher parts here means the fixed stars rather than the Solar System objects near the ecliptic. It is important, as it presents the first exposition published by Kapteyn in which he describes his ideas for the scientific programs he would like to undertake. Evidently, he also delivered it to support his request for a *photographic telescope* (although without mentioning it). Again I follow it in some detail.

Kapteyn started by reminding his audience that until relatively recently astronomy had almost exclusively been concerned with the study of objects in the Solar System, and that knowledge of the 'higher parts of the Heavens', concerning fixed stars and nebulae, could be summarized in a few words. He attributed this lack of interest in the fixed stars to the absence of changes, since 'science starts only when there is change, evolution, life'. He referred to Giordano Bruno (1548–1600), who assumed motions among the fixed stars, Edmund Halley (1656–1742), Giovanni Domenico Cassini (1625–1712) and Christian Mayer (1719–1783) who had proved their existence, while some stars were found to be variable in brightness and there had even been the appearance of new stars sometimes. The distances to stars had been unknown for a long time, but it was now clear, Kapteyn continued, that motion and change were the rule in the sky of fixed stars, that the Sun was an ordinary star among the others and that gravity worked everywhere in the universe, as was convincingly demonstrated by the motions of binary stars. Next he summarized the 'great cosmic problem' in the words of Simon Newcomb (1835–1909). Although

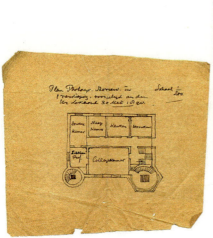
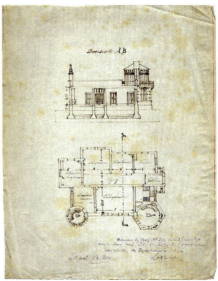

Fig. 7.10 Drawings for Kapteyn's observatory with his plans of 1890. On the left Kapteyn's sketch, dated May 30, 1890; on the *right* the plan that the architect van Lokhorst drew on June 3, 1890, on the basis of this sketch. The top figure on the left is a crosscut along the line 'AB' in the lower panel (Kapteyn Astronomical Institute)

he did not tell us where Newcomb's words were to be found, the quote was easy to trace back, viz., to page 408 in Newcomb's book *Popular Astronomy* of 1878 [11] and the quote reads in full:

> 'When we have bound all the stars, nebulae, and clusters which our telescopes reveal into a single system, and shown in what manner each stands related to all the others, we shall have solved the problem of the material universe, considered, not in its details, but in its widest scope.'

To get anywhere close to solving this problem, Kapteyn listed as requisites *time*, *completeness* and *accuracy*. 'Time', obviously, since changes are slow, and 'completeness' not only because the positions of stars must be charted, but also 'their motion, form, brightness, chemical composition' need to be measured as far as this is practical. 'Accuracy' requires the detailed study of as many individual objects as possible. Now the new 'weapon' that observational astronomers had recently obtained was the *photographic camera*. After reviewing the development of the photographic process, Kapteyn showed some remarkable photographs of astronomical objects, especially of the Pleiades star cluster. This cluster of rather young stars is also sometimes referred to as the Seven Sisters. Its brightest members can be seen with the naked eye; the group is at a distance of about 420 light years. Kapteyn referred to photographs taken by the Henry brothers (which are not reproduced at any easily accessible place) and Isaac Roberts, whom we have met on page 189 as a participant in the Carte du Ciel meeting. Roberts' picture is reproduced in Fig. 7.11. In relation to this image Kapteyn notes: '…on which new, linear nebular structures

7.6 One Last Attempt at an Observatory

Fig. 7.11 Photograph of the Pleiades star cluster by Isaac Roberts in 1888 that Kapteyn referred to in his lecture as departing rector magnificus in 1891; see text (DSpace@Cambridge [12])

can be seen – nebulae that, at least in all likelihood, form bridges of cosmic dust between star and star, sun and sun'. The nebulous structures are indeed formed by reflection of the light from the stars on the dust surrounding them.

Naturally, Kapteyn spent some time describing the advantages of photography in the production of star catalogs and his work with Gill on the Cape Photographic Durchmusterung. The text mentions that he became aware of the fact that Gill could not possibly undertake the measurement of all the plates and that he therefore offered his help. Nor did he forget to acknowledge the support he received from Huizinga in the form of the provision of working space in his laboratory and to thank him profusely. Of course, at the time the CPD work was still in the stage of measurements of the plates. He described the Carte du Ciel effort that had been started a short while before and praised it as an important record of the sky at the time of the late nineteenth century, fundamental for further studies, in particular to find changes of stars in brightness or position.

Kapteyn went on to describe what else photography could mean for astronomy, obviously with a view to what he would plan to do if he succeeded in obtaining a photographic observatory. He mentioned the discovery of minor planets and the tracing out of orbits in binary stars, but concentrated on positional work that would lead to more accurate parallaxes from measurements of relative positions. But then he turned to photometry and spectroscopy. Photometry would help determine the color of the light from stars. He had already noted that the CPD had demonstrated that stars near the Milky Way were systematically bluer and therefore hotter. But most progress he expected from spectroscopy. This had already shown that stars, just like our own Sun, were made up of the same chemical substances as found on Earth. The field was developing rapidly; Kapteyn described the possibility of classifying stars in three types: the white stars where the temperature is so high that metal vapors are not visible in the spectra, yellow stars that resemble our Sun, and red stars that are sufficiently cool to allow chemical compounds (he used the Dutch term 'chemische verbindingen') to exist. Photography should make it possible to do spectroscopic studies of stars (and nebulae) on a grant scale. He must have had precisely this in mind when thinking of his scientific program once the CPD would be finished and he would have access to his own photographic telescope. He continued to describe how spectroscopy could provide information on the radial velocities of stars and tell whether or not they were binary stars of which the individual components cannot be separated.

On this occasion he *did* actually address the question of his own observatory; he probably felt it to be more appropriate for this lecture rather than for the presentation in which as rector magnificus he outlined the state of affairs of the university as a whole. I quote the paragraphs in full.

'I hope that many of you who have honored this ceremony with your presence for many years, and all who have listened to the complaints articulated year after year by the departing rector when he discussed the needs for astronomy at this university, will be struck by this contradiction:

There, at the Cape, an observatory, as a result of an excess of work, from which it cannot extract itself, without the possibility to bring a great work that is needed urgently to a quick completion;

Here in Groningen, no observatory, but an astronomer doomed to relative idleness as a result of the failure of all his attempts to acquire one.

You may not have noticed this contradiction, but it certainly struck me, when exactly during that period, making use of the inestimable hospitality of my colleague, I was spending my vacations at Leiden Observatory to devote myself, albeit under unfavorable circumstances, to my obligations to science, which has made heavy demands on my strength and energy.

But it must also be clear that here, of all places, science and the astronomer concerned could both profit in this way.

So I felt I should not hesitate, only a few months after Gill had started his work, to propose to him a collaboration, of course with the proviso that I would succeed in finding the rather modest means that would enable me to do my bit in terms of calculations and discussions of the plate material.'

This description, incidentally, is fully consistent with the assumption that Kapteyn had not read any paper in a journal inducing him to offer his services to Gill. But obviously he was fully aware of Gill's plans from the correspondence they had been having concerning the latter's need for help.

However, all this did not make Kapteyn backtrack from his desire to own a telescope and an observatory. He ended his presentation by listing all the nations that were in possession of a photographic facility (Britain and colonies at the Cape and in Australia, France and Algiers, Belgium, Prussia, Baden, Austria-Hungary, Sweden, Denmark, Russia, Italy, the Vatican, Spain, the United States, Mexico, Brazil, the Argentine Republic, Chile). 'And the Netherlands? Still I cannot include the Netherlands in the list. That is sad, because now the first, usually not the worst, fruits of the new method will be to the benefit of others. But the list will not be without the name of the Netherlands forever. Because the excellent name that the Netherlands has established for itself in the area of astronomy since the days

7.6 One Last Attempt at an Observatory

of Kaiser may not be lost.' He did not mention Groningen, already resigned to the fact, perhaps, that once again he might fail to obtain his observatory.

To Kapteyn's disappointment van de Sande Bakhuyzen did in fact succeed in getting an astrophotographic telescope for Leiden. In the annual report for the Observatory of 1888, [13], he reported his participation in the Carte du Ciel meeting and the resolution to proceed to produce a photographic atlas of the skies.

> 'At the congress six directors of observatories already announced their preparedness to take part of the tasks upon them, while later ten others joined in. [...] The observatories that participate in this work need to have at their disposal a photographic telescope, mounted on an isolated pier and in a dome, the costs of which will amount to about 50,000 Guilders (in purchasing power about 1.4 million € today), so that it was unthinkable that I would agree to offer the cooperation of Leiden Observatory. In addition, I hesitated to make a proposal after my return to obtain funds for the observatory to acquire the instruments to photograph the lights in the sky, since I felt that the instruments we have at present could be used more profitably. Recent experience, however, has not confirmed this opinion, but convinced me, on the contrary, that the use of photography would make important changes in our current methods of observing.'

Van de Sande Bakhuyzen did mention, however, that he had offered to help, determining the distortions of the photographic images as a result of optical defects in the objective lenses.

In subsequent annual reports he mentions this effort and other involvement in the Carte du Ciel project, but only in the reports for 1891 and 1892, [14], did he write – in connection with his attendance of a meeting with the Permanent Commission of the Carte du Ciel in April 1891 – that he had become more and more convinced of the 'great significance of photography for the accurate determination of star positions. In this area of astronomy in particular the efforts of Leiden have been extensive, and if Leiden Observatory wants to keep its name and fame, it will be necessary to add a photographic facility to it. This does not necessarily have to be devoted to all aspects of astronomical photography, but just for the determination of relative positions of stars that are not too far apart on the sky ...'. In the archives of Leiden Observatory there is a drawing of the design for the building by the very same architect as in Kapteyn's case, Jacobus van Lokhorst, the Chief Government Architect. This, of course, was a much simpler building than Kapteyn's, since after all Leiden was already in possession of the infrastructure necessary for an Observatory and the plan only included the structure of the dome and a single office room. The funds to build an astrophotographic telescope were included in the government's budget for 1895, to be allocated to Leiden Observatory. At this stage Kapteyn's last chance to obtain his own observatory had come to a final and definitive end.

7.7 An Astronomical Laboratory

After his last and unsuccessful attempt to obtain his own observatory, Kapteyn seems to have given up. But he still had no other place for his measurements apart from the two workrooms, temporarily made available to him in the physiological laboratory of his friend Dirk Huizinga. He set out to improve his facilities for the measuring of photographic plates taken elsewhere and planned to extend his own *Astronomical Laboratory* (as he already used to describe his temporary offices) with more apparatus and locating it in a laboratory building of his own. First he needed funds for a machine for improved measuring of photographic plates. In May 1892 he appealed to the Curators of the university for funds:

'The Groningen astronomical collection is already in possession of
1. The measuring instrument with which the almost completed Photographic Survey of the Southern Hemisphere has been measured.
2. The measuring apparatus that has been allocated to me by the *Comité Permanent de la Carte du Ciel* for my use and that has been almost completed, built according to the principles of the previous model, but designed to obtain the highest precision currently possible.
Furthermore, Prof. Haga has offered me a room in his laboratory for the operation of the instruments that is in all respects suitable for this purpose. If in addition a Repsold measuring instrument is purchased for the price of about 2400 Guilders [currently a purchasing power of about 65,000€], the astronomical institution here would have in its possession a unique set of photographic measuring instruments [...], and we will have in Groningen the possibility to study photographic plates of star fields by a variety of different methods and measurements that would be better than elsewhere, even abroad.'

Hermanus Haga (1852–1936) was a professor of physics and meteorology in Groningen from 1886 to 1922. The request was granted, but the instrument did not go to Haga's laboratory. Instead Kapteyn succeeded in securing a laboratory of his own. He was allowed the temporary use of the unoccupied official residence of the Queen's Commissioner. Remember that King Willem III of the Netherlands had died in 1890; his three sons from his first marriage had died earlier. His daughter from his second marriage, the later Queen Wilhelmina (1880–1962) was only ten years old, so her mother, Emma van Waldeck-Pyrmont (1858–1934), acted as regent until Wilhelmina reached the age of 18. So, in 1890 the title *King's* Commissioner was changed into *Queen's* Commissioner. Carel Coenraad Geertsema (1843–1928) had been Queen's Commissioner since 1893, but he had also been a member of the Board of Curators of Groningen University since 1894. He seems not to have used the Commissioner's official residence, which is not surprising as Geertsema was a banker in Groningen and had his private dwelling in town. The house assigned to Kapteyn for his Laboratory is located in Oude Boteringestraat (see Figs. 4.7 and 7.12) Currently this building, no 44, is used for the offices of the Governing Board of the University.

7.7 An Astronomical Laboratory

Fig. 7.12 Oude Boteringestraat in Groningen. The main building on this picture is the court house. The first location of the Astronomical Institute is the third building to the right with the monumental three-floor facade (From SkyscraperCity [15])

The Astronomical Laboratory in Oude Boteringestraat was officially opened on January 16, 1896, with a Public Lecture, Kapteyn (1896b), with no specific title (see Fig. 7.13). To begin with, Kapteyn stressed that it concerned a temporary location and not a new laboratory. Nothing to the building had been changed or adapted – his current instruments had only been moved there. He felt that he had to explain the 'curious fact that [he] has been so bold as to recommend an astronomical workshop, where one will not, at least not for the moment, have an opportunity to observe the stars, an observatory that is not an observatory. Would I have been able to point out analogous cases, nationally or abroad, I had refrained from the duty to give an explanation,' In any case, he seemed to have given up his hopes for an observatory (or maybe not, as he added 'at least for the moment'?), but he nevertheless went into a detailed exposition of the background of his attempts to get one.

> 'And when I do, let me first recall how I have insisted year after year on the founding of an observatory, which *would* really be an observatory, an observatory with domes and telescopes, with pillars and multiple foundations, with meridian slits and flaps, movable roofs and what not. Although at some stage the realization of such a facility seemed forthcoming, and at a later stage immeasurably far away, in the end this goal has never been reached, since the costs are high, the number of astronomy students in the Netherlands small and our country is already in the possession of two observatories.

Those costs have not diminished with the large-scale application of photography in astronomy. Because in addition to the old observatory, where the classical methods are being employed, in order to realize such a *complete* facility a photographic facility needs to be constructed; and in addition to the old personnel that already has its hands more than full, new personnel is required.

To be absolutely complete, the current observatories require substantial extension. Could it be expected that the government would decide in favor of such an extension and then immediately think of Groningen for a completely new facility?

But why, then, attach the condition of *absolute completeness* of each observatory? Why not have the fundamental astronomy in Leiden with all its traditions following the classical methods, photography in Groningen, and in Utrecht spectroscopy and photometry, for example? [...] Would it not be beneficial for science if the interests of all its parts are not entrusted to only a few hands, when one has more of them? Is it not particularly in the interest of the teaching [of astronomy], which has to be offered at *all* universities according to the law and which *cannot* be missed, that no complete faculty of mathematics and physics should be closed at some of our universities? [...]

When the establishment of a photographic facility was promised to Leiden, it was very easy to predict that the government would not fund a dedicated photographic observatory in Groningen. It was clear that the government was not prepared to provide much funding for the astronomical education and research in Groningen, even if those costs could be recovered, in their entirety or at least to a large extent, from lesser expenditure elsewhere?'

This long introduction reveals in clear terms the frustration that Kapteyn suffered throughout the years that he was not given the opportunity to establish a scientific environment that was in all respects equivalent to what others had. He asked himself whether it would not be possible to salvage his ambitions with no large costs, by restricting his plans by eliminating all things that make an observatory really expensive.

'In order to answer this question we need to look at the work that needs to be carried out at a photographic observatory. It consists of two parts:
1°. The work of the photographer. His part ends with the completed photographic plate.
2°. The work of the astronomer. His part starts with the completed photographic plate.
What is expensive is clear: everything I called typical for an observatory above. [...]

Eliminate the photographer and all you need is a building with some six rooms, for which no very special requirements exist. Still we cut out what seems to us the least vital, the lack of which would present an eminent advantage to most astronomers if they prefer research to teaching,. [...]

7.7 An Astronomical Laboratory

Fig. 7.13 Cover of the printed version of Kapteyn's public lecture on the occasion of the opening of his Astronomical Laboratory (Kapteyn Astronomical Institute)

No laborious efforts on the part of the photographer, only concentration on the real astronomical work! What pleasure, what alleviation!'

So, his unsuccessful bids became a blessing in disguise. But then Kapteyn qualified this statement. Was it really possible to make this division? After all, one needed to know how observations had been taken. Which parts of the sky needed to be photographed? For how long and when? Which parts of the year could contribute the most to the particular question asked? What about just looking for perceptible changes, in nebulous objects, for example? Photographs for such purposes did not demand much analysis, generally not more than a very careful look at the plates. And what about trying to find new planets or comets? This necessitated a telescope nearby and the two could not really be separated. But usually research required very careful measurement, to such an extent that 'one hour's work of the photographer provides days, weeks, months of work for the astronomer' and there the division between the two became much less problematic and even preferable. After a few more examples, Kapteyn made a reference to his lecture as departing rector magnificus, when he listed countries that possessed photographic facilities. He noted that at many of these the amount of photographic material was consciously limited. He mentioned the Harvard College Observatory with its 'legion of assistants' as an

Fig. 7.14 Logos designed by Kapteyn for his Astronomical Laboratory. These pictures are kept in the Kapteyn Room at the Kapteyn Astronomical Institute. They have faded somewhat, especially the red ink. The structure of the letters in the picture on the left is not due to ink spreading, but results from design efforts by Kapteyn (Kapteyn Astronomical Institute)

exception to the rule. But there they were able to keep up with the rapidly growing number of plates since they concentrated on discovery. What if they would actually proceed to measure up all those plates?

Kapteyn referred to the Carte du Ciel for which many plates for the catalog have been taken in the meantime (i.e., ever since his lecture as departing rector magnificus).

> 'What has actually been measured? An insignificant fraction. Will we have finished the catalog in *fifty* years? Who knows? As far as we can see, this is unlikely. There are some astronomers who are in doubt, not whether the living generation of astronomers will ever see the catalog, no realistic person even cherishes such hopes; no, they doubt whether the work will actually be completed at all. [...]
>
> I just want to draw the following conclusions: As early as the days of the old methods, the curse of most observatories was, and *particularly where people worked the hardest*, the threatening accumulation of arrears of work. [...] Photography has increased this curse tenfold, and soon it will

be hundredfold. [...] Is it not time that the balance between photographer and astronomer be at least partly restored?'

The question to be asked, of course, was whether the photographers should stop providing more plates than could be measured by the astronomers. Would that be the solution? To answer this, Kapteyn first listed the contributions by the giants of his age. First, he mentioned the large number of observations collected in Bonn by Argelander and Schönfeld. Next he referred to the fewer, but far more accurate, observations at Pulkovo, Greenwich and Leiden, and the enormous efforts of many astronomers (including Gill). He went on to list the number of '*accurate* star positions': from 'the Observatory at Bonn, 81,000; Córdoba, 106,000; the Astronomische Gesellschaft Katalog will contain 130,000 stars. And greater numbers of less accurate positions at the Observatory at Bonn under Argelander, 324,000, and Schönfeld, 134,000; once more Córdoba, 340,000; while in Groningen with *intermediate* accuracy, 450,000.' But, he asked, would the future researcher, in spite of all this work, be able to find what he was looking for in the enormous body of data?

After more such examples, Kapteyn came to his real conclusion:

'In summary, the urgent question that has become even more urgent as a result of photography, is whether a larger part of the work of astronomers should not be theoretical research. What Darwin, in one of his letters, so characteristically calls the *grinding of huge masses of facts into law*. I do not speak of research into the motions of bodies in the Solar System, the part of science that up till now has exclusively been designated theoretical astronomy. [...] I speak of the Universe of the fixed stars. Have we not reached the point already in that respect, where efforts are justified to try to lift the laws of the structure of the universe out of the available material? Even if the results were not as definite, as a result of the incompleteness of the data, as would be desirable, so what? It is the very manner that will tell us how the observations should be planned, and that will provide a guideline for future work. [...]

For a long time we have been in the possession of enormous numbers of observations of positions, i.e., of directions where stars are located, and of at least approximate estimates of apparent brightness. But from the two properties so little can be deduced, and even such simple deductions require rather bold assumptions. It is because of this that the studies in this area by the Herschels, the older von Struve, Argelander, Proctor, von Seeliger, not to speak of others, have resulted in so few fruitful, and even then mutually contradictory results.

More is to be expected from the studies of *motions* and *distances*.

As far as the latter subject is concerned, the available number is small. To work on the collection of distances using new, promising ways seems to me to be one of the most useful things with which practical astronomy should occupy itself and already here in Groningen a considerable share of our effort is being devoted to this, but more about this later. As far as *motions* are concerned, what is needed first and foremost is that they become measurable. Over a few years interval they are too small to be measurable with any precision, and precise astronomy is still young, was *too* young in the times of the Herschels and von Struve to produce results that are satisfactory for these aims. But I believe that the time is near and will soon be here that such investigations can be undertaken with the promise of at least some fruits.

And this will get even better from year to year, now that spectroscopy is beginning to generate significant information.

If that is the case, then such studies are *the primary duty*, because only this way may we hope to be no longer forced to restrict ourselves to the collection of observations, *completely at random*, in the future. Since the measurements for the southern *Durchmusterung* have been completed here, [...] I have integrated regular research on the structure of the universe, on the distribution of stars in space, into the research program of the astronomical establishment in Groningen.'

Obviously, after all these years of the drudgery of measuring positions, Kapteyn was looking forward to doing real astronomy, which in his case meant trying to understand the arrangement of stars in space. What he definitely did not have in mind was that his laboratory should only be involved in measuring plates and providing more properties of stars in large numbers. Neither did he aim to do more positional astronomy or produce more star catalogs. What he was interested in, was to measure distances of more stars, their proper motions, and chemical composition and radial velocities from spectra, and subsequently use those to find out more about the sidereal universe. And in most of this he was fully occupied already, as we will see below.

The point about the growing backlog of observations and photographic exposures in observatories was well taken. Had he had his own photographic telescope, he would have run the same risk. Even in his own days in Leiden, not much more was produced than the astrometric work. There simply was no more manpower – the professor, two observers and some computational support – available than was essential to execute the program of accurate positional astronomy, in which Leiden University had a long tradition and an excellent reputation. Indeed, the available literature suggests that during Kapteyn's time in Leiden it seemed as if the work was often considered finished with just the provision of accurate positions and contributions to high precision stellar catalogs. It would be highly unfair, it seems to me, to state that van de Sande Bakhuyzen was satisfied with this situation. Surely he would have liked to go further and produce studies of celestial phenomena based on these observations. But he had too little time. When we look at his publication list in the NASA Astronomy Data Base, almost all his papers (excluding those involving administrative matters) concern or are related to positional astronomy. There are only a few exceptions of a different nature, notably three on different subjects (and in different languages) concerning the rotation of Mars, the motion of the Solar System in space and the distribution of stars in space (*On the rotation period of Mars* [16], *Mémoires et observations. Mouvement du système solaire déduit de différents groupes d'étoiles* [17], *Über die Vertheilung der Sterne im Raume nach der Grösse*

7.7 An Astronomical Laboratory 233

der Eigenbewegungen [18]). For decades this was the only interpretative astronomy produced in Leiden. Of course, this was to change later, what with the appointments of Willem de Sitter, Ejnar Hertzsprung and Jan Oort, not surprisingly all protégés of Kapteyn.

In the rest of the lecture Kapteyn gave a few examples of ongoing research elsewhere, or in which he was involved himself. Spectroscopy had shown that two types of stars could be distinguished, with spectra of either, like the Sun or Sirius. Kapteyn was of the opinion that the evidence was mounting that on average the solar types were closer to us than the Sirius type stars. But parallaxes are needed to study that in more detail, and he outlined the work he was doing together with the Finnish astronomer Donner, and which I will describe later in this chapter. But that was only a beginning; for a significant development of our understanding of the structure of the universe, Kapteyn stated, we did not need the one-and-a-half million accurate positions that the Carte du Ciel would determine, but a census of a few thousands of carefully chosen stars for which parallaxes were determined, together with simple counts. Also measuring the proper motions of all or most of the stars accurately, down to magnitude 6, he considered 'infinitely more urgent' than the same for the stars with the highest proper motion.

He concluded by pointing out that he now had excellent facilities to teach astronomy to his students, who could gain experience in measuring photographic plates and take part in theoretical studies. The lecture ended with a profound display of thanks for the Curators, from whom he expected further support in terms of funding instruments and a library, before he could call himself one of the *professores contenti*. He again expressed his thanks to 'My dear friend' Dirk Huizinga, for placing his own laboratory at Kapteyn's disposal for ten years. As in the Introduction to the CPD, he seized the opportunity to note that 'this major work, on which I have worked all those years and that can be considered finished – after all, one third has been printed, one third is being printed and the third part is almost ready to go to the printer – is entirely the product of the physiological laboratory.'

So far Kapteyn's lecture. I quote from the *HHK biography*:

> 'There was still much to be wished for. The laboratory was only a temporary accommodation not allowed to erect the pillars necessary to mount some instruments on. Many other instruments were necessary, there was no housing room for observational instruments and no library. 'An astronomical institute without a library is impossible', he said, 'especially in an institute as this one where the work program concerns theoretical investigations based on observational data obtained elsewhere.'
>
> But the first step had been taken, the laboratory was elevated to an official establishment with its own premises and annual budget, and with his beloved motto: *'Le mieux est l'ennemi du bien'* ['the perfect is the enemy of the good']. Quoted from Voltaire (François-Marie Arouet; 1694–1778)], he accepted his astronomical laboratory.
>
> He received appreciation from all sides. Oudemans, the astronomer from Utrecht, who had failed to support him in realizing an observatory, wrote later: 'In hindsight it is possible that things worked out better (not having

Fig. 7.15 Second home of the Astronomical Laboratory at the corner of Oude Boteringestraat and Spilsluizen (Kapteyn Astronomical Institute)

an observatory). When I was an observator, it must have been 1853 or '54, I had to write to Airy [who was England's most famous astronomer at the time]. I told him that there was a plan to build a new observatory in Leiden and what was his reaction? 'Your message that there will be a new observatory in Leiden, does not make me particularly happy. There are more than enough observatories, but it is a computational institution that we need.' The computational institutions at Berlin and Groningen have shown that this judgment was not that bad after all. But for real progress the tenacity and perseverance like yours is essential.'

In 1904 the building was needed again for its original functions and the Astronomical Laboratory had to move to another – again temporary – location at the corner of Oude Boteringestraat and Spilsluizen (see Figs. 7.15 and 7.16). Before this it was home to the meteorological laboratory. It is currently the location of the 'Hotel Corps de Garde'. This building, which dates from 1634, was originally the home of the military who defended the city. The laboratory was housed here for 8 years.

7.8 Refraction and Aberration

I now turn to the scientific work that Kapteyn was also carrying out during the years that he was primarily occupied with the preparation of the Cape Photographic Durchmusterung. In addition to this activity, other scientific research went on, most

7.8 Refraction and Aberration

Fig. 7.16 Oude Boteringestraat seen from the north across the Boteringe bridge around 1899. The building on the right was the second home of the Astronomical Laboratory. The first location, at the home of the Commissioner of the Queen, was down the street on the right, just beyond where it can be seen from this vantage point (From Beeldbank Groningen [19])

but not all of which was related to the ongoing work on the CPD. We have already seen that he did a fair amount of work associated with the Carte du Ciel, as a member of committees and the author of long papers for the publications of the Permanent Committee. These were mostly on his parallactic method of measurements, but as late as 1896 he published a paper on the theory of corrections for refraction and aberration for measurements of rectangular coordinates of stars on photographic plates, Kapteyn (1896a). At the same time he was preparing his work on parallaxes with the meridian circle in Leiden for publication.

In 1890, a short note appeared in the *Astronomische Nachrichten*, written by Friedrich Robert Helmert (1843–1917), the director of the Geodetic Institute and a professor of geodesy at the University of Berlin. It was entitled *Starke Änderung der geographischen Breite in der zweiten Hälfte des Jahres 1889 zu Berlin, Potsdam, Prag und Strassburg* [20]). Helmert reported that with regard to the orientation of the Earth's axis (from the geographical latitude), no perceptible change had been observed in Berlin and Potsdam during the first half of 1889, but a strong change had occurred in the second half of that year, amounting to no less than $0\rlap{.}''5$ to $0\rlap{.}''6$. This was confirmed by observations from Prague and Strasbourg. Although he had already reported this during a geophysical conference, he also wanted to draw attention to his findings by alerting astronomers and ask them if their observations were consistent with his. Kapteyn read this note and sent a letter to Helmert, who offered it to the Astronomische Nachrichten for publication.

In the letter, Kapteyn (1890a) did not discuss the effect itself. He pointed out, however, that there was a very accurate method to measure one's geographical latitude using photography of stars that pass through the zenith, through an instrument that used the principle of the 'reflex zenith-tube'. This had been designed around 1850 by Sir George Biddell Airy (1801–1892), Astronomer Royal and director of the Royal Greenwich Observatory. It had been used extensively to monitor the star γ Draconis to determine the constant of aberration (see Box 3.2) and its parallax. Kapteyn suggested the following set-up, using that principle: Take an objective lens and point it to the zenith; light from a star going through the zenith is then reflected below the lens by a basin filled with mercury. The mercury surface is located at a little more than half the focal length from the objective, so that an image of the star is formed just below it. Then mount a photographic plate at this position. It will block the lens so the plate should have a rather small size. This can be used to measure the zenith distance of a star with great accuracy by measuring the angle between the star on the plate and the position of the optical axis (also on the plate).

Now Kapteyn proposed not to measure absolute zenith distances, but to use the instrument to measure *differences* in zenith distance. A star that passes close to the zenith, will leave a trail on the photographic plate; this trail will have a slight curvature, since the daily path of a star on the sky is a small circle centered on the pole. Then take a second star that also goes through the zenith, but on the other side from it with respect to the first star; then rotate the objective and plate by exactly 180° and observe that second star. Then there are two trails; the one from the second star will also be curved but in the other direction, so that the trails can be

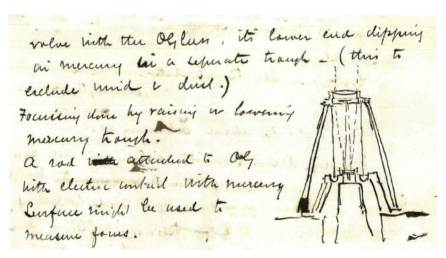

Fig. 7.17 Sketch of a zenith telescope by David Gill based on the suggestions of Kapteyn, designed to measure changes in the orientation of the Earth's rotation axis. This is part of the letter from Gill to Kapteyn of December 27, 1890 (Kapteyn Astronomical Institute)

distinguished. Now if the polar height changes, for example such that the first star crosses at a larger zenith angle, the second star – being on the other side of the zenith – will cross at a smaller zenith angle. But since the plate has been rotated between the measurements, the distance between the two trails has changed and the method can be used for accurate measurement of *changes* in the positions of the poles. The mercury 'mirror' ensures exact orientation towards the zenith. Also note that this method does not suffer from effects of light deflection in the atmosphere and flexure in a telescope tube.

This was a bright idea. Kapteyn went on to discuss the possible sources of error and evaluated their seriousness and how this could be corrected. He contacted the Henry brothers in Paris on the performance of their objectives, to judge down to what magnitude stars could be recorded. Based on a Henry objective of 33 cm diameter, 343 cm focal length and allowing for blocking, loss of light in reflection, etc., Kapteyn estimated that stars of magnitude 7.6 would still leave a useful trail on the photographic plate, and he therefore concluded that the method was feasible. As far as I could find out, there was no published response from Helmert. But the method is definitely ingenious and demonstrates Kapteyn's insight and skill when it came to designing observational procedures.

One person who did react to this paper was David Gill, in his correspondence to Kapteyn. He promised to give the method a try and in his letter of December 27, 1890, he even sketched an alternative design with a mercury trough that could move vertically with a view to focusing (see Fig. 7.17). Whatever came of this is not further recorded in Gill's letters.

7.9 Kapteyn's Star

Other publications are related to Kapteyn's ongoing measuring of the CPD plates. Kapteyn (1890b) focused the attention of the astronomical world on the fact that during his reduction of the plates and the preparation of the Durchmusterung, numerous variable stars were being found. As it would take years before the work was completed, he reported nine objects found to be variable for the benefit of other astronomers. He had as a matter of fact been corresponding with Gill on how such results should be published and Gill had authorized Kapteyn 'to publish any notes about variables' in his letter of April 2, 1890. In a follow-up a few years later, Kapteyn (1896c) reported another eight variable stars. Remarkable about this paper for the Astronomische Nachrichten is that it was written in English, whereas he reverted to German in following papers.

Kapteyn (1893c) reacted to the discovery of a 'new star' in the constellation Norma. This star had been discovered by Williamina Paton Stevens Fleming (1857–1911) on October 26, 1893, and reported by her and Edward Charles Pickering, director of Harvard College Observatory. The star reached a brightness of magnitude 7.0 or so. In those days, nobody had any idea what these 'new stars' or 'novae' were, so it was important to know what the progenitor star looked like. So Kapteyn immediately went to the CPD plates. He did not see the star on three plates taken in 1887, but it was visible at about magnitude 9 in two plates in April and May of 1890. In 1887 it must have been fainter. No further follow-up was recorded, but it shows that the CPD plates were a useful archive of the sky for time-dependent phenomena.

Another occasion that attracted special attention, was the discovery that came to be known as *Kapteyn's Star*; the discovery paper, Kapteyn (1897b), bore the title *Stern mit grösster bislang bekannter Eigenbewegung* (Star with the largest currently known proper motion) and is only five lines of text and a table with five entries (see Fig. 7.18). Kapteyn reported five positional measurements. Two were from the CPD plates taken in 1890 and 1893. The star was also part of the Córdoba Durchmusterung, in which the position had been obtained in 1873. The information was extended with new visual measurements with an equatorial and a meridian telescope at the Cape Observatory, by Robert Innes (Innes can be identified in the photograph in Fig. 7.30). In 1897 he determined the position of the star twice. The proper motion was an unprecedented 8″.7 per year (the 0ˢ.621 in Right Ascension translates into 6″.60 in angular measure 'along the great circle'). That is no less than about 3′.5 over the 24 years that Kapteyn listed measurements. Since it is important with a view to what follows later, I quote Kapteyn's final sentence, translated into English: 'The discovery is the result of the work on the C.P.D., as well as that of the Cape astronomers (Innes) and myself.'

The oldest reference to it by the name 'Kapteyn's Star in Pictor' is by Agnes M. Clerke in a paper *Recent determinations of the Sun's movement in space* of 1902 [21]. Since its discovery there has been only one star, Barnard's Star, that has surpassed the proper motion of Kapteyn's star; in that case it is a little over 10 arcsec per year. It was discovered in 1916 by the American astronomer Edward Emerson Barnard (1857–1923).

There is an interesting incident related to the discovery of this large proper motion star, which throws some light on Kapteyn's character and his strong sense of justice. It started with the publication of a 'note' by the editors in the February 1898 issue of the *The Observatory*, in which they drew attention to Kapteyn's publication and the record proper motion of the star [22]:

'In Astr. Nachr. No 3466, Dr Kapteyn publishes a note stating that in examining a plate of the Cape Photographic Durchmusterung he discovered a star which apparently had a large proper motion. [...] From comparison with the places found from reductions of a Cape photographic plate, and with equatorial and meridian observations of the star made last year, Dr Kapteyn deduces a proper motion of +0ˢ.621 in R.A., and -5″.70 in declination.'

So, no mention of Innes' role in the discovery, even though Kapteyn in his note had made sure to make mention of Innes' contribution. In the section 'Observatories' of the November issue, there was a report by the Royal Observatory, Cape of Good Hope concerning the year 1897, presented by its director, David Gill [23]. Gill noted that 'Mr Innes has done a great deal of work observing stars to detect variability or to verify the reality of stars given in some catalogues of precision.' The editors of the journal added the following footnote:

7.9 Kapteyn's Star

Stern mit grösster bislang bekannter Eigenbewegung.

Der Stern Cordoba Zone Catalogue 5^h243 hat eine Eigenbewegung von $8''\!.7$ im grössten Kreise, wie dies aus folgenden Beobachtungen hervorgeht :

	Grösse	Epoche	α 1875	δ 1875
Cord. ZC. (2 Beob.)	8	1873.04	$5^h 6^m 40^s\!.61$	$-44° 58'\!.176$
Cape Phot. DM.	9.2	1890.1	50.8	59.9
Cap Catalog Platte (geschätzt)	—	1893.9	53.8	60.2
Innes, Equatorial	8.2	1897.1	55.8	60.4
Cap, Merid. Beob.	—	1897.81	56.0	60.530

welche alle gut stimmen zu einer Eigenbewegung von $+0''\!.621$ in gerader Aufsteigung, und von $-5''\!.70$ in Declination. Innes findet den Stern orange-gelb.

Die Entdeckung ist aus den Arbeiten für die C P. D. hervorgegangen, also aus den Arbeiten der Cap-Astronomen (Innes) und meinen eigenen.

Groningen 1897 Dec. 14.   *J. C. Kapteyn.*

Fig. 7.18 The note that Kapteyn published in the *Astronomische Nachrichten*, Kapteyn (1897b), on the discovery of the high proper motion star now known as Kapteyn's Star (From Kapteyn (1897b))

'In connection with this work Dr Gill gives in detail the facts as to the discovery of the large proper motion of Cordoba Zone V. 241, which was reported on p. 106 of our February number. It seems that Prof. Kapteyn, in co-operation with whom Mr Innes is working, did not find the star on the Cape Durchmusterung plates, and told this to Mr Innes, who looked for it in the telescope. He also did not find a star in the Cordoba place, but found one about 15 sec. distant in R.A. and suggested that this was the star, and that it has a large proper motion. Prof. Kapteyn again examined the Durchmusterung plates and found the star in a position which corroborates the proper motion hypothesis. The *kudos* therefore seems to belong to Mr Innes. We are glad to make this statement, as our note of February last might give a wrong impression.'

The credit was taken away from Kapteyn and he was furious. On November 10, 1898, he wrote to Gill on routine matters: eight pages of tight and compact handwriting. After concluding the letter, he started a P.S. referring to the footnote in 'The Observatory', and this postscript ran another six(!) pages.

'I just now receive the November Number of 'Observatory' and I am considerably vexed by what is said there in connection with the discovery of the PM of CZ5^{h}243, the conclusion being reached, in consequence of the detailed facts communicated by You in the report of the Cape Observatory for 1897, that 'The Kudos therefore seems to belong to Mr Innes' etc.

Now I certainly do not think such a conclusion fair, but would not mind the matter very much if the fact that I published and the way in which I published the discovery did not show pretty clearly that I laid claim at least to half the honour of it. It is pretty evident, by what the editors of Observatory are pleased to say, that they think me guilty of appropriating wholly or in part what of right belongs to another.

Now this I consider to be a glaring injustice and I certainly would be grieved beyond anything if You or Mr Innes agreed with the gentlemen named. I think it due to myself therefore to say again explicitly that I really think myself entitled to half at least of whatever honour may belong

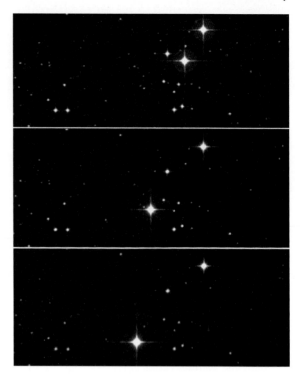

Fig. 7.19 The motion of Kapteyn's Star against the background of 'fixed stars'. The pictures are from the United Kingdom Schmidt Telescope Unit Survey taken with different photographic emulsions in 1975 (top, blue sensitive), 1990 (middle, red) and 1996 (bottom, infrared). The star moved by about 3 arc minutes in this time interval (Produced with the Digital Sky Survey DSS2 [24])

to the discovery of this PM and that I still think the terms in which I expressed myself in the Astr. Nachr. perfectly fair to anybody concerned in the matter.'

Kapteyn explained that he had written to Gill as early as 1896 how to go about credit for discoveries (of variable stars) during the CPD measurement. He also documented the course of events resulting in the finding of the high proper motion. It started when Kapteyn noted that a star in the CPD was missing in the Córdoba Durchmusterung. Then Innes found a bright star nearby and suspected large proper motion. Next Kapteyn 'strengthened' the suspicion on the basis of his measurement of the Right Ascension, and Innes found a correction to the Córdoba declination that proved the proper motion. Kapteyn's conclusion was that the credit should at least be shared. He reminded Gill of the large amount of work he had devoted to measuring the plates and making comparisons to existing catalogues. He then asked: 'Shall the man who leads the thing up to the point that discovery can hardly escape, have no right at all to the discovery, and only the man who looks for the stars pointed out?' He suggested that there should be a publication with a 'list of errors of catalogues and that of the proper motions [...] in the latter of which our star will of course appear', to be written under both Kapteyn's and Innes' authorship. He wound up by saying:

'This supposes that it will be Mr Innes who takes it upon himself to further elaborate these lists. – If this will not be the case, or if Mr Innes or You

7.9 Kapteyn's Star

have any other objection to this plan, then I leave the matter gladly to Your judgment in which I trust absolutely.

I hope not that in all what I have written You will see anything in the least offending to Mr Innes. If so, please impute it to a necessity on my part of working off my steam at once. – As my letter is to be dispatched in a few moments to catch the mail, I have no time to cool off thoroughly. But let me say this at least: I am really extremely pleased with the thorough way in which Mr Innes has taken up the work of revision and most candidly hope that he will be the man to take up the work of the PM's.'

Gill replied on December 1, 1898:

'Now for your postscript.

I posted a hurried pencil line, which I hope caught the mail – so soon as I read, or rather glanced over your letter.

I had been in Cape Town [...] – my mail had been posted before I left for town – and now Mr Innes had gone off for dinner.

I found on his return that Mr Innes had written in the mail that had just left privately to the Editor of the Observatory to say that the note in the Observatory about his share in the work was unfair to you. What he had felt was that a lot of things had been published by others about Kapteyn's new PM star without any mention of Innes' share in the matter. He had no complaint to make – quite the contrary – of your note in the Astron. Nachr. What I put in my report was a simple statement of the facts of the case, which I thought the fairest and most dignified way of soothing Mr Innes without raising any question on the matter. At the same time Innes was fully sensible of the injustice of the Observatory note, and had written, without either my knowledge or suggestion to the Editor (& privately) to say that it was not fair to you.'

After reviewing his recollection of the course of events (not unlike Kapteyn's account, actually) he continued:

'Everything was fully and fairly stated by you in the Astr. Nachr. – and we were both quite pleased that it should be so. The first steam was when poor Innes found himself ignored by the popular writers and then when the Observatory people tried to ignore you – both he and I were equally indignant with yourself. What Innes complains of rather is that you did not let off any steam at the paragraph on page 106 of the Febr. number of the Observatory, in which the entire credit is given you and no mention whatever is made of his name.

This however is certain, that of all discreditable and undignified things, the worst is the washing of scientific dirty linen in public and I will right gladly give my adhesion to any cause that will prevent such a thing.'

Gill recounted how he reacted when 'our good friend Müller', presumably Gustav (in full Karl Hermann Gustav) Müller, the director of Potsdam Observatory (1851–1925), in his review of the C.P.D., tried to give all the credit for the Durchmusterung, including the original conception, to Kapteyn. It is not clear where this

		Mag.	α. h m s	° ′	
Gould Z.C.	1873·0	8	5 6 40·6	−44 58·2	2 obs.
C.P.D.	1890·1	9·2	6 50·8	59·5	4 obs.
Innes	1897·0	8·3	6 56·0	58·0	

the R.A. agreeing very well on the supposition of a proper motion of 0^s·64, but not so the Declinations. Could there be a mistake in Mr. Innes' Declination? On September 22nd I replied that there was no doubt that the star C.Z. V. 243 is the star of greatest known proper motion; that Innes' observation of 1897·0 had not been an observation in Declination, because, finding no star in the required R.A., he had simply noted the neighbouring star, differing 15^s in R.A., and having *about* the same Declination as the missing star. But on Feb. 15 he had re-observed the star, and found a 5^h 6^m 56^s·0, δ −45° 0′·4, and that the estimated position of the star on one of the Catalogue plates made a proper motion of about +0^s·64 and −0′·1 in Declination certain. On October 27 I forwarded results of two meridian observations of the star made on October 23 and 24. A preliminary note has been communicated to the Astron. Nach. 3466, by Prof. Kapteyn. A complete investigation of the proper motion and parallax of this remarkable object will be made at the Cape.

Fig. 7.20 The detailed text published by Gill in *The Observatory* of the sequence of events concerning the discovery of Kapteyn's Star (From Gill (1899); see text)

review was published, maybe it was just in draft form. Anyhow, Gill reminded Kapteyn that he wrote to Müller thanking him for the review and 'the kind way you speak of my comparatively small share in the work. There is no praise too high for what Kapteyn has done'. 'I think', Gill continued his letter, 'that was far better than writing to say that part of what he implied was wrong. At the same time I think you might quietly put him right. Were I to do so it might lead to the idea that I grudge you some part of the credit due to you – which, God knows is the reverse of the case.' Here Gill seems to be imply that Kapteyn would be wise not to be too headstrong.

Gill expressed his wish that 'there should be a loyal CPD fraternity – all resolved to make the CPD the best and most complete thing possible, all assured that each will be fairly dealt with, and all content that the work of each is mentioned, and all taking a legitimate pride in their common labours.' He added that he agreed with Kapteyn's plan and assured him that Innes would cooperate. 'I think you can leave that safely to me. I am quite ready to consider any other suggestion you may have to make. Meanwhile, I have sent a little note to the Observatory of which I enclose a press copy.'

The note appeared in the February 1899 issue [25] together with another note by the Editors. Gill explained that the credit should be shared between Kapteyn and Innes. He reproduced in full the course of events (see Fig. 7.20 for the details). The Editors seemed to feel unjustly criticized: Gill's note 'accuses us of some want of judgment', taking the point of view that the problem arose from the fact that Kapteyn was allowed to publish by himself.

'The suggestion of a possible proper motion occurs, perhaps, once a day in a hard-working observatory; the verification of such is rarer. This is mentioned to explain that no one has shown unusual genius; but because it is the largest proper motion hitherto known, some name will be attached to it in sensational astronomical articles, and the owner of that name will acquire some notoriety. We should fancy the staff of the Cape Observatory will feel aggrieved if the name is Kapteyn, although Dr Gill does say they are satisfied with the credit given for their share in the work. The trouble seems to have arisen because Dr Kapteyn was allowed to publish, in his own name, results of work that emanates from the Cape Observatory. [...] no member of the staff of an observatory should, in justice to his colleagues, make a spurious reputation from circumstances more or less accidental.'

The editors also noted that the word *kudos* was used because it was 'less emphatic than *honour*', but refused to retract anything they said: 'these are our views, and we think the wording of our notes is consistent with them.'

Kapteyn, for his part, appeared to have understood Gill's lesson. On January 3, 1899, he wrote to Gill:

'Now that I know that neither You nor Mr Innes complain of my note in the Astr. Nachr. I am completely satisfied. I have not the slightest desire, and never had, to be considered the sole or the principal discoverer of the PM of this star and have certainly felt the injustice done to Mr Innes by some writers. I now feel that I ought perhaps to have protested; but, I am ashamed to say, the thought had not occurred to me before, probably because I felt, as I do now, that the announcement in the Astr. Nachr. was perfectly clear and fair to all concerned. [...]

That I was indignant and showed it, as soon as my share in the discovery was questioned, is true. But what especially galled me was not this. It was the implication that in my note to the Astr. Nachr. I had tried to rob Mr Innes of his part of it, [...]

In Müller's review of the C.P.D. I think there are more points than one which are not quite fair. [...] Will it not be the best thing that in my Introduction to Vol. III I say a few words about that matter? In my opinion that will be better than a private communication.'

Gill replied on February 22, 1899 with the words: 'Their [the editors] object and desire is apparently to make bad blood between us – and to sow mischief. Thank God I do not think the latter crop will flourish in any of our fields'. Kapteyn responded on March 15, 1899: 'I would certainly not like you to write anything in reply to the note of the Editors of 'Observatory' (the nastiness of which is the only thing which is perfectly clear to me). I feel already ashamed to have been the cause of so much ado about nothing. Be sure that not all the proper motions in the world would compensate me in the slightest degree for the loss of your friendship'.

In the Preface to the final volume of the CPD, Gill wrote in very complimentary terms about Kapteyn (see page 182). Kapteyn, in his Preface in the same volume, wrote extensively about Gill (without mentioning him by name, but describing him as 'the man who originally planned the whole work, but who feared that the measurement and the cataloguing must be relegated to his old age', paraphrasing

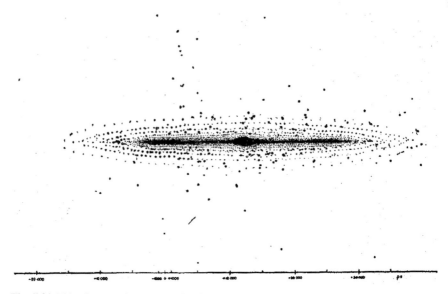

Fig. 7.21 This diagram shows a sketch of the Milky Way Galaxy's structure. The Population I disk is the flat structure, the Population II halo is spherical and consists of many individual stars, but some of these are concentrated in so-called globular clusters. The diagram was published by W. de Sitter in his book *Kosmos* of 1934. This is a Dutch translation of his *Kosmos: A course of six lectures on the development of our insight into the structure of the Universe, delivered for the Lowell Institute in Boston, in November 1931* (The figure is credited to Jan H. Oort)

Gill's Preface), giving the full credit of the idea of a photographic Durchmusterung to him, where it belonged, but made no further mention of Müller.

All in all I feel that at least during this episode, Gill seems to have been the wiser and the more tactful of the two.

Kapteyn's star has a proper motion of about 8.7 arc seconds per year (see Fig. 7.19). It has a parallax of about 0.26 arcsec and is thus at a distance of 12.8 light-years (3.9 parsecs). The proper motion, then, corresponds to a transverse velocity of about 160 km/s. But is also has a radial motion of some 245 km/s with respect to the Sun, so the total space velocity relative to us is over 290 km/s. Only 25 known stars are closer to us than 4 parsec, so it is 'in our backyard', in a manner of speaking. Its mass is about one quarter of the Sun's and its radius about 0.3 times that of the Sun. With reference to the discussion at the beginning of this chapter, I note that Kapteyn's Star is the nearest example of a halo or Population II star, and it is currently crossing the disk of the Milky Way (see also page 209 and Fig. 7.21). So, it is part of the ancient, more or less spherical halo of our Galaxy, in which stars have very small amounts of chemical elements apart from hydrogen and helium. In terms of those elements, Kapteyn's star has only a few tenths of one per cent of its mass or ten times less than our Sun. In the halo, stars move on criss-cross orbits, but the whole of the halo has little organized (rotational) motion. The stars in the halo can have very high velocities with respect to each other, since as a consequence

7.9 Kapteyn's Star

Fig. 7.22 The large globular cluster ω Centauri, from which Kapteyn's Star may have originated (The Hubble European Space Agency Information Centre [28])

of little organized motion their orbits have a wide variety of orientations in space. Kapteyn's star is currently moving at a modest velocity of some 20 km/s, in the direction of the center of the Milky Way and at about 50 km/s perpendicular to the disk of the Galaxy. Its high velocity is due the fact that in its orbit it now moves in the opposite direction with respect to the rotation of the disk of the Galaxy (which is 220 km/s). With respect to the Sun it moves at almost 290 km/s. The corresponding 'retrograde' orbit opposite Galactic rotation has been derived in a paper by Kotoneva *et al.* in 2005 [26] This orbit takes it in towards the center of the Galaxy from its current 8 kiloparsec or so (similar to the Sun, of course) to as close as almost 1 kpc. In the direction perpendicular to the Milky Way disk it oscillates up and down to at most 0.8 kpc from the plane. Since it is a halo star its age must be high, roughly that of the Milky Way Galaxy itself, i.e. at least 10 billion years.

In the 1960's it was discovered that Kapteyn's star is in fact part of a collection of stars that move together through space. This has come to be known as the 'Kapteyn (moving) Group' and at present has about 33 known members (see Olin J. Eggen, *Star Streams and Galactic Structure* (1996) [27]). These stars share the large retrograde motion in the direction of Galactic rotation to within 19 km/sec or so with Kapteyn's Star, but have a much larger deviations in the direction to/from the Galactic Center ($+13 \pm 82$ km/sec) and perpendicular to the disk of the Galaxy (-16 ± 67 km/sec). As far as is known they also have low amounts of chemical elements apart from hydrogen and helium. Other such 'moving groups' are known with far less extreme systematic retrograde motion or no such motion at all, and are thought to arise from either resonance effects in the disk (a collective response to some gravitational disturbance) or from small satellite galaxies that have fallen into the Galaxy and are now dissolving and merging with the general stellar population.

There is an interesting suggestion made recently with regard to the chemical properties of Kapteyn's Star and other stars in the Kapteyn Group. The Population II stars in the Galactic halo are mostly distributed smoothly throughout space, but a small fraction concentrates in aggregates that are called 'globular cluster' because of their spherical shapes. The distribution of globular clusters is shown in Fig. 7.21, taken from de Sitter's book *Kosmos* [29]. Now, the stars in the Kapteyn Group show a distribution of chemical properties that very much resembles one of the populations in the largest globular cluster by far, which is called ω Centauri. Unlike the smaller ones, this cluster, which can be seen with the naked eye in the southern hemisphere (see Fig. 7.22), seems to have at least three populations of stars of a different chemical content. One of these is very similar to that of the Kapteyn Group. Its extraordinary size and number of stars of ω Centauri has given rise to the suggestion that it is actually the remnant of a dwarf galaxy that has fallen into our Galaxy and moved through the disk of the Milky Way. In that process it would have lost its most loosely bound stars through so-called tidal stripping. The hypothesis is that this is the origin of the Kapteyn Group and Kapteyn's Star (see Wiley-de Boer et al.: *Evidence of Tidal Debris from ω Cen in the Kapteyn Group* (2010) [30]). This makes Kapteyn's star a truly remarkable one.

Very recently two planets have been discovered around Kapteyn's star, both a few times the mass of Earth and one orbiting in the 'Habitable zone' where the presence of liquid water is possible [31]. It is remarkable that a star with that little heavy elements can still harbor planets.

7.10 Gill Visits Groningen

While the production of the Cape Photographic Durchmusterung was progressing, Gill visited Kapteyn a few times in Groningen (Kapteyn's only return visit to the Cape took place in 1905). The first visit, which has been mentioned before, was just before the first Carte du Ciel meeting in Paris in the spring of 1887, and it must have been the first time for a very important astronomer to visit Kapteyn. It made a strong impression on the family. Henriette Hertzsprung-Kapteyn, then a little girl of only five years old, described the event as follows:

> [...] And when Gill was planning to visit Groningen in 1887, the complete Kapteyn family lived in great expectation. Everybody prepared themselves. Mrs Kapteyn's English was excellent and in a hurry she taught her daughters, 7 and 5 years old by that time, some rudimentary English they could use as words of welcome. Gill came, and everyone was instantly fond of him. He had the talent to win all hearts in no time through his open, honest and warm enthusiasm. The first evening, Gill and the Kapteyns were sitting in front of the fireplace and Gill told his life's story, which left a strong impression because of his modesty and dignity. It taught the two young people a beautiful lesson for life that not many will understand: trust and you will be trusted, and your life will be enriched.

Mother looked at them with joy, as she was standing in front of the house when they left arm in arm for the Laboratory (the two small rooms had this grandiose name in the Kapteyn family). Gill talking loudly and gesticulating, stared at by the Groningen citizens, Kapteyn, thin and modest, quietly happy next to him. This first visit was a great success, and also the children were delighted with this big playmate, who had won their hearts notwithstanding his incomprehensible language. From then on he was 'Oom Gill' [Uncle Gill] and remained that forever.'

Gill's correspondence tells us that he made another visit to Groningen while traveling through Europe in 1891. Kapteyn and Gill had both been at the March meeting in Paris of the Permanent Committee of the Carte du Ciel and on his way back to the UK via Hamburg late April, Gill visited Groningen once again. From Berlin, where he was before stopping in Hamburg, he announced his arrival by train, also stating 'Remember me most kindly to your wife and the little ones'. In many later letters he sent his regards to 'the lassies'.

7.11 Parallax Again

Despite all the time required for the CPD and related research, Kapteyn started some research into general matters concerning the distribution of stars in space and the structure of the universe, as he had already indicated in his lecture at the opening of his laboratory. The problem that had had his attention for a long time was that of the distances of stars. But even before the CPD was finished, he began to investigate patterns in proper motions of stars and distributions on the sky of intrinsic properties. It is possible to estimate the colors of the stars by comparing their magnitudes on photographic plates with those determined visually. This is so, because the photographic plate, at least the ones used early in astronomy, were more sensitive to bluer wavelengths and less sensitive to redder ones that the human eye is. So, if stars have different colors, their brightness on the photographic plates will be different from that to the eye.

I defer mentioning his work on proper motions and colors to the next chapter, and concentrate here on what had been his longstanding interest (already the title of his inaugural lecture), the determination of stellar parallaxes on a large scale. Having proved that parallax determinations with timing observations on a meridian circle were feasible, and having published his comprehensive paper on the subject in the Leiden Annals Kapteyn (1891a), he kept looking for new opportunities. Obviously, photography offered new possibilities in this area. Kapteyn had made suggestions to Gill on how to proceed here. As we will see below, this must have been before 1889 and possibly as far back as the Carte du Ciel meeting in Paris in 1887. The method proposed by Kapteyn was published in a short note by Gill in the Bulletin of the Permanent Commission, in 1889. He said this was at the request of Kapteyn, and he added his own remarks, likewise at Kapteyn's request. The full title of the paper is

Fig. 7.23 Anders Donner as he appears in the *Album Amicorum* presented at the retirement of H.G. van de Sande Bakhuyzen as professor of astronomy and director of Leiden Observatory in 1908 (Archives Leiden Observatory; see caption Fig. 3.7)

Exposition of a project by Mr J. C. Kapteyn, in relation to the determination of the proper motions and parallaxes of stars. For this reason I have included it in the list of Kapteyn publications, in Appendix A as Gill & Kapteyn (1889).

Basically, the method is straightforward. Every plate taken 'for the series' (so the proposal seems to be all plates for the Astrographic Catalogue) should be taken in three steps. Start with an exposure at the time the parallax for the stars in that field is maximum. Do not develop it, but take a second exposure 6 months later, but with the pointing of the telescope slightly different. Repeat this another six months later and develop the plate then. The difference between the first and third position (after correcting for the shift in the pointings) is the annual proper motion and similar difference between the mean of that and the middle position then is twice the parallax. This seems straightforward, and possibly others had thought of it as well, but Kapteyn was the first to present it to the Permanent Committee. If it had been adopted it would in principle yield accurate values or otherwise upper limits to parallaxes for an enormous number of stars. In practice it would only have worked for the brightest stars, but the amount of work in measuring the plates would have been enormous, not to speak of the trouble in taking plates that had to be stored and conserved for a full year. The obvious two problems that Gill saw in his comments are the following: Firstly, the possible deformation of the emulsion cannot be measured now with respect to the reseau, since the reseau only gives that at the time the reseau has been exposed on the plate and that is not necessarily the case at all three instants the plate has been exposed. Secondly, photographic plates, at least the ones Gill had used, have a diminishing sensitivity with time. He advised the Permanent Commission to discuss this possibility.

7.11 Parallax Again

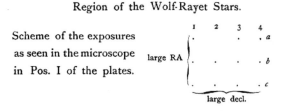

Fig. 7.24 Lay-out for the observations of stars for the determination of parallaxes. 'Large' in association with RA or Decl. should be read as 'increasing'. For more detail see the text (From Kapteyn & Donner (1900))

When the Permanent Commission met in 1891, it did not adopt the scheme. Not surprising, perhaps, in view of the enormous extra effort required (not to mention the difficulties pointed out above and Gill's remarks). As it was, some good arose from it. Kapteyn had become good friends with Anders Severin Donner (1854–1938) of Helsingfors Observatory. Helsingfors is Swedish for Helsinki. Finland was autonomous duchy within the Russian Empire since 1809; it was bilingual (Swedish and Finnish) so the name Helsingfors Observatory was not unusual and Donner used Helsingfors for his residence throughout his career. They must have met at the original Carte du Ciel meeting in 1887 (see Fig. 6.20 where they both figured in the group photograph and a later photograph in Fig. 7.23). Donner was Kapteyn's junior by three years, and since they were among the minority of less senior astronomers at the first Carte du Ciel meeting, this may have been drawn them towards each other. Donner had been professor of astronomy and director of the Helsingfors Observatory (see Fig. 7.26) since 1883. His observatory took part in the Carte du Ciel project and therefore was in possession of a photographic telescope (astrograph). As it turned out, the Helsingfors part of the Astrographic Catalogue was among the first to be completed.

Donner's correspondence with Kapteyn has been added to the archives in the Kapteyn Astronomical Institute in paper copies, as part of the effort to restore Kapteyn's correspondence (see the article by Petra van der Heijden in the *Legacy*), but the Astronomy Department of the University of Helsinki has provided the full set of Kapteyn letters on their Website [32]. The correspondence is in German and stretches in time from 1889 until 1922, but apart from an occasional exception it consists of letters by Kapteyn. Apparently, Donner did not keep copies of his letters to Kapteyn (or any drafts). Curiously, a set of letters from 1891 and 1892 is missing on the Website, but have been included in the paper copies now at the Kapteyn Astronomical Institute. These have been provided in electronic form by the University of Helsinki.

During the 1891 meeting of the Permanent Committee of the Carte du Ciel, when Kapteyn failed to get his scheme adopted, Donner and he decided to try the scheme with the Helsingfors telescope. And indeed, in the final publication Kapteyn & Donner (1900). Donner wrote that he *offered* to help securing the plate material, and in an early letter to Donner on June 25, 1891, (see Fig. 7.25), Kapteyn wrote:

'I have reflected often on the plan that we discussed in Paris (and which has also been described already in Bulletin du Com. Perm.. 4me Facs. S. 262) and discussed it with Gill, who visited me in Groningen for a few days. Not much new has of course come out of that. Only an experiment will show definitely to what extent the method is practical and expedient and will also tell us about the single element that matters most.

More important than anything else is the question: will there be a measurable shift between the various exposures? It seems highly unlikely to me and also Mr Scheiner [Julius Scheiner (1858–1913) of the Potsdam Astrophysikalisches Observatorium, who also was heavily involved in the Carte du Ciel] thinks it is hardly imaginable. It remains to prove this irrefutably by experimentation.'

Kapteyn mentioned that he had asked Gill to take some special plates on which the Carte du Ciel reseau was exposed over intervals of a few months, in order to investigate the stability of the photographic emulsion. He went on to outline the sequence of observations (see Fig. 7.24) for the parallax measurements. The horizontal lines of images correspond to the first observation when the parallax was maximum (1), two separate exposures 6 months later (2 and 3), and the final one (4) another 6 months later, always with the position of the telescope displaced by the same amount in right ascension. In some cases, to increase accuracy, at each instant the exposures were repeated (a, b and c) by shifting the telescope by a certain amount in declination. Not one plate per position on the sky was taken, but rather 4 or 5, partly to increase accuracy by independent measurement and to check the stability of plates, but also in case something happened to the plates while they were stored for months.

On August 18, 1891 Donner replied (see Fig. 7.25):

'In summer, when the bright sky in the north [meaning northern latitudes] inhibits the astronomer, and especially the photographic astronomer, to observe and therefore gives a few months of vacation, I have been on a number of trips. Unfortunately your friendly letter arrived here during one of such trips and has been forwarded so that I only recently had the pleasure of receiving it. I am afraid that maybe this lack of reply, for which in view of the reasons mentioned I offer my apologies, has made you so impatient, that you turned to someone else. I hope that from our discussions in Paris you noticed that I was very interested in the plan you proposed, and that you will understand from this that it was not for lack of interest but some other reason caused the delay.'

He promised to take the plates as soon as possible and this marked the start of a long and fruitful collaboration.

7.11 Parallax Again

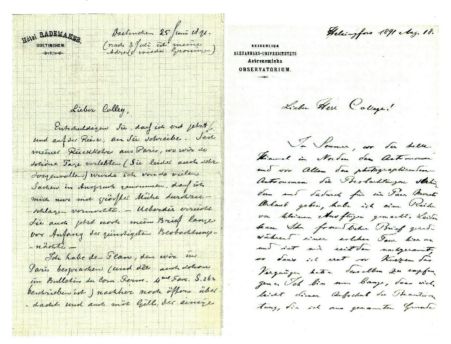

Fig. 7.25 Left: The first page of Kapteyn's letter to Donner of June 25, 1891. Right: The first page of the letter Donner sent in reply on August 18 of the same year (Provided by the Central Archives of the University of Helsinki)

In all, Donner produced 21 plates centered on five stars that Kapteyn had chosen. These all have high declinations so that they could be photographed in these positions while the sky was dark. The first exposures were obtained between early November 1891 and March 1892. The whole set was completed by March 1893. The actual publication of the project did not take place until 1900. Kapteyn had decided that, now that he had an laboratory, he would start a series of publications, as all major establishments did. He had apparently succeeded in persuading the University to provide the necessary funds. This series was launched with the publication of the results of this work as no 1 of the *Publications of the Astronomical Laboratory at Groningen* Kapteyn & Donner (1900). The title page of this publication has been reproduced in Fig. 7.27. The series was used for publication of extensive material (including complete PhD theses in some cases) produced at the Laboratory, and was only terminated in 1953 with the publication of No 55. In contrast to Leiden Observatory, where the title of the 'Annalen' was in Dutch and the publications (certainly up to this point in time) were usually in German, Kapteyn decided that both the title of the Series and the text of the papers should be in English.

The reduction of the plate material like this proceeds as follows: For all stars in the field one measures the distances between positions 1 and 2 and between 3 and 4. This can then be turned into a large set of equations, of which the solutions are the parallaxes of all stars. [For mathematically inclined readers: if these distances

Fig. 7.26 Helsingfors Observatory in the *Album Amicorum* presented at the retirement of H.G. van de Sande Bakhuyzen as professor of astronomy and director of Leiden Observatory in 1908 (Archives Leiden Observatory; see caption Fig. 3.7)

for each star are p and q and the time intervals t_1 between 1 and 2 and t_2 between 3 and 4, Kapteyn calculated the properties $v = (t_2/t_1)p - q$, which are independent of proper motion and depend only on parallax. Now various effects, such as differential diffraction and aberration, changes in tilt and orientation of the plates, will result in corrections, depending in a predictable manner on the position of each star on the plates. For n stars, this gives n equations with n unknowns.] Kapteyn gave a long and involved discussion on how to solve this set of equations, on how to correct for systematic errors in the observations and in the measuring procedures, the details of which are of no relevance in this context.

Kapteyn explained in his introduction that although Donner was quick to provide the plates, the measurement took much longer than anticipated. First he measured some plates (centered in the star 61 Cygni, which is one the first stars in which a parallax was measured; see page 139) using an instrument in Leiden, but he was not satisfied with that. Only after he was given funds by the government to purchase his own measuring apparatus and after he had installed it in the laboratory of Prof. Haga, measurements could start, in September 1895. Note that this contradicts the statements by Henriette Hertzsprung-Kapteyn in her biography, that the instrument did *not* go to Haga's laboratory. Kapteyn: 'Even then could we only devote to it the time that could be spared from the work connected to the Durchmusterung and some investigations in progress, so that both the measuring and the reduction advanced very slowly indeed.' One set of plates was rejected, and for one other region which was very rich in stars, three of the four plates were measured in detail (the fourth was rejected). Donner provided a new set of plates, centered on the Hyades cluster (see page 384), which is one of the nearest star clusters at about 48 parsecs (roughly 150 light years). However, 'seeing that the progress of the work was perpetually interrupted or rather that, what with the very small

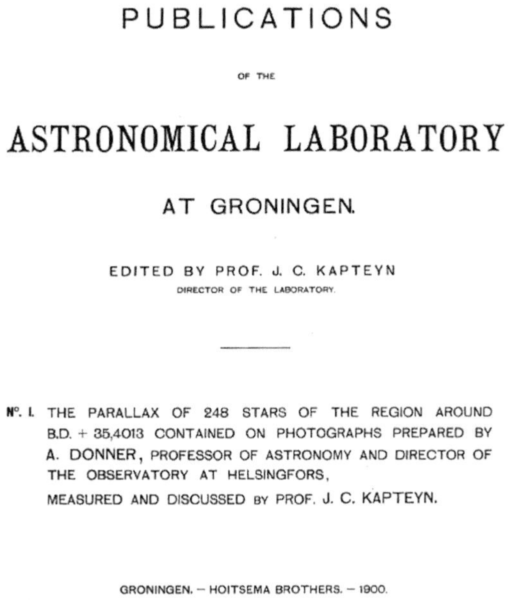

Fig. 7.27 Title page of the first volume in series *Publications of the Astronomical Laboratory at Groningen*, published in 1900 (Kapteyn Astronomical Institute)

PUBLICATIONS

OF THE

ASTRONOMICAL LABORATORY

AT GRONINGEN.

EDITED BY PROF. J. C. KAPTEYN
DIRECTOR OF THE LABORATORY.

N°. I. THE PARALLAX OF 248 STARS OF THE REGION AROUND B.D. + 35,4013 CONTAINED ON PHOTOGRAPHS PREPARED BY A. DONNER, PROFESSOR OF ASTRONOMY AND DIRECTOR OF THE OBSERVATORY AT HELSINGFORS, MEASURED AND DISCUSSED BY PROF. J. C. KAPTEYN.

GRONINGEN. – HOITSEMA BROTHERS. – 1900.

resources at my disposal, the continued prosecution of any work other than the Photographic Durchmusterung was nearly impossible, I resolved to stop further measurement altogether'. But although no new measurements had been done by 1900, he felt he needed to publish the results that he had, because of interest in the method elsewhere.

The only finished reduction, therefore, was on three plates centered on DM+35°, 4013, designated as 'Region of the Wolf-Rayet stars'. This region was given that name because it was the region where in 1867 Charles Joseph Étienne Wolf (1827–1918) and Georges Antoine Pons Rayet (1839–1906) of Paris Observatory found three stars with unusual spectra. These had spectral lines in emission, whereas in ordinary stars there are only absorption lines. They are massive stars in the latest stages of their lives, when they are very bright and, as we know now, are losing their outer regions in strong stellar winds. Although the discussion in the publication, Kapteyn & Donner (1900), was restricted to these plates, there was some mention of the parallax of 61 Cygni also. The actual publication had been preceded by preliminary discussions on the method at the fourth 'Nederlandsch Natuur- en Geneeskundig Congres', a national organization dedicated to physical and medical sciences (see page 284), which was held in Groningen, Kapteyn (1893d). and Kapteyn published in addition a preliminary list of results in the Astronomische Nachrichten, Kapteyn (1898), under the title *Determination of 250 parallaxes*.

Fig. 7.28 Table 15 from Kapteyn & Donner (1900) (Kapteyn Astronomical Institute)

TABLE 15.

	Theor. numb.	Observ.	Obs. — comp.
$< -0''105$	0	0	0
$-0''105$ to $-0''085$	1	0	-1
$-.085$ „ $-.065$	3^5	6	$+2^5$
$-.065$ „ $-.045$	12	15	$+3$
$-.045$ „ $-.025$	27	36	$+9$
$-.025$ „ $.000$	55	49	-6
$.000$ „ $+.025$	55	47	-8
$+.025$ „ $+.045$	27	20	-7
$+.045$ „ $+.065$	12	8	-4
$+.065$ „ $+.085$	3^5	9	$+5^5$
$+.085$ „ $+.105$	1	4	$+3$
$> +0''105$	0	2	$+2$

The aim of the publication was not 'so much to derive a certain number of parallaxes as to advance the method'. After many pages of discussion, Kapteyn concluded that the accuracy of the parallaxes was 0''.035 from a single plate and 0''.020 from a combined measurement on three plates. From the 196 stars that he felt delivered the best results, he found only a dozen stars or so with measured parallax (see Fig. 7.28). The left column is a range of formally determined parallaxes, which can be negative due to observational error; next the number expected from his analysis of the accuracies; then the actual observations, and on the right the difference between observed and predicted. The three numbers at the bottom show the excess due to 'determined parallaxes'. Kapteyn (1898) actually added that the parallax of the two components of 61 Cygni (a binary star with a period of about 722 years) is 0''.30±0''.031 and 0''.36±0''.034. The current adopted parallax of the system is 0''.28588±0''.00054.

7.12 A Durchmusterung for Parallax

Kapteyn came to the conclusion that parallaxes could be measured with this method with an accuracy of 0.''020, but also that the number of stars having such a large parallax (or small distance from us) was still small compared to the number of stars measured. The aim of the investigation, however, was to 'prove by experiment the truth of the thesis presented by myself as probable in 1889 (i.e., in the papers

7.12 A Durchmusterung for Parallax

Kapteyn (1889a) and Kapteyn (1891a) on the determination of parallaxes by timing measurements), 'that a systematic determination of the parallaxes of all the stars, perhaps even all those of magnitude 8.5 or 9, with a probable error of $\pm 0\rlap{.}''05$ for every individual value, is not any longer to be considered a task exceeding the powers of modern Astronomy.' Now that Kapteyn felt he had proved this, he proceeded to discuss in detail the feasibility of such a general 'Durchmusterung for Parallax'.

First the telescopes to be used. 'It will be safe [...] to assume the work was executed with such telescopes as that which furnished the plates discussed in this paper. Already the time is not far off that many of these telescopes will have finished their tasks for the 'Carte du Ciel' and that they may be used for other purposes.' Exposure times were to the order of one to one-and-a-half minutes, and a complete plate would then take an effort at the telescope of about one hour's work. The corresponding labor 'will slightly exceed that demanded for the *catalogue* plates for the *Carte d Ciel*'. Kapteyn did not go into further detail, but his assumption that the international community could be rallied for another such undertaking sounds unrealistically hopeful and confident.

Kapteyn was equally optimistic about the rest of the work required for such a large-scale Durchmusterung: 'The *measures* and the *reduction* will demand a labour very probably not exceeding that required for the measures of the catalogue-series and their reduction to *corrected rectangular coordinates*. So that finally we may come in the possession of the parallaxes of 800,000 stars, with a probable error of $\pm 0\rlap{.}''025$ for each value, investing an effort comparable to what is involved in the work now jointly undertaken by a series of observatories, on behalf of the *Carte du Ciel*.' This sounds to the modern ear as totally unrealistic, as the history of the Carte du Ciel would seem to rule out another of such an undertaking. Yet, Kapteyn proceeded to discuss what one could expect to learn from a Parallax Durchmusterung.

He was certainly honest. There was every reason to expect that *mean* parallaxes of groups of stars might be much easier to obtain, such as for 'stars of different magnitude, different proper motion, different spectral type, of clusters, of stars at different galactic latitudes, etc.' As to the parallax of individual stars, Kapteyn gave the results of unpublished calculations that ultimately showed that he could expect 658 stars to have a measured parallax exceeding $0\rlap{.}''05$, but 74% would indeed have such a parallax in reality. Separating the real and spurious parallaxes might require heliometer or meridian circle observations. Still, the conclusion was that only for a very small fraction of stars in the sky the parallax was large enough to be measurable with current techniques.

Finally he asked: '*Is there no shorter way to obtain results of equal value?*'. He rejected the option of restricting surveys for parallaxes to the brighter stars only, stating that better prospects were offered by restricting oneself to 'stars of great proper motion'. But this suffered from the fact that there was no completeness yet in a census of stars of large proper motion. The paper ended on an ambivalent note. On the one hand, Kapteyn urged for a parallax Durchmusterung, although he seemed unconvinced that it was realistic to expect that this could be organized. Between the lines he expressed the view that although more individual parallaxes would be determined in the near future, the realistic expectation was that determining mean distances of groups of stars was the more fruitful course of action.

7.13 Enter de Sitter

What happened to the other plates Donner took? They were indeed measured, reduced and published, albeit not by Kapteyn himself, but by his assistant Willem de Sitter (1872–1934) (see Fig. 7.29). De Sitter had been born in the Frisian city of Sneek, in a family that had produced many lawyers, and had come to Groningen to study mathematics. During his undergraduate studies he measured photographic plates in Kapteyn's laboratory and also followed his lectures. Then he met Gill, during the latter's third visit to Groningen. The circumstances of this event were described in a well-known obituary of de Sitter written in 1935 by Harald Spencer Jones (1890–1960) [33], then Astronomer Royal. It was in fact based on a letter that de Sitter wrote to Gill (on May 12, 1911), apparently in reply to a request to record the story for the history of Cape Observatory that Gill was writing at that time. This letter is kept as part of a collection of Gill letters at the Royal Geographical Society in London (transcriptions of Gill's letters are being prepared by Mr Paul Haley of the Share Initiative of Hereford, UK [34] and kindly made available to me). The relevant part reads as follows:

'My dear Sir David,
It is only today that I find time to reply to your letter. I have not been able to hunt up the exact date of your visit to Groningen, when you and I first met. In must have been in the autumn of 1896. I was then a student at Groningen, and was working in Kapteyn's laboratory, making the measures which we discussed in Gron. Publ. 2 & 3. You came to pay a visit to Kapteyn. Although I cannot hunt up the precise date, I remember the circumstances very well. One afternoon, you came with Kapteyn to look at the plates and the measuring microscope at which I was working. I tried to speak a few words of English to you, but I am afraid you never understood what I intended to say! The next morning I was having breakfast in my rooms, when a message came from the laboratory that you wanted to speak to me. I remember quite well how you sat down in Kapteyn's room – he was lecturing in the lecture room – and I stood listening, trying to understand. Then at last Mrs Kapteyn came in and acted as an interpreter. You offered me to come to the Cape as a computer, and thereby complete my astronomical education – or rather to begin it, for up to that time I had never made a specialty of astronomy, and intended to become a mathematician. It was agreed, after consulting my parents, that I should first pass my examinations preparatory for the doctor's degree, and then come out to the Cape. I reached the Cape on 1897 Aug 27 and left it on 1899 Dec 6.

7.13 Enter de Sitter

I came with the intention of making photographic parallax observations with the McClean telescope, and brought with me a complete working programme for these observations. The telescope was, however, as you know, not complete in time. Beyond a very few occasional observations of meteors, comets, occultations, etc. the observational work I did at the Cape consisted in the photometric work, described in Gron. Publ. 12, and the Heliometer observations: parallax of four stars, observations of red stars (appendix to parallax Cape VIII, 2) and a part of the polar triangulations. The reduction of the parallax observations was entrusted to me in my function as a computer. And finally I began under your auspices my work on the satellites of Jupiter. Papers relating to my work at the Cape are: [and here follows a list of these publications].
I hope these few informations will be sufficient. I was so glad to hear that Lady Gill and yourself were benefiting by your stay in Wales. I hope the fine air of Scotland will also do you good. With many kind remembrances to you both from my wife and myself.
Ever yours,

W. de Sitter

P.S. After having written this letter I have succeeded in finding out the date of our first meeting at Groningen. It is 1896 October 2.'

This meeting between Gill and de Sitter can be seen as a crucial event in de Sitter's life and career, but also in Dutch astronomy. Had Gill not impulsively invited de Sitter to the Cape, the latter would probably have become a mathematician and the course of Dutch astronomy (and physics!) might have been completely different.

In his chapter on *Kapteyn and South Africa* in the *Legacy*, Michael Feast quoted from a letter from Gill to Bryan Cookson (1874–1909), who was to come to the Cape and who eventually became an astronomer at Cambridge. I also include Feast's introduction to Gill's words:

'The training that Gill thought suitable for a young astronomer is given in a letter he wrote to Bryan Cookson at that time [see Forbes, G., *David Gill, Man and Astronomer*, 1916, John Murray, London, p. 234]. After telling him that there was nowhere suitable to study in England and that he should come to the Cape, he goes on:
'I have a very nice young fellow here, de Sitter, a young Dutchman who has passed his PhD examinations [Here Gill was mistaken. De Sitter came to South Africa to work on his PhD thesis, which he completed only after returning to the Netherlands. De Sitter had completed his 'doctoraal' (Masters degree) before going to Cape Town.] in pure mathematics at Groningen <u>cum laude</u> and has come out to learn practical astronomy. He is engaged from 9 to 3 just now in reducing my Heliometer observations for stellar parallax at a table near me. At night he is learning the use of the Geodetic Theodolite and Transit Circle. From these he will go to the Heliometer – then to the Equatorial with the filar micrometer, the pho-

Fig. 7.29 Portrait of Willem de Sitter as he appears in the *Album Amicorum* presented at the retirement of H.G. van de Sande Bakhuyzen as professor of astronomy and director of Leiden Observatory in 1908 (Archives Leiden Observatory; see caption Fig. 3.7)

tometer and the spectroscope, and before he returns to Holland – some two years hence – will have done some independent work of his own.'

De Sitter made extensive studies of stellar astronomy at the Cape. He published a paper (*On the use of electric light for the artificial star of a Zöllner photometer* in 1899 [35] and was involved in heliometric measurements of parallaxes (*Researches on stellar parallax made with the Cape heliometer* by D. Gill, W.H. Finlay, W. de Sitter & V.A. Lowinger (1900) [36]). Figure 7.30 shows a group picture of the persons at the Cape Observatory in 1897.

'De Sitter took up, particularly, a discussion of the heliometer observations of Jupiter's satellites made by Gill and his assistant W.H. Finlay […]. As a result the study of the satellites of Jupiter became a major part of de Sitter's scientific work, extending over the next thirty years and not entirely completed at the time of his death. De Sitter's time at the Cape was fruitful in two other ways; his friendship with Innes […] and the fact that he met for the first time the lady who was to become his wife.'

De Sitter married Eleonora Suermondt (1870–unknown) in 1898. After returning to Groningen, where he was appointed assistant to Kapteyn, he defended his PhD thesis *Discussion of heliometer observations of Jupiter's satellites made by Sir David Gill, K.C.B. and W.H. Finlay* in 1901 [37]. The satellite system referred to here consists of only the four major ones, Io, Europa, Ganymede and Callisto, which are similar in mass to our Moon and had been discovered by Galileo Galilei in 1610. The Galilean satellite system of Jupiter remained of major interest to de Sitter throughout his life. He eventually published a comprehensive study of its structure. An interesting account of this is in his George Darwin Lecture *Jupiter's*

7.13 Enter de Sitter

Fig. 7.30 The staff of the Royal Observatory, Cape of Good Hope, in 1897, photographed in front of the main building of the Observatory. De Sitter is on the extreme right; next to him is Gill and next to Gill and slightly behind him (in a cap) is Innes (From Michael Feast in the *Legacy*; South African Astronomical Observatory, Cape Town, archives)

Galilean satellites [38]. He ascribed the stability to the peculiar characteristics of the inner three satellites that dominate it. These are very close to a system of three bodies in circular orbits and in one orbital plane in exact resonance. The periods of these satellites around Jupiter are in a ratio of 1 : 2 : 4 (Ganymede, Europa and Io respectively), so that the mutual attractions and distortions of the orbits are far from random but return systematically in time and result in orbits locked in resonances. This provides a long-term stability to the whole system. The actual situation is complicated by the effects of the somewhat different planes and shapes of the orbits of the satellites and the influence of the fourth moon, Callisto, which is more or less in a 3 : 8 resonance with Ganymede (the outer one of the major three).

One of the assignments given to de Sitter by Kapteyn was to further reduce the plate material that Donner took. The results were published in the *Publications of the Astronomical Laboratory*, Kapteyn, de Sitter & Donner (1902), and are easy to summarize. Some plates were centered on one of the double clusters h and χ Persei (see Fig. 7.31). These are two rather young clusters (aged only about 3 and 5 million years respectively) in the Milky Way disk, at a distance of the order of 2.3 kiloparsec (some 7500 light years). The parallaxes of their stars are less than one-thousandth of an arcsec and could not be measured by de Sitter and Kapteyn. For the star Groombridge 745, on which some other plates were centered, they found a parallax of $0''.083 \pm 0''.024$. The designation 'Groombridge' refers to the British astronomer Stephen Groombridge (1755–1832), who compiled a list of circumpolar stars (as seen from England), which was published posthumously in 1838. The star is now known as HD24451, an 8th magnitude high proper motion (0.6 arcsec/year)

Fig. 7.31 The double cluster h and χ Persei (National Optical Astronomy Observatory [39])

star with a parallax of $0\rlap{.}''06154\pm0\rlap{.}''00070$. For 61 Cygni the parallax is in the end found to be $0\rlap{.}''326\pm0\rlap{.}''035$. It is now known to be a binary with a system parallax of $0\rlap{.}''285888\pm0\rlap{.}''00054$.

7.14 Statistical Parallaxes

At this point in time, Kapteyn seems to have given up trying to measure parallaxes directly on a grand scale. Stellar parallaxes, possibly with the exception of a small number of nearby stars, were simply too small to measure accurately. Even concentrating on stars with large proper motions produced just a few reliable distances. He concluded that the motion in space of the Solar System provided a much better baseline than the diameter of the Earth's orbit did. That baseline reflects itself in proper motions and it increases with time. That is an advantage, but then the problem, as we will see in more detail in the next chapter, is that it is only possible to use proper motions to provide parallaxes on a statistical basis.

With the help of Donner, Kapteyn set up a program to measure proper motions of a large number of stars, allowing for the option to measure direct parallaxes of that same sample of stars. To achieve this, Donner took a number of plates of some areas in the sky around 1897/8, and repeated this on the same fields around 1904/6. He complemented this, however, with plates of the same field to measure parallaxes directly. This constituted Kapteyn's last attempt to provide parallaxes on a wholesale basis. Again de Sitter was asked to supervise the measurement of these plates and reduce and discuss the results. Much of the measuring was entrusted, as usual, to the faithful member of staff T.W. de Vries. The result was published in the Groningen Publications in 1908, separately for the proper motions, Kapteyn, de Sitter & Donner (1908a), and for the parallaxes, Kapteyn, de Sitter & Donner

7.14 Statistical Parallaxes

(1908b). The resulting parallaxes were very accurate indeed, considering the difficulty of measurement, and de Sitter and Kapteyn concluded that whenever three sets of images were available, the final parallax had a probable error of as small as 0″.023. A major accomplishment, but a quick count from the published tables shows that exactly 100 of 3650 stars have a measured parallax in excess of 0″.1 and should be reliable. However, this is quickly qualified when another count is made, which shows that 68 stars have a measured, but spurious, parallax less than -0″.1. However, the authors did not draw attention to this. The paper concluded with the presentation of tables useful for others to reduce such measurements; e.g., parallaxes are most easily measured in right ascension, and depending on the circumstances of the observations of the photographic plates these should be corrected to the full value in the direction of astronomical or ecliptic longitude. Tables to do this quickly were provided. In an earlier publication in the Groningen Series (*Tables for photographic parallax-observations* [40]), de Sitter had actually provided tables that could be used to select stars suitable for parallax measurements at a given geographical latitude and the times at which plates then should be taken. This would facilitate other observers in planning such observations; see de Sitter (1906).

Presumably, at the time of the publication in 1900 of the paper reviewed above, Kapteyn had already given up on wholesale measurement of stellar parallaxes and had extensively involved himself in systematic studies of motions and colors of stars. This is in line with his statement at the opening of his laboratory, that in order to tackle the problem of the structure of the sidereal universe with any chance of success, positions and apparent brightnesses are not the useful properties by themselves; motions are the key to further understanding, while classifying stars on the basis of their spectra is an essential tool as well. Spectroscopy was beyond his reach; it required dedicated observations of individual stars in large numbers. But properties of the stars are also revealed in their colors. Having exhausted the issue of direct measurements of large numbers of distances to individual stars, Kapteyn turned to systematic studies of colors of stars, of distributions of stellar motions and *statistical* determinations of parallaxes on the basis of their proper motions.

Fig. 7.32 Kapteyn drawn by Cornelis Easton in 1906 (Kapteyn Astronomical Institute)

Chapter 8
Colors and Motions

For it is the duty of an astronomer to compose the history of the celestial motions through careful and expert study. Then he must conceive and devise the causes of these motions or hypotheses about them.
Andreas Osiander (1498–1552).[1]

'Messy', concluded Mr Charles C. M. Carlier. 'And sloppy!'
It all was, if you considered it carefully, really quite messy, sloppy actually, and unfinished – the creation and all that, the starry sky and all that, even though it was the summer sky of his beloved Provence.
'Typical rush job', said Mr Charles C. M. Carlier.
Havank (1904–1964).[2]

The Cape Photographic Durchmusterung appeared in print in its entirety in the course of the year 1900, the last of the nineteenth century. Upon entering the twentieth century on January 1, 1901, Kapteyn was getting on for fifty. He was recognized as an accomplished scientist, and at last found himself in a position to devote much time to the work he had always wanted to do, i.e., study the distribution of stars in space and the structure of the sidereal universe. As we have seen in the previous chapter, he published papers on results directly related to or resulting from his Durchmusterung work. However, it turned out that during most of the 1890s he performed research that was a prelude to his tackling the question of what Herschel had called the 'Construction of the Heavens'. Before discussing the work Kapteyn embarked upon when the completion of the CPD gave him time to address other astronomical issues, I will first have a look at his current private life. To this end, I quote from the *HHK biography*:

---

[1] Osiander supervised the printing of *De revolutionibus orbium coelestium* by Nicolaus Copernicus (1473–1543) and published this in an unsigned letter preceding the Preface.

[2] Hendrikus Frederikus (Hans) van der Kallen was a Dutch writer of detective stories. This quote is from *Lijk Halfstok* (1948).

Fig. 8.1 Jacoba, Kapteyn's oldest daughter, is the person on the right. She is photographed here with two classmates at the HBS highschool; the caps are 'Burgerschoolmutsen', caps worn at school as a variation of the caps that female university students used to wear (Universiteitsmuseum Utrecht and Dr. Inge de Wilde)

'In the meantime the children were growing up. The girls attended the 'Hoogere-Burgerschool' for boys and the Gymnasium, which was unusual in those days. Kapteyn did not for a moment consider making a difference between his daughters and his son, who also went to the HBS.'

The HBS, see page 62, for boys had apparently opened for girls as well at that time. The Kapteyn girls were not the only female students, as Fig. 8.1 shows. In the gymnasium, unlike the HBS, much of the curriculum was devoted to Greek and Latin. Since the son and the oldest girl attended the HBS, the younger daughter must be the one who went to the gymnasium.

'The same opportunities and the same rights for all, to him that was the way it was and the way it should be. The problems of males and females were of equal interest to him, he thought deeply about the future of his children and wanted to provide all that was necessary to attain what they felt was their mission in life. No course that he deemed useful for them to prepare them for later life was too much or too expensive for him. He and his wife gave up all luxury if need be. He was greatly interested in the schoolwork of the children, but he also made sure that it would not play too large a role in their lives. [...] When his oldest daughter came home one day from school and told him that she was number one of the class, for which she had worked very hard, his reply was: 'Don't do that to me again!' [...] He was also opposed to all sorts of competition and did not want his children to participate in them, pointing out the lack of depth and true value in applause.

As an example of the beauty of his wise fatherly love, which went far beyond the usual views of upbringing, the following anecdote shines as a light between all things forgotten from the past. One day a large bunch of grapes was on the table, which was a rare event in this simple family. The youngest daughter looked at it with delight and sighed: 'Oh, if only I could have the whole bunch for myself; then I would be perfectly happy!' The righteous pedagogue would have pointed out that this would be selfish and that everything had to be shared. However, he said: 'Well, in that case, my little child, then for once you shall experience the perfect happiness that is so rare'. [...]

But he was not interested exclusively in his own children, as many parents are; other children, too, enjoyed talking to him and discussing their things of interest. For example, there was a small girl, who had learned in school that the earth is moving around the Sun, which she found difficult to believe or understand. When she asked her father for an explanation, he said: 'Why don't you ask Prof. Kapteyn; he knows these things better than I do'. She actually went to see him and he explained everything in such a clear way that she returned home satisfied and convinced. Many parents came to Kapteyn to obtain his advice about how to raise their sons, which profession they should choose, etc., and they always went home with the right advice and warm consolation if there were difficulties. His interest concerned all that lived and flourished, equally for all. His children he taught to respect opinions of other people and not to criticize them immediately. He himself was the clearest and purest example of this and he lived for this every day. We were very surprised when one day we heard him criticize a colleague sharply. We were so used to his humane, understanding judgment, that this made a very deep impression. Such criticism is valuable and has its effect.

These are obviously remarks by a loving daughter who has nothing unfavorable to say about her father. The picture Henriette Hertzsprung-Kapteyn paints of him makes him larger than life, and in many respects he must have been no more interesting and special than any ordinary person.

8.1 Vries

At a bargain price he bought a second house with a large garden in Vries, in the northern part of the province of Drenthe, some fifteen kilometers from Groningen and this was where the family spent their summers. Mrs Kapteyn worked in the vegetable garden all day, which was a great joy to her, and the daughters did the housekeeping. Many guests came and went and everyone was welcome in this family where a simple and pleasant hospitality was offered. There was a big study, where anyone who felt so inclined, could work in the mornings, and Kapteyn himself also worked there in the morning as his work never left him alone. In the afternoon he took his bicycle, roaming around the neighborhood and getting to know it very well. He was always watching and listening with an open ear for what nature was presenting. He listened to the birds and knew their particular calls. It became a favorite pastime; his face would lighten up when he heard an unexpected call and he would silently move closer to observe the bird from nearby. Bird-trips, a full day with binoculars and an inexhaustible enthusiasm were a never-ending joy to him. He went over ditches and trenches and right through thick weeds, if he had some goal in mind. Prof. Boissevain, who often accompanied him on his trips, felt on one occasion that became a bit too much. 'But', he said later, 'I was running behind him and I thought; what he can do, should also be possible for me' and with such enthusiastic guidance the impossible became possible.

Figure 8.2 shows a map of Vries and surroundings around 1868. The thick, slightly curved line down the middle to the right of Vries, is a major canal called 'Willemskanaal'. To the east lies Tynaarlo and further to the right the railroad to Groningen. Vries is located about 15 kilometers directly south of Groningen. At that time it was a village with approximately 3,000 inhabitants and could be reached easily by train; there was a station 'Vries-Zuidlaren' directly to the east, just beyond Tynaarlo.

The Kapteyns had their second home along the road between Vries and Tynaarlo, about 400 meters from the center of Vries. The Kapteyns bought the house, that had been built in 1876, in September 1899 for 5500 Guilders (165,000€ nowadays in terms of purchasing power) and sold it again in March 1909 (116,000€), according to records of the notary who drew up and kept the formal ownership certificates. The house still stands today. It is known as 'De Burcht' (The Castle), but even the present owners have not been able to ascertain how or why this name originated; in formal documents it first appeared in 1947. The current address is Tynaarlosestraat 58, but at the time the Kapteyns owned it, it was 'wijk A 22a'. The Kapteyns held on to their formal domicile in Groningen, except for a short period between May 31, 1906 and October 19, 1906. Presumably this was during their removal from their home on Heerestraat to Eemskanaal ZZ(=South Side) 16a, later renamed Oosterhaven ZZ 16a (see Fig. 4.7). It appears from correspondence and the *HHK biography* that the Kapteyns usually lived in Vries during the summer months.'

'Kapteyn would give anything to be allowed to watch this feast of color and motion, with full abandon: getting up early in the morning, making long trips over unpaved roads, with endless patience – and he came home with a very special glow in his eyes and with growing amazement over and respect for nature.

Also stones started to tell him their stories. His long geology hammer became a trusted companion during his wanderings over heather and sand drifts. Megalithic tombs and grooves, earlier dead and meaningless to him, now came to life and told him stories of old times and the wonders of nature. The children sometimes became very impatient when during their bicycle trips together their father got off at every heap of stones, [. . .],

8.1 Vries

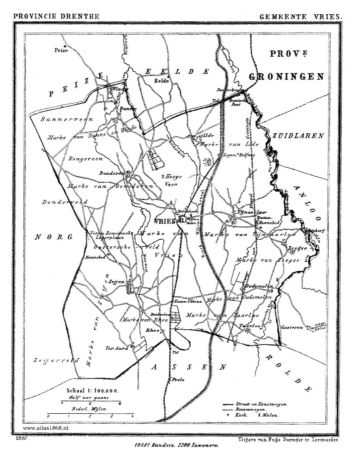

Fig. 8.2 Map of Vries around 1868. The 'Nederlandse mijl' (Dutch mile) in the bottom indicator equals 1 kilometer. The map covers about 22 × 15 km. The Kapteyn home was located roughly halfway the center of Vries (indicated by the name) and the canal that runs with a long curve just to the east of Vries itself. The eastward road passes though the village of Tynaarlo and just beyond that crosses the railway, where a station was in use at the time (Gemeente Atlas Jacob Kuyper (1865–1970) [1])

but his enthusiasm was getting a hold on them too, and after some time they searched and hammered along, shouting out when they found an interesting stone that could be added to the collection. Geology books were read and discussed, bringing up new ideas. Kapteyn discovered a few elevations near Vries, which in their regularity made the impression of having been made by men. From the owner of the land he purchased a plot with one of the mounds for a few guilders, and the Kapteyn family and friends, all armed with pickaxes and shovels, went to try and uncover the secrets of the mound. They dug and worked, but nothing of relevance was ever discovered. The diligence started to wane and in the end the work

Fig. 8.3 'De Burcht', the house in the village of Vries where the Kapteyns spent their summers (Photograph by the author)

was abandoned, but their interest in geology never abated. [...] He asked Prof. Molengraaff, the geologist from Delft, who undertook excursions with his students every year, if he could join in these trips and this was gladly granted. So, a few times he went along to far-away places as if he were the youngest and most enthusiastic student.

Through his modest friendliness he immediately pulled all young men to him, was untiring, and during long marches, his spirit, with only a few of the others, remained high.' This is what Prof. Brouwer wrote after participating in one of those trips. And: 'He often asked the leader of the trip for a further explanation, which showed his interest in the general findings of geology, and since we would forget the main theme for all the details, it was also very instructive to listen to those exchanges. I remember vividly that during the dinner on the last day of the trip, when various speeches were made as was customary, Kapteyn turned to the students and emphasized that they should never be too serious about things. He said it rather casually, but everyone felt that here was someone expressing himself exactly the way he was and it made a lasting impression.'

Gustaaf Adolf Frederik Molengraaff (1860–1942) was a professor of geology at the current technical university of Delft and Hendrik Albertus Brouwer (1996–1973) was a professor of geology and paleontology, also at Delft at first and later at the University of Amsterdam.

8.2 Visits by Gill, Donner and Newcomb

In the astronomical world his star was rising rapidly. With his work on the CPD, his participation in the planning of the Carte du Ciel, and his other publications, Kapteyn had undeniably established himself as an important contributor to astronomy. Astronomers began to visit him in Groningen. During the 1890s Gill visited three times (April 1887, March 1891 and October 1896). Donner, it seems, was not a very avid traveler. In 1896, however, he too visited Kapteyn. Donner had taken some plates for the Hyades cluster (see page 384), and Kapteyn was anxious to discuss the project with him. On March 15, 1896, he wrote:

'I would be delighted if it would be possible to examine the results of these plates together. Unfortunately I cannot come to Paris. [This was for a meeting of the Permanent Committee of the Carte du Ciel] Would it be possible for you to travel via Groningen either before or after the congress. It will be a great pleasure for me and my wife to have you as a guest for a short time. You would then be able to see my laboratory and the measuring apparatus at leisure, tell my about your extensive experience with the Repsold machine and we can discuss during a few days what would take a month by letter. Even if I were to come to Paris, it would be more difficult to discuss our work than would be possible in the laboratory itself.' [...] 'Hopefully Gill will be here also for a few days, but when I do not yet know.'

Donner did indeed agree to come. On May 13, 1896, Kapteyn sent him the schedule for his train trip from Paris (after the Carte du Ciel congress) to Groningen. He was to depart from Paris at 8:10 (a.m.) and had to change trains in Rotterdam, Utrecht and Zwolle, arriving in Groningen at 10:08 in the evening, a trip of almost 14 hours (currently with a large part of the trip by high-speed rail, the duration is a little over $5\frac{1}{2}$ hours), a long time for those hating travel. Donner's visit started on May 18, but the correspondence gives no clue as to how long he stayed. In subsequent letters there is no reference to the visit, except that on June 22 Kapteyn thanked Donner for his 'portrait', presumably a picture of himself, and Kapteyn encloses one of himself in that letter.

Henriette Hertzsprung-Kapteyn refers only briefly to Donner's visits to Groningen: 'Donner visited Groningen a few times and their collaboration developed into a warm friendship, although their correspondence was infrequent, as they were both poor letter writers. Their letters usually started with an earnest lamentation and excuses for the long interval it had taken. Both had busy, hard-working lives from which their correspondence suffered.' On Gill she actually wrote two complete, albeit short chapters, and in view of the significance of her father's collaboration with Gill this comes as no surprise. However, it hardly did justice to the importance of Donner's plates taken for Kapteyn. To another important visitor she again devoted a full chapter, viz., Simon Newcomb (1835–1909).

Newcomb (see Fig. 8.4) and Kapteyn were not close collaborators, but shared many interests; in the 1890s, for example, they both worked on the solar motion in space and its use to measure distances of stars statistically. Newcomb, who was born in Canada, had been an undergraduate at Harvard, where he received a bachelor's degree. He joined the US Naval Observatory in Washington D.C., where he became director of the Nautical Almanac Office in 1877. In 1884 he was appointed professor at Johns Hopkins University in Baltimore. His interests were in positional astronomy, but he also concentrated much of his effort on planetary motions, speed of light and fundamental constants.

Kapteyn had corresponded with Newcomb on matters related to the distribution of stars in space and of their motions since 1897.The Kapteyn Institute has copies of letters between Kapteyn and Newcomb from the latter's archive at the Library of Congress; see the article by Petra van der Heijden in the *Legacy*. The correspondence seemed somewhat one-sided at first. Kapteyn sent at least two letters in 1897 on his publications on motions of stars, actually offering to 'gladly undertake to translate the paper on the distribution of the stellar velocities for you in such English as I can command'. The paper referred to could be Kapteyn (1895), but is more likely Kapteyn (1897a), a paper in Dutch on the distribution of cosmic velocities, which he presented for the National Academy in Amsterdam (see below). There is only a short note by Kapteyn, dated May 31, 1898, in which he congratulates Newcomb on his appointment as a foreign member of the Royal Netherlands Academy, but no reaction from Newcomb was forthcoming. There must have been one, however, since on January 9, 1899, Kapteyn wrote: 'My cordial thanks for your kind letter. Of course I will be only too happy if my work embodied in the MS sent to you two years ago, can be of any use to you.'

Kapteyn had not gone unnoticed by Newcomb. In May 1897, Newcomb delivered an address on the occasion of the dedication of the Flower Observatory of the University of Pennsylvania. Reese Wall Flower (?–1891) of Philadelphia was an amateur astronomer, who had left funds in his will (contested by his relatives, with some success) for an observatory; it no longer exists. Newcomb's address concerned *the Problems of Astronomy* [2] and he reviewed some the most pressing problems of the time, such as the extent of the universe in space and time. The problems with time were the difficulties of reconciling the age of the Sun, derived from estimating the time it can shine as a result of gravitational energy being released from contraction, compared to the geological timescale as it was known at the time. The first is the solar contraction timescale, now known as the Kelvin-Helmholtz contraction time, named after William Thomson, also known as Lord Kelvin, (1824–1907) and Hermann Ludwig Ferdinand von Helmholtz (1821–1894), which amounts to no more than twenty or thirty millions of years, whereas geological time scale rather covers hundreds of millions of years. This was only solved in 1938, when Hans Albrecht Bethe (1906–2005) proved that the Sun derives its energy from nuclear reactions in its central parts. The spacial dimension of the Universe is related to the measurement of parallax and proper motion, of which Newcomb – referring to the Carte du Ciel project – said: 'A photographic chart of the whole heavens is now being constructed by an association of observatories in some leading countries of the world. I cannot say all leading countries, because then we would have to exclude our own, which, unhappily, has taken no part in this work.' But then he mentions 'the new astronomy' using the spectrograph. To illustrate this he men-

8.2 Visits by Gill, Donner and Newcomb

tions Kapteyn: 'How useful it may become has been recently shown by a Dutch astronomer, who finds that the stars having one type of spectrum belong to the Milky Way, and are farther from us than the others.' This work of Kapteyn, which will be discussed in more detail later in this chapter, attracted Newcomb's attention, so he was very much aware of Kapteyn and his work.

But then in a letter in July, Kapteyn stated that 'my wife and I will be extremely pleased to have you in Groningen.' [...] 'I need certainly not tell you that the longer you stay with us, the happier we will be.' Mrs Kapteyn added a short note: 'Do let me tell you that it will be a great pride and pleasure for me, to have you for my guest' (signed Elise Kapteyn, while Kapteyn signs with J.C. Kapteyn). Particularly interesting is the note she wrote on 14 September 1899, when the visit was approaching: 'Dear Mr Newcomb, Thanks very much for your kind letter, which hardly wants any answering, as it is a matter of course that you shall be just the same welcome by me in whatever apparel you choose to make your appearance in. Were it not that I must tell you that we live in an upper story. I hope sincerely that this will not be any inconvenience to you. I am looking forward to your visit as a great treat, and I hope you will not find me too forward if I shall try to have as much of your company as possible. My husband has always been indulging me in this respect when we had our dear friend Gill as a visitor and so I hope will you. Respectfully yours, Elise Kapteyn Kalshoven.' The Kapteyns were planning to receive them at their apartment in Groningen rather than their summer home in Vries.

It seems that the visit was almost canceled, as Kapteyn wrote Newcomb on September 27 that 'I perfectly realize how very difficult it will be for you to come to Groningen'. And on October 6: 'Happy as I would have been to have you at Groningen I do not think it will be possible.' On October 13: 'I now realise how absolutely egoistical were my endeavours of meeting you here or elsewhere;' [...] 'and I can only express the hope that your health may promptly be restored and that now at some future time you will see occasion to indemnify me for what I had to miss now.'

In the end, however, Newcomb did come. Henriette Hertzsprung-Kapteyn wrote in the *HHK biography*:

'When his visit was announced, it caused Kapteyn some concern as he feared he did not have much of importance to show. When it seemed for a moment that the visit was going to be cancelled, Kapteyn wrote him: 'You will not see my instruments in this case, but this will only save you so much disappointment, as there is certainly no observatory in Europe, not to speak of America, so scantly equipped.'

Newcomb came to Groningen and turned out to be a remarkably stern man, which became manifest through his appearance and his economy of speech. Kapteyn was not much at ease with him and was sometimes upset with the visit, even though he appreciated it and felt it was informative and interesting for him. Not so Mrs Kapteyn. She had an easy, open manner of associating with dignitaries that was both very amusing and grand. She was not easily impressed by others, since she never thought much about

Fig. 8.4 Simon Newcomb as he appears in *the HHK biography*

the impression she would make, but was always keen to make her guests feel at ease and comfortable. She liked him immediately; his quietness did not bother her, since she had enough subjects for conversation herself and had a good command of English. 'He is a king among people', she used to say and her children noted to their surprise that he was allowed to put his tired, dusty feet up on the beautiful golden chairs that were the centerpieces of their mother's salon. She baked buckwheat cookies for him in the morning, because he liked them so much, and was the nicest hostess the stern American could imagine. The admiration came from both sides. For example, Newcomb wrote to Gill around that time: 'I have read a letter of Mrs Kapteyn', apologizing. 'It is hard to keep anything from so delightful a woman.' Gill wrote about this remark to Kapteyn, who was very pleased to hear it. Newcomb had a special dead-pan sense of humor that surprised the children in so imposing a man. He was tall and robust with thick and wavy white hair that was remarkably beautiful. He mentioned that his hair was his wife's pride. Later his name will be entered in the lexicon as 'the man who had the most beautiful head of hair; seems to have been an astronomer'.

8.2 Visits by Gill, Donner and Newcomb

Due to a neural disorder in his leg he was temporarily forced to use crutches, which made him even more interesting in the eyes of the children. He left them alone, which they only felt to be natural, and they behaved quietly and restrained in his commanding presence. Still he was shy and had difficulty expressing himself. His appreciation and interest he expressed in a strangely dry and yet to-the-point manner. In 1907 he wrote to Kapteyn: 'Next year I pass through the Hague on my return from Rome. If I do this, I hope you will be able to place yourself near my line of movement'. And later: 'I will spend a week or two at the Hague. Perhaps you and Mrs Kapteyn can also come to that region which I believe is very pleasant in the early autumn, when I shall probably be there.' Once he even wrote: 'I think it is nearly a year since I have heard from you personally. And I now write rather from a general desire to hear how you are doing than from having anything important to say.' Which from this very withdrawn man might be called remarkable.'

There is a delightful and amusing letter from Mrs Kapteyn to Newcomb that illustrates the different characters involved as well as the fact that she was much at ease with Newcomb, probably more than Kapteyn was. Newcomb had sent a book, which he must have promised to do during his visit to Groningen. It is referred to as 'Autocrat' and undoubtedly is the book *The Autocrat of the Breakfast–Table* by Oliver Wendell Holmes, Sr. (1809–1894), published in 1858 [3]. It is an amusing set of essays, recording conversations taking place during breakfast among a group of persons in a boarding house. It was first published as a series in a literary magazine, the 'The New–England Magazine'. Its witty, but at the same time serious, tone and the philosophical issues it addressed made it very popular, to such an extent that it was followed by two sequels, *The Professor at the Breakfast–Table* and *The Poet at the Breakfast–Table* [4].

'Groningen, 20 November 1899

Dear Mr Newcomb,

We were glad to hear that you had reached home safely. Your family will be pleased to have you back after so long an absence.

I thank you very much for the sending of the 'Autocrat' – Do you know that this present of yours has given rise to some gentle sparrings between my husband and myself. I thought you had promised the book to, or rather that you had told me you would order it for me. No, says Kapteyn, Mr 'Newcomb said he would send it to me'.

Well, I thought, the address may perhaps put that right. Alas, it was addressed to Prof. Kapteyn. So, as it more happens in this world of ours, the weaker sex had to give way to the stronger. When your letter came inquiring whether I had received the 'Autocrat', you can imagine the look of triumph that would came to my face. Well, Kapteyn wrote my name on the flyleaf and I thank you heartily for it.

Will you please tell Mrs Newcomb that I am in her debt for the photograph that Dody [nickname of the oldest Kapteyn daughter Jacoba] has not finished

yet. I will send it to her [Mrs Newcomb] with one of Kapteyn's which she would perhaps like to see.
With my sincere regards both for you and Mrs Newcomb, believe me dear sir,
Yours,
Elise Kapteyn Kalshoven'

Mrs Newcomb was Mary Caroline Hassler (1840–1921); she had married Simon Newcomb in 1863. Mrs Kapteyn sent Mrs Newcomb the picture 'of your husband that my eldest girl took' on March 15, 1900. In the fall of 1902 Mr and Mrs Newcomb visited the Kapteyns in their home at Vries. On September 5, Mrs Kapteyn (always signing her letters with Elise Kapteyn, whereas Kapteyn himself, rather formally, used his initials) wrote to Mrs Newcomb how 'very pleased and much honoured' she was that they would visit, telling her that 'a quarter of an hour by train and another quarter driving will take us there'. I presume that would be by taxi. On December 25 she wrote to Newcomb, thanking him again for sending a book (not stating a title this time, saying that she 'will try and read it and Kapteyn shall explain when I can't fully understand it') and sending her regards to 'your dear wife', who has stimulated the girls to come to concerts and play music. Mrs Kapteyn also mentioned that 'both [girls] have become engaged. The eldest with an assistant under whom she worked at the hospital the last three years, and who wants to become a surgeon afterwards. We are very pleased with both the young men. And in January Hetty shall pass her examination for Cadidate and after that it will be made formal'. And a little further: 'I am glad to be able to tell you that my husband's eyes are getting better and that he is in general good health.' In 1901 she had written out a letter of Kapteyn to Newcomb because he could not write himself as a result of severe pains in the eyes. The relation between the ladies Newcomb and Kapteyn remained very friendly throughout. As we will see later, the Kapteyns visited the Newcombs in Washington in 1904, when Kapteyn delivered a crucial presentation at the Louisiana Purchase Exposition (the fourteenth World Fair) at St. Louis, Missouri.

Gill returned to Groningen again for a visit in 1900. This was at a tense period in history regarding the relation between the Netherlands and Great Britain as a result of the 'Boer Wars', also called the South African Wars. The most violent period was between 1899 and 1902, but had been preceded by an earlier war around 1880. It was fought between Great Britain and two Boer republics – the South African Republic, also known as Transvaal Republic with the capital Pretoria, where it had started in 1880, and the Orange Free State of which Bloemfontein was the capital. The Boers, originally known as 'voortrekkers' or pioneers, were Calvinists and Huguenots, the majority from Dutch descent, and had settled in these two republics earlier in the century as a result of the 'Great Trek' away from the British rule and the drought in the Cape provinces. The two republics had gained independence from the British in the 1850s. The influx of other settlers after the 1886 Gold Rush at Witwatersrand made the political situation unstable. The British eventually defeated the Boers in the second war, after much bloodshed, and officially annexed the two provinces in

8.2 Visits by Gill, Donner and Newcomb

1902. The war was controversial both in Britain and in the Netherlands, because of the alleged (and later confirmed) 'scorched earth' policy of the British army and internment of civilians in concentration camps. Henriette Hertzsprung-Kapteyn refers to this visit by Gill as his 'third' visit, but Gill's letters indicate that his third visit to Groningen had already taken place in 1896.

'In 1900 Gill visited Groningen for the third time. He was welcomed as an old trusted friend. The Boer War that had outraged many among the Dutch, was in its last phase, and the British were not very popular in those days. Their greedy, unjust policies made them much hated by the Dutch, who enthusiastically took sides with their own descendants in this far-away country. The Kapteyn family decided not to bring up this delicate subject, as it would be uncomfortable to the English guest and so the situation was not discussed. Gill's visit was as nice and pleasant as always. His warm interest and spontaneous, lively spirit had a major influence on Kapteyn's work. [...] He had himself photographed with the two girls, who he called his 'lassies', teased and spoiled them and was the most admirable uncle they could imagine. One day Gill and Kapteyn visited the Physics laboratory in Groningen, where over each door a wise quote had been painted. One said: 'Wisdom is more valuable than rubies.' (Wijsheid is beter dan robijnen) 'I know what that means', Gill called out. 'Whiskey is better than red wine.' He enjoyed his own jokes as much as others did. Typical for Gill was his joking translation of the Latin saying 'Experientia docet'. 'You know: Experience does it.'

'I have obtained all my wisdom from Gill', Kapteyn used to say. In his letter to Gill on the occasion of the latter's 70th birthday he wrote: 'I think I picked up some of your great 'Lebensweisheit', of your capacity of making life a joy to yourself and to others.' Kapteyn's own wisdom certainly was no less; his inner life was a serene confidence of spirit. He radiated a still strength that could not be explained, since he was so modest and calm without pretension, contrary to what we see around us nowadays. One of his clerks, who had worked at his laboratory for twenty years, said of him: 'He had a very deep wisdom in life, which was often explained in such simple terms that many failed to appreciate it. But I was always aware of the wisdom in it. He gave stability in my life, and in difficult moments, when I thought of him, he was a great source of inspiration.' He told me much more about him and suddenly said: 'Maybe you feel it is disrespectful that I always speak about 'him'. But it isn't, he always was 'He' with a capital 'H'. [...]

After Gill had returned to the Cape the correspondence was continued. Kapteyn had always found it hard to force himself to write letters, his life being arduous, busy and full. In response to an urgent question of Gill, he answered: 'The fact is that I am and have always been a bad correspondent, wanting some direct stimulus for writing a letter. Add to this the feeling that I had better avoid writing on the subject of the disastrous war, a subject which as soon as we Dutchmen write to an Englishman will come uppermost in our mind, and you will see, how it is that I don't let you hear for ever so long.'

Gill was very much moved by this and replied: 'You certainly are a bad correspondent, but there is this about you, when you do write a letter, there is always something or rather there are many things well worth reading in it. Now first of all I really did not realize fully before, what a perfect gentleman you are — I did not realize till now that you feel so strongly about this miserable war. I only wonder, how you had the power to keep so completely away from the subject during the time I was with you in Holland. As your guest of course I did not open the subject,' — was he not just as well a perfect gentleman? — 'but I had felt as keenly as you do about it, I do not think, I could have refrained so perfectly as you did. I only feel about the war that we had to fight or to make up our minds to submit to a Boer Republic throughout South-Africa. But I have no feeling of animosity against Boers. Blood is thicker than water, so if you really want to let off steam about the war, you will find me an excellent safety valve, and don't let in the future the fear of speaking your mind interfere with your writing to me on that or any other subject. Indeed it would interest me greatly, if you do so.' So this friendship became stronger and more complete.'

8.3 Natural Sciences Society

Another activity of Kapteyn that must have taken much of his time over many years was his involvement with the 'Natuurkundig Genootschap' or Natural Sciences Society of Groningen. This is the current Royal Natural Sciences Society or KNG [5]. It is a learned society that was founded more than two centuries ago and it still leads a very active life. I have had the honor of being its chairman since 1998. In 1976, on the occasion of its 175th anniversary, Queen Juliana of the Netherlands bestowed on it the honorary title 'Royal' (so that it is now 'Koninklijk' Natuurkundig Genootschap or KNG). This was renewed in 2012. The celebration of its bicentennial in 2001 included a day-long symposium on 'Evolution from large to small', attended by 650 members and guests. This is not the place to present a comprehensive history of the Society, but some background information to appreciate Kapteyn's role in it seems called for.

It was the ever-increasing broad interest in science among the educated citizens and academicians during the last part of the seventeenth century and the first part of the eighteenth that gave rise to the society. Many other societies similar to the one in Groningen were founded in this period, their activities usually being centered on presentations concerning recent developments in scientific research, execution and repetition of experiments and the building up of collections of scientific specimens. This particular society in Groningen was founded on February 28, 1801 as an initiative of Theodorus van Swinderen (1784–1851). Van Swinderen was a

8.3 Natural Sciences Society

Fig. 8.5 Commemorative medal, issued on the occasion of the fiftieth anniversary of the 'Society for Natural Sciences at Groningen' (From *Spiegel der Wetenschap*)

law student, but became interested in natural science when he attended the lectures of Jacob Baart de la Faille, professor of mathematics, physics and astronomy at the University of Groningen (see Fig. 4.3). Van Swinderen eventually became professor of Natural History, which is a collective term for natural science, although usually referring to biology and sometimes also geological and paleontological sciences. Among the founders of the society were three more students of law, as well as the later prominent Groningen professor of chemistry, pharmacy and technology Sibrandus Stratingh (1785–1841). On the occasion of its fiftieth anniversary in 1851, a commemorative medal was issued (see Fig. 8.5).

The Natural Sciences Society flourished particularly when during the late nineteenth century, the city of Groningen and its university increased significantly in size (the number of inhabitants rose from 35,000 or so around 1850 to about 75,000 in 1910). The expenditure on education increased with the founding of the Hoogere Burgerschool (HBS) (see page 62). Not only university professors, but also teachers and headmasters of the Groningen HBS, played an important role in the activities of the Society. One important member was the teacher of physics and headmaster Florentius Goswin Groneman (1838–1929), who joined the board in 1867 and remained a member of the board, apart from a few years interlude, until 1907 (see Fig. 8.6). He served as chairman between 1895 and 1907, was the driving force behind the Society's favorable development and presided over the major celebration of the first centenary on March 1 and 2, 1901. The key-note celebratory speech was delivered by Rudolph Sicco Tjaden Modderman (1831–1924), a professor of organic, anorganic and pharmaceutical chemistry. In some references, in particular in the records of the Society, his third given name Tjaden was made part of his surname (so he was in fact R.S. Tjaden Modderman). His father, however, a well-known local lawyer and politician, was H.J.M. (Hendrik Jacob Herman) Modderman [6] (1796–1859). The centennial speech contained a detailed summary of the history of

Fig. 8.6 Florentius Goswinus Groneman, chairman of the Natural Sciences Society 1867–1882 and 1893–1903 (From *Spiegel der Wetenschap*)

the Society. The proceedings of this celebration with all the speeches was published by (Tjaden) Modderman *et al.* as *Het honderdjarig bestaan van het Natuurkundig Genootschap te Groningen, gevierd op 1 en 2 maart 1901* [7]).

Another account of the Society's history of the (by then 'Koninklijk') Natuurkundig Genootschap is part of a book published on the occasion of the *second* centenary in 2001. This historical sketch, written by Franck Smit, and introduction to a collection of descriptions of scientific progress during the two-hundred years since the founding of the Society (Kees Wiese (ed.): *Een Spiegel der Wetenschap: 200 jaar Koninklijk Natuurkundig Genootschap te Groningen* [8]). There is a third historical introduction – also in Dutch – in a book on science in the Netherlands around 1900 (Jan Guichelaar, George B. Huitema en Hylkje de Jong: *Zekerheden in Waarnemingen: Natuurwetenschappelijke ontwikkelingen in Nederland rond 1900* [9]). This is not the place to discuss the history of the KNG Society in much further detail.

Throughout his years in Groningen, Kapteyn gave much attention to the Natural Sciences Society. Other prominent Groningen scientists likewise made efforts on

8.3 Natural Sciences Society

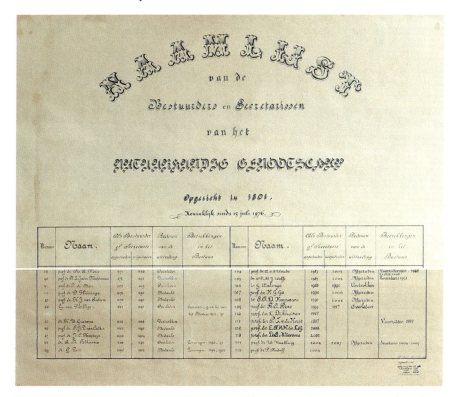

Fig. 8.7 Top and bottom part of the 'Name-list of board members and secretaries of the Natural Sciences Society' (Natuurkundig Genootschap), 'Founded in 1801, Royal since July 15, 1976'. Kapteyn is third from the bottom on the left. This list and a recent second one are on display in the Kapteynzaal (Lounge) in the Concerthuis, the home of the Society (Archives Koninklijk Natuurkundig Genootschap)

its behalf, e.g., and particularly, Hartog Jacob Hamburger (1859–1924), professor of physiology and histology (board member from 1905, chairman 1907–1924) and Gerardus Heymans (1857–1930), professor of philosophy and psychology (board member 1907–1915). From 1840 onwards the home of the Society has been the 'Concerthuis' on Poelestraat in the center of Groningen (see Figs. 8.8 and 8.9). Van Swinderen and his wife provided a substantial amount of cash, and it was bought for 16,000 Guilders (equivalent to some 58 yearly wages of an unskilled worker and a purchasing power of about 335,000€ today). In exchange the van Swinderens obtained a life annuity from the Society. In 1928, the board decided to name the room on the top floor (with the windows just above the name images in Fig. 8.9) 'Kapteyn Lounge' (Kapteyn Zaal), in recognition of Kapteyn's service and merits. This room should not be confused with the Kapteyn Room in the Kapteyn Astronomical Institute. On the walls of this Kapteyn Lounge hang a copy of Kapteyn's portrait (a photograph of the painting of Kapteyn in academic robe in the University's Senate Room (see Fig. 16.4) and a list of all board members (Fig. 8.7). One noteworthy

Fig. 8.8 The Concerthuis at Poelestraat 30 in Groningen, the home of the Koninklijk Natuurkundig Genootschap. Here we see a poster of around Kapteyn's time advertising it as a location for parties and to perform plays (Archives Koninklijk Natuurkundig Genootschap)

name is that of the secretary of the Society between 1870 and 1881 – which is around the time Kapteyn came to Groningen – one Dr. W. Gleuns Jr.; this was in fact the father of the person of the same name, who together with Kapteyn had been considered for the job of observator in Leiden. The elder Gleuns was a mathematician, teacher in secondary schools and author of many books for schools. He spoke regularly for the Society on many different subjects, but often on astronomy in which he appeared to have a special interest. In those days the secretary was not a member of the board.

I have mentioned above (page 104) that Kapteyn already gave a lecture for the society in 1878, not long after his arrival in Groningen. That presentation had a long title and seemed an attempt to cover almost all aspects of astronomy: 'About the relation between different stellar systems and particularly the clusters of stars, the binary and multiple stars, the Milky Way system and the nebulae' (November 28, 1878). Kapteyn delivered more lectures, e.g., on 'The history of the comets' (Febr. 7, 1881), and on 'Astronomical accuracy' (Febr. 23, 1882). On January 24, 1884, just before his lecture on 'Shooting stars' (held on Febr. 21, 1884), Kapteyn was nominated to be an honorary member of the Society as a 'regular speaker'. He must have been a regular member prior to this, since he gave his first presentation shortly after he arrived in Groningen. He became a member of the Board in 1886;

Fig. 8.9 The Concerthuis at Poelestraat 30 in Groningen in 2010 (Archives Koninklijk Natuurkundig Genootschap)

the physiologist Dirk Huizinga (see Fig. 8.10), who was a close friend and who gave Kapteyn working space in his laboratory to do his work on the CPD, had been a board member since 1876, and may have been a party to this. Kapteyn remained a regular, though not too frequent (compared to some others) speaker: 'The problem of weighing the Earth' (Febr. 17, 1887), 'Determining one's geographical position at sea' (Oct. 20, 1887), 'The various planetary systems' (Jan. 22, 1889), 'The binary stars' (March 3, 1891), 'How the heavenly bodies are being moved' (Oct. 13, 1993), 'The history of the development of our understanding of the structure of the solar system' (three lectures on Jan. 14/21/28, 1996).

The Natural Sciences Society reorganized itself in the time that Kapteyn was a board member. In 1897 it instituted the 'central bureau for the knowledge of the province of Groningen and surroundings' to collect and discuss information on the city and province of Groningen and the surrounding region. This bureau flourished under the chairmanship of Pieter Roelf (Roelof) Bos (1847–1904), a teacher of geography at the Groningen HBS and the author of the famous 'Schoolatlas der geheele Aarde', the atlas of the world that he developed for the teaching in secondary schools in 1877. Later editions of this 'Bos-Atlas' are still being used in many schools in the Netherlands. In addition to this activity, the chemist Johan Frederik Eijkman, (1851–1915) proposed to found a 'scientific chapter' to start a

Fig. 8.10 Dirk Huizinga was a professor of physiology at the University of Groningen from 1870 to 1901 (University Museum Groningen [10])

regular series of lectures (two, on every evening of the first Saturday of the winter months) to stimulate contacts among scientists in Groningen and discuss scientific issues. Eventually this became the major activity of the Society and its most successful department, the other activities slowly fading into the background. Basically, scientific lectures are essentially the sole function of the KNG at present. The major driving force behind the success of the scientific chapter would prove to be Kapteyn.

This came about as follows: In 1900, the Society adopted the formal bylaws of the scientific chapter, but their definition of the relation between the Society and this department was not quite what Eijkman had envisaged. He resigned as a member of the chapter's board (and even as a member of the Society), and Kapteyn took his place. The reports do not elaborate on the reason of Eijkman's action, but the most plausible is that Eijkman wanted more autonomy for his department. The new bylaws required that the board of the scientific chapter had at least two members of the Society's board, and Eijkman was not a member of that body. Eijkman probably wanted to organize meetings with in-depth scientific discussions, in any case more profound and less superficial than the ones the Society's board used to organize. He opposed the continuing influence by the board on the chapter, and resigned when his ideas were not adopted. In 1901, Kapteyn's first student, Willem de Sitter, became secretary of the scientific chapter, possibly at the instigation of Kapteyn. The chapter's board was chaired at the time by the chemist Arnold Frederik Holleman (1859–1953), who had come to Groningen to assume the position of director

8.3 Natural Sciences Society

of the 'Rijks-landbouwproefstation' (State Agricultural Laboratory). However, he was soon appointed professor of anorganic and organic chemistry. When Holleman resigned to take up a professorship in Amsterdam in 1904, Kapteyn was elected chairman of the scientific chapter, which he remained until his retirement in 1921. He retired from the board of the Society in 1904, quoting his heavy workload as the reason (he was a board member of the 'Central Bureau' as well). When de Sitter was appointed professor of astronomy in Leiden in 1908, he resigned as secretary and was replaced by Joost Hudig (1880–1967),. Hudig was a chemical engineer, specializing in pedology (soil study); he worked at the State Agricultural Laboratory in Groningen. In her biography Henriette Hertzsprung-Kapteyn writes:

'[The Society] had as its primary aim to organize lectures and other presentations to keep its members up-to-date with developments in the rapidly changing scientific developments and views. To this end they invited prominent scientists to explain the outcomes of their research. Kapteyn and his energetic secretary Hudig managed to find many important speakers to come and give presentations. For example, they were able to organize lectures by, among others, Eugène Dubois, the discoverer of the skull of *Pithecanthropus erectus,* and Dr. Tienemann, the director of the Vogelwarte of Rossiten in the Kurische Nehrung, whose interesting studies were of much interest to the bird watcher Kapteyn.'

An interesting incidental circumstance of this story is that this 'energetic Joost Hudig' eventually became Henriette's second husband, after she had divorced astronomer Ejnar Hertzsprung and Hudig had become a widower For more details see Appendix B. The Hudigs and the Kapteyns were neighbors, as Henriette notes elsewhere in her biography: 'Also Hudig, a young neighbor who quickly became a friend, read to him regularly from geological and historical publications [...]' (see page 426). It is not clear from the context when this took place, but if it was shortly after Hudig moved to Groningen (around 1905), it would have been in Heerestraat, or else Eemshaven. Eugéne Dubois (1858–1941) was a professor of anthropology at the University of Amsterdam and Johannes Tienemann (1863–1938) a German ornithologist and pioneer in studying the migration of birds by banding or ringing them; he established the Rossitten Bird Observatory in Eastern Prussia.

'During his tenure of office the Society came to full prosperity. He held many presentations himself and never tired to keep scientific life in Groningen at a high level. After the lectures came the most important part: the gatherings at Restaurant Willems, where they continued their discussions and exchanged ideas about the subjects they had been presented.'

Restaurant Willems was located in the center of Groningen at the address Heerestraat 52; this is still one of the city's busiest shopping areas. Until recently, these informal post-lecture gatherings were being held in the 'Kapteyn Lounge' in Concerthuis, where speakers and board members enjoyed a glass of wine. Since 2012, until the completion of the construction of the major 'cultural center 'Forum', planned for 2017, the meetings and the drinking of wine took place elsewhere. The KNG board, however, still holds its meetings in the Kapteyn Lounge.

For some time the regular lectures of the Society took place along those of the scientific chapter. The annual report made a distinction between the two: those of the Society itself were referred to as 'non-specialist', i.e., meant for a general audience. Kapteyn was a very active lecturer for the scientific chapter: 'Growth of trees and the weather' (Jan. 6, 1899), 'About the luminosity of stars' (Apr. 27, 1901), 'About the apparent motion of the nebulae of Nova Persei' (Febr. 1, 1902), 'Skew distributions' (March 14, 1903). But he also held a 'non-specialist' lecture on 'The Doppler Principle' on Dec. 14, 1904. This was not a novel development – the effect of the shifting of wave frequencies emitted from a moving object had already been proposed in 1842 by Christian Andreas Doppler (1803–1853). He applied it to variable stars, which he suggested were binary systems in which the white light of each component is shifted towards the infrared when it is moving away from us, the components becoming periodically invisible. This explanation did not hold up, but in 1845 Doppler's principle was proven to be correct by Buys Ballot in the case of sound waves. The lecture on Nova Persei was very timely. The nova had been discovered in February 1901 as a third magnitude star that reached a peak brightness of magnitude 0.2, so for a while it was one of the brightest stars in the sky. That was unusual, but then in August of the same year astronomers detected a developing nebulosity around the star that showed extremely high proper motions on the sky. This very energetic outburst seemed associated with the ejection of material with extremely high velocities. Kapteyn became greatly interested in this phenomenon and wrote a paper about it some years later. I will return to this in the next chapter (see page 312).

For many years, Kapteyn was also a member of the Netherlands Natural and Medical Science Congress, a nationwide organization with aims similar to those of the Groningen Natural Sciences Society. This 'Nederlandsch Natuur- en Geneeskundig Congres' (NNGC) was founded in 1887 and had members all over the country. It organized congresses on a regular basis to bring scientists in these disciplines together with a view to elucidating and discussing progress in scientific research. It is active to date, although it is now aiming for a broader audience, holding congresses on specific subjects twice a year [11]. Originally it had specialized sessions and plenary ones. Kapteyn had joined the NNGC early on and remained a member throughout his life. At the meeting of 1893, which was held in Groningen, he made a presentation to the section on physics and chemistry on his method to determine distances to stars by transit timing, Kapteyn (1893d). Later he was a key-note speaker, Kapteyn (1911d), of the NNGC when it met in Groningen again.

8.4 Colors of Stars

In 1888, Kapteyn was elected a member of the 'Koninklijke Nederlandse Academie van Wetenschappen' (Royal Netherlands Academy of Arts and Sciences) (KNAW). Reports show that he attended most of its meetings, in spite of the long journey from Groningen to Amsterdam, and he started delivering reports on astronomy and

8.4 Colors of Stars

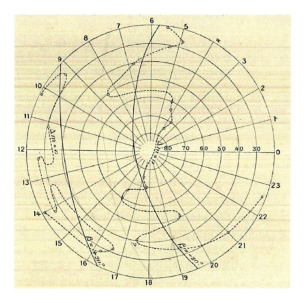

Fig. 8.11 This diagram shows the positions on the southern sky, where stars are found to be systematically bluer and redder in comparison with photographic CPD determinations and visual catalogs. For further explanation see the text (From Kapteyn, 1892c)

his research. These were published in Dutch, but in his early days summarized in English in foreign journals. From about 1900 onwards the Academy also published English versions. Kapteyn used this forum to present his views to a wider audience, restricting his publications in astronomical periodicals to technical presentations and discussions. By and large, the Academy was very important to him.

His first contribution, Kapteyn (1890c), involved explanatory remarks on an as yet unpublished paper on the determination of stellar parallaxes using timing observations at the Leiden meridian circle. His next contribution, Kapteyn (1892d), concerned an issue that had been on his mind for quite some time. In his second report to the Astronomische Gesellschaft on the progress of the CPD, Kapteyn (1890d), he had already reported on a strange anomaly that he had noted when comparing the plates from the Cape with stellar catalogs, such as the Bonner *Südliche Durchmusterung* and the Argentinian *Cordoba Durchmusterung*: the relative number of stars on his plates compared to visual star catalogs was significantly higher in the areas near the Milky Way than away from it. If his assumption was correct, it would mean that on average the stars in the Milky Way are bluer than those away from it. This is because the photographic emulsions that were used were relatively more sensitive to bluer wavelengths than the human eye.

Kapteyn first worked this out in more detail in a paper in the Bulletin of the Carte du Ciel, Kapteyn (1892c). For various parts of the southern sky he calculated the mean difference between the visual magnitudes in star catalogs and those in his photographic Durchmusterung. The definition of magnitudes is such that the redder the light of a star, the smaller (or more negative) this magnitude difference Δm. Such 'color indices' are still used to date; however, the modern convention is to always subtract the redder wavelength band from the bluer one, so that a small or negative index indicates blueness. Kapteyn's definition is opposite to this, unfortunately.

Figure 8.11 has been taken from the 1892 Carte du Ciel paper. It was not published in his KNAW paper, which was more of a summary, but it is likely he showed it when he presented the paper.

It shows a mapping of the southern sky onto a flat plane, such that the south pole is at the center and circles are lines of equal declination, the outer one corresponding to $-20°$ (roughly the extent of the CPD). The smooth curved lines correspond to Galactic latitudes $+20°$ and $-20°$ and the intervening area represents a strip of $40°$ width following the Milky Way. The irregular lines separate the areas where the mean magnitude difference of the stars Δm is positive and the color relatively red and where it is negative. The separation of these two occurs 'without any exception'. According to the KNAW minutes, Kapteyn's presentation at the Amsterdam Academy was received with considerable reserve on the part of his astronomy colleagues van de Sande Bakhuyzen and Oudemans; it is recorded that these 'gentlemen point to the possibility of absorption by nebulae or the difference in richness of star fields in different parts of the Milky Way'. Such dissenting comments were not usually printed in these proceedings, so the 'gentlemen' seem to have given Kapteyn a hard time.

It did take not very long, though, before Kapteyn's results were confirmed. The support came from the Potsdam astronomer Julius Scheiner, who based himself on a comparison of the Carte du Ciel plates that were taken at Potsdam, with the visual observations of the Bonner Durchmusterung, (*Über die Abhängigkeit der Grössenangaben der Bonner Durchmusterung von der Sternfülle* [12]). Kapteyn wrote a further note, Kapteyn (1898b), commenting on Scheiner's work. Although he was pleased to see his work confirmed, he warned that a really independent determination was required before it could be accepted that the effect was a property of the starlight. In Kapteyn (1899), he commented on the fact that Newcomb had found that there was no unequivocal evidence for the reality of the effect; such conclusions were premature and based on too few data, and needed further foundations.

Kapteyn was well aware of the questionable aspects of his analysis. In his paper for the Carte du Ciel he had worried about effects of different zenith angles during observations, different seasons at the times of the exposures, various possible effects of the atmosphere and many other effects. He proposed that a series of special plates should be taken to study the reality of the effect and David Gill actually proceeded to take such exposures. Seasonal effects on the photographic magnitudes can be eliminated by making exposures on the same plate of two different regions and during the same night. One region is close to the Milky Way and the other far away from it. The exposures are made as near to each other in time as possible, and with the regions selected in such a way that the zenith angles are virtually the same. The conditions should be absolutely without cloud or moonlight. The question of the colors can be investigated by not only taking such plates with regular emulsions, but also with 'isochromatic' ones. These are emulsions designed to have a more even response ('actinic effect' or ability of the light to act on the silver in the emulsion) over all (optical) wavelengths; hence 'isochromatic' (or 'panchromatic', nowadays). These emulsions have lower sensitivity, though, and require longer exposure times. Unfortunately, Gill had taken both types of plates with the same exposure times,

8.4 Colors of Stars

which was the cause of great regret afterwards, in the ensuing analysis. Kapteyn put de Sitter on the tasks to measure and analyze the plates, and his results were published in the Groningen Publications as de Sitter (1900a). The paper was written by de Sitter, but as the work was done at the suggestion and under the guidance and responsibility of Kapteyn, I include Kapteyn in the reference lists as the second author. The study involves ten areas, and on the regular emulsion plates there are 454 'galactic' and 336 'polar' stars. Numbers that are sufficiently large for a statistical analysis.

The photographic exposures do not affirm the color variations, but a more thorough comparison of de Sitter's result with visual *catalogs* confirms the trends. What to make of this? The conclusion is that the effect is indeed real when photographic and visual data are compared, but the fact that it does not show up in the comparison of the two photographic studies leaves open the possibility that it is due to systematic errors in the visual observations rather than in the actual mean colors of the stars. So the result was not conclusive and more work was required. In the course of his further studies, de Sitter continued with another paper in the Groningen Publications, de Sitter (1900b), *On isochromatic plates* [13]. His assertion was that 'the isochromatic plates are not sensibly different from the ordinary ones in their behaviour with respect to such differences of colour as exist among stars'. He proved this by means of a detailed comparison of photographic measurements and visual ones. So, these emulsions were not able to detect the color differences he was after. So why is that? The fact is that de Sitter claimed that in point of fact the sensitivity ('sensibility') curves of the isochromatic emulsion have a 'secondary maximum in the less tangible part' (redder wavelengths) of the spectrum contrary to the ordinary emulsions, but a primary maximum that is the same as that of the ordinary emulsions. Furthermore, the focus position of the telescope differs with wavelength, whereas all plates were focused similarly. Contrary to what had been done during the procedure, the isochromatic plates needed to be exposed through a filter blocking the blue light and with its own appropriate focus. Hence the results cannot be trusted.

De Sitter published his final study into this issue in a paper in the Groningen Publications (1904) on *Investigation of the systematic difference between the photographic and visual magnitudes of stars depending on galactic latitude, based on photometric observations by W. de Sitter, visual estimates by R.T.A. Innes, and photographs taken at the Cape Observatory, together with catalogues of the photometric and photographic magnitudes of 791 stars*, de Sitter (1904). When de Sitter left for the Cape, the idea was that he would obtain photometric observations of stars that had been measured on the CPD plates. This was in fact the reason why he brought along the Zöllner photometer of the Astronomical Laboratory at Groningen 'to perform these observations'. Kapteyn seems to have managed to acquire such a photometer. 'Some time was lost', as de Sitter wrote in the paper, 'in practice and preliminary results, before the instrument was in satisfactory working order'. However, since the weather was not helpful during de Sitter's stay at the Cape, Innes offered to help by observing all his stars visually.

Now, this work did again confirm the systematic difference found earlier by Kapteyn, but also showed that nature was more complicated than had been assumed. The variation of this mean color between the Milky Way the poles was found to be two tenths of a magnitude, but if one looked *along* the Milky Way in detail, the mean color there varied irregularly with longitude and to an even larger extent. De Sitter concluded: 'If the apparent variations in that average colour, as we proceed along the Milky Way, are real, and the present investigation hardly leaves any room for doubting their reality, it is a problem of great importance to investigate them more closely and find their relation, if any, to the general structure of the Milky Way.'

8.5 Spectral Types of Stars

Obviously, the matter did not rest here, but was resolved when the same effect was found when spectral types rather than colors were used. In order to follow that I have to review the state of spectroscopy at the time. I have already referred to spectral types above in passing, but have deliberately postponed the presentation of a more comprehensive discussion until now. Spectra of stars, which are recordings of their light distribution as a function of wavelength, have been taken since the middle of the nineteenth century and are still a highly useful approach to the study of stellar structure and evolution. Spectral distributions of light show a broad continuum of light at all wavelengths, intersected by 'spectral lines' which are dark areas of very narrow wavelength range due to absorption of light by chemical elements in the outer layers of the stars. Each spectral line corresponds to an energy transition in the atoms/ions of a specific element. It thus carries information on the chemical composition, but also on the physical conditions (primarily temperature, density and pressure) of the emitting layers. The modern spectral classes, which I will introduce shortly, are one way of defining the horizontal axis in the Hertzsprung–Russell diagram of Fig. 7.1. They are indicated on the top axis. The bottom axis gives the color of the stars, which is an indication of the temperature at the surface as we see it from outside the star. We see that the spectral classes correlate well with temperature, the hottest (bluest) stars being towards the left. Remember that the vertical axis represents the absolute magnitude or equivalently total luminosity (see page 206).

The earliest systematic work on stellar spectra was performed by Father Pietro Angelo Secchi SJ (1818–1878), who was a Jesuit astronomer. He was director of what is now the Vatican Observatory, originally established in 1774 as the observatory of the Roman College. This observatory was confiscated by the Italians after Secchi's death, but was reactivated at the Vatican around 1890 by the Holy See and still exists today. Although commonly known as the Vatican Observatory, the official name is Specola Vaticana. Secchi did pioneering work in stellar spectroscopy and in the 1860s designed the first classification scheme of stellar spectra. His earliest set of 'Secchi classes' encompassed three types, which also correlate with the colors of the stars (and with modern classes), and were used by Kapteyn and his contemporaries.

8.5 Spectral Types of Stars

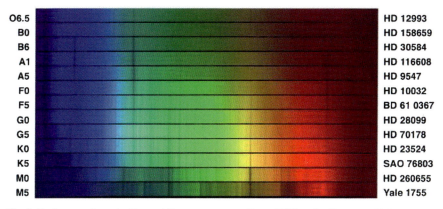

Fig. 8.12 Spectra along the Harvard/Draper sequence (From Wikimedia Commons [14])

Secchi's Class I were stars with broad heavy spectral lines of hydrogen. Their color is blue, which indicates that they are hot. In Fig. 7.1, this means from the modern class A (see top axis) to about the start of F. He later included a sub-class, which he called Orion sub-type and in which the lines are narrower. This is now class B. His Class II has weaker hydrogen lines, but more lines of elements such as calcium, as we also find in the Sun. Today this corresponds to the rest of the classes F, G and K. Class III has the reddest stars with complex systems of spectral line and that is the modern class M. In his work of around 1890 to 1900, Kapteyn referred to the types by their numerical designations I, II and III only (without referring to Secchi, as we would do today).

For the modern spectral classification we are indebted to Edward Charles Pickering (1846–1919) of Harvard College Observatory (see Fig. 10.8) and some of his female staff (known as Pickering's harem). It is named after Henry Draper (1837–1882), who was a physician as well as an accomplished amateur astronomer. Draper was one of the first to photographically record the spectrum of a star at his private observatory and to recognize the similarities of the many stellar spectra he obtained to that of the Sun. After his death his widow supported the spectroscopic work of Pickering through a fund called the Henry Draper Memorial. Pickering obtained spectra of hundreds of stars and these were classified in a new system, mainly by Williamina Fleming (1857–1911). This was achieved by first subdividing Secchi's type I up into types A through D, type II into E through L, using the designation M for type III and introducing a new type O. The result was published in 1890 as the *Draper catalogue of stellar spectra* (*The Draper Catalogue of stellar spectra photographed with the 8-inch Bache telescope as a part of the Henry Draper Memorial* [15]). Pickering, in conjunction with Antonia Caetana de Paiva Pereira Maury (1866–1952) and Annie Jump Cannon (1863–1941), among others, refined and revised it and applied it to about 400,000 stars. This resulted in the *Henry Draper Catalogue* and its extension, published between 1918 and 1936. The spectra types were restricted in the process to a smaller number and the rest was rearranged to some extent: 'O' was put before type I and only a few of the letters were retained, leaving the sequence OBAFGKM. This final sequence has been illustrated in Fig. 8.12; note the systematic change in the brightest parts from blue to red (going from O to M or hotter to cooler), and the systematic variation in appearance and strength of spectral lines. For reasons having to do with previous incorrect understanding of stellar evolution through the spectral type range, the blue (O-star) end of the range is often called 'early type' spectra and, moving through the sequence towards M-stars, 'later' types.

8.6 Parallactic Motion

In 1892, Kapteyn gave a presentation on the distribution of stars in space at a meeting of the Royal Academy, elucidating his efforts to study the systematic differences of the distributions of spectral type. The paper is Kapteyn (1892e); it has been summarized in English in Kapteyn (1892f). He used proper motions as a statistical means to derive distances of stars. Before going into this, we need to understand some of the background, since it is more than a simple assumption that large proper motion might be a sign of proximity.

All stars move through space at different velocities, both in magnitude and in direction. The Sun also moves with respect to the collective of stars in its neighborhood, which is indicated as the Local Standard of Rest. This motion of the Sun reflects in a systematic pattern of the proper motions of stars in the sky. Think of the simple case that all stars are stationary, but that the Sun moves with respect to the stars. This will result in proper motions of the stars as seen from the Earth. When we direct our eyes in the general direction (the point in the sky) where the Sun is heading, we will see the stars on the sky move, i.e. have proper motions away from that point. When we look in a direction perpendicular to the motions, stars will show the largest proper motions, since our motion with respect to them is projected onto the sky without any foreshortening. Finally, when we look in the direction on the sky the Sun comes from, we will see stars having proper motions directed towards that point. This point is called the Solar Antapex and the point opposite (where stars would move away from and which is the direction in which the Sun moves) is called the Solar Apex. The effect has been illustrated in Fig. 8.13, taken from a publication by Kapteyn in 1906; it was written in French, which is the reason for the French names of the constellations Hercules and Canis Major.

Another way of seeing how this comes about is to consider the situation where you sit in a car with trees on either side of the road. Focus on two trees in the distance that are on opposite sides of the road. When you are far away, they will be seen at only a small angle from each other, but as you approach them they will move apart. When you pass them they are directly to your left and right and when they disappear in the distance you will again see them approaching each other with a smaller and smaller angle between them.

8.6 Parallactic Motion

Fig. 8.13 Illustration of the effect of the motion of the Sun with respect to the common stars on the distribution of proper motions (From Kapteyn, 1906e)

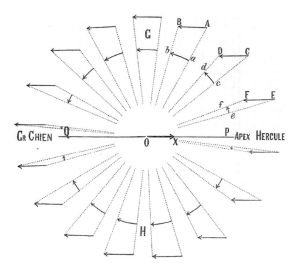

In actual practice the stars are not stationary, of course, and will have a random pattern (in principle at least) of proper motions on the sky. However, that will have the systematic pattern just described superposed on that. This was determined for the first time by William Herschel in proper motions of very bright stars and published in a paper *On the proper motion of the Sun and Solar System; with an account of several changes that have happened among the fixed stars since the time of Mr Flamsteed* [16]. The Solar Apex lies in the constellation Hercules. Herschel's position was based on proper motions of a dozen or so bright stars in catalogs (by others) that he had available, and only some ten to twenty degrees away from the currently adopted position. Figure 8.14 shows the constellation Hercules and the position adopted by Kapteyn in 1901 (large black square) compared to a recent determination (small square) [17]. The Apex lies some 20° mostly south-west of the bright star Vega (the brightest in the northern sky) in the constellation Lyra. The Antapex is the position opposite on the sky and this lies some 15° south-west of Sirius, the brightest star in the sky. The best current value for the velocity of the Sun with respect to the Local Standard of Rest is about 18 km/s, which is almost four times the distance between the Earth and the Sun per year. Kapteyn thought the Suns's velocity was about 17 km/sec.

Proper motions, certainly at Kapteyn's time, suffered from a major uncertainty that was a constant source of concern of positional astronomy. Remember (see e.g. Box 3.1) that positions of stars are measured by determining the altitude with a meridian circle and timing the meridian passage – or comparable measurements that relate the position of the star on the sky to that of the pole. Measuring positions at one instant and noting how they differ at a later point in time, gives the proper motion. The problem, however, is that the position of the pole changes over time as a result of precession. This is the effect that the direction of the rotation axis of the Earth is not constant in space. It can be compared to the motion of a gyroscope,

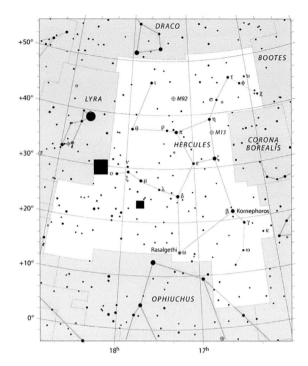

Fig. 8.14 The Apex is the direction on the sky towards which the Sun is moving with respect to the mean of stars in its neighborhood. It lies in the constellation Hercules. The larger black square is the position found by Kapteyn (1901b), the smaller one a more recent determination based on proper motions from the Hipparcos satellite. North is at the top, west to the right (when one looks at the northern sky towards the south, west is to the right) (From International Astronomical Union constellations Website [18])

whose rotation axis is not vertical with respect to the gravitational field of the Earth. It then precesses slowly, keeping the angle with the vertical constant. The Earth does the same as a result of asymmetric pull of the Sun on the Earth, which its has a not quite spherical, somewhat flattened shape. The Earth's rotation axis then precesses while keeping the angle with the orbit (the ecliptic) constant. A precession cycle is about 26,000 years. This means that even if a star did not have any proper motion, its right ascension and declination would change. The problem now was, that in Kapteyn's time the speed of precession, expressed as the precession constant, was not known with the precision necessary for comparing positions measured at different times.

8.7 Secular Parallax

Knowing the velocity of the Sun through space allows another important application, which is called 'secular parallax'. In 1843, Auguste Bravais (1811–1863) demonstrated that the motion of the Sun in space can be used to estimate distances of groups of stars from their mean annual proper motion. Bravais actually worked and taught as an astronomer, but is particularly known – together with his brother Louis François Bravais (1801–1843), a physician and botanist – for Bravais lattices in crystallography. The method can be used to estimate the mean distance for any

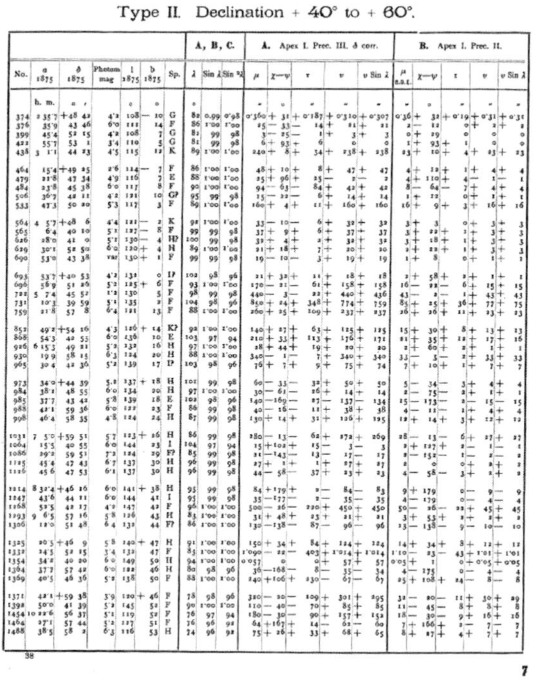

8.7 Secular Parallax

Fig. 8.15 A page from the publication of components of the proper motions of Bradley stars. From Kapteyn (1900d); see text for explanation (Reproduced from *Publications of the Astronomical Laboratory Groningen*)

group of stars in the sky for which one suspects the distances to be about the same (such as stars of similar spectral type and apparent magnitude). He showed that for a group of stars that move together in space (as in a star cluster) the proper motions define an apex corresponding to the direction of the Sun's motion relative to the center of gravity of that group, and this can be used to find the mean distance of the group to the Sun. Since the relative velocity of the Sun with respect to the group is unknown, the unit in which this distance is expressed is the annual distance covered by the Sun relative to the center of gravity of the group.

Kapteyn concluded that the measuring of trigonometric parallaxes on a wholesale basis was unlikely to be realized any time soon. The baseline of the diameter of the Earth's orbit was simply too small to produce measurable effects in the positions of stars, except for the few nearest ones. As he pointed out repeatedly, the Sun moved through space with such a velocity that it covered a distance four times the radius of the orbit of Earth in one year, and this provided a larger baseline that would furthermore grow with time. Parallaxes measured on this basis were of necessity statistical.

In two presentations at the Academy, Kapteyn (1892d) and Kapteyn (1892e), he reported on a new approach he was going to pursue, viz., the possible systematic relation between spectral types of stars and their proper motions. For reviews in English of these presentations see also Kapteyn (1892f) and a review of some years later by the authority on historical aspects of astronomy Agnes M. Clerke on *The distribution of the stars* [19]. Kapteyn found that among the stars with low proper

Fig. 8.16 The Andromeda Nebula (M31) is the spiral galaxy nearest to our own Milky Way Galaxy. It is number 31 in the list of Messier. The small galaxy bottom-center is one of its companions (M32) (From spacetelescope.org/ [20])

motion, (spectral) Type I predominated, while stars with high proper motions were predominantly of Type II. This is in line with what he found for colors and it points at a distribution in space where the relatively red stars of Type II were local, whereas those of blue Type I stars were concentrated towards the Milky Way and more distant. He concluded as follows: 'If what has been found here will be confirmed, it seems that the observations are best represented by assuming that our stellar system more or less has the form of a ball, surrounded by a ring, resembling to some extent the form of the Andromeda Nebula on photographs...' (see Fig. 8.16 for a recent picture)'. This may sound prophetic, since our galaxy indeed resembles the Andromeda Nebula, except that we now know that the Sun is not located in the central 'ball', but in the disk that Kapteyn describes as a 'ring'. It should not be left unmentioned that the effect was also found independently, but using a much smaller sample, by William Henry Stanley Monck (1839–1915) from Dublin, also in 1892.

Having established this, Kapteyn embarked on a series of studies related to the use of proper motions. Again he reported on this in presentations before the Amsterdam Academy, first using the distribution of the *directions* of the proper motions on the sky in Kapteyn (1895). In a second presentation he also made use of the *amount* of proper motion, Kapteyn (1897a). Both presentations are entitled *Distribution of*

Tabelle 4.

Photom. Grösse	Grösse BD.	π Typus I	π Typus II	π alle Sterne
2.0	1.3	0."0315	0."0715	0."0530
3.0	2.5	0.0223	0.0505	0.0375
4.0	3.7	0.0157	0.0357	0.0265
5.0	4.9	0.0111	0.0253	0.0187
6.0	6.0	0.0079	0.0179	0.0132
7.0	7.0	0.0056	0.0126	0.0094
8.0	8.0	0.0039	0.0089	0.0066
9.0	9.0	0.0028	0.0063	0.0047

Fig. 8.17 Table 4 from Kapteyn (1898a) on mean stellar motions, the absolute value of the Sun's velocity in space and the mean parallax of stars of various magnitudes; see text for explanation (Kapteyn Astronomical Institute)

cosmic velocities. In a paper in the Astronomische Nachrichten, Kapteyn (1898a), he outlined the approach in full technical details and gave a first application. His basic, novel idea was that the proper motion of any star can be resolved into two components, one along the direction towards the Apex (or Antapex) and one perpendicular to that. The latter is entirely due to the 'peculiar' motion of the star, but the first contains a component due to the motion of the Sun with respect to the Local Standard of Rest (LSR) and he named this component the *'parallactic'* proper motion. Its magnitude then depends on the distance of the star, at least in a statistical sense. Now Kapteyn made the fundamental assumption that the motions of the stars in space relative to the LSR were distributed without preferred direction. In more modern terms we would denote that as 'isotropic'. He further assumed that the distribution of the magnitude of these velocities (the percentage of stars that have a space velocity between two particular values) was independent of the distance from the Sun; in other words, the mean velocity was the same everywhere. This would mean that there was no preferred direction of these motions and the 'law' of the distribution of velocities had the same shape everywhere. With these assumptions he outlined how the observed distribution of the directions and magnitude of proper motions on the sky are related to the mean velocity of stars in space. With his fundamental assumptions of isotropic velocities and homogeneous mean motions he found that he could express the mean linear velocity on the sky and that along the line of sight in terms of the space velocity.

The data Kapteyn used were the 'Auwers-Bradley' stars. Referring back to Box 3.3, I point out that the first catalog of stars in the modern era was that by James Bradley, based on 60,000 observations of 3000 stars from Greenwich between 1750 and 1765. He never reduced his data, but Friedrich Bessel turned these observations into a catalog of fundamental positions, named *Fundamenta Astronomiae* of 1818. In 1888, Georg von Auwers produced a fundamental catalog from observations made at Greenwich and Berlin between 1836 and 1880. The stars in that catalogue, 3268 in number, which included the Bradley stars, had proper motions derived from the change in positions over the roughly one century between Bradley's and von Auwers' measurements.

The title of Kapteyn's work was *Components τ and υ of the proper motions and other quantities for the stars of Bradley*, Kapteyn (1900d). In this publication he went through the laborious work of calculating, for each star, the angle λ between the position of the star on the sky and that of the Solar Apex and the angle that the proper motions make with the direction towards the Apex (which he indicated as $\chi - \psi$). Next he calculated the projections of the proper motions *onto* (ν) and *perpendicular to* (τ) the great circle through the star and the Apex/Antapex. In the end he did this for 2,640 suitable stars. The results were eventually published in the Groningen Publications Kapteyn (1900d). Not shunning laborious computations, he proceeded to do all of this for *five* separate assumptions, which were different combinations of three values for the precession corrections and three for the position of the Apex! A sample page is reproduced in Fig. 8.15. The publication had 108 of such pages. As if this were not enough, he performed a sixth calculation, Kapteyn (1902b), using what he considered the most likely combination of these parameters.

Kapteyn complemented this information with the spectral types from the Draper Catalogue. The important next step was that he used the observed distributions, together with his basic assumptions, to estimate a value for the mean linear velocity of the stars on the sky divided by the space velocity of the Sun. He then looked into the radial velocities, for which he used published data on 51 stars observed by Hermann Carl Vogel (1841–1907) [21] from the Potsdam Astrophysikalisches Observatorium. and discussed at his request by Paul Friedrich Ferdinand Kempf (1856–1920) [22]. Of course, these radial velocities also have a component related to the Sun's motion and one from the star's peculiar motion with respect to the LSR. Taking everything together, Kapteyn arrived at a mean velocity of stars in space of 31.1 ± 2.2 km/s and a space velocity of the Sun of 16.7 ± 1.2 km/s.

But that was not all. Having a value for the motion of the Sun in space, he was also in a position to estimate the mean parallax of various groups of stars. The final result of all this work can be summarized in a single, small table, which I reproduce in Fig. 8.17. It shows the mean parallax of stars as a function of their apparent magnitude and spectral type. So, for any bright stars for which the spectral type was determined, Kapteyn had a *statistical* distance.

8.8 Praise by Newcomb

Obviously, Kapteyn was not the only one to get on to this idea. Indeed, Simon Newcomb worked along much the same lines. But Kapteyn's results were much more reliable, and Newcomb, by recognizing this, was instrumental in drawing the attention of a wide audience to Kapteyn's work. Presumably, the fact that he praised him extensively was also helpful. In 1896, Newcomb published a paper *On the Solar motion as a gauge of stellar distance* [23], in which he also laid out methods to use the motion of the Sun through space with a view to obtaining statistical information on the distances to stars. In a sequel paper in 1899, *Some points relating to the solar motion and the mean parallax of stars of different orders of magnitude* [24] he notes:

'In this journal [...] I published a paper [...]. Soon after it appeared I learned that its methods were not new, Kapteyn having treated the subject from the same point of view in a communication to the Amsterdam Academy in 1893. In some points this author had not only anticipated the methods of my paper, but carried his investigation farther than I had contemplated doing, especially in investigating the relative proper motions of stars having different types of spectra. More recently, in Kapteyn (1898a), he has applied the method to determining the mean parallax of stars of different magnitudes, as a preliminary datum for this, deriving the absolute amount of the solar motion from Vogel's observations of motions in the line of sight. Under these circumstances it would seem still less necessary that I should continue my unfinished work on the same lines, were it not that the subject is of such breadth and interest that it is not easily exhausted.'

In his book *The Stars: A study of the Universe* [25], Newcomb notes at the beginning of his chapter on *Statistical studies of proper motions*: 'The number of stars now found to have a proper motion is sufficiently great to apply a statistical method to their study. The principle steps in this study have been taken by Kapteyn, who, in several papers published during the past ten years, has shown how important conclusions may be drawn this way.' This is the same book in which Newcomb, when discussing the Cape Photographic Durchmusterung, writes the footnote (quoted on page xiii) that has been taken as the title of this book. Obviously, Kapteyn's fame was spreading.

8.9 A Resume for Gill

This work, using statistical methods, was setting the scene for Kapteyn's further work on the structure of the stellar system, requiring a study of the distribution of stellar velocities as a means to finding statistically 'secular' parallaxes. It is interesting to see how he himself looked upon the development of his scientific studies and we are fortunate that some of his notes on the subject have survived. In Kapteyn's papers there is a handwritten note of 16 pages, dated 12 April, 1907. The cover page says *Resumé van K's Onderzoekingen over de 'Structure of the Universe'* (Summary of Kapteyn's studies on the Structure of the Universe) to which he adds between brackets: 'written at the request of D. Gill and sent to him on the same date' (see Figs. 8.19 and 8.18).

The request was made by David Gill in a letter of March 19, 1907, from London, where he had returned after his directorship at the Cape had ended. He was President of the British Association in 1907–1908. The British Association, currently the British Association for the Advancement of Science, is a learned society in the UK that was founded in 1831. It should not be confused with the 'Royal Society' (of

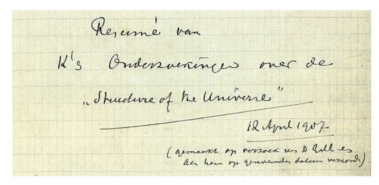

Fig. 8.18 The cover page of Kapteyn summary prepared for Gill (Kapteyn Astronomical Institute)

London for the improvement of Natural Knowledge), founded in 1660. Gill wrote: 'I am very anxious in my Presidential address to make reference at considerable length to your researches on the structure and motions in the Sidereal System –; and I write, if you will be kind enough to spare the time, to assist me by putting together a statement in regard to your work on the subject. Can you do this; say in course of the next month?'. Kapteyn complied (on April 12), after Gill urged him once again, on April 2, to deliver the summary as soon as possible: 'Do this for me like a good friend.'

I will briefly have to go ahead of the chronological order in this book in introducing the manuscript.

In his letter of March 19, 1907, Gill invited Kapteyn and his wife to attend this meeting. 'I will see to it that a good host is provided to entertain you there so that the cost to you would only be that of your traveling expenses to and from Leicester'. In reply Kapteyn wrote that they were not in a position to come to Leicester: 'It's not the money question only, though this counts heavily, now that we have three of our children on our hands out of door, but also of time.' His elder daughter Jacoba Cornelia was 26 at the time and had married 31-year old Willem Cornelis Noordenbos, but the other two children, Henriette Mariette Augustine Albertine and Gerrit Jacobus, were 25 and 23 (see Table 5.2) and probably still pursuing their studies at university. It seems that the Kapteyns were still supporting their elder daughter. Actually, Jacoba Cornelia was pregnant at the time, and gave birth to a daughter, Greta Noordenbos (1907–1925) on July 7 of the same year in Groningen. According to the birth certificate the father was a medical doctor in Groningen and the mother without occupation. Further correspondence shows that the Kapteyns had been to London to visit the Gills in January of the same year. Although Gill, in his letters of April 10 and April 24, repeatedly asked Kapteyn to reconsider his decision, the latter did not come to Leicester. Gill expressed his regret about this in further letters (which also dealt with the proofs of the printed version of his address) right up to the time of the meeting itself, which took place from July 29 to August 6.

8.9 A Resume for Gill

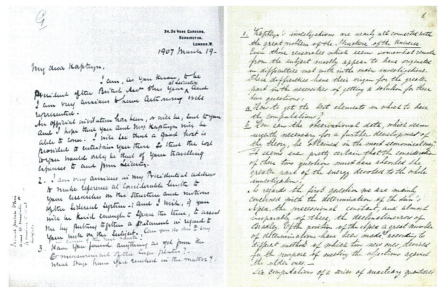

Fig. 8.19 On the left the first page of Gill's letter requesting Kapteyn to write a summary of his work and the first page of Kapteyn's reply on the right (Kapteyn Astronomical Institute)

The first part of the document relates to Kapteyn's work and ideas dating back to the period covered in this chapter. I will restrict myself to what is relevant up to this point in Kapteyn's career. I have left out most notes and references Kapteyn wrote on the left-hand pages. The full text of Kapteyn's resumé is available at my Kapteyn-website. It starts as follows:

1. Kapteyn's investigations are nearly all connected with the great problem of the <u>Structure of the Universe</u>. Even those researches which seem somewhat remote from the subject mostly appear to have originated in difficulties met within the main investigations. These difficulties have their origin for the greater part in the necessities of getting a solution for these two questions:

<u>a.</u> How to get the best elements on which to base the computations?

<u>b.</u> How can the observational data, which seem urgently necessary for a further development of the theory, be obtained in the most economical way?

As regards the first question we are mainly concerned with the determination of the Sun's Apex, the precessional constant, and almost inseparably of these, the declination errors of Bradley. Of the position of the Apex a great number of determinations have been made [Kapteyn gives a note with some references], according to different methods of which two new ones, devised for the purpose of meeting the objections against the older ones.–

Six computations of a series of auxiliary quantities have been made, starting from different assumptions in regard to Apex precession and declinations errors. The process is laborious but several advantages are gained lacking in other methods. The errors of the proper motions in declination of Auwers Bradley were derived independently of any fundamental system; the precession constant from stars with insignificant proper motion alone. [For a modern explanation, see W. Fricke, *Fundamental Catalogues. Past, present and future* [27].]

Part of these investigations nearly coincide with almost simultaneous ones of Newcomb. Notwithstanding the great difference of the methods the agreement of the results is, with a single exception,[3] most gratifying. In regard to the observational data most urgently wanted, K.'s work relates mainly to Wholesale determinations of Parallax and Proper Motion.

Already his parallax determinations at the Leyden Meridian Circle made between 1885–87 prosecuted the aim of obtaining numerous data. In publishing his results he sketched a plan which promised to yield the parallax of all the stars down to the 5th magnitude in a few years. Such determinations have later been taken up by Albert Stowell Flint at Washburn [This is Albert S. Flint (1853–1923) of Washburn Observatory of the University of Wisconsin, Madison. and by Jost at Karlsruhe [Ernst Heinrich Rudolph Jost of the Sternwarte Karlsruhe] and have furnished previous results.

Subsequently K. devised a photographic method both for parallax and proper motions, capable of dealing, without sacrifice of precision, with thousands of stars. Already work done at Groningen, Helsingfors, Pulkovo,[4] Bonn, shows that we need not despair of obtaining thousands of eminently trustworthy parallaxes and proper motions. There is one other subject to which attention might be here directed.–

One of the most difficult elements in investigations of the structure of the universe is no doubt that of extinction of the light in its course through space.– One of the means proposed by Kapteyn is that based on observations of nebulae. If there is no extinction then the surface brightness of a nebula will not diminish if it is viewed from a greater distance. If there is extinction it will. The application of this idea to the question of the determination of extinction is obvious. [...]

2. Parallactic motions as a measure of the distance. Let us come to the considerations of the way followed by K. in the treatment of his subject proper.–

Practical astronomy furnishes the directions in which the stars are seen; in order to get at the distribution in space we want the distances. Direct parallax determinations furnish only a few exceptionally small distances. Even if tenfold more refined it could do no more; for it is almost of necessity <u>differential</u> and so can only lead to <u>differences</u> of parallax. Instead of such direct determinations, or rather along with them, K. uses extensively the parallactic motion as a measure of distance. If the stars were at rest, the Sun's motion through space would furnish a yearly increasing basis for the measurements of their distances.[...]

As early as 1843, Auguste Bravais[5] showed that whatever be the motion of the stars comprising an arbitrary group, we can determine the direction of the Sun's

---

[3] The exception obtains for the correction of the Auwers/Bradley declinations for the stars between declination $+20°$ and $+40°$.

[4] <u>Kostinsky</u>, Pulkovo Obs. Vol. - - - . [Footnote by JCK with his underlining] This is Sergej K. Kostinsky (1867–1936); Pulkovo Observatory is at St. Petersburg, Russia. [PCK]

[5] Journ. de Liouville 8 (1843), p. 435. [JCK] Actually this is the 'Journal de mathématiques pures et appliquées, founded in 1836 by famous mathematician Joseph Liouville (1809–1882). [PCK]

motion relative to the center of gravity of the group and the mean distance[6] of the group to the Sun. The unit in which this distance is expressed being the yearly way covered by the Sun's motion relative to the center of gravity of the group.[...]

3. Velocity of Sun's motion As the distribution of any group of stars found by the parallactic method is expressed in the yearly Sun's motion through space as a unit, the velocity of its motion is one of the fundamental quantities to be determined.–

K.'s determination in 1897 is the first which furnishes an approximately correct value of this element. It is based on the radial velocity of 51 stars determined at Potsdam. It is true that Kempf, some years before, had already determined the Sun's velocity from this material, but the result, 12 kilometers per second, was evidently still extremely uncertain. K. showed how another independent result of far greater accuracy can be derived from the same material. His method is as follows: spectral observations give the total approach or recess of star and Sun. Now, the direction of the Sun's motion being known within some precision, one can free the observed velocity from the part which is due to the Sun, as soon as we have a first approximation of the Sun's velocity. As such we may take Kempf's or any other determination.– We then obtain a first approximation for the true radial velocity of the stars. – Now, by a method presently to be considered, K. had already found that the mean radial velocity of the stars must be equal to 0.93 times the Sun's velocity in space. From the mean star-velocity one can obtain at once a second approximation to the Sun's velocity. By continuing the process we very soon reach a final result. It is 17.6 kilom. per second, or, combined with the result obtained by the more direct method 17 kilom. per second with a probable error of 1.2 kilom. six years afterwards. Campbell,[7] repeating the process, using a number of stars more than 5 times greater, finds the 20 kilometer by the direct, 18 Kil. by the indirect way.

4. Mean velocity of the stars. Adapting a velocity of 19 kilometers we find that the yearly motion of the Sun through space is four times the Sun's distance. To see how the proportion of the mean velocity of the stars, so that of the Sun can be found, imagine a group of stars at a distance of $90°$ from the Apex. The mean parallactic motion, is evidently free from any admixture of the Sun's motion and solely owing to the motions of the stars themselves. True – it is but a component of the star's real motion, which already is seen by us projected on the sphere, but a simple mathematical consideration shows that, on an average, this component must be just one half the motion we would have observed if we saw the star's motion unforeshortened at the distance of the star from the Sun. Therefore the proportion of the parallactic motion with the mean of the components in question gives at once the proportion of the Sun's motion with

---

[6] Rather the distance corresponding to the mean parallax. [JCK]

[7] Astrophysic. Journ. ...... [This seems incorrect; presumably, the reference is Publications of the Astronomical Society of the Pacific, 13, p.51, 1901. The author is William Wallace Campbell. PCK]

the mean of the components of the Sun's motion to half the average motion of the stars. It is thus found that this average velocity is about 1.86 times that of the Sun.[...]

5. Mean parallax of given magnitude and proper motion. Reverting now to the general problem we see that, the Sun's yearly motion being 4 times the Sun's distance, the parallactic motion for stars which see the motion unforeshortened must be 4 times their parallax. How this number varies with the degree of foreshortening needs no explanation. The point is that from the mean parallactic motion of a group of stars we are now enabled the derive at once its mean parallax.

So much for these excerpts. The next steps would be to use this information to derive the two 'laws' of the stellar distribution; the first one, which he called the *luminosity law*, would give the relative number of stars in any volume element of a particular luminosity. This law was supposed to be uniform or the same everywhere in space. This, then, could be used to derive the second law, the total number of stars as a function of distance from the Sun. But just after the turn of the century this was still something of the future.

By this time, Kapteyn seemed to have settled in the position of director of an astronomical laboratory, where he could work without the burden of running an observatory with a heavy load of observational work, maintenance of telescopes and related heavy duties. He used data from others, often photographic material produced for his purpose and at his request. The most extensive suppliers up till then had been David Gill at the Cape and Anders Donner at Helsingfors. Later on, as we will see, Edward Pickering of Harvard College Observatory, and later still George Hale of Mount Wilson Observatory would be added to the list. He also made extensive use of published catalogs, especially those with large numbers of proper motions and spectral types. Now that he had decided on secular parallaxes as his main weapon, his attempt to tackle the problem of the determination of the structure of the Sidereal System was ready to start.

8.10 Personal Matters

Before continuing my description of how Kapteyn developed his scientific program, this seems a good moment to have a look at some of his personal traits. Apart from what we notice when reading the love-letters (keeping in mind these were written at a relatively early age) and the HHK-biography (colored by the sentiments of a loving daughter), there are various other anecdotes and stories about Kapteyn that illustrate his personality. Some of them can be found in the chapter by Wessel Krul in the *Legacy*, but there is also a useful article on Kapteyn by Cornelis Easton (see Fig. 11.12). I have mentioned this article before, in connection with Easton's remark that Kapteyn was rather tall (see page 56). Easton wrote his *Personal Memories of J.C. Kapteyn* shortly after Kapteyn's death and published them, together with an obituary by Pieter van Rhijn, in the Dutch periodical *Hemel & Dampkring*. To make

8.10 Personal Matters

Fig. 8.20 Gerardus Heymans 1857–1930), professor of philosophy and psychology (University Museum Groningen)

it accessible to the English speaking world I have translated it and made it available through my dedicated Kapteyn Website. In Appendix C, I have provided excerpts of this text, leaving out all passages that describe his professional life and his research.

Kapteyn was unconventional in some respects, in spite of – or possibly because of – his religious upbringing in Barneveld. We have seen that his daughter described him as a firm believer in equal treatment of boys and girls. Indeed, his daughters went through the same secondary school as his son and all three children eventually entered university to pursue academic studies. The elder daughter Jacoba took up medicine, the younger daughter Henriette read law, later English in Amsterdam, and the son Gerrit went to Germany to become a mining engineer. That the daughters took up university studies was rather exceptional in those days. The first female student at any university in this country had been Aletta Jacobs (1854–1929), who enrolled in the University of Groningen in 1871 to study medicine.

In this connection Kapteyn's friendship with psychologist and philosopher Gerardus Heymans (1857–1930) (see Fig. 8.20) should be mentioned. Heymans had started research in what he called 'special psychology', in which he made extensive use of surveys based on questionnaires. He had set up a psychological laboratory in Groningen to conduct these studies. One of them was based on more than a hundred biographies from which he deduced a number of personal traits. These had to be correlated and it is believed that Kapteyn's human calculators in his laboratory provided extensive help in calculating the significance level of the many correlations that were possible (see Jan Bank and Maarten van Buuren, *1900. Hoogtij van burgerlijke cultuur* [28]). Heymans furthermore conducted a study on the differences between men and women and was the author of a widely read (and translated) book *Die Psychologie der Frauen* [29]. The book was partly based on a survey of professors and lecturers at the four major Dutch universities of the day, asking them to provide a list of personality traits (ability for abstraction, individuality, memory, adroitness, studiousness, etc.), and also to describe themselves as either predominantly male or female. It was Heymans' view that the difference between the sexes is gradual rather than discrete, can vary statistically and are not one-to-one, and that the sexes are in principle of equal value. The precise outcome of this is of no relevance to the present discussion, but it was much along the lines of expected and prejudiced opinions (in general terms it acknowledges generally held notions, e.g., that men are more rational and women more emotional). Actually, the low response and the suspicion that the answers were by and large following prejudices, invited much criticism. This led him to enable the critical student Anna Wisse (1887–1968) to conduct follow-up studies involving students and results of academic studies.

It is sometimes claimed that Kapteyn and his staff also contributed to this effort, but there is no provable reason to assume so. It is true that Kapteyn and Heymans were very good friends, but they often disagreed on philosophical matters. One of the things they apparently agreed on was the position of women in society. Heymans was actively involved in securing the voting right for women (which was only granted formally in the Netherlands after the revision of the Constitution in 1917 and women were first elected into elective office in 1922). Unlike Heymans, Kapteyn was not a member of the Association for Women Voting Rights (or suffrage), but seems to have been part of 'Heymans' inner circle in Groningen that looked favorably upon women's emancipation' (see Inge de Wilde, *op. cit.*).

Between Heymans and Kapteyn there were also some major differences and on some points they had even opposite opinions. Again I quote Wessel Krul from his chapter in the *Legacy*.

> 'In the early 1900s, a group of professors initiated a campaign to improve the level of cultural life in Groningen by staging exhibitions, organizing concerts and inviting speakers on art and literature-related subjects (see W. Moll: *Persoonlijke Herinneringen aan de Stad Groningen rond de Eeuwwisseling* [30]). Heymans, who had his private house built by H.P. Berlage [Hendrik Petrus Berlage (1856–1934)], the most prominent modern architect in [the Netherlands] at the time, was one of the most energetic among them. It seems that Kapteyn followed these activities only from a distance. There are no indications that he was touched by the vogue for modern art that captivated certain academic circles in Groningen. The movement appealed to a vague spirituality and to a sense of luxury, both of which were entirely foreign to him. Heymans, basing himself on his studies of human sense perceptions, gradually came to believe in a conception of the Universe as a coherent organism. For him, this also entailed the possibility of life after death and of spiritual messages from the Beyond.

8.10 Personal Matters

Again, it seems that Kapteyn was not prepared to follow him in this. He showed more interest in the revival of musical life under the new conductor of the local orchestra, Peter van Anrooy [Peter Gijsbert van Anrooij (1879–1954)]. Personal sympathy may also have played a large part here. Kapteyn's appreciation of music always remained limited. To the amusement of his acquaintances, when concentrating on his work he often began to produce strange and decidedly unmusical hums, grunts and other noises.'

Kapteyn's own marriage was rather conventional in that his wife looked after the household, while Kapteyn kept financial affairs as his specific responsibility. They were not religious and church attendance was not part of their moral routine. All the same, Kapteyn adhered to some rather conservative or formal habits. Take, for example, the way he signed his letters. From the very beginning of their correspondence, he opened his letters to Gill with 'My dear Gill' or sometimes 'dear Gill' and signed them with variations of '(very) truly/sincerely/faithfully yours, J.C. Kapteyn', without writing out his first name. He stuck to this habit throughout their correspondence. On the other hand, Gill usually wrote 'my dear Kapteyn' and kept using that phrase, but his early closing phrase 'sincerely yours' rapidly became 'your sincere friend, David Gill' or variations thereof. Similarly, Kapteyn opened his letters to Donner with 'Lieber Herr College' or 'Lieber Donner', and invariably signed them 'Ihr (ganz) ergebener, J.C. Kapteyn'. Newcomb was addressed by Kapteyn as 'dear Sir' and the letters ended with 'sincerely yours' or 'yours very truly, J.C. Kapteyn'. The few letters from Newcomb to Kapteyn that we have in our possession, on the other hand, usually started with 'My Dear Friend', but were signed 'Sincerely yours, S. Newcomb'. In this respect Kapteyn was rather formal, in spite of the obvious fact that some deep friendships developed with these colleagues in particular, with Gill and especially Newcomb becoming less formal and actually referring to their friendship as time progressed. On May 2, 1906, for example, Gill wrote: 'My dear Kapteyn, I have just received your letter dated 11th July 1906. Whether it is an "astral" development of what you will in future write on that date I don't know, but I think it probable you intended to write "April" instead of "July".' Such frivolous passages are difficult to find in Kapteyn's letters. To people like de Sitter, and other students of his, and to the younger van de Sande Bakhuyzen, Ernst Kapteyn opened letters with the much less formal 'Amice', but remained formal when he addressed the older brother Hendricus, who had been his superior.

From a letter by Gill to Kapteyn on December 22, 1897, I quote one particular passage. After informing Kapteyn that he has little time to write and will come back to the issues at hand soon, he adds: 'So I only send you my warmest good wishes and kind love to you and to your wife and my friends the dear lassies; and Mrs Gill, who I am thankful to say is much stronger, sends the same. God bless you all and may your new year be a happy and prosperous one. Your ever sincere friend, David Gill.' This reveals more than just a casual friendliness. The letters between Gill and Kapteyn regularly referred to their illnesses or those of their wives, and increasingly so with advancing age. This is much less the case in other correspondence of Kapteyn, which demonstrates the close personal friendship that the two men must have felt.

I seize the opportunity to bring up another rather puzzling point. Gill refers to the two daughters (the 'lassies') with some regularity, but there is no mention at all of the son Gerrit Jacobus, who was called 'Rob' in the intimacy of the family. The boy had just turned fourteen and must often have been around. Why was he not mentioned? What kind of person was Kapteyn's son in the first place? Actually, in the HHK biography, he is mentioned only a few times by his sister and each time only in passing. The first time she wrote that the daughters and the son were attending the HBS and the second time she mentioned that the son was studying in Freiburg, Saxony, to be a mining engineer. Another mention of the son is, when she told about the removal in 1905 of the Kapteyn family to a smaller place ('now that the elder daughter was married and the second was a student in Amsterdam', after which she added that the son still studied in Freiburg. And later we learn from the *HHK biography* that in 1908 the son had moved to the United States, where Kapteyn saw him on his way to and from his first Pasadena visit, and that later he and Mrs Kapteyn would visit the son whenever they traveled to America. But that is the full extent to which Gerrit Jacobus is being brought to the fore. Except for these remarks we know very little of him.

The reason for this may have had to do with the fact that Gerrit Jacobus left home and the Netherlands at a young age, going to Freiburg for his academic studies rather than to a Dutch university. Kapteyn's great-grandson Jan Willem Noordenbos has provided some more information on Gerrit Jacobus Kapteyn, collected from the latter's offspring. The reason he went to Freiburg for his studies in mining engineering was the fact that this was much less expensive at the time than studies at Dutch universities. Gerrit Jacobus also was apparently a rather adventurous man. After his studies he spent some time in the USA, living at Boulder in Colorado. But he must have returned to the Netherlands at some time during World War I, since he got married in Groningen on January 30, 1918. The marriage certificate describes him as living in Groningen at that time, as was his bride Wilhelmina Henriette van Gorkom. According to a genealogy on the Web [31], he worked in the Dutch East Indies from 1917 to 1921. He and his wife had a daughter Wilhelmina Elisabeth, born in Batavia, Java in 1919, a son Henri Carel Marie, also born in Batavia, in 1920, and a son Jacobus Cornelius, who was born in The Hague in 1923. This is in agreement with the years he reportedly spent in the Dutch East Indies. He worked there in mines on the island of Sumatra and was involved also in the building of bridges. The story goes that he spent quite some time down in the mines (which formally was forbidden for engineers), where it was so cold that he regularly caught colds returning to the surface and the tropical air. This may be the cause of his later health problems.

There is a story that at some stage (apparently before the coast of South America) he was involved in a shipwreck and survived only by clinging to a leather suitcase, that still is kept at the attic of one of his grandchildren. He and his wife lost al their financial capital in the stockmarket crash of 1929. It appears he was most of the time away from home, working on temporary contracts also in places like Rumenia. Later he developed tuberculosis (possibly related to his mining activities in the East Indies) and he spent his last years at a sanatorium in Switzerland, where he died in

8.10 Personal Matters

1937 (see Table 5.2). After so many years in different countries, and being away from home so often, he probably had lost contact with his sisters, at any rate with Henriette, but not his parents, one would think.

Another characteristic of Kapteyn was that he seemed to have been a good story teller. Cornelius Easton offers a few examples in his recollections of Kapteyn, published shortly after Kapteyn's death. I quote a single paragraph as an illustration and refer the reader to Appendix C. for more.

'Sometimes he told us about his youth. Of the prowess of one of his brothers, who was posing as 'little Mercury' on the ridge of the house's roof, standing on one leg, while his father looked up and was not able to move being filled with fear. – Or that on one occasion he had sat down on the hanging frame that house painters were using, after which the frame gradually started to come down – how he escaped from it he did not remember. Or how he and one of his brothers in the times before the 'Rover safety' [The 'Rover safety' is the model for our modern bicycle with the diamond-shaped frame and two wheels in the same line, the rear one driven by a chain] exultantly cycled into their hometown Barneveld; nobody had ever seen a bicycle and the people all gathered in great numbers to see them so that the two cyclists fell from their two-wheelers 'like ripe pears'.'

The following quote is again from Wessel Krul's chapter in the *Legacy*.

'In the relatively small and still very class-conscious community of Groningen, someone like Kapteyn could not fail to attract attention. He was often entirely absorbed in thought, which caused him to behave oddly in the streets. To his fellow townspeople he became the archetype of the absent-minded professor. This gave rise to many anecdotes concerning his person, some of which have survived. One evening, as Kapteyn returned home from his laboratory, an apparently inexperienced policeman noticed something unusual about him. He did not seem to know what he was doing, and the disorderly way he was dressed aroused suspicion. Was he drunk? A madman? Or a criminal? Kapteyn soon became aware that he was being followed, and decided to play the game. Instead of walking directly to his house, he took his persecutor on a tour of the town, leading him around all sorts of alleys and byways. Great was the disappointment of the loyal police officer (or: upholder of justice), when finally the suspect took his key, bade him farewell and disappeared behind a front door bearing the name of a well-known university professor (see W. Moll, 1957 [30]).

Inevitably, Kapteyn belonged to the unlucky category of people who tend to leave their luggage on the train. Once a friend went to meet him at the station in Rotterdam; Kapteyn stood waiting for him in the pouring rain, without an umbrella, his head only covered by an old straw hat. When asked why he didn't carry something to protect himself, he said: *'My wife refuses to buy any more umbrellas. I always seem to lose them along the way'* (see Easton, *Persoonlijke herinneringen*, p. 151.). Kapteyn's decision to travel second class only often brought him into contact with traveling salesmen, who then were the most frequent users of this railway class at the time. As he was an agreeable conversationalist, and as he did not distinguish himself in any way by his appearance, traveling with a large bag like all the others, they sometimes forgot he was not one of them. One day, upon arrival at Groningen station, he left the compartment without taking his bag with him. One of the salesmen ran after him to return it, saying: 'Professor! You forgot your bag of samples!' Kapteyn answered: 'Thank you very much! But these are not samples, you know'. At which the salesman retorted: 'Well then, if you wish, your bag of stars!' (see Perdok, W.G., 1964, *Leven en Werken van J.C. Kapteyn en F. Zernike*, [32] p. 181; the original conversation is in the Groningen dialect).

Everybody agreed that Kapteyn was an excellent and inspiring teacher. From his students he required very little ready knowledge, and he abhorred pedantry. To his successor, P.J. van Rhijn, he once said: 'Give me the failures, then I will allow you all the bright ones' (van Rhijn, 1922). He was interested in method, not in learning. But for the very same reason he also had very little patience with every kind of amateurism [Easton, p.112]. He was not a gifted orator in the traditional sense, and he had little literary talent [Easton, the *HHK biography*]. His publications are written in a loosely organized, exclamatory style, which looks slightly awkward on paper, but which must have been very effective in the classroom. What he lacked in rhetorical refinement, he compensated by spontaneity. He often underlined his statements by expansive gestures. Someone remembered him pushing the tip of a gigantic pair of compasses firmly into the blackboard, saying: 'You see, gentlemen, I need a fixed point!' [Perdok, p. 181].

Instead of isolating himself, Kapteyn always reminded his more conventional colleagues of the artificiality of social distinctions. Johan Huizinga, who usually traveled first class, remembered a trip with Kapteyn as a lesson in democratic behavior. 'I once made the journey with him from Groningen to Zwolle, second class, accompanied by an indiscriminate group of people, in whose presence I bluntly decided, according to modern manners, to keep silent. Not so Kapteyn. Without making any attempt at showing himself off, he was in no time in conversation with the whole company. At once, they all proved to be agreeable persons; they told him about their business and their opinions; he had suddenly brought the discussion to a much higher level than the common railway talk of salesmen, or whatever they were. I can still see how pleased and interested they looked at him, and it seemed as if they regretted not to be able to bid him farewell in a more cordial manner, when he descended the train' (see J. Huizinga, 1922, *J.C. Kapteyn* [33]).

8.11 German to English

At about the time he completed the CPD, the language in which Kapteyn used to write his scientific papers became a different one. As German was the lingua franca of science in the nineteenth century, it is no surprise that he started out writing in German. His early papers appeared in the German journal 'Astronomische Nachrichten'. As I noted above, his two early papers in the Irish journal 'Copernicus' were strangely enough in German as well. From the *Love Letters* we observe that he felt much less at ease with English (contrary to his fiancee) and he offered his apologies for his poor English in his early letters to Gill. His papers in relation to the Carte du Ciel are all in French, but so were all publications from the Central Bureau. Henriette Hertzsprung-Kapteyn mentions that at the boarding school in Barneveld, the boys spoke French during conversations. Kapteyn's texts in the Cape Photographic Durchmusterung were unavoidably in English and must have provided some training ground and incentive for the development of his command of this language. However, his papers in Astronomische Nachrichten continued to be in German, with one notable exception in 1896, when a paper bore the title *New southern variable stars*. He kept publishing in the German journal until 1903 and continued to do so in German, even though this was not strictly required. But after that he started to use English almost exclusively; from this time onward, the bulk of his papers appeared in his Groningen Publications (where all publications were in English) and later on, increasingly, in US journals and observatory publication series. After 1903, he published just one other paper in the Astronomische Nachrichten; that was in 1910 and it was in English. The turning point was around 1904, the year when he visited the United States for the first time, followed in 1905 by his first (and only) visit to South Africa.

The Annals of Leiden Observatory were in German up to Volume 10, which appeared in 1916, after which the series was titled in Dutch. Actually it had the German name 'Annalen der Sternwarte in Leiden' up to Volume 9 in 1915. Volume 11 (on an instrument after the French physicist Jean Bernard Léon Foucault (1819–1868) and dedicated to him) was in French, but from 1918 (Volume 12) onward, when de Sitter had taken over, the series was exclusively in English, except for a posthumous set of two papers by H.G. van de Sande Bakhuyzen in Volume 13. De Sitter was Kapteyn's first student. Kapteyn, being a Dutch astronomer, was at least a decade ahead of his time in using English for his publications.

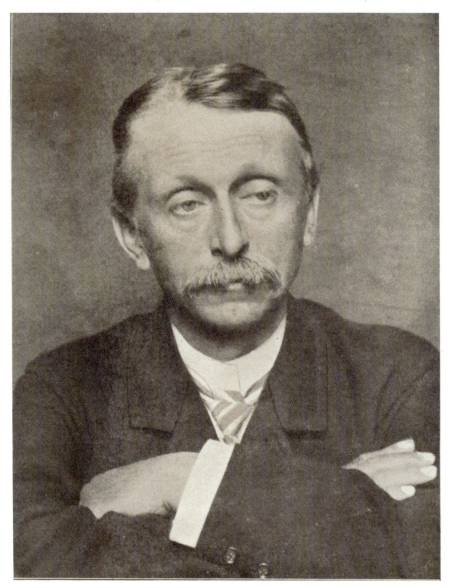

Fig. 8.21 Kapteyn at age 45 (Picture reproduced from the *HHK biography*)

Chapter 9
Star Streams

> *To ancient Chinese fancy, the Milky Way was a luminous river,*
> *– the River of Heaven, – the Silver Stream.*
> Lafcadio Hearn (1850–1904).[1]

> *[...] we are reminded that everything is flowing – going somewhere, [...]*
> *While the stars go streaming through space pulsed on and on forever*
> *like blood globules in Nature's warm heart.*
> John Muir (1838–1914).[2]

9.1 Nova Persei 1901

In February, 1901, a bright star appeared in the constellation Perseus, which quickly became known as Nova Persei 1901. It was discovered on February 21 by Thomas David Anderson (1853–1932). He was an amateur astronomer in Edinburgh, who apparently was sufficiently rich to pursue astronomy full-time and who has been credited with the discovery of many variable stars. The nova quickly reached a magnitude of 0.2, becoming one of the brightest stars in the sky. Such phenomena had been observed before – between 1890 and 1900 there had been three with a peak magnitude of 4 to 5 and therefore fainter, but visible to the naked eye– but in comparison this one was exceptionally bright so presumably rather close. These objects

---

[1] Patrick Lafcadio Hearn, Japanese name Koizumi Yakumo, was a writer especially on Japan.
[2] John Muir, Scottish-American writer and naturalist. From 'My first summer in the Sierra'.

were called novae and usually, in a few weeks' or months' time, they faded to levels that could only be seen with telescopes.

A nova, we now know, is an example of a so-called cataclysmic variable, consisting of two stars rotating around each other in a binary configuration, of which one is a white dwarf and the other a regular Main Sequence star. The white dwarf was a heavier main sequence star than the remaining one, but has progressed through its more rapid evolution into the white dwarf phase. As the second star also evolves into the stage of red giant, it expands and the result of this is that the outer layers become so close to the white dwarf that this material gets pulled off the main sequence star onto the white dwarf. What happens then is that this accreted gas is accelerated to high speeds in the strong gravitational field of the very compact white dwarf, so that when it hits the surface of the latter, it is violently compressed and heated. This gives rise to nuclear reactions in the hydrogen in that gas, to some extent similar to what happens in stellar interiors, except that here elements heavier than helium are also being produced. The system brightens considerably and only fades when the accreted gas is blown away again from the surface of the white dwarf.

The nova in Perseus (originally called Nova Persei, more recently with the addition 1901, but formally designated as GK Persei) happened to be a spectacular nova. Not only was it very bright in the sky – this, to some extent, reflects its proximity – but it had also been brightening very rapidly. As it happened, the region on the sky where it resides, was photographed two days before discovery and at that time it was thirteenth magnitude. But when it was discovered it was third magnitude, so it had brightened by ten magnitudes or a factor ten thousand in just two days. It remained visible to the naked eye for about a year, slowly fading until it reached a level of about magnitude 12 by 1904. From then on it showed irregular fluctuations and some outbursts after 1948. Observations of the radial velocities show that the binary has an orbital period of about 1.9 days. The light curve up to 1983 of GK Persei has been published by F. Sabbadin, F. & A. Bianchini and the orbital parameters by A. Bianchini, F. Sabbadin & E. Hamzaoglu [1].

Within a year's time a luminous shell or nebula was detected around the nova. This was unusual, especially that it happened so quickly. In most cases a nebula or shell is actually observed, but then these take years to become bright enough to be visible, even with the best telescopes. At Lick Observatory, Charles Dillon Perrine (1867–1951) (see Figs. 9.1 and 9.2) and at Yerkes Observatory George Willis Ritchey (1864–1945), a famous telescope designer, discovered on photographs that there were features in the nebula that were moving on the sky at an astonishing rate of 11 minutes of arc per year. This speed was spectacularly high, since – unless the star is closer than about 10 parsec – this would mean an expansion velocity higher than the speed of light! We now know that the distance to GK Persei is about 460 parsec, so this apparent expansion would exceed the speed of light by an enormous factor.

Nowadays people grow up with the notion that nothing can move faster than the speed of light. But this was posed by Albert Einstein (1879–1955) in his theory of special relativity of 1905, and although in 1901 possibly superluminal speed was not considered necessarily unphysical, it was completely unexpected. Kapteyn was also puzzled by these observations, but he quickly came up with an idea how this

9.1 Nova Persei 1901

Fig. 9.1 Nova Persei (GK Persei) photographed during a 10 hour exposure at Lick Observatory on November 12/13, 1901 (left), and one of 9 hours and 45 minutes on January 31/February 2, 1902 (right) (Lick Observatory, University of California, Bulletin No. 23 [2])

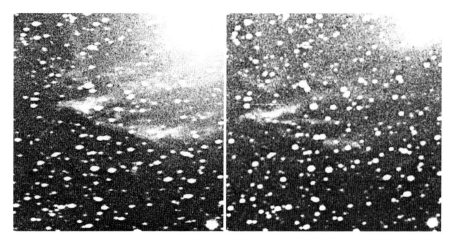

Fig. 9.2 Nova Persei (GK Persei). Details and contrast in enhanced versions of the pictures in Fig. 9.1 (Lick Observatory, University of California, Bulletin No 23, see Fig. 9.1)

could be used to estimate the distance of Nova Persei. He published it in German, Kapteyn (1901d), in a short article, which he submitted on December 8, 1901, only months after the first announcements of these motions. Fortunately, the paper was translated by the Journal *Popular Astronomy* into English, Kapteyn (1902e), and made available to circles of American astronomers in particular, where it certainly drew attention. As it turned out, the basic idea was correct, but the geometry and the distance were not. Kapteyn presumed that what was seen as moving nebulae were in fact *not* motions of nebular material, but of spots of reflected light. What Kapteyn envisaged was the following:

First he noted that the assumption that the 'nebulae' are much closer to us than the nova gives rise to other problems, which is not very helpful. He assumed that the space around the nova was filled with material that reflects the light from the nova. This nebula that surrounds the star does not emit much light itself, so we only see the reflections of the nova light. Think first of the ideal case that the surrounding nebula is uniform and that the nova phenomenon is a relatively short flash. Of course, here on Earth we see the flash as soon as the light reaches us along a straight line from the nova to us. But now think of light that takes off in another direction and some part of which gets reflected into our direction at some distance from the nova. Usually this is projected for us at a small distance on the sky from the nova, but since the light has to travel a longer path length it arrives later than the direct light. Kapteyn showed that for a particular time after we have seen the nova, there is a three-dimensional figure for which for every point on it the distance from the nova plus the distance from that point to us is the same. If light is reflected by any material in the surface of this figure, it would reach us at the same time. He then showed that this figure is an ellipsoid of revolution (think of an ellipse rotating along its long axis) with the nova and the Earth in the two foci. This is easy to appreciate from the fact that in an ellipse the length of any line from one focus via the figure itself to the other focus is constant, a property often used for the construction of ellipses. In fact, the figure is so stretched that it can be approximated by a parabola. At a later time, this became a larger paraboloid. If all reflecting matter were uniformly distributed, we would at any time see all material lighting up along the paraboloid corresponding to that moment. If the nebula were a sphere around the nova, we would, as time progresses, see subsequent slices of the form of the intersection of the paraboloid and sphere. Kapteyn compares this to the way 'the biologist learns the structure of an organism by a series of consecutive cross-sections'.

But there is no uniform nebula, as the photographs show. Kapteyn inferred from these that there was a 'set of streamers, which is enclosed in a plane', which makes an angle of about 79° with the line of sight; this angle he derived from the shape of a prominent streamer which should be the circumference of the circle that corresponds to the intersection of the paraboloid and this plane. It seems to me a far from unique description, but is not inconsistent with the observations. Anyway, because of this angle the illuminated part of the streamer travels along it at about the speed of light. And from this geometry Kapteyn deduced from the changes on the sky that the distance to Nova Persei is 9 parsec (since he did not use the parsec, he actually quoted a parallax of $0''\!.011$).

The basic idea of regarding the expanding features as light echoes from the nova eruption was correct, but the geometry was not. Further refinements were published, e.g., by Arthur Robert Hinks (1873–1945) in 1902, in *The movements in the nebula surrounding Nova Persei* [3]. Hinks considers a geometry in which the 'nebulous matter is disposed in a circle with the star as center'. He shows that the apparent motion could then even surpass the speed of light. Indeed these 'superluminal velocities' can easily be explained by Kapteyn's fundamental idea of reflection of light on dust structures. A model with apparent superluminal motion was eventually developed successfully by Paul Couderc (1899–1981) in 1939 (see *Les*

9.2 Cosmic Velocities

Fig. 9.3 A recent picture of Nova Persei (GK Persei). The nebula around it is sometimes referred to as the Firework Nebula (National Optical Astronomy Observatories, [4])

auréoles lumineuses des novæ, [5]) The apparent motion (from further analysis of photographs) is about 4 to 6 times the speed of light and results from light echoes on a sheet in front of the nova. Note that special relativity forbids matter, energy and information to move faster than light. But here the expansion is not some physical piece of material; the situation can be compared to the light of a rotating lighthouse, which throws a spot of light on a large, distant projection screen. As we move the screen further from the lighthouse, the speed of the spot on the screen will increase and there is no fundamental principle of physics that prevents this from exceeding the speed of light.

For the sake of completeness I note that the nebular material observed *today* around GK Persei *is* remnant material thrown out from the cataclysmic binary (see Fig. 9.3).

9.2 Cosmic Velocities

We have seen in the previous chapter that even before his work on the CPD was completed, Kapteyn had already focused his attention on the problem of the distribution of stars in space, the 'Construction of the Heavens' as William Herschel had called it. Having rejected the option of obtaining stellar parallaxes in a wholesale manner as a matter of routine, he decided to turn to motions of stars as a (statistical) guide to distances of stars, using the motion of the Sun with respect to the general system of stars (the Local Standard of Rest) as a base for the distance determination.

Knowing the direction of the Sun's motion (the position of the Apex) and its magnitude makes it possible to calculate for every star what part of its proper motion is the reflection of the Sun's motion: the 'parallactic' motion. Improved determinations for the Apex and solar motion were presented in Kapteyn (1898a).

Kapteyn presented his way of tackling the problem of the structure of the Sidereal System, using motion of stars, in a presentation he gave before the Royal Academy on May 31, 1895, Kapteyn (1895). This was followed by some additions on June 10, 1897, in Kapteyn (1897a). Both presentations were titled *Distribution of cosmic velocities*. The printed reports, which appeared in the 'Zittingsverslagen', were at that time published in Dutch only. They became considerably longer. Kapteyn's first contributions, up to 1892, were no more than a few pages, whereas the one in 1893

Box 9.1 Density distribution of stars from proper motions
In the notation of Kapteyn (1895) we have:
r, distance expressed in units of the distance at which the annual motion of the Sun is seen under an angle of $1''$,
s, the linear velocity of the star,
v, projection of s onto the line perpendicular to the plane through Sun, star and Apex,
τ, projection of the proper motion μ perpendicular to the direction the Apex/Antapex,
$\Omega(v)$, probability that the projection of the linear proper motion onto a given line has the value v,
$f(s)$, the probability that the linear proper motion has the value s,
Δ_r^m, the number of stars per unit volume of apparent magnitude m at distance r,
N_τ^m, the number of stars of apparent magnitude m that have a projected proper motion of magnitude τ.
The distribution of projected linear proper motions relates to the distribution of space velocities through the relation

$$\Omega(v) = \int_v^\infty \frac{f(s)}{s} ds.$$

Now, if we have a solid angle ω on the sky, then between distances r and $= dr$ the number of stars in ω of apparent magnitude m is $\Delta_r^m R^2 \omega dr$. Of these the number of stars with linear, projected proper motion v will be $\Delta_r^m \Omega(v) R^2 \omega dr$. Then it follows that (since $v = r\tau$):

$$N_\tau^m = \omega \int_0^\infty \Delta_r^m \Omega(r\tau) r^2 dr.$$

And, using the equation above, this becomes

$$N_\tau^m = \omega \int_0^\infty \Delta_r^m dr \int_{r\tau}^\infty \frac{f(s)}{s} ds.$$

9.2 Cosmic Velocities

was 16 pages long and subsequent ones usually covered 10 to 20 pages. Starting with the meeting of May 1898, the Academy published English versions; Volume I of the 'Proceedings of the Section of Sciences' was published in 1899, [6]. The presentations reported here are therefore only available in Dutch.

In the first contributions, Kapteyn (1895), he started by outlining his aims: to determine (**1**) the law according to which the absolute linear velocities were distributed; (**2**) the law according to which the stellar density changes with distance from the Sun; and (**3**) the law of the distribution of absolute luminosities. He went on to explain that to achieve this he needed to adopt three hypotheses: (**a**) among the directions in which the motions of the stars take place, there is no preferred one; (**b**) the law of the distribution of the velocities does not change with distance from the Sun; and (**c**) the form of the distribution of the linear velocities has only one maximum. The last two were necessary for the mathematical theory he was developing to derive these laws from the observed distributions. Basically, the second one can easily be checked by examining the *mean* linear velocity as a function of distance from the Sun. At the time there were some conflicting opinions on this issue, but Kapteyn showed in his contribution that his new analysis supported this assumption.

Kapteyn then developed the mathematical tools to derive information from the observed distribution of the proper motions, on that of the *linear* proper motions, i.e., the velocities perpendicular to the line of sight corrected for distance (so in linear measure such as km/sec). All this under the assumption of the hypotheses posed. This was not easy; it involved very complicated calculus, including evaluation of triple integrals. It turned out that he had so much trouble getting this done, that he enlisted the help of his mathematician brother Willem. Eventually this theory was published in detail in Groningen Publications 5, Kapteyn & Kapteyn (1900). The article is almost 90 pages long and full of mathematical formulae and tables. I will return to this in more detail in the next chapter.

The problem with deriving the space distribution of stars from observed distributions of proper motion and apparent magnitude is that it involves the inversion of an integral. For the mathematically inclined, Box 9.1 shows some of the details. For others I will try to explain what is involved with the help of a simple example.

Take a sports competition such as in national leagues in soccer, baseball, field or ice hockey, American football, etc., e.g., the premier league ('Eredivisie') in Dutch soccer. Here 18 teams compete during a 'full' competition, which means that each team plays every other team twice: once at home, once away. After 34 matches, the final ranking is arrived at based on the points earned. But from the final standings (the integral) it is not possible to derive for each team how many times they won, lost, and played a draw at home and away; that information is not preserved in the integral and it cannot be converted. You need additional information if you would like to analyze the individual performances of teams. Such additional information could, for example, be the statistical result that half the games end in a win by the home team, one quarter in a win of the visiting team and one quarter in a draw (see my article *Home advantage and tied games in soccer*, [7]). Another piece of information would be that the top teams usually win most of their home games and the bottom ones almost always lose their away games. This may not be sufficient to uniquely trace back how the scores came about, but at least it may give some more insight.

Kapteyn then went on to make the following inference first. One would expect that for stars that have small velocities compared to that of the Sun, the directions of the proper motions need to be more concentrated towards the direction of the Apex. After all, for these stars the 'parallactic' component of the proper motion (the part of it resulting from the motion of the Sun) should in general be larger than that from the peculiar motions of the star itself. Using the stars of Bradley's catalogue, Kapteyn found that the direction of their proper motions on the sky is consistent with the mean velocity of stars being constant with distance from the Sun. This was contradictory to a claim by Frederich Ristenpart (1868–1913) from the Sternwarte Karlsruhe (*Veröffentlichungen der Grossherzoglichen Sternwarte zu Karlsruhe*, IV., pp. 286 and 287), but Kapteyn argued that his result was superior, both methodologically and in the data used.

Up till now he had only used the *directions* of the proper motions and not their magnitude. Kapteyn then calculated the proper motion *in* the direction of the Apex and that *perpendicular* to it. Next he calculated their ratio. He had to exclude stars belonging to two nearby clusters of stars, as the motions in space of their stars are not uncorrelated. These two clusters are the nearby ones that we have discussed before, the Hyades (see page 384) at a distance of about 48 parsecs (roughly 150 light years) and the Pleiades (see Fig. 7.11), which have a distance of about 130 parsecs (roughly 400 light years). Now, the component of proper motion in the direction Apex/Antapex contains a part due to the Sun's motion, the amount of which depends in a known way on the star's position in the sky relative to that of the Apex, while the component perpendicular to it does not. In principle, the ratio of the two should then provide statistical information on the stellar velocities. It is possible to estimate from this what the mean velocity of stars is, expressed in terms of that of the Sun. From this, Kapteyn found that stars move in space with a mean velocity of 1.86 times that of the Sun.

That is an important finding. However, Kapteyn was not satisfied and repeated the analysis not using the ratio above itself, but its square or cubic root. Of course, the final outcome then needs to be squared or cubed. The reason he did this was to make sure it was less sensitive to extreme values, by making dominant contributions of large proper motions over small ones less pronounced. He did not go through this in detail, as the outcome did not affect his conclusion to a high extent.

In the same contribution he showed how, when the distribution law of the velocities of the stars was known, his approach could be used to derive from the observed proper motions what the density of stars (number per unit of volume) was as a function of distance for stars of a certain apparent magnitude. In other words, for the stars of a particular apparent magnitude you would know how many of them are closer than 10 parsecs, between 10 and 20 parsecs, etc. – and that for a range of apparent magnitudes. This makes use of the fact that the proper motions have a systematic component (the parallactic proper motion due to the motion of the Sun) and a random one that should average out in large samples.

9.2 Cosmic Velocities

Once one has determined the number of stars per unit volume as a function of apparent magnitude and distance from the Sun, it is possible to turn this into a distribution of the density of stars as a function of distance. Kapteyn had done such calculations before, for the Bradley stars, but had encountered a serious problem. It concerned the distribution of the angles that the proper motions made with the direction to the Apex, in particular for stars in and around the Galactic south pole at Galactic latitudes between $-40°$ and $-90°$. These seemed to indicate that the Sun moved in the direction of the Apex rather than the Antapex! And this was true for stars of all spectral types. The effect cannot be attributed to errors in the assumed direction of the Apex or in the value of the precessional constant. Although it could mean that the assumption that the distribution of stellar velocities did not vary with the location in space (Kapteyn's assumption **b**), he attributed it to errors in the Bradley proper motions. But it remained a serious problem.

Kapteyn returned to the Royal Academy with a sequel presentation, Kapteyn (1895), and began by presenting a diagram that I reproduce in Fig. 9.4. Here he showed how he used both the direction and the magnitude of proper motions to determine the 'law of cosmic velocities'. To understand what Kapteyn did, consider the following: We designate the angle between the proper motion of a star and the direction towards the Solar Apex with p and consider intervals in p of $15°$. First look at stars at a particular distance from the Sun. The line SB has a length proportional to the mean proper motion of these stars in the interval centered at $p = 15°$, the line SC for $p = 45°$, etc. The points B, C, etc. will then show a smooth curve that points towards the Apex (A) and has a form that depends on the distribution of space velocities n of the stars involved. Now, if we take a set of stars at a larger distance and make the same plot, the resulting curve will have the same shape, but will have shrunk by an amount equal to the ratio between the two distances. That is, if assumption (**b**) is correct and the velocity law does not depend on distance from the Sun. Also the number of stars contributing to the points B, C, etc. will be in the same proportion. This (unique) dependence can of course be expressed in – as it turns out – a somewhat complicated set of mathematical formulae.

In Kapteyn (1897a), he then collected as much material as possible, excluding again stars from the Pleiades and Hyades, but also excluding all stars with proper motions so small that their directions were uncertain, faint components of binary stars, etc. At that stage he was left with no fewer than 2355 stars. He then started testing the randomness of the sample by considering the distribution of the angles p and he found a strong systematic deviation from axial symmetry. He convinced himself first that this was not due to some systematic motion. Closer examination left him three options to explain the asymmetry (and I quote verbatim from the paper):

1. 'There is a systematic motion in the direction of the south pole for all stars with large proper motion with respect to stars of the rest.' He means the Galactic south pole.

2. 'A negative correction to the assumed declination of the Apex.'

3. 'A negative correction to all proper motion in declination.'

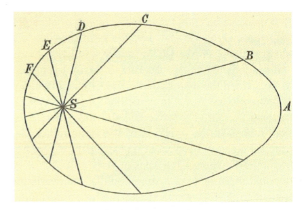

Fig. 9.4 Figure from Kapteyn's presentation for the Royal Academy of Arts and Sciences on June 10, 1897. See text for explanation (From Kapteyn (1897a))

Kapteyn regarded the first option as '*a priori* not likely'. Option **2** is more difficult to exclude, but he concluded that the main cause probably was option **3**: 'It leaves after all this little doubt that we should look for the cause of the observed asymmetry chiefly in a correction of the proper motion in declination.' We should read 'systematic error' instead of 'correction'. This correction, he argued, should be the same for stars of the same declination. As support he quoted a result by van de Sande Bakhuyzen, who had found that the Bradley stars with small proper motion indicate a more southern declination of the Apex than stars of larger proper motion.

9.3 The Solar Apex Revisited

This was a very unsatisfactory state of affairs. The theory of the approach of analyzing the data that I have just described, was published in full in the paper Kapteyn & Kapteyn (1900). It appeared as a Groningen Publication **5** under the title *On the distribution of cosmic velocities. Part I. Theory.*, and Groningen Publications **6** was going to be part II, dealing with the application of the theory. That volume has *never* appeared. Kapteyn became gradually convinced that the discrepancies were not due to problems in the proper motions, but that in fact the assumption of absence of preferred motion of the stars in space had to be dropped. The systematic differences were simply too large to account for in the framework of isotropic velocities, in spite of the simplicity and attractiveness of this assumption. This led to his most impressive discovery – that of the two Star Streams.

Before going into this in more detail, I first go back to his work in the late 1890s where he tackled the sidereal problem. Three things were foremost in his mind: firstly the improvement of the accuracy of the proper motions that were known at the time, especially the issue of systematic errors in declination. Secondly, the improvement of the accuracy of the position of the Apex and the velocity of the Sun in space. He must have hoped that this would eventually solve the problems. And thirdly, he was in the middle of developing the mathematical machinery, already

9.3 The Solar Apex Revisited

mentioned above, to turn the observed proper motions into a statistical distribution of mean parallax of stars as a function of observed proper motion and apparent magnitude. He published a torrent of papers on these subjects around 1900.

Some of these, especially the ones that were published in 1900, seem to be a backlog that had been building up for a number of years. This year marked the start of the new series of the *Publications of the Astronomical Laboratory at Groningen*. It must have been a matter of great satisfaction to Kapteyn, as it established his laboratory as a full-fledged scientific institution. Obtaining the funds for this was not easy, of course, and that must have been the reason that there was a substantial backlog of papers waiting to be printed and the numbers 1 through 8 (except of course the notorious number 6) all appeared in 1900, almost 400 pages on large format. After that they appeared at regular intervals, nos 9–11 in 1902, 12–14 in 1904, 15 and 16 in 1906, etc. No 7, *Components τ and υ of the Proper Motions and other quantities for the stars of Bradley*. involved a careful consideration of the available proper motions to base the statistical studies on, using five different assumptions for some parameters, while No 9, Kapteyn (1902b), provided a sixth calculation using his most probable value thereof. This was discussed on page 296. The result was his final and most accurate set of proper motions that he felt was available.

The first of Kapteyn's concerns was the task to improve the position of the Apex as much as possible. This was not only affected by the accuracy and treatment of available proper motions, but in addition there was still uncertainty in the value of the precessional constant that had to be used to compare stellar positions at substantially different epochs (see page 291). He published an extensive paper, Kapteyn (1901b), with the title *Der Apex der Sonnenbewegung, die Constante der Praecession und die Correctionen der Eigenbewegungen in Declination von Auwers-Bradley*. He also reported the result at the Royal Academy, Kapteyn (1900b; in the English translation as Kapteyn, 1900c). The principle of the determination of the Apex is simple: find the point on the sky where the extensions of the directions of proper motions of stars all over the sky cross. The matter is much more complicated in practice. Stars have their own (*secular*) motions superimposed, so the directions only cross approximately. Some proper motions are large and well-determined, others are small and have large uncertainties, others again are far from the Apex and extrapolation of their proper motions produces large uncertainties near the Apex. Therefore the result depends on the actual distribution of available stars on the sky (ideally one uses a distribution that is as uniform as possible). The procedure is a statistical one and for its execution assumptions have to be made with regard to the actual distributions of the space velocities of stars. Usually the assumption in this context is the one that we have already seen a few times, which Kapteyn – in the introduction of the paper – called the 'Fundamental Hypothesis', designated *Hypothesis H.*: 'The peculiar proper motions of the fixed stars have no preference for any particular direction'.

Kapteyn criticized the previous determinations by George Airy and Friedrich Argelander, which were stated to be based on Hypothesis H. He found that they were 'not entirely based' on that hypothesis (meaning that the mathematical treatment was not in full accord with its full exploitation). He further noted that there was a determination by Hermann Albert Kobold (1858–1942) of Hungary, who stated that the hypothesis should not be used as there were indications that it did not represent the actual situation in nature. Kapteyn, however, was not yet convinced that this was the case, but found the method proposed by Kobold likewise unsuitable. Kapteyn indicated at the end of the paper, as Simon Newcomb pointed out as early as 1843, that the physicist Auguste Bravais (1811–1863) from Lyon had determined the direction of the Apex without making use of any assumption such as Hypothesis H (see also page 292).

After developing the necessary mathematics for a correct treatment of proper motions that was fully consistent with Hypothesis H., Kapteyn applied it to the best data he felt were available, the proper motions of the Auwers-Bradley stars, but supplemented them with a new set of proper motions determined by Russell William Porter (1871–1949) from the Cincinnati Observatory. This way he ended up with no fewer than 699 stars with sufficiently accurately determined proper motions. He carried out the analysis for the two samples separately. To do this, he proceeded as follows: First he took an educated guess at the position of the Apex. Next he calculated for each star the angle between that direction and that of its proper motion. These angles together should add up to zero if the assumed position of the Apex is the correct one; actually this should hold for various parts of the sky separately. In a second approach he used the property that if there were no random motions of the star the pattern on the sky for stars of a certain distance, the proper motion in the direction of the Apex would be largest 90° away from the Apex (actually it would scale with the sine of the angle to the Apex). So he then corrected the proper motions for this pattern. Now, for the correct Apex position the sum of these corrected proper motions should be maximum. In a third and fourth method he calculated the squares of the differences between the expected pattern (zero angles in method I and zero deviation from the expected pattern in method II) and found the correct Apex by selecting the smallest of the sums of the squares of these differences.

This sounds laborious, since you would have to do the calculations again and again for many assumed positions of the Apex. However, one can develop mathematical techniques – which is what Kapteyn did – where 'all' you have to do is enter the observed data (in this case the two coordinates of the stars and the direction and magnitude of their proper motions) into a set of equations and then solve them, again using mathematical techniques. This is the 'least squares technique'. Who designed this technique is not a matter of consensus; it is usually credited to Johann Carl Friedrich Gauss (1777–1855), but also frequently to Adrien-Marie Legendre (1752–1833).

In any case, the basic idea is the following (skip the next paragraph if this is not your prime interest): Suppose you have some points in a plane more or less along a straight line and you wish to find the straight line that fits it best. Often this involves a measurement of a property y for a given x. By 'best' is meant the following: Take any line and calculate for each point how far it is away (in y since the x is given) from this line. Square all these deviations in y and add them up. For the best fitting line in a 'least squares sense' this produces the smallest number. In this case it is straightforward and well-known how to find the best fit: add up all the values x and y, and also x^2, y^2 and xy. Then there is a mathematical formula that makes it possible to calculate from these 5 sums the slope of the line and the 'intersect' (where it crosses the y-axis). There are also formulae to estimate the uncertainties in these properties.

Not having too much faith in the precession constant and the declinations, Kapteyn not only made fits of the data to derive a solution for the position of the Apex, but also for the precession constant and corrections of the Auwers declinations in five declination zones. So, in the end his analysis gave the best fit of the data using eight unknowns. Kapteyn converged on a value of the Solar Apex with uncertainties of order $1°$. These were formal errors and even today astronomers would not claim to know that position to better than a five degrees or so. Kapteyn's result was Right Ascension 18^h 16^m and declination $+29°$; the current best values are 17^h 46^m and $+23°$ (see Fig. 8.14).

Naturally, he also produced best values for corrections on the precession constant and the declinations in the Auwers' catalogue.

9.4 Reactions to the Apex Determination

Johannes Wilhelmus Jacobus Antonius Stein (1871–1951), a Jesuit priest who had obtained a doctor's degree in Leiden in 1901 under Hendricus van de Sande Bakhuyzen, communicated a reaction to Kapteyn's results to the Academy through his supervisor. He strongly disagreed with Kapteyn's criticism of the Airy treatment of proper motions (the English version has the title *On J.C. Kapteyn's criticism of Airy's method to determine the Apex of the solar motion.*, [8]). Stein, who was to become director of the Vatican Observatory in 1930 and was responsible for its move away from the Rome city lights to the papal summer residence and retreat Castel Gandolfo, argued that actually the Airy method *was* fully consistent with Hypothesis H. Kapteyn insisted Stein was wrong and published a reply in Kapteyn (1901c; English version Kapteyn, 1902a]. 'It appears', he opened the paper, 'that Dr Stein has not completely understood my paper. [...] This fact, and the fear that on the other hand I may also have misunderstood Stein's reasoning (for one part at least this is certain) have led me to make my reply more circumstantial and elementary than might otherwise seem necessary'.

The disagreement focused on how to minimize the squares of the deviations in the two celestial coordinates (together or separately). For any possible position of the Antapex, one can calculate what – at the position of any star – the components of the parallactic proper motion should be and derive the deviations of these from the observed components. The squares of these should then be minimized. Now Airy used the *equatorial* coordinates for the components to be minimized, so he projected observed proper motions onto great circles through the pole (giving declination) and small circles perpendicular to that (giving right ascension). For any assumed position of the Apex he then calculated the resulting components of the parallactic proper motion (due to the motion of the Sun in space) projected onto these axes. Next he minimized the deviations from the observed proper motions by shifting the assumed position of the Apex. Stein did not use the celestial pole, but a *fixed* point near the probable Antapex and thus he projected the proper motions onto great circles through this point and small ones perpendicular to it. When he changed the assumed Apex coordinates, he kept this same point for reference. Kapteyn, on the other hand, *varied* the 'pole', so that it coincided with possible positions of the Apex. He believed that this was more in accord with hypothesis H. Kapteyn in the end defended his own approach, stating that it was based 'on the following: the equations which are derived [...] are identically the same as those which Bravais derived, long before Airy, from a quite different, mechanical, principle.' Stein published an adapted version of his critique in the astronomical literature, *Der Charakter der Airy'schen Methode zur Bestimmung des Apex der Sonnenbewegung*, [9]. Kapteyn followed suit and likewise published an article in the same journal with almost the same title, Kapteyn (1902f). It seems to me that the difference between the two methods is smaller than the intrinsic uncertainty in the position of the Apex as a result of the peculiar stellar motions themselves.

A further follow-up was called for when, shortly after Kapteyn (1901b) had submitted his paper, Lewis Boss of the Dudley Observatory in Schenectady, New York, published another determination of the Solar Apex (*Tentative researches upon precession and solar motion*, [10]), in which he found the same right ascension, but a wildly different declination of 45° (compared to Kapteyn's 30°). Neither of them was aware of the other's papers when they submitted theirs. Kapteyn published a long and detailed comparison showing that the difference is due to incorrect treatment of systematic errors, inclusion of stars with small proper motions and therefore relatively large errors and methodological ones, Kapteyn (1903a). Boss did not stick to his results; some years later, in 1910 he published a study in which he found a declination for the Apex of about 35° [11].

9.5 Mean Parallaxes and Luminosities

One of Kapteyn's primary aims in his quest to unravel the structure of the Sidereal Universe was to find ways to estimate the luminosities of stars and determine their relative frequency in space. A first step would be to see to what extent proper motion and apparent magnitude were indications of the distance. A first paper towards this aim was contained in No 8 of the Groningen Publications, *On the mean parallax of stars of determined proper motion and magnitude*, Kapteyn (1900e). In the preface he stressed that these results were only provisional, since new data were becoming available rapidly and further studies on the distribution of stars in space would further constrain the distributions. He based his study on the one hand on the available measured parallaxes, and on the other hand on the outcome of the study reported in Kapteyn (1898a). In addition he made use of the spectral types as available in the Draper Catalogue.

8. *Summary of results. Tables.*

The following formulae may be considered as the final results of the preceding investigation:

(50) Type I $\quad \bar{\pi}_{\mu.m} = (0\cdot905)^m - 5\cdot5\sqrt[1\cdot11]{0\cdot116\,\mu}$

(51) Type II $\quad \bar{\pi}_{\mu.m} = (0\cdot905)^m - 5\cdot5\sqrt[1\cdot54]{0\cdot0262\,\mu}$

(52) All the stars together $\bar{\pi}_{\mu.m} = (0\cdot905)^m - 5\cdot5\sqrt[1\cdot405]{0\cdot0387\,\mu}$

} mean parallax of stars with determined proper motion and magnitude,

obtained by a combination of the formulae (16), (17), (18) with (21).

(53) . . . Type I . . . . . . $\pi_m = 0''0098\,(0\cdot75)^m - 5\cdot5$

(54) . . . Type II . . . . . $\pi_m = 0''0223\,(0\cdot75)^m - 5\cdot5$

(55) . . All the stars together $\pi_m = 0''0160^5\,(0\cdot75)^m - 5\cdot5$

} Mean parallax of *all* the stars of magnitude m,

by formula (26).

(56) . . . . . $\varrho = $ Probable amount of $\log \dfrac{\pi}{\pi_0} = 0\cdot19$

(57) . . . . $\pi_0 = 0\cdot810\,\bar{\pi}$

for which consult article 7.

Fig. 9.5 Final summary of the result reported in Groningen Publication **8**, *On the mean parallax of stars of determined proper motion and magnitude*, Kapteyn (1900e). The formulae in the top part give the *mean* parallax of stars of given proper motion, and of apparent magnitude. The bottom formulae give the *most probable* parallax as 0.810 times that of the mean one, based on an assumed distribution of parallaxes (Kapteyn Astronomical Institute)

The number of directly measured parallaxes that Kapteyn regarded as reliable is 58, although it must be stated that some are unrealistically small for an accurate measurement and some are negative. He seems to treat those as stars with upper limits to their parallaxes. To this he adds the parallaxes obtained by Albert Stowell Flint, which were unpublished at the time. At the Washburn Observatory, Flint had used the meridian passage timing technique that Kapteyn had pioneered in himself, Boss & Kapteyn (1891). Remember that Flint was the person who had written so favorably about Kapteyn's publication when it first appeared (see page 146). The results of his use of proper motions to determine distances from the parallactic method produced information on the mean parallaxes of stars of various different apparent magnitudes. Kapteyn's final outcome was a set of equations where one could calculate the mean parallax of stars of a given apparent magnitude and proper motion for each spectral type separately or for all stars together. By summing over the proper motions one could also find the mean parallax of stars of a given apparent magnitude. I reproduce this important result directly from the paper in Fig. 9.5.

If Kapteyn's results would hold strictly, they would be very powerful instruments to study the space distribution of stars as a function of their spectral type and luminosity. But Kapteyn realized he needed a much more solid basis to establish a reliable guide towards stellar distances and luminosities. His next study was published as No 11 of the Groningen Publications: *On the luminosity of the fixed stars*, Kapteyn (1902d). This paper is actually the same (except for slight changes in the wording here and there) as the English translation, Kapteyn (1901g), of a presenta-

> **Box 9.2 Kapteyn's definition of absolute magnitude**
> The following is directly quoted from *Groningen Publications* 11, 1902:
> 'If we put further:
>
> $L =$ luminosity, or total illuminating power of a star of apparent magnitude m and parallax $= \pi$,
>
> we can easily find by [the the definition of the magnitude scale]:
>
> $$\log L = 0.2000 - 0.4m - 2\log\pi.$$
>
> We further define the *absolute magnitude (M)* of a star, of which the parallax is π and the distance r, as the apparent magnitude which that star would have if it was transferred to a distance from the Sun corresponding to a parallax of 0″.1. It is easily seen that
>
> $$M = m - 5\log r = m + 5 + 5\log\pi = 5.5 - 2.5\log L.$$
>
> For the Sun $L = 1$; the formula thus gives for the absolute magnitude of the Sun $M = 5.5$, [...].'

tion before the Royal Academy, Kapteyn (1901a). He started by summarizing the result of Fig. 9.5.

He then made a compilation of the number of stars within narrow ranges of proper motion as a function of apparent magnitude from the most reliable data he could lay his hands on. For the interval magnitude 3.5 to 6.5 he used his own improved proper motions of the Auwers-Bradley stars and for fainter stars other determinations of proper motions. This could be combined with his earlier work on the distributions of proper motions as a function of apparent magnitude. He then determined the distribution of parallaxes as a function of proper motion and apparent magnitude. But what he needed was the luminosities or absolute magnitudes. Here he used a definition that is worth pausing for, as it is the very same one that is still in use today.

9.6 Absolute Magnitude

Kapteyn defined *absolute magnitude*, which he designated (as we do today) with a capital M, as the apparent magnitude a star would have if its parallax were 0″.1. Because of its historical significance I quote that section in full (including some extremely simple equations).

9.6 Absolute Magnitude

'As unit of luminosity I will adopt the total luminosity of the Sun. It is true that our knowledge of the relation between the quantities of light which we receive from the Sun and from certain fixed stars is still very imperfect. This is however of little importance, because, when this relation will be better known, it will only be necessary to multiply all our results by a certain constant in order to bring them into accordance with the new determination.

I will here adopt: light of the Sun = 40,000,000,000 × light of Vega [Kapteyn refers to the authoritative *A Text-book of General Astronomy for Colleges and Scientific Schools* by Charles A. Young (1834–1908), which was first published in 1888, followed by a revised version in 1898. Kapteyn refers to page 231, but it is not on this page, in neither edition. In the 1888 version it is found on page 213, in the 1898 version on page 233 [12].] According to the Potsdam measures the apparent magnitude of Vega is 0.41. From these data it can be easily derived that the Sun, when transferred to a distance corresponding to the parallax $\pi = 0\rlap{.}''10$, would have the apparent magnitude 5.048. I will adopt 5.5, which accidentally agrees exactly with the mean magnitude of the Bradley stars.'

The rest of the text can be found in Box 9.2, where we see some formulae that should be very familiar to anyone who has attended a course on elementary astronomy. In astronomy the absolute magnitude of an object is still defined as the apparent magnitude it would have at a distance of 10 parsecs; Kapteyn's convention is universally adopted and used. Kapteyn does not really explain why he chose a parallax of $0\rlap{.}''1$ as the standard distance. But the last sentence indicates why; the absolute magnitude of the Sun is conveniently the same as the mean apparent magnitude of the Auwers-Bradley stars that he uses so extensively.

In the *Legacy*, Owen Gingerich added this about Kapteyn's definition of absolute magnitudes:

'Kapteyn had originally defined absolute magnitude based on a standard distance corresponding to a parallax of $0\rlap{.}''1$ (in *Groningen Publications* 11, 1902), and used this measure as a standard distance without giving the unit a name. After H.H. Turner [this is Herbert Hall Turner (1861–1930) from Oxford] defined the *parsec* in 1913, Kapteyn apparently felt obliged to use a magnitude scale based on a distance of only 1 parsec, as he did in this letter to Hale [this letter, referred to a little earlier in Gingerich's text, is dated March 3, 1920], although he believed that the parsec was *'barbaric'*. [In Kapteyn & van Rhijn (1920b) we read about the parsec: 'For the sake of uniformity we have resolved henceforth not only to use this unit but also use the name, which is very convenient (though very ugly).']
[...] Kapteyn's original definition of absolute magnitude was adopted by the International Astronomical Union in its first meeting, in 1922.' [See Box 16.1 and discussion on page 605].

Kapteyn did not always adhere to this definition himself. In a landmark lecture he held in St. Louis, USA, in 1904, that I will discuss next in this chapter, Kapteyn (1904b), he defined absolute magnitude as the apparent magnitude of a star at a parallax of $0\rlap{.}''01$ or a distance of 100 parsecs. And in the seminal papers in which he presented his final attack on the Sidereal Problem, Kapteyn & van Rhijn (1920b) and Kapteyn (1922a), he finally accepted the parsec as the unit for distance, but redefined absolute magnitude to a distance of *1* parsec (see page 559). Actually, van Rhijn for some years stuck to this convention in spite of the IAU resolution. But let us have another look at Kapteyn's paper on the mean parallax of stars as a function of proper motion and magnitude, Kapteyn (1900e).

9.7 Stellar Densities as a Function of Distance

One can see that Kapteyn was now in a position to derive the space densities of stars (Fig. 9.6) and the luminosity curves (the relative frequencies of stars in any part of space) for stars of both primary spectral types (Fig. 9.7). Solutions A. and B. differ only in adopting slightly different numerical values for the functional form assumed (see Fig. 9.5) when used to estimate the parallax from apparent magnitudes and proper motions, but that is of no consequence here. Let us have a closer look at these results.

The table in Fig. 9.7 is his fundamental result. It shows the number of stars per unit volume, where the unit of volume is a cube with sides equal to the distance unit he used, i.e., that corresponding to a parallax of $0.''1$. In modern terms that would be 1000 cubic parsec. The columns are of constant stellar absolute magnitude or the logarithm of the luminosity. The values run from just over 5, which means a bit more than 10^5 or 100,000 times that of the Sun, to about -2 (10^{-2} or 0.01 times the Sun). There is a curious notation issue. If the log is a negative number (say -1 for $\log 0.1$) Kapteyn adds 10. This was rather common in those days. Where this fear for negative numbers comes from, I am not sure. The (horizontal) lines are of constant parallax or distance (again in units that would be 10 parsec today), the larger distances (850 pc) at the top, down to about 5 pc at the bottom. Obviously, the stars used have a limited range of apparent magnitudes, so that at larger distances the stars included are brighter than at smaller distances. This has been indicated by the thick, step-wise lines, the left one corresponding to an apparent magnitude of 3.1, the one on the right to 7.1. The total densities on the right are therefore very incomplete.

One could, of course, use these results with an adopted constant luminosity law (as in Fig. 9.5) to extrapolate over the whole range of absolute magnitudes and derive a total density of stars. This is very sensitive to what the adopted parameters in that 'law' are. Kapteyn provided some examples, but the results ranged from a significant increase (up to a factor 3) from about 50 to 100 pc locally, followed by a factor 2 to 4 at 300 pc to a relatively constant density out to 50 pc followed by a decline by a factor 6 from that peak at 700 pc.

The numbers in Fig. 9.7 show two possible solutions for the density distribution and the luminosity law (number of stars per unit volume between two absolute magnitudes) for stars of the two principal spectral types separately. In the case of the densities he now used linear numbers (i.e., not logarithms). The interval in absolute magnitude in the luminosity law he chose here is 1 magnitude (about a factor 2.5 in

TABLE 3. LOG. NUMBER OF STARS PER UNIT OF VOLUME.

Mean π	r \ M. / log. L	-7.55	-6.55	-5.55	-4.55	-3.55	-2.55	-1.55	-0.55	0.45	1.45	2.45	3.45	4.45	5.45	6.45	7.45	8.45	9.45	10.45	Density.
		5.22	4.82	4.42	4.02	3.62	3.22	2.82	2.42	2.02	1.62	1.22	0.82	0.42	0.02	9.62	9.22	8.82	8.42	8.02	
0″,00118	84.7	3.51	3.81	4.40	5.033	5.685	6.326	7.068	7.713												0,122
,00187	53.5	4.11	4.38	4.53	5.286	5.984	6.589	7.188	7.838	8.446											0.234
,00296	33.8			5.18	5.60	6.167	6.852	7.428	7.992	8.543	9.111										0.418
,00469	21.3				5.88	6.477	6.991	7.628	8.190	8.726	9.201	9.719									0.656
.00743	13.5		5.90	5.90	5.90	6.613	7.209	7.714	8.310	8.858	9.366	9.769	0.225								0.869
,0118	8.47						7.28	7.878	8.377	8.899	9.423	9.884	0.225	0.651							0.985
,0187	5.35							7.97	8.441	8.975	9.423	9.892	0.295	0.580	0.935						1.031
.0296	3.38							8.00	8.52	8.949	9.476	9.857	0.269	0.596	0.830	1,113					1.000
.0469	2.13							8.32	8.60	9.021	9.364	9.907	0.200	0.552	0.812	0.999	1.202				0.917
.0743	1.35								8.90	8.90	9.422	9.709	0.270	0.479	0.742	0.962	1.100	1.220			0.829
,118	0.85									9.51	9.51	9.71	9.980	0.560	0.688	0.863	1.097	1.164	1.201		0.742
,204	0.49									9.98	9.98	0.219	0.004	0.173	0.663	0.864	1.041	1.210	1.266	1.257	0.648
apparent mag.											0.1	1.1	2.1	3.1	4.1	5.1	6.1	7.1	8.1	9.1	

Fig. 9.6 Kapteyn's 1901 determination of the space densities of stars as a function of parallax π (or distance r) and absolute magnitude M (or logarithm of Solar luminosities $\log L$) (Kapteyn Astronomical Institute)

TABLE 8. DENSITIES Δ.

π		Type I.		Type II.	
Limits.	Mean.	Sol. A	Sol. B	Sol. A	Sol. B
0″00100 — 0″00158	0″00118	0.280	0.278	0.070	0.102
.00158 — .00251	.00187	.470	0.478	0.156	0.190
.00251 — .00398	.00296	.738	0.726	0.254	0.314
.00398 — .00631	.00469	1.006	0.986	0.440	0.474
.00631 — .0100	.00743	1.202	1.172	0.622	0.655
.0100 — .0158	.0118	1.215	1.171	0.802	0.790
.0158 — .0251	.0187	1.189	1.283	0.960	0.933
.0251 — .0398	.0296	1.000	1.000	1.000	1.000
.0398 — .0631	0.469	0.822	0.826	0.993	1.186
.0631 — .100	.0743	0.669	0.669	0.940	1.083
.100 — .158	.118	0.338 } 0.583	0.505 } 0.619	0.883 } 0.908	1.059 } 1.072
> 0.158	.204	0.290	0.368	0.598	0.981

TABLE 9. LUMINOSITY-CURVE.

(Log. number per unit of volume for π = 0″0296)

Log. L.	M.	Type I		Type II.	
		Sol. A −0.039	Sol. B	Sol. A −0.078	Sol. B
4.82	−6.55	4.433	4.382	4.182	4.192
4.42	−5.55	4.951	5.000	4.702	4.678
4.02	−4.55	5.640	5.636	5.375	5.371
3.62	−3.55	6.223	6.262	6.104	6.131
3.22	−2.55	6.804	6.854	6.767	6.849
2.82	−1.55	7.358	7.422	7.443	7.491
2.42	−0.55	7.902	7.972	8.046	8.116
2.02	0.45	8.413	8.477	8.634	8.703
1.62	1.45	8.902	8.907	9.170	9.232
1.22	2.45	9.308	9.362	9.635	9.691
0.82	3.45	9.644	9.632	0.062	0.054
0.42	4.45	9.927	9.843	0.401	0.422
0.02	5.45	0.093	0.002	0.670	0.640
9.62	6.45	0.297	0.139	0.937	0.818
9.22	7.45	(0.50)	(0.08)	1.080	1.004
8.82	8.45			1.306	1.115

Fig. 9.7 The total density of stars as a function of parallax (top) and the 'luminosity curves' or relative density of stars of different luminosities L or absolute magnitude M for the two principal spectral types. The meaning of the two solutions A and B are explained in the text (Kapteyn Astronomical Institute)

9.7 Stellar Densities as a Function of Distance

luminosity) and the actual densities are normalized such that they all correspond to a distance of 34 pc (parallax 0''.0296).

Kapteyn described his results as follows:

'[For] Sol. A. we [...] find, for both types, a strong decrease of the density with diminishing distance. By the alteration[s ...] in Sol. B. this decrease disappears practically entirely for type II. For type I the decrease has become somewhat less rapid, but it has not disappeared. The weight of this result is but very small however. The number of stars of type I whose parallax is $> 0''.063$, is so small that any conclusion based thereon is of necessity hardly reliable, especially in a case like the present where, as has been shown above, the adopted number of stars with large proper motions may be very materially in error. For reasons which have already been mentioned, it must be considered as probable that, as soon as more reliable data will be available, we will, for this type also, find the density not far from constant for parallaxes larger than 0''.02.

As a consequence of this result some of the conclusions, at which I had previously arrived (Proceedings Jan. 1893 [this is Kapteyn (1890c)]), must be withdrawn, or at least considerably altered.

These conclusions were based on the result, derived by Stumpe, Ristenpart, and others, *viz.* that, if the stars are arranged in groups according to their proper motions, the mean parallaxes of these groups are approximately proportional to the mean proper motions. It is only subsequently that I found that this result was arrived at by an illegitimate reasoning and is certainly not in accordance with the facts.

For the stars with large proper motions (say larger than 0''.10) it follows from the above that the variation of the quantity Q [the ratio of stars of two types] in the paper quoted, is, either entirely or at least to a large extent, a consequence, not of a condensation of the stars of type II in the neighborhood of the sun, but of the fact that the number of faint stars of the first spectral type, as compared to the number of bright stars of the same type, is not so large as in the case of the second type.'

In other words, his earlier conclusions on the distributions of the two types as derived from distributions relative to the Milky Way were no longer valid. The method to obtain insight into the problem of the Sidereal System had been worked out. But the result was still unreliable, partly due to the number of stars available, partly due to the assumptions that had been made. One such assumption was that previously it seemed safe to assume that smaller average proper motions mean larger average distance. Kapteyn no longer found evidence to support the fundamental assumptions that the space motions of stars had no preferred direction and that the luminosity curve was the same for different distances from the Sun. So, he had three worries. In the first place there was growing evidence that the fundamental assumption of no preferred direction of motion among the stars was wrong. Secondly, the different distributions of the principal spectral types cast doubt on the assumption of a similar 'luminosity curve' throughout space. And finally, the distribution of stars in space might not simply depend on distance from the Sun, but also on direction. In Kapteyn

(1890c), and extended in Kapteyn (1900e), he had earlier found that there are in fact two classes of stars. The brighter stars with larger proper motion were distributed evenly across the sky, but 'the total mass of stars of a determined magnitude, is condensed towards the Milky Way, the more strongly, the higher (fainter) the magnitude'. Things were becoming more and more complicated.

> **Box 9.3 The Fundamental equation of statistical astronomy**
> Statistics astronomy is the study that relates the distribution of stars on the sky as a function of apparent magnitude to the density distribution in space. The Fundamental Equation of Statistical Astronomy is easily derived as follows: See also box 11.1.
> Assume that the luminosity curve, the relative number of stars in any volume element as a function of absolute magnitude M, is $\varphi(M)$ and that the density distribution as a function of distance r is $\Delta(r)$. The number of stars in the sky of apparent magnitude between $m - 0.5$ and $m + 0.5$ is then
>
> $$N_m = \text{constant} \int \varphi(M) \Delta(r) r^2 dr.$$
>
> Of course $M = m + 5 - 5\log r$ from the definition of absolute magnitude. Interpretation of star counts relies on solving this equation. Note that N_m is easily measured by counting stars, but that $\varphi(M)$ needs large numbers of parallaxes (distances). Derivation of $\Delta(r)$ requires the inversion of the integral.

Before continuing, I would like to note that in the *Legacy*, Maarten Schmidt (b. 1929) has commented on the importance of Kapteyn's working method using the table reproduced in Fig. 9.6. Remember that the table lists space densities as functions of absolute magnitude (horizontal) and parallax (vertical). These had been derived from a listing of number of stars of apparent magnitude and parallax. The diagonals were lines in the table of constant apparent magnitude (as indicated at the bottom). This representation makes it possible to sum in the horizontal direction and find the density as a function of distance. Or in the vertical direction to find the 'luminosity curve', the number of stars in any volume element of various absolute magnitudes. Of course Kapteyn's data were incomplete, since closer to us stars were included that were intrinsically fainter than were used at larger distances. It is an incomplete, tabular representation of the fundamental equation of stellar statistics (see Box 9.3). Schmidt was an undergraduate at Groningen University, where he was taught the principles of stellar statistics by van Rhijn. Later he discovered the quasars. These 'quasi-stellar radio sources' had been noted in the 1950s and early 1960s as very strong radio sources that were optically associated with starlike objects. Schmidt noted in 1963 that these had very high redshifts and therefore large radial velocities away from us, as a result of the expansion of the Universe. They are now known to correspond to the centers of massive galaxies at cosmological distances, where enormous amounts of energy are liberated by black holes. Schmidt related in *the Legacy*, how he devised the so-called $V/V_{\max}$ test to study their cosmological evolution as a natural extension of Kapteyn's approach (see also Schmidt's paper *Space distribution and luminosity functions of quasi-stellar radio sources* [13]).

9.8 The Louisiana Purchase Exposition

In 1904 the Louisiana Purchase Exposition, also known as the St. Louis World's Fair, was held in St. Louis, Missouri (see Fig. 9.8). Such major expositions had been held in various countries and various forms. Famous predecessors of great international events on a grand scale – to mention but a few – were the Great Exhibition of the Works of Industry of all Nations, in London (1851), for which the Crystal Palace had been built, and the Exposition Universelle of 1889 in Paris. The latter coincided with the 100th anniversary of the storming of the Bastille and featured the Eiffel Tower as its main symbol. In the 1803 Louisiana Purchase, the United States had acquired more than two million square kilometers from France, stretching from the New Orleans to and slightly beyond the current Canadian border and involving all or part of fifteen or so current states. The 1904 World Fair, which had originally been planned for 1903, marked the centennial of that event.

It took place in St. Louis, Missouri, and lasted for the most of the year 1904, from April 30 to December 1. It also hosted the 1904 Olympic Games, although attendance by teams from outside the USA was limited by the costs of transport. The exposition is reported to have been visited by some 20 million people. In the course of 1904 there were a number of conferences or congresses in addition to the many exhibits. By far the largest of these was the 'International Congress of Arts and Sciences', which took place in the week of 19 to 24 September. This was reportedly attended by one thousand participants. It took place on the campus of Washington University (founded in 1853), where all available lecture halls were required for the purpose. The overarching theme of the ICAS was 'Progress of Man since the Louisiana Purchase' and it featured a large number of lectures on scientific, technical, literary and industrial topics. The ICAS was organized by a group of officials and its president was the astronomer Simon Newcomb.

Newcomb wrote to Kapteyn about this on July 15, 1903:

'I wish to write you in a purely personal and friendly way, on a subject on which I hope, in a few weeks, to address you officially. It is that of the International Congress of Sciences, etc., to be held at St. Louis next year, of which I send you a programme, under separate cover. What I wish to know is whether there is any chance that you would be able to accept an invitation to attend the Congress and deliver one of the astronomical addresses, less than an hour in length. In case of your coming, the sum of five hundred dollars would be given to you, towards paying the expenses of your journey [today that would be of order thirteen thousand dollars, [14], or a little over ten thousand Euros. Lowest ship fares for immigrants were about 30$, one way [15], but the Kapteyns must have traveled for a multiple of that.] If you came alone, this would more than suffice. I think it might almost,

Fig. 9.8 Frontispiece and title page of the official guide to the 1904 Louisiana Purchase Exposition. The guide is a volume of over 200 pages, describing all features and aspects of the exposition (St. Louis, The Official Guide Co [16])

but not quite suffice, should you bring Mrs Kapteyn with you, which I, for one, would decidedly want you to do. I am sorry that I am not yet in a position to send a formal invitation, but I hope to be able to do so, at no distant day. I write now, in order that you may have as much time as possible to think the matter over, and also, in order that, if your coming should be out of the question, I may know it at as early a date as possible, but I hope such will not be the case.'

Kapteyn's reaction on July 23, 1903, was:

'You honour me far above my deserts by your invitation to deliver one of the astronomical addresses at St. Louis. Still, however, as you place this confidence in me, I think, after careful consideration, that I may accept it, on the condition that the intention is (as I have little doubt it is) that I shall lecture before the section 13a. and not on the relation to other sciences but on problems of the day. If this condition is granted I shall be exceedingly glad to come and I thank you cordially for having thus planned a splendid holiday for me.

I have long hoped to see America and the Americans, and to meet several of the American astronomers. I have wished to see the St. Louis exposition for especially the aeronautics department. If there are no serious money difficulties in the way I will certainly bring my wife too.'

Newcomb then wrote on May 3, 1904:

'[...] The fact is, I have had great difficulties in arranging a presentation of the subjects in the department of astronomy so as to be a connected whole of the kind required by the general plan. The latter calls for two addresses in each section, of which one shall treat the relations of the subject to other branches of science, and the other of the present problems. But it did not seem possible to divide up astronomy exclusively on these lines. The outcome of the matter is that I have put you down in the section of Astrometry, but shall ask you to treat the subject in which you have been most eminent, namely, the Problems and Relations of Stellar Statistics; the bearing of this branch on other branches of science, and what we may expect from it in enlarging our knowledge of the universe at large. [...] Meanwhile, I shall be glad if you will let me know whether Mrs Kapteyn will accompany you, as I hope she will, and what time you will probably sail. I need hardly add that the earlier you can secure passage, the more likely you will be to get satisfactory accommodation on the crowded steamers.'

On June 8, 1904, Kapteyn sent a reply:

'I am doing the best I can in treating the subject of which you put me down; I foresee some difficulties as to the 'bearing of this branch on other branches of science' but will earnestly try what I can do.

I am happy to write that after much consideration, my wife has now made up her mind to come with me to America. She cordially thanks you for the interest shown in her coming.

We intend to be in Washington on Sept. 27 (Tuesday), but I am not quite sure whether we can remain till Friday.'

This visit to Washington took place after the St. Louis congress (it was held in the week of September 19 to 24, 1904). There is a short note from New York by Kapteyn, when they were already on their way back, dated October 4, 1904:

'Dear Mr Newcomb,

The cheque arrived in due time. I found even occasion to convert it into money. Thank you for your interest. We remember the times spent with Mrs Newcomb and you as one of the brightest of our stay in America. Hope you will enjoy your earned vacation.

With kind regards to Mrs N. and yourself,

Very sincerely yours,

J.C. Kapteyn.'

9.9 International Congress of Arts and Sciences

The proceedings of the ICAS extended over fifteen printed volumes. Volume VIII (see Fig. 9.9), which according to the copyright statement was not printed until 1908, also contains Earth Sciences. It is described as 'comprising lectures on Astronomy,

Fig. 9.9 Logo and frontispiece in the proceedings on Astronomy and Earth Sciences of the International Congress of Arts and Sciences in St. Louis in 1904. The text accompanying the picture on the right reads: 'PHOEBE – Photogravure from the painting by Louis Perrey – In Greek mythology Phoebe was the special name given to Artemis as moon goddess. She was the twin sister of Apollo who was called Phoebus, the sun god. No other goddess surpassed Phoebe in beauty except Venus. The painting, reproduced here, presents the goddess of the moon with all her accredited attributes of grace and loveliness. Above her is her chosen luminary, while she is surrounded by a galaxy of glittering stars' (From International Congress of Arts and Science [17])

Astrophysics, Sciences of the Earth, Geophysics, Geology, Paleontology, Petrology and Mineralogy, Geography, Oceanography, Physiography and Cosmical Physics'. The latter is in fact meteorology and terrestrial magnetism. There are three astronomical sessions, each consisting of two lectures.

The first session was on General Astronomy and took place in the afternoon of September 20. It was chaired by George C. Comstock, director of the Washburn Observatory at Madison, Wisconsin. The two lectures were given by Lewis Boss, director of the Dudley Observatory on *Fundamental conceptions and methods in astronomical science* and by Edward C. Pickering, director of Harvard College Observatory on *The light of the stars*. The second session (morning of September 21) was on Astrometry and Kapteyn was the second speaker there. The first speaker was Jöns Oskar Backlund (1846–1916), director of Pulkovo Observatory, who addressed the issue of *The development of celestial mechanics during the nineteenth century*. It was chaired by Ormond Stone (1847–1933) of the University of Virginia. The final astronomy session in the afternoon of September 21 was chaired by George Ellery Hale, the then director of Yerkes Observatory. Speakers were Herbert H. Turner on

9.9 International Congress of Arts and Sciences

Fig. 9.10 The Hall of International Congress at Washington University in St. Louis, where the International Congress of Arts and Sciences was held in September 1904. It is now named the Stephen Ridgley Hall (From *Celebrating the Louisiana Purchase*, St. Louis Public Library [18])

The relations of photography to astrophysics from Oxford University and William Wallace Campbell (1862–1938), director of Lick Observatory on *The problems of astrophysics*. These chairs and speakers were among the most important and influential astronomers of the day (we have met most of them before in this book) and Kapteyn must have been thrilled to meet the Americans, whose work and papers were very familiar to him, face to face. Figure 9.11 shows pictures of Campbell and Turner. The fact that Kapteyn met George Hale in St. Louis and that they must have had extensive conversations, would prove to be of vital importance to Kapteyn's future, comparable to his first contact with David Gill in 1884, which had been a decisive turning point in his astronomical development and career.

The text of the St. Louis lecture, Kapteyn (1904b), is difficult to find. In the Kapteyn Room, in the Kapteyn Astronomical Institute, we have no printed copy, but there are three (identical) versions on carbon paper, the thin, semi-transparent paper used for making carbon copies of typed letters. Indeed they are in typed form with the figures drawn in manually for both copies. Two of these copies are in the bound volumes in which Kapteyn collected his papers (see Fig. A.2), one in the relevant blue volume and another one in the red volume which contains papers by various authors on the subject of Star Streams. Then there is a single copy, which has Van Rhijn's name on it and was apparently presented by Kapteyn to his (prospective) successor in Groningen. It seems that not even Kapteyn himself owned a printed copy of the book with the proceedings or a reprint of his own lecture. All of the volumes of proceedings of the International Congress of Arts and Sciences in St. Louis in 1904, including volume VIII on Astronomy and Earth Sciences (see Fig. 9.9), are also lacking in the central library of Groningen University, of which the former library of the Kapteyn Laboratory and Institute is now part. The book in which the lecture was eventually printed with much delay (it did not appear in print until 1908)

Fig. 9.11 William W. Campbell and Herbert H. Turner as they appear in the *Album Amicorum* presented at the retirement of H.G. van de Sande Bakhuyzen as professor of astronomy and director of Leiden Observatory in 1908 (Archives Leiden Observatory; see caption Fig. 3.7)

is available on the Web though [19]. A shorter version was published after Kapteyn also presented his ideas at the meeting of the British Association in South Africa in 1905. It can also be found on the Web [20].

Kapteyn started with a long introduction of what questions sidereal astronomy was trying to answer, leading to the statement that in general the distances to stars, absolutely necessary for a comprehensive study of the subject, were too small to measure. Parallaxes were only accessible for *selected* objects, not representative of the whole. But since Bradley's time the Sun had moved in space more than 300 times the diameter of the orbit of the Earth, and this baseline presented a suitable basis, albeit that in this way parallaxes could only be measured in a statistical sense. Kapteyn then outlined the fundamental hypothesis: 'The peculiar notions of the stars are directed at random, that is they show no preference for any particular direction.' He then summarized his work on the parallaxes of stars as functions of their apparent magnitude and proper motion, and showed how this could be used to find information on the number of stars as a function of distance from the Sun and the distribution according to luminosity or absolute magnitude. Strangely, he introduced here the absolute magnitude of a star as its apparent magnitude if the parallax were $0''\!.01$, whereas before (and after) he generally used $0''\!.1$.

The results are presented in a few tables. The first is a table which for stars out to a particular radius showed their number as a function of luminosity, the 'luminosity function'. Then he derived the relative total density of stars in space, the relative number (for unity near the Sun) of stars per unit volume, as a function of parallax or

9.9 International Congress of Arts and Sciences

distance. And finally he tabulated the fractional distribution of the absolute values for the total velocity of stars in space. This must have been impressive; he showed how we could find how stars were distributed in space and what the distributions of their luminosities and space velocities were.

He then discussed two more hypotheses that were necessary if progress was to be made. The first was the one stating that the distribution of stars as a function of their luminosity is the same everywhere. The second concerned the question of the extinction or absorption of starlight in space. This was timely, as George C. Comstock, who should have been present as he had chaired the previous session, had recently published a paper on proper motions of faint stars. I will come back to this later on in more detail, but what it comes down to is that Comstock repeated micrometer measures done by the von Struves at Pulkovo of positions of faint (about ninth magnitude) stars relative to nearby bright ones that in the meantime had well determined proper motions. Using Kapteyn's determination of corresponding parallactic proper motions, he then found that these stars had rather high space velocities, if they conformed to Kapteyn's extrapolated relations between mean parallax and proper motion and apparent magnitude. This meant in Kapteyn's terms that 'hypothesis A' –' the fainter stars are of equal intrinsic luminosity with the brighter stars'– must be replaced by 'its antithesis: The intrinsic luminosity of stars diminishes with their apparent brightness in such a ratio that a star of the tenth magnitude possesses only one-tenth of the luminosity of a star of the fifth magnitude'. Comstock postulated that this is due to an 'absorbing medium' producing considerable extinction of starlight. Kapteyn announced that he tended to disagree for the time being and would elsewhere present a method to test this. I will return to this in a later chapter.

Kapteyn also refrained from further discussions and turned to the question of the fundamental hypothesis that there was in space no preferred direction in the motions of the stars. For this he pointed to a figure that he actually presented earlier on in his lecture and which is reproduced in Fig. 9.12. In part **P** he illustrated the fundamental hypothesis. On average, proper motions were equal in size and distributed evenly over all possible directions. But we observe them from a vantage point moving through space with the Sun, which will modify the distribution as in **Q**; the distribution of proper motions would be drawn out in the direction of the Solar Antapex, the horizontal line labeled x. 'Near to this line', Kapteyn noted, 'on the Antapex side, the proper motions will be most numerous, and they will be greater in amount.' He did not stress this, but the illustration was for stars at the same distance for which the parallactic proper motion is equal in size. If there were stars at different distances the actual figure would be a superposition of such diagrams. The point was, of course, that whatever the distribution of distances or even real space velocities, the symmetry properties were conserved into the end result. In part **R** he had divided the directions of the proper motions up in 12 sectors of 30° degrees and shaded four of these. 'If our fundamental hypothesis were satisfied, and if, in consequence thereof, the symmetry of our figures were complete, the blackened parts of the figure would have been equal to the corresponding lighter-tinted parts. (This ideal case is represented in Fig. [9.12], **R**).'

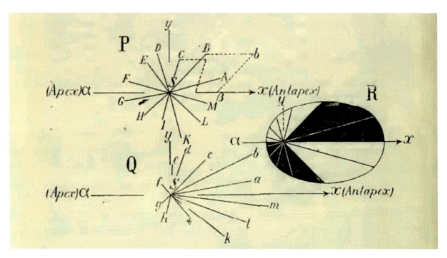

Fig. 9.12 Figure 1 from the 1904 lecture of Kapteyn 1904 on *Statistical methods in stellar astronomy* at the Louisiana Purchase Exposition (also known as the fourteenth World Fair), held at St. Louis, Missouri (Report of the seventy-fifth meeting of the British Association for the Advancement of Science, South Africa (see text))

Next he produced such figures using the (over 2,400) Bradley stars, which covered over two-thirds of the whole sky. This area he divided up into 28 sub-areas. The resulting picture would be 'too overburdened', so he only showed the results for a subset of ten figures, where the 'phenomenon to which I wish to draw your attention is most marked' (Fig. 9.13). 'The real state of things is something quite different' (from the ideal case of Fig. [9.12], **R**), 'and, what is all-important, we see at once that the divergences are strikingly systematic. The figures at each pole of the Milky Way show them in nearly every particular of the same character. Near the North Pole the blackened parts are invariably much greater; at the South Pole the case is reversed.'

Kapteyn did not leave at this, but went on to perform some more quantitative analysis, making use of symmetry properties. It (see Fig. 9.12) was clear that if every proper motion is decomposed in one in the (horizontal) direction x and vertical direction y we should have two conditions. The sum of all components y should be zero. They are not. And the sum of all components x on one side of the x-axis should be the same as the sum of these components on the other side. With regard to the latter Kapteyn demonstrated that there was a clear and systematic difference as a function of galactic latitude and when different assumptions were made for the direction of the Apex; the result was strikingly systematical. What may be more difficult to see in Fig. 9.13 is that for each sub-diagram he had drawn four arrows (vectors); two had open heads, two had filled ones. The filled arrows concerned 'direct' proper motions, that is only those that were by and large directed away from the Apex. The open ones were directed generally towards the Apex and were called retrograde. Then for each there was the direction of the symmetry line of maximum

9.9 International Congress of Arts and Sciences

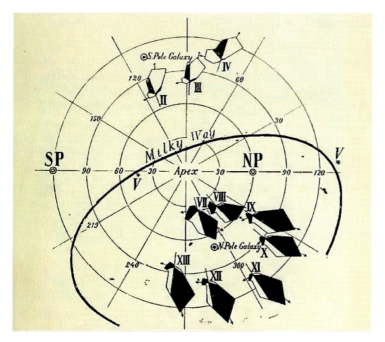

Fig. 9.13 Figure 3 from the 1904 lecture of Kapteyn 1904 on *Statistical methods in stellar astronomy* at the Louisiana Purchase Exposition (also known as the fourteenth World Fair), held at St. Louis, Missouri (Report of the seventy-fifth meeting of the British Association for the Advancement of Science, South Africa (see text))

proper motion and the one of symmetry in minimum motion. Now the direct ones converged very well for a general streaming towards a point on the sky some 20° from the Apex. For the retrograde ones the effect was less clearly defined but there appeared to be two streaming directions about 125° apart. In a sense, these were Apex determinations for the direct and retrograde stars separately.

Obviously, these are not, as Kapteyn stressed, directions in which stars moved exclusively, but 'there only is a decided preference for these directions'. The fact that they were not diametrically opposite was because of the motion of the Sun with respect to the system of stars in general. The mean velocities of the stars had to be known to correct for this. Doing that in an approximate manner resulted in two opposite points on the sky indicated by the letters **V** in Fig. 9.13. Kapteyn used the word 'vertex' for these points to distinguish them from the Apex. They were lying in the Milky Way, so obviously had something to do with this.

Kapteyn had a look at measured radial velocities and concluded that, although the evidence was limited by the small number of stars available for use in this context, it confirmed the Star Streams and their vertices. Kapteyn said little about the amount of streaming. From his discussion we see that the velocities involved were of order 10 to 20 km/s, so this would be the difference in mean space velocity in the direction of a stream compared to that perpendicular to it.

9.10 Aftermath of the Announcement of Star Streams

This announcement of the Star Streams must have made an enormous impact. As indicated, the result appeared in print only years later. I speculate that the carbon copy versions of which we have three in Groningen, might have been produced to meet demands for requests of the lecture. Fortunately, as we will see in the next chapter, Kapteyn traveled to South Africa in 1905, where he also presented the lecture at the meeting of the British Association for the Advancement of Science. An abridged version of his St. Louis lecture appeared in the Report of that meeting, Kapteyn (1905a). Apparently Kapteyn managed to convince some of the most prominent astronomers present and his discovery did not fall into oblivion in spite of the lack of a printed version of the presentation. I quote a few lines from two astronomers looking back at the event. The first comes from a lecture in 1923 by Hector MacPherson, entitled *Astronomy in the twentieth century* [21]. Hector Carsewell Macpherson (1851–1924) was a Scottish journalist and science writer. In this well-publicized paper read before the British Astronomical Association he stated:

'In stellar astronomy, however, the rate of advance has been most phenomenal.

(i) The first outstanding discovery of the century in this field was announced casually in 1904 at the St. Louis Congress. Before that year, it was generally held that the proper motions of the stars were at random. But the late Professor Kapteyn of Groningen had been working for years at the problem of statistical astronomy, and by the time of the St. Louis Congress he was able to announce his discovery that the proper motions of the brighter stars in the Bradley-Auwers catalogue showed a strong preference for two opposite directions in the galactic plane. He interpreted this result as meaning that the stars – or in all events nearer and brighter stars – are to be divided into two streams moving in opposite directions. At first Kapteyn's discovery did not attract the attention it deserved; for as Hinks has paradoxically said (see for the actual full text page 354) – 'To announce a great discovery to a scientific congress may render it inaccessible for a long time afterwards.' But the discovery was brought to the test by two distinguished English astronomers, Professor Eddington and Sir Frank W. Dyson, who independently carried out similar investigations on stars in different catalogues, and the results were confirmatory. The observed fact of star-streaming is fundamental to any dynamical theory of the stellar system.'

Arthur Eddington said the following in his introductory address on the occasion of the Royal Astronomical Society's centenary celebration in 1922 [22]. He listed what he felt were the 'outstanding landmarks in these hundred years'. There were six, in chronological order, of which number (5) was: '1904 – Kapteyn's discovery of the two star-streams, the beginning of the modern era of investigations of the sidereal system.' And: 'But I think the great impetus on sidereal astronomy came from Kapteyn's discovery, which I have mentioned along with the six landmarks of the century. The two star-streams were the first taste of the many amazing results obtained in the statistics collected or being collected. They were the first indication to us of something like organization among the myriad of stars. Paradoxical as it may seem, the duality of the stellar universe was the first clear indication of its unity.' And in his obituary of Kapteyn [23] Eddington wrote:

'The fine series of *Groningen Publications* containing the researches of Kapteyn and his students on the sidereal system form one of the most often consulted works in an astronomic library. But perhaps the most famous of them is No 6 – the one which was never written. The neighbouring numbers, growing each year more ragged, stand waiting to be bound up with it when it arrives; but the gap will never be filled, for Nature took an unexpected turn, and the stellar universe would not fit into the scheme which No 6 was promised to elaborate. The previous number was entitled *The distribution of cosmic velocities: part I, Theory*; it was a study of how the motions of the stars, all pursuing their courses independently, would turn out statistically when the solar motion and its effects of varying distance were allowed for. Meanwhile, the observed proper motions of Auwers-Bradley were being prepared for comparison, so as to determine the numerical constants of the formulae. But the theory, although it represented the most unquestioned views of the time, turned out to be so wide of the mark that not even the beginnings of a comparison could be made. Kapteyn's greatest discovery – the two star-streams – revealed for the first time a definite organization of the system of the stars, and started a new era in the study of the relationships binding the most widely separated individuals. It was in 1904 that Kapteyn first discovered the striking phenomenon that there were two favoured directions of motion. When once pointed out, this was as conspicuous a feature of the proper motions that the results rapidly won acceptance. The last great opponent, Lewis Boss, surrendered in 1911, converted by his own most excellent proper motions; since then no survey of the scheme of the stellar system could disregard the existence of this peculiarity.'

9.11 Impressions of America

The journey to the United States made a strong impression on Kapteyn and his wife. In the *HHK biography* the trip was described as follows:

'Mrs Kapteyn accompanied him. Their departure was characteristic for these unassuming people. They left towards the end of August, traveling together by bike from their second home in Vries to the train station in Groningen to take the long-distance train to Rotterdam. The children were sitting on the fence and waved them goodbye. That way their first major trip to America started. Kapteyn to conquer the astronomical world, Mrs Kapteyn to make new friends in these remote places where she would find like-minded people. Upon arrival in St. Louis they were welcomed by the

Fig. 9.14 The administrative building of Washington University at St. Louis, where the International Congress of Arts and Sciences was held in September 1904 (From *Popular Science Monthly*, Volume 66, November 1904 [24])

consul of the Netherlands, who had to put them up in primitive timber buildings, since nothing else was available. They had arrived just in time to be present at the opening of the astronomical conference and they sat quietly at the back of the auditorium. Newcomb, the president, had noted them and came immediately to Kapteyn to take him to the table with the Board members where the important astronomers were gathered. He was welcomed warmly and respectfully by everyone and immediately felt accepted as a welcome and honoured guest. Newcomb, who did not know yet about his new discovery, introduced him as follows: 'Prof. Kapteyn will tell us something about his interesting Durchmusterung work'. So it was a surprise when it turned out that his lecture was about a completely different subject. They were very impressed and he was happy with the interest in his discovery, but the most important part of his trip to America was the fact that he made the acquaintance of Prof. George Ellery Hale, at that time the director of Yerkes Observatory. [...]

On their way back they visited Newcomb in Washington, where they also attended the big reception at the White House. They arrived in Newcomb's carriage with two horses and a black groom, and were introduced to President Roosevelt together with Hugo de Vries, who happened to be in Washington at the same time. H. de Vries (1848–1935) was a plant physiologist and geneticist of the University of Amsterdam. The President had a suitable word and a handshake for everybody. 'Ah Mr de Vries, I suppose we are cousins, because the name de Vries is in my family.' This, of course, with pride as Dutch names meant distinction in America. Kapteyn was greeted with a friendly remark about astronomy, and this remarkable moment was history.'

9.11 Impressions of America

They returned towards the end of October, so had been away for about two months. Not long after their return, on November 8, 1904, Kapteyn wrote to Newcomb:

'Now that I begin to get some way through accumulated arrears, it is a pleasure to sit down and revert in thought to the happy days spent in America and in your home. We will never forget these days, so full of new impressions and of many kindnesses. The cordial welcome Mrs Newcomb and yourself gave us, at a time when you had so many important things to care for and to think of, has deeply impressed us. We often remember the time and I can assure you that the interest you have shown me these years has really given me 'an infusion of increasing vigor in pursuing my labors'. Our journey has been a very pleasant one. Before sailing we spent a couple of days on the Hudson and in Albany. Splendid Hudson, seen on a glorious autumn day. How we enjoyed the trip! Backlund was with us on the boat and was no less enthusiastic than we were. [...]
The ocean was mostly prettily smooth and my wife was only sea-sick for half a day. We arrived in due time at Groningen, nearly rested from our intensely interesting but at the same time fatiguing trip.
Will you kindly remember me to Mrs Newcomb,
Yours very truly,

J.C. Kapteyn'

This was mostly business, but Mrs Kapteyn had already written to Mrs Newcomb on October 31, 1904 on a much more personal note and with different news. The relationship between the two women had become very cordial indeed. Before I quote that letter I first offer some biographical details of Mrs Newcomb. She had been born Mary Caroline Hassler (1840–1921), granddaughter of the surveyor and head of the US Coast Survey and the Bureau of Weights and Measures, Ferdinand Rudolph Hassler (1770–1843). She had married Simon Newcomb in 1863. They had three daughters, the eldest of which was Anita Newcomb McGee (1864–1940), who became a well-known military physician and first female acting Assistant Surgeon of the United States Army.

'My dear Mrs Newcomb,
We are home now and we begin to get somewhat used again to the old kind of life. And were it not for a great sorrow that has befallen us, we should be as happy as possible. To tell you what we feel about our visit to America and the thankfulness we owe to all the dear friends that made our stay one never to be forgotten, a better pen than mine could only do that. Especially your dear self and Professor Newcomb were so good and kind to us that we feel very much indebted to you. And now I am going to tell you, dear Mrs Newcomb, what is the matter that made such a sad ending to our delightful holiday.
Coming from Bologna we found a letter from our eldest girl to tell us that Hetty, the second, had broken off her long engagement and at Rotterdam we heard from Hetty herself to ask us to go and [crossed out 'take leave'] say goodbye to the dear boy who was with his parents at The Hague. As we really love him as one of our own already this was a very hard thing

to do. – And coming home and finding it so full of memories of him and knowing him to be lost for us was harder still. However, all this has to be got through and now something happened that made it somewhat easier. On the boat going home, we made the acquaintance of a young lady teacher at one of the university schools at Chicago, who went to Europe for a couple of years to restore her health. She was bound for Bologna, so to go on at Paris, but after a few days we asked her to accompany us to Holland in which she was just very much interested and go to Paris afterwards.

This she decided to do and after having got our letters we told her everything about and begged her to come with us all the same, leaving her free at the same time to go her own way should she so feel inclined. However, she did accompany us and proved to be a great boon to us, as she directly took very much to both of the girls, especially to Hetty and gave her daily some lessons that proved so interesting that Hetty has taken up the study of English and French literature and is working as hard as ever she can. Do, the eldest is hard at her study too and the boy will leave us in a couple of weeks to go to Germany for his engineering studies.

Dear Mrs Newcomb, coming home and looking over my bills I found that you had made me a present of the tablecentres that I took home for the girls. Thank you very, very much for them. I do not think I shall ever forget the great kindness you so liberally bestowed on us. The pretty butterdisk you gave me with the N engraved on it, found great favour in Do's eyes as her future name will be Noordenbos. But I am not going to let her have it. And now I must take leave of you. I have still one wish. May the future bring me a real good photograph of a certain lady who I noted as being with her husband the most beautiful couple I ever saw together. Will you tell Mr Newcomb for me that I am very proud of having been your guest. I shall be prouder still when you will permit me to sign myself,
Your affectionate friend,

Elise Kapteyn Kalshoven.'

A tablecenter is probably some kind of stand for ornaments or decorations in the centers of a dinner table. Mrs Kapteyn apparently found out that she had not been billed for one of those that she had brought with her.

9.12 Kapteyn Homes in Groningen

I conclude this chapter with a short description, taken from the *HHK biography*, of the Kapteyns in Groningen and the places they lived. I refer back to Fig. 4.7 for a map of Groningen and the locations of these homes and the years they occupied these. First Henriette Hertzsprung-Kapteyn takes us back to the period around 1900, when they lived in Heerestraat.

9.12 Kapteyn Homes in Groningen

'The children had grown up by now and all three of them were doing academic studies. The two girls were among the first female students. The elder girl had chosen for medicine and the younger law as a subject, which gave rise to much criticism in these turbulent days of the fight for woman rights, but Kapteyn felt that females studying at universities was so natural and self-evident, that one did not get far with counter arguments. The son went to Freiburg in Saxony to study mining engineering.

It was comfortable and colourful in the large upper level house they occupied in Heerestraat, where Mr and Mrs Kapteyn were a prime example of hospitality. The atmosphere was pleasant and free, people were received simply but well; they were surrounded by youth, music and singing. Plans were made for the future, problems youngsters encountered were discussed, counsel and comfort were sought and given. The father was ever present when there was a need to solve a problem or a judgment required. He was the counsellor and friend, versatile and familiar with all matters. He was not like other great scientists, whose special studies came to stand in the way of their interests in other persons. 'What an unusual man', a simple woman, who had visited the Kapteyn family with some timidity, remarked, 'He was so normal and had real interest in me, while I am not anything special.' Every human being was special to him and nobody ever was given this onerous feeling of inferiority which makes many great people so unpleasant. And in the end this does most harm to themselves.

In 1905 a revolution took place in the Kapteyn household. When the elder daughter had married, the second one studied in Amsterdam and the son studied in Freiburg, the house became too large and they decided

Fig. 9.15 Heerestraat in Groningen in 1902, looking north. The house on the left is where the Kapteyns lived on the first floor from 1891 to 1906. There was a cigar-shop on the ground floor. The incomplete sign reads 'Apotheek [pharmacy] H. Rinsma', of which the address is Zuiderbinnensingel 5 (now Coehoornsingel), the street to the right perpendicular to the Heerestraat. In the distance the tower of the Martini Church in the center of town (Beeldbank Groningen [37])

Fig. 9.16 Ossenmarkt, where the Kapteyns occupied the upper part of the house seen just to the right of the large tree in the middle of the photograph. Their entrance was on the right-hand side of the house. They lived here from 1910 to 1920. This picture must have been taken from the building that was the second home of the Astronomical Laboratory (see Fig. 7.16). Kapteyn lived here well after the Laboratory had been moved to its final location (From Groningeninbeeld.nl [26])

to look for a smaller place to live. Mrs Kapteyn, who had developed a lot of sympathy for her American 'sisters' and their regiment of self-help during her short visit to America, proposed to her husband to break with the Dutch convention and stop hiring a maid and having a cleaning lady for help during the morning hours. She was the first in her circles to take this bold step, and her husband fully agreed with it, as long as the work would not be too much for her. They rented a small first-floor house on Eemskanaal on the outer edge of town, the maids were sent away and a cleaning lady was hired. Mrs Kapteyn did the cooking herself and joined in with the household work. The Groningers found this not-done, but she enjoyed her ability to do the work and the financial help this provided for her husband. What would they care about the opinion of others? In all respects they were ahead of their times; what is common today was then still out of the question. And their lives flourished in fullness and happiness.

They lived in this small place for 5 years, very happy with their simple but active lives. Mrs Kapteyn hated washing the dishes most of all, a necessary evil, until she found a magic formula that made it a pleasant activity. Every evening she put the problem before her to find a way to do it in the quickest way possible, and every day she became more skillful at it and quicker and happier. 'Les grand esprits se rencontrent' [Great minds think alike], since

9.12 Kapteyn Homes in Groningen

these were Kapteyn's own words: 'Fortunately, nature has been designed in such a manner, that every job, even the most unpleasant one will eventually develop its own charm. The most boring routine calculations (and nothing in the world is more boring than routine computing) will have its appeal when one has developed a certain aptitude in it.' In this manner everyone did his boring, routine computing with a happy cheerfulness.

After 5 years they moved to a larger, old upper house at Ossenmarkt. This was a house 'with a soul'. People who visited it never forgot it. Not particularly beautiful, but very special was the entrance; a narrow alley like a cleft between two tall houses and a very humble door with an old-fashioned doorbell, which was seldom used, however, since the doors were seldom locked and friends and acquaintances were always able to let themselves in. The beautiful, old oak staircase led to a lobby unto which large rooms opened, with a serenity and peacefulness that felt beneficent. The spacious living room with its old carpentry, antique wallpaper and deep windowsills had a view through the high windows with the orthodox Dutch half-drawn curtains of the quiet, grand square. Kapteyn's second laboratory, the former meteorological institute, which had been given to him when the old Commissioner's house had been restored to its proper use, could be seen lying on the other side of the bridge with its friendly old style and bricked-up archway. They lived in this house, which was so fitting for them and was adored by everyone, for 8 happy years.'

Fig. 9.17 Photograph of Kapteyn accompanying the article of C. Easton in Hemel & Dampkring (see Appendix C)

Chapter 10
Selected Areas

The wonder is not that the field of the stars is so vast,
but that man has measured it.
Anatole France (1844–1924).[1]

Stars everywhere.
So many stars that I could not for the life of me understand
how the sky could contain them all, yet be so black.
Peter Watts (1958–present).[2]

10.1 R.A.S. Gold Medal

In 1902, Kapteyn was awarded the prestigious Gold Medal of the Royal Astronomical Society 'for his work in connection with the Cape Photographic *Durchmusterung* and his researches on stellar distribution and parallax'. The address by the President of the Society [1], James W.L. Glaisher (1848–1928), contains an interesting comparison, referring to the goal of the CPD 'to extend the Argelander *Durchmusterung* to the south pole'.

'Argelander's work, which included stars to the 9th or 10th magnitude, contained 324,188 stars; Schönfeld, who continued the catalogue from $-2°$ to $-23°$, included 133,659 stars between these limits. The Cape *Durchmusterung* contains 454,875 stars between $-18°$ and the south pole.

---

[1] Anatole France, born François-Anatole Thibault, was a French poet and novelist. This quote is from *Le Jardin d'Épicure*, translation Alfred Allinson in *The Works of Anatole France in an English Translation* (1920).

[2] Peter Watts, Canadian science fiction writer and biologist.

The catalogues of Argelander and Schönfeld contain 431,760 stars for the rest of the sky. The average number of stars for each square degree in the Cape, Argelander and Schönfeld is 32.66, 15.19 and 18.21 respectively, so that the star density in the Cape catalogue is double that of the density mentioned in Argelander's catalogue. The latter catalogue sets a good example as regards freedom from errors, and there is every reason to believe that Kapteyn has fully maintained this high standard in the present volumes. It is difficult to speak in too high terms of the zeal and ability displayed by Kapteyn during the years that he devoted to this catalogue.'
The address finishes with:

'It will have been seen that one spirit runs through all Kapteyn's work, viz. an effort to treat the great cosmical problems comprehensively, and to apply to them general methods which – though not exact – may be expected to give information which would be quite unattainable by any procedure resting upon an absolutely sure foundation with respect to every individual star. Always occupied with the most effective means of attacking the great problems of the universe, he has never flinched from the heavy labour required to carry his ideas into practice; and it is very fitting that one whose mind has been from the first so much attracted by the more difficult questions presented by the statistics of the heavens should himself have contributed such invaluable material for the study of the subject which he has so much at heart.'

The Gold Medal of the R.A.S. is not only its own highest award, it is also one of the most prestigious awards worldwide. Although early on – the Medal was first awarded in 1824 – there were often two recipients per year (and none in some years), between 1833 and 1963 there was only one (if any). Currently there is one available annually for achievement in geophysics, solar physics, solar-terrestrial physics, or planetary sciences (the 'G' award), and one for achievement in astronomy, cosmology, astroparticle physics, cosmochemistry, etc. (the 'A' award). A select number of persons have received the medal twice, among whom John Herschel in 1826 and 1836, David Gill (1882 and 1908) and Edward Charles Pickering (1881 and 1901). The Society was founded in 1820 and William Herschel was its first president. The logo of the R.A.S., which is also depicted on the Gold Medal, shows Herschel's 40-foot telescope and his motto *quicquid nitet notandum* (whatever shines should be observed). Kapteyn's medal, along with other awards and decorations, is reproduced in Fig. A.6.

The award certainly put Kapteyn in the spotlight. Among the laureates immediately preceding him were the famous French mathematician, astronomer, physicist and philosopher Jules Henri Poincaré (1854–1912) and Edward Charles Pickering. In the USA an announcement of the award was published together with a summary of the President's address (*Kapteyn's contributions to our knowledge of the stars* by J.D. Galloway [2]). Kapteyn's star had risen to great heights and his stature was all the more enhanced by his presentation in St. Louis and his discovery of the Star Streams. In 1913 Gill published a history of the Cape Observatory [3] and in this he wrote:

'In 1902 the Royal Astronomical Society marked its high appreciation of the value of Kapteyn's work by awarding him its Gold Medal.

But probably the most valuable result of the C.P.D. to science is the fact that its preparation first directed Kapteyn's mind to the study of the problem of cosmical astronomy, and thus led him to the brilliant researches and discoveries with which his name is now and will ever be associated.'

Kapteyn owned a private copy of this book on the Cape Observatory, presented to him by Gill himself, who wrote on the first page: 'To his dear old friend Kapteyn, from David Gill'.

10.2 Visit to the Cape

As mentioned in the previous chapter, the publication of his St. Louis paper on the Star Streams was much delayed by the publishing process of the congress. However, Kapteyn traveled to South Africa in 1905, where the British Association for the Advancement of Science held its annual meeting and a summary of his lecture appeared in the report of that meeting Kapteyn (1905a). In 1906 a French summary, even shorter, was published [4]. As it turned out, this remained Kapteyn's only visit to South Africa and to the Cape Observatory. Henriette Hertzsprung-Kapteyn describes it as follows:

'In the summer of that year Kapteyn traveled to South Africa, at the invitation of the British Association, which held its large international meeting there. De Sitter, who was his assistant at the time, accompanied him. A few other astronomers also traveled with him: Backlund, the director of the Pulkovo Observatory (Russia), Hinks and Cookson both English astronomers. They discussed many things during the voyage and for its duration established the 'Astronomical Society of the Atlantic'. [...]

It had been an old wish for him to see Gill in his own observatory, since you only get to know a man completely when you see him in his working quarters. With admiration he saw the observatory that Gill had brought to fruition, and he saw his friend as the all-important person, as President of the British Association and host. He was the radiant centre of attention, flexible and sociable, full of thoughtfulness and untiring. Gill found it difficult to fulfill his high position with the dignity associated with such positions. One of his friends said to him that he should not forget he was the president and 'preserve his dignity'. 'That is just what my brother said to me', was his reply. 'Davie', he said, 'you've no more dignity than a duck.' And he remained his amiable and jovial self.

Kapteyn returned home happy and satisfied. He had seen and admired Gill's observatory, had seen his friend in full glory and had made major progress with his own plans. He had seen the land with which the Dutch felt so related to in all its beauty, since the Association had organized a

Fig. 10.1 The constellation Crux or the Southern Cross. It is the kite-like (see quotation from Mark Twain on page 153) combination of four stars on the right. The two bright stars on the lower left are the 'Pointers', guiding the eye towards the Southern Cross. These are α and β Centauri; the first is also the nearest star to the Sun at about 4 light years. In the background is the Milky Way. The dark structure below and to the left of the Cross is the 'Coalsack Nebula', the most prominent dust cloud in interstellar space that the naked eye can see. It is of order 600 light years from us. For comparison the brightest star in Crux is at about half that distance (European Southern Observatory ESO [5])

few excursions that aroused his interest. He was delighted by the abundant flora. Fields full of arums had made a strong impression. The Zambezi Falls he felt were less impressive, but more mysterious than Niagara Falls. The Southern Cross, the much praised constellation, had disappointed him and did not live up to his expectations. In general he felt that the starry sky of the northern hemisphere was more beautiful, although seeing Scorpius in the zenith was magnificent. He was most touched by the grave of Rhodes [Sir Cecil Rhodes (1853–1902), businessman and politician, founded the state of Rhodesia, the present Zimbabwe], which he found extremely impressive. Only a large, flat stone with the name Rhodes chiseled in it, among the majestic emptiness of the Matoppo hills in Matabeleland, an impressive reminder of Rhodes' victory over the Matabele king Lobengula. He met a few South Afrikaners of distinction and heard many interesting facts about the country after the war. So in all respects the trip had been interesting and successful.'

Arthur Hinks wrote a report of the Cape meeting of the British Association [6], from which the paragraphs on Kapteyn are worth reprinting here.

10.2 Visit to the Cape

Fig. 10.2 Sir David Gill as he appears in the *Album Amicorum* presented at the retirement of H.G. van de Sande Bakhuyzen as professor of astronomy and director of Leiden Observatory in 1908 (Archives Leiden Observatory; see caption Fig. 3.7)

'Most astronomers have heard somewhat vague reports of a great address delivered by Prof. Kapteyn at the St. Louis Congress of 1904, which has not yet been published. One or two have had the privilege of reading the manuscript; but it is curious – and perhaps a warning to others – that to announce a great discovery to a scientific congress may render it inaccessible for a long time afterwards. Even to repeat the substance of it twelve months later, as Prof. Kapteyn did at Cape Town, does not bring it very much nearer publication; for although the Sectional Committee resolved to ask the Council to print the address in full in the Association Report, and the Council agreed, that Report will probably not appear for some months. The consequence of this delay is that, eighteen months after it was first announced, comparatively few people know what 'starstreaming' means. The members of Section A at Cape Town are a happy minority, and the writer will be so bold as to set down his recollections of an intensely interesting but rather complicated subject.'

I skip the paragraph in which Hinks summed up the procedures followed by Kapteyn to show the evidence for two streams among the stars. He then continued:

'And now for the interpretation. Practically the whole of the stars, according to Prof. Kapteyn, belong to one or other of two great streams, and the *apparent* vertices of their motions lie 140° apart, at points about 7° south of α Orionis and 2° south of η Sagittarii. The true vertices of their motion are at ξ Orionis: and the point diametrically opposite it – that is to say, the streams are moving past one another along a diameter of the Milky Way.

In the course of the discussion that followed, Prof. Darwin expressed his admiration of the development which the science of stellar statistics had received through the research of Prof. Kapteyn, but admitted a grudge against these results, which furnished no confirmation of the theory put forward in his address. Sir David Gill expressed the amazement of the practical astronomer that the systematic character of proper motions should have escaped notice for so long, and was sure that the steps which Prof. Kapteyn advised for the elucidation of this matter would command the best efforts of the best observatories of the world.

There can be no question of the revolutionary character of this discovery, and it will take some reflection to make sure of what is left of the supposed established facts of stellar astronomy. What will happen to the precession constant may easily be guessed. An apex of the Sun's way becomes indefinable when two systems are involved. But perhaps the most interesting point is this: What relation do the stars which have up to the present shown sensible proper motion bear to the stars in general and to the star-clouds of the Milky Way? Kapteyn says that the brighter Bradley stars with proper motion are nearly equally distributed, but that the fainter are condensed on the plane of the Milky Way. Newcomb considers the latter part of this result unreliable, and prefers to say that the distribution of stars whose proper motion is certainly determined bears no relation to that plane. The difference is clearly of extreme importance – just the difference between having found out something of the structure of the whole visible universe and having analyzed into two separate parts the particular star-cloud in which our Sun is involved. For the solution of this and many other points in the theory of star-streaming which are at present obscure, we await with lively impatience the appearance of Prof. Kapteyn's detailed discussion.'

10.3 Acceptance of the Star Streams Concept

In the next three years (1906 to 1908), three papers appeared on the subject of Star Streams that quickly established the phenomenon as generally accepted. This must to a large extent have been due to that fact that these were papers by astronomers whose opinions carried a lot of weight. The first was by Arthur S. Eddington in 1906, who had just become chief assistant at the Royal Greenwich Observatory

10.3 Acceptance of the Star Streams Concept

Fig. 10.3 Arthur Stanley Eddington as he appears in the *Album Amicorum* presented at the retirement of H.G. van de Sande Bakhuyzen as professor of astronomy and director of Leiden Observatory in 1908 (Archives Leiden Observatory; see caption Fig. 3.7)

after having been one of the brightest students in Cambridge (see Fig. 10.3). In the paper, *The systematic motions of the stars* [7], Eddington devised a quantitative test and applied this to a set of stars in the Groombridge Catalogue for which new proper motions had been determined by Frank Watson Dyson (1868–1939) and William Grasett Thackeray (1853–1936) at Greenwich. I remind the reader that the Groombridge stars constituted all circumpolar stars as seen from England, collected in a catalogue (and published posthumously) by Stephen Groombridge. In fact, the roughly 4,500 stars are all less than 52° away from the north pole. Dyson and Thackeray had used this for a new determination of the position of the Solar Apex (*A determination of the constant of precession and the direction of the Sun's motion from comparison of Groombridge's Catalogue (1810) with modern Greenwich observations (1905)* [8]). Kapteyn used brighter stars all over the sky, but Eddington's material constituted more stars over a limited part of the sky, which were also fainter (between tenth and ninth magnitude). His results 'strongly support Kapteyn's hypothesis of two star-drifts', the number of stars belonging to reach drift being approximately equal.

The next person was Karl Schwarzschild (1873–1916), director of Sternwarte Göttingen at the time. His paper was in German, *Über die Eigenbewegungen der Fixsterne* [9]. Schwarzschild had problems with the idea of dividing up of the stel-

Fig. 10.4 The distribution of space velocities of stars according to Karl Schwarzschild (1907) (From: *The systematic motions of the stars* by A.E. Eddington (1905))

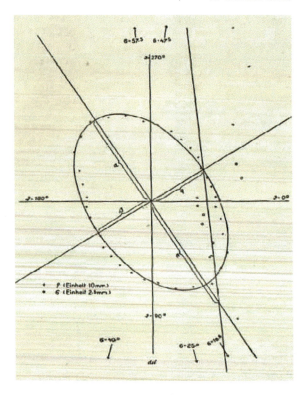

lar system into two distinct swarms and tried to combine everything into a single picture. He noted that Eddington had described the system as two streams each with a 'Maxwellian' distribution. 'Maxwellian' here derives from the motions of molecules in a gas as described by James Clerk Maxwell (1831–1879). This means that in each stream the random velocities of the stars are superposed onto the systematic motion according to a Gaussian ('bell-shaped') distribution, which then has a mean value that is the same in all directions. Schwarzschild now proposed that there was really only one system, and that onto the systematic velocity that a star had relative to the Sun, the random velocities were indeed distributed in a Gaussian fashion, but that the mean velocity had two preferred directions. In fact, he described the distribution as 'ellipsoidal', so that the curve of mean motion as a function of direction traced an ellipse. Such a distribution has a projection onto the sky that should match the distribution of directions and magnitudes of proper motions that can be compared to observations. The data of Eddington indeed resulted in a distribution of velocities that was roughly ellipsoidal (see Fig. 10.4). The axis ratio of the ellipse was 0.55 and the largest axis had a value 1/0.70 (about 1.43) times the velocity of the Sun towards the Apex.

Although conceptually different, Schwarzschild's approach is not distinguishable from two separate streams in observational terms. It actually represents the current view. When the mean motion is largest in two opposite directions, the result is that there appear two opposite, preferred directions or vertices in the distribution of proper motions on the sky. The concept of an ellipsoidal velocity distribution was eventually incorporated in the dynamical theory of the Milky Way of Jan Hendrik Oort (1900–1992). In this concept, the Milky Way rotates differentially, that is to say the outer parts take progressively longer to make a full rotation than regions closer to the center, and the axis ratio of the velocity ellipsoid is determined by the amount of this differential rotation.

The third person was Frank Dyson, who analyzed the motions on the sky of stars with large proper motion only (between $30''$ and $80''$ per century) in a paper also with the title *The systematic motions of the stars* [10], and confirmed both the results of Kapteyn and those of Eddington and Schwarzschild.

Kapteyn's Star Streams quickly won acceptance. However, at the same time this meant that the whole edifice that he had built to go from observations of proper motions as a statistical means towards determining the structure of the sidereal system was useless; it relied on the 'fundamental assumption' of no preferred direction of motion for stars in space. Still the concept of parallactic motions would work for groups of stars that shared a common general motion in space, such as a star cluster, in which there was a mean motion of the cluster as a whole relative to the Sun. The most promising group of stars was the very nearby star cluster the Hyades. We have seen (page 384) that with the help of plates taken by Anders Donner, Kapteyn had attempted to measure parallaxes in that cluster. And, of course, the determination of as many parallaxes as possible was still as urgent as ever. We will see that Kapteyn researched all these issues extensively.

10.4 Plan of Selected Areas

The idea that Kapteyn contemplated now, or had in fact been contemplating for some time, was to make a census of the sky by comprehensive observing of stars to faint levels in carefully chosen, selected regions. This resulted in his 'Plan of Selected Areas', which was published in 1906. In its Introduction, Kapteyn (1906d) (see Fig. 10.5 for the first page of the handwritten manuscript) wrote:

'In the following pages a plan is outlined, which may be realized by the cooperation of a few astronomers in a, relatively speaking, moderate time.

The aim of it is to bring together, as far as possible with such an effort, all the elements which at the present time must seem most necessary for a successful solution to the sidereal problem, that is: the problem of the structure of the sidereal world. [...]

After much consideration the most promising plan seems to me to be some plan analogous to that of the gauges of the Herschels.

In accordance with the progress of science, however, such a plan will not only have to include numbers of stars, but all the data which it will be possible to obtain in a reasonable time.'

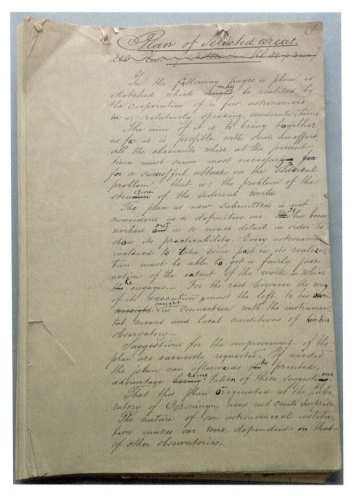

Fig. 10.5 First page of the handwritten introduction of the Plan of Selected Areas in Kapteyn's final manuscript (Kapteyn Astronomical Institute)

What he proposed to do was to bring together comprehensive data sets of stars all over the sky to fainter levels than had been possible till then from observations by a number of observatories and with the use of the measuring capabilities of the Groningen Astronomical Laboratory. Since it is impossible to do this for the whole sky, he proposed to define a set of some two hundred areas, strategically chosen so as to sample the sky in a systematic way, yet steering away from areas that for one reason or another would be problematic. He noted that his early proposal to do a Durchmusterung for parallaxes was not met with much enthusiasm and that at the time Gill suggested to restrict it to certain small areas to make it manageable. He now used this idea for his Plan.

10.4 Plan of Selected Areas

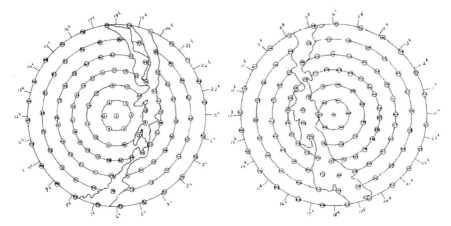

Fig. 10.6 The Selected Areas (systematic plan) in both hemispheres. The poles are in the middle, the equator at the outer circle. The irregular band is the Milky Way (Kapteyn Astronomical Institute)

Kapteyn set out the Plan as follows: 'It consists simply in this: For 206 areas, regularly distributed over the sky and for another, little extensive series of particularly interesting regions, to obtain astronomical data of every kind, for stars down to such faintness as it will be possible to get in a reasonable time'. This is the *'Systematic Plan'*; the fields are distributed along lines of constant declination of 15° apart (see Fig. 10.6). Kapteyn was looking for 200,000 stars in all (all stars down to the limiting magnitude within areas of accurately known extent had to be included), for which should be measured rough positions, but sharply defined photographic and visual magnitudes. For some 20,000 stars, 'fitly distributed over magnitude' he proposed to obtain proper motions down to $0\rlap{.}''01$ per annum in each coordinate, parallaxes accurate to $0\rlap{.}''02$, and for as many as feasible the class of spectrum and the radial velocity. He added that it would be important to determine the total amount of light received from different parts of the sky. This is a bit out of line with the rest, but the total contribution of starlight from faint stars would constitute an important and useful measurement. Kapteyn based this suggestion on a paper by Simon Newcomb, *A rude attempt to determine the total light of all the stars* [11], in which he devised various ways to estimate the surface brightness of parts of the sky by comparing through eye vision the sky compared to the 'spread-out light from a star of known magnitude'. Newcomb's result was very inaccurate. Kapteyn was also experimenting with this; in 1909 he had a PhD thesis prepared under his supervision by Lambertus Yntema (see page 437). Later van Rhijn also performed a detailed study of the same property.

10.5 Finding Support for the Plan

Kapteyn had prepared the idea extensively and had consulted and talked to a great number of astronomers, for the first time, apparently, while at the 1904 St. Louis congress, then during his voyage to and stay in Cape Town in 1905 (there were apparently 'copies of proceedings of the Astronomical Society of the Atlantic') and by correspondence. By 1906 he was in a position to publish and distribute the Plan, having enlisted the support of several key astronomers. To this end he took extreme care that with the publication of the Plan he could present excerpts from letters by a long list of astronomers who endorsed it and committed themselves to provide help. To accomplish that he had to adapt his own ideas, if he wished to keep certain players on board. For example, Henriette Hertzsprung-Kapteyn wrote:

'Many had their remarks or suggestions. For example Pickering, the director of Harvard Observatory, proposed to investigate 46 additional special areas that showed peculiarities, such as unusual lumpiness or emptiness, mostly in the Milky Way. These were added to the Plan. Others had other proposals.'

It is believed that Kapteyn was not too much charmed by the idea of these extra fields, which threatened to delay the main project. But Pickering's help was essential, so he gave in. I first look at the appendix, which presents excerpts from letters of endorsements.

He starts with the correspondence with Pickering, who pointed out that in the Milky Way

'points of greatest and least density [of stars] will necessarily occur at irregular intervals. There are certain centers of cosmical disturbance, like the Nebula in Orion [see Fig. 11.8], which should surely be included. [...] At least one region should include portions of where the sky, as in Herschel's [this must have been John Herschel] Coalsack in Crux [see Fig. 10.1] and the great nebulous region in Scorpius, stars are almost entirely wanting, apparently owing to large adjacent nebulae.'

So, Kapteyn reluctantly included 46 special areas in the Milky Way.

Edwin B. Frost (1866–1935), director of Yerkes Observatory committed himself for one night per week on the 40-inch telescope ('it might turn out later that one and a half nights could for a while be devoted to the work'). Karl Friedrich Küstner of Bonn (1856–1936) considered his climate and latitude less ideal for photographic parallax work, but committed meridian observations of 1500 stars. George C. Comstock of Washburn Observatory in Madison offered help with radial velocities. David Gill wrote from the Cape that they were ready to help with parallax and proper motion plates and the necessary additional meridian observations in the southern hemisphere. Also Pulkovo Observatory would take part and in a postscriptum added in proof, there is support from Oxford's Radcliffe Observatory.

In the proposal Kapteyn first reviewed what was known about the brighter stars and concluded that

'summing up, I think that we may say, in a rough way, that the state of our knowledge of the various elements of the sidereal problem is, or will shortly

10.5 Finding Support for the Plan

be fairly satisfactory for the stars visible to the naked eye. For most of the elements it will still allow a thorough statistical treatment for stars as faint as magnitude 7 or 7.5. Below this limit data become decidedly fragmentary and less accurate. They are nearly wanting below the 9th magnitude.' [The problem is of course that] 'the aspect of the Milky Way is mainly dependent on the distribution of the faint stars'.

Kapteyn stressed that the need to know these *numbers* of faint stars, was the reason for Herschel's star gauges. But he also quoted 'an eminent investigator' to illustrate the point of the necessity of 'such systematic counts of faint stars of determinate brightness'.

This 'eminent investigator' is Hugo von Seeliger (1849–1924), whom we met as a candidate for Kapteyn's chair in Groningen. Independently of Kapteyn, von Seeliger had done much fundamental work in the area of the study of the distribution of stars in space and the structure of the sidereal system. Before continuing it seems opportune to have a look at the relevant work of von Seeliger, which is not discussed in great detail in this book on Kapteyn. By and large, the two men worked on the same subject, but hardly corresponded with each other and seem to have had completely different characters. Their contributions to the development of statistical astronomy and its applications happened in almost compete isolation from each other. This point is the subject of a thorough and very enlightening study by E. Robert Paul, reported in a number of important studies and an excellent scholarly book *The Milky Way Galaxy and Statistical Cosmology, 1890–1924* [12]. Paul shows how different the styles and approaches of the two were, Kapteyn gaining much international praise through his observational approach, whereas von Seeliger's was much more mathematical and much more abstract. In addition, von Seeliger hardly engaged in international projects and collaborations. The development of Kapteyn's work was largely independent of von Seeliger's. The issue of what lies at the base of this lack of communication has received a full and comprehensive treatment in Robert Paul's, book to which little substantially new material can be added. I therefore restrict myself to mentioning von Seeliger where appropriate and necessary.

Of course it was wise that Kapteyn mentioned von Seeliger in his Plan of Selected Areas, so as to ensure that people knew he was aware of this work. He was also careful to mention the 'Carte du Ciel' effort and acknowledge its importance. He noted his 'perfect concurrence with the opinion of those who see in this work a *duty* to posterity' (italics by Kapteyn), but also stressed that even the catalogue part would take a long time to complete, while magnitudes would be subject to inaccuracy, but that proper motions were not going to be part of it. What was urgently needed was 'elements' as data about parallax, spectra, visual magnitudes, radial velocities.

'The gauges of the Herschels have been found extremely useful in studies about stellar distributions. That they no longer fully satisfy our needs, is *not* a consequence of the fact that they only cover 145 square degrees of the 41,000 contained in the firmament, but that they are restricted to this *one* element: *total number of stars*. Did we possess these numbers for *each separate, sharply defined magnitude*, then already an immense step forward would be possible, as is well shown in Seeliger's investigation:

Fig. 10.7 George Ellery Hale (From the *HHK biography*)

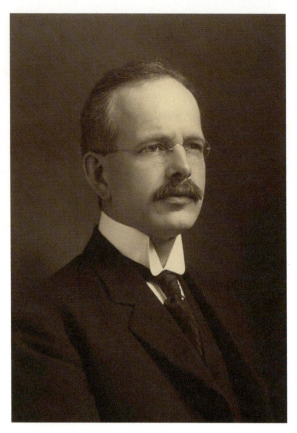

Betrachtungen über de räuml. Verteilung der Fixsterne. Still our need for the other elements is certainly not less urgent.'

10.6 Enter George Ellery Hale

Certainly it was important for Kapteyn to get as much support as he could to get his plans adopted. Most important, if not vital, was the fact that he gained the support of George Ellery Hale (1868–1938). In Kapteyn's career, it seems to me, three persons were essential by providing opportunities at a critical time. The first was of course David Gill, who 'invited' Kapteyn to work on the Cape Photographic Durchmusterung. The second was Anders Donner, who first – when the Carte du Ciel did not adopt Kapteyn's idea of measuring parallaxes from multiple exposures – provided photographic material to test the method and later contributed much more photographic material for Kapteyn's studies. The third was George Hale, who provided not only support for the Selected Areas project but also, as we will see shortly, offered Kapteyn to spend an important part of his time at Mount Wilson for

10.6 Enter George Ellery Hale

a number of years. Some of the later correspondence between Hale and Kapteyn (documented by Petra van der Heijden in the *Legacy*) has been sent on microfiche to Groningen. I have consulted it extensively, but for a complete picture it was necessary to examine the rest at the Cal Tech Archives and the Huntington Library. A very helpful document is a list that Hale prepared of all the letters he received from Kapteyn between 1905 and 1922. This list resides in the Hale papers at Cal Tech Archives.

Hale had started out as a solar astronomer and physicist. As a professor of astronomy at the University of Chicago, he founded the Yerkes Observatory in Williams Bay, Wisconsin, in 1897, and built the largest refracting telescope (still in operation) with a 40-inch (102 cm) lens as objective. Hale next founded the Mount Wilson Observatory near Pasadena in California with funds from the Carnegie Institution of Washington, which was founded by Andrew Carnegie (1835–1919) in 1902. Carnegie had made his enormous fortune in the steel industry, which boomed in the later part of the nineteenth century. Hale had become a leading figure in American astronomy at the time of the St. Louis meeting. He actually organized a meeting on solar research at the same time the 'International Congress of Arts and Sciences' was held there. Kapteyn was invited to participate there in a letter Hale wrote to Kapteyn on May 27, 1904. Obviously they did not know each other then; Hale addressed Kapteyn as 'My dear Sir'. He started out by telling Kapteyn of a Committee on Solar Research, set up by the National Academy of Sciences, of which Hale was chairman. It aimed at improving coordination in solar research. The Academy in Amsterdam had been asked to appoint a similar committee, and Hale asked Kapteyn to be kind enough to interest himself in the matter.

> 'The various societies to which requests for the appointment of committees have been addressed are invited to send one or two delegates to a conference of the various committees, which will be held in St. Louis at the time of the International Congress of Science. It gave me great pleasure to learn from Professor Newcomb, that you expect to attend the congress, and to give one of the principal addresses. I trust that you may also come as a representative of the committee to be appointed by the Royal Academy of Sciences, and that you will join in the discussions regarding a plan of organization which will insure satisfactory representation of all interests. [...]
>
> I hope that while you are in this country you will also visit the Yerkes Observatory at Williams Bay, Wisconsin, where you will be given a most cordial welcome by Professor Frost, who is acting Director in my absence. I regret much that I shall not be there at the time of your visit, unless you can conveniently go to Williams Bay immediately after the St. Louis Congress. I am in charge of an expedition for solar research on Mt. Wilson, California, and shall spend most of my time here for several years to come.'

Interestingly, Kapteyn had taken the initiative to write to Hale on June 8, 1904. In this letter he informs Hale that he is interested in measuring parallaxes and is keen to find out what the focal length would be to obtain high precision determinations of parallax and proper motion. He says he has plates from various other observatories as well and has heard that Hale is trying 'to get plates for parallax with the big telescope of your observatory'. He is interested in getting data from Hale as well and asks for 'some negative of good or middling quality', adding that 'even a broke negative or one which for some reason or other you do not care for would be welcome'.

> 'I have an earnest hope that next autumn it will be my good fortune to be able to thank you personally. I can arrange to be in Chicago on Sept. 17 if you will allow me to come and see the observatory where so much splendid work is being done.'

Obviously the two letters crossed. Hale had scribbled on Kapteyn's letter: 'I had already invited Kapteyn to visit us, but he had not received the letter when he wrote this'. There was also correspondence between Frost and Kapteyn, which survives in the Yerkes archives (see van der Heijden in the *Legacy*), and of which copies have been made available to the Kapteyn Astronomical Institute. Frost had followed up Hale's letter and sent a letter to Kapteyn on June 13 inviting him to visit Yerkes and on June 28 he sent a reply to the letter of June 8. Kapteyn accepted the invitation on July 3, saying he would be in Chicago 'about September 16'. Frost and Kapteyn (then at Harvard) exchanged telegrams involving his travel details some days before Kapteyn arrived. Kapteyn actually visited Yerkes just before he met Hale in St. Louis.

Kapteyn was also present at the Conference on Solar Research, which was held on September 23. The proceedings are available on the Web (*Minutes of the Meeting of Delegates to the Conference on Solar Research*. It was held in the Hall of Congresses, St. Louis, September 23, 1904 [13]). Hale was president, the vice-presidency being in the hands of no less a person than Henri Poincaré, while Austria was represented by the famous Ludwig Eduard Boltzmann (1844–1906). The national committee for the Netherlands was recorded as consisting of Kapteyn and Julius; the latter was Willem Henri Julius (1860–1925), who was a professor of physics at the University of Utrecht and is generally regarded as the founder of solar research in the Netherlands. Julius was not recorded among those present in St. Louis.

There are some interesting observations referring to Hale and Kapteyn meeting each other in the *HHK biography* (note that below Henriette Hertzsprung-Kapteyn is wrong on the presence of Julius in St. Louis), including some recollections Hale wrote at Henriette's request (see page 646):

> 'At the same time there was an international conference in St. Louis 'on solar research' at which Kapteyn and Prof. Julius from Utrecht, were the delegates from the Netherlands. Hale presided over the conference. Kapteyn showed him his Plan of Selected Areas. The meeting took place on the grounds of the exhibition in the Tyrolean Alps, an artificial scenery set up on a grandiose American scale. They talked all afternoon, absorbed by their deliberations and unconcerned with the natural beauty(!) around them. Hale was enthusiastic and would later prove to provide important support to make the Plan reality. Kapteyn was deeply impressed by the wide views and the sharp mind of this American, who was about to become the leader of the biggest and best equipped observatory in the world, the Mount

10.6 Enter George Ellery Hale

Wilson Observatory in California, which was nearing completion. Support from this man was of the utmost importance to him.'

Hale wrote later about this first meeting in a letter with 'Reflections of Kapteyn' [and attached it to a letter to Henriette Hertzsprung-Kapteyn of August 13, 1926, as a 'contribution for your book']:

'As chairman of the Academy's Committee I was deeply interested in the possibilities of international cooperation, and convinced that much might be accomplished through joint effort. I was therefore greatly impressed by Kapteyn's scheme of Selected Areas, which he presented at the International Scientific Congress, held in conjunction with the St. Louis Exposition. My own experience had been in the field of astrophysical research, and my plans for the then nascent Mount Wilson observatory were chiefly confined to an attack upon the physical problems, involved in the study of the Sun as a typical star. Researches on the distribution of stars in space did not then enter into the scheme. However as I listened to Kapteyn's masterly paper and realized the wide scope of his plans and the skill with which he availed himself of international cooperation in assuring their execution, I was deeply impressed by his appeal. Could we not help him to secure the data, needed for the fainter stars and at the same time broaden and strengthen the attack in our own problem of stellar evolution? The answer is obvious today, but at that time, approaching the subject along the path marked out by Huggins, Lockyer and other pioneers of solar and stellar physics, and seriously hampered by lack of funds, the case was not so clear. Nevertheless the genius of Kapteyn and the personal charm which brought him the unqualified support of astronomers the world over, convinced me at once that the Mount Wilson Observatory ought to profit by his cooperation as soon as circumstances might permit.'

William Huggins (1824–1910) was a spectroscopist, who was the first to realize that some nebulae like the Orion Nebula (see Fig. 11.8) were gaseous. Joseph Norman Lockyer (1836–1920) was a solar spectroscopist. And further:

'A powerfully creative imagination, glowing with optimism and enthusiasm is prone to set itself too vast a task. But Kapteyn, though he would gladly have measured all the stars of heaven, recognized the necessity of limiting his endeavor. Hence the Plan of Selected Areas, and the successful appeal for international cooperation.'

However, the undertaking was so gigantic, that it was impossible for Kapteyn to undertake the leadership and coordination all by himself. He proposed to set up a committee and wrote about this, among other things, to Prof. Küstner, the director of the Observatory in Bonn on July 22, 1907:

'The coordination in my Plan of Selected Areas is now so much that the responsibility threatens to become so heavy for me that it is no longer possible for a single human being to bear the full oversight of the whole project and the coordination of the arrangements. Gill has insisted for a long time that a conference be organized. I have declared myself strongly against that. Almost all matters, yes, really all of them, are in good hands.

Fig. 10.8 Edward Charles Pickering, director of the Harvard College Observatory (Courtesy of the Harvard College Observatory [14])

Maybe a conference will be too unwieldy, and will in the end not result in better plans. We have extensive experience with this.'

So he asked Sir David Gill, Prof. Edward C. Pickering, Prof. George E. Hale, Prof. Karl Küstner, Prof. Karl Schwarzschild, Sir Frank Dyson and Walter S. Adams, men who together represented the astronomical world of those days, to serve on a committee, an invitation that each of them accepted.'

10.7 Early Progress of the Plan

This committee was in fact formed in the summer of 1907, but consisting initially of Gill (Cape), Pickering (Harvard), Hale (Mt. Wilson), Küstner (Bonn) and Kapteyn. In 1910 it was extended by appointing Schwarzschild (Göttingen), Dyson (Edinburgh, and Greenwich from 1910 onwards) and Adams. Walter Sydney Adams (1876–1956) was an assistant at Mount Wilson Observatory, where he became director in 1923, after Hale's retirement. The committee met twice in the early stages: for the first time in Paris in 1909 on the occasion of a meeting of the permanent Commission of the Carte du Ciel, and a second time with all members present at Mount Wilson in 1910 during a meeting of the Solar Union. Figure 13.18 shows some of the participants of the 1910 meeting and some members of the committee for the Selected Areas. Kapteyn began to publish 'reports on the progress'; the first and second of these appeared in 1911, although the first had actually been written in 1909, Kapteyn (1911c). However, the third report was not published until 1923, by Pieter van Rhijn (*Third report on the progress of the Plan of Selected Areas, together with some remarks concerning the future investigations of the plan* [15]), and the fourth in 1930 in the *Bulletin of the Astronomical Institutes of the Netherlands* [16]. These were not records of actual meetings but collections of contributions from various members. From then on the role of the committee was taken over by Commission 32 (Selected Areas) of the International Astronomical Union (see Box 16.1), later the Subcommittee on Selected Areas of IAU Commission 33 (Structure and Dynamics of the Galactic System). I would like to point out to the interested reader two more general publications on the Plan of Selected Areas (both with that title) of later years, viz., by Beverly T. Lynds in 1963 and by A. Blaauw & T. Elvius in 1965 respectively [17].

10.8 The Harvard-Groningen Durchmusterung

Especially in the first decades after its inception, the Plan of Selected Areas was very actively pursued in various countries and at various observatories. An important share of the work involved the collaboration with Edward Pickering of Harvard (see Fig. 10.8), with whom Kapteyn completed a Durchmusterung of the Selected Areas, referred to as the Harvard-Groningen Durchmusterung. The first thing needed was an inventory of stars in the Selected Areas with relatively, but not extremely, accurate positions and magnitudes. This was done using plates taken at the Harvard College Observatory in the north and at its Boyden Station in Arequipa, Peru, in the south, and measured at the Groningen Astronomical Laboratory. The Boyden Station of Harvard Observatory had originally been erected near Lima, but had been relocated to Arequipa (also in Peru) in 1890. It was named after Uriah Atherton Boyden (1804–1879), engineer and inventor, who left a large sum of money in his will to be used by Harvard for astronomy. In 1927 it was moved to Bloemfontein in South Africa. The parts on the 'Systematic Plan' were published in three installments, Pickering & Kapteyn (1918), Pickering, Kapteyn & van Rhijn (1923) and Pickering, Kapteyn & van Rhijn (1924), in the Annals of the Astronomical Observatory of Harvard College. The corresponding part on the 'Special Plan', on which Pickering had been insisting so much, was only published much later as van Rhijn & Kapteyn (1952).

These publications provided positions and photographic magnitudes (the 1963 review by Beverly T. Lynds incorrectly states that both photographic and visual magnitudes were provided) of stars in areas of $80' \times 80'$ at the higher latitudes, but smaller ones in parts of the sky with larger densities of stars, down to $20' \times 20'$ in the Milky Way. The faintest stars were of about magnitude 16. The plates were measured by Kapteyn, later van Rhijn, and his staff in Groningen. The production

Fig. 10.9 The 16-inch Metcalf Telescope of Harvard College Observatory, which was used for the northern part of the Durchmusterung of Selected Areas, Pickering & Kapteyn (1918) (From Harvard College Observatory [18])

of the results was described in detail in the first publication, Pickering & Kapteyn (1918). Pickering discussed how the plates were taken with the 16-inch Metcalf telescope at Harvard and in the south with the 24-inch Bruce telescope in Arequipa. The plate scales were different, respectively 93″.3 (later after regrinding the telescope objective 98″.2) and 59″.7 to the millimeter. Also the exposures had to be different to produce more or less uniform depth, allowing for the different focal lengths of 221 (later 210) versus 345 cm; they lasted 60 minutes in the north and 120 minutes in the south. The Arequipa observations (for the full Durchmusterung) were taken between 1906 and 1916 and at Harvard between 1910 and 1912. The procedure was to take a short (1, respectively 2 minute) exposure with the telescope slightly moved in right ascension for the brightest stars to be recorded well, in addition to the deep observations.

The magnitudes were calibrated with a set of special observations to accurately measure those of a series of stars in each field (a so-called 'sequence of standard stars'). The procedure to obtain these was very careful. First it is referred to a set of adopted magnitudes for stars near the North Pole. This was the result of a plan that had been drawn up by the Permanent Committee for the Carte du Ciel with extensive international representation. Using this North Polar Sequence and Harvard Standard Regions related to it, all other magnitudes were put on a uniform scale. Consequently the resulting magnitudes were designated the 'International Scale of Magnitudes'.

For this purpose, exposures were obtained, using plates from the same supplier box, of the North Polar Sequence and one or two Harvard Standard Regions and

10.8 The Harvard-Groningen Durchmusterung

Fig. 10.10 The apparatus that Kapteyn and his staff used to measure the Harvard plates for the Selected Areas Durchmusterung. Picture taken from the *Album Amicorum* presented at the retirement of H.G. van de Sande Bakhuyzen as professor of astronomy and director of Leiden Observatory in 1908 (Archives Leiden Observatory; see caption Fig. 3.7)

Fig. 10.11 The computers and technical staff of the Kapteyn Astronomical Laboratory around 1926. The measurements of the Selected Areas Harvard plates was performed by T.W. de Vries (see also Fig. 6.11) in the middle, with the assistance of J.M. de Zoute (front-left) and E.J. Rondeel (not shown) (Kapteyn Astronomical Institute [19])

some of the Kapteyn Selected Areas. All these plates were then developed together and measured. Images on these plates had been obtained by exposing with a whole range of exposure times (and telescope shifts in between), ranging from 1 second to about 17 minutes. This resulted in a set of a dozen to twenty Standard Stars with 'known' magnitudes in each of the Selected Areas 1 through 115. In Pickering's Introduction this work is credited to Henrietta Swan Leavitt and assistants. Henrietta Leavitt (1868–1921) was part of the group of female astronomers at Harvard ('Pickering's harem'); she became famous for her discovery of the period-luminosity relation of Cepheids. Briefly, these are pulsating stars in the later stages of evolution, where the luminosity changes periodically during the pulsation. Since the period of pulsation depends on the physical condition in the insides of the star, the pulsation period, much like a standing wave, scales with overall properties such as luminosity or absolute magnitude. Since the absolute magnitude of a Cepheid can be derived from the period of its brightness variability, these stars have proven to be crucial for distance determinations in astronomy.

In contrast to the CPD, Kapteyn now employed a method of measuring rectangular coordinates for the stars on the plates. To facilitate conversion to right ascension and declination, he provided very simple formulae to convert the x and y measures listed in the catalogue, with only a few parameters, which varied from plate to plate. The apparatus (see Fig. 10.10) was really meant for accurate *differential* measurement, but at Kapteyn's request the producers (the Repsold brothers) had provided options of more accurate measuring, which were required here. According to Kapteyn's acknowledgments the measuring was performed under the supervision of the 'assistants Dr. W. de Sitter, Dr. H.A. Weersma, Dr. F. Zernike [later Nobel laureate in physics] and Dr. P.J. van Rhijn'. His trusted senior technical assistant, computer and observer T.W. de Vries, together with E.J. Rondeel and J.M. de Zoute performed the measurements and the reductions (see Fig. 10.11).

The care with which Kapteyn set up these measurements, the details taken into consideration in their preparation and execution and the use of his experience with the CPD, have been eloquently illustrated by Adriaan Blaauw in the *Legacy* in his chapter on the Kapteyn Room, in which he presents the 'Kladboeken'. From KB no 104 inverted, pp. 6/7, he has taken a draft for a letter to Pickering. The draft is dated March, 13, 1907 (see Fig. 10.12); the chapter shows reproductions of the second page and a transcription of the full text. Kapteyn reported to Pickering that he had just received the first batch of plates and then continued (in draft form so Kapteyn's English is not always very clear):

> '1st. I am most positively of the opinion that we ought to measure the originals and not the copies. For the CPD we always had two plates of every region, which gives an enormous advantage in the certainty with which real stars can be distinguished from defects in the plates. [...] Now that we have but one plate of each region the danger is enormously greater. Using copies it would be doubled. I think this alone is more than sufficient reason to stick to the originals.
>
> The dangers to which they are exposed are not great. I got perhaps some 2000 photographs for the CPD. Not one of them was broken.

10.8 The Harvard-Groningen Durchmusterung

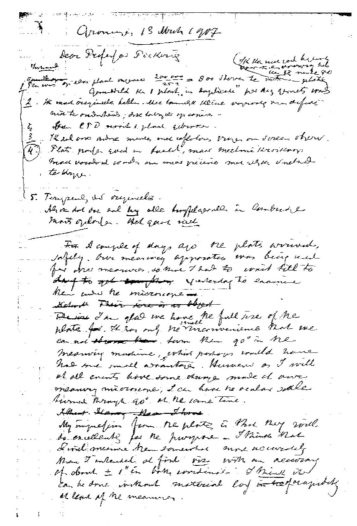

Fig. 10.12 First page of the draft letter of Kapteyn to Pickering on the execution of the Harvard-Groningen Durchmusterung (Kapteyn Astronomical Institute)

2nd. As to sending the plates back as soon as measured, I fear this will land us in the greatest difficulties. As soon as other parts of the work come to be handled (parallaxes, proper motions, spectra etc.) we are sure to have every day some doubt to be settled, which requires a direct reference to the plates. [...] For this reason Gill let all the negatives for the CPD remain at Groningen. They are even now in my possession. – Still I have no desire whatever to keep them forever. On the contrary, they are far safer with you in your fire-proof buildings. Still I think that it would be advisable to let them remain here at least for a few years to come. – [...].

3rd. Of course you will easily be able to let the production of plates keep pace with our measurements. Still, nearly the whole of my working power must be used for a time in the execution of the measures and their reduction. Any stoppage in the regular progress of the work means to me a very serious loss of labour. I will rather still wait for a couple of months than risk the danger of being stopped in the work once started. In fact, as I said above, some things have to be altered at the measuring microscope. Also some other changes have to be made for which it was better to wait till I had some specimen plates. So it would be difficult for me to begin in good earnest for 6 or 8 weeks to come. Everything considered, I think we ought not to begin our work before you are some 40 or 50 plates in advance.

My original intention has been to measure on an average some 800 stars on each plate. See 'Plan [of Selected Areas]' p.9. With the enormous richness of your plates it may be difficult to confine ourselves [... so] that we do not somewhat surpass this number. Still I am rather unwilling to measure more than say 300,000 stars in all (instead of 200,000 as proposed in Plan). This would mean an average of 1200 stars pro plate. We have, very often, measured a plate with this number of stars, in duplicate, in one day. In two days it can be done with the greatest possible care.

Considering all this I think that the best plan will be: that we wait in making a beginning till you are about 40 or 50 plates ahead. If from that time you try to get the whole of the plates in the course of one year, we are pretty sure that everything will run smoothly.

4th. I am very anxious to get the best possible determinations of photographic magnitude. For the 26 regions of 'Plan p. 31' at least I would like to have as many different determinations as I am able to get. I therefore am trying to interest astronomers provided with good reflectors in the work.
[...]
Yours very sincerely,
JCK'

Kapteyn and Pickering authored the first installment of the resulting 'Harvard-Groningen Durchmusterung', and it was published in 1918. This encompassed all areas of the northern sky, those from the North Pole to the Areas on the equator (1 through 115). The southern part was published under P.J. van Rhijn, after both Pickering (1919) and Kapteyn (1922) had died. These publications were Pickering, Kapteyn & van Rhijn (1923) for the Areas at Declination $-15°$ and $-30°$ (116 through 165) and Pickering, Kapteyn & van Rhijn (1924) for the remaining ones (164 through 206 at the South Pole). In the latter two publications, the acknowledgments included a new assistant, J.H. Oort, in addition to de Sitter, Weersma and Zernike.

The numbers of stars in the catalogues were 83,489, 65,763 and 81,866 respectively, for a total of 231,118. The number per square degree runs between just over 200 (SA's 54–58 near the North Galactic Pole) and 21,656 (SA 193 right in the

Galactic Plane; in SA 157 close to the Galactic Center it is 11,554). These extensive counts were of course available long before they were actually published and were – at least in part, as we will see – the basis for studies that Kapteyn performed towards the end of his life on the statistical distribution of stars and the structure of the Sidereal System as deduced from star counts all over the sky.

10.9 Further Progress of the Plan

The limiting magnitude of 16 was too bright, in Kapteyn's opinion. We will see in Chap. 12 how he got involved with George Hale at Mount Wilson Observatory, where he went each year for an extended period. While there, he organized a Durchmusterung of the Selected Areas that can be seen from California with the then newly built 60-inch telescope (see page 458). This Mount Wilson Catalogue, which was only published in 1930, will be discussed in the next chapter.

It was also necessary to have a reliable and uniform inventory of the brighter stars, since the Selected Areas covered only a small fraction of the sky and their centers were chosen such that the field was not dominated by bright stellar images. To this end, Kapteyn collected star counts from all catalogs and sources available and published the results in Groningen Publication 18 *On the number of stars of determined magnitude and determined Galactic latitude*, Kapteyn (1908d). He starts out by writing: 'In the investigation of the arrangement of the stars in space, there is no element which has played so prominent a role as the number of stars of a determined magnitude and the variation of this number with Galactic latitude. These investigations all suffer more or less from one or both of the following causes: *a.* that they do not rest on a trustworthy *photometric* scale; *b.* that they are restricted to too small a range of magnitude.' So, he sets out to provide such data on the 'Harvard scale' of magnitudes, defined using the North Polar Sequence. This scale is somewhat different from the International scale that we have seen above, but they could be corrected (and actually were, at a later stage) in relation to each other. Of course such data would be a product of the Carte du Ciel, but Kapteyn noted that 'the charts for the Carte du Ciel make slow progress. Most of the cooperating observatories have not yet made a beginning with their publication. It seems certain that the completion of the chart is still very far off.'

The provision of Durchmusterungs of the Selected Area progressed relatively rapidly. This was partly due to the strong support of Pickering and Hale, and partly thanks to the personal involvement of Kapteyn and his Laboratory in Groningen. However, it was largely because this was the easiest and least time-consuming part of the work and obviously the most urgent. Accurate measurements of positions, magnitudes and colors – and the provision of standard stars for such measurements – were much more time-consuming and were performed at a much slower pace. Also, as this was done by various observatories at a wide variety of locations, the work obviously took much longer to be completed. Proper motions in particular required long timescales. Trigonometric parallaxes proved very difficult to obtain as these were often too small to be measured and in his later studies Kapteyn had to rely to a large extent on secular parallaxes from proper motions. Parallaxes and proper motions were the contributions where the Cape Observatory did much work (albeit with limited success). Radial velocities and classifications of spectra were likewise slow in forthcoming, in spite of the great efforts by Milton Lasell Humason (1891–1972) and Walter S. Adams (1876–1956) at Mount Wilson (Solar) Observatory.

It is interesting to read the *Third report on the progress of the Plan of Selected Areas*, which van Rhijn published in 1923, the year after Kapteyn's death. Except for the Durchmusterungs, progress was indeed slow, scattered and not always very systematic. For example, work on standards for photographic and photo-visual magnitudes and colors was performed at Mount Wilson and Yerkes, Harvard, Greenwich, Córdoba and Pulkovo; on proper motions at Radcliff, Pulkovo, Yerkes and Yale; trigonometric parallaxes at Yerkes, Yale and the Cape; position standards at Babelsberg (Berlin), Bonn, Leiden, Paris, Strasbourg, Bordeaux, Lick and La Plata near Buenos Aires, Argentina; spectral classification at Mount Wilson, Hamburg (Bergedorf) and Potsdam; radial velocities at Mount Wilson.

10.10 The Bergedorfer Spectral Durchmusterungs

The Astronomical Laboratory played the central role that Kapteyn had envisaged. In addition to the Harvard and Mount Wilson Durchmusterung/Catalogue and measurements on plates for parallaxes and proper motions, the Laboratory was involved in one other major project, viz., the spectral classifications. Some work was done at Mount Wilson, where Humason determined spectral types for 4066 stars in the first 115 Selected Areas. In the *Legacy*, Tom D. Kinman writes: 'It was demanding work. He used a slitless spectrograph at the Newtonian focus of the Mount Wilson 60-inch telescope with exposures that averaged 4.5 hours; this must have involved great patience as well as a lot of observing time.' But this was only a small selection of stars. A more comprehensive survey was performed in collaboration between Groningen and the Hamburg Observatory at Bergedorf, but this was not taken on in earnest until after Kapteyn's death. On fainter stars work had been done in some Selected Areas at Mount Wilson, but in his second report, Kapteyn (1911c), little progress could be reported. It was only in the 'Third Report' of 1923 that the work was described as having started. Actually, van Rhijn had proposed the Durchmusterung to Friedrich Karl Arnold Schwassmann (1870–1964) of the Bergedorfer Observatory. This survey aimed at determining spectral classes of stars in the magnitude range 8.5 to 11. The method was to take photographic images of the sky with a prism in front of the primary lens in the telescope, a so-called 'objective prism', so that each stellar image is replaced with a small spectrum (see Fig. 10.13) on the photographic plate. The telescope used is a double astrograph: two parallel telescopes designed to make photographic images. One of them is equipped with the objective prism, so that the resulting plate with its small spectra can immediately be

10.10 The Bergedorfer Spectral Durchmusterungs

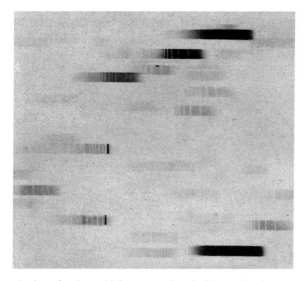

Fig. 10.13 Reproduction of a plate with images produced with an objective-prism for the Bergedorfer Spektral-Durchmusterung (Kapteyn Astronomical Institute [20])

compared to a 'direct' one that only shows stellar images as dots and to plates in the same manner of stars with known spectral type. The spectral classification was made possible with a reasonable accuracy.

The work took a long time to complete. The spectral classification was done in Hamburg, the photometric magnitude determinations in Groningen from extensive plate material collected at Harvard with the cooperation of director Harlow Shapley. Most of the work in Groningen was done by van Rhijn's assistants Bart J. Bok (1906–1983), Adriaan Blaauw, Lukas Plaut (1910–1984) and others. Eventually, the results were published in five volumes: *Bergedorfer Spektral-Durchmusterung der 115 nördlichen Kapteynschen Eichfelder* by F.K.A. Schwassmann and P.J. van Rhijn [21]. A similar exercise was performed at the Potsdam Observatory in the *Spektral-Durchmusterung der Kapteyn-Eichfelder des Südhimmels*, again in five volumes by Wilhelm Becker (1907–1996), partly in collaboration with Hermann Alexander Brück (1905–2000) [22].

So this part of the Plan of Selected Areas took a long time to be executed and the question arises: did the Plan of Selected Areas ever get completed? The answer must be negative, but then the Plan had no specified final goal other than collecting as much data of as many stars as possible in the designated areas. The impact of it, however, by no means depended on its 'completion'. First of all, the relatively quick execution of the (in Adriaan Blaauw's words in the *Legacy* 'impressive') Harvard-Groningen Durchmusterung provided a very important census of the stellar distribution on the sky. And the stars that were used as magnitude calibrators, determined at Harvard and Mount Wilson, formed a uniform, all-sky system of calibration standards. A major accomplishment of the Plan of Selected Areas was that it got astronomers from all over the world and at many observatories to cooperate and agree on a research program. The fact that for the larger part of the research effort, observations were taken in coordination of the same parts of the sky and of the same stars, turned out to be of crucial importance on various occasions.

For instance, Frederick Seares at Mount Wilson had been working on a program of photometric observations during the 1910s with the 60-inch telescope (see page 468). 'It involves the determination of the photographic magnitudes of the fainter stars immediately surrounding the central star of each of the 115 Selected Areas on and north of the equator.' [From Sears (1914) [23]]. In a paper on globular clusters in 1917, Mount Wilson astronomer Harlow Shapley (1885–1972) noted [24]: 'The derivation of photographic and photo-visual magnitudes, and hence the colors of stars in clusters, has been made possible through two important factors. The first is the 60-inch reflector at the Mount Wilson Observatory; the second and more necessary one is the fundamental work by Seares in establishing photographic and photo-visual magnitude scales for faint stars'.

I quote from the chapter by Tom Kinman on the Selected Areas in the *Legacy*:

'Shapley's tribute to Seares is important. It is not easy today for us to realize how difficult it was to set up an accurate magnitude sequence for faint stars using photography. Much labour and ingenuity was employed to avoid systematic errors and it was not until 1930 that the *Mount Wilson Catalogue of the Northern Selected Areas* was published by Seares, Kapteyn and van Rhijn. This photometry was at the foundation of stellar and galaxy work in the first half of this century. It became increasingly obvious however, that even the best photographic photometry had substantial errors in spite of all the care that had been taken. This led Hubble and Baade to ask Stebbins to use the Mount Wilson telescopes to re-observe some of the stars in three Selected Areas with his photoelectric photometer. Stebbins started his observations in 1947, these were continued by Whitford in 1948 and by Whitford and Johnson in 1949. In this important work, Stebbins, Whitford and Johnson (1950) reached 18th magnitude. [...]

Substantial errors are present at quite bright magnitudes in the muchused Harvard Selected Area magnitudes; this led to significant systematic errors in (for example) the magnitudes of variable stars. If, on the other hand, there had been no Selected Areas with which observers could standardize their photometry, the situation would undoubtedly have been worse.

Few people can have been more aware of the importance of photometry and of the difficulty in obtaining the needed accuracy than Baade. When, however, the question of whether the Radcliffe Observatory would extend the Mount Wilson work on the Selected Areas to the Southern Hemisphere, he was strongly opposed to the idea. In a letter to Thackeray (Director of the Radcliffe Observatory) dated 1949 December 13, he wrote:

'The work involved in setting up a long precision photometric sequence is so formidable — even with the assistance of the photocell — that the extension to the total of S.A.s is out of the question. I hope that you in the southern hemisphere will not repeat the detours and mistakes which have been made in the north. Kapteyn's program, excellent in the light of his time, is dead and nobody will be able to revive it. Even those of my older colleagues here who devoted a large part of their life to this program — and with enthusiasm — admit freely that in retrospect the returns did not come up to their expectations and the best contributions of the observatory came from those men who were not bound by long range programs. After Adams' retirement — and on his strong advice — all work on the Kapteyn program was brought to an end and it is doubtful if someone can ever induce this observatory to go on a similar binge. I hope very much that you can keep the radial velocity work in Pretoria within reasonable limits because such routine programs have a deadening effect once they get the upper hand. Worst of all, they attract the wrong kind of people —the so-called hard workers — and once they are established the fate of an observatory until its obsolescence is sealed.' (this quotation is taken from Feast, 2000 [25]).

We can understand why Baade objected to setting up photometric sequences in *all* the southern Selected Areas. [...] Baade's comments on *'hard workers'* are less easy to understand in the light of the Palomar–Groningen survey which he proposed only a few years after this letter was written.'

The Palomar-Groningen survey concerned a particular type of pulsating variable stars (RR Lyrae stars, related to Cepheids), that occupied Groningen astronomer Lukas Plaut for many years. The passage just quoted illustrates both the strengths and weaknesses of the Plan.

10.11 The Unofficial End of the Plan

The work on the Bergedorfer Spectral Durchmusterung was definitely not a good use of time and resources. Adriaan Blaauw often complained that it had been so much of an effort that very little else could be done at the Laboratory. Now, it is true that van Rhijn's determination of the 'luminosity curve' in a paper in 1925, entitled *On the frequency of the absolute magnitudes of the stars* [26], was a major accomplishment. It survived for a very long time as the standard 'van Rhijn Luminosity Function', albeit in revised and improved versions, but the fact that it survived so long also demonstrates that not many new, improved results appeared. In the meantime, astronomy had taken a turn towards astrophysics and interpretative research rather than cataloging and data collecting. Leiden, largely through the influence of Kapteyn, as we will see, reorganized itself into a research center in a broad sense, no longer being only occupied with positional astronomy as it had been under Kaiser and van de Sande Bakhuyzen. Leading positions in Leiden were soon filled by protégés of Kapteyn, in particular Willem de Sitter,, Ejnar Hertzsprung and Jan Oort, while Kapteyn himself returned briefly to Leiden as deputy director after his formal retirement. The Laboratory in Groningen went into a deep decline until

Fig. 10.14 Attendants at IAU Symposium No 1 at Vosbergen near Groningen in June 1953. Some notable persons are: Walter Baade, fifth from the left in the front row, four persons to his right Jan Oort, to the right and in front of Oort Lukas Plaut, behind Plaut Adriaan Blaauw and sixth from the right Pieter van Rhijn (Kapteyn Astronomical Institute)

the appointment of Adriaan Blaauw as director in 1957. Blaauw also used to complain [27] that the work on the Bergedorfer Durchmusterung was not even honored with the Groningen appearing in the name. Would it not have been appropriate to designate it the Bergedorf–Groningen Durchmusterung? Or maybe even the other way around? Woody (Woodruff T.) Sullivan put it even more harshly in the *Legacy*:

'Finally, at Groningen after the death of Kapteyn in 1922, his eponymous Laboratory went into a period of decline coinciding with the long directorship (1921–1956) of his protégé Pieter van Rhijn. Van Rhijn did some excellent work in the 1920s in developing the standard stellar luminosity distribution, but both his research and leadership for the rest of his career were unimaginative and stuck in a rut [See also *Sterrenkijken bekeken, (op. cit.)*]. Kapteyn's influence on his own institution ironically was negative [...]

One direct result [...] was Kapteyn's Plan of Selected Areas, the study and organization of which occupied a large portion of his successor van Rhijn's career. As previously stated, this work was done in a routine and unimaginative manner, and contributed to Groningen's period of decline. Van Rhijn also aped his mentor unprofitably by teaming up with another observatory (Bergedorf) and spending twenty years reducing data on

10.11 The Unofficial End of the Plan

170,000 stars for the Bergedorfer Spektral-Durchmusterung, a survey whose results turned out to be of no great import.'

In June 1953, The Kapteyn Astronomical Laboratory organized the International Astronomical Union Symposium no 1 on *Co-ordination of Galactic research*, held at the estate Vosbergen near Groningen, which was owned by the University. The 'Purpose and character of the conference', quoted from a circular letter sent to all participants, was described as follows:

'During the first third of this century an important concentration of work on Galactic structure and motions has been promoted by the Plan of the Selected Areas, initiated in 1906 by Kapteyn. Although the scheme outlined in 1906 has not lost its significance, it is widely felt that further research into structure and dynamics of the Galaxy should be extended beyond the original Plan. At the same time it is felt by many that some kind of co-ordination of effort remains highly desirable, if only because the value of many observations is greatly enhanced if the data can be combined with other data for the same stars.

Several observatories have asked for suggestions with regard to future work on galactic structure. The recent advent of several large Schmidt telescopes furnished with objective prisms, red-sensitive plates, etc., brings forward with some urgency for each of the observatories concerned the problem of how to organize and restrict the work with telescopes that can produce much more than can be measured and discussed by the existing observatories. The question arises, whether it is desirable to formulate a new plan of attack, and whether such a plan should also include recommendations for concentrated work on particular regions. If the answer to the latter is affirmative, which regions should be selected and what data would be the most urgent.'

This is the unofficial end of the Plan of Selected Areas. As described above, Kapteyn had instituted a committee to oversee the Plan of Selected Areas, which later became Commission 32 of the International Astronomical Union and later a sub-committee of Commission 33 (Structure and Dynamics of the Galactic System). The latter action took place in 1958; at the time van Rhijn had been chairman of that sub-committee for two years into his retirement. This constituted a final landmark in terms of finishing of the systematic work on Kapteyn's Plan. The reports for 1954 and 1957, prepared by van Rhijn, summarized the publication of data obtained on the Selected Areas. There were 79 such publications, some of which were of course major catalogs constituting years of labor. In the end 28 observatories from 11 different countries contributed to the Plan of Selected Areas.

Fig. 10.15 Kapteyn in 1908 (Centraal Bureau voor de Genealogie, collectie Veenhuizen [28])

Chapter 11
Extinction

> *In all this, no question is raised in regard to the integrity of the record, nor whether in its long journey, any planet, sun, comet, meteorite or nebula has interfered to modify or in any way to corrupt the story it was commissioned to tell.*
> *What Faith!*
> De Volson Wood (1832–1897)[1]

> *Why after the dust settles, someone has to come by and blow at it, stirring it up into the air again?*
> Anthony Liccione (1968–present).[2]

The discovery of the Star Streams had disrupted Kapteyn's grand plan to turn a distribution of apparent magnitudes and proper motions across the sky into a three-dimensional distribution of stars in space. After all, it was based on three assumptions (no preferred space velocities, a 'luminosity curve' independent of position in space and an absence of absorption of light by an interstellar medium). In any event, the first assumption had now been proven wrong. The second was still thought to be reasonable, but the third, as we will see in more detail in this chapter, remained a constant worry. Before going into extinction of light of stars on its way to us, I first return to some other work Kapteyn had done in the first few years after his discovery of the Star Streams.

An important part of Kapteyn's attack on the sidereal problem was determining the luminosity curve. His first approximation to this distribution of stellar

---
[1] De Volson Wood, American civil engineer: *Faith in the integrity of the interstellar medium* [2].
[2] Anthony Liccione is an American poet.

luminosities had been published in 1901 (see Fig. 9.7), but obviously much improvement was required. For this he needed parallaxes of many stars. But these had proven to be very difficult to collect on a wholesale basis on account of the large stellar distances. So he concentrated on deriving distances of stars using statistical methods involving the use of proper motions, allowing for the systematic effects of the Star Streams.

11.1 Weersma and the Distance to the Hyades

Kapteyn first continued his work on an on-going program of determining proper motions and parallaxes of stars in the Hyades cluster, based on plates obtained by his old collaborator Anders Donner. The Hyades cluster still plays a vital role in astronomy; as the nearest star cluster it is a stepping stone of studies of distances and fundamental properties of stars. Since distance determinations historically depend on calibration of stellar properties through stars in the Hyades, any revision of the adopted distance to the cluster results in a revision of the full distance scale in astronomy. In the later part of the twentieth century such revisions up to 15% occurred. As we will see, the distance is about 150 lightyears, which corresponds to a parallax of about 0″.02. Even today, this is at the limits or just beyond what can be measured accurately with ground-based telescopes. The measurement of an accurate and reliable value for the distance to the Hyades becomes possible only from space, with the European Space Agency ESA's astrometric satellite Hipparcos (see *The Hyades: distance, structure, dynamics, and age* by Perryman *et al.* (1998) [3]).

The Hyades cluster covers somewhat more than the area of sky outlined by the constellation Taurus. This is illustrated in Fig. 11.1. The main part of the constellation, which is easily seen at night with the naked eye, is the slanted V-shape set of stars just below the center. Of these, the brightest star, (α Tauri), is called Aldebaran, the red eye of the bull. It is a red giant star at about 68 lightyears and one of the brightest stars in the northern sky. It does not belong to the Hyades, but most of the other brighter stars in the Taurus constellation do. They form the cluster's central part; actually most of the stars seen projected on the slanted V are Hyades members. Because of their proximity, the Hyades are not easily recognized as a star cluster simply by looking at sky images. However, this did not prevent astronomers as far back as the ancient Greeks to list them as a cluster (along with the Pleiades), although of its members obviously only the naked-eye stars of the Taurus constellation could be seen. In Greek mythology, incidentally, the Hyades were the daughters of Atlas and sisters to the Pleiades.

The cluster does extend over a much larger part of the sky, at least the part shown in the figure, but further away from the densest central regions only a smaller fraction of the stars are members. The group of stars near he top-right are the Pleiades (Fig. 7.11), which is further from us than the Hyades.

11.1 Weersma and the Distance to the Hyades

Fig. 11.1 Constellation Taurus. The reddish bright star is Aldebaran and the cluster near the top-right is the Pleiades (Fig. 7.11). Most of the brighter stars in this photograph, except Aldebaran and those of the Pleiades are members of the Hyades cluster, but the cluster extends beyond the boundaries of this picture (From the Hubble European Space Agency Information Centre [4])

Kapteyn realized that the Hyades offered a unique possibility. It was very likely to be close, since so many apparently bright stars belonged to it and it was also known that the Hyades stars had a relatively large proper motion. So if one knew the distance to the Hyades, one would immediately have obtained parallaxes of many different stars. The cluster was also known to contain stars of various colors and spectral types and offered a chance to study differences among stars. The first task was to determine which stars in the field were members and proper motions seemed to be the best way to discriminate members from background stars. The proper motions of stars in the Pleiades were known to be almost parallel, indicating a common motion in space, and this property could be taken as a signature of a physical cluster. Thus determining proper motions of Hyades stars might be used for such discrimination. In addition, trying to measure parallaxes directly would be also useful. The plan Kapteyn drew up was to use two sets of plates. The first were parallax plates, as he had been using in Kapteyn & Donner (1900) and in Kapteyn, de Sitter & Donner (1902) for the h and χ Persei clusters, and in on-going work in other fields in Kapteyn, de Sitter & Donner (1908b), utilizing a set of three exposures separated by half a year.. A second set of plates involved multiple exposures separated by a few years, to determine proper motions, quite similar to the work he was doing in Kapteyn, de Sitter & Donner (1908a; see page 260). Of course, it was Donner

Fig. 11.2 H.A. Weersma as he appears in the *Album Amicorum* presented at the retirement of H.G. van de Sande Bakhuyzen as professor of astronomy and director of Leiden Observatory in 1908 (Archives Leiden Observatory; see caption Fig. 3.7)

who was found interested in providing the necessary plate material for the Hyades; he took the parallax plates in 1893 and 1894 and the proper motion plates between 1895 and 1901. The amount of photographic material provided through the years by Donner and the Helsingfors Observatory was very substantial indeed, and the fact that Kapteyn met Donner at the Carte du Ciel meetings would prove to be a considerable stimulus in his work.

For this project Kapteyn enlisted his second PhD student, Herman Albertus Weersma. Weersma (1877–1961) (see Fig. 11.2) obtained his PhD in 1908 (see page 434). He remained at the Laboratory as an assistant and in 1909 he was appointed 'privaatdocent' in the 'determination of orbits and theory of perturbations of celestial bodies' [5]. Weersma left astronomy and Groningen University in 1912, but his name kept appearing in publications of the projects that he had been working on, such as Kapteyn & Weersma (1914); in a paper in 1918, Kapteyn, van Rhijn & Weersma (1918), he is described as 'formerly assistant'.

There is a short *in memoriam* written about Weersma after his death in 1961 (*In memoriam Dr. H.A. Weersma, Groningen 1877 – Heelsum 1961* by D. Bartling [6]). From this we learn that he had a strong interest in philosophy and the 'search for meaning'. He had hoped to find answers to his philosophical questions through a study of biology, but after his Candidaats exam (comparable to a Bachelors degree) he switched to astronomy because of his clumsiness with conducting experimental work. In the end, astronomy did not provide what he was looking for and after his

PhD thesis, while he was still employed as assistant to Kapteyn, he studied philosophy with Heymans, becoming particularly interested in the writings of German philosopher Georg Wilhelm Friedrich Hegel (1770–1831). Weersma also developed a keen interest in music, especially Wilhelm Richard Wagner's (1813–1881). He became a mathematics teacher, so that he could devote more time to these interests. Later he became a fervent socialist and he wrote a number of books on philosophical issues related to Marxism, dialectics and logic.

Kapteyn started with proper motions, since these could be used to decide on cluster membership. Weersma's first assignment was to collect as much information on proper motions of Hyades stars as was available using existing catalogues. It resulted in No. 13 of the Groningen Publications in 1904: *The proper motions of 66 stars of the Hyades derived from the observations of 34 catalogues between 1755 and 1900*, Weersma (1904). These were necessary, as Weersma explained, for the determination of the plate constants taken by Prof. Donner for the purpose of deriving proper motions of stars in the Hyades. The results of the proper motion measurements themselves were presented in the next issue of the Groningen Publications, Kapteyn, de Sitter & Donner (1904).

For Kapteyn, the work also constituted a study on the feasibility of obtaining wholesale proper motions of faint stars. After all, having realized that trigonometric parallaxes of stars were beyond the abilities of the telescopes of his time, with the exception of some nearby ones, he had turned to proper motions. Kapteyn concluded in Groningen Publications 14 (1904): 'With a hundred plates for each hemisphere, on each of which 40 or 50 stars would be measured, together with two or three meridian observations of 12 stars for each plate, we might have, in the lapse of a dozen years, as good materials for the discussion of the proper motions of the stars of the 10th to the 14th magnitude, as we have now for the stars as bright as 5th or 6th magnitude. [...] For the study of the structure of the stellar system such data would be of inestimable value.'

The measurement of the parallaxes took much longer than expected and was not published until 1909. The actual work had to a large extent been done by de Sitter. In the Introduction to that paper, Kapteyn, de Sitter, Donner & Küstner (1909), Kapteyn brought up an issue that had been gaining attention, viz. the possibility that the cluster contained stars of 'widely different spectral classes', and 'numerous spectroscopic doubles', (stars that appear single but that have two independently moving sets of spectral lines in their spectra, pointing at a binary star with two components revolving around each other). This makes the cluster useful for studying 'internal motions, or for the study of evolution'. Kapteyn's program, which he said was set up in 1892, when he discussed it with Donner, aimed to determine the membership of the Hyades as well as an accurate parallax in order 'get clear notions not only of the linear dimensions of the system, but also about the linear velocities and absolute magnitudes'.

The work is not only based on data from Helsingfors, but also on plates taken by Karl Friedrich Küstner of Bonn Observatory. The Bonn Observatory had a telescope with a relatively long 5-meter focal length. This ensures a large plate scale, which is very suitable for measurements of parallaxes (and proper motions). Unfortunately due to 'our uncertain climate' the exposures were shorter than Kapteyn had wished. Küstner took plates in three Hyades fields during 1903 and 1904.

The lists of measured parallaxes show the property noted before, viz. that for many stars formal parallaxes were too small to be trusted or even negative. However, the results translated in a parallax for the Hyades as a whole, though. For Helsingfors the result is +0″.024±0″.010 and for Bonn +0″.023±0″.0025. Since the two values were essentially the same, except that the Bonn errors were smaller, Kapteyn adopted the latter as his best determination of the Hyades parallax. The Hipparcos parallax of the Hyades is 0″.0215±0.0028. Kapteyn's value was wrong by only 7%! Kapteyn calculated the absolute magnitudes (at a parallax of 0″.1) and luminosities (compared to the Sun, which has an absolute magnitude of 5.5 in the photographic band) of 82 stars, the absolute magnitude ranging from 0.5 to 7.2 (luminosity from 98 to 0.2 Suns).

11.2 Comparison to the Results of Boss

In the Introduction to the paper just referred to, Kapteyn mentioned that his final value for the Hyades parallax was also very lose to the one determined only shortly before by Lewis Boss, who used a different method, of Dudley Observatory in Schenectady, New York. His determination (see also page 292) was based on a method designed in 1843 by Auguste Bravais (1811–1863). In brief, the method works as follows. For a group of stars that move essentially parallel in space with respect to the Sun, the proper motions will point to a point on the sky, quite similar to the proper motions of all stars do towards the Solar Apex. However, this point on the sky is different from the Solar Apex, and now it gives us the direction of the velocity of the *group* (or cluster) with respect to the Sun. Now, if for a number of these stars the radial velocity is also measured, we also know the projection of the velocity of the cluster onto the line of sight. The angle on the sky between the cluster's position and the convergence point gives us the angle over which this projection has taken place. This means that we can calculate the space velocity of the cluster with respect to the Sun. But it also means that we can calculate the projection of the space velocity onto the plane of the sky (known as the tangential velocity) and therefore the proper motion in kilometers per second. Comparing this to the proper motion in seconds of arc per year then yields the distance. This is called the 'moving cluster method'.

Figure 11.3 shows the determination of the convergence point by Boss. Kapteyn was aware of this method; actually it is underlies his resolve to use the motion of the Sun through space as the basis for distance determinations rather than that of the Earth around the Sun. It seems likely that he already knew about this and the possibility to measure secular parallaxes from proper motions already at the time of his PhD thesis, as I pointed out in relation to his Proposition 15 (see page 48). And he had directed Weersma to write a thesis on the subject. This thesis, published in the Groningen Publications in 1908, was entitled *A determination of the Apex of*

11.2 Comparison to the Results of Boss

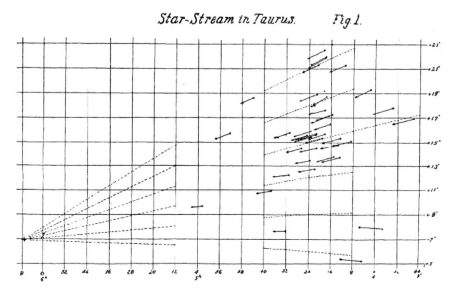

Fig. 11.3 Proper motions of the Hyades stars as published by Lewis Boss in 1908. The lengths of the arrows correspond to 50,000 years of proper motion (From *Convergent of a moving cluster in Taurus* by Lewis Boss [7])

the Solar motion according to the method of Bravais, Weersma (1908). So, why had Kapteyn not used it already himself?

The answer is that Boss' work had shown that the Hyades extended over a much larger area of sky than Kapteyn was aware of when he started his work on the cluster. The problem was that when the angular extent of the cluster was small (and the convergence point far from the cluster on the sky), the determination of the convergence point was not possible with any accuracy. There was the case for the Pleiades where the proper motions were almost parallel (and the convergence point very distant on the sky). The same was the case for the Hyades as Kapteyn knew it. But now that Boss had extended the membership of the Hyades over a considerably larger area of sky, the situation was different. Figure 11.3 indicates that it actually extends over more than 15°, while the convergence point is only some 25° away; that point can then be determined very accurately from the observations.

As it happened, Kürstner had reported radial velocities of three(!) Hyades stars, giving almost exactly 40 km/sec. This would mean a space velocity of about 45 km/sec of the cluster. Boss calculated what the proper motion of each star was and arrived by taking the means of all individually determined parallaxes at a final value of 0."0253. The path of the cluster through space was now also known and Boss actually mentioned that the cluster was closest to the Sun 7,600 years ago, when the central parallax was 0."05.

The parallax work on the Hyades constituted Kapteyn's last effort to measure trigonometric parallaxes on a wholesale basis, and from then on he concentrated on proper notions and star counts. That does not mean that parallaxes became second priority, but it was obvious that the progress in getting more parallaxes would be slow. As we have seen, progress in parallax determinations in the Plan of Selected Areas was slow as well. Actually, programs of wholesale parallax determinations have been possible only since the 1990s from space with the Hipparcos satellite. Well within a century after Kapteyn's death, there now is a GAIA satellite[3], which was launched by the European Space Agency ESA in late 2013. Among others things, it measures the positions, parallaxes and proper motions of 1 billion stars with an accuracy of about 20 μarcsec (microarcsecond or a millionth of an arcsec) at magnitude 15, and 200 μarcsec at magnitude 20. It will also measure the radial velocities for the brighter stars from spectra. To Kapteyn this would have been unthinkable, although it may have been his ultimate dream.

Before turning his attention definitely to star counts, Kapteyn – with the help of Weersma – produced a compilation of parallax determinations, Kapteyn & Weersma (1910), containing 360 stars. As before in such lists, some stars have negative parallaxes and these (and some more with small parallaxes) should, according to current practice, have been quoted with upper limits. As an illustration, I count only 156 parallaxes in excess of 0″.05.

11.3 Star Counts, Extinction and the Star Ratio

I already mentioned that Kapteyn made an exhaustive investigation, published in in 1908 in the Groningen Publications *On the number of stars of determined magnitude and determined Galactic latitude*. This was a major effort with which he hoped to arrive at a better determination of the space density of stars. It was as far as he could hope to go before the results from the Plan of Selected Areas would become available. That grand undertaking would take years to get anywhere near completion. However, Kapteyn was too impatient not to try to make progress on the understanding of the distribution of the stars in space. Also, his interest had been aroused with regard to the issue of determining the stellar distribution from star counts by a paper *Distribution of stars* by Edward Pickering [1].

There were two issues in that paper that Kapteyn was uncomfortable with. The first was the data themselves. In Pickering's work the conclusion was reached that: 'the number of stars for a given area in the Milky Way is about twice as great

---

[3]This looks like and indeed is an acronym, which stood for Global Astrometric Interferometer for Astrophysics, but the design changed and this full name is no longer appropriate and not used anymore.

11.3 Star Counts, Extinction and the Star Ratio

as in other regions and this ratio does not increase for faint stars.' Kapteyn (1908d) stated: 'This conclusion has been to me extremely startling. It cannot well be reconciled with the results of the Herschels and I think that we have to admit, either a very serious systematic error in the work of the Herschels, or some mistake in the work of Prof. Pickering.' As we will see below, Kapteyn concluded eventually that the latter alternative had to be accepted.

The other conclusion by Pickering is that he found 'the ratio of the increase in number of stars of diminishing apparent brightness to deviate from the 'theoretical' ratio, and seems also inclined to adopt the hypothesis of a considerable absorption of light in space' [quoted from Kapteyn (1904a)]. Now, this alarming conclusion was reinforced in a paper by George C. Comstock, director of the Washburn Observatory at Madison. In this paper, on *Provisional results of an examination of the proper motions of certain faint stars* [8], Comstock used proper motions of a set of 68 stars to estimate statistically their parallax, following methods he attributed (rightly) to Kapteyn. That is he decomposed the proper motions in a component towards or away from the Solar Apex and perpendicular to it. The first can be used to statistically derive a secular parallax for an assumed velocity of the Sun though space and the latter can be used to estimate a statistical distance from an assumed mean random motion of the stars. Comstock first found that 'the intrinsic luminosity of stars diminishes with their apparent brightness in such a ratio that a star of the tenth magnitude possesses only one-tenth of the luminosity of a star of the fifth magnitude.' The alternative would be that 'the conditions for infinite transparency are not satisfied in the celestial spaces'. He stated he was aware of several studies that failed to establish the existence of this medium, but he also emphasized that they 'have equally failed to demonstrate its absence'. Comstock clearly leans toward the second explanation.

Furthermore, Comstock found a second property, namely that the mean proper motion of stars at a fixed apparent magnitude depended on Galactic latitude, being smallest in the Milky Way. 'If this relation shall hereafter be established by more abundant data [he had 'only 68 stars' so it was a small sample] its explanation will doubtless be that in the plane of the galaxy, stars of a given brightness are more remote than in the direction perpendicular to that plane. This conclusion tends to fortify the supposition of an absorption by sparsely distributed gross matter [...]'. Note the modern sounding use of 'plane of the Galaxy'.

Full transparency of space was a vital assumption in Kapteyn's analyses and these developments were cause for grave concern. He had conscientiously always stated that he explicitly assumed transparency and had not been aware of any evidence to the contrary, but now the issue seemed open. He addressed the matter in a paper in an American journal, entitled *Remarks on the determination of the number and mean parallax of stars of different magnitude and the absorption of light in space*, Kapteyn (1904a). The submission was on May 1904 and it was published August 10, 1904, just before his address at St. Louis. Both Comstock and Pickering were present in St. Louis, the first chairing the first astronomy session and the latter making a presentation in that session on September 20 (Kapteyn lectured in the second session the next day). They must surely have discussed the issue of absorption (heatedly?), but no record of that is known to me.

I will explain the 'theoretical' value of the ratio of the number stars between subsequent apparent magnitudes. Take the case where all stars have the same luminosity and are distributed uniformly in space. Stars within half a magnitude brighter or fainter than some particular magnitude then reside within a well-defined shell around the Sun. Stars one magnitude fainter (a factor of 2.512; see Box 2.4) are $\sqrt{2.512}$ more distant, since their absolute luminosities are the same. The corresponding shell then has a surface area that is 2.512 times larger and since that shell is also $\sqrt{2.512}$ thicker, the total volume contained in that shell is $2.512^{3/2} = 3.981$ times larger. But, since the number of stars in space per unit volume is constant, there are also that many more stars in the shell. The 'theoretical' value of the number of stars in one magnitude intervals in subsequent mean magnitudes, the 'Star Ratio', is 3.981. It assumes that all stars are intrinsically equally luminous and uniformly distributed in space, and might serve as a benchmark against which to compare observed ratios. Actually, Pickering did not use that number of 3.981, but the logarithm of it, which is exactly 0.6.

This model must be unrealistic, because the predicted number of stars grows very rapidly with increasing magnitude; actually between magnitude 0 and 15 (well within observational possibilities in Kapteyn's day) it would correspond to an increase by a factor of 1 billion. This is in contradiction with observations. Actually, the predicted number of stars from the Star Ratio eventually grows to infinity. This is reminiscent of Olbers' paradox and, indeed, it is equivalent to it. Olbers' (or the 'dark night sky') paradox makes the same assumptions. But then the argument goes slightly differently. Take shells of equal thickness. These have volumes that grow with the square of their radius. So does the number of stars in it, but the apparent brightness of each star decreases with the square of the distance and therefore the total light from each shell is equal. In an infinite Universe, the total light from the sky would be infinite. This is a bit extreme, since stars have a finite dimension and shield stars behind them, but allowing for this still gives the result that the sky should be ablaze like the solar disk. The paradox is named after Heinrich Wilhelm Olbers (1758–1860), but was probably already know to Kepler. In the case of the stellar distributions the solution is that the stellar system is finite, for Olbers' paradox the solution is the finite age of the Universe. An excellent introduction has been written in 1987 by Edward R. Harrison [9] This paradox and the star count problem have in common that interstellar extinction is not the final explanation.

11.4 Luminosity and Density Laws

On page 326 and further I discussed Kapteyn's early attempt (1902) at a description of the distribution of stars, using information on stellar proper motions as a function of apparent magnitude, Kapteyn (1902d). He derived from this the mean parallax of stars as a function of the apparent magnitude and then related that to two fundamental 'laws', the luminosity curve, which gave the distribution of stars as a function of their luminosity, and a density law, that specified the total number

Box 11.1 Star counts

The number of stars in the sky (or any particular direction) depends on the density distribution (designated Δ) and that of stellar luminosities (the 'luminosity curve' φ). Δ is a function only of distance r ($r=1$ for a parallax of $0''\!.1$, so the unit is 10 current parsecs), and φ only of stellar luminosity L. The latter is turned into $\psi(L)$, which is the probability of the absolute luminosity L being within ± 0.5 magnitudes from L.

Now write h_m for the apparent brightness of a star of $m-5.5$ (the absolute magnitude of the Sun); then Kapteyn shows that the number of stars in the sky between apparent magnitudes $m-\frac{1}{2}$ and $m+\frac{1}{2}$, N_m is

$$N_m = 4\pi \int_0^\infty r^2 \Delta(r) \psi(h_m r^2) dr.$$

With $z = h_m r^2$, this can be rewritten as

$$N_m = 2\pi h_m^{-3/2} \int_0^\infty \sqrt{z} \Delta\left(\frac{z}{h_m}\right) \psi(z) dz.$$

For a constant density Δ, the integral in the second equation is easily seen to be constant. Then N_m is proportional to $h_m^{-3/2}$, independent of the luminosity function, as long as this function is the same everywhere. A change in apparent brightness by one magnitude is 0.4 in the log and therefore the change in N_m is 0.6 in the log or a factor 3.981.

Kapteyn similarly derived a formula for the mean parallax of stars at a particular apparent magnitude, but I will not use that in the present discussion.

---

of stars as a function of distance from the Sun. In Kapteyn (1904a), he extended that approach to investigate the matter of the predicted number of stars as a function of apparent magnitude. He used a simple approximation, in which the star density depended only on the distance from the Sun (not on the direction) and the luminosity curve was the same everywhere. All one needed to do is specify these two distributions as mathematical functions and then calculate directly the distribution of stellar magnitudes on the sky. The mathematical details of all of this are in Boxes 11.1 and 9.3.

In his paper, Kapteyn tabulated the results of his calculations. It is instructive to turn these into graphs as would be customary today. In Fig. 11.4, I show his assumed starting points, derived (and somewhat improved) in his earlier work in Groningen Publications 8, Kapteyn (1900e), and 11, Kapteyn (1902d). The dots in these figures represent his best, most reliable determinations and the lines fits by mathematical functions. The luminosity curve on the left shows that for fainter absolute magnitudes the number of stars per unit volume (cube with sides corresponding to a parallax of $0''\!.01$) increases. The values on the vertical axis are logarithmic, so that a change by 1 means a factor 10. On the right the densities as a function of distance from the Sun (again units corresponding to a parallax of $0''\!.01$, or 10 parsec ≈ 32.6

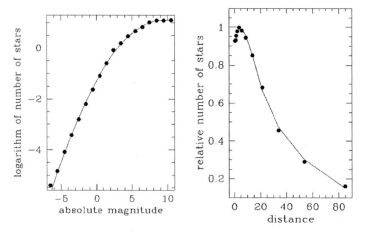

Fig. 11.4 Graphical representation of the fundamental curves Kapteyn assumed in his discussion of star counts. On the left Kapteyn's luminosity curve, which gave the relative frequency of occurrence of stars as a function of their intrinsic luminosity. On the right the distribution of the space density of stars as a function of distance from the Sun. The dots were Kapteyn's best determinations, the lines his adopted mathematical function. For the units see the discussion in the text (Produced from tables I and II in Kapteyn, 1904a)

lightyears in modern units). They have been normalized to a value 1.0, which Kapteyn assumed corresponded to 136.9 stars per unit volume in his calculations.

In Fig. 11.5 I show the outcome of Kapteyn's analysis also in graphical form. On the left the star counts of Pickering have been compared to a model using the curves of Fig. 11.4 and assuming completely transparent space. The slope of this line should in the theoretical case correspond to the factor 3.981 between subsequent magnitudes or a shift of 0.60 in the logarithm per magnitude. In fact the ratio varies (in the observations) from 3.28 at the brightest magnitudes to 2.31 at the faintest. On the right we have the mean parallax as a function of apparent magnitude; the dots are Kapteyn's 1904 values, the line lowest on the left an earlier determination. The two other lines are for either Kapteyn's transparent space and for Comstock's interstellar absorption.

The outcome is not completely surprising: the star counts could be represented by either a significant drop in density with distance from the Sun (as Kapteyn did) or a substantial interstellar extinction of star light. Or, of course, a combination of the two. Moreover, there was no way the two could be distinguished without further, independent information. Kapteyn wrote:

'In conclusion, I think it will be admitted that, neither in the number of stars of various magnitudes, nor in their mean parallax, we have the slightest indication of an appreciable absorption of light. Must we conclude that there is no sensible extinction? I think not, even if the agreement found just now between theory and observation were far more perfect. [...] If, therefore, we may represent the N_m [the star counts] and the $\pi_m{'}$ [parallaxes] nearly equally well with largely different values of the star density, merely

11.4 Luminosity and Density Laws

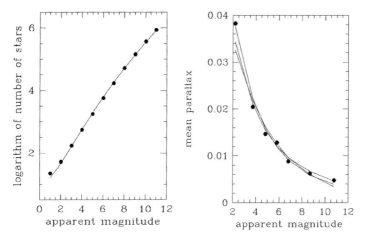

Fig. 11.5 Graphical representation of the results of Kapteyn's computations On the number and mean parallax of stars of differing magnitude'. On the left the number of stars in the sky as a function of apparent magnitude according to Pickering (dots) compared to Kapteyn's model (line). On the right the dots are Kapteyn's mean parallax as a function of apparent magnitude according to Kapteyn (1902d); one of the almost identical curves represents Kapteyn's modeling assuming transparent space (the highest on the left) and another (middle on the left) Comstock's model that includes interstellar extinction. The lower one is from Kapteyn (1900e) (Produced from tables IV and V in Kapteyn, 1904a)

by introducing a corresponding value of the extinction, *vice versa*, it is evident that we cannot hope to determine the law of densities from these quantities alone.'

Before continuing, let me comment on the actual curve of the star density and Comstock's amount of extinction. Kapteyn's drop in star density (Fig. 11.4-right) was quite significant. Away from the Sun, it amounted to a halving at 1000 lightyears and a factor 5 or so at 2500 lightyears. Still, a Sidereal System with a diameter of a few thousands of lightyears would have seemed not unreasonable to him. It would be a very large system indeed and the number of stars in it enormous. On the other hand, Comstock's absorption was very large, even for that time. Kapteyn (1904a) expressed it with a property a (in later publications that I quote below, Kapteyn sometimes used a different symbol, but I will keep using the a), which is the extinction in magnitudes when the light traveled his unit of distance (the current 10 parsec), and he quoted Comstock's amount as 0.18. The currently adopted value is of order 0.8 to 1.0 magnitudes per *kiloparsec* at visual wavelengths, which would mean $a \approx 0.01$. Indeed, as Kapteyn also pointed out, Comstock's absorption value would translate into a very rapid increase in stellar densities; he gave a table in which he did the calculation and found that in round numbers the star density at a parallax of 0."00469 was 120 times the local one. His table extended to a parallax of 0."00118, where Comstock's extinction would imply a star density of 110 million the local one. And he found that a value of $a = 0.016$ was already sufficient to make the star density constant with distance from the Sun. Actually, this is only slightly larger than what we now know it to be. But then, we know that the star density is indeed changing very slowly with distance from the Sun.

So we see that the assumption of transparent space, that Kapteyn used here and in his earlier work – and would continue to use – translated into a substantially decreasing space density of stars with the Sun located near the densest point (see Fig. 11.4-right). The latter may seem very suspicious to us, but it did not seem to have bothered him at all. What *did* bother him, was that Comstock's absorption implied that the Sun was near some pronounced *minimum*: 'But even if we stop at a parallax of 0.″003, it must be admitted that we have already to do with densities which are in the highest degree improbable [in his table we read a value of 1600 for this distance], especially as they require us to assume for the sun a very exceptional position in space'.

11.5 Star Density or Extinction?

In this 1904 paper, Kapteyn discusses how one may escape from the problem of the degeneracy of the decreasing star density and the allowance of interstellar absorption in the interpretation of star counts. He proposes to look at the 'luminosity curve' as can be determined, at least in principle, at various distances. The method would be along the following lines. Assume, for a moment, that the luminosity curve is not a continuously rising curve as in Fig. 11.4(left), but has a well-defined maximum. If it is independent of distance from the Sun, this maximum will move out to fainter apparent magnitude when we determine it at larger and larger distances. The amount of shift of magnitude is geometrically determined: using stars at twice the distance the peak in the luminosity curve should appear for stars four times fainter, which means shifted by 1.51 magnitudes. If there is absorption, the shift appears larger in apparent magnitudes.

There were two problems and one assumption with this. The assumption obviously was that the luminosity curve was the same everywhere. One problem was that the luminosity curve in reality had no maximum. But its slope changed and, although in that case it was more difficult to perform the method accurately, it was still possible. The other problem was that Kapteyn did not have many parallaxes at his disposal (and at distances where the absorption would be significant the parallax measurements were also less accurate). In his earlier studies, particularly Kapteyn (1902d; see page 325), he had analyzed the observations he had at his disposal into tables that gave the distribution of stars as a function of proper motion and apparent magnitude. This could then statistically be turned into a table giving the number of stars in bins of parallax and apparent magnitude. Each line (at a constant parallax) in such a table corresponded to the luminosity curve at that particular parallax. So, as there was no peak in the luminosity curve, Kapteyn had to use slopes.

Not surprisingly, the results were not very accurate. In simple terms, the shifts in magnitude that he found for various parts of the luminosity curves were generally somewhat larger than expected from a model with no absorption, but definitely

11.5 Star Density or Extinction?

smaller than Comstock's amount of extinction. 'As matters stand', Kapteyn wrote, 'it seems unsafe to draw any more definite conclusions than the following: Our numbers do not support the theory of an absorption of the amount found by Comstock. They are well reconcilable with an absorption of, say, one-third the amount. I am not inclined, however, to admit even an absorption of the latter amount, at least, before we possess much more cogent reasons.'

His reluctance to accept interstellar absorption stems from his argument that even a relatively small amount results in the star density increasing substantially when moving away from the Sun. Even for a third of Comstock's absorption the stellar density increases by a factor of 5 or so at a parallax of 0.″003. Clearly, the solution to the problem has to be sought in obtaining reliable proper motions for large numbers of faint stars. But why he seemed to have rejected the option that the star density may actually be constant and remained insisting on a declining star density (with the Sun in a special position), is not made clear.

Kapteyn for many years did not try to improve on his determination of the distribution and density of stars in space. One thing he did do was to improve the determinations for the star counts. In Kapteyn (1908d), *On the number of stars of determined magnitude and determined Galactic latitude*, he compiled the star counts into a uniform system. The main motivation was that the existing counts were not on a 'trustworthy photometric scale', while the range in apparent magnitude was too small. He tried to improve on this by combining all information that was available into a single product. Some of this was work undertaken primarily for other purposes, such as the catalogues of stars in regions around variable stars by Father Johann Georg Hagen, S.J. (who at that time has just moved from Georgetown to the Vatican Observatory). In addition this included in particular data for faint stars from the ongoing efforts to produce the Astrographic Catalogue, but also extensive counts, many of which by Kapteyn himself, of stars on photographic plates (or copies thereof), together with estimates of their limiting magnitudes. Some of these were Carte du Ciel plates, but also many from other sources.

The work was timely in view of the ongoing surveys, as Kapteyn argued in the Introduction. 'The charts for the *Carte du Ciel* make slow progress. Most of the cooperating observatories have not yet made a beginning with their publication. It seems certain that the completion of the charts is still very far off.' But also work on the Plan of Selected Areas 'has as yet hardly begun.'

There was another reason as well. In his 1903 paper referred to on page 390, Pickering did make a statement [10] that 'the number of stars for a given area of the Milky Way is about twice as great as in other regions and this ratio does not increase for faint stars down to twelfth magnitude'. This conclusion Kapteyn found 'extremely startling', as it was irreconcilable with the Herschel gauges. Some of Pickering's work was based on some Hagen counts and these turned out to be incomplete, while Pickering had assumed the opposite.

Kapteyn used various inter-and extrapolations to arrive at a final table with numbers of stars as a function of both magnitude and Galactic latitude. Kapteyn felt confident to extrapolate the counts to magnitude 19, while the actual observations extended only to magnitude 15 or 16. Because of the laborious work to convert

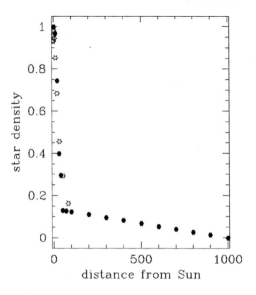

Fig. 11.6 This star density distribution was derived by Kapteyn using his determined star counts in Groningen Publications 18. The open symbols represent his 1904 determination (see Fig. 11.4-right) (Produced from Tafel V in Kapteyn, 1908b)

right ascension and declination into Galactic latitude, the paper included also tables to make this conversion. There are two interesting conclusions Kapteyn drew from this. In the first place he found that the ratio of light from stars brighter than 15-th magnitude varied with Galactic latitude. Above latitude 40° the bright stars contributed about 90% of the total light, while in the Milky Way it was less than half. Secondly, adding up all contributions, he quoted that the total amount of light we see on the whole sky was equivalent to that of 2384 stars of apparent magnitude 1.0. He did not discuss that any further, except noting that it would be interesting to compare it to the total brightness of the night sky (which was an objective in the Plan of Selected Areas). The latitude dependence could have been used to infer some general properties of the structure of the Sidereal System, but Kapteyn refrained from that.

That would not mean Kapteyn ignored that issue of the distribution of stars in space. On two occasions in 1908 he lectured on the matter of the distribution of stars in space. In February he delivered a presentation before the Royal Academy in Amsterdam; the content is published in Kapteyn (1908a; English and French translations were published as Kapteyn (1908b) and Kapteyn (1908c) respectively). He quoted his result in Publication 18 as very reliable down to magnitude 11.5 and reasonably reliable down to magnitude 15. The main purpose of this paper was to present the star density to larger distances from the Sun than was possible previously. Before that Kapteyn noted, but did not discuss in any detail, that he assumed that there was no absorption of light in space. Also he noted that an analysis at a few different Galactic latitudes would be interesting, but that he considered that at this stage not expedient in view of the fact that such a study should address the

11.5 Star Density or Extinction?

Fig. 11.7 Color image of the constellation Orion. The brightest stars, that are easily visible to the naked eye are the three on a row in the middle (Orion's Belt) and the two near the top and two near the bottom. Below the Belt, in what is sometimes referred to as Orion's sword, is a nebulous structure, which is the Orion Nebula. The red, extended, circular material is known as Barnard's Loop (From AlltheSky.com [11])

issue of possible changes in the luminosity curve in these different directions. This is not surprising since he after all was aware of changes of stellar mix with latitude. Anyway, he stuck with a single (average) run of stellar density with distance.

Kapteyn's analysis showed that the fall-off of the density at larger distances was much slower than a simple extrapolation of the formula for his 1904 determination (see Fig. 11.4-right), so he adopted a linear, straight line. The tabular result is reproduced in Fig. 11.6 as a graph. In this solution the density declines steadily reaching zero at 1000 units (10 kiloparsec or 32,000 about lightyears). There was no discussion by Kapteyn of this result in his presentation.

In May, 1908, Kapteyn gave a discourse at the Royal Institution of Great Britain in London on *Recent researches in the structure of the Universe*, Kapteyn (1908e). A summary has been published as Kapteyn (1908f). He presented the same results, and added that studies of this at various latitudes might address the question of the constancy of the luminosity curve. About the assumption of no absorption of light in space, he said, that 'the truth of [this] is doubtful; its elucidation will be difficult, but we can even now see ways in which it may be overcome'.

Fig. 11.8 The Orion Nebula is a region of hot, ionized hydrogen gas around a few very massive, hot and young stars. It is number 42 in the Messier Catalogue of extended objects. The picture has been taken with the 2.2 meter telescope at of the Max Planck Gesellschaft and the European Southern Observatory ESO at the La Silla Observatory in Chile (European Southern Observatory [12])

11.6 Interstellar Extinction and Dust

Before continuing with Kapteyn's studies of interstellar absorption, I first present a brief summary of current knowledge. Interstellar space is not empty, but permeated with a very tenuous medium consisting of gas atoms and molecules and of dust particles. Most of the gas is invisible in the optical, consisting for a large part of cold, neutral hydrogen en helium (in the same ratio as in stars as this has been determined in the Big Bang). The cool hydrogen can only be seen with radio telescopes. But there are also molecules, such as for example molecular hydrogen (H_2) and carbon monoxide (CO), but also much more complex ones. At optical wavelengths the gas, but particularly the dust, can be seen through various mechanisms. The area around the constellation Orion is an excellent place to illustrate what I discuss here (see Fig. 11.7). It is situated almost in the Milky Way and therefore is certainly not typical for any part of the sky. As it is also very close to the equator (the stars in the Belt have declinations of a few degrees negative), it can be seen from both hemispheres. People on the southern hemisphere are of course used to see it upside down from what their antipodes do and when I use 'below' in what follows it would for them be 'above'. The figure shows it as seen from the north.

The bright blob below the central three stars (the Belt) is the Orion Nebula. In Fig. 11.8 it seen in more detail. This is a region where gas becomes visible when it is exited by starlight. Most effective in this is the ultraviolet light from hot, massive and (since their lifetime is short) young stars. This radiation separates electrons from atoms so that these become electrically charged ions. Some electrons 'recombine' with the ions and there will be an equilibrium between neutral atoms and charged ions. The light of the nebula is to a large extent in narrow spectral lines in emission, corresponding with the energy that is released when the electrons and ions

11.6 Interstellar Extinction and Dust

Fig. 11.9 Image of a region near the 'belt' in the constellation Orion. somewhat to the right of the middle we see the Horsehead nebula, which is a dust cloud, obscuring the light of the extended reddish glow behind it. The latter is an emission nebula of glowing gas, ionized and excited by hot, young stars. Some stars, such as the one just below and to the left of the Horsehead, have blue hues around them, which is reflected starlight by the dust particles in the interstellar medium. The brighter stars also have regular blue structures around them, consisting of circles and spokes. This is an artifact due to light reflection in the telescope. The picture, which has north to the left, has been produced from photographic survey plates taken in the 1950s with the Schmidt Telescope at the Palomar Observatory in California (Space Telescope Science Institute [14])

recombine. The shape of the nebula depends only partly on the density distribution of the gas and is mainly determined by the dust absorption. The Orion Nebula has a distance of about 410 parsec (some 1300 lightyears) from us.

The big circular structure in Fig. 11.7 is called Barnard's Loop. It is somewhat more distant than the Orion Nebula and is estimated to be at 500 parsec (1600 lightyears). It is the image of the shell of expanding material left over from a supernova explosion, probably some 2 million years ago. When the expanding material hits the interstellar gas, it also is heated and becomes ionized.

Dust, which by mass overall makes up only a few percent of the interstellar material, can be seen in two ways. A prominent example of a dust cloud in this area is the Horsehead Nebula. In Fig. 11.9 we see an area near the Belt (it is just to the left and below the left-most star in Fig. 11.7), but rotated 90° anti-clockwise with respect to the large view. The faint background glow is another emission nebula (known as IC434), but the dust grains in the Horsehead Nebula absorb the light and scatter it out of our line of sight (interstellar absorption is not really the right word, although surely absorption is part of the process; extinction is used often too and is a more proper term). The absorbed light heats the grains, but the energy is emitted again and the dust is in that manner kept at a temperature of tens of Kelvins. Therefore, when one observes in the infrared, dust can be seen in emission by its heat ('black-body') radiation.

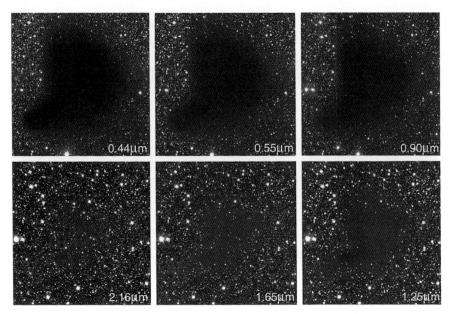

Fig. 11.10 The effects of interstellar 'absorption'. The extinction of starlight occurs mostly through scattering and absorption of light and is most effective at shorter wavelength. This picture shows the 'Bok globule' Barnard 68 photographed at various wavelengths: at the top from left to right blue, visual and red, on the bottom three images in the near-infrared with increasing wavelengths from right to left. Each of the upper pictures have been taken with one of the four 8.1-meter unit telescope of the Very Large Telescope at the Paranal Observatory of the European Southern Observatory ESO in Chile, and the bottom ones by the New Technology Telescope at the La Silla Observatory, also of ESO (European Southern Observatory [15])

Finally there are the reflection nebulae. An example is the light to the lower left of the Horsehead Nebula in Fig. 11.9. It is reflected light from the star on dust grains in the interstellar medium. We have seen that also in the Pleiades in Fig. 7.11.

Dust clouds can easily be recognized in pictures of the Milky Way such as in Fig. 7.2. Actually the naked eye can see dust clouds also, such as in the case of the southern Coalsack in Fig. 10.1. It is instructive the quote the description of Sir John Herschel in his influential book *Outlines of Astronomy* (first published in 1849) [13]:

'In the midst of this bright mass, surrounded by it on all sides, and occupying about half its breadth, occurs a singular dark pear-shaped vacancy, so conspicuous and remarkable as to attract the notice of the most superficial gazer, and to have acquired among the early southern navigators the uncouth but expressive appellation of the coal-sack. In this vacancy which is about 8 in length, and 5 broad, only one very small star visible to the naked eye occurs, though it is far from devoid of telescopic stars, so that its striking blackness is simply due to the effect of contrast with the brilliant ground with which it is on all sides surrounded.[...]

Thus when we see, as in the coal-sack, a sharply defined oval space free from stars, insulated in the midst of a uniform band of not much more than twice its breadth, it would seem much less probable that a conical or tubular hollow traverses the whole of a starry stratum, continuously extended from the eye outwards, than that a distant mass of comparatively moderate thickness should be simply perforated from side to side, or that an oval vacuity should be seen foreshortened in a distant foreshortened area, not really exceeding two or three times its own breadth.'

Herschel interpreted the dark regions as areas where there was a real deficiency of stars, but it was sometimes assumed that such regions were actually exceptional concentrations of absorbing material, although this not necessarily implied that there was an all pervasive medium of dust particles in interstellar space. It is was Friedrich Georg Wilhelm von Struve, who in 1847 first speculated on the presence of absorption of starlight (see page 216). In external galaxies (not yet known as such in Kapteyn's days) the dust is seen very well, especially for galaxies that are seen edge-on, such as our Milky Way look-alike NGC891 illustrated in Fig. 7.3.

The extinction of light by dust is not the same at all wavelengths. It is most effective at shorter wavelengths, so blue light is affected more than red light. We can see that happening in the Earth's atmosphere, which is the reason the setting Sun appears red. The extinction is to some extent the result of scattering of light on small dust particles that are smaller than the wavelength of the light, which is known as 'Rayleigh scattering', first described by Lord Rayleigh (see below). The dust particles have a broad range of sizes, peaking around a tenth of a micron (a micron is one thousands of a millimeter) or so, which is comparable to the wavelength of optical light. Red light has a wavelength larger than many dust particles and passes the medium more easily than bluer light with wavelengths comparable or shorter than the size of the dust grains. This effect can be appreciated by looking at Fig. 11.10. The dark cloud shields more distant stars from our view, but when we move to infrared wavelengths we can see through the cloud. The wavelength effect of extinction is also referred to as interstellar reddening. This makes it possible in principle to detect extinction by comparing colors of stars of similar spectral type (and therefore temperature and color) at various distances. Rayleigh scattering should not be confused with Thomson scattering, named after Joseph John Thomson (1856–1940), which is scattering of light by charged particles. Both effects were known at Kapteyn's time.

It may be helpful for what follows to briefly summarize what was known about nebulae and dust absorption around the beginning of the twentieth century. Nebulae had traditionally been closely interwoven with star clusters. The first famous catalogue by French astronomer Charles Messier (1730–1817), meant as a source list of diffuse objects not to be confused with comets during searches for the latter, contained star clusters as well as nebulous objects. The Catalogue of Nebulae and Clusters of Stars that William Herschel published between 1786 and 1802 and that was extended by John Herschel in 1864, likewise contained both clusters and nebulae. William Herschel even changed his opinion more than once on the question of whether all nebulae were actually star clusters that could not be recognized as such because of poor image quality or were completely different objects (see page 215).

An important finding had been the one by pioneer spectroscopist William Huggins, who found as far back as the 1860s that there were two types of spectra of nebulae. Some, like for example the Orion Nebula, had their light concentrated in a few (originally four) restricted wavelengths, spectral lines in emission of unknown origin, others had spectra like stars with more continuous wavelength distributions. A prominent example of the latter was the Andromeda Nebula (see Fig. 8.16). This led to a prevailing view that the former were gaseous (described e.g. in prominent works as Simon Newcomb's *The Stars: A study of the Universe* or Karl Valentiner's *Handwörterbuch der Astronomie in vier Bänden*, Vol. 3-II [16], both published in 1901). Remarkable was that these 'gaseous' objects like the Orion Nebula were concentrated towards the Milky Way and the ones with continuous spectra as the Andromeda Nebula, prominently present away from it even up to the Galactic poles. This discovery is usually credited to W. Stratonoff, working around the turn of the century at Tachkent Observatory in Russia (currently Uzbekistan). We now know that the latter are predominantly external galaxies like our own and that this is the reason for this dichotomy.

The nebulae with continuous spectra were often of the type 'spiral nebula' because of their obvious spiral structure. A most prominent of this was the 'Whirlpool Nebula' or M51 (number 51 in the Catalogue of Messier; see Fig. 11.11, left). It had been William Parson, the third Earl of Rosse (1800–1867), who had discovered this spiral nature with his giant 72-inch (183 cm) home-telescope at his Birr Castle in Parsonstown, Ireland. Another example is the 'Pinwheel Nebula' M101 (Fig. 11.11, right). When astronomical photography took off, many more were discovered. Some scientists at the time around the *fin de siécle* believed these were stellar conglomerations and not part of the Milky Way. In a lecture before the Société Astronomique de France in 1906 on *The Milky Way and the theory of gases* [17], Henri Poincaré even developed a theory of the motions of the stars in the Milky Way by analogy of molecules in gases. Discussing the two types of nebulae he noted: '... the spiral nebulae are generally considered as independent of the Milky Way; it is admitted that they are, like it formed of a multitude of stars, that they are, in a word, other Milky Ways very distant from our own. The recent works of Stratonoff tend to make us consider the Milky Way as a spiral nebula ...'. This view was not shared by most other astronomers, though.

11.7 Cornelis Easton and Spiral Structure

What was Kapteyn's point of view on the matter? We do not know for sure, as he has not recorded much in writing on the subject at least up to the first years of the twentieth century. But we do know what Cornelis Easton (see Fig. 11.12) has said about it. Easton (1864–1929) was a Dutch journalist and newspaper editor, but also an accomplished amateur astronomer. Since the 1890s he had been publishing

11.7 Cornelis Easton and Spiral Structure

Fig. 11.11 The spiral galaxies M51 (whirlpool galaxy) and M101 (pinwheel galaxy). Note the dust lanes on the 'inside' of the spiral arms (HubbleSite/Space Telescope Science Institute [18])

every now and then in professional journals, such as the Astrophysical Journal, on astronomical subjects in general, but mostly on Milky Way related issues. In 1900 he had written a paper *A new theory of the Milky Way* [19], in which he used the precise morphology of the Milky Way on the sky to infer information on the three-dimensional structure of the system. He took the dark spots and band on the sky as indications of structure in the system rather than absorption features, mostly on the authority of John Hershel. The structures on the sky are then interpreted as features such as branches, streams, rings and appendages in the distribution in space and Easton describes after his analysis the view of the Milky Way from a distance as similar to spiral nebulae as in Fig. 11.11. The Sun is not in the center. It is known (see also Appendix C.) that Kapteyn and Easton discussed astronomical matters regularly, including the nature of the nebulae, so that Easton's opinions, although not necessarily the same as Kapteyn's, at least were influenced or formed by those of Kapteyn and vice versa. In 1903 Kapteyn proposed that Easton should receive an doctorate *honoris causa* from the University of Groningen, which was conferred upon him subsequently with Kapteyn as honorary 'supervisor'.

Easton communicated in 1906 to the Royal Academy through van de Sande Bakhuyzen two contributions, printed back to back, on the matter of the nebulae (*On the apparent distribution of the nebulae* and *The nebulae considered in relation to the galactic system* [20]) as a guide to the structure of the Milky Way. He first addressed the issue of the apparent distribution of the nebulae and recalled the two types of spectra. He used the term 'green' nebulae for the one with emission

Fig. 11.12 Cornelius Easton as he appears in the *Album Amicorum* presented at the retirement of H.G. van de Sande Bakhuyzen as professor of astronomy and director of Leiden Observatory in 1908 (Archives Leiden Observatory; see caption Fig. 3.7)

lines (like oxygen has at green wavelengths) and 'white' ones that have continuous spectra. The effect that the latter are seen away from the Milky Way and their frequent spiral nature was of course known. But then Easton addressed the suggestions that these are 'distant galactic systems'. He rejected that view on the basis of the fact that their correlation on the sky (avoiding the Milky Way) pointed towards 'an unmistakable organic connection between the great mass of the nebulae and our system of stars'. He then proceeded to use catalogues of nebulae to study their density on the sky as a guide towards the system's structure and ended up with a picture in which the Sun was at a region somewhat poor in stars and nebulae. The main structure beyond this displayed spiral features running out from a center in the constellation Cygnus.

In 1913, Easton published an even more elaborate picture of the stellar system (see Fig. 11.13) in a paper in the Astrophysical Journal: *A photographic chart of the Milky way and the spiral theory of the Galactic System* [22]. The spiral structure he had drawn was something resembling both the Whirlpool galaxy M51 and the Pinwheel Nebula M101 (Fig. 11.11). In Fig. 11.13, the Sun is to the lower left of the center. The band around the figure is not part of a model, but shows the actual appearance of the Milky Way on the sky; the irregular structure due to the dust was interpreted as indicating structure in the stellar distribution. As seen from the Sun the center is located in the constellation Cygnus, which is more or less the brightest

11.7 Cornelis Easton and Spiral Structure

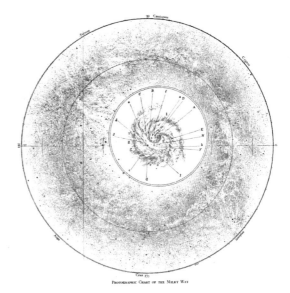

Fig. 11.13 Schematic model for the Milky Way by Cornelius Easton, published in 1913 (Kapteyn Astronomical Institute)

part of the Milky Way from the northern hemisphere. Easton believed firmly that the other spiral galaxies are part of our Stellar System.

It must have been less than convincing to Kapteyn. He wrote (see Appendix C for the full text): 'I have put this map on the wall next to my work table, but for the moment I feel that the profound implication works *de*pressing. While working with papers I sometimes feel a certain satisfaction and especially hope – the appearance of what all needs to be explained, subsequently reduces this to a proper feeling of insignificance.'

Easton believed that the dark structure in the Milky Way is not due to absorbing material and that the spiral nebulae with their continuous spectra are part of the Milky Way system. We do not know whether Kapteyn concurred or not. The view, that dark regions such as the Coalsack, indicated real sparsity of stars rather than absorbing material, was held by others also. Prominent American astronomer Edward Barnard summarized his views in a paper in the much read journal Popular Astronomy in 1906. It was called *On the vacant regions of the sky* [21] and in it he showed how the dark regions bore a relationship with the nebulae. He was particularly struck by the fact that the vacant regions were 'not dependent on the Milky Way entirely, for we find them also in the nebulae, and in or connected with the clusters'. Of individual features he makes statements like 'No one would suspect for a moment that this lane is anything but an actual vacancy among the stars' and 'My own opinion is that it is a true hole through which we look out into space beyond which there are no more stars'. This argumentation was not very strong and convincing, but it showed that the view that dust is *not* ubiquitous in interstellar space and that even the Coalsack and that such features are not strong evidence for interstellar absorption, was widely held.

11.8 Kapteyn and Nebulae

In 1906, the year he published his Plan of Selected Areas, Kapteyn made an interesting excursion into the study of the nebulae. In a presentation before the Royal Academy in 1906, he addressed the issue of the *'parallax of the nebulae'* Kapteyn (1906a; English and French translations appeared in Kapteyn, 1906b, 1906c). He made use of an extensive publication *Beobachtungen von Nebelflecken* by Carl Otto Louis Mönnichmeyer (1860–??) of Bonn Observatory [23], in which positional work is reported on 208 nebulae. Kapteyn took the 168 ones from this, which he believed to be sufficiently accurate. The copy of the publication from 1895 in the library of the Kapteyn Astronomical Institute is full of penciled notes by Kapteyn, some of which clearly show his selection process. Mönnichmeyer collects data from other sources, mostly from the 1860s. The study provides estimates of the proper motions of nebulae (or bright features therein) relative to stars nearby in the field.

The analysis Kapteyn performed was aimed at deducing a secular parallax for the nebulae as a whole. He started with summarizing the technique (see page 292), based on the method of Auguste Bravais to find the distance of a group of stars that move together in space from their proper motions. The secular parallax, as employed here by Kapteyn, used the motion of the Sun to statistically deduce the distance of a set of objects. Briefly, all proper motions were projected onto the direction to the Solar Apex; assuming the velocity of the Sun through space was known, the mean of these components of proper motion of the objects would give their mean distance. This was so because in the mean proper motion statistically all peculiar components were averaged out, so what was left was the annual motion of the Sun seen from the mean distance of the objects. Kapteyn performed two steps. As a first step he deduced that the mean parallax of the nebulae compared to the mean parallax of the comparison stars is $-0''.0017 \pm 0''.0012$ (both the Dutch and the English version show an incorrect error of $0''.012$, but the French version is correct). In other words, the mean distance of the nebulae was somewhat larger than that of the comparison stars. The latter were of mean magnitude 8.75, and the mean parallax according to Kapteyn (1900e) was $0''.0063$. So the end-result was $0''.0046 \pm 0''.0012$, which was about the mean parallax of stars of magnitude 10. In Kapteyn's own reprint of the French text he corrected this to $0''.0031 \pm 0''.0012$, but this was based on Groningen Publication 29, Kapteyn, van Rhijn & Weersma (1918), so this must have been added in 1918 or later. As he did on other occasions, Kapteyn insisted that the result is not very definite, but that his analysis also showed that it could be done better in, say ten years.

The reason why Kapteyn got interested in this, was not revealed in this publication. However, it can be deduced from the text of the Plan of Selected Areas and indeed had to do with interstellar absorption. On page 23 of the Plan he proposed to identify nebulae, especially in the areas of the 'Special Plan', and obtain proper motion plates so that their distances can be derived. On page 57 he elaborated:

> 'The all important question, however, the solution of which might eventually be found by these measures was not mentioned. I consider this question to be the question of the absorption of light in space. – In Astr. Journ. No. 566 (1904) I tried to show, both the paramount importance of

the investigation of this element, and at the same time its exceeding difficulty. – After that article was written the idea struck me that the nebulae offer the best means of attacking the problem. For it is evident that in the case that there is no absorption of light, the surface brilliancy of the nebulae will not vary with distance, whereas if there is absorption it will do so.'

And Kapteyn proceeded to explain that he had tried this already using the diameter of nebulae as an distance indicator, without significant conclusions as a result of the difficulty of judging surface brightness differences from the material he had available. With special plates and distances based upon 'parallactic motion' the task would not be hopeless and he referred to the study just reviewed.

11.9 On the Absorption of Light in Space

In 1909 two papers appeared by Kapteyn in the Astrophysical Journal. These were the first publications of his after his (part-time) appointment as Research Associate at Mount Wilson (see next chapter) and were also part of the 'Contributions of the Mount Wilson (Solar) Observatory'. The papers, Kapteyn (1909c) and Kapteyn (1909d), both titled *On the absorption of light in space*, were a major effort on Kapteyn's side to address this urgent matter. He stated quite decidedly at the outset: 'Now there can be little doubt, in my opinion, about the existence of absorption in space'. To the second paper a 'Correction' was published, Kapteyn (1909e), in the section 'Minor Contributions and Notes'. The correction, however, is far from minor, since it replaces the final result with a significantly different one (see below).

The evidence for absorption, according to Kapteyn, was twofold. Firstly he stated that 'we know that space contains an enormous mass of meteoric matter', but did not review the evidence for that. Secondly, he referred to his 1904-paper, in which he had shown that the star counts could be equally well represented by a declining density with distance from the Sun or a constant one with a relatively small amount of absorption. 'The choice', he wrote, 'between the two probabilities does not seem to be very difficult'. He quoted as support for this that the distribution of stars on the sky showed very little difference between diametrically opposite regions. 'From this we may conclude that, if there is a thinning-out of the stars for increasing distance from the sun, it must be so in whatever direction from the sun we proceed. This would assign to our sun a very exceptional place in the stellar system, viz. the place of maximum density. On the other hand, if we assume that the thinning-out of the stars is simply apparent and due to absorption of light, the apparent thinning-out in any arbitrary direction is perfectly natural. The latter alternative is therefore undoubtedly the most probable one, [...]' This was definitely a much firmer point of view than the one expressed in his 1904 paper.

Kapteyn mentioned his proposal, as worked out in the 1904-paper, to use the 'luminosity curve' and the independent possible method of using nebulae. But then he offers another possibility, that would work if the 'absorption in interstellar space

Fig. 11.14 The UBV or two-color diagram. In total slightly more than 46,000 stars, most brighter than magnitude 10, have been included (From Nicolet (1980) [25])

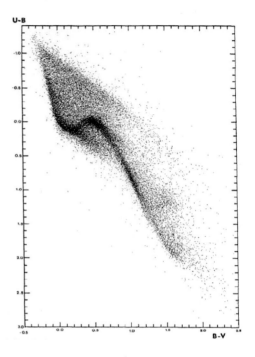

is more or less *selective*. 'If the absorption and scattering of light by meteoric material is really sensible, then there can be no reasonable doubt but that the violet end of the spectrum must be more strongly affected than the less refrangible rays.' and gas would leave evidence for its existence in the form of absorption lines in stellar spectra.

Kapteyn's first paper treated general absorption as indicated in spectra of stars. He relied on the work by Antonia Maury, who we met on page 289 as one of the female astronomers working at Harvard under Edward Pickering on classifications of stars. The relevant paper Kapteyn used presented classifications of stellar spectra, viz. *Spectra of bright stars photographed with the 11-inch Draper Telescope as part of the Henry Draper Memorial* by A.C. Maury & E.C. Pickering (1897) [24]. Many of these stars (in fact 355 out of 681) had been classified as XVa, which would be K0 in later and current notation. Most of these are red giant stars in a later stage of their evolution (see page 206) at the top-right in the Hertzsprung–Russell Diagram (Fig. 7.1). Now, Antonia Maury had noted that some stars show a slight 'general absorption in the region having wave-length shorter than 4307', such as α Bootes (the bright star Arcturus), or a 'more conspicuous' absorption, such as α Cassiopeiae. Actually – Kapteyn did not make this remark – Arcturus is about 20° from the Galactic pole, while the second (also known as Schedar) is in the Milky Way (latitude 2°).

Kapteyn used their proper motions as an indication of distance. Proper motions of many such bright stars were very reliable and he had respectively 45 stars with spectra of the α Cas type and 25 of α Boo type. And indeed, he found that stars in

11.9 On the Absorption of Light in Space

the α Cassiopeiae division to have 'exclusively small' proper motions and those of the α Bootis stars to be 'invariably considerable'. For the first set it was 11.″4 per century and for the second 47.″1. This strongly supported the existence of absorption although the actual amount could not be determined here of course.

In the second paper Kapteyn looked at the colors of stars. If there were significant absorption (and if it were selective), stars at larger distances should appear redder. The principle of using colors to study the 'reddening' of stars by interstellar absorption can be illustrated using Fig. 11.14. This is the so-called 'two-color diagram.'[4] The horizontal axis has the difference of the apparent magnitude of a star in a blue band **B** and a visual band **V**. Since the magnitude scale is a logarithmic one this gives the ratio of the brightness of the star in blue and visual wavelengths. The definition of the magnitude scale translates into a relatively blue color corresponding to a small value for **B-V** and a red color a high value. The vertical axis is similar, but now compares an ultraviolet magnitude **U** with the blue one. Both color indices are by (arbitrary) definition zero for stars of spectral type A. The plot is such that a bluest stars are at the top-left and a reddest at the bottom-right.

The S-shaped band show the location of unreddened stars. These run from the hot O-stars at the top-left to the cool M-stars at the bottom-right. The range simple indicates that the hotter a star is at its surface, the bluer the light that it radiates. It has an S-shape rather than being a smooth line as a result of the presence of very strong absorption lines of hydrogen around spectral type A (at the minimum near the values 0.0 for both color indices). Most stars in the figure are relatively nearby and show little reddening. However, there are also many stars to the right of the curve. These are stars reddened by interstellar extinction. The slope of the upper boundary (also one emanating from the location of the type A) shows the direction extinction takes stars in this diagram. It is steeper than 45°, which indicates that the effect is stronger in **U-B** than it is in **B-V**.

Kapteyn did not have such detailed information at his disposal; in his case the color is measured by comparing the photographic and the visual magnitudes. The sensitivity of photographic plate is bluer, Kapteyn quoted a central wavelength of 0.431 micron (μ) or 4310 Å, than for the visual (about 0.54 μ). He used two catalogues. The first is the *Draper Catalogue*, published by Edward Pickering in 1890. The history of the Draper Memorial was described on page 289. The second was the *Harvard Revision*, published in 1908, also by Pickering [26].

Kapteyn divided the stars available into groups of similar spectral type (from the Draper Catalogue) and 3 or 4 groups of proper motion (which would be an indication of distance). He also distinguished galactic stars less than 20° away from the Milky Way from 'extra-galactic' stars at higher galactic latitudes. The analysis has been summarized in Box 11.2. Indeed Kapteyn found as expected that the intrinsic color of a star depends on its spectral type; actually over his range of types from B to M the color changed quite similarly as the **B-V** in Fig. 11.14, as should be the case. And the stars at larger distances did appear to have been affected by reddening due to interstellar extinction.

---

[4]This is a misnomer; it really is a 'three-color diagram' or the 'two-color index diagram'.

The presence of reddening itself did not tell of course what the amount of total extinction of the starlight was. In order to arrive at this property Kapteyn made two assumptions. The first was that 'practically all the loss of light is due to scattering' and that therefore his results were lower limits. He assumed it to be 'Rayleigh scattering', named after Lord Rayleigh, John William Strutt, 3rd Baron Rayleigh ((1842–1919), and deduced that the scattering should be inversely proportional to the fourth power of the wavelength. Rayleigh scattering is due to particles that are much smaller than the wavelength. In reality the distribution of sizes of dust particles (some comparable to the wavelength) has to be taken into account and interstellar extinction is in first approximation linearly, inversely proportional to the wavelength. Therefore Kapteyn had assumed much too steep a wavelength dependence. The conclusion reached by him was that (in round numbers) the loss of light over a unit distance of 32.6 light years is 0.01 magnitude in the photographic band and 0.005 in the visual. Current best values are respectively 0.013 and 0.010. Curiously, Kapteyn did not comment on the nearness of his values to that of 0.016 required to turn the observed star counts into a space distribution constant with distance from the center.

However, he very soon found an important error in his paper and published an erratum only a few months later Kapteyn (1909e). The problem was that he had made an error in turning *average* parallaxes into *average* distances. This needed to be done taking the distribution of distances into account, which he had not done properly. The effect was substantial, cutting the derived absorption in magnitudes roughly in half. Again he did not comment, but it would mean that the apparent decline of stars as we move away from the Sun was partly due to extinction and partly to the actual distribution of stars in space.

This must have left Kapteyn with a dilemma. Either there was no extinction and the Sun would end up in a very special place (his own words), or there was absorption and then it appeared to have too small an effect in the sense that the Sun would still be in the special position near the maximum density. The special position of the Sun near the center of the stellar system thus had to be accepted independent of the effects of extinction.

Kapteyn returned to the issue in Kapteyn (1914d). This presented no new research but discussed what has been published in the literature; the subject was defined by the title of the paper: *On the change of spectrum and color index with distance and absolute brightness. Present state of the question.* There were two 'observed phenomena' that Kapteyn accepted. The first was that on average stars of fainter apparent magnitude were redder. The second was that stars of the same apparent magnitude and the same spectrum were redder at larger distances than nearby. The first made sense with what we now know. Indeed, when we look at very bright stars they are predominantly nearby and intrinsically bright. Some of these are massive stars burning hydrogen as Main Sequence stars, others are red giants in a later stage of their evolution. When we move to fainter stars in the sky the relative percentage of stars that intrinsically are less bright increases, and those would be redder stars. Indeed Kapteyn gave as one possible explanation that the relative proportion of 'later' spectra types (remember that the usual jargon in astronomy calls spectral types at the G, K and M end of the range 'late'; see page 289) assures that fainter stars are on average redder. Kapteyn assumed, by the way, that stars of different spectral type were mixed equally in space, but also noted that Kapteyn (1902d) showed this to be probably not true.

Kapteyn cited other investigations than his own in 1904 and 1909 that have addressed the question of changes in color of stars with apparent magnitude, such as H.H. Turner: *Diminution of light in its passage through interstellar space* (1908) or E.S. King: *Photographic magnitudes of 153 stars* (1912) [27], and others. An important part was a preliminary discussion of a study by his student Pieter van Rhijn (it would become his PhD thesis), which gave a mean change in color of 0.0050 ± 0.0019 per unit distance. This would be 'confirmatory' of Kapteyn's value of 0.0031 ± 0.0006.

In the end Kapteyn concluded that there is a real effect of increasingly redder colors of stars with distance. But he remained undecided between the possibility of it resulting from a correlation of color with absolute magnitude or 'a selective absorption (or scattering) of light in space'. Work in which it had been tried to separate the two effects (e.g. by studying the distribution of spectra types for fainter stars) remained inconclusive.

11.10 The Thesis Work of van Rhijn

Kapteyn in his further studies on the structure of the stellar system never applied explicit corrections for extinction or took the effects formally into account in his modeling, although he did always remark that the effects could mean important changes and any model he presented was in that sense preliminary. One wonders

---

Box 11.2 Interstellar reddening
The analysis in Kapteyn (1909d) of the stellar colors proceeded as follows: If the parallax is π, then the distance Δ is $0''.1/\pi$, as usual in units of 32.6 lightyears. If the apparent magnitude is m and the absolute magnitude $M = m + 5 + 5\log\pi$, then he fits the colors to

$$Phot. - Vis. = a + bm + cM + d\Delta.$$

The constants b, c and d are the change in redness for a change in one unit of m, M and Δ.
Contrary to expectations b is different for different spectral types. This, however, has only a small effect in the final outcome for c and d, which are the properties with physical meaning. The value for c is close to zero, as expected: $+0.0058 \pm 0.0112$ magnitudes. The value for d is found to be $+0.0066 \pm 0.0031$ magnitudes, later corrected to $+0.0031 \pm 0.0006$.

what had happened had he decided to take these outcomes at face value and applied them consistently. One problem is of course that not only absorption exists, but is in space confined in its most severe consequences to a rather limited layer in the Milky Way disk. Anyhow, the question is what caused Kapteyn to assume that the option of completely transparent space was still viable, although there were some, albeit not fully compelling, indications to the contrary?

To address this issue I must first turn to the PhD thesis that van Rhijn wrote under Kapteyn in 1915. The thesis was entitled *Derivation of the change of colour with distance and apparent magnitude together with a new determination of the mean parallaxes of the stars with given magnitude and proper motion*. The thesis was never fully published; a short paper appeared as van Rhijn (1916b) – with the same title as the thesis itself. But the work eventually led to three Groningen publications in which van Rhijn was involved, notably van Rhijn (1917), Kapteyn, van Rhijn & Weersma (1918) and Kapteyn & van Rhijn (1920a). Van Rhijn had as a graduate student spent significant time at Mount Wilson (1912–1913) and the 1916 publication was also number 110 in the Contributions from the Mount Wilson Observatory. The paper, van Rhijn (1916b), hardly treated the first part of the thesis (*On the mean parallaxes ...*), but it did summarize in some detail the second part, which concerned the question of reddening of starlight with distance from the Sun. Van Rhijn used a study called the *Yerkes Actinometry*, which was published in Parkhurst (1912) [28]. In this John Adelbert Parkhurst (1861–1925) used a double-lens system which also included an objective prism (see page 376). It could be used without prism to measure magnitudes in two bands simultaneously and with a prism to determine spectral types. Parkhurst used 'color-sensitive plates with a 'visual-luminosity' filter'; in other words, in addition to photographic magnitudes he determined visual ones as well using photographic exposures adjusted to the wavelength response of visual observations. The name Actinometry refers to an instrument John Herschel had designed to measure the radiation flux from the Sun by recording the increase in temperature in an enclosed space after exposure to sunlight, which he called an actinometer. This refers to the 'actinic' effect light has on a silver-based photographic emulsion. An entertaining article has been written on an experiment to recreate an actinometer by A.C. Voskuhl: *Recreating Herschel's Actinometry: An essay in the historiography of experimental practice* [29]. Parkhurst applied this to 124 stars north of declination $+73°$ and brighter than visual magnitude 8.0; these could be seen everywhere from the northern hemisphere and therefore could serve as calibration objects. Among other things Parkhurst found a very tight correlation between spectral type and color.

With this dataset, together with proper motions from other places, van Rhijn determined the change in colors of stars with distance. He found that it amounted to '*ceteris paribus* [all other things being equal] only 0.000195 ± 0.00003 magnitudes per parsec'. Van Rhijn used the parsec, which had recently been introduced (see Box 5.6), so we have to multiply by ten to compare to Kapteyn's value. Still it was only about half that what Kapteyn had found (and also smaller than the preliminary value Kapteyn had quoted). Interestingly, there was absolutely no discussion of this difference, neither in the paper, nor in the thesis itself. In any case, the inferred amount of extinction was again smaller than Kapteyn had assumed and maybe this made him question the seriousness of the extinction issue. Van Rhijn's value was much lower than the currently adopted roughly one magnitude per kiloparsec. However, this region of Parkhurst Actinometry is centered on the celestial pole and that is almost 30° away from the Milky Way. No surprise in hindsight that the inferred extinction was smaller than *in* the Milky Way.

11.11 Shapley's Absence of Extinction

At about the same time, Harlow Shapley, whose star was rising rapidly and who had joined the staff of Mount Wilson Observatory after completing a PhD thesis under Henry Russell at Princeton in 1913, was publishing an impressive and groundbreaking series of publications on globular clusters. These are roughly round clusters of stars situated in the spherical halo of the Milky Way and part of the (old) Population II (see page 246). Shapley (*Studies of magnitudes in star clusters. I. On the absorption of light in space* (1916) [30]) had obtained color indices (photographic minus visual) for stars in the globular cluster Messier 13 in Hercules (he measured 1300 stars!) and found none to be very red. Actually their color distribution was very similar to Parkhurst's Actinometry. From the apparent magnitudes of the individual stars Shapley inferred that the distance of the 'Hercules cluster, we may be sure, is not less than 1000 units of stellar distance'. This is 10,000 parsecs or 10 kiloparsec (kpc; the M13 distance corresponds to 32,600 light years). The currently accepted distance is not very different; it is 6.8 kpc. Shapley noted that if Kapteyn's value for the reddening were correct, the smallest (bluest) color index he would have observed should be 2.5 (magnitudes). For M-stars, which were presumably present in globular clusters, he should have observed a color index of 5! This was magnitudes redder than Shapley actually observed. With van Rhijn's value for the extinction all observed colors should have exceeded one magnitude. Shapley concluded from this that the absorption of starlight in interstellar space could not exceed 0.0001 magnitudes per unit distance of 10 parsecs. '...an amount', he concluded, 'completely negligible in dealing with the ordinary isolated stars. In the light of this result we are probably justified in assuming that the non-selective absorption in space (obstruction) is also negligible.'

M13 is about 40° from the Milky Way and even now we know that the absorption towards it is very small indeed. Shapley's generalization that it also applied to 'ordinary isolated stars' was in hindsight not justified, but it was not known at the time that extinction is only significant in the Milky Way. Therefore Shapley's extremely low upper limit for the reddening was widely accepted and an important reason for the prevailing view of a reasonable transparency of interstellar space.

Von Seeliger, who worked parallel to Kapteyn on the problem of the distribution of stars in space, is reported to have taken a point of view (see Paul's *Statistical Astronomy*, page 75) 'invoking the principle of parsimony'. [Von] 'Seeliger argues that given the equally arbitrary assumptions the least problematic alternative – that of no absorption – was certainly the most desirable.' Whether Kapteyn also

Fig. 11.15 The Astronomical Laboratory after Kapteyn had moved into this building in 1911 (Kapteyn Astronomical Institute)

was led by such an 'Occam's Razor' (this *lex parsimoniae* or law of thrift is often connected to Franciscan monk William of Ockham or Occam (1287–1347) type of approach) is not documented, but possible.

Shapley's result was very widely accepted still in the years after Kapteyn's death. In his 1924 Bakerian Lecture *Diffuse Matter in Interstellar Space* [31] Arthur Eddington [(1882–1944)] said: 'The assumption that interstellar space is nearly transparent has usually been based on Shapley's observation that the light of stars from the most distant globular clusters is not appreciably reddened. It is held that dimming without reddening is impossible (unless the obstruction is by solid meteoric particles of considerable size).' Eddington believes that interstellar gas is ionized and electron scattering (on free electrons that have been removed for the atoms) is significant and independent of optical wavelengths. But he does believe dark nebulae to be obscuring stars behind them. 'The only way of obtaining the obscuration with the mass at our disposal is by taking it to be in the form of fine solid particles. I have great reluctance (which is perhaps a prejudice) to admit meteoric matters of this kind in interstellar regions, but I cannot suggest an alternative.' Even by that time Shapley's result still provided the evidence for mostly transparent space. That dust might be responsible even outside dark nebulae and that is was confined more or less to the Milky Way, was not appreciated.

Actually, Eddington's lecture prompted van Rhijn to estimate the extinction towards globular clusters from studying their radii (determined in a uniform manner from counts of numbers of star counts with distance from their centers) as a function of distance in *On the absorption of light in space derived from the diameter-parallax curve of globular clusters* (1928) [32]. The result was in Kapteyns' terms

11.11 Shapley's Absence of Extinction

Fig. 11.16 The Academy Building of the University of Groningen (center) with the Astronomical Institute to the left. The church opposite the Academy is the Roman Catholic Sint-Martinus Church, not to be confused with the Martini Church near the Grote Markt. It was demolished in 1982 to be replaced with the Central University Library (Beeldbank Groningen [35])

0.000355 ± 0.000208 magnitudes per unit distance of 10 parsecs, so formally barely significantly different from zero.

In 1908, Kapteyn gave an important lecture before the Royal Institution of Great Britain, an organization established in 1799 and devoted to research and education of the general public [33]. I did refer to it already above in connection with the constancy of the luminosity curve. It has been published as Kapteyn (1908e), and in summary in Kapteyn (1908f). Kapteyn only devoted one paragraph to the matter of extinction: 'Last, not least. Is the universe really absolutely transparent? There are reasons which make this seem very doubtful. A couple of years ago I obtained some evidence in the matter, which shows that the absorption of light in space, it exists to an appreciable amount, must at least be so small that over a distance of a hundred light-years not more than a few percent of the light can be lost. To determine so small an amount to within a small fraction of its total value will be a difficult task indeed. Still we can even now see definite ways, which, given the necessary data for very faint stars and nebulae, will probably enable us to overcome this last difficulty.'

The existence of appreciable interstellar extinction was established beyond any doubt only in 1930 by Swiss-born American astronomer Robert Julius Trumpler (1886–1956). He used distances derived from the apparent magnitudes and spectral types of stars in Galactic clusters to calculate their linear diameters and found these to be decreasing with increasing distance if space was assumed to be transparent (*Absorption of light in the Galactic System* [34]).

Kapteyn's worry about the implied special position of the Sun in space near the maximum in the stellar density and a corresponding rather central position in

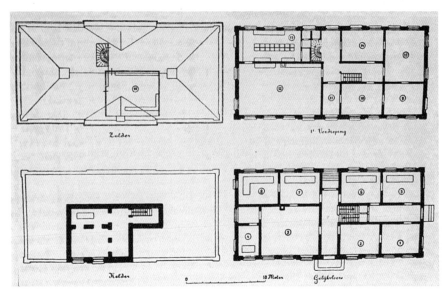

Fig. 11.17 The lay-out of the Astronomical Laboratory. From top-left and clockwise: loft, top-floor, ground-floor and cellar. The Academy building is towards the bottom-right (Kapteyn Astronomical Institute; Kapteyn, 1914e)

the Sidereal System was put to rest. On September 23, 1915 he wrote to George Hale (see also Paul's *Statistical Astronomy*, page 106):

'[...] One of the somewhat startling consequences is, that we have to admit that our solar system must be in or near to the center of the universe, or at least some local center. Twenty years ago this would have made me very skeptical... Now it is not so. [Von] Seeliger, Schwarzschild, Eddington and myself have found that the number of stars is greater near the Sun. I have sometimes felt uneasy in my mind about this result, because in its derivation the consideration of the scattering of light in space has been neglected. Still it appears more and more that the scattering must be too small, and also somewhat different in character from what would explain the change in apparent density. The change is therefore pretty surely real.'

11.12 The Laboratory's Permanent Location

In 1911 an important change occurred. Kapteyn and his Astronomical Laboratory were assigned a permanent location. Ironically, it turned out the place where he had started the measurements for the CPD, when he was given the use of two small rooms by his good friend Dirk Huizinga in his physiological laboratory. Huizinga's successor Hartog Jacob Hamburger (1859–1924) moved into a new and larger physiological laboratory in 1911 and the building was assigned to Kapteyn (Henriette

11.12 The Laboratory's Permanent Location

Fig. 11.18 Two of the rooms in the Astronomical Laboratory that were used for the measuring machines. Kapteyn writes that the one in the back is one of the two rooms that were used to measure the CPD plates (Kapteyn Astronomical Institute; see Fig. 11.17)

Hertzsprung-Kapteyn in her biography incorrectly dates this occasion in the year 1903). Finally, his own laboratory (see Fig. 11.15). It was located next to and somewhat behind the central University Academy building; Fig. 11.16 shows an aerial photograph of the center of Groningen. The street in front of the Academy Building is called the 'Broerstraat' and the address of the Astronomical Laboratory was Broerstraat 7. This picture is reported to have been taken 'around 1920', but that cannot be correct. We see on top of the Laboratory the telescope dome that was installed by Kapteyn's successor Pieter van Rhijn and this occurred in 1931.

The Astronomical Institute, after his retirement named after Kapteyn, has been used as such until 1968, when the University expanded into a campus with laboratories on the north side of Groningen (see Fig. A.10). The building on the Broerstraat was used for various administrative departments of the University until a fire destroyed part of the loft and the upper floor in 1982; rather than restore it, the University decided to have it demolished and replaced with an unimaginative, modern building.

The interior has been described by Kapteyn himself in a short article in the Commemorative 'Gedenkboek' on the occasion of the 300-th anniversary of Groningen University in 1914, Kapteyn (1914e). The accompanying plan has been reproduced in Fig. 11.17. The rooms numbered 4 through 8 were designated for the measuring machines; they had foundation independent of that of the rest of the building to ensure isolation from vibrations.

Fig. 11.19 The Kapteyn Astronomical Institute maintains a 'Kapteyn Room', where some of Kapteyn's materials are being kept. We see his desk and 'blackboard globes', on which he could draw and write with blackboard chalk, together with some of his books (see also Fig. A.1). The volumes on the left are Kapteyn's copies of the Draper Catalogue. We see also his portrait painted by Jan Veth (Fig. 16.3) and his teenage star map (Fig. 1.9) (Kapteyn Astronomical Institute)

'The students find all they need on the ground floor [...]: lecture room [this was number 3 on the plan], rooms practical work and for reducing their observations. For this the two rooms (plan numbers 1 and 2) have been designated. [...] The upper floor has the office of the professor (10), of the assistant (9), the library (15) and three rooms for calculators (12), (13), (14).'

Kapteyn's office had a nice view of the tower of the Academy building and part of the square in front of it (see Fig. 11.16). The lay-out of the Laboratory probably did not change after Kapteyn's retirement, and was in van Rhijn's days very likely still the same. I have asked some colleagues that have seen the insides of the Astronomical Laboratory in the 1940s and 1950s up to the time Adriaan Blaauw took over the professorship in 1957. They mentioned only minor differences; there was a workshop on the ground floor, probably since the telescope at the roof had been erected after Kapteyn's days. Van Rhijn used room number 9, at least in his later days, as did Blaauw after him. Figure 11.18 shows two of the rooms on the ground floor used for measuring plates.

Having established his own Laboratory in its own building must have been extremely gratifying for Kapteyn. It had been a long path to get to this recognition. Also is had not been straightforward. Yet, on November 10, 1907 Kapteyn

11.12 The Laboratory's Permanent Location

Fig. 11.20 The Oosterstraat around 1913. Kapteyn had lived here between 1885 and 1891 in the upper part of the sixth house from the right edge. Between 1906 and 1910, when he lived on the Eemskanaal/Oosterhaven he probably walked through this street on his way to and from work (Beeldbank Groningen [36])

asked Gill for a copy of the latter's lecture for the British Association in which Kapteyn was mentioned extensively. 'A sum on the budget for the acquisition of a new measuring apparatus (a stereo-comparator) has given rise in our parliament to the question, just by 'some' members whether it would not be better to abolish the Groningen Astronomical Laboratory altogether. In connection with this question the 'curators' of our university have asked me to provide them with the necessary data to prove the efficacy of the Laboratory and the opinion held about it by other astronomers. As one of the main of these I would be very glad if I could send him a copy of your speech before the British Association.' Gill replied on November 19, 1907: 'It would surely be an act of brutal absurdity on part of your Government to abolish the Groningen Astronomical Laboratory, from which such splendid work has emanated. May I address a letter to the Curators of the University on the subject?' Kapteyn wrote on December 8, 1907: 'A thousand thanks for your cordial warm letter. I do not think that there is any real danger for our Laboratory and our curators are doing whatever they can for me. So kind thanks for your offer of writing to them; for the present it is surely not necessary. If in future I should want a helping hand I will know where to look for it. Meanwhile I have taken the liberty (knowing that you will not mind) of showing your letter to one of our curators, one of the most influential of the members of our house of commons, who has asked for the data of defending the laboratory if it was openly attacked.' Gill replied on December 10, 1907: 'I am quite sure that your country can never be so unpatriotic as not to support you and an institution that has done such great things for science as the Groningen Astronomical Laboratory has done.' He did have the support of his university, but his position had been vulnerable. With a permanent housing a more secure future would be reached.

Some of the furnishings Kapteyn used in his office have been preserved and are now located in the Kapteyn Astronomical Institute in the Kapteyn Room, which is shown in Fig. 11.19.

11.13 Some Routines

I conclude this chapter with some more parts of the *HHK biography*. It starts with some daily routines. Figures 11.20 and 11.21 show two streets that must at various times have been parts of his daily route to and from the Laboratory.

'Groningers could see Kapteyn walking in the streets every morning at 9 o'clock, on his walk to the laboratory. Everyone recognized his characteristic, boyish walk, with small shuffling steps, which he himself called a 'heath-step', and that he explained as due to his many walking trips while he was young through the heath. Between nine and noon no-one was allowed to disturb him and in those daily three hours with concentrated work behind his desk, his mind wandered far and his major scientific problems were addressed. He was very good at concentrating whatever was at hand. Nothing would distract him. At home he worked preferably in the living room, where the warm atmosphere was a stimulus.

'I learned concentration as a youth', he said laughingly. 'Once when the chatter of the birds hindered my working and I closed the window, my mother became so cross with me about this sensitivity, that she slapped me and threw the window open again. That slap was more useful to me than all well-intended words.' Also to have to work surrounded with all the boys had taught him to close himself off from the rest of the world whenever necessary. The other side of the coin of his ability of intense concentration was his real professorial absentmindedness. Fortunately there were always loving and caring persons around to help him. 'Professor, you forgot to put on your hat', his clerks often reminded him, when he was about to leave the laboratory, or 'the leg of your trouser is folded up'. Or they called him on his phone to remind him that he was expected at a thesis defense ceremony. The old housekeeper, who had a tremendous memory, was of immeasurable worth in reminding him of exams and other important matters. It did happen for example, that he was on his way to America one day and got to Rotterdam when he noted he had forgotten his wallet with his money. He was only just able to return to Groningen and come back in time to catch his ship. Another time in America he has lost his wallet

11.14 Philosophy, Music and Poetry

in a train; after a frantic search the black attendant was consulted. 'Just look under your pillow, Sir', was his advice. And indeed, that was where it was found.

The clerks at the laboratory, who increased in number all the time, were much dedicated to him. He possessed a respectful love and an unshakable faith in them. No wonder, since he shared in their joys and personal troubles, supporting them and giving advice where life brought them ill luck, had an infinite trust and was, as all felt, like a father for them. In his laboratory there was an optimistic, serene atmosphere, brought about by Kapteyn wherever he was and that made the work which was often endlessly mechanical and boring, lighter for all. His radiating optimism helped him to overcome many difficulties. 'Come, let us do first things first and then we will find out what comes next. We then will get it done.' And when a problem was tackled with determination, the difficulties resolved themselves automatically.

The afternoons were for the contacts with the rest of the world: his lectures, discussions, exams, his walks, among which his usual walks on Monday afternoon with his trusted friends Heymans and Boissevain. The three of them could be seen every Monday walking along the Harenschen weg. Heymans in the middle, the tall imposing figure in a pelarine coat with distant view in his eyes and his mind rising above humanity. Boissevain, who was small, but always moving and full of a lively interest. Kapteyn at the other side, slim and entertaining, enjoying the open air and the interesting things that each of them told about their work, while at the same time they noted the songs of the birds, recognizing them. They walked to Haren, a little village at an hour's walk from, Groningen, and they never failed to continue this habit. They used to do this for twenty years, and it was remarkable, they noted themselves, how infrequently they had to terminate their trip because of the notorious weather in the Netherlands. It did happen, especially in later years that they first went to the Café de Passage and kept talking there, being just as satisfied when they returned home. It was a beautiful friendship, full of mutual admiration and respect.'

The walk from the center of Groningen to Haren is almost 6 kilometers. The Café 'De Passage' (see Fig. 11.22) is in what was then the little village Helpman. Now it simply is the southern part of the city of Groningen. It is about 2.5 kilometers of the way from Groningen to Haren, not quite halfway. In addition to a pub and teahouse, it also was a restaurant and hotel. The building no longer exists. The location is on the current 'Verlengde Hereweg' east side between the 'Helperbrink' and the 'Helper Oostsingel'.

Fig. 11.21 The Heerestraat in Groningen around 1900. Kapteyn must often have walked here at that time. The horse-drawn streetcar is typical for the times (From Groninger Archieven [37])

11.14 Philosophy, Music and Poetry

'Kapteyn had major problems of nervous pain in his eyes, for which he was unable for many years to find a solution, until in the ends the problem solved itself and disappeared. But for a long time he was forced to stop reading at night. That was a big sacrifice, but good friends came to help regularly and read out to him at night. Heymans did this for many years and both enjoyed these evenings. In the beginning these were articles from Kapteyn's area of research, but later it concerned things of mutual interest, mostly articles or brochures of contemporary scientific problems. They discussed these problems, and since they were very different in character and temperament, there were always many things to discuss, which gave rise to new insights and new questions. Both had a great respect for science and held a firm belief in its sovereignty.[...]

Heymans and Kapteyn studied other books as well. They read the *Novum Organum* by Bacon and even Boswell's *Life of Johnson*; this bulky volume that everyone knows but almost nobody has read. They read about Einstein's relativity theory, that interested them much, but for which he [Kapteyn] felt it did not 'become fully clear to him'. Prof. Ehrenfest, the physicist from Leiden, offered to give a few lectures, for which he came to Groningen. Also philosophical articles were discussed, which Kapteyn did not appreciate much as a result of his strong taste for graphical realism,

11.14 Philosophy, Music and Poetry

Fig. 11.22 Café-restaurant 'De Passage', where Kapteyn and his companions Heymans and Boissevain would pause (or stop) during their Monday afternoon walks to Haren and back. It is located in a village called Helpman on the main street from Groningen southward to Haren (Beeldbank Groningen [38])

which prevented him from 'coming into this'. An exception on this was exact psychological research, for which he held a lively interest.'

Sir Francis Bacon (1561–1626) was a British philosopher, scientist and politician and James Boswell (1740–1795) was a Scottish lawyer and writer, especially known for his biography of the British author Samuel Johnson (1709–1784). Paul Ehrenfest (1880–1933) was an Austrian-born professor of theoretical physics at the University of Leiden. For what follows: Gerardus Johannes Petrus Josephus Bolland (1854–1922) was a professor of philosophy in Leiden.

'He also always was keen to have correspondence with teachers and others, who had imagined to have found errors in Newton's theories or to have done important discoveries, although this always led to nothing but never made him impatient.

Although Kapteyn did not have a real interest in pure philosophy as his nature was not one of deep reflection, he did whatever was needed to keep himself informed. For a year he followed a course by Prof. Bolland, the famous philosopher from Leiden, not because his ideas appealed to him –

to the contrary he rejected them since he felt they were unbalanced and wild in his opinion – but he wanted to know and try to understand why so many got carried away with it and admired it. For a while he was captured by Bolland's eloquent speech, but too soon they became unmotivated outbursts and illogical turns, and his reckless dislike of those who had different convictions resulted in Kapteyn's indignation. Level-headed as he was, this hot-headed and unconstrained temperament, this absolute subjectivity did strike him as unsympathetic. Scientific research should according to him, being used to think as a natural scientist from his early days on, be in the first place objective; without objectivity, he felt, science would become a ship without helm, from which nothing good and lasting would result, but could only confuse and blur. Lessing's philosophy and classic tranquility appealed to him more and more. His wisdom resonated with him and he resolved that after his retirement, when he hoped to have more time to devote to literature, he would study his work in detail. A few days before his death he ordered Lessing's complete works. He still did see the beautiful volumes and he looked forward to reading them, but they were left unopened. In this way many other things that he wanted to do after retirement remained undone.

Also Hudig [Henriette's later second husband], a young neighbor who quickly became a friend, read to him regularly from geological and historical publications and those evening were a great pleasure to both of them. Mrs Kapteyn read him daily from novels that they had chosen together, and stories in bright and colorful magazines that amused them both. They also worked together often, since in the course of time it had become a normal routine for her to write his letters as he dictated them. And since his correspondence was very extensive that was a help of enormous value. And in this way the difficult years passed, until his eye problems disappeared and he himself became again master of his life.[...]

It was around his 60-th birthday that he started to attend a course about music in order to better appreciate this art. He was not gifted as a musician, did not play an instrument and did not have a good singing voice. He had his own manner of expressing himself using music, which at home was referred to as 'trumpeting'. As soon as he started producing this sound, Mrs Kapteyn could not resist going to the piano and accompanying him, which was a very original effect. You could hear him coming home while singing and sometimes while working he would suddenly start singing, usually parts of sonatas or symphonies that he was familiar with. He wanted to hear the same old pieces again and again, the new unknown ones having little appeal to him. But the art in itself, the depth of it, was a mystery for him; it filled him with a quiet, respectful awe and for an artist he felt the deepest admiration. He followed a course, given by Peter van Anrooy, at that time the conductor of the Groningen orchestra, which interested him enormously. Every Wednesday evening he attended the concert of the orchestra in the 'Harmonie', concentrating on the beauty of music and

11.14 Philosophy, Music and Poetry

always felt enriched by it. His acquaintance with van Anrooy, which soon became a close friendship, took him closer to art. He found many parallels between science and art. Isn't it true that both are in their ideal form unselfish and striving towards truth and purest expression? Oblivious to earthly fame and prosperity in order to give the highest that a person has to give? In that way he regarded art as the sister of science.'

Peter van Anrooy (1879–1954) was appointed director of the local orchestra at Groningen (1905); later he was conductor in Arnhem (1910) and of the 'Residentieorkest' in the Hague (1917). The clubhouse 'Harmonie' was housed in a major building in the center of Groningen, which contained a well-known concert hall, famous for its acoustics.

'One piece of beauty actually remained a mystery for him and that is poetry. Not that he was looking down on it with pity as so many intellectuals do, but he did not have the ability to appreciate poems. Make up your own mind about his only poem as his present to the world. It was the poem he wrote when initiated in the student society upon entering university. Now it has to be admitted that this period really is not the right time for high-level poetic expression, but his poem was undoubtedly the most un-poetical among those of his fellow students that went through initiation:[5]

'The initiation poem is obligatory
for students to be initiated, I notice.
So it seems to be the wisest thing
To simply accept this custom.'

' Is anything less poetic possible? Still it is interesting, since it shows characteristically his logical philosophical mind!

Epic poems did appeal to him, however. His favorite one was the poem of Waltharius [A Latin poem of the ninth or tenth century about a German heroic personality.] with delightful primitiveness, of fighting and blood, of primitive people with primitive instincts. It moved him and roused his enthusiasm. Many found this incomprehensible about him, but it was his simplicity, this little piece of primitive character that he certainly possessed, that found satisfaction and expression in this thundering song of the superman with unrestrained passion.

The beauty of life revealed itself to him in many ways: he had an unlimited ability to learn and enjoy. The first flowers and birds of the spring were a source of great happiness for him. When he had made a walk he always brought some daisies with short stems, that he put on the table, partly shyly and partly overjoyed. Every bird that he heard for the first time exalted him and the appearance of the first swallows was noted in the diary. 'In my next incarnation I would like to be a swallow', was his often heard wish. They seemed to be to him the personification of carefree joy. Some other time he wished to be a dandy in a new life. The admiration for

---

[5]Het groenenversch is een eerstvereischte/ voor groenen, zo merk ik/ Het is dus zeker wel het wijste/ dat ik mij naar die gewoonte schik.

a neat appearance remained part of him. 'There is a lot of self-confidence in an elegant appearance', he often said. He never reached that ideal; the best he managed was to put on a clean collar every day.

He was able to enjoy and laugh unrestrained and irresistibly, especially when he was telling an anecdote – it made no difference whether he told for the first time or for the twentieth – he laughed so that tears came in his eyes and even the most gloomy among them forgot his dark thoughts and laughed along, taken over by his talent and his genuine joy. A wise Frenchman once said: 'The prudent one does not laugh, but only smiles'[6]. Kapteyn knew better.[...]

He also did not disapprove of movie theaters, unlike the norm in intellectual circles. He liked to go there with an open mind, looking for relaxation without looking for deeper values. Why always this destructive criticism of hopelessly superficiality and inferiority in life? Why did one have to cling so desperately to a life of superior values? Much more harmful is the deadly criticism that withers everything that it touches. One can after all leave the unbearable and sensational to the side, like one does with theater or literature. And then enjoy the unlimited technical and artistic possibilities that modern art offer.

At a time there was funfair in Groningen where a tent had been erected on the Ossenmarkt, close to our home. Every evening we heard the most interesting noises: pistol shots, joyful yelling, outrageous clapping and we could not resist to go and have a look and join the excitement. Seated in the first row my father and I watched a funny show of stupid detectives that were always making a fool of themselves, creepy bandits that started to sing a long song 'the birds of the night' at the moment of supreme tension, and went through the most ridiculous and unlikely situations. And we laughed and clapped our hands over so much silliness and did sing this song of the birds at night for a long time afterward. Also in this we were shown the divine rhythm of the grand, versatile life for whoever was open to it.

Life in Groningen went on calmly and evenly. Someone uninitiated and unfamiliar would not suspect that behind these quiet going-ons of things a slowly developing discovery was hiding, as had not occurred in its area since the days of Herschel. Young people that look at their own development as the only and most important thing, fail to notice the great things that are happening in their immediate surroundings. 'Oh, for a life of emotions rather than of thought!', I once cited Keats [John Keats (1795–1821) was an English poet from the Romantic period] when I was still a young and enthusiastic child that has little experience. 'My child, do you think that a life of thought does not know any emotions?' was his grave answer. No argumentation or presenting different points of view, just this calm assuredness, which made a lasting impression.'

---

[6]Le sage ne rit pas, il sourit.

Chapter 12
Students

> *There is no recipe to be a great teacher,*
> *that's what is unique about them.*
> Robert Sternberg (b. 1949)[1]

> *The truth is, when all is said and done,*
> *one does not teach a subject,*
> *one teaches a student how to learn it.*
> Jacques M. Barzun (1907–2012)[2]

12.1 Kapteyn as an Educator

Before continuing with Kapteyn's research I would like to have a closer look at the subject of Kapteyn's teaching. The educational tasks of university professors were a major part of their duties and this probably was even more true (in the case of astronomy) in Kapteyn's days. Teaching was a major investment of effort and took the lion's share of the time. I start by quoting part again from the *HHK biography*.

'He did not agree much with the methods of teaching in his time. The aim of just preparing oneself for exams and collect learned facts did not appear to him to be a way of creating independent scientists. The most learned men often were unfamiliar and awkward when it concerned reality,

---

[1] Robert Sternberg is a psychologist at Cornell University.
[2] Jacques Martin Barzun was a French-American historian and philosopher.

and were not able to solve even the simplest of problems. To substantiate his opinion he used to take examples from his own experience. Many years he was 'gecommitteerde' [This is a person who at an exam has been appointed by the government as an inspector to oversee the proceedings] at final exams for the 'gymnasium' [grammar school]. He usually had a different opinion on the mathematical skills of the candidates than the teachers had. At one time there was a boy who had solved a problem concerning the calculation of an interest, but as a result of a small error in his calculations had ended up with an enormous number, so large that he had to turn the paper horizontally to find enough space for all the figures. The teacher felt that the calculation was acceptable except for the small error. But Kapteyn felt the boy should not pass the exam as he had shown a clear failure of understanding what it was all about, revealing a complete lack of mathematical insight. In another specialty he took the same position. Once a candidate had to make a French translation about a tired pilgrim in a desert. 'Comme son coeur rit quand il s'approche d'un gîte.' Apparently the student did not know the word 'gîte' and translated it as: 'How his heart laughs when he sees a wild animal', probably thinking of 'gibier' (game). A strange pilgrim indeed! His opinion was that in such a case one can better leave that word untranslated rather than provide illogical absurdities that showed a lack of common sense. 'Give me the candidates who fail', he said, 'you may keep the ones who pass', with the usual overstatement of a sound theory. [...]

Van Rhijn, his assistant and successor, described his lectures as follows: 'When students explained their ideas ostentatiously demonstrating their erudition, Kapteyn used to say: 'That is fine, but I would like to see some more fundamental understanding from you gentlemen, not to approach the matters first as a mathematician, but as a physicist.' And then he embarked on an explanation so transparent and understandable and so fantastically graphic... That is', van Rhijn said, 'what your students are grateful for above anything else, namely that you rescued them from being pedantic learned men.'

He did not demand too much actual knowledge from his students and in his lectures he used little factual matters; he preferred to show his students how the questions that we today consider to have been answered indisputably, were once unsolved scientific problems and how earlier scientists had attacked and solved them. That was the secret of his fascinating lectures: one had the impression that the student and the professor worked together to solve a scientific problem rather than having the professor explain and the student listening.'

His lectures were clear through and through and well-organized. The aims of studies and theories became crystal clear for the students. 'I have never met anyone who was so able to do this and I believe that this is the best way to do scientific research', wrote one of his former students, J. Oort, the current conservator at Leiden Observatory. And he continued:

12.1 Kapteyn as an Educator

Fig. 12.1 Jan H. Oort and his future wife Mieke at the dinner after his PhD defense in Groningen in 1926 (Leids Fotoarchief, Sterrewacht Leiden [1])

'Being in his company, during lectures or a colloquium, with something unusually stimulating, most of the time I left with more beautiful and happier thoughts than when I arrived. This is because he saw the beauty of nature and of science, the interrelationship of which he saw more clearly than anyone else... Whereas other lectures would tend to make a first-year student lose his confidence, the astronomy course restored and enhanced that.'

Jan Hendrik Oort (1900–1992) studied under Kapteyn as an undergraduate. Oort later wrote a PhD thesis under van Rhijn (see Fig. 12.1, and also Fig. A.5 for van Rhijn on this occasion), but always regarded Kapteyn as his real teacher. In his inaugural lecture as professor of astronomy in Leiden in 1931 [2], Oort referred to Kapteyn as 'mijn inspireerenden leermeester' (my inspiring teacher). Eventually Oort became director of Leiden Observatory and one of the most prominent and influential astronomers of the twentieth century. In *Some notes on my life as an astronomer* in 1981, Oort writes [3]:

'When I began my studies in Groningen in 1917, at the age of 17, I was almost immediately inspired by Kapteyn's lectures on elementary astronomy. Although I had been greatly interested in astronomy since my high school years, a fact that influenced my choice of the University of Groningen because Kapteyn was there, I was still undecided in 1917 whether to choose physics or astronomy as my major subject; I remember that I was so impressed by the way he taught elementary celestial mechanics that I tried to convey my new insights to friends who had likewise entered the University, but were studying humanities. But I do not believe that I succeeded in conferring to them a full appreciation of the fascination of celestial mechanics.

Perhaps the most significant thing I learned – mainly, I believe, from Kapteyn's discussion of Kepler's method of studying nature – was to tie interpretation directly to observations, and to be extremely wary of hypotheses and speculations. In the first part of his course, Kapteyn refrained, for instance, from introducing the notion of 'force' to replace the measurable quantity 'acceleration'. His disliked intricate mathematical formulations which prevented one from 'seeing through' a theory; he feared the danger that the formulae might make one lose sight of the essentials. This was, of course, before quantum mechanics brought home the fact that one's insight is insufficiently developed to 'look through' the deeper domains of physical science without the aid of mathematics.'

Kapteyn must have thought highly of Oort as well. In the Strömgren archives there is a letter from Mrs Kapteyn, dated December 28, 1936, to Mrs Strömgren, wife of Elis Strömgren, director of Copenhagen Observatory. It was a reply to a card the Strömgrens had sent Mrs Kapteyn and they mentioned apparently in it that their son Bengt Georg Daniel Strömgren (1908–1987), later a famous astronomer, had been invited by Otto von Struve to spend some time at Chicago. Mrs Kapteyn wrote:

'I do hope that Bengt is not remaining for good in Chicago; much that I love our American friends but I have the feeling that each country has the right to keep their big man to herself. Prof. Oort is now in Leiden at the Observatory, I know he was my Darling's best beloved student and assistant and I felt glad that he did not accept the positions they offered him in America, but came back to Holland. And O? let him choose his wife from your country. We have found the international marriage a great drawback.'

This was written weeks before her daughter Henriette Kapteyn was formally divorced from Ejnar Hertzsprung and months before she married Joost Hudig. Actually, Bengt Strömgren had married Danish Sigrid Caja Hartz already in 1931.

12.2 Courses Kapteyn Taught

What courses did Kapteyn teach? Some information can be found in the *Groningsche Studentenalmanak*, published annually by the student fraternity 'Vindicat Atque Polit' (Maintains and Refines). In the years prior to 1900 in particular there were reports from most Faculties containing short presentations and discussions of lecture

12.2 Courses Kapteyn Taught

courses. Kapteyn's Faculty (then called 'Philosophical'), even before 1900, often omitted to file such a report. I quote from the ones available:

1879: 'Kapteyn gave his first courses on general and spherical astronomy. [...] As the academic year was already progressing he could only cover a part of astronomy. He chose for the subject of the fixed stars. We hope that he will soon have the required instruments at his disposal, as even the most necessary tools are still lacking.'

1883: 'From Prof. Kapteyn we learned how to have a look at the depths of the universe; astronomy always captivates the layman, but one may be convinced that after his fascinating lectures, even if it would concern a less popular subject, we would still practice this science with zeal and pleasure. We feel, however, that [he] should judge our mathematical skills as somewhat more advanced, especially in view of the available time.'

1887:'Prof. Kapteyn lectured this year again in a clear and interesting manner about general and spherical astronomy. The fact that one night [he] showed us the position of the most prominent constellations in the sky and continuously tried to compensate for the lack of an observatory, is highly appreciated.'

1889: 'We confirm the positive opinion expressed in earlier reports about the lectures of the professors Kapteyn and F. de Boer' [Floris de Boer (1846–1908) was a professor of mathematics].

1894: 'The lectures of Professors Kapteyn, Schoute [Pieter Hendrik Schoute (1846–1913) was professor of mathematics] and de Boer were very clear, as always.'

1895: 'Following the history of astronomy up to the present, in which the logical course of the development was followed, Prof. Kapteyn lectured on perturbations in the solar system. After that he presented some smaller subjects, 'precession and nutation', 'theory of the tides', 'stability in the solar system' and 'shooting stars'. The enjoyable presentation and the clear exposition of the subjects make Prof. Kapteyn's lectures some of the most pleasant among our many lessons. The lectures on mechanics for candidates [advanced students] were also very clear.'

1907: Lectures were on general astronomy, spherical astronomy and parallaxes and eclipses. 'A word of gratitude for his clear and impassioned presentation'.

There is no doubt that Kapteyn was an excellent educator indeed. Physics professor Wiepko G. Perdok (1914–2005), in the *Festschrift* on the occasion of the University's 350th anniversary wrote about Kapteyn [4]:

'As a professor, Kapteyn showed himself to be a worthy descendant of a lineage that boasts a century-old tradition of teachers. Understanding was more important for him than knowledge and he liked to show in his lectures how currently accepted wisdom were once unsolved problems and how these were solved by his predecessors. All Kapteyn's students testify to his exceptionally clear presentations and experienced the inspiring influence he had on others. Even the memory of the manner in which their esteemed teacher constructed a circle is unforgettable: Kapteyn stood at some distance from the blackboard with a large pair of compasses under his arm, stormed towards it with the well-sharpened point and drove it deep into the blackboard exclaiming: 'I need a fixed solid point'!'

12.3 Thesis Projects in Groningen: Weersma

I now take up the issue of Kapteyn's early students. For a complete list of PhD theses written under his supervision, see Appendix A.3. Although he had been a professor in Groningen since 1878, there had been only one student that had completed a thesis with him (Willem de Sitter in 1901), when he celebrated his thirtieth anniversary as a professor in 1908. It is true that Kapteyn had initiated the award of a doctorate *honoris causa* to Cornelis Easton in 1903, but no others had received an doctors degree with him. Was that exceptional? In 1900 there were 9 professors in Groningen in the Faculty of Science ('Wis- en Natuurkunde'), two each in mathematics and biology and one each in astronomy, physics, chemistry, geology and pharmacy. In the years 1890 to 1910, there 58 doctorates were awarded, or 2.9 per year. Among 9 professors this would correspond to one doctor per professor every 3.1 years. So, Kapteyn was very much behind such a rate at the start of 1908. Eventually, he did make up for the difference, but only to some extent. Not counting Easton's doctorate *honoris causa*, but including Adriaan van Maanen (who graduated from the University of Utrecht with A.A. Nijland as formal supervisor; however there is good reason the count van Maanen as a student of Kapteyn since there had been considerable involvement of Kapteyn; see page 490), he produced $8\frac{1}{2}$ PhD's over a period of 43 years, averaging one per 5 years or so. His contemporary Herman Haga (1852–1936) was a professor of physics and meteorology from 1886 to 1922 and produced at least 10 PhD's during these 36 years. On the other hand, Pieter Hendrik Schoute (1846–1913), professor of analytical, descriptive and higher (further) geometry from 1881 to 1913 has had only four students (at least as listed in the Mathematics Genealogy Project [5]). In the MGP, Kapteyn has all his 8 students listed and Haga 9 out of the 10.

In 1908 Weersma completed a thesis on *A determination of the apex of the Solar motion according to the method of Bravais*, Weersma (1908), and the following year Lambertus Yntema one *On the brightness of the sky and the total amount of starlight*, Yntema (1909). Both were published in the Groningen Publications.

We have met Weersma before (see page 386). The thesis concerned a precise and new determination of the Solar Apex following 'the method of Bravais'. We have seen that the apex determinations assumed that the distribution of stellar velocities had no preferred direction (except for groups of stars that obviously belong together as in the Hyades; but then this is replaced with the assumption that the stars in the Hyades move together with a small, random component superimposed on it). Kapteyn had discovered that there were significant systematic motions which he had identified as the two Star Streams. He proposed to use the original approach of

Bravais. Weersma's work built on the Apex determination as discussed in Kapteyn (1901b; see page 321), where Kapteyn had introduced the *hypothesis H.*: 'The peculiar proper motions of the fixed stars have no preference for any particular direction'.

In his 1907 summary of his work for Gill (see page 293), Kapteyn had introduced the problem as follows: 'As early as 1843, Auguste Bravais (footnote: Journ. de

---

Box 12.1 Bravais's method for determining the solar Apex
Auguste Bravais derived in 1843 three equations for the motion of the Sun with respect to the center of gravity of the stars in its neighborhood, free of any assumptions. These can be written as follows.
Define a coordinate system with x towards the vernal equinox (the zero point of right ascension) and z towards the north pole (y is then towards right ascension 6^h) and the Solar velocity components as V_x, V_y and V_z.
For each star in the solar neighborhood we need its right ascension α, declination δ, distance r, proper motions μ_α and μ_δ, radial velocity v_r and mass m. Then the following three equations can be used to solve for V_x, V_y and V_z.

$$V_x \Sigma m(1-\cos^2\delta\cos^2\alpha) - V_y \Sigma m \cos^2\delta \sin\alpha\cos\alpha - V_z \Sigma m \cos\delta\sin\delta\cos\alpha =$$
$$\Sigma mr(\mu_\alpha \sin\alpha\cos\delta + \mu_\delta \sin\delta\cos\alpha) - \Sigma mv_r \cos\delta\cos\alpha.$$

$$V_y \Sigma m(1-\cos^2\delta\sin^2\alpha) - V_x \Sigma m \cos^2\delta \sin\alpha\cos\alpha - V_z \Sigma m \cos\delta\sin\delta\sin\alpha =$$
$$\Sigma mr(-\mu_\alpha \cos\alpha\cos\delta + \mu_\delta \sin\delta\sin\alpha) - \Sigma mv_r \cos\delta\sin\alpha.$$

$$V_z \Sigma m \cos^2\delta - V_x \Sigma m \cos\delta\sin\delta\cos\alpha - V_y \Sigma m \cos\delta\sin\delta\sin\alpha =$$
$$\Sigma mr(-\mu_\delta \cos\delta) - \Sigma mv_r \sin\delta.$$

The summations obviously are over all stars.

Liouville 8 (1843), p. 435.) showed that whatever be the motion of the stars comprising an arbitrary group, we can determine the direction of the Sun's motion relative to the center of gravity of the group and the mean distance of the group to the Sun. The unit in which this distance is expressed being the yearly way covered by the Sun's motion relative to the center of gravity of the group.' The reference to Bravais' publication is the 'Journal de mathématiques pures et appliquées', founded in 1836 by famous mathematician Joseph Liouville (1809–1882). The important point to note is that Bravais had realized that the fundamental approach should be to use the center of gravity. Now the determination of this center and the Sun's velocity with respect to it was only possible if the distances of all stars, their masses and their three-dimensional velocities in space were known. Bravais had derived a set of equations that could be used to solve for the Sun's motion in a rigorous manner. I summarize these in the form Weersma's presented them in Box 12.1 for mathematically inclined readers.

In Bravais' days (i.e. in 1843) very few distances to stars were available, no radial velocities had been measured and stellar masses were completely unknown, except for the Sun. What Bravais had done then was to assume that all stars had the same mass and distance (later the apparent magnitude was used as an indication of distance) and he ignored the radial velocities. Kapteyn (and before him Argelander in 1838; *Über die eigene Bewegung des Sonnensystems* [6]) used the hypothesis of random motions so that only symmetry arguments were necessary to find the position of the Apex. The difficulty remained how to combine the proper motions (see also the disagreement of Kapteyn with Stein on Airy's work of on the Apex determination; page 323). Weersma found the following way to proceed.

He had to ignore the differences in mass, or more precisely he had to assume that they all were identical. This was not a problem. It simple meant that the center of gravity, which he called the 'mechanical' center of gravity, was replaced with that for the case where all stars had equal mass and this he named the 'geometrical' center of gravity. He argued that stars had at least for 'many centuries to come' straight, uncurved motions and therefore the geometrical center of gravity moved in a uniform and rectilinear manner. Distances were estimated from proper motions. For this Weersma used a provisional position of the Apex and using mean proper motions in different parts of the sky, he corrected the observed ones to 'reduced' proper motions. The radial velocities were initially ignored, just as Bravais had done originally, but Weersma later solved for the value of the Sun's velocity by using the available ones in a second approximation. He concluded that in his approach he had replaced Hypothesis H of no preferred direction of motion to *'Hypothesis B'* (B from Bravais): 'The resultant of the radial peculiar velocities of all stars relative to the geometrical center of gravity is zero'.

Weersma's detailed analysis need not be discussed here. Suffice to say that indeed he had followed a different approach than had been customary at the time (undoubtedly with guidance from Kapteyn). The final result was a position of the Apex that compared well with that of Kapteyn (1900e), and somewhat less well with those of Argelander and of Airy. Weersma found no significant dependence of the Apex on the Galactic latitude of the stars used nor on the spectral type. The velocity of the Sun in space was 0.75 times that of the average space velocity of stars, but that depended on use of the very incomplete measurements of radial velocities.

Weersma's work did constitute an important step for Kapteyn, who relied heavily on an accurate knowledge of the Sun's velocity in space for his statistical determinations of (secular) parallaxes and the 'luminosity curve', necessary for Kapteyn's ultimate aim of determining the arrangement of the stars in space.

12.4 Yntema and the Background Starlight Brightness

Another important item was the extrapolation of the number of stars in the sky to faint magnitudes. For that question already in the Plan of Selected Areas, Kapteyn had proposed to perform measurements on the integrated light of all the stars on

12.4 Yntema and the Background Starlight Brightness

the sky. This would set limits on the total amount of background starlight from stars too faint to be visible individually. and on the extrapolation of star counts to fainter magnitudes than actually observed and was therefore very important for studies of the arrangement of stars in space. Kapteyn's next PhD student, Lambertus Yntema, was commissioned to work on this. He produced a thesis with the title *On the brightness of the sky and the total amount of starlight – an experimental study* in 1909. It has been published in the Groningen Publications as Yntema (1909). Yntema was born in 1879 in a very small village on the west coast of Friesland called Cornwerd, belonging to the municipality of Wonseradeel. After obtaining his PhD, Yntema left the university and astronomy. Not too much is traceable about him. However, I did find that he was registered in Leeuwarden, the Frisian capital, as director of a christian (i.e. protestant as opposed to catholic or public) high school of the type HBS (see page 62). The records state that he and his family moved to Leeuwarden in 1918 from Breda in the south of the country and away to Bussum near Amsterdam in 1931. These records incorrectly give his year of birth as 1870 (day and month are OK) and list him as having three children. The oldest of this was a daughter Anna Elisabeth, born in Neuilly sur Seine in France in 1904, well before his marriage in 1907 with Grietje van der Molen and his PhD defense in 1909. This thesis was dedicated to his mother (his father had died) and his wife.

Before going into the research in Yntema's thesis, I will first summarize what we currently know about the sky background light in order to appreciate what is involved. For technical details and an analysis of the starlight background in terms of structure of the Galaxy I refer to a paper by myself in 1986 (*Surface photometry of edge-on spiral galaxies. V – The distribution of luminosity in the disk of the Galaxy derived from the Pioneer 10 background experiment* [9], see also my chapter in the

Fig. 12.2 The Pioneer 10 spacecraft measured accurately the background starlight distribution from the Milky Way on its way to Jupiter (From Yoshiki Matsuoka [7])

Legacy). When one refers to the sky background this obviously means the sky as it appears away from human light pollution and the city lights that are responsible for many current inhabitants of urban areas never having seen the Milky Way. Also this refers to the sky when there is absolutely no moonlight, so when it is around New Moon. It turns out that even then the night sky is dominated by the so-called airglow, which originates from various processes in the upper layers of the Earth's atmosphere. These include chemical reactions involving the recombination of atoms that were photo-ionized during the day by Sunlight and interactions with high-energy cosmic ray particles that strike the atmosphere. At a very dark site, this airglow has a surface brightness of order 22 magnitudes per square arcsecond (mag arcsec^{-2}), which means that the light coming from a piece of sky (a 'solid angle') of one by one second of arc is the same as that of a star of apparent magnitude 22. For comparison, we know now that the integrated light of all stars is about one magnitude (thus a factor about 2.5) brighter than the airglow in the brightest parts of the Milky Way, but in the Galactic poles it amounts to 24 mag arcsec^{-2}, so it is more than six times fainter than airglow.

The problem in addition to the airglow is zodiacal light, which is reflected sunlight from dust in the Solar System (at night of course beyond our planet). This is concentrated towards the plane of the Solar System, so on the sky to the ecliptic. In gross terms, it has a similar surface brightness as the stars (22 mag arcsec^{-2}) in the ecliptic and one magnitude fainter in the ecliptic poles (actually that statement is false for the south ecliptic pole, which happens to be the direction towards the Small Magellanic Cloud, a small companion system to the Milky Way Galaxy). In addition to starlight, there also is a component of diffuse Galactic light, mostly but not entirely due to scattered starlight by interstellar dust. This is at least a magnitude fainter than the background from direct starlight. Finally, there is a very small contribution of background light from the Universe at large distances (the 'cosmic background'). So, the background starlight is significantly affected by airglow and zodiacal light. The can be avoided now by going to space, but even there zodiacal light remains a problem, unless the probe used is moving far out into the Solar System.

Nowadays the distribution of background starlight has been mapped accurately with the NASA probes Pioneer 10 and 11. These were launched in 1972 and 1973 to study Jupiter and its satellites (and Saturn in Pioneer 11's case) and have now left the Solar System after monitoring also the solar wind and cosmic rays out to large distances from the Sun. Now, while traveling between Mars and Jupiter, where there is less dust and the zodiacal light much fainter, the spacecrafts were used to map the light from the sky at red and blue wavelengths. The primary aim for this was to correct local measurements of the zodiacal light for that from the Milky Way, since beyond Mars the dust density in the Solar System is much lower and there is essentially no zodiacal light left. But, although not primarily obtained for this; the Pioneer measurements can also be used to study the structure of the Galaxy. This was my interest in 1986 when I used these data. In Fig. 12.3 I show 'isophote' maps (lines of equal surface brightness) from the Pioneer 10 measurements compared to

12.4 Yntema and the Background Starlight Brightness

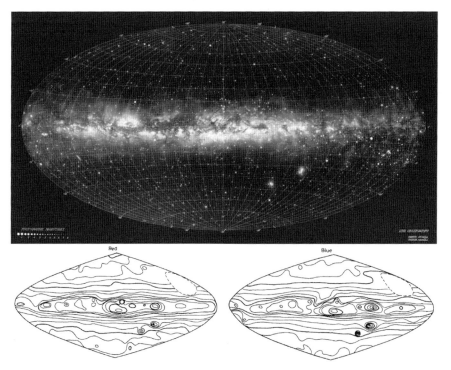

Fig. 12.3 At the *top* a famous drawing of the Milky Way produced in the 1950s at the Lund Observatory in Sweden. At the bottom two drawings of the Pioneer 10 surface brightness distribution at red en blue wavelengths. The Galactic center is in the middle; the two blobs to the right and below the Milky Way are the Magellanic Clouds. In both cases the faintest isophotes are at 24 mag arcsec^{-2} (Top: Lund Observatory, bottom: van der Kruit (1986) [8])

a famous drawing of the Milky Way produced in the 1950s at the Lund Observatory in Sweden. In a recent paper, (Matsuoku *et al.*, 2011), the measurements of Pioneer 10 and 11 were presented in a half-tone form and I have reproduced some of these in Fig. 12.4. The figure shows the two Galactic hemispheres. The dark area on the left corresponds to the general area on the sky where the Sun was located as seen from the Pioneer probes while on their way to Jupiter (see also the dashed outlines in the bottom part of Fig. 12.3). On the rims of the two hemispheres in Fig. 12.4 we see the light from the Milky Way.

Yntema started his thesis with an overview of previous attempts to measure the integrated starlight, going back to 1847, when F.G.W. von Struve called attention to the importance of this property. The most relevant work done up until Yntema's thesis was for a large part due to Newcomb (*A rude attempt to determine the light of all the stars* (1901) [10]), who used spy glasses (hand-held, usually monocular telescopes) with the lenses covered with varying diaphragms produced from cardboard, and mirrors and dark glasses. The method always involved comparison of the amount of light between an area of the sky with another region of the sky or the

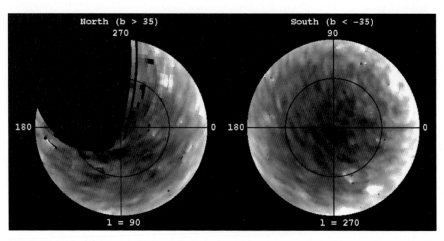

Fig. 12.4 Optical background of star light measured by the spacecrafts Pioneer 10 and 11. The two galactic hemispheres are for latitudes over 35° (From Matsuoka *et al.* (2011) [11])

spread-out light of a bright star. Yntema quoted Newcomb's final result in the poles of the Milky Way as 0.9 in units of the light of a star of magnitude 5.0 spread out over a circular area with diameter 1 degree. This is 0.79 square degrees and the unit then corresponds to 22.53 mag arcsec^{-2}. That is not bad at all. In the Milky Way it is another 1 to 1.5 units brighter, corresponding to 21.8 to 21.6 mag arcsec^{-2}. Again pretty good. Others followed, some using photographic methods. Yntema mentioned that Kapteyn had stated in his Plan of Selected Areas that he had experimented as well on measurements of the sky brightness, both in the Netherlands and in South Africa, and described an instrument that he had used involving a tube with diaphragms and opal glasses.

But as starting point Yntema chose experiments by Kapteyn, also while in South Africa, in which he took an 'ordinary visiting-card', held up towards the sky, and illuminated from behind him, diminishing the illumination until the card turned from standing out brighter than the sky into a dark spot. Yntema devised a more sophisticated instrument, that is reproduced in Fig. 12.5. Undoubtedly this was done together with Kapteyn; in the Introduction Yntema writes about Kapteyn: 'It is a great honor for me to have been able to make a small contribution to Your grand work plan. The results of this investigation are for no small part due to Your encouragement and ever present interest'. A technician from the physics department did the actual construction. The main structure was 1.7 meters long. A screen Q at the far (left) end was illuminated with a lamp M, which could be moved along the pole. The observer looked at the screen Q through the hole S. Everything was coated black. The screen had a hole in its middle through which the sky could be seen. Since the sky illuminated the screen Q as well, the box T was placed on the tube. The larger hole was for the lamp to shine on the screen and the smaller hole for the observer to look at the screen. It was covered with blackened cardboard. As the lamp-holder moved along the pole, its position could be read off on a scale provided

12.4 Yntema and the Background Starlight Brightness

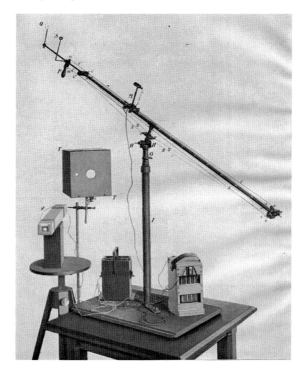

Fig. 12.5 Photograph of Yntema's apparatus that he and Kapteyn designed for the measurement of the integrated background light from the sky (Kapteyn Astronomical Institute)

on the surface of the pole (see Fig. 12.6). The observations were performed from a village called Borger, 36 km from Groningen (to the south-east), where almost no human illumination occurred. It was performed from within a blackened fence on the roof of a house. Readings and notes were recorded by his wife, 'who has faithfully assisted me in carrying out this work.'

Such an experiment could provide only relative measurements. In order to find the absolute brightness, Weersma decided against Newcomb's spreading out of starlight. Instead he devised the method, in which a box V in Fig. 12.5 was used as an 'artificial star'; it contained an electric lamp and an opal glass in its diaphragm. When observed from nearby on the roof is appeared as a surface, but when observed as it stood in the garden (at 53 meters) as a star. The magnitude of the artificial star is determined by comparison with real stars with a Zöllner photometer (see page 46), owned by the Astronomical Laboratory, in which the 'usual oil-lamp is replaced by an electric lamp'.

Finally, Yntema performed photographic studies. For this he used tin boxes with equal size circular openings in the top and bottom and with photographic plates attached to the bottom one. He exposed different parts of the sky by attaching these to a 'hemisphere of iron hoops', shown in Fig. 12.7. That figure has only three boxes but he used in practice more of them. Calibration this time was done with exposures of the full Moon.

Fig. 12.6 Detail of the Yntema's apparatus in Fig. 12.5 (Kapteyn Astronomical Institute)

Fig. 12.7 Mounting used by Yntema to make photographic exposures to measure the background light of the sky (Kapteyn Astronomical Institute)

12.4 Yntema and the Background Starlight Brightness

The full details and complications need not be mentioned here. But Yntema noted that the brightness of the sky was variable throughout the night and from night to night. Yntemma's final result was that the difference between the Milky Way and the poles was 'astonishingly small' and the relative brightness did not increase towards the Milky Way. And contrary to expectations the brightness increased towards the horizon. The starlight should have been more reduced there due to the longer pathway through the atmosphere. From this Weersma concluded that 'the illumination of the background of the sky is not entirely due to direct light of telescopic stars.' The other component he called 'Earth-light'. He speculated that part of that would be diffuse starlight, but another part had to be due to a 'permanent aurora'. But he remained optimistic and proceeded to derive the difference of starlight between various parts of the sky, in particular in the Milky Way and the Galactic poles. Further, using Kapteyn's extrapolation of star counts in Kapteyn (1908d), he proceeded to make estimates of the actual starlight, the diffuse starlight and the Earth-light. The diffuse starlight he estimated from the sky brightness during the day as diffuse Sunlight compared to direct Sunlight and the same for Moonlight (all measurements were done by others and available in the literature).

Expressed in modern units (Yntema used number of stars of magnitude 1 per square degree), he arrived at a surface brightness in the Milky Way of 20.4 mag arcsec^{-2} to 22.6 in the poles. The diffuse starlight was 60% of the direct starlight. The Earth-light at the zenith varied among his various night between 20.3 and 20.6 mag arcsec^{-2}, so comparable to the brightest parts of the Milky Way. This is still a very good results compared to present knowledge. He concluded that improvement should be possible by observing from the top of a high mountain.

I have gone into this work in some detail to illustrate the stubborn adherence to a chosen way, that is typical for Kapteyn and probably many other astronomers. Never give up unless you have proved the impossibility beyond doubt. It is fascinating to see how Kapteyn and his student worked under extremely challenging conditions with enormous care and perseverance and actually arrived at a result that was not far from the truth as we know it now. Kapteyn knew that knowing the integrated starlight (as we would call it today) would put very useful constraints on models for the distribution of stars. Earth-light prevented them from obtaining independent, trustworthy results. Progress has to be sought elsewhere.

A final remark concerns one of the propositions of Yntema. His proposition VIII reads: *The ellipsoidal velocity distribution of Schwarzschild is identical to the two-starstream theory of Prof. Kapteyn.* Indeed, without other, independent evidence the two cannot be distinguished easily. Kapteyn held on to his interpretation of Star Streams (but for good reasons as we will see later). His student took the correct position that there were two equally plausible interpretations.

Fig. 12.8 H.G. van de Sande Bakhuyzen (University of Leiden [12])

12.5 Directorship of Leiden Observatory

On November 30, 1907, Kapteyn received an important letter from Leiden. It was an offer to be appointed at the position that would become vacant in the summer of 1908 when H.G. van de Sande Bakhuyzen (see Fig. 12.8) at the age of seventy would have to retire from his professorship in astronomy and as director of the Observatory. It was a generous offer (the letter is available in the very few pieces of correspondence, except that of Gill, in the archives of the Kapteyn Astronomical Institute):

> 'With this offer we do not ignore the fact that in view of the large scope, into which the astronomical discipline has expanded and branched out, and also in view of the directions that your own researches have opened, a positive decision according to our wishes can only be taken by you if the arrangement of the teaching or in the managing of the Observatory, or maybe in both, is reconsidered, together with increases in personnel and instruments. What these would have to be is not for us to answer at this time, as we would prefer, in case you would find yourself willing to consider our offer, to define the details of our proposals in consultation with you. We therefore restrict ourselves to declare that any request you would have will be taken into very serious consideration, and that the realization of any suggestions that would come out of our joint deliberations, will be supported as far as it is in out power.'

12.5 Directorship of Leiden Observatory

Fig. 12.9 Willem de Sitter around 1898, as he appears in the *Album Amicorum* presented at the retirement of H.G. van de Sande Bakhuyzen as professor of astronomy and director of Leiden Observatory in 1908 (Archives Leiden Observatory; see caption Fig. 3.7)

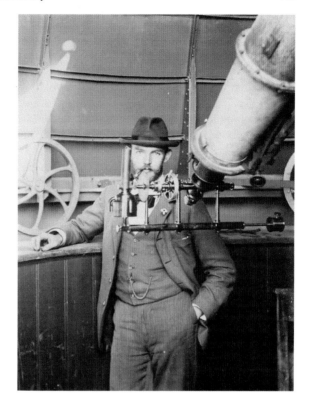

On behalf of the Faculty of Mathematics and natural Sciences of the State University of Leiden,

J.C. Kluyver, chairman
J.P. Kuenen, secretary.'

Jan Cornelis Kluyver (1860–1932) was a professor of mathematics and Johannes Petrus Kuenen a professor of physics. This offer was indeed generous (although no definite promises were offered) and must have been tempting, but certainly very gratifying to Kapteyn. It would have given him direct access to the Leiden telescopes, among which the photographic one that he so disappointingly had seen being allocated to Leiden. And the workforce in Leiden and the support from the University were far from insignificant. But still he declined. The professional reasons for this are not too difficult to imagine. At that time the work on the Plan of Selected Areas had just been started and much effort on Kapteyn's part must have been required to coordinate things not only on a global basis, but also in relation to the measurements of the plates and the reduction of the data. It would have been the wrong time to leave Groningen. The telescopes in Leiden, even the photographic one, was not much suited to the work related to the Plan, which required photography and spectroscopy from large telescopes at dark, excellent sites. The fact that Kapteyn had at that time only just been appointed as part-time Research Associate at Mount Wilson (see next chapter) undoubtedly played an important role as well. His annual absence for four or five months would not be compatible with the position of director of the observatory, at least would have made an effective leadership difficult to say the least. In Groningen, he did have dedicated hands to take over his responsibilities during his long period of travel to America. One may wonder what would have happened had Hale not made the offer. Would Kapteyn have moved to Leiden and left Groningen to de Sitter? Would the execution of the Plan of Selected Areas, as far as the Harvard and Mount Wilson Durchmusterungs, have proceeded as expeditiously as it did now?

The directorship of the Observatory, furthermore, was very demanding, especially running a research program involving nighttime observing and much numerical work during the day. Maybe Kapteyn was too much emotionally attached to his laboratory in Groningen, which had become recognized as an important research center, to make the move. And possibly he, or his wife, were not prepared at this stage of their lives to give up their home and social life in the north.

Leiden was not anymore the attractive place for him that it had been before. I will come back to this in more detail when I discuss the 1918 reorganization of Leiden Observatory and Kapteyn's role in that (see page 581), but the events of 1908 are a precursor to that and some further discussion is relevant. On the one hand, astronomy had changed indeed and Leiden Observatory was lagging behind developments elsewhere. On an international scale, the focus had changed from positional astronomy to astrophysics, from measurements of positions to studies of physical properties and structure. Measuring the cosmos was replaced with understanding the cosmos. Much of this had to do with the enormous rise of astronomy in the United States, where observational studies of astrophysical issues were facilitated with new telescopes, techniques, methods, approaches and ideas. Contrary to this in Europe much of the astronomical research effort remained focused on astrometry, on traditional work with existing methods and techniques.

This is sometimes ascribed to the Carte du Ciel effort, which took up a considerable fraction of the resources, both in terms of manpower as well as in telescope time and budgets, while its effort was almost entirely concentrated to European observatories. A very clear exposition of this is given in *American astronomy: Community, careers and power, 1859–1940* by John Lankford [13], especially his final chapter 'Astronomy compared'. The often expressed opinion, nicely discussed by Lankford, is that the cause of all of this has been not only the enormous efforts required for the Carte du Ciel and the fact that American Observatories did not participate in this and gave them the option to develop observational astrophysics next to astrometry. Also the uniform approach to the Carte du Ciel's execution and the rigid definition of the observational procedures left little room for innovation and progress. In addition, other factors contributed also, as pointed out by Jones (2003), (*Was the Carte du Ciel an obstruction to the development of astrophysics in Europe?* [14]), concluding that 'there would have been not enough people with necessary interests, skills and training' in Europe anyway.

Leiden was not taking active part in the Carte du Ciel, but still shared in the malaise. Antonie Pannekoek (1873–1960), who had done a PhD thesis in Leiden in

1902, compared the atmosphere in Leiden around 1905 to the 'smell like in catacombs, of deadly rigidity and boredom' (this is quoted here a bit out of context; for the complete quote see page 582). In 1902, volume 8 of the Annals of Leiden Observatory had appeared, but volume 9 appeared only in 1915, thirteen years later(!); and it contained only 5 papers. The number of Leiden articles in the Astronomische Nachrichten similarly was minimal. Between 1900 and 1908, H.G. van de Sande Bakhuyzen published two contributions in that journal; he did publish typically twice per year in the proceedings of the Royal Academy and annually produced his report on the proceedings of the Observatory. Obviously, the production of significant scientific results had not completely vanished, but fallen to a minimum.

12.6 De Sitter's Appointment to Leiden

In any case, Kapteyn declined the offer to come to Leiden. But he did put forward and strongly promoted de Sitter. The latter was only 36 years of age, but had been and was very productive scientifically. Taking over Kapteyn's position would have to wait until 1921, when Kapteyn would turn seventy, but by that time de Sitter would be approaching fifty himself. With the limited number of professors in the university, it was essentially impossible to be appointed to a second chair in astronomy in Groningen. The position in Leiden was his chance, and undoubtedly Kapteyn realized this as well. There were two problems though. The first was that de Sitter was interested in theoretical problems. True, he had done observational work during his period at the Cape, but his interests (the satellites of Jupiter and at that stage increasingly gravity theory, in addition to contributing to Kapteyn's program to determine the constitution of the Heavens) were very different from running an observatory and directing an active research program involving telescopes. Secondly, Leiden did still have van de Sande Bakhuyzen's younger brother Ernst Frederik (see Fig. 12.10) on the staff as the senior observator. And he was at age 60 much more senior than de Sitter. In the end, de Sitter was appointed as professor of astronomy and E.F. van de Sande Bakhuyzen as director of the Observatory. This was a good choice; it left the observatory in the hands of an observer (although the younger van de Sande Bakhuyzen proved to be an uninspiring leader) and added a very talented theoretician to the university to set up a strong research and educational program. As it turns out, de Sitter himself has described this period in some detail in two letters to David Gill around the time, that are kept in the Gill correspondence at the Royal Geographical Society archives.

De Sitter's first letter on the issue was written on August 8, 1908. At the top the letter has been qualified as 'Confidential'.

'My dear Sir David,
At last my fate has been decided. I am appointed as professor and director of the observatory at Leiden. The trustees of the University wished the two officers to be separated and the younger Bakhuyzen to be director, but the University would not hear of it. At least that is the course of

Fig. 12.10 E.F. van de Sande Bakhuyzen as he appears in the *Album Amicorum* presented at the retirement of H.G. van de Sande Bakhuyzen as professor of astronomy and director of Leiden Observatory in 1908 (Archives Leiden Observatory; see caption Fig. 3.7)

events as I suppose it has been – officially I know nothing. However – here I am, a young theoretical astronomer, appointed to be director of an observatory with assistants much older than I, and who know far more practical astronomy. So my position may not be without difficulties. [...] Of course I shall have to make a programme for the several instruments – meridian circle, photographic spectra, refractor – and I should like you to suggest suitable subjects for investigations, as well as to advise me on whatever points of general policy as may occur to you. I am so suddenly placed before the responsibility of managing this observatory and making it do good work – having during the last few years always looked forward to a purely theoretical professorate – that for the moment I feel quite unprepared for the task. [...]'

De Sitter asked Gill for advice, noting that he will also consult with Kapteyn, and wrote again on November 30, 1908. He started by thanking Gill for his letters and apologized for having not written earlier.

'[...] Then in early September my dear father died suddenly after only a few days of illness, which entirely upset me for a time. He enjoyed my appointment more than I did myself, and it was a great grief that he could not be present at my inaugural address. About my position here I will tell you all I know, but, of course, strictly confidentially.

Bakhuyzen, as you know, and the majority of the faculty, wished to have the office separated. There was, however, a strong minority in the Faculty – the majority was only one vote I think – who opposed this and insisted on the appointment of a complete successor to Bakhuyzen. The minister has followed the advice of this minority and appointed me. So far I had not been consulted in the matter. As soon as I was appointed I wrote to you,

12.6 De Sitter's Appointment to Leiden

and I spoke to Kapteyn and to Bakhuyzen. Your letter of Aug 8 arrived at Groningen while I was in Leiden. Having found out how the matter stood I suspected, and this afterwards proved true, that the appointment of a separate director would not be made unless I wished it. I confess I have not been entirely averse of the idea of taking up the directorate and facing all the difficulties connected with it – though they are very great – and trying to make the Leiden Observatory do as good work as it can. But I clearly saw that in that case it would be impossible for me to do any work myself, at least of a theoretical kind, until I should have trained a staff which worked after my ideas. On the other hand the directorate would of course give me the disposal of the computing force of the observatory, and I could use one or two computers for the initial work, and leave the practical and administration direction to E.F. Bakhuyzen, whom I could appoint sub-director.

In fact this was the course suggested by several persons, and also by the University. However I saw from the beginning that this would not do. If I were nominally director, I must be also actually director, and take the whole responsibility, and therefore also the whole work, on my shoulders. Finally, I made up my mind to give up the directorate, but I asked of Bakhuyzen to grant me the use of two computers for my own work, to which he at once agreed.

I then went to the Hague and had a long talk with the minister –more than an hour, and at last succeeded in persuading him to appoint E F Bakhuyzen as 'professor extraordinarius' and director of the observatory. He made, however, several restrictions, the gist of which is that from the bureaucratic point of view I am to have precedence over Bakhuyzen, I have certain privileges as to the use of the resources of the Observatory, I must be consulted in the nomination of the director, etc. All these things may be useful in case difficulties arise, but in practice Bakhuyzen is the only director of the Observatory, and I have nothing whatsoever to do with it, except that I give my lectures in the lecture room of the Observatory, and that I am to have two computers for my own work. These I would never have got if the division had been carried out at once.

The younger Bakhuyzen is most agreeable and kind. He has offered me a room in the Observatory, he has offered me to publish my theoretical work in his Annals, etc. Our intercourse is as friendly as it can possibly be. So the end of it is that I have now a position better than I could ever have dreamed. [...]'

This concludes this episode, and I will return to the situation in Leiden later. Kapteyn lost de Sitter as the senior assistant in his Laboratory and important collaborator, just at the time he went for his first visit to Mount Wilson. But he must have been happy for de Sitter and with the arrangement of a dual appointment. On August 10, 1908, when it still seemed that de Sitter would be appointed both professor and director, he wrote to Gill: 'You will probably have heard already from de Sitter, that he has been appointed in Leiden as a successor to Bakhuyzen. I am exceedingly happy, for if he had lost this chance, he might have had no other for an indefinite time. Only I would have liked that there had been appointed two successors, one as a director to the Observatory, one as a professor. Only Government would not hear of such an arrangement.'

H.A. Weersma completed his PhD at about the same time (July 1908) and seemed to have been appointed in the vacancy that de Sitter's departure created.

12.6 De Sitter's Appointment to Leiden 451

Fig. 12.11 Undated photograph of Kapteyn (University Museum Groningen)

Chapter 13
Mount Wilson

> *If I have seen further...*
> *it is because I have a bigger telescope than you.*
> Anonymous.[1]

> *[...], the driving public was discouraged*
> *from using the toll road. [...] there were the sharp turns*
> *that required most cars to negotiate dangerously precipitous road edges.*
> *Mr L. L. Whitman of Pasadena who made the ascent [...] said,*
> *'Not for five hundred dollars would I make the trip again.'*
> Wikipedia.[2]

13.1 George Ellery Hale

George Ellery Hale (see Fig. 13.1) was director of the Mount Wilson Observatory of the Carnegie Institution of Washington, located near Pasadena, not far from Los Angeles. He invited Kapteyn to become a Research Associate to his Observatory and spend a few months in California every year. This invitation was extended in June of 1907, and Kapteyn went to Mount Wilson for the first time in the fall of the next year. George Hale had established a solar observatory at Mount Wilson

---

[1] Persiflage of Isaac Newton's quote *'If I have seen further it is by standing on the shoulders of giants'*. Another is the anonymous *'If I have not seen as far as others, it is because giants are standing on my shoulders'*.
[2] Article on the Mount Wilson Toll Road [1].

Fig. 13.1 George Ellery Hale, painting by Seymour Thomas in 1929. This portrait hangs in the Hale Library of the Carnegie Observatories, Pasadena (From the Huntington Digital Library [2])

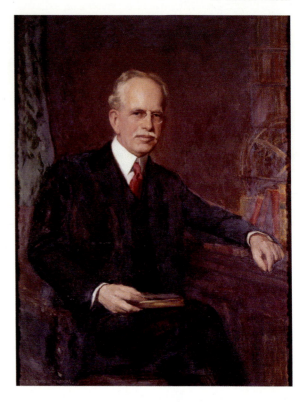

(elevation about 5,700 feet or ∼ 1,750 meters) in 1904. Its history has been very well described in two books by Helen Wright on Hale and the Mount Wilson Observatory (*The Legacy of George Ellery Hale* and *Explorer of the Universe: A biography of George Ellery Hale* [3]). But the most comprehensive, detailed and authoritative account is *Sandage's Mt. Wilson History*, written by longtime staff member Allan Sandage. Very good, but older and much shorter accounts on Hale and Mount Wilson can be found in two excellent obituaries, written by his successor Walter S. Adams (1876–1956) and longtime collaborator Harold D. Babcock (1882–1968) (*George Ellery Hale, 1868–1938* [4]).

Hale had been director of Yerkes Observatory near Chicago, which he had founded in 1897 and where under his supervision the largest refracting telescope, with an objective lens of 102 cm, had been built (it had been commissioned in 1893). While in California in 1903, Hale visited Mount Wilson, which had already been considered by others (e.g., Edward Pickering of Harvard College Observatory) as an observatory site. Hale's plans were made possible after the foundation of the Carnegie Institution of Washington (now of Science) in 1902, by steel magnate and philanthropist Andrew Carnegie (1835–1919). In the words of Carnegie himself, this Institution 'shall in the broadest and most liberal manner encourage investigation,

13.1 George Ellery Hale 455

Fig. 13.2 The Mount Wilson Observatory as illustrated in Henriette Hertzsprung-Kapteyn's biography of her father (Picture reproduced from the *HHK biography*)

research, and discovery [... (and)] show the application of knowledge to the improvement of mankind.' Hale succeeded in convincing the Institution to establish a (solar) observatory at Mount Wilson. This led to the lease of a piece of land atop Mount Wilson by the Carnegie Institution in 1905 and the observatory was quickly established. In 1908 the 60-inch telescope was taken into operation. I will return to the story of the building of the 60-inch later (see page 458). Figure 13.2 shows Mount Wilson at about the time Kapteyn was a regular visitor.

Hale was a solar astronomer/physicist with a keen interest in spectroscopic studies of the Sun. The Mount Wilson Observatory was, however, not to be just a solar installation and Hale set out to build a 60-inch (152 cm) reflecting telescope for stellar astronomy. This telescope was completed in 1908, which coincided with Kapteyn's first visit to Mount Wilson (see page 458 and Fig. 13.4). At the time it was the largest working telescope in the world and would become one of the most productive ones ever. It also was to play an important role in Kapteyn's Plan of Selected Areas. In 1910, a 150-foot (46 meter) tall Solar Tower had been erected at Mount Wilson, and in 1917, when Kapteyn had already ceased to visit Pasadena, the 100-inch (2.5 meter) giant Hooker telescope was inaugurated. Mount Wilson quickly became the center of observational astronomy. Hale did not rest here, but planned the design and building of the 200-inch (5 meter) telescope that was eventually erected

Fig. 13.3 The Mount Wilson Observatory around 1910. The solar telescopes (the Snow Telescope and the two Solar Towers) are on the left. The 60-inch dome is just left of the middle. The Kapteyn cottage (see later) would be near the right border, but hidden by trees (From the Huntington Digital Library [8])

at Palomar Mountain, some 150 km towards San Diego, away from the encroaching city lights of Los Angeles; it became operational in 1948, well after Hale had died in 1938.

13.2 Mount Wilson and the Selected Areas

Most of Hale's correspondence is kept at the Cal Tech Archives in Pasadena, but there is also a small part at the Huntington Library in nearby San Marino. The letters to (usually carbon copies of typed letters) and from Kapteyn start in 1904 pertaining to Hale's invitation to the Dutch Academy to appoint Kapteyn on the national committee to the Conference on Solar Research in St. Louis that year (see page 366). Somewhat later on the correspondence turned to Kapteyn's Plan of Selected Areas as it was taking shape. Much concerned the use of the Mount Wilson 60-inch telescope that was under construction, with a view to the execution of the Plan. Hale first wrote about this on January 9, 1907: 'I am afraid you may have come to the conclusion, that I am not interested in your 'Plan of Selected Areas'. As a matter of fact, I am a great believer in this plan and consider it of the highest importance.' He described in detail the stellar work performed at that time comparing the spectra of stars such as Arcturus to that of the Sun in the sunspots, and the

13.2 Mount Wilson and the Selected Areas

Fig. 13.4 The 60-inch Telescope at the Mount Wilson Observatory (From the Huntington Digital Library [9])

expected contributions to stellar spectroscopy that the 60-inch would bring within reach. The selection of stars for spectroscopic measurement could be guided by the Selected Areas and also Hale hoped to start some proper motion and parallax work. He invited Kapteyn to give his 'views as to the usefulness of the 60-inch reflector'. He said that he regretted that, due to earthquakes in San Francisco (where the mountings were produced), the 60-inch was delayed, but was expected to be operational by the summer of 1908. And Hale looked forward to the 100-inch reflector being completed 'four or five years hence' and their joint work being extended in many ways. In passing he also suggested a visit by Kapteyn to Mount Wilson not long after the completion of the 60-inch to 'go over the whole question'.

Kapteyn replied on February 17, 1907, and detailed what he thought the 60-inch could contribute. Hale answered on March 14, 1907: '... I hasten to assure you that much of the work you mention will be done on Mt. Wilson. The more I think of the matter the more I perceive the great possibilities of a close cooperation with you. Many of the data you require can be easily furnished by us and, on the other hand, many of the results obtained in the course of your investigations will be indispensable, if we are to draw any joint conclusions regarding stellar development.' He committed to photometric work, radial velocities, parallaxes, proper motions and nebulae. He stated he hoped to attend the meeting of the Solar Union (in Meudon near Paris in 1907) and suggested that maybe they could meet. Indeed, Hale's aim with the 60-inch telescope at Mount Wilson had been to study stars and stellar spectra and elucidate the development (we now would say evolution) of stars. He saw Kapteyn's program as an important way towards that goal.

Before continuing the story of Hale and Kapteyn's Selected Areas, I turn to the development of the Mount Wilson Observatory, but refer the interested reader for more details and a better understanding to *Sandage's Mt. Wilson History*. As early as 1894, George Hale had persuaded his father, American elevator tycoon William Ellery Hale (1836–1898), to buy him a 60-inch mirror for a large reflecting telescope. It was to be used for a telescope at Yerkes, where George Hale became director in 1895, but the University of Chicago never provided the funds to build the telescope itself. When the Carnegie Institution of Washington decided to fund the Mount Wilson (Solar) Observatory in 1905, Hale donated the mirror to the Institution and the 60-inch was built and completed in 1908. This was possible because of the outstanding work in optics and telescope design by George Willis Ritchey (1864–1945). Ritchey had already been hired by Hale to lead the optics workshop at Yerkes and moved with Hale to Mount Wilson, where he oversaw the construction of the optics for the solar telescopes. For the spectroscopic work at very high resolution on the Sun, a telescope and a spectrograph with a long focal length was required and this meant large physical size. The first instrument erected for that purpose was a so-called coelostat telescope, in which a set of two flat mirrors directed the sunlight along a horizontal path towards a focal plane. This 'Snow Telescope' belonged to Yerkes Observatory and had been financed by Helen Snow, daughter of building contractor George Washington Snow (1797–1879). Then there were two Solar Towers, in which the sunlight was directed by a mirror along a vertical, open structure into the focal station underground. The 60-foot Solar Tower was completed in 1907 and the 150-foot one in 1911. The 60-inch telescope, completed in 1908, was the largest light collecting instrument in the world. In 1906 Hale became acquainted with John Daggett Hooker (1838–1911), a local businessman and a hardware and steel-pipe millionaire. Hooker donated enough funds for a 100-inch mirror to be cast. Eventually this became the Mount Wilson 100-inch 'Hooker'-telescope (see Fig. 13.5), which had a profound influence on the development of astronomy in the first half of the twentieth century. It was completed in 1917.

Hale had been spending quite some time defining an observing program for the 60-inch Telescope while it was being constructed and – as mentioned above – settled on Kapteyn's Plan of Selected Areas. Sandage, in his *Mt. Wilson History* quotes small passages from Hale's annual report for 1907 in the Carnegie Institution Year Book for 1907 [11], primarily aimed at the Carnegie trustees; I also quote them below in somewhat more extended form. After having described his work on the Sun (in the framework of which he reports on an extended visit by Willem Julius from Utrecht), he turns to 'a different phase of the problem of stellar evolution.'

13.2 Mount Wilson and the Selected Areas

Fig. 13.5 The 100-inch Hooker Telescope at the Mount Wilson Observatory (From the Cal Tech Archives [10])

'The closeness of the bond that unites the astronomy of position with astrophysics is well illustrated here, for one of the greatest of present needs is the determination of the velocities of motion in the line of sight of a very large number of stars. These spectrographic results would settle the question of the Sun's motion in space, add greatly to our knowledge of the two groups or streams into which Kapteyn and Eddington have separated the stars, and contribute in other important ways to the solution of the problem of sidereal distribution.

[...] Since faint stars, as well as bright ones, must be included in any investigation of the general laws of stellar distribution, it is evident that an attempt to cover the entire sky, which would involve the minute study of many millions of stars, is out of the question, at least within any reasonable period of time.

Accordingly, Kapteyn has selected a certain number of areas, uniformly distributed over the heavens. The brightness, proper motion, parallax, spectral type, and radial velocity of all stars within these areas will be determined, through the joint efforts of many observatories. There has been little difficulty in providing for the work on the bright stars, but the instrumental means available do not suffice for the fainter ones. For this reason I have long felt that the large reflecting telescopes of the Observatory should provide some of the necessary observations of the fainter stars, if this could be done in such a way as to contribute effectively toward the solution of our own problem of stellar evolution.'

So, he goes through great pains to show that collaborating with Kapteyn fits naturally into the work envisaged for Mount Wilson.

'During the past year I have devoted much time to the preparation of a working program for the 60-inch reflector. Since all the lines of research with this instrument should converge on our principal problem, cooperation with Professor Kapteyn, who is concerned with the very different question of stellar distribution, might appear to be excluded. A single illustration will show, however, that this is not the case. Kapteyn believes that the stars of the Hyades may belong to a true physical system. since their proper motions, and probably their parallaxes, are in good general agreement. Hitherto, however, he has been unable to find whether their radial velocities also correspond, since no large telescope could be employed to measure them. Should the 60-inch reflector prove them to do so, a common origin may with confidence be assigned to the group. If these stars could be shown to have commenced their evolutionary career at the same period and under similar conditions, a minute study of their spectra might throw light on the possible relationship between a star's mass and its spectral type –a question of fundamental importance. In many such cases the measurement of radial velocities, parallaxes and stellar magnitudes will be essential in connection with our spectroscopic work. [...]

[...] Although these instruments must be devoted, for the most part, to spectroscopic observations and the photography of nebulae, I believe that they may also yield important contributions to Kapteyn's cooperative undertaking.'

Hale's efforts extended beyond support of the Plan of Selected Areas. He adopted it as a spearhead in his use of the 60-inch and 100-inch reflectors. In Sandage's words in his *Mt. Wilson History*:: 'Hale bought the Kapteyn plan lock, stock and barrel. He proposed that it become a major focus of nighttime observing at Mount Wilson.' How did he arrange the agreement with Kapteyn?

13.3 Hale Invites Kapteyn to Mount Wilson

On June 12, 1907, Kapteyn sent a letter to Hale with a reaction to the proposal Hale had made to him 'yesterday', and which 'took me absolutely by surprise', to come to work at Mount Wilson in 1908 for a few months. There is no letter of invitation preceding that. However, further on in the Cal Tech archives there is a telegram from Kapteyn to Hale, dated 9/6/1907 and inserted as if the date was September 6. There actually is an official stamp with the month quoted in words, revealing that in reality the date was June 9, 1907. (the notation being day/month/year as is customary in

13.3 Hale Invites Kapteyn to Mount Wilson

the Netherlands and many other European countries). It was addressed to Hale at the Hotel d'Europe in Amsterdam and reads: 'I will be at the hotel Tuesday at one afternoon = Kapteyn'. Obviously, Hale was in Amsterdam and Kapteyn visited him. A quick check shows that the 'yesterday' in Kapteyn's letter, June 11, 1907, was indeed a Tuesday. So, Hale's invitation to Kapteyn to become a research associate at Mount Wilson was extended in person in Amsterdam. It seems likely that Hale especially came to Amsterdam, while traveling in Europe, and had asked Kapteyn to come and see him in a note or letter of which he did not keep a copy (of course the original would have been part of Kapteyn's lost papers). Indeed, Hale attended the meeting of the International Solar Union [12], which was held in Meudon near Paris from May 20 to May 23, 1907, and he must have been traveling in Europe for the intervening two weeks. Hotel de l'Europe is still a luxury 5-star hotel in the center of Amsterdam. Hale apparently asked Kapteyn to think the proposal over and reply by mail as soon as possible. Kapteyn wasted no time and – as mentioned – wrote Hale the next day (June 12):

'Dear Sir,

Today I have much reflected on Your proposal of yesterday, which absolutely took me by surprise.

I expect you will draw up a sort of contract and as I think I rightly understood what you expect from me I have not a doubt but that I will agree with my full heart. For I see in it the best way for the realisation of [a] great part of the work I planned for the remainder of my life.

Your idea to make the arrangements 'elastic' seems to me just the way to make our cooperation truly productive. I understand it to mean that the requirements of the work will decide about the exertions expected on my part. You may be sure that in order to make the work succeed I will spare no effort, as well in America as at home; but I would be sorry to lose time at epochs when the work would not seem to demand it and merely to execute some written arrangement.

About the amount of the 'salary' I would propose to put it, not at 1500, but at 1800 Dollar – Not that, if you have fundamental objections I will not accept, and gladly accept the former amount. If I were rich enough I would undoubtedly prefer not to receive any salary at all. – Now that I am unable to indulge in such a luxury, it would be a great thing for me to be free from any money cares; which means: free from the necessity of delivering extra lectures, going out on examinations etc. in order to meet the demands which three out of door children make on a professor's income. In short the increase proposed must enable me to give the whole of my leisure to astronomical research.'

For a better understanding it is helpful to convert the amount of 1800$ to current currency. The exchange rate in around 1910 was about 2.50 Guilders to the Dollar [13]. The resulting 720 Guilders in 1910 correspond to about a full year's salary of an unskilled worker and a purchasing power of 17000€. Kapteyn's request would be equal to about a few months' salary for a full professor.

This must have come as a godsend to Kapteyn; he would be able to devote all his time to astronomy, not having to do all the extra things he mentioned that kept him from research, to provide the income required for his family. He must have felt extremely lucky and privileged. The letter ended with some remarks that he was confident they would work together without disagreements, as he had never had problems with Gill. 'I feel proud to work with him for a great end, I feel proud that, now again, I may work with a man as you, for still greater ends. With kindest regards to Mrs Hale. Very sincerely Yours, J.C. Kapteyn.'

One can imagine how Kapteyn must have felt on his trip back by train from Amsterdam to Groningen. His grand plan had the support of arguably the most influential astronomer in the United States, who was director of the observatory with the most powerful telescopes. Not only would Hale support him, he was prepared to make his Plan an important part of the work to be done with the (two) large telescopes in the first years of their operation. And Kapteyn would be directing the work from the site itself, made possible by a generous financial grant. Kapteyn was 56 years old at the time, with 14 years to go until retirement, normally speaking, so the timing was perfect too.

Things moved fast. Only three days later (on June 15) Hale already replied from Monley's Hotel, Trafalgar Square, London. The letter is in the Cal Tech archives, in the form of a copy by Hale himself, obviously written in haste.

'Dear Professor Kapteyn,

I was greatly pleased to receive your letter, and heartily agree that you should receive a salary sufficient to relieve you from outside work. Let us begin with 1800$ per year, and increase the amount to 2000$ per year if you find that the expense of annual trips to America renders this change necessary. [...].

I hope the arrangement may go into effect January 1, 1908, provided there is good reason to believe that the 60-inch reflector will be in regular operation during the following summer so as to make your presence in California desirable.

It may be advisable to work out a more formal contract later, but my views are expressed in this letter, and you may therefore be willing to favor me with your acceptance before I leave England.

[...] You have made all the arrangements in the most admirable way, and I have never heard a word of criticism from anyone. So I fully believe you will succeed.

Assuring you of the great pleasure I feel in anticipation of the work before us, and with warm regards, I am

Yours very sincerely,'

The parts I left out contain matters concerning the Solar Union. Kapteyn wrote back on June 21, 1907, again leaving out matters relating to the Solar Union.

'Dear Prof. Hale,

I accept the proposals made in your letter of the 18th very gladly and without reserve.

They are to me what I think Englishmen call a godsend. They open to me the prospect of contributing far more efficaciously to the solution of the stellar problem than I had ever dared to hope and at the same time relieve me of much care and worry. – The way too in which you propose to establish our co-operations on a footing of absolute confidence has my fullest sympathy. Generally there is not a word in your letter with which I do not cordially agree and the similarity of our views gives me great confidence in the future of our labour. [...]

Before finishing let me thank you with my whole heart for your broadminded s[y]mpathy with my work and for all that it means to me personally.

Yours very truly,

J.C. Kapteyn'

13.4 Kapteyn's First Visit to Mount Wilson

Kapteyn sailed for America on September 12, according to a letter to Gill. Unfortunately it turned out that the construction of the 60-inch had been delayed. On December 11, 1908 Kapteyn wrote from Pasadena to Newcomb, whom he had planned to visit on his return trip:

'The plans for my return to Holland could only be settled now. The completion of the 60 inch has taken somewhat longer than was expected, so that it will be only possible to take some plates tomorrow. I am of course unwilling to leave here before I have seen as much of its performance as I possibly can. So I will stay here nearly to the last possible moment.

Still I am very unwilling to leave America without having at least tried to see Mrs Newcomb and yourself. – I can be in Washington on Sunday the 27th of this month, quite early and if that suits you I will come to your house at about 11 a.m.

I have to start for New York in the afternoon, somewhere about 4 or 5 o'clock and have to make a short visit to Prof. Woodward.'

Physicist and mathematician Robert Simpson Woodward (1849–1924) was at that time the President of the Carnegie Institution. It was a matter of courtesy and politeness that Kapteyn should meet him before leaving the US.

Kapteyn mentioned to Gill what he did at Mount Wilson, while waiting for the 60-inch to become available. On January 17, 1909, after having returned to Groningen he wrote: '[...] the Mount Wilson affairs gave me little to do, in fact far less than I would have desired, because the telescope was not finished. I can only plead that I have not been idle, notwithstanding. I feel sure that I am on the right scent for determining the amount of loss of light in space, and this is so fundamental a question in my studies, that I gave all my working power to the subject.' When he stayed here, Kapteyn actually lived on the mountain, not in Pasadena. This meant that he had to travel up the so-called Mount Wilson Toll Road. This was usually done by carriage drawn by mules or horses (see Figs. 13.6 and 13.7). The road started north of Pasadena at what is now Altadena. There was also an older

Fig. 13.6 Lower part of the toll road up to Mount Wilson around 1904 (From the Huntington Digital Library [14])

Fig. 13.7 The tube of the 60-inch telescope being transported up the Mount Wilson trail. Kapteyn had to travel up this road also (From the Huntington Digital Library [15])

13.4 Kapteyn's First Visit to Mount Wilson

trail starting at nearby Sierra Madre (about 10 km to the east from Pasadena, directly below Mount Wilson), used at the time the Harvard College Observatory had operated a 13-inch telescope at Mount Wilson between 1889 and 1890, before it was moved to Arequipa in Peru (see page 369). The Toll Road was opened in 1991, and when Hale established the Mount Wilson Solar Observatory, the observatory had been granted free use of it. It was ten miles long, started at an altitude of just under 1,000 feet (about 300 meters) and ended at Mount Wilson Observatory at 5,700 feet (about 1,740 meters). It had been broadened from the original four feet a number of times, until it was a ten feet (about three meters) wide roadway by 1907. This was necessary to allow the transport of the parts of the large solar towers and the 60-inch telescope.

In the beginning Kapteyn camped in a tent and was probably assigned some space in the Monastery, the living quarters built in 1905 for visiting astronomers who came to use the telescopes. Sandage in his *Mt. Wilson History* notes that there was office space in the early Monastery, before it was destroyed by a fire in 1909. The rebuilt Monastery only had bedrooms. Otherwise Kapteyn would have had some place to work, for example in the 60-inch dome, but that is unlikely. The pictures of Kapteyn in Figs. 13.23, 13.8 and D.1 were taken in the Monastery during his 1908 visit.

At this point it is also of interest how his daughter described the first visit to Mount Wilson in the *HHK biography*:

Fig. 13.8 Kapteyn in the Mount Wilson Monastery in 1908 (Kapteyn Astronomical Institute)

'In October 1908 Kapteyn traveled to Mount Wilson for the first time with a three months' leave-of-absence from the government. Shortly before this, his son, who had finished his studies, had left for America to try his luck as a mining engineer. Kapteyn saw him in Denver on his way to and on the way back. On many later travels, too, the parents would see their son in distant America, a piece of luck not many parents had.

This first trip, however, did not give him satisfaction in all respects. Without his wife he felt a half person. She was the radiant and easy one, who took advantage of every opportunity with her carefree self-confidence and her playful openness. He was as timid as when he was a youth, and admired the easygoing manners of others. Characteristically he wrote his daughter about this in 1913, when he once more made a trip by himself: 'On board ship. How do I go about getting to know people? I don't have that ability. Mother is much better at that and many times she was the one who broke the ice. Now that I am by myself, I feel this weakness twice. I just had a hard lesson. I was walking on the deck with the safety vessels. I saw a man who had drawn my attention because of his pleasant appearance. When he saw me, he came straight to me and said: 'Mister Kapteyn, may I introduce myself, my name is Biesterman' (or something like that), a grain merchant from Rotterdam. 'Simple comme bonjour'. I think I always worry about a 'rebuff'. 'It won't do to be too sensitive in life.' We people of the wisdom from books always run that risk. People who are always among other people, on the exchange, at their office or travelling, lose that unnecessarily deep sensitivity and obtain an expansive readiness and become easygoing, which I always envy. Although it seems to me that in order to get far in arts, like van Dijk [(professor of theology in Groningen at that time; footnote by HHK); this was Isaac van Dijk (1847–1922)] says, one has to take position at the window of the soul. He that lives much in himself and who always keeps himself occupied with things that are far from every-day life, it seems to me, would never feel at home in this every-day life like others. Now I would miss this inner life, but still... instinctively my heart is longing for the other way. If, later, we would travel the world for a full year, daily meet other people and see other things, every day need to adapt to a new environment, then we will become real 'people of the world' ['Weltmenschen'], but then would be 70 years old.'

The basis of course goes deeper and he was certainly aware of that. One does not become a 'Weltmensch' simply on the basis of external circumstances. After all, Mrs Kapteyn was the easiest of persons and did not have other opportunities in her quiet, simple life. Now they complemented each other in a beautiful way: together they experienced the fullness of life. She cleared obstacles out of the way for him, taking the first step by introducing themselves and in this way they made good friends everywhere. Further steps were taken by others, who got to know him better and immediately started to like him. In that way their travels were full of interesting experiences and lasting friendships.

13.4 Kapteyn's First Visit to Mount Wilson

To his disappointment during this first year of his stay at Mount Wilson the large instrument was not completed, and Hale was too much preoccupied with his research on the Sun to spend much time on that of the stars. 'That way I have little to keep myself busy. I do more or less the same work I would have done at home. Not very satisfying, but that will be better next time.'

But then it was very nice on the mountain. Sunshine every day during November, which is the most chilly and unpleasant month in the Dutch year. He loved the beauty of the landscape with his constant interest in nature. The 2000 meter high mountain was like an enormous castle high above the Pacific, the clouds being a sea of fog at his feet.

'The view in the evening over the valley and behind that the great Pacific is often idyllic in its beauty. All features in the wrinkled valley become softer in a indescribable manner in the evening fog, the colours above that being in magnificent warm harmony with all this', he wrote full of marvel.'

I already mentioned earlier (see page 122) that in December 1908, so during this first visit to Mount Wilson, he gave a public lecture on tree rings which was published in the 'Pasadena Star', the local newspaper. On the way back home after this first visit to Mount Wilson, Kapteyn wrote a farewell letter from the ship, presumably before it left America.

'My dear Hale,

A last farewell from the boat, where I found your letter. Thank you kindly both for the news and, most, for your appreciation. And on the former – let me cordially congratulate you on the success of the 60-inch. I feel that with such power and such perfection, to quote the words of an eminent man - - - 'there must be lots of fun ahead'.

My journey did not begin under the best auspices, for, hardly were we underway for some hours, or we were stopped by a train with gravel which had got out of the tracks. We lost 4 or 5 hours.

There has been, however, no serious delay in my reaching Colorado Springs and my son. I found him in good spirits and things seem to go pretty well with him.

In Washington I saw Profs. Newcomb, Woordward and Abbot. I enjoyed these visits very much but I am sorry to say that I found Newcomb not at all well. He will probably have to submit (submit is not the right word for he is rather eager for it himself) to an operation ere long. It seems to be an internal abscess which troubles him exceedingly and gives very much anxiety to Mrs N.

In New York I found that Prof. Osborne had started for Baltimore. He had kindly taken care, however, that another gentleman should show me around and I enjoyed my morning exceedingly. I will certainly come back at some future time and hope to see Prof. Osborne himself on that occasion. And now, before leaving American soil (or rather water) let me thank you and Mrs Hale once more for many kindnesses and many pleasant days.

I learned much and I enjoyed much and take with me a bright hope that our joint work may lead to some real good things.

Yours very truly,   J.C. Kapteyn'

Charles Greeley Abbot (1872–1973) was a solar astronomer from the Smithsonian Astrophysical Observatory in Washington. Prof. Osborne might be Henry Fairfield Osborn (1857–1935), president of the American Museum of Natural History in New York.

When Kapteyn first visited Mount Wilson and when the 60-inch telescope became operational, the available staff was made up of Walter S. Adams and Francis Gladhelm Pease (1881–1938). Adams had been with Hale at Yerkes since 1902 and moved to Pasadena in 1904. He originally worked on spectroscopy of sunspots and such matters, but with the advent of the 60-inch and later the 100-inch telescopes he increasingly studied stellar spectra. Pease joined the Yerkes staff in 1901 and also moved to California in 1904, where he was associated with the construction of the solar towers and after 1906 with that of the 60-inch. But to ensure an effective use of the 60-inch, Hale needed to expand the staff for night-time astronomy and hired three men in 1909: Edward Arthur Fath (1880–1959), Frederick Hanley Seares (1873–1964) and Harold Delos Babcock (1882–1968). Their roles are described by Sandage in his *Mt. Wilson History* as follows: 'Fath was assigned to carry out spectroscopy and direct photography in those Selected Areas reachable from Mount Wilson. Babcock was also to devote part of his time [Babcock was actually a solar astronomer] to the Kapteyn program. Seares was appointed chief photometrist. Apart from his other duties, he was also made responsible for developing the library at the Santa Barbara Street offices.' The latter were (and still are) the offices that accommodated the Carnegie Institution in Pasadena. Seares in particular has been very important to Kapteyn's program of Selected Areas (see e.g., page 378). Between him and Kapteyn a very deep friendship developed. Their correspondence was extensive and their collaboration very fruitful. Initially Seares (see Fig. 13.9) was in charge during the months when Kapteyn was not at Mount Wilson, and later on also when the latter no longer visited California. Most of the letters concern the ongoing work on what came to be the 'Mount Wilson Durchmusterung' (see below), but there was always a friendly word at the end referring to personal matters. There were many requests from Seares to Kapteyn to assist in obtaining books or other publications that originated from the Netherlands or Europe, for the Pasadena library and sometimes also for the smaller one on Mount Wilson. During WWI it must have been difficult for US citizens to obtain German publications and apparently Kapteyn mediated when it came to providing papers published in the Astronomische Nachrichten to Mount Wilson Observatory. Seares, incidentally, was largely responsible for developing the library in Pasadena into one of the most extensive and complete in astronomy.

Fig. 13.9 Frederick Seares as he appears in the *Album Amicorum* presented at the retirement of H.G. van de Sande Bakhuyzen as professor of astronomy and director of Leiden Observatory in 1908 (Archives Leiden Observatory; see caption Fig. 3.7)

13.5 The Kapteyn Cottage

Kapteyn's second visited took place in 1909, this time together with Mrs Kapteyn. They still camped out on the top of the mountain near the Observatory (see 13.11). During their stay at Mount Wilson, Simon Newcomb died on July 11, 1909, at the age of 74. Some correspondence on this between Mrs Kapteyn and Mrs Newcomb has survived. Still in Cambridge, Mass., where they must have been visiting Pickering and Harvard, Mrs Kapteyn wrote on July 4, 1909:

'Dear Mrs Newcomb,

I have come over to America with my husband and I should so much like to come and see you and Professor Newcomb. But we find it impossible to do so now. Would it be too free of me to dare ask you to send me a little word to tell how Mr Newcomb is doing? I told you once he is one of my heroes and I feel very bad in not being able to go and see you.'

She mentioned that they would visit their son in Colorado on the way to California, where they would see him for only one day. She also said that her elder daughter now lived in Utrecht and had two children. This was followed by a letter of July 24, 1909, written at Mount Wilson. She wrote that 'coming in the United States, the

Fig. 13.10 Kapteyn and Mrs Kapteyn on board ship underway to America in a photograph taken from *the HHK biography*

Fig. 13.11 The Kapteyns in front of their tent on Mount Wilson (Kapteyn Astronomical Institute)

first news we heard was 'Newcomb is dead' ....' This must mean the Kapteyns learned of his passing away upon arrival in Pasadena. Newcomb had died of cancer of the bladder.

> '... and it has been and is a cloud on my happiness to accompany my husband on his trip to Mount Wilson. I wrote you some time ago that he was my great hero and <u>that</u> he will always remain. I am very thankful that it has been my lot to know so great a man and doubly thankful that I have been so privileged to feel as if he were in some way our friend too.
> I now thank you most heartily for the visit that you both paid us at our home in Vries – I shall never forget him –.
> Thank you also for the lines that you sent me in his last time with you. Those few minutes taken from him, seem holy to me.
> Going back in September we want very much to come and see you. You do not know how I longed on the steamer to arrive and to see Mr Newcomb. It was not to be.–
> My husband is just overpoweringly busy. He will write you soon.
> I remain, with high regards, your grateful and affectionate friend,
>
> Elise Kapteyn Kalshoven'

On the way back the Kapteyns visited Edwin B. Frost at Yerkes (Williams Bay near Chicago), as we learn from a letter written by Kapteyn from Mount Wilson in August. He expected then that they would be arriving on September 27 or 28. On the way to Williams Bay Mrs Kapteyn wrote to Mrs Newcomb that they planned to be in Washington on the 29th or 30th. On October 9, 1909 Kapteyn wrote to Hale a 'few words before leaving American soil', and on November 7 that they were due to return home on October 25. From websites on immigration to the United States during the first years of the twentieth century, for example [16], it can be inferred that a direct crossing between Rotterdam and New York took about 10 to 11 days at the time. A train trip between New York and Los Angeles took close to a week in regular service. The Transcontinental Express, between New York and San Francisco, set a record in June 1876 of 83 hours and 39 minutes, but that was not a regular service [17].

I continue with another fragment from the *HHK biography*, starting with the second visit to Mount Wilson:

> 'He enjoyed life up there much more when the following year Mrs Kapteyn accompanied him to America. On their way back and forth they visited Gill in London, who would use the opportunity to organize an astronomical gathering, so that Kapteyn kept in touch with his friend and his English colleagues.
>
> 'We rejoiced to hear again the familiar guttural exclamations and quaint expressions, as with youthful mind and enthusiasm he unfolded his latest ideas', Eddington wrote about these meetings with Kapteyn.
>
> Everywhere on his travels through America he was greeted as a very welcome guest. They visited most observatories: Harvard, Yerkes, Princeton, Albany, Newhaven (Yale), Allegheny, etc. and their trip was a real triumph. For Kapteyn this annual meeting with American astronomers was of the utmost importance. America had for some time been the leading nation

in astronomy, as a result of the extent of its resources, of the enormous work spirit and energy of its astronomers and also of its openness towards problems, free of the limitations imposed by traditions. This young nation developed an energy and fresh strength that was unknown to the older Europe and which captivated the ever young and enthusiastic Kapteyn. [...]

Now that the Kapteyns went to America every summer the house in Vries had to be given up and the wonderful summers near the heather fields of the province of Drenthe came to an end. The family said goodbye with much sorrow to have to leave such a beautiful thing behind. The mayor and his wife had come to say farewell and had expressed the hope that they would see them many more times. 'My wife and I have discussed and concluded', he said, 'that we do not know anyone else who puts a modest income to such good use and who gets so much happiness for it as you.' This remark made them very happy, since they had talked much with him about their lives and aspirations. Kapteyn indeed did know how to live in such a manner; he did not know how to increase his financial wealth, but he did know how to put it to use for things that had real value and those fruits were more important than increasing their wealth.

The first summer at Mount Wilson they stayed in a tent (see Fig. 13.11), as the observatory did not provide any lodging facilities for couples. The house where astronomers stayed was called the 'Monastery'. This was so, because it only had small rooms for astronomers who were working at the telescopes, which they did in turns, while those not observing worked in Pasadena at the Solar Office. Their families lived in Pasadena. [...] For the Kapteyns however it was not very practical. Living in a tent had many disadvantages, although they accepted these with a wise optimism. It was small and not very comfortable. They spent the days in the open air, which was easily possible with the warm and reliable climate. Tables, chairs, books, everything was outside. And there they worked, although they were bothered by small flies that sometimes disturbed the idyllic setting; these two human beings, being relocated from a comfortable upper house in a city to live in pure natural circumstances, quickly adapted to be happy.

Next year a major surprise was awaiting them. Upon arrival they found a small wooden house that Hale had had furnished with all sorts of comfort. Everyone at Mount Wilson had contributed. And they had asked themselves: 'Now, if it were to be yours, how would you like this or that to be done?'. And so with the help of all, this became a small jewel in all its simplicity.'

The story tends to be somewhat over-romantic here. Kapteyn surely knew that the Cottage was being built and it certainly did not come as a complete surprise upon their arrival. On May 5, 1910, Hale wrote in a letter to Kapteyn that the work on 'your cottage' was progressing well, so he definitely had advance knowledge of its construction. Kapteyn in turn wrote on July 18, 1910 from Mount Wilson to Hale in Pasadena, that the cottage was 'ideal'.

13.5 The Kapteyn Cottage

Fig. 13.12 The Kapteyn Cottage at Mount Wilson Observatory at the time the Kapteyns visited there (This photograph is taken from *the HHK biography*)

Of course, the cottage (see Fig. 13.12) was small and facilities on the mountain top limited. A description on the Web reads: 'Used for decades as a guest house for visiting astronomers and as a weekend retreat for observatory staff. Originally [it had] just a living room, one bedroom, and a kitchen, it had no shower or bath. Heating was by wood-burning stove. Acquired by the CHARA project in 1995 to be renovated and used again as living quarters.' [18].

The cottage is owned by the Center for High Angular Resolution Astronomy of Georgia State University, which operates an array of six 1-meter telescopes, whose signals can be combined to provide very high angular resolution images. This organization restored and partly rebuilt the cottage after they acquired it (see Fig. 13.13), and currently uses it as a residence for visiting staff. However, some parts of it are still intact and in the original condition, so it is easy to form a mental image of what it must have looked like. The HHK biography contains a photograph of the Kapteyn Cottage as it was when the Kapteyns lived there for a few months each year; it is reproduced in Fig. 13.12. A more recent picture of the Kapteyn Cottage appears in the *Legacy* (on page 366). It is located close to the 60-inch dome. The porch faces west and the view from there on clear days is indeed spectacular (see Fig. 13.14). Sitting on that porch the Kapteyns would have been able to see the valley stretching out from Sierra Madre at the bottom (Pasadena is hidden behind some nearby lower ranges) to Los Angeles and beyond, to the Pacific coast.

I include the following passage from the *HHK biography* on the periods the Kapteyns stayed at Mount Wilson.

'The Kapteyn Cottage, as it is known up to the present day, became their American home for many years and it became as dear to them as their home in the Netherlands. It was erected at a beautiful spot, in the shadows of knotted oaks and centuries-old pine trees, and large yuccas were growing around the house as enormous bouquets. The view was magnificent,

Fig. 13.13 The plaquette at the Kapteyn Cottage at Mount Wilson Observatory (Photograph by the author)

Fig. 13.14 The (current) view from the porch of the Kapteyn Cottage at Mount Wilson on a relatively foggy day (Photograph by the author)

13.5 The Kapteyn Cottage

stretching far over the mountains and the canyon, the deepest ravines of Western America, over the valley with its many towns that enhanced the feeling of peace high up there at night with brilliant star clusters, but which also gave the feeling that the richness of human life was close at hand. And in the distance they could see the majestic waters of the Pacific Ocean. The peaceful quiet that reigned there far from the hectic world, was doubly dear to them after the exhausting and emotional trip. Often deer came to look at them with curiosity, squirrels ran silently from tree to tree, and the chickadees, the tits of the West, were singing all day. The cool nights were wonderful and Mrs Kapteyn put her bed on the porch that stretched the length of the house. One night she saw something dark on the bed, and it turned out to be a nest of young squirrels for which the mother had chosen a soft place to give birth. Idyllic but also a little bit creepy. She took up another place to sleep that night.

Kapteyn missed the heather fields that he loved so much and he decided to import these. The next year he traveled with a small box filled with heather as if it were his most valuable piece of luggage, which held his attention most closely. One time it was left behind in a carriage that was uncoupled unexpectedly from the train and he telegraphed everywhere and it was returned to him. It reached the mountain unharmed and was planted near the house, but the heather did not catch on in spite of all love and care. So this wish had to be given up.[...]

The mountain air was for him like a elixir of life that gave him double the strength and desire to work. He had no constitution for the climate of the Low Countries, which made him tired and depressive. It was a suffering for him that he fought by the use of quinine. In the mountain air, however, he felt a different person with a much larger vitality and unlimited power.

'I feel a youthful desire awakening to clamber around in the mountains, descend into the canyons, and since I cannot give in to that as *pater familias*, I have recently tried my strength on a trip well-known as strenuous, following steep, poorly kept mountain paths', he wrote from there.

With the Americans he felt at home. The lack of conventions and the simplicity of the Kapteyns fitted in well with the open-minded, uncritical people of the far West, who were not burdened with centuries-old commitments, conventions and sentimentalities. So they were always received with much joy when they arrived. 'We are glad to see you folks again!', all said with happiness in their eyes and warmth in their handshake.

In 1913 Kapteyn wrote to his friend Boissevain: 'I have had a wonderful, very moving time in America. Parties and parties and parties. Speeches and trumpet sounds. Vanitas vanitatum.[3] And, yes, the American is much more unprejudiced and cheerful than we are. And I have found cordiality here.'

No wonder everyone was impressed by his personality. Barnard, astronomer at Yerkes, wrote him in September 1913: 'I simply wish to tell you

---

[3]From *Vanitas vanitatum omnia vanitas*; Vanity of vanities; all is vanity. (Ecclesiastes 1:2)

how much we enjoyed the short stay you made at the Yerkes Observatory. It is a great pleasure to see you here, and it always leaves behind a recollection that makes me feel good for a long time afterward.' [...]

However, his visits to Mount Wilson were not satisfactory in all respects. He wrote about this in 1913: 'I don't think I can continue with these Californian trips very much longer. It would be the nicest and most beautiful position one can think of – as long as it were not paid for. Now that I am being paid for it, I sometimes feel: do I deliver my money's worth? I would probably answer that with a yes when I could just put these people here to work on the things that are necessary to accomplish my plans and ideas. But of course I only have an advisory position – a voice that is listened to, but still ... the institution is in reality mostly a 'Sonnenwarte', an observatory for solar physics – and I represent the other part. The employees are all persons from the other part. But then they have done very much according to my ideas. That is the 'illness of the doubt' [Maladie du doute] that I have deep in my soul.' [...]

Sometimes, but that was rare, he would feel exhausted under the pressure of the problems that gathered around him, and that took up so much of his attention to leave any time for other things, to which he longed. He wrote from Mount Wilson to his daughter: 'My life is almost a constant struggle with scientific problems – and so seldom do I bring any of these to a satisfactory conclusion, and then I look out for something better. I sometimes think of Cauchy[4]:

'O, what sad occupation
What a humiliating weakness
To have to calculate all day
and to integrate without pause.'

But things don't change and that is it, even though I have such a strong desire for the time of my retirement, when I shall put down the pencil to perform calculations and will enjoy more like a human being all that is human. Will I ever be able to do that?' [...]

Mrs Kapteyn did all she could to make the stays up there homely and cozy. She did all the work on the house herself, which was according to her character. But that work was not all that easy. Everything had to be ordered from Pasadena by telephone and all the time the wrong things were brought up the mountain, which then had to be sent back and exchanged. Everything had to be negotiated over a distance of a dozen kilometers and a kilometer or so in height. But she enjoyed overcoming all obstacles and remained cheerful.

Already on one of the first days of the stay she would travel to Los Angeles, the large business and harbour city that bordered the villa town of Pasadena, with a long list of items to buy that she wanted to take back. The shops offered a range of interesting but unknown household items and

---

[4] 'Oh mais quel triste emploi/ Quelle humiliante faiblesse/ Que de chiffrer toujours/ Que d'integrer sans cesse.'

13.5 The Kapteyn Cottage

Fig. 13.15 A piece of artistic paper cutting found on the wall in the Kapteyn Cottage. The word 'Nachts' is German and means 'at night'. Although it superficially sounds as if it were Dutch, it was not left there by the Kapteyns (see page 477) (Photograph by the author)

elegant new things that irresistibly attracted a real housewife and loving mother. And when our mother returned home, she brought a suitcase full of surprises, those great American wonders, ingenious, practical and graceful – in one word tempting. Together they had selected and bought the more monumental presents, beautiful books, a typically American rocking chair, an enormous swing for the garden for the use of their grandchildren that was confiscated at Liverpool for it looked like a camouflaged airplane and that gave them much trouble during transport. But nothing was too much for them, as long as it would bring joy.'

The picture in Fig. 13.15 can be found on a wall in the Kapteyn cottage. One might think it was put there by the Kapteyns. The black paper cutting has the title 'Nachts', which would mean nights in Dutch or 'at night' in German. However, on the back it says that it was 'given to the Kapteyn Cottage by Mrs Theodore Boveri of Würzburg, Germany after a happy time here in July 1924.' Theodore Heinrich Boveri (1862–1915) was a German biologist and cytologist, who died in Würzburg. His wife was the American biologist Marcella O'Grady (1863–1950), who returned to the US in 1925. She had apparently stayed in the Kapteyn Cottage.

'Kapteyn came back from America with his head full of new plans and ideas, with renewed health and work drive, longing for his laboratory and his peaceful life of concentrated work. It was always a joy to go to America and a happy returning back home. Both valued this unusual, happy life to the full.'

13.6 The Luminosity Curve Revisited

I now return to the astronomical research that Kapteyn and his Laboratory were occupied with at the time when his association with Mount Wilson began. Starting with his first paper on interstellar absorption in 1909, Kapteyn published almost all of his papers in the (American) Astrophysical Journal. Subsequently they were included in the Contributions of the Mount Wilson Observatory), footnoting his name with 'Research Associate of the Mount Wilson Observatory' (in the earliest of these papers the annotation Solar was used in the name of the observatory. Often, Kapteyn also added the Carnegie Institution of Washington). Exceptions to this rule were presentations of extensive material, which were published in the Publications of the Astronomical Laboratory at Groningen, and in this context Kapteyn had written in English from the start.

His last paper, in which no association with Mount Wilson was mentioned, was in the Astronomische Nachrichten and appeared in 1910. It was also in English and concerned a vital ingredient of a model for the distribution of stars in space, viz., *The luminosity curve*, Kapteyn, (1910a). The reason for publishing in the Astronomische Nachrichten is explained in a footnote to the title. Kapteyn had sent this paper to the other American journal, the Astronomical Journal 'almost two years ago' (this he wrote in December 1909), '... but it seems that the paper and some letters afterwards were lost on the journey. Shortly after that the Journal ceased to appear.' So he had sent the paper early in 1908. What happened? The editor was Seth Carlo Chandler, Jr. (1846–1913) The story seems to be that Chandler, who worked as an unemployed affiliate of Harvard College Observatory (before that he had worked as an insurance actuary and called himself an 'amateur astronomer' [5]) severely neglected the journal when his health deteriorated. In 1909 the number of pages that were published had decreased from 65–85 per year, to a mere 16. Now, Kapteyn submitted his paper the year prior to this. The archives of the journal are kept at the Dudley Observatory in Schenectady, New York. Here we learn [6]:

> 'In 1909, in ill health, Chandler persuaded assistant editor, Lewis Boss to become editor of the journal. The Dudley Observatory agreed to become the publisher. When Boss took over the helm, transferred to him were boxes of mail which had been unopened for 4–5 years. Much of the first few years as editor were spent sorting out the subscription records of the journal and either publishing or returning manuscripts which had never been seen. When Benjamin Boss became editor upon his father's death, in 1912, he too continued to have to sort out the problems which had been forced upon the periodical by the unfortunate ill health of Seth Chandler. After a few years, the journal was restored to its original status and Benjamin Boss continued as editor until 1941, when he arranged for the transfer of responsibility of the Astronomical Journal to the American Astronomical Society, where it rests today.'

Lewis Boss (1846–1912) was director of the Dudley Observatory until his death, when his son Benjamin (1880–1970) took over. Evidently Kapteyn's letters had never been opened and his paper was simply ignored. But the Astronomical Journal

13.6 The Luminosity Curve Revisited

did not cease to exist, as Kapteyn wrote (in December 1909), but rallied again and is in fact one of the main professional journals in astronomy to date.

Kapteyn's paper concerned a reply to a paper by Comstock, *The luminosity of the fixed stars* [7], who had determined the distribution of absolute magnitudes of stars, using only those which were felt to have reliable parallaxes. The result differed significantly from the determination by Kapteyn (1902d). His footnote continues as follows: 'After the long delay caused by these circumstances I felt that I might perhaps well keep back the contents still somewhat longer. Opportunity would offer at a presumably not too far off time to embody the main parts in a paper in a further study on the arrangement of stars in space, which I had long in contemplation. Now that it appears, however, that Professor Seeliger in his recent paper in the Astronomische Nachrichten No 4359 adopts Prof. Comstock's luminosity curve and uses it in his derivation of the arrangement of the stars in space, I feel that it would not be right further to hold back my reply.'

Kapteyn's criticism concerned a number of points. First he noted that, when Comstock used Kapteyn's curve to compare his results with, he had made a fundamental mistake. Kapteyn listed in a table the absolute magnitude M and the number of stars corresponding to that value. But that must be within some interval, say between $M - \frac{1}{2}$ and $M + \frac{1}{2}$. Now, Kapteyn had listed the corresponding luminosity L and gave the number of stars with that luminosity without correcting for the interval. Let us assume that we have $L = 2$ in some unit. The magnitude scale is logarithmic, so the corresponding M is related to the logarithm of L, and $\log 2 = 0.30$. Say, the interval in $\log M$ is from 1.5 to 2.5; if then the interval in $\log L$ is from 0.2 to 0.4, this gives about the same interval in L, namely from 1.59 to 2.51. But now look at $L = 5$. If the intervals are the same we get in L the interval 4 to 6, while in $\log L$ it is from 4.5 to 5.5, which in L is from 3.98 to 6.31. This is elementary, but Kapteyn made a valid point that Comstock had made an error here, which he illustrated with a somewhat more elaborate but in essence similar argument as I have just presented.

Further criticisms were that Comstock used some unreliable data, and on the other hand ignored some sets of data that were reliable, and made a few other systematic errors (that are more technical and need not be presented here). Kapteyn proceeded then to redo Comstock's analysis, ending up with a luminosity curve not all that different from his earlier one. He spent quite some time on the incompleteness question. After all, even among brighter stars a sizable fraction had no determined parallax and he estimated correction factors as a function of apparent magnitude to allow for this ('magnitude-factors'). To this end, Kapteyn made the (reasonable) assumption that the parallax of a star of given magnitude and proper motion did not depend on the question whether or not it had been on an observing list of some observatory. As usual he assumed the luminosity curve to be independent of distance and the absence of absorption. In the end he used his more recent publication on the distribution of stars as a function of apparent magnitude and Galactic latitude, Kapteyn (1908d), to find improved values for the mean parallax and mean luminosity as a function of apparent magnitude.

Kapteyn's discussion was summed up by stating that Comstock had used distribution tables and average values incorrectly. Luminosity curves listed numbers of stars per interval of luminosity and it was important to take proper notice of whether this interval was in linear luminosity or its logarithm. Also Comstock failed to take into account that values for the mean parallax depend on the underlying space distribution.

Kapteyn must have felt that this discussion belonged in an American journal. Why he then did not send it to the Astrophysical Journal, which had become –and would remain– his preferred journal until the end, does not become clear.

13.7 The Lecture Before the Royal Institution of Great Britain

Before leaving for his first visit to Mount Wilson, Kapteyn traveled to London to deliver a lecture at the Royal Institution of Great Britain. From his correspondence with Gill we know that Kapteyn stayed with the Gills and that he had brought his daughter Henriette along. Gill had a few times urged him to bring his wife as well, but in the end Mrs Kapteyn stayed at home. Kapteyn hoped that Lady Gill could introduce his daughter to Octavia Hill (1838–1912), the great social reformer, but from the correspondence there is no further indication whether or not this happened. Often Gill ended his letters with Lady Gill's best wishes for Mrs Kapteyn and the promise she would write soon. Kapteyn often concluded his letters on a similar note. Surely there must have been a certain amount of correspondence between the two ladies, but these letters are not to be found in any of the archives.

More background from correspondence between Mrs Kapteyn and Mrs Newcomb. On February 8, 1908, Mrs Kapteyn wrote a letter to Mrs Newcomb with a summary of the private circumstances of the Kapteyns. The letter was a somewhat belated reaction to Christmas greetings from the Newcombs. Mrs Kapteyn started by apologizing for the delay, pleading busy times.

'I do not think that you know that I have become a grandmother as in July a baby girl was born in the little house next door. [Apparently the elder daughter Jacoba and her husband Willem Noordenbos lived next door. It is not clear, however, whether this would be at Eemskanaal/Oosterhaven (see Fig. 4.7, where they lived at the time, or whether she was referring to their second home in Vries. Genealogies on the Web give the child's place of birth as Groningen]. All went well, with the exception of a three-weeks' ailing of the young mother by her breast being infected. However, the baby began to thrive wonderfully well, as she was able to feed it all through. In autumn however, he [this must be Willem Noordenbos] was appointed lector in surgery in Utrecht, and in January they moved away. He hopes to get in practice too so everything seems fair for the future.

Another thing that made us very happy was our boy passing his examinations for mining engineer and surveyor. True to his father's instincts he took a fortnight's rest and was off again to the mines to get a practical insight in his future work. He has to serve [offer his services?] for a month and then he will start for Spain to learn the language and thence to America.

13.7 The Lecture Before the Royal Institution of Great Britain 481

You will certainly have heard from Mr Newcomb that Kapteyn has got an appointment to go to America this year, and if all goes well will return there every summer. This year he will go alone, next year I hope to accompany him. If it were possible I should like very much to come and see you, but beforehand I must not make plans.

Hetty, our second daughter is living in the Hague. She studied in Amsterdam for two years and has become a house inspector. In May she will go to England with her father and pay a visit to Octavia Hill, whom perhaps you have heard of.

I have followed the example of Mrs Moulton [hardly legible; possibly the wife of Forest Ray Moulton in Chicago (1871–1952), a staff member at the University of Chicago in 1904. He worked in celestial mechanics and had written a well-known book about the subject] whom we visited, being in America. I have [a] nice woman who helps me a few hours daily and the rest I do myself. It gives a quiet in the house that I love, and besides it is very economical, –as we are still in our expensive years. [...]'

In the letters to Gill, Kapteyn noted that just before coming to London, he had been on a geological trip to Thuringia and Jena (where the Zeiss company was producing a measuring microscope for him). The lecture was delivered on Friday May 22, 1908. It was published by the Royal Institution as Kapteyn (1908e), but also in the form of a summary in Kapteyn (1908f), and as a full reprint by the Smithsonian Institution, Kapteyn (1909a). The copy in Kapteyn's bound volumes is the one from the Royal Institution and is full of notes. These are textual changes and corrections, plus indications where the illustrations should appear (these appeared in the printing of the Royal Institution on separate pages of better quality paper, but in the Smithsonian reprint they were positioned in the text).

The lecture is an exposé for a general public on Kapteyn's program of determining the distribution of stars in space. He did not mention the determination of stellar distances by methods of secular parallax or to individual clusters. He described his sorting of stars in rough distance intervals using their proper motions, but made no mention of his Star Streams. Also under his necessary assumptions he mentioned that of the same 'mixture law' (the luminosity curve) with distance and latitude and that of no absorption, but there was no reference to the assumption of no preferred direction among motions in space. However, at the end he did refer to his Plan of Selected Areas, without putting himself forward as the initiator and coordinator. His conclusion is amusing to the modern ear:

'If, at the end of this lecture, somebody summarizes what has been discussed by saying that the results about the structure of the universe are still very limited and not yet free from hypothetical elements, I feel little inclined to contradict him. But I would answer him by summing up in another way, viz.:

Methods are not wanting which, given the necessary observational data obtainable in a moderate time, may lead us to a true, be it provisionally still not very detailed, insight into the real distribution of stars in space.

I think this time need not exceed fifteen years. They to whom such a time may still seem somewhat long may be reminded of the fact that that time will be elapsed, that we shall have finished our work, before any but the very few of our nearest neighbours in space can be aware of the fact that we have begun.'

In the meantime, Kapteyn's appointment to MW increasingly became common knowledge. Kapteyn was absolutely delighted. He wrote to Gill on June 24, 1907, about fixing dates for meetings of the committee to oversee the progress of the Plan of Selected Areas. Since there were members in Europe and in the U.S., and it might be difficult for all members to attend an annual meeting, he wrote: 'Of course it will probably be somewhat difficult for Prof. Hale to attend the meeting every year. [...] In years he would be unable to come, we would communicate with him beforehand by letter. Possibly also I could act as a sort of intermedium for you know, as Prof. Hale will tell you, that he has proposed to me, under singularly favourable conditions, to come to Mount Wilson every year.' Gill replied on June 29: 'I rejoiced to hear of the arrangements made about your going to Mount Wilson –they are in every way satisfactory– and now I hope we may see you in London on your way back and forth to America'.

Edwin Frost at Yerkes Observatory also tried to arrange a get-together with Kapteyn. On September 1907 he noted: 'We had a pleasant visit from Julius early this summer.' [This must have been Willem Henri Julius, the solar physicist from Utrecht.] 'I frequently asked him if there was any possibility that you could get a leave of absence and come over and spend some time here. We have the stereocomparator already, and I could give you the equivalent of a night per week with the 40-inch for photographing fields while you were here. I later found that Dr. Hale made an arrangement with you for coming to Pasadena later when his apparatus [the 60-inch telescope] is ready. That I assume will not be for a couple of years from now, at least not for next summer. Will you not seriously consider this? I am sorry to say we have no funds that would be available for salary or traveling expenses [...]' Kapteyn replied on October 23, 1907: 'What a splendid thing it would be to be able to accept your delightful invitation. My wife too would be so happy. But such things are not to be for poor astronomers like us. Besides, I hope and trust that you are too skeptical as to the time required at Pasadena for the erection of the 60-inch. Prof. Hale rather confidently hoped to have the thing done before next summer and in that case expects me to come at once. If so, then I will certainly make a pilgrimage to Williams Bay to see you all again.' And Pickering from Harvard on August 10, 1907: 'I am very glad that there is a prospect of seeing you soon again in this country, and I hope that, as before, Mrs Kapteyn may be able to accompany you.'

13.8 The 1910 Solar Union Meeting

In 1910 the Mount Wilson Solar Observatory hosted the fourth meeting of the International Union for Cooperation in Solar Research. This organization had been founded on the occasion of the 1904 St. Louis meeting and Kapteyn had been present as the representative from the Netherlands (see page 366). Additional meetings had been held in Oxford, UK, in 1905, and in Meudon, France, in 1907. The meeting took place from August 31 to September 2, 1910. The Kapteyns were at Mount Wilson at the time. Many astronomers, not just solar but also stellar, attended the meeting. This was in keeping with the interest of Hale, who although primarily a solar astronomer, was in fact interested in the more general problem of the evolution of stars. Consequently he felt that the Sun needed to be studied as well, apart from stars, and in particular spectroscopically. Together, solar and stellar studies would give insight in stellar evolution. Kapteyn was again the delegate of the Netherlands, the responsible organization in the Netherlands being the Royal Academy. The other Dutch members of the Solar Union were not present. These were Willem H. Julius from Utrecht and physicist Pieter Zeeman (1865–1943), discoverer of the Zeeman effect of the splitting of spectral lines the presence in magnetic fields and co-laureate for the 1902 Nobel Prize in physics.

The attendance was large, with almost half of the just over 80 participants coming from Europe (see Figs. 13.16 and 13.17). Many saw the two solar towers and the 60-inch telescope on Mount Wilson for the first time. There are two enjoyable accounts of the meeting, both published in 1910 [20]. The *Transactions of the International Union for Cooperation in Solar Research, vol. 3* were published in 1911 [21].

In those days, traveling was far more exhausting and time-consuming than today. The European astronomers first had to cross the Atlantic by steamer and then the American continent by train. Many of them, however, like their American colleagues, had been at Harvard College Observatory in Cambridge, Mass., to attend a meeting of the Astronomical and Astrophysical Society of America from August 17 to 19. They traveled to the West Coast by special train, which arrived in Pasadena on August 28.

The meeting started in the morning of Monday August 29, with a visit to the offices, laboratory and workshops of the Observatory in Pasadena. Here the participants were shown the glass disk that was to become the 100-inch mirror. At that time the disk was of unacceptably poor quality, but later it had to be used all the same. In the afternoon there was a garden party hosted by Prof. and Mrs Hale. The Kapteyns, who were up at Mount Wilson, had excused themselves in a letter to Hale from Mount Wilson on August 22. 'We would certainly like to be present, but going down now, would mean for me giving up the idea of bringing my work here to some sort of a close.' Kapteyn apologized and hoped that Hale would understand.

On Tuesday everybody went up to Mount Wilson, either using the old trail by electric car to the foot and then on horseback or on foot(!), or by six-passenger carriages along the Mt. Wilson Toll Road. The whole procedure is said to have taken over eight hours. Those on foot as well as the horses pulling the carriages had to rest

Fig. 13.16 The participants of the 1910 meeting of the International Union for the Cooperation of Solar Research at Mount Wilson Observatory (From the Huntington Digital Library [19])

Fig. 13.17 In the centre of the picture in Fig. 13.16 there are a few persons who were of particular importance to Kapteyn. From the left we have Walter S. Adams, Edward C. Pickering, George E. Hale and Mrs Kapteyn. Kapteyn himself is on the right in the second row. The lady at the front-right is Williamina P.S. Fleming (From the Huntington Digital Library)

13.8 The 1910 Solar Union Meeting

Fig. 13.18 Some participants of the 1910 Mount Wilson meeting. The person Kapteyn is talking to has been identified as a Mr McBrids [22]; I have been unable to find any data with regard to him. The persons to the right are Karl Schwarzschild and spectroscopist Vesto Melvin Slipher (1875–1969) (From the Huntington Digital Library [23])

frequently. Most participants stayed in the Mount Wilson hotel that had been opened in 1905 offering accommodation to tourists. But its capacity was insufficient and some people had to spend the night in tents. The meeting was held at the Observatory in a building that was called the museum, a one-story wooden building. To provide a cool environment a white canvas tent was erected over it (see Fig. 13.16).

Hale increasingly suffered from nervous breakdowns and depressions. According to many these were brought about by the enormous strain of carrying out major projects such as the Mount Wilson Observatory, including the start of the construction of the 100-inch (completed in 1917), even before the 60-inch was operational. Eventually, he relinquished the directorship to Walter Adams in 1923, who had been assistant director since 1913. Hale's condition caused him to leave the meeting after the first day, but not after having given an extensive welcome address, focusing on the spectroscopy of sunspots.

On the Wednesday and Thursday evenings there were special addresses. The first one was by Charles G. Abbot of the Smithsonian Astrophysical Observatory in Washington and concerned the solar constant, the amount of energy received from the Sun in the form of radiation per surface area (square centimeter for example) perpendicular to the line towards the Sun. The address on Thursday was delivered by Kapteyn. Hale had invited him to speak and Kapteyn had agreed, but not without expressing some reluctance at addressing a meeting on solar astronomy. The presentation was entitled *On the systematics proper motions of the Orion stars*, Kapteyn (1910d) (the ADS-link given in Appendix A for this paper provides the texts of both evening addresses; Kapteyn's starts on page 215).

'Orion stars', also called 'helium stars' at the time, are stars that are of current spectral type B (or sometimes O). In these lines of neutral helium are prominent. Bright stars in the constellation Orion (except the brightest one, the red colored Betelgeuse) are predominantly of this type, but they can be seen all over the sky, mostly concentrated towards the Milky Way. This is so because they are bright, young stars. Kapteyn had been associated with a detailed study with Edwin B. Frost of Yerkes Observatory, in which they determined the Sun's velocity in space with respect to these stars, Kapteyn & Frost (1910). Kapteyn was one of the collaborators for an even more extensive study of Helium stars that resulted in a number of papers up to 1918, which will be discussed later. In this contribution he illustrated the existence of the two Star Streams by deriving their properties from observations of proper motions and radial velocities. What is important here is that in this presentation to predominantly spectroscopic (both solar and stellar) astronomers, he tried to link their field – the evolution of stars and the Sun – to his own, viz., the arrangement of stars in space. The opening paragraphs read as follows:

'I have been much surprised at being invited to give an informal talk before the present company. My studies are not those of the Solar Union, and I had expected to sit down in a quiet corner and enjoy many things that would be new and interesting to me, and to hear discussions about matters that I have long wanted to understand better. But thinking the matter over my surprise has been lessened.

You have seen our spectrographs and spectroheliographs, our Snow telescope, and our towers [being a (part-time) staff member, Kapteyn felt justified to refer to *our* telescopes], the 60-inch reflector, with its focal plane spectrograph, and a hundred other things, and you have had explained to you the work that is being done with these instruments and the aims in view. I have tried to explain many things myself. It remains, however, for the explainer to explain himself. What is the use of a man of the astrometric branch of astronomy, whose main study is the arrangement of stars in space, at an astrophysical observatory, where the study prosecuted is that of the evolution of the stars, and where the Sun is studied most, because it is the nearest of the stars and consequently can be more thoroughly investigated than the rest? I take it that the meaning of the investigation is that I should offer you such an explanation.'

The other interesting aspect of the presentation is that it was based on preliminary results. Kapteyn eventually published the paper as he presented it, but inserted notes whenever he felt that his point of view had changed, in particular where the results of Kapteyn & Frost (1910) had to be revised. 'I might have suppressed the paper, but as I have myself always taken a particular pleasure in those cases in which it is possible to follow the writer on the way, howsoever crooked, that he has actually followed, I have supposed that some of my readers might not dislike the idea of seeing this method applied in the present case. To prevent misunderstandings, I have added footnotes, indicating the views I hold at the present time (March, 1911).' I will return to the science case later.

The meeting produced one significant outcome, and that was the appointment of a 'Committee to consider and report on the question of stellar classification'. This was preceded by the adoption of an important resolution, that stipulated 'that the Solar Union extends its activity so as to include general astrophysics, and that a committee be appointed to consider and report on the question of the classification of stellar spectra'. In this the Union became the predecessor of the IAU. The committee had as its chairman Edward Pickering and as secretary Frank Schlesinger (1871–1943), the then director of the Allegheny Observatory of the University of Pittsburgh. Among the members we find Hale and Kapteyn, and also Walter S. Adams, William W. Campbell, Edwin B. Frost, Karl F. Küstner, Henri N. Russell and Karl Schwarzschild. A very powerful composition indeed. An immediate consequence, as described in some more detail by Sandage in his *Mt. Wilson history* (Sandage writes erroneously that Schlesinger was chairman of the committee; actually he was the secretary), was the adoption of the Draper system of classification of stellar spectra, first informally, later officially by the IAU during its first meeting in Rome in 1922.

The return trip to Pasadena was on Saturday morning (going down taking only four hours), after which a closing dinner was offered by Prof. and Mrs Hale. It was attended by over one hundred participants and guests. After the meeting, nearly all participants went to visit Lick Observatory near San Francisco before traveling back home. Europeans must have left their homes no later than early August, and returned early October at the earliest. But a lot was accomplished at this very important meeting. It brought together astrometrists, solar physicists, astrophysicists and sidereal astronomers. Kapteyn had been part of it and there can be doubt as to the significance of his role.

13.9 The Mount Wilson Catalog

In the meantime the Mount Wilson work related to the Plan of Selected Areas was progressing more and more rapidly. Adams was obtaining radial velocities of stars at an unprecedented rate. For example, on February 26, 1910 Hale mentioned in a letter to Kapteyn that Adams had taken spectra of 25 stars and on May 5 of the same year that he 'has more'. The important photometric part, determining the positions and magnitudes of stars in the Selected Areas with the 60-inch telescope in all Selected Areas that could be seen from Mount Wilson, took place under the direction of Seares. It would take quite some time, however, to complete the program. The paper *Mount Wilson Catalogue of photographic magnitudes in Selected Areas 1–139* by Seares, Kapteyn, van Rhijn, Joyner & Richmond (1930), that eventually grew out of the effort, required the development of new methods not only related to establishing magnitude scales at these fainter levels, but also the technique of measuring plates taken with such large reflectors. The field of view of the 60-inch telescope is rather small (a circle with a diameter of about 23 arc minutes) and few stars for overlap with brighter magnitudes were available. In fact, most of the fields had no stars brighter than 14-th magnitude, while reliable magnitude sequences got no fainter than magnitude 12 or so. Furthermore, the stars away from the center of the field had distorted images due to the parabolic shape of the mirror. This did ensure sharp images on the axis, which a spherical mirror would not provide, but off-axis the images are severely compromised. Measuring positions and particularly magnitudes was far from straightforward. Seares, with the help of Kapteyn and later also Pieter van Rhijn carefully developed methods to overcome all difficulties.

The plates at the 60-inch were exposed for one hour for each of the 139 Selected Areas north of declination $-15°$ and were taken by Harold D. Babcock and Edward A. Fath from 1910 to 1912. Copies were made to be left in Pasadena and the originals were sent to Groningen. This was complemented with a set of plates with shorter exposure time and/or reduced aperture (reducing or 'stopping' the primary mirror by a diaphragm) in order to image brighter stars necessary for establishing a scale of magnitudes for each Selected Area. This series of plates was taken from 1913 to 1919, the observers being Seares himself and new staff astronomer Harlow Shapley (1885–1972). Then a set of exposures of the Selected Areas and a field with stars in the North Polar Sequence on the same plate were taken and similar plates to connect adjacent Selected Areas. This was done with a 10-inch telescope, mostly by staff member Milton Lasell Humason (1891–1972). Overall more than one thousand plates were taken. The long exposure plates were measured at the Astronomical Laboratory in Groningen, the other plates were measured and reduced at the Mount Wilson offices in Pasadena. Measurements included coordinates on the plates and diameters of the images for magnitudes. For the effects of distortions, off-axis corrections were determined, taking into account that these were dependent upon the aperture used (whether a diaphragm was used to stop the aperture). Technically, these corrections depended on the f-ratio between aperture size and focal distance.

The computers at Groningen were largely paid by Hale and the Carnegie Institution. The support of the University of Groningen remained ungenerous, in a way; not when it concerned the status and success of Kapteyn, but in a financial sense its support was marginal. Kapteyn was constantly looking for sources of money to employ his computers. For example, in a letter of February 2, 1910, Kapteyn asked Hale for 100$ for a computer to perform the measurements of the Selected Areas plates, to which Hale reacted by sending a check on the 18th of that month. Incidentally, this cannot have been with a full-time position in mind. On September 2, 1912, Kapteyn sent a letter from Mount Wilson requesting two computers in Groningen for a longer period. He estimated the required salaries at '350 to 400$ per year, increasing to 500$ in 5 years' to start with. Using the estimated exchange rate for 1910 as applied on page 678, this corresponds to 5500€ today; so here again this seems to be indicate part-time employments. *How much did you say?* quotes a salary of an unskilled worker in 1910 of 750 guilders per year. Maybe the University wanted Kapteyn to match each appointment with part of the costs.

The Mount Wilson Catalogue was finally published as a special publication of the Carnegie Institution, Seares, Kapteyn, van Rhijn, Joyner & Richmond (1930), eight years after Kapteyn had died. The last two persons were in fact not presented as the primary authors, but were credited on the title page as assistants. Seares actu-

13.9 The Mount Wilson Catalog

ally married his assistant Mary Joyner in 1942 after his first wife had died in 1940. A very important result obtained with the Mount Wilson Catalogue was published as early as 1925, Seares, van Rhijn, Joyner & Richmond (1925). Kapteyn was not a posthumous author on this paper, as he would be in the Catalogue itself five years later, but the paper certainly did what Kapteyn had meant to do with the data all along. I have therefore included it in the Kapteyn publication list of Appendix A. The paper provided the fundamental star count data needed to derive a model for the distribution of stars in space, i.e., the number of stars as a function of magnitude and Galactic latitude. For the brightest stars (between magnitude 4 and 9), the 1930 catalog itself is based on van Rhijn (1917), a study in the Groningen Publications Series with the best counts available in 1917. For the interval magnitude 9 to 13.5, counts were based on the Astrographic Catalogue in the Carte du Ciel effort (see page 194), and for magnitude 13.5 to 18.5 on the Mount Wilson Catalogue. Figure 13.19 shows the results as the solid lines. The dots are based on the Mount Wilson Catalogue, the open circles on the Harvard-Groningen Durchmusterung. Down to magnitude 18.5 there are 13,300 stars per square degree in the Milky Way, decreasing to 660 in the poles. As I mentioned, Kapteyn did not live to see this monumental result of his Plan.

Now that Kapteyn had a long-term association with the Mount Wilson Observatory, he reflected extensively on programs to be performed after completion of the

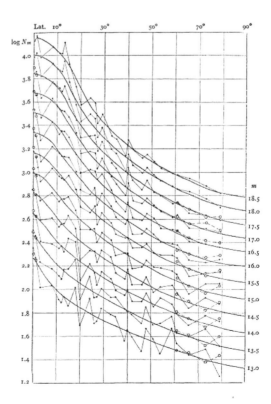

Fig. 13.19 The distribution of (the logarithm of) stellar surface density N_m, the number of stars per square degree brighter than magnitude m as a function of Galactic latitude and m (From Seares, Kapteyn, van Rhijn, Joyner & Richmond; 1930)

100-inch telescope, at the request of Hale. He made estimates of how many magnitudes could be gained by the larger aperture, what use could be made of the prime focus and in particular the extent to which spectroscopy could be employed for all aspects of studies. A brief word is required about the mention of 'prime focus' and the design for the 100-inch telescope. The 60-inch telescope had three observing stations or foci. The first one is called 'Newtonian'. In this set-up there is a mirror near the top of the telescope that reflects the light by about 90° and the observer is positioned in a 'cage' that is attached to the telescope structure. The second is called 'Cassegrain' and here the light is reflected by a hyperbolic mirror through a hole in the middle of the primary mirror, where a camera or spectrograph can be attached to the back of the telescope structure. In the 'Coudé', a few mirrors reflect the light into the dome where a large, non-moving spectrograph can be placed to provide spectra with very high spectral (wavelength) resolution.

The prime focus is the 'real' focus of the primary mirror, a point in the middle and near the end of the telescope tube. The advantage is that there is no need for a second reflection of the light, which is a potential source of loss of light and image distortion. But the drawback is that in such a case the observer must also be at prime focus and a cage blocks a larger part of the mirror than a secondary mirror would. Eventually the 100-inch telescope was built with a Newtonian focus (and like the 60-inch the other two mentioned above). Only the 200-inch telescope erected on Palomar Mountain, which was developed and designed under Hale, but not completed until after his death, in 1948, has such a prime focus cage. It is about a meter wide and blocks only a few per cent of the area of the 5 meter primary.

13.10 The 'Pipeline'

Kapteyn also started what David DeVorkin in the *Legacy* called *the Dutch Pipeline*, the introduction of promising young and talented astronomers to Mount Wilson to gain experience. The first were Ernst Arnold Kohlschütter (1883–1969), Adriaan van Maanen (1884–1946), Ejnar Hertzsprung (1873–1967), and Pieter Johannes van Rhijn (1886–1960). Kohlschütter had studied under Karl Schwarzschild and obtained his PhD in 1908 at Göttingen. By 1909 Schwarzschild had moved to Potsdam and Kohlschütter worked at the Bergedorfer Sternwarte near Hamburg. Schwarzschild had recommended Kohlschütter for a stay at Mount Wilson and Kapteyn wrote a letter much in favor of the request to Hale on March 4, 1911. Kohlschütter came that same year. Van Maanen had obtained his degree in Utrecht in 1911 under A.A. Nijland on a thesis on proper motions, *The proper motions of 1418 stars in and near the clusters h and χ Persei*, van Maanen (1911). Part of this work had been completed during an extended period when he worked with Kapteyn in Groningen between 1908 and 1910 and it is not at all unlikely that Kapteyn had been behind the definition of the research project. The plate material for the study was obtained from Anders Donner at Helsingfors and Sergej Kosinsky at Pulkova, undoubtedly through Kapteyn. So, there is good reason to regard van Maanen as

13.10 The 'Pipeline'

a disciple of Kapteyn. It is not unlikely that Kapteyn had already mentioned him to Hale before the name of Kohlschütter came up. In any case, van Maanen, supported by a generous gift from a wealthy cousin, went to Yerkes Observatory as a volunteer observer. He traveled with Kapteyn when the latter was on his way to his annual Mount Wilson visit in 1911. Klaas van Berkel in his contribution to the *Legacy* shows and quotes a letter of June 23, 1911, from Kapteyn to Edwin B. Frost, written on the steamer, in which he announced his arrival for a short stay at Yerkes: 'If my calculations are correct, we will be at Williams Bay on the 28th or the 29th and will then deliver into your hands Dr. van Maanen'. Van Maanen, undoubtedly on the recommendation of Kapteyn, moved to Mount Wilson in 1912, where he stayed for the rest of his career.

In the same letter of March 1911, in which Kapteyn recommends Kohlschütter, he mentions that Gill will have written Hale about Voûte. This is Joan George Erardus Gijsbertus Voûte (1879–1963), who was originally a civil engineer, but who had turned to astronomy and was an observator in Leiden. At about that time, Voûte had been offered a position at the Cape Observatory, which he took up in 1912. In the archives in Leiden there are a few letters between the then director Ernst van de Sande Bakhuyzen and van Maanen, in which the latter applies for the position of observator that would become available when Voûte would have left (June 17, 1911). This letter was also written on board the ship (M.S. 'Baltic') from which Kapteyn wrote a few days later. On April 26, 1912, van Maanen wrote again saying that he had also had an offer to come to Allegheny Observatory, and awaited advice from 'his highly valued teacher [hooggewaardeerde leermeester] professor Kapteyn'. This seems to indicate that Kapteyn had played a major role in van Maanen's studies. On December 30, 1911, van Maanen wrote to van de Sande Bakhuyzen, that he declined his offer and chose for Allegheny. However, he came to Mount Wilson Observatory instead, in July 1912.

Hertzsprung (see also Appendix B.), was Danish by birth and trained as a chemical engineer. After a variety of jobs, including that of a volunteer observer in Copenhagen Observatory under director Svante E. Strömgren (1870–1947), he was appointed in Potsdam in 1909, by Schwarzschild (the story how Hertzsprung managed to shatter the objective lens of the Copenhagen Observatory's 12-inch refractor is told entertainingly in Sandage's *Mt. Wilson History*). Hertzsprung was much interested in gaining access to the 60-inch telescope at Mount Wilson, and Schwarzschild suggested that he contact Kapteyn for a recommendation to Hale. Hertzsprung came to Groningen, where Kapteyn suggested he accompany him on his next Mount Wilson visit. In a letter to Hale dated December 6, 1911, Kapteyn actually recommended both Hertzsprung and his own student van Rhijn. Hale answered (December 26) that Hertzsprung was welcome and suggested van Rhijn accompany Kapteyn on his next visit. Kapteyn wrote on January 21, 1912, that Hertzsprung would actually come with him and that van Rhijn would also come. Of van Rhijn he said: 'Whether he will be a good observer remains to be seen.' Hertzsprung actually traveled to America with Kapteyn in June of 1912. But during his stay in Groningen he had also become interested in Kapteyn's daughter Henri-

Fig. 13.20 Van Maanen (extreme right) in a part of a group photograph of the Mount Wilson Observatory staff, apparently on the occasion of a visit by H.A. Lorentz (second from the left). The picture is dated 'circa 1923'. Fifth from the left is Adams (From the Huntington Digital Library [24])

ette and by the time he left the two were engaged. A picture of Hertzsprung as a young man has been reproduced in Fig. 13.21.

Van Rhijn was still a student working on his PhD thesis, which he completed in 1915. There is a letter from Hale to Kapteyn (February 18, 1913) that after the receipt of a telegram (from Kapteyn?), van Rhijn had been informed, while on Mount Wilson, that his father had died. Van Rhijn was devastated and Hale advised him to take a short vacation in the Alleghenies. As a matter of fact, many young Dutch astronomers have followed in the footsteps of their predecessors in later years, spending some time abroad (usually, but not necessarily in the USA) at an early stage of their careers. Sometimes they remained abroad or went back on a regular basis. It remained and still is a (sometimes even formal) requirement for a tenure track position in astronomy at universities in the Netherlands to have had a foreign postdoctoral position first. Essentially all professional astronomers currently working in the Netherlands have held postdoctoral fellowships abroad after obtaining their PhDs. For example, after graduating from Leiden University, I was a Carnegie Fellow at the Mount Wilson and Palomar Observatories before I came to Groningen to join the institution where Kapteyn started the tradition. In his chapter *Growing*

Fig. 13.21 Ejnar Hertzsprung as a young man (Aarhus University [25])

astronomers for export: Dutch astronomers in the United States before World War II in the *Legacy*, Klaas van Berkel has documented the visits that occurred up to 1940.

Hale was certainly happy with the persons Kapteyn recommended, although not always from the very start. On December 3, 1912, he wrote: 'I am very much better pleased with van Rhijn than I was at first and feel that the main difficulty lay in the problem of getting acquainted with him. He has made excellent preparations for his work with the 60-inch, and he will commence observations in a few days. I think we may expect him to obtain valuable results and prove to be a very good man. Van Maanen is as satisfactory as ever – a great addition to our staff.'

13.11 The Structure of the Universe

On January 11, 1913, Kapteyn – through Hale – received an invitation to deliver an important address to the US National Academy on *The Structure of the Universe*. The date would be April 15 of that year and the fee a generous 500$. Kapteyn accepted in a letter of January 25, 1913, to Hale, albeit somewhat reluctantly, as one can appreciate from the tone, as if he felt he could not decline in all decency. By that time he had also been notified that he would be awarded the Catherine Wolfe Bruce Gold Medal, [26] awarded for lifetime contributions to astronomy by the Astronomical Society of the Pacific.

'The letter which informed me about the Award of the Bruce Medal contained the names of the former medalists. I think I quite feel what it means to have your name added to such a list. If I were not I, but another, would I have voted for such an addition? Let me not try to answer the question, but let me say, as I feel strongly that I know how much I owe to the good –too good– [here the letter contains a footnote: 'No, no, E.K.', obviously added by Mrs Kapteyn, who must have read the letter before it was sent] opinion of such men as yourself, a good opinion shown again by asking me to address the National Academy. I hope my cablegram has duly reached you. I cannot deny that I had a time of hesitation in accepting. I never had so many and so troublesome things to settle as since my last return from America.'

The troubles were with his work on 'Helium stars', which will be discussed later. He was reluctant to 'stop the work again to prepare an address'. However, he had to speak for only half an hour and the preparation should not take too much time. But he would have to travel to and from America especially for this address. He stated that unfortunately he could not combine this with his visit of that year to Mount Wilson, as he would have to return to the Netherlands immediately because of his daughter's marriage. This was the marriage of Henriette Kapteyn and Ejnar Hertzsprung, which took place on May 16, 1913. He wrote that the last date he could leave from New York by boat was on April 22. To make it even more inconvenient for Kapteyn, the Academy changed the date of the meeting to April 22 (letter of Hale of February 28). Whatever was arranged, Kapteyn made it for the ceremony, as appears from the presence of his signature on the wedding certificate in Fig. B.4.

The lecture was published in 'Science', Kapteyn (1913a), and reprinted in an extended form in the Italian journal 'Scientia: Rivista di Scienza' (Review of Science; Kapteyn 1913b), and the same version also by the Royal Astronomical Society of Canada, Kapteyn (1914a). The Scientia publication is in English and contains a general introduction on measurements of stellar distances for a wide audience. In the 'Science' version, Kapteyn explained that he could have chosen to describe what was known about the structure of the Universe in those days, but instead decided to talk about 'the history and evolution of the system'. The article is therefore rather speculative. There were some general issues that he took as relatively certain facts. The first was the notion of the Star Streams; he reported that these were directed about 100° from each other. However, this was relative to the Earth, and they were in effect opposite and had, when corrected for the motion of the Sun, vertices that were 180° apart. Moreover these vertices were located very close to the Milky Way.

Secondly, stellar spectra can be arranged in a sequence from the 'Helium stars' (now 'B-stars') followed by respectively the Secchi types one, two and three. Now recall (see page 288) that this sequence is now understood as the Main Sequence of stars, ranging from hot, massive and short-lived O-stars to cool, light, long-lived M-stars. Kapteyn interpreted this, as did many at that time, as a sequence of age, preferring to assume that the Helium stars were the youngest. Thirdly, he inferred from observations that one Stream (the one he had called Stream II) contained older stars, while the other Stream is almost devoid of the young helium stars. Also the random motions of older stars within the Streams were larger than those of young ones. Indeed, in groups (clusters) like the Hyades, Pleiades, etc. there were stars of the three Secchi types, but no helium stars, and they showed the usual change of spectrum with brightness. So, the sequence in the Hyades starts with 'the second stages of a star's life'. Kapteyn even inferred (correctly) that the Pleiades were somewhat younger than the Hyades.

This difference between the two streams (and the fact that they contained unequal numbers of stars) was very strong evidence that the two-stream interpretation should be preferred to the Schwarzschild interpretation of a local elliptical velocity distribution, as described on page 358. We now know that the latter interpretation is correct, but Kapteyn assumed that the phenomenon indicated a streaming of two distinct star clouds that moved through one another. Kapteyn regarded the matter that made up the nebulae as the source of material for the formation of stars, and therefore he had to assume that the near absence of helium stars in the second Stream was due to the fact that 'since some time nebulous matter must have been exhausted in this cloud', whereas in the first Stream it 'must not yet have been exhausted, or if so, only at a very recent period'. The two clouds, 'owing to their initial velocity, have come to meet and intermingle in space'. The increase of random motions of the stars was ascribed to their mutual attraction.

There is one question that springs to mind immediately, viz., to which of the two Streams our Sun belongs. Kapteyn did not enter into this. In other places he sometimes made the statement that the 'two Streams cross and our Sun is situated at a point in space where the two Streams are mixed to the extent that wherever one looks one sees everywhere stars belonging to both Streams. [...] It is therefore in many cases impossible to decide to which of the two groups a particular star belongs' (Quoted from Kapteyn (1911d). But he always denoted Stream I as the one in which stars moved towards a true vertex at (R.A.$\sim 6^h 30^{min}$, $\delta \sim +10°$) and Stream II (R.A.$\sim 18^h 30^{min}$, $\delta \sim -10°$). As the Sun moved towards its Antapex at (R.A.$\sim 6^h$, $\delta \sim -30°$ it would seem that it was likely that it was part of Stream I.

13.12 Pickering's Harvard Northern Durchmusterung

In 1912, Pickering of Harvard proposed to undertake as a project, separate from but in support of the Selected Areas, a new photographic Durchmusterung of the northern hemisphere, using the Harvard telescopes.. This catalog would contain all stars in the northern sky from declination 20° to the North Pole, down to about 14th magnitude. Pickering suggested that Kapteyn and his laboratory would take part in this. But Hale was not very happy with this idea. On December 3, 1912, he wrote:

> 'Both Adams and I are a little inclined to regret that you have committed yourself to Pickering's plan, as it seems inevitable that it will make a heavy tax upon your energies, apparently without sufficient return. The value of the work is of course obvious enough, but a man of your ability ought not to be compelled to devote time and attention to such a piece of routine. The

more opportunity you have for thought on the larger phases of astronomical work, the more will astronomy benefit through the extra-ordinary range of your imaginative power. Routine work may not do you any harm, but it will certainly prevent you from dwelling on the larger theoretical aspects of the subject, and insofar as it does this it will handicap you.'

Interestingly, when Henriette Hertzsprung-Kapteyn recited these sentences she added a second quotation: 'It seems such a pity that you should be merely collecting material for somebody else, incalculably less able than yourself.' That must have been in another letter. Kapteyn replied extensively, explaining that there were too few stars between magnitude 6.5 and 8.5 in the Selected Areas (he estimated one to one-and-a-half per area) and measurements of color index (as a substitute for spectral type) would be extremely helpful to study the distribution of stars in space. After all, a full understanding requires the determination of this distribution for spectral classes separately. '... with external data for colour index, say for stars down to 11th magnitude we can at once make a very important step forward in the discussion of the problem of the arrangement of stars'. Kapteyn wrote that the letters from Hale and Adams (who apparently also wrote to Kapteyn) 'have strengthened my resolve not to commit myself to more than somewhat over one <u>fourth</u> of the whole work. The whole thing in my mind stands about thus. Here is an enormous piece of work for which Pickering will pay.' He planned to educate an assistant who would eventually do almost all the work. The additional argument that Kapteyn brought forward came at the end of the letter. 'I have hope that after the completion of the whole work, I will be able to persuade our Government to extend our Laboratory permanently by appointing all the men who made the observations as regular employés.' Kapteyn feels he will be too old for that but 'I have the future success of the Laboratory much at heart.'

> 'There is a sort of fate which makes me do all my life long just what I want to do least of all. – The making of the Durchmusterung has no attraction whatever for me, but in what I have tried to do in the direction that has a true attraction for me, I have always been hindered by want of suitable material. So if – after I hope not too small a number of years – I come to die, I will probably leave behind me more Durchmusterung work and bringing together of material than almost anybody – leaving it to the next generation to do the real work that I hoped and longed to do. Well, when you don't have what you love, you have to love what you have.'[5]

In the end, the Durchmusterung never came about. In the *HHK biography* we read:

> 'The collaboration proved to be very difficult. After the first few weeks which had to be spent on organizing the project, Kapteyn did not have to spend much time on it, since it could be done by his assistants, but Pickering was not like Gill or Hale, and after a year he regretted that he had agreed to the undertaking. The problems, resulting from differences of opinion in 'matters astronomical', became after a while so great that they had to abandon the project before it led to any conclusion.'

---

[5] *'Quand on n'a pas ce qu'on aime, il faut aimer ce qu'on a'*. This has been attributed to Roger de Rabertin, Comte de Bussy (1618–1693).

13.13 End of the Mount Wilson Visits

World War I made an abrupt end to the annual visits. The year 1914 had started sadly. The following quotation is from the *HHK biography*:

'In the beginning of 1914 Kapteyn was hit with a severe blow. After a short, not very severe illness Gill passed away. Deeply saddened he and his wife went to London to see their friend for the last time. They found Lady Gill in deep mourning. She was lying on her bed, completely exhausted, waiting for Kapteyn as the most trusted friend of her husband. He knelt next to her bed, and solemnly putting her hand on his head, she blessed him and thanked him for all he had been for her husband. All the love and gratitude, all the grief and mourning of a great human heart were expressed in this impressive gesture.

Gill was buried in his place of birth, Aberdeen, where he was taken by his wife and a few friends. In London at the same time a memorial service was held at St. Mary Abbot Kensington, where the Kapteyns and many sad friends attended. On the organ the beautiful song of Tennyson, 'Crossing the Bar' was played, and a pure boy's voice as if heavenly music sang the comforting words. Silently they returned to Gill's house, packed their suitcases and left the house on 34 de Vere Gardens, this hospitable house that they would never return to, as Lady Gill would leave it as well. They returned home poorer. And every Sunday morning Mrs Kapteyn would play this sad death song, of which we heard the sounds rustling through our home in the quiet morning. We were quiet for a while and remembered this trusted friend.'

Alfred Tennyson, 1st Baron Tennyson, FRS (1809–1892) was a well-known and very popular British poet. He was poet laureate (the official poet of the King or Queen) during much of Queen Victoria's reign.

Kapteyn (1914c) wrote an extensive obituary about Gill in the Astrophysical Journal.

'On January 24 died in London, at the age of seventy years, Sir David Gill, formerly Her Majesty's astronomer at the Observatory at the Cape. In him science loses the foremost practical astronomer of the age. [...]

Only a few words must be said about the *Astrographic Chart and Catalogue*. The initiative for this great undertaking is due to the joint action of Gill and Admiral Mouchez, the director of the Paris Observatory, aided by the brothers Henry. What the whole undertaking, not only at starting, but during the whole of its progress, owes to Gill's untiring energy, all will know who attended the meetings of the Comité Permanent. Up to the last, his was the great driving force. [...]

No man could be long with him without feeling that here was a man to whom the real interest of astronomy was paramount, a man who was always ready to sacrifice any pet plan of his own to the real interest of astronomy. A favourite expression of his, in giving up his opinion, would be: 'The man who never makes a mistake is he who does nothing.' I cannot help thinking

Fig. 13.22 The Kapteyns in Groningen on an undated photograph from an album in the Kapteyn family (Courtesy of Jan Willem Noordenbos and other Kapteyn descendants)

that such personal qualities –his indomitable energy, his broad-mindedness, love of his work, kindness –his manliness in the best sense of the word, in short the charm in his strong personality, had almost as much to do with his achievements as his qualities as a scientist. [...]

As a scientist Gill is best compared in my opinion to F.G.W. Struve. [...] And might not the following words be applied to Gill's *History of the Cape Observatory*: 'There is inspiration to be found in nearly every page of it, for its author had the true genius and spirit of the practical astronomer –the love of refined and precise methods of observation and the inventive and engineering capability'. As a matter of fact they were written *by* Gill *about* Struve. Even in the details of their careers there is the greatest parallelism.

In the annals of astronomy, Gill's name will take place with those of Bradley, Bessel and Struve. In many a human heart his image will last as long as life itself.'

Kapteyn did not dwell on the Cape Photographic Durcmusterung, only listing it along with other major projects Gill was involved in.

The return from Mount Wilson by the Kapteyns in 1914 was not without difficulty. I continue with Henriette Hertzsprung-Kapteyn's biography.

'At the outbreak of the war Kapteyn and his wife were at Mount Wilson. In one instant the world had changed completely. Everybody lived in fear and worry about distant friends and family, and who was able to return home did so as quickly as possible. For the Kapteyns the voyage across the ocean was impossible, however, due to the threat of mines; they stayed until January and then came home safely, expected with much trepidation.

13.13 End of the Mount Wilson Visits 499

Fig. 13.23 Kapteyn in 1908 in the 'Monastery' (the residence of astronomers during their observing sessions) at Mount Wilson Observatory (Kapteyn Astronomical Institute)

It was their last journey across the ocean; they would never return to America.'

Kapteyn described the journey back to the Netherlands in a letter to Seares of January 23, 1915.

'Our journey has been without any difficulty. We were stopped by a man of war (English) when we were no more than a few hours out of New York. A couple of officers came on board, but they only examined the ship's papers. After that we came unmolested to Portsmouth, where, after some

hours waiting, we got a pilot, who had to bring us safely from here up to near Harwich. We wanted him less on account of the mines, the situation of which (as far as they had not broken loose) was well known to our captain, but on account of the lighthouses, which partly were kept dark, partly had been displaced on purpose.
In Dover harbour we spent a night. The ship's papers had been taken to the shore, but even if they had not, our captain did not care to go by night. The permit to leave reached us somewhat late the next morning, so that we could not land at Rotterdam that same night, but had to lie 'on stream' off Hook of Holland. Altogether we lost about 2 days, but that was all.
Mrs Kohlschütter had no trouble at all up to Rotterdam. We have not yet any news of her and cannot say therefore how she fared at the border.'

Apparently Kohlschütter's wife had been traveling with the Kapteyns. Kohlschütter himself had left the USA 'shortly after the outbreak of the war' (according to H. Schmidt in the obituary *Arnold Kohlschütter 6. 7. 1883–28. 5. 1969* [27]) and had attempted to reach Germany via Gibraltar. There he was arrested by the British and kept there as a prisoner of war for the rest of WWI. He was later transported to England, where Eddington gave him opportunities to do some scientific work. After the war he returned to Germany.

Kapteyn's visits to Mount Wilson had come to an end, but his association continued. His presence, especially in the early stages of the Observatory and the operation of the 60-inch telescope, had been very beneficial on both sides. In his 1923 obituary in the German 'Vierteljahrsschrift der Astronomischen Gesellschaft', de Sitter (1923) quoted from a letter sent to him by George Hale 'a few years ago' (so before 1922, the year Paul incorrectly mentioned in his *Statistical Astronomy*). This letter is not to be found in the de Sitter papers in Leiden.

> 'Nothing has been so valuable to the Mount Wilson Observatory as the inspiration and guidance of Kapteyn. His splendid imagination and fine optimism have stimulated us to our best efforts and encouraged us to attack problems of wide scope.'

Chapter 14
Tides, Statistics and the Art of Discovery

> *I couldn't claim that I was smarter than sixty-five other guys*
> *– but the average of sixty-five other guys, certainly!*
> Richard P. Feynman[1]

> *There is a tradition of opposition between adherents of*
> *induction and of deduction. In my view it would be just as sensible*
> *for the two ends of a worm to quarrel.*
> Alfred North Whitehead[2]

14.1 Origin of the Tides

At different points of time Kapteyn was involved in scientific issues outside astronomy. We already encountered his work on growth-rings in trees (see page 122 and further), which he finally published after having presented it during one of his visits to Pasadena and Mount Wilson – first in the local newspaper, the Pasadena Star, and later privately. Another issue that was only indirectly related to astronomy and in which he apparently had some interest, was the origin of the tides on Earth. The work that he published, Kapteyn 1909b, was based on a lecture he had held before the 'Mijnbouwkundige Vereeniging', the Association for Mining Engineering, in Delft. This must undoubtedly have had links with what is now the Technical

---
[1]Richard Phillips Feynman (1918–1988) was a physicist and Nobel laureate. The quote is from *Surely you're joking, Mr Feynman!*.
[2]Alfred North Whitehead (1861–1947) was a British mathematician and philosopher [1].

University of Delft, where mining engineering was being lectured. Although his son Gerrit Jacobus studied mining engineering (in Freiburg), Kapteyn had no known affiliation with Delft. So, how this lecture came about is not clear. The published paper, which was a contribution to the Annual Report of the Association, consists of a set of notes by one 'F.T. Mesdag'. This must have been Ferdinand Taco Mesdag (1886–1961), who was indeed a mining engineer. The paper itself, which summarizes Kapteyn's lecture, had been annotated with footnotes, presumably constituting Mesdag's comments. There was no introduction to what Kapteyn had set out to tell his audience, nor a set of conclusions.

Kapteyn's presentation addressed the response of the oceans for the ideal case that the Earth was completely covered by water and the tides only arose from the gravitational attraction of the Moon. Present insight is that on the Moon's side the attraction is greater at the surface of the ocean than it is at its bottom (and the surface of the solid Earth) and greater on the opposite side at the bottom of the ocean than at the surface. This results in a bulge in the surface of the ocean where it faces the Moon and at the antipode. This then gives rise to a periodic high and low tide at any position on Earth, with two high and two low tides in the time the Moon takes to arrive above the same place on Earth. This is the rotational period of the Earth as seen from the Moon and amounts to 24 hours and about 50 minutes. So the tides should return (on average) every 12 hours 25 minutes.

Kapteyn presented a completely different approach. The published account had no mathematics in it but relied heavily on a set of figures that I have reproduced in Fig. 14.1. He started with the simple case that the Moon's orbit would be in the plane of the Earth's equator (so that it would always be directly above the equator). He then looked at the gravitational attraction of the Moon at a point on the equator. This force had a component in the direction parallel to the line Earth-Moon, but according to Kapteyn this force did not result in any displacement of the water relative to the Earth since the force of the Moon on a small volume of water was the same as the attraction on a piece of the Earth of the same mass (or weight). This force could be decomposed in a force perpendicular to the surface and a component parallel to it (all of this is illustrated in the top-left drawing in Fig. 14.1). He then argued that the component parallel to the surface was zero at the point below the Moon (which would from there be seen in the zenith), but maximal at positions on the equator 90° away from this (where the Moon was seen on the horizon). This then sped up the rotation in one of these spots and diminished it at the others.

The right-hand part of Fig. 14.1 served to illustrate this. In the orientation used here the lower one (designated Fig. 14.2) showed the forces as one goes around the equator. The Moon was thought to be beyond the top of the page. The upper one showed the resulting changes in the rotation of the oceans. At the bottom and top the rotation sped up, at the left and right it slowed down. So, Kapteyn said, water was moving away from the spots of highest rotation and collecting in those with slower rotation. His first conclusion was: 'Under the Moon we have low tide and 90° away from it high tide'.

In a footnote Mesdag wrote: 'This conclusion is different by 90° from what is the most widely adopted opinion.' And indeed, Kapteyn was dead wrong! He

14.1 Origin of the Tides

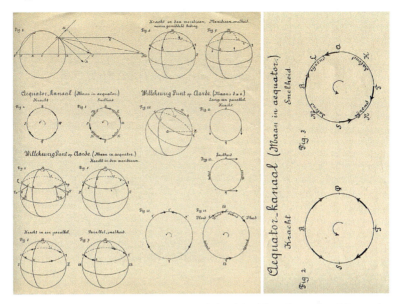

Fig. 14.1 Figures from Kapteyn (1909b) on the origin of the tides. Figures 2 and 3 are enlarged at the right (and rotated) (Kapteyn Astronomical Institute)

seemed to have completely missed the point that what counts was *differences* in attraction in the direction towards the Moon, which he dismissed as 'not producing any movement of the water'. Furthermore, if it were the direct gravitational forces that produced the tides by causing streaming motions in the oceans, the influence of the Moon would be completely overwhelmed by that of the Sun and there would be no periodicity corresponding to the orbital period of the Moon. The gravitational force of the Sun on a piece of matter on Earth is almost 200 times *larger* than that of the Moon on the same piece of matter. If we look at *differences* in forces, the situation is different. Newtonian gravity on a piece of matter is proportional to the attracting mass and inversely to the square of the distance. But the tidal force is differential and therefore inversely proportional to the *cube* of the distance. In contrast to the gravitational force itself, the tidal force of the Sun is about 45% of that of the Moon, so the Moon is the dominant and primary cause of the tides, although the Solar influence is non-negligible. I should point out that the attraction on the Earth on water at the surface of the oceans is of course (and fortunately) even greater than that of the Moon and the Sun; the Earth's gravity at its surface is more than 1,650 times larger than that of the Sun (and more than 300,000 times that of the Moon). The origin of the tides is often described as 'lunar-solar'; indeed, when the relative positions of the Sun and Moon are such that their effects add up, we have a stronger high tide – which we call spring tide – and when they are opposite the other way around (neap tide). It is the more surprising that Kapteyn came to this explanation, as the origin of tides had been discussed long before him by Isaac Newton, Laplace (Pierre-Simon, Marquis de Laplace; 1749–1827 and others.

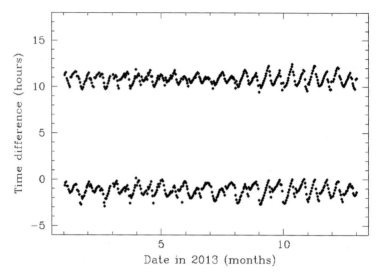

Fig. 14.2 The difference in passage of the Moon through the meridian and the times of high tide in Scheveningen (predictions for 2013) (Using the Website of Rijkswaterstaat; Ministry of Infrastructure and Environment [2])

One might also look at what nature shows in practice. After all, in principle it would be easy to decide what is the correct result by looking what happens to the tides when the Moon crosses the meridian. But unfortunately, this does not help since in reality the effects of tides are extremely complicated due to the presence of continents, etc. The tidal wave originates primarily in the Pacific Ocean, and travels around Cape of Good Hope and South America into the Atlantic Ocean. After a considerable delay (about two days) it reaches the North Sea and the Dutch coast from one of two directions, either directly through the Strait of Dover (Pas de Calais) or all the way around the British Isles. Any correlation between Lunar passage of the meridian and the times of high tide is therefore completely lost. To illustrate this, I have plotted this difference in Fig. 14.2 for Scheveningen near The Hague, which is close to Delft on the Dutch coast, for the year 2013. We see then that there is about a 2 hour difference between the high tide and the time the Moon crosses the meridian (and about 10 hours for the second high tide). This varies considerably moving along the Dutch coast. At Vlissingen (just north of Belgium) it is somewhat less than 1 hour, while it amounts to a little over 6 hours at Den Helder, where the Dutch coast starts to turn towards the east, and almost $10\frac{1}{2}$ hours at Eemshaven just west of the German border.

In his Delft lecture Kapteyn next discussed the more complicated cases when the Lunar orbit does not coincide with that of the equator and of tides at other geographical latitudes. It seems hardly opportune to go more deeply into this somewhat unfortunate episode.

Box 14.1 The log-normal probability distribution function
This is a skew distribution that is related to the usual 'normal' distribution in that it really is normal distribution plotted with a logarithmic horizontal scale. The equation for the normal probability function is

$$P(x) = \frac{1}{\sigma\sqrt{2\pi}} \exp\left\{-\frac{1}{2\sigma^2}[x-M]^2\right\}.$$

Here σ is called the standard deviation. The log-normal distribution is

$$P'(x) = \frac{1}{x\sigma\sqrt{2\pi}} \exp\left\{-\frac{1}{2\sigma^2}[\ln(x)-M]^2\right\}.$$

An example of an log-normal curve has been plotted in Fig. 14.3-left for $M = 100$ and $\sigma = 2$; on the right is the same distribution plotted using a logarithmic x-axis (now the parameters are $M = 4.61$ $\sigma = 0.693$). Note that the horizontal scale on the right has the common (base 10) logarithm, not the natural one. The colors change at multiples of σ (median to 1σ 34.1% of the total, 1σ to 2σ 13.6%, etc.).
In the normal distribution the mode (the maximum in the curve) and the median are the same and by definition the skewness (the third moment divided by the second moment to the power 3/2) is zero. For the log-normal distribution, the mode is at $10^{M-\sigma^2}$ and the median at 10^M and the skewness is $\sqrt{10^{\sigma^2}-1}(2+10^{\sigma^2})$. The family of formulae Kapteyn proposed for the general skew probability density function is

$$P_{JCK}(x) = \frac{1}{F'(x)\sigma\sqrt{2\pi}} \exp\left\{-\frac{1}{2\sigma^2}[F(x)-M]^2\right\},$$

where $F'(x)$ is the derivative $dF(x)/dx$. Kapteyn particularly studied the case that $F(x) = (x+\kappa)^q$. The special case $q = -1$, which he uses extensively, is the log-normal distribution, as can be seen from the fact that then $F' = (x+\kappa)^{-1}$.

14.2 Skew Frequency Distributions

Kapteyn spent much time on theoretical issues concerning the use of statistics. Before we look at Kapteyn's work in this area, I will briefly explain what a skew distribution function and in particular the log-normal distribution are. More details are in Box 14.1. Many distributions in nature, such as the lengths of people, follow the 'normal' distribution that we see on the right in Fig. 14.3. But there are also distributions that cannot be fit so well, because they are not symmetric but skew. For these the log-normal distribution is well suited. Figure 14.3 shows the distribution that needs to be used then on the left. This distribution becomes the normal distribution when the logarithm of the parameter that is being studied is plotted on the

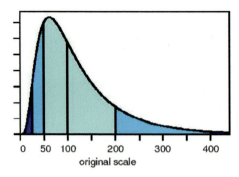 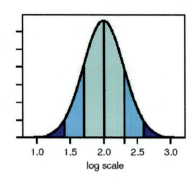

Fig. 14.3 On the left a log-normal distribution. On the right the same distribution but plotted on a logarithmic scale. The colors change at integer standard deviations σ (Adapted from Limpert et al. (2001))

horizontal scale (on the right in the Figure). The book *The lognormal distribution* (1996) by Aitchison and Brown [6] includes a thorough discussion of Kapteyn's work. A more technical presentation of the log-normal distribution is the paper *Lognormal distributions across the sciences: Keys and clues*, Limpert et al. (2001) [5]. This paper points out that the work of Kapteyn on skew distributions has been very important; it concludes the acknowledgment section with the statement: 'Because of his fundamental and comprehensive contribution to the understanding of skewed distributions close to 100 years ago, our paper is dedicated to the Dutch astronomer Jacobus Cornelius Kapteyn.'

So, how had Kapteyn become interested in statistics and skew distribution in particular ? It was the result of a request by 'by several biology students and by some other persons interested in statistical methods of Quetelet, Galton and Pearson' [...], to deliver a few lectures in which these methods would be explained in a popular way', Kapteyn (1903b). The course of events in this subject has been studied and discussed in detail in studies by Ida H. Stamhuis from the Free University in Amsterdam, who was trained as a mathematician and statistician in Groningen before she started working on the history of statistics. In what follows, I rely heavily on the authoritative papers by her and her collaborators (See Ida H. Stamhuis and Eugene Seneta: *Pearson's statistics in the Netherlands and the Astronomer Kapteyn* (2009) [3] and references therein).

Lambert Adolphe Jacques Quetelet (1796–1874), a Belgian mathematician, had been interested in statistical application in astronomy (he was the founder and one-time director of the Royal Observatory of Belgium in Brussels). He promoted the use of statistics by stimulating international collaborations and introduced statistics and the 'normal distribution' in social sciences. Sir Francis Galton (1822–1911) had introduced the concept of statistical correlation and had developed statistical methods to study heredity of human qualities. Karl Pearson (1857–1936) was a British mathematician and statistician, who is being credited with the establishment of 'mathematical statistics'. These scientists, together with Walter Frank Raphael Weldon (1860–1906), who had had applied statistical methods in zoology and stud-

14.2 Skew Frequency Distributions

Fig. 14.4 Jan Willem Moll (1851–1933) and Jantina Tammes (1871–1947) were biologists in Groningen. Moll was a professor of botany and Tammes of botany and genetics (University Museum Groningen [4])

ies of evolutionary biology, had made a great impression on plant physiologist and geneticist Hugo de Vries (1848–1935) from the University of Amsterdam (de Vries was the other Dutchman attending the congress in St. Louis in 1904). De Vries had started to use statistics in his studies in botany.

Jan Willem Moll (1851–1933) had studied botany in Amsterdam and was very good friends with de Vries. In 1890 Moll had been appointed professor of botany in Groningen and de Vries must have influenced him to a high degree when it came to applying statistics in his work. Jantina ('Tine') Tammes (1871–1947) had entered Groningen University in 1890 as one of the few female students and had obtained first-level teaching qualifications in a number of disciplines, ranging from physics and chemistry to biology. She was appointed Moll's assistant in 1897. She was awarded an honorary doctorate in 1911 and in 1919 became the first female professor of genetics. She had become interested in using statistical methods and probability theory. It seems quite likely that Moll and Tammes (see Fig. 14.4) were the scientists Kapteyn referred to, or were among them, and who asked him to lecture on the subject. Statistics based on the normal (binomial or Gaussian) distribution and the technique of least squares (see page 322) had been in extensive use, especially in astronomy, for the reduction of measurements of stellar positions and other studies. This must have been one reason why Kapteyn was approached. The other problem at the time was that often distributions in nature were not conforming to the symmetrical normal distribution but were skew.

Fig. 14.5 The Skew Curve Machine, produced by Jan Willem Moll after design by Kapteyn. It used to be kept at the biology department of the University of Groningen, but is now lost (From Kapteyn & van Uven, 1916)

In Kapteyn (1903b), the skew distribution was introduced by taking the example of berries picked from a tree, having sizes distributed symmetrically around some mean value. Kapteyn concluded that in a realistic case in nature, distributions had a large number of causes which usually resulted in the symmetric distribution corresponding to the 'normal' (Gaussian) bell-shape. The distribution function in the case of the size of berries would also be close to that. But maybe the distribution of their *volumes* rather than their diameter (after all corresponding to weight) was more relevant. This is proportional to the cube of the diameter and the resulting distribution function is no longer symmetric. The diameters result from a number of possible causes, such as whether the position in the tree was more or less in the shade, how easily nutrients could reach it, etc. If a large number of causes produced the normal distribution of diameters, the distribution of the volumes would therefore no longer in general be symmetric. Now, Pearson had published extensive studies of skew distribution in a widely known paper in 1895, *Contributions to the mathematical theory of evolution. II. Skew variation in homogeneous material* [8], but had regarded skew distributions as originating in a process in which the probability to arrive at a certain

14.2 Skew Frequency Distributions

Fig. 14.6 Another machine, this time producing the 'normal' probability distribution curve, possibly also designed by Kapteyn. This is still kept in the Groningen biology department (University Museum Groningen)

result was not symmetrical with respect to some 'central' value. That would result in a form of the distribution function that would be unrelated to the causes and nothing about these causes could be deduced from the form of that distribution. Kapteyn disagreed and felt that as a matter of principle the distribution function should in one way or the other be related to the causes and therefore contain at least some information on these. So, in preparation for his lectures for non-mathematically schooled botanists, he developed a different theory of skew distribution functions, that I have summarized in Box 14.1.

Kapteyn also presented a picture of an instrument, (the 'Skew Curve Machine'), that had been built after the example of a rather similar machine designed by Francis Galton, which was generally known as the 'quintux'. The machine Kapteyn showed (Fig. 14.5) had been designed by himself and built by Moll. It was constructed such that when sand was poured into it at the top, this collected at the bottom in a form that illustrated the skew, log-normal distribution. This apparatus has gone missing in the University of Groningen. In the *Legacy* (page 78). it was noted that 'The [then] curator of Groningen University Museum, F.R.H. Smit, could not find this apparatus in a recent inventory of the University's buildings, but did locate the copied machine [in Fig. 14.6]. It [was] in the Genetics Department [...]. It was probably also built for Kapteyn with a similar purpose as the log-normal machine, but uses small balls instead of sand. It illustrates the normal bi-modal distribution rather than the log-normal one. It dates from 1901 and measures 30 × 40 × 3 cm. The machine works by re-distributing the balls at each point into two equal parts.'

The text on the apparatus at the top-left says: 'Mechanical illustration of the cause of the probability curve. According to Galton. National Inheritance. Constructed by W.A. Weening 1903.' *National Inheritance* (1889) was a book by Galton, in which he summarized his statistical work on correlations and regressions. Galton was involved in human statistics, but also (he was a cousin of Charles Darwin), highly interested in heredity and supported the ideas of eugenics, the improvement of the pool of genetic material in a population. In order to keep the genetic pool of as high a quality as possible, he promoted marriage between higher class citizens and encouraged having children at a relatively young age. This idea was reflected in the title of his book, although it was to a large extent a treatise on correlation and statistics.

Kapteyn's paper of 1903 is 43 pages long plus another 7 with figures and tables. The one in Kapteyn's bound volumes of his publications (see Fig. A.2) is full of handwritten notes, corrections and remarks, probably related to his further publications on the subject that will be discussed below. After a first non-mathematical description and an involved exposition of the mathematical theory he applied his skew functions by fitting to three data sets. The first was a set of observations of thresholds of sensations by persons, conducted by his friend Gerardus Heymans, professor of philosophy and psychology. The second data set was taken from a publication of Pearson on prices of properties (homes) in Wales and England, and the third on the sizes of foreheads in Naples crabs, which was published by Weldon. Although he initially used his full family of curves, he eventually refitted the data with a special case which he identified and which we now know as the log-normal distribution.

Pearson published a very sharp reaction in 1905 with the title *Das Fehlergesetz und seine Verallgemeinerungen durch Fechner und Pearson. A rejoinder* [9] (the law of errors and its generalizations by Fechner and Pearson). Gustav Theodor Fechner (1801–1887) had developed and introduced statistical methods invoking probabilities in psychology. In this long article Pearson addressed Kapteyn's work on just over 5 pages, and the tone was rather rude and harsh. He started out by stating that much of what Kapteyn said had already been published by Edgeworth and blamed Kapteyn for not acknowledging that. Indeed, the work of Francis Ysidro Edgeworth (1845–1926), especially in a paper *On the representation of statistics by mathematical formulae* in 1898 [10] is relevant. Likewise, Pearson objected to Kapteyn's use of the log-normal and his perceived impossibility to calculate concepts like means, standard deviations and the like. Finally Pearson objected to Kapteyn's principle to relate the form of distribution functions to causes.

This critique was unnecessarily harsh and Kapteyn felt he had to respond. He did so in the Dutch journal 'Recueil des Travaux Botaniques Néerlandais', Kapteyn (1905b). It was relatively short with only just over 6 pages. He started by frankly

14.2 Skew Frequency Distributions

admitting his neglect to acknowledge Edgeworth: 'I must plead guilty *in part* and offer Prof Edgeworth my apologies. I confess to have overlooked his papers. I may perhaps adduce as an attenuating circumstance that these papers have been also overlooked in the bibliographies of both Prof. Ludwig and of Davenport, the only bibliographies on the subject with which I am acquainted.' It was not so nice to put part of the blame on others, though.

His main point was that he felt that 'If I still feel justified in claiming my part in the ownership of this formula, it is on the ground that Prof. Edgeworth's remark, correct though it be, is still not equivalent to a general theory'. Further Kapteyn attacks Pearson by asserting that '(I) that Prof. Pearson actually adopts my theory (which he refutes) as the only rigorous and general one; (II) that Person's formulae, even now that he has tried to derive them from *our* equation (1) may, at the very best, be accepted as *empirical* representations.' So, Kapteyn accused Pearson of stealing his general formula and ridiculed to some extent the latter's work by dismissing it as empirical. Indeed, Stamhuis & Seneta (2009) state 'But, on the whole, Kapteyn (1906) is not a carefully considered response to Pearson (1905).'

Pearson took this up and wrote in 1906 an agitated response, *A rejoinder to Professor Kapteyn* [11]. He mainly addressed the issue of priority of the general theory and the curves proposed by Kapteyn, claiming that the formula concerned 'has been habitually discussed in my lectures on statistics for at least five or six years, if not longer!'

On the whole, Kapteyn's claim to priority seems overstated. He may have worked independently of Edgeworth and have improved significantly on earlier work by Pearson, but at least some of the priority should go to these persons as well. Kapteyn was rather sensitive to being refused priority and credit that he felt he was at least partly entitled to. It reminds one of the incident related to the discovery of Kapteyn's Star, which I discussed at length above (see page 238). This somewhat resentful touchiness was evidently one of Kapteyn's less admirable character traits. The sensitivity, it seems, was deeply rooted. Some insight comes from the following episode in Henriette Hertzsprung-Kapteyn's biography. She refers to the American physician and explorer Frederick Albert Cook (1895–1940), whose claims to have reached the North Pole were challenged by Robert Edwin Peary (1855–1920), who said that he and not Cook was the first person to reach the North Pole, in 1909.

'It was typical for Kapteyn to have a strong sense of honesty. When in 1907 Frederick Cook said he had reached the North Pole and when later his scientific statements of how he climbed Mount McKinley proved to be false, many made fun of the gullibility with which the scientific world had accepted Cook's story. Kapteyn opposed that attitude fiercely; he felt that it was only natural to believe a scientist on his word of honour and not to consider the possibility of deception. To him the fraud itself was a cause of deep indignation. After many years he would spark into anger about a physicist, who had proposed a theory, but when Kapteyn recommended him to read certain books, replied: 'But you cannot expect me to collect data disproving my own theory!' The same would happen with a well-known

English biologist, whose articles had been found to be fraudulent, and Kapteyn said; 'This is the only man I hate'. Twice in his life he personally came into contact with men who had been dishonest in science. He was very upset about that, as it was totally impossible for him to comprehend how one may come to such a thing. He never wanted to have anything to do with them, since they had failed the elementary rules of the code of honour and had lost the right to carry the torch of scientific research.'

We cannot be sure who that physicist might be, but – as Stamhuis & Seneta also suggest – it seems quite possible or even likely that the biologist that is referred to here is in fact Karl Pearson.

14.3 Work with van Uven

In 1916 Kapteyn returned to the subject, or at any event he published on it again. The first publication was together with Marie Johan van Uven (1878–1959), who was a mathematician originally working in Utrecht as a grammar school and university teacher, but who became a professor of mathematics in Wageningen in 1913. He was awarded the title professor on a personal basis, even though Wageningen was not in fact a university, since he had had an offer of a professorship at the Polytechnical School in Delft (the current Technical University of Delft) at the same time when he was offered a position at Wageningen (see *In Memoriam Prof. Dr. M. J. van Uven 1878–1959* by N.H. Kuiper [12]). Van Uven was a passionate musician and a well-known collector of musical manuscripts, which is still evident from the van Uven Foundation, which is named after him and owns a large collection of sheet music [13]. Together, Kapteyn and van Uven published an article of 69 pages plus figures, entitled *Skew frequency curves in biology and statistics – second paper*, Kapteyn & van Uven (1916). Both papers expressed Kapteyn's opinion that the use of frequency curves would lead to some understanding of the underlying causes of the observed distributions. Kapteyn said in the introduction that he had earlier resolved to return to the subject only after his formal retirement due to lack of time before that. But van Uven had offered his help and that had prompted Kapteyn to return to the subject at that time.

The joint publication consisted of three separate articles. The first two both had the title *Development of the theory*, the first article having been written by Kapteyn and the second by van Uven. They developed the principle of skew frequency curves with as little algebra as possible. In the third article they discussed some applications. For a technical discussion I once again refer to the paper by Stamhuis & Seneta (2009). The publication is very technical; the purpose is 'to describe the causal origins' [3] of the distributions and I refer to that paper by Stamhuis and Seneta for a detailed and expert discussion.

Kapteyn published another paper on the matter, Kapteyn (1916b), which is really a printed version of a lecture he had presented to the scientific chapter of the Natural Sciences Society of Groningen (see page 276), of which he was

Box 14.2 The correlation coefficient

The correlation (in terms of linear regression) of two variables with n measurements x_i and y_i ($i = 1$ to n) is often described by the **Pearson product-moment correlation coefficient** r. It uses the second moments

$$M_{xx} = \sum_{i=1}^{n}(x_i - \bar{x})^2 \; ; \; M_{yy} = \sum_{i=1}^{n}(y_i - \bar{y})^2 \; ; \; M_{xy} = \sum_{i=1}^{n}(x_i - \bar{x})(y_i - \bar{y}).$$

The full definition is

$$r = \frac{M_{xy}}{\sqrt{M_{xx}M_{yy}}} = \frac{1}{n-1}\sum_{i=1}^{n}\left(\frac{x_i - \bar{x}}{\sigma_x}\right)\left(\frac{x_i - \bar{y}}{\sigma_y}\right),$$

where

$$\bar{x} = \frac{1}{n}\sum_{i=1}^{n}x_i \; ; \; \sigma_x = \sqrt{\frac{1}{n-1}\sum_{i=1}^{n}(x_i - \bar{x})^2}$$

and

$$\bar{y} = \frac{1}{n}\sum_{i=1}^{n}y_i \; ; \; \sigma_y = \sqrt{\frac{1}{n-1}\sum_{i=1}^{n}(y_i - \bar{y})^2}.$$

The absolute value of the correlation coefficient $|r|$ can adopt values between 0 (no correlation) to 1 (total correlation).

chairman. That paper contained very few mathematical equations, but was intended to help biologists apply statistics of skew distributions to their data. Probably the best example of a scientist using Kapteyn's efforts was Jantina Tammes (see *Statistiek en waarschijnlijkheidsrekening in het werk van Tine Tammes (1871–1947)* by I.H. Stamhuis [14]), although her last work in genetics using statistics dates from 1915.

14.4 The Correlation Coefficient

There is another example of a paper on statistical issues by Kapteyn; it concerned the *Definition of the correlation coefficient*, Kapteyn (1912), published in the Monthly Notices of the Royal Astronomical Society, one of the main international astronomical journals. The mathematical definition of the correlation coefficient is summarized in Box 14.2. It had been introduced by Pearson based on work by Galton. Interestingly, Kapteyn did not mention the name Pearson, referring only to the 'usual formula for the practical computation of the correlation coefficient', which he assigned the notation r as usual. This could take values between 0 and 1 for cases between completely uncorrelated ($r = 0$) to perfectly correlated ($r = 1$).

Kapteyn's problem that made him write this paper, was the question how to interpret the value of r. Why this parameter and why not define the correlation coefficient as the square root or the square or any other root or power of r, such as $\sqrt{r}$, r^2, etc. After all, all these could vary between 0 and 1 just as well. Kapteyn presented the case of two variables that were both the sum of four quantities. Assume that one of the quantities in the first variable was perfectly correlated with one in the second one and that all the others are completely uncorrelated. Then it would be reasonable to find that the correlation coefficient is 0.25, since only one quarter of the quantities are correlated. 'But is $r = 0.25$?', Kapteyn asked. In other words, in order to arrive at a interpretable value, does one have to use the 'usual' (Pearson) r or its square, or square root, or etc.? Kapteyn continued: 'On those rare occasions that I was led to peruse some articles on correlation I have asked myself the question, but have been unable to find a definition that would settle my doubt. [...] Finally I requested a friend, well versed in these matters, to look up the literature on the subject. The result has been negative. We may thus, I think, assume that any acceptable definition does not exist, or at least that it has escaped nearly every worker in this branch of science.'

I stated above that Kapteyn did not mention Pearson. Instead he referred to the 'theory of Bravais, which seems to have been taken as the base of the study of correlation'. We have met Auguste Bravais (1811–1863) before in relation to determinations of the Solar Apex from proper motions (see page !292). Kapteyn used the 'beautiful paper' (Kapteyn's words) of Bravais *Analyse mathématique: Sur les probabilités des erreurs de situation d'un point* (the reference has the year 1846, but it apparently had been published earlier in another form in 1844 [15]). Kapteyn explained that this was not 'really concerned with the quality of correlation and for that reason no definition of the coefficient is given. I will simply try, in what follows, to give Bravais' theory the *very small* extension needed for getting at such a definition.'

It would lead too far afield to follow the technical details of Kapteyn's paper. In the end he concluded: 'The quantity r *itself*, therefore, and not such a function as $\sqrt{r}$, r^2, ... must be considered as the natural measure of correlation.' So, in the end, it was Pearson's coefficient that Kapteyn concluded was the one to use, but he did not feel the need to mention Pearson. This may be a streak of some narrow-mindedness in his character, but it should also be remembered that he felt poorly treated by Pearson. In his paper of 1920 *Notes on the history of correlation* [16], Pearson treated the contributions of Galton in an extensive manner (Pearson had written an authoritative biography of Galton, *The Life, letters and labours of Francis Galton* (1914–1930) [17]), and in his turn omitted to mention Kapteyn. However, Kapteyn published in a professional *astronomical* journal, and it is very likely that Pearson had never heard of Kapteyn's 1912 paper.

Kapteyn must have considered his work on statistics and correlation a duty among friends he could not refuse, yet on the other hand he considered it a severe loss of precious time. On November 30, 1903, he wrote to Donner: 'In a few days I hope to send you a short paper on a statistical method, from which you will see how little time I spent on astronomy lately. I have now, thank Heaven, returned

14.4 The Correlation Coefficient

to astronomy. It will take a few months, however, before I have anything to tell you about this.'

After the first paper, Kapteyn (1903b), and the heated rebuttals by both Pearson and himself in 1905 and 1906, Kapteyn apparently considered to publish a paper that would reach the British biologists. He wrote to Gill on November 2, 1906.

'.... Might I ask whether you can help me in the following. For months I have been working on a non-astronomical subject, viz on frequency curves. I would not have done so, were it not that I am strongly convinced that biologists under the guidance of Karl Pearson are on the wrong way with their study of these curves. Now this study is mainly in the hands of English investigators and in order that my investigation might reach them it seems absolutely necessary that it be published in one of the English publications concerned with this matter. Of such I know but two, viz. 1st Biometrika; 2nd The Philosophical Transactions. Would you do me the favour to present it? There is nothing polemical in it. It is merely a study of frequency curves from quite another point of view as that of Pearson or anybody else. I think I can finish it in another 6 weeks. I think there are a few results in it which will not merely interest biologists and statisticians. So for instance: that the arithmetical mean cannot be regarded as the most plausible value, but that what usually is called the median is preferable in all cases where the two differ. I would gladly make my publication in Biometrika, but I am sure that Pearson, who is its principal editor, will see his way towards refusing it. I have had an abominable experience in a similar matter. I will tell you about it another time, but it has led me to downright contempt of the man's character. The question therefore is: Is there any possibility of getting it published in the Philosophical Transactions? If such a thing is against the rules, there is of course an end of the matter and it will then be published (in English) in a Dutch Journal. If not ....'

Now, Pearson's second critique had appeared in the October issue of 'Biometrika', so apparently Kapteyn was considering not to let the matter rest, even after two publications by Pearson and himself, and publish a further paper in a British journal to seek support among UK biologists in his crusade against Pearson. He certainly did not easily let matters of perceived injustice rest. On November 5, 1906, Gill wrote: 'I shall be delighted to communicate any paper of yours to the Royal Society.' However, it seems the paper was never written.

Kapteyn came back to the issue a few months later. In his letter of March 22, 1907, in which he told Gill he was unable to come to Leicester for Gill's address as President of the British Association, he gave an explanation (see page 298; I here quote from the same paragraph in that letter):

'My wife and I cannot well come to Leicester. It's not the money question only, though this counts heavily, now that we have three of our children on our hands out of four, but also of time. What evil spirit possessed me to go on and on I know not. Once deeply in a question it gets such a tight hold on you. And just now I find that the neglect of a higher order term

Fig. 14.7 Procession of professors, at the celebration of the tercentennial of the University of Groningen in 1914. Kapteyn is third from the right (University Museum Groningen)

or something of that sort at the very beginning of the work may perhaps have spoiled all. However that may be I am sadly in arrears with matters astronomical and ought not to think of taking a holiday.'

Gill reacted to this on March 27, 1907 in a letter that is also (for a smaller part) quoted in the HHK-biography. He started by expressing his regret that Kapteyn is not coming to his Leicester address (see page 298):

'I am very sorry indeed to hear that you cannot come to Leicester – and still more sorry for the reason. Why a man of special astronomical gifts like yourself could waste his days in abstract mathematical work which so many men are capable of working at – whilst they are so free to do what you can so well do – I don't know. After all – what is the value or interest in a frequency curve compared with the structure of the universe? I am glad to hear that you confess to a temporary possession by an evil spirit [Henriette Hertzsprung-Kapteyn in the biography cited this letter also, but she wrote 'are coil spirit', which is not understandable English; the original letter, which resides at the Kapteyn Astronomical Institute clearly shows 'an evil spirit'] – Some form of exorcism is necessary – and I wish to administer it, if I can. I do think that in astronomy at the present time there is nothing comparable in interest with your work and the place you propose for its accomplishment. I want to put these before the B.A. as strongly as I can.'

In spite of this, Kapteyn kept studying statistics at least up to his joint work with van Uven in 1916. Although this was probably largely based on a genuine wish to help his biologist friends, it is not impossible that he was unable to let the disagreement with Pearson rest and stubbornly kept contrasting his view with his in the expectation to be proven right in the end and receive the credit he felt he deserved.

14.5 Tercentennial of the University of Groningen

Meanwhile, Groningen University remained the smallest of the three state-funded universities. In the years prior its 300th anniversary in 1914, it had only about 450 students (see G. Jensma & H. de Vries, *(op. cit.)*.), while those in Leiden, Utrecht and Amsterdam (a Municipal University since 1877, funded by the city of Amsterdam) had around 1100. Only the Free ('Liberated') University of Amsterdam, which was founded in 1880 as a orthodox-protestant or Calvinist university, had fewer students: between 150 and 200. Only about 50 of the 500 students in mathematics and natural sciences studied in Groningen (but the Free University had none). Today the University of Groningen ranks third among 13 universities in total number of Master students, and fourth among 10 universities offering Mathematics and Natural Sciences. The number of ('ordinary') professors in 1914 in Groningen was 38; 4 of Theology, 7 of Law, 8 of Medical Sciences, 9 of Mathematics and Natural Sciences and 10 of Philosophy and Humanities.

Being such a small university, and in view of the large emphasis on education, it would seem natural (and reasonable) that the funding for research in Groningen was small, and that in spite of Kapteyn's standing expansion of his Laboratory seemed out of the question. I alluded above already to the difficulties to hire computers and technicians for his Laboratory and many were funded with outside support, such as from Pickering and Hale. The future for astronomy in Groningen was reasonably secure, but funding would not reach the level of the Observatory in Leiden, which had then two professors in astronomy, van de Sande Bakhuyzen and de Sitter.

During the celebrations of the tercentennial of the University of Groningen, Kapteyn was asked by the US National Academy of Sciences to be its formal representative. Kapteyn had been elected a Fellow of the Academy in 1907. The invitation came through Hale. A statement was delivered through Kapteyn, which along with the invitation itself is among the Hale papers at Cal Tech. The content of the congratulatory text is interesting, where it refers to the remote location of Groningen (which must have been meant and should be understood as a compliment):

'The National Academy of Sciences, rejoicing in the completion of three centuries of educational progress and scholarly research by the University of Groningen, sends greetings and congratulations through its distinguished Foreign Associate, Jacobus Cornelius Kapteyn. In searching for the secret of Holland's great achievements in science we may best look to the example set by this University. Removed from the haste and confusion of the world, and steadfast in their devotion to the search for truth, her scholars have quietly pursued their way for centuries. Thus have they built up a center of learning whose foremost spirits inspire and guide the men of science

Fig. 14.8 Nieuwe Academiegebouw 1909. (Beeldbank Groningen [18])

of all nations. For their wise counsel and high example we express the appreciation and gratitude of the National Academy of Sciences. We can wish the University of Groningen no brighter future or more lasting renown than the production and support of such great leaders of research will ensure.'

It was signed by William H. Welch, president, Charles D. Walcott, Vice President, Arthur L. Day, Home Secretary, and George E. Hale, Foreign Secretary.

Some years before the tercentennial, in 1906, the central Academy Building was completely destroyed by fire. Fortunately the painting of the professors, some of which I used in this book, were saved. As early as 1909, a new Academy Building was completed (See Fig. 14.8) and opened (Fig. 14.9). In view of the number of professors quoted for 1914 (five years later) the procession seen on the latter photograph would show essentially the entire professorial staff.

It is of interest to mention that the windows above the central entrance and balcony, behind which the main auditorium or 'Aula' of the University is situated, have been decorated in stained glass between 1930 and 1953. The (from the inside) rightmost window depicts famous scholars, in particular Petrus Camper (1722–1789) (physician and anatomist) and Kapteyn (see Fig. 14.10).

Fig. 14.9 Opening Academiegebouw 1909 (Beeldbank Groningen [19])

14.6 World War I

In the first years after Kapteyn's last visit to Mount Wilson, his rather extensive correspondence with Hale focused on a few issues that are interesting enough to discuss in more detail, as they throw light on Kapteyn as a person and as a scientist. Sometimes we encounter remarks on the ongoing World War I. There are some interesting observations by Kapteyn on that issue, related to a visit the Kapteyns had made to the Hertzsprungs, their daughter and son-in-law. Hertzsprung after all worked at the Astrophysical Observatory at Potsdam. Traveling to Germany was a complicated matter in wartime. On May 27, 1915, Kapteyn and his wife obtained passports to travel to Germany from the Queen's Commissioner of Groningen province, and had to pay 6.75 Guilders each, which was the equivalent to 2.8 daily wages of an unskilled worker, and this roughly corresponds to 185€ today. In a letter to Hale of July 15, 1915, Kapteyn wrote:

> 'He [Kohlschütter] wrote that the commander of his camp had given permission to complete the theoretical astronomical work on the luminosity curve of the K stars. So I sent him the MS, which was in my keeping. But I hear nothing of it and I begin to fear it may not have reached him.
> My wife and myself have lately been in Potsdam for about a fortnight. As the position of our children, who are said to be of a 'unwohlwollende Neutralität' [malevolent neutrality] made it desirable in our discourse with the

Fig. 14.10 In the Main Auditorium ('Aula') of the Academy Building there are stained glass windows. This one depict famous physician and anatomist Petrus Camper (left) and Kapteyn (right)

astronomers not to touch on sore points connected with the war. I have had little discussions on the topic. Still I noticed a considerable change in the minds of the intelligent people. Those to whom I talked were pretty well unanimous in talking about the war as 'dieser blödsinnige Krieg' [stupid war], which is far away from the 'erfrischende Krieg' [refreshing war] they were talking of so freely at the beginning.'

A little earlier (May 13, 1915), Kapteyn wrote in a letter to Seares:

'The war too is getting on our nerves. I imagine that in America too anti-teutonic feeling must rapidly increase. The sinking of the Lusitania is such a crowning piece of barbarism. It must open the eyes of whoever still felt something for Germany. It is no war, it is simply murder'

The sinking of the British ocean liner 'Lusitania' on May 7, 1915, by a German submarine played a role in the declaration of war on Germany by the USA. And on August 18, 1915, Kapteyn sent a letter to Seares containing pretty much the same information as he had sent to Hale, as quoted above. He was somewhat concerned for the well-being of his daughter and son-in-law and and took the position that Germany should be blamed for starting the war.

'Of poverty or lack of food you can see little or nothing. as yet. The 'Kriegsbrot' [war bread] is not by a long way of fine quality, but then the quantity is sufficient. How it will be in the future is another question. In the whole of North Germany there has been little or no rain between the beginning of April and the time we were there, i.e. the latter half of June, and there was a general anxiety about the harvest. In fact in the whole of Brandenburg I was assured that even if then good moist weather set in, it could not save the wheat harvest

One other thing which was told by a correspondent of one of our Dutch newspapers, surprised me. He said that the officials in the ministry talked a great deal more sensibly about the war then did for example your professors. They acknowledged pretty frankly that the war was not a war forced on Germany.'

14.7 Induction Versus Deduction

The letter of July 15, 1915 by Kapteyn to Hale, from which I quoted a part above, opens with an expression of thanks for a book that Hale had sent him. It dealt with the history of the National Academy of the US and it had been written by Hale. This rather surprised Kapteyn, who realized it must have taken a great deal of time, which also for Hale was a scarce commodity. Next, Kapteyn turned to Hale's recent work:

'Your experiments in the laboratory on vortices interest me greatly. Better even than to be convinced by subtle mathematics, it is to see things before your eyes. I have often wished that my life had taken another turn and had made me an experimenter. Fate, however, has willed otherwise. My education as a boy naturally led me to study mathematics. Experimenting was practically impossible. So I went to the university as a student for mathematics. Owing to the fact that our physical professor was himself the poorest of experimenters, the chances were very much against my being drawn away from the science of my choice.

Then my reluctance to become a teacher brought me to Leyden, where I fell at once under the charm of the art of observing. I think I would have felt just the same, if chance had led me to a physical laboratory and I felt at once that I had in me what is needed for a good observer. I think I might have also made a pretty good experimenter, with a sufficient admixture of the mathematician. But fate decided that it should not be so. My appointment in Groningen soon put a stop to my observing and as nothing came of the promised and hotly desired observatory there was an end to my observing and by and by, in the lapse of years I gradually lost my inclination for the matter.

Still probably fate has known better what is good for me, than I would have known myself. In all events I cannot complain that my life is not full of interest. I might more easily have lost than gained. If but the theoretical work would not be so very long and what is infinitely more, be so continually marred by lack of data. Decidedly, your contempt for the 'catalogue' is not altogether sound. According to my view, astronomy needs nothing so much as good catalogues of everything. With our catalogues what could we not do! But the misery is that so many of our catalogues were made without any clear notion about the purpose for which they were wanted, and I feel pretty sure that it is just this particular that has led you to your contempt. The past six months have not been good for my astronomical work. Though I was not really sick, I felt little inclination for work. Moreover I had all sorts of trouble. I had to deliver about a double number of lectures, in order to make up for the time spend in America. Then my assistant left me, who was appointed lecturer of mathematical physics at our university. Then circumstances compelled me to finish a second paper of 'Skew curves in Biology and Statistics', a work which has weighed heavily on my mind for a long time.

But fortunately that is all past now. My paper is finished and we have our vacation. The was kept us at home and so I may have a chance of doing some astronomical work with almost as little disturbance as in other years I have at this time of the year on your unrivaled mountain.'

The assistant that left him was later Nobel laureate in physics Frits Zernike (for his discovery of the phase contrast microscope).

In a further letter from Kapteyn to Hale (September 23, 1915), we learn something about how the discovery of the Star Streams took place. After agreeing with Hale's point of view on catalogues, Kapteyn continued:

'The random way in which data have been collected in astronomy is astonishing. Take star positions, in a certain way the strong point because for over one hundred and fifty years such positions have been accumulating.

14.7 Induction Versus Deduction

> Still, as soon as in stellar research you want particular positions you are pretty sure not to find what you want. So for instance data for the proper motions of stars fainter than 6th magnitude. So in many other instances. The trouble, I think, is that work was undertaken without having in view any particular problem for the solution of which the work is required. I know that many astronomers saw no other purpose in the colossal undertakings of the Astron. Gesellsch. Zone Catalogue and the Carte du Ciel than the providing of fixed points for eventual observations of comets and minor planets. Of course I knew that you hated 'catalogues' just on account of this, but I may have feared sometimes that your hate had led you too far. I am extremely glad to find that it is not so. My studies have made me of more and more of a statistician and for statistics we must have great masses of data of course.
>
> The question you put in your last letter interests me highly. I certainly believe in an influence on the man of the science he has taken up. The science in which every error committed in reasoning or assumption is pretty sure sooner or later to lead to evident falsehoods, must influence a man's honesty and self-criticism. I also believe, as you do, that we neglect the 'Art of discovery' too much. My impression is that we are still not sufficiently imbued with the sense of absolute necessity of proceeding by induction. Deduction sets in too soon and too much is still expected from it.
>
> To illustrate what I mean take the Star-Streams as an example. I wish I could think of another example, but do not find anything nearly as well suited for my present purpose. For I really was within a hair's breadth of it for a couple of years. So excuse my introducing myself as an actor.'

Here Kapteyn hits on an issue that was taken up quite extensively by Hale, who disagreed. This issue, of inductive methods versus a deductive approach became the subject of extensive correspondence. In connection with the following, I may remind you that Eduard Schönfeld (1828–1891) was assistant to Argelander in the construction of the Bonner Durchmusterung and succeeded him as director. Schönfeld subsequently carried out the southern extension (Südliche Durchmusterung).

> 'Schönfeld was led, I think by analogy, to consider the question: May there not be a rotating motion of the Milky Way as a whole. He made the necessary computations, but found practically nothing.
>
> Other men tried a rotating motion of all the stars in orbits in the Milky Way, not necessarily all with the same period. Some, I believe, tried to adhere to a common direction of motion, others assumed both direct and retrograde motions. I myself tried: is there at all a preference for motion parallel with the Milky Way but for the rest quite arbitrary. I found nothing.
>
> Now all this seems to me too much deductive. We begin by making a wild guess, deduce its consequences and see whether it agrees with the observations. How long might we have guessed before we come to put the question: are there two Star Streams?
>
> I blundered along for a long time in the same mistaken way, till one day I swore to go along as inductively as I could.

I made drawings showing at a glance the observed data for each point of the sky. These showed very decided deviations from what was to be expected according to existing theory. Considering these deviations as perturbations I tried to isolate these perturbations.

I superposed all the drawings belonging to zones in which, according to existing theory there ought to be equiformity and took averages. The result was a figure pretty well in conformity with existing theory. This drawing I then took to represent the undisturbed form and subtraction from the individual figures then gave the isolated perturbations. There showed all at once a great regularity, which regularity was almost at once seen to consist in a convergence of the lines of symmetry to a single point of the sphere. From this to the recognition of 2 Star Streams 'il n'y a qu'un pas' [is only one step].

Thus the inductive process led in a very short time to a result which others, myself included, had tried in vain to bring out in a more deductive way, for ever so long. From the standpoint of a finished theory the method must be conceded to be very little rigorous. I have a notion that it will be so in most cases. Therefore it has to be supplemented by a deductive treatment of the problem: given that there are two streams, show rigorously that the observations are well represented. I think that it is on account of the mathematical rigour that the latter problem (the deductive one) appeals to so many even of the very best men and that it is this part of the investigation which finds its way into the textbooks. Still, in my mind the inductive part of the investigation, with all its defects, is incomparably the more important part in research. The deductive problem can be solved by any well skilled mathematician. Still the textbooks teach only the least important part.

Your way of putting your questions immediately brought to my mind Bacon's <u>Novum Organum</u> which I read a couple of years ago. Bacon in it tries to show how you can arrange set of rules, by the aid of which even the man in the street could, almost mechanically, work for the progress of science. My friend Heymans, the psychologist and philosopher, gave me some titles of books such as you desire. He particularly recommends the first book: 'de la méthode dans les sciences'.'

Sir Francis Bacon (1561–1626) in his *Novum Organum Scientiarum*, published in 1620 and promoted in it the use of inductive reasoning. The title Heymans recommended is very likely *Discours de la méthode pour bien conduire sa raison, et chercher la vérité dans les sciences*, the famous 'Discourse on the Method' by René Descartes (1596–1650). Kapteyn's letter continued with his recent thoughts on the Star Streams, to which I will turn in the next chapter.

Hale did not fully agree with Kapteyn. He replied extensively in a letter of November 4, 1915.

'The account of your discovery of the two star streams is a splendid illustration of the value of the inductive method, which doubtless serves best in a large class of investigations. And yet I cannot help feeling, that in many

14.7 Induction Versus Deduction

other cases a combination of deduction and induction is more likely to be successful. In fact, I constantly find myself instinctively framing hypotheses as guides to research, always bearing J.J. Thomson's definition in mind, and therefore endeavoring to construct multiple hypotheses to account for obscure phenomenon. Each hypothesis suggests the application of a series of criteria, and it usually becomes possible to eliminate some of them very soon.'

Hale wrote 'phenomenon', while the plural seems better suited to the sentence. The reference to Joseph John Thomson, Nobel Prize winner for his discovery of the electron, probably relates to his putting forward hypotheses for the explanation of his observation of 'rays' that flowed between two electrodes in a vacuum tube when a voltage was applied over it. He found that the rays were reflected by electrically charged plates and by magnets and proposed that the rays consisted of small, electrically particles, corpuscles, that are smaller than the atom. It took some time before this hypothesis, that the atom is not the smallest constituent of matter, was accepted by the scientific community.

Hale gave a set of examples, relating to his work on spectra of sunspots and other solar phenomena to illustrate his point. He then wrote:

'In my own experience, therefore. deductive methods are almost invariably applied. Frequently they are the merest guess, without a substantial theoretical basis. But each suggests experiments or tests which would hardly occur to me otherwise, and thus almost any hypothesis may prove useful. Is it not true that you also employ deductive methods in many instances? Sometimes they may enter only tacitly, and nevertheless play an important part in your investigations. In fact, on a basis of pure induction your imagination would have little opportunity to serve as a guide.'

In a follow-up letter (November 11, 1915) Hale corrected the first sentence. 'Of course I meant, as the context shows, that I employ a combination of deduction and induction.' He further noted that he had ordered all books on the list Kapteyn sent him, except Bacon's, 'which I reread recently', and asked him to thank Professor Heymans for it. The letter ended as follows.

'Pardon this long letter. If I have given too many illustrations of the use of deductive methods it is in the hope that you will examine your own investigations again to see whether deduction does not play a considerable part, directly or indirectly. [And in a footnote:] It is even possible that you would not have undertaken the investigation which led to the discovery of the star streams if you had not had started in this direction by an (incorrect) hypothesis? [end footnote] I wish you would get Lorentz's view on this subject.'

It seems that Kapteyn had the more rigid and Hale the more subtle view in these matters. Kapteyn returned to the issue in a letter of March 26, 1916. After some pages on work by van Maanen (his well-known erroneous result of having measured proper motions in the Pinwheel galaxy M101; see page 591) and van Rhijn, he wrote:

'I am sorry to say that I cannot report so favourably on my own work. I feel farther from the end than at the beginning.. I find a good deal of facts to be explained but every time something will turn up that will not fit my explanations. This brings me back to the question of inductive vs. the deductive method. I have talked a good deal on the matter both with our philosopher Heymans and with Ehrenfest, Lorentz's successor. They take more nearly your view of the matter than mine. After all, I think that the difference is more a difference of degree. We cannot pass from the observations to the laws underlying them, than by making certain jumps, certain hypotheses and applying them to the observations in hand, and modifying them till they fit. If however we take but very small jumps, we will call our methods inductive, if we take bolder jumps, we call it deductive. Well, my feeling again is, that I tried too big a jump, and I have come back to the very long and weary method which proceeds by insensible steps. I find that we can find the distributions in space of nearly all the helium stars.'

So, Kapteyn found Heymans and Ehrenfest not on his side, but refrained from fully conceding that Hale was right, merely qualifying his own point of view.

14.8 Continuation as a Research Associate

In a letter of March 9, 1916, Hale had urged Kapteyn not to come to Mount Wilson that year in view of the 'unnecessary risk' in wartime. That letter presumably had not reached Kapteyn yet when he wrote the following on March 26.

'I hope to come this year at Mount Wilson about the middle of July. It will be somewhat later than usual, because our daughter Hetty is staying with us; she expects her baby in the beginning of June and her mother wants of course to be with her till all is well over. Meanwhile the war is taking a turn that might easily prevent us from coming at all. Some of our passenger companies (to England and East India) are stopping service partly or wholly; others send out their ship accompanied by a relief steamer which carries all sorts of saving apparatus. Heaven knows whether there will be any boat sailing at all in July. If I cannot or judge better not to come, I have in my mind I ought to resign as a research assistant.

In any circumstance I think I ought not continue this position much longer. In about 5 years time I will have to resign as a professor. I cannot well bear to leave many of the works undertaken at the Laboratory unfinished and there are a great number of these. They will require pretty well all my thoughts and leisure. If in addition to this I come to feel – more even than I did sometimes in the past – that my services to the M W Observatory are not what they ought to be, it will certainly be more satisfactory to go now, than next year or the year after this. I feel that after all you did for me I ought not to go away before consulting with you. What I will regret more

14.8 Continuation as a Research Associate

than anything else in disconnecting myself from the observatory, will be that owing to the state of your health our personal intercourse has been so limited. Meanwhile I still hope and expect to be able to come this year and that I may devote my strength in helping to make plans for the 100-inch.'

The child born to Henriette and her husband Ejnar Hertzsprung was a daughter that was named Rigel after the bright star (see also Fig. 16.14 and page 646). On June 6, 1916, Hale replied:

'From what you say, to my great concern, about giving up your position as Research Associate of the Carnegie Institution, I am afraid I may have given you a false impression of the duties of Research Associates. It would be extremely desirable from the standpoint of the Observatory and the Carnegie Institution, for you to retain the position even if you never returned to Mount Wilson and never did a single piece of work for us. In other words, the purpose of the grant is to facilitate your own investigations, and not to secure any direct return to the Institution. This is done in the case of many other men, such as Professor A.A. Noyes of Boston, and various others I could name if I had the Year Book here. So don't think of resigning under any circumstances. We want your close interest and counsel, but you need not do any work for us unless you have plenty of time for it.'

Arthur Amos Noyes (1866–1936) was a chemist at the Massachusetts Institute of Technology (MIT) in Cambridge, Mass. near Boston and later professor at Cal Tech. Kapteyn's reply, dated July 23, 1916, was brief.

'I thank you cordially for your letter of June 2 (which arrived only a few days ago). After what you write it would be very foolish in me to insist on giving up my position as a Research Associate of the Carnegie Institution. The position has given me so much happiness that I could not do so – believe me – with a light heart.

Moreover I really think that it has resulted in some good astronomy both through work done by myself and by work done by others.

If at times I have felt somewhat discouraged, pray excuse me. It would perhaps not have been had circumstances permitted me to serve the Astronomy without any money considerations.'

So Kapteyn did not resign, but nor did he return to Mount Wilson, neither that year 1916, nor any other year after that.

Fig. 14.11 Kapteyn probably some time around 1914 (Kapteyn Astronomical Institute)

Chapter 15
First Attempt

One thing is sure. We have to do something.
We have to do the best we know how at the moment.
If it doesn't turn out right, we can modify it as we go along.
Franklin Delano Roosevelt (1882–1945).

Take the first step in faith.
You don't have to see the whole staircase, just take the first step.
Martin Luther King, Jr. (1929–1968).

Kapteyn's efforts towards a better understanding of the distribution of stars in space had been directed first to star counts and parallaxes and when he had discovered the Star Streams towards proper motions. By the time the Plan of Selected Areas began being executed with the help of Hale at Mount Wilson and Pickering at Harvard, a stream of plates started to arrive in Groningen to be measured at his Laboratory for the Durchmusterung parts of the Plan. This routine work was to go on for many years to come. With the advent of spectroscopy of faint stars, spectral types could be established, but also determinations of radial velocities had become possible and in fact became an important part of the Plan. This significantly broadened Kapteyn's interests. But the scope of the program to understand the Construction of the Heavens grew accordingly.

15.1 Radial Velocities

Observations of radial velocities with the Mount Wilson 60-inch telescope started in 1909. It was executed in two ways. In the first place by using the so-called Coudé

Fig. 15.1 Seven astronomers during the 1910 Mount Wilson meeting of the Solar Union. Edwin B. Frost is the middle one. To the right of him William W. Campbell, Karl Schwarzschild and Frank Schlesinger. Second from the left John S. Plaskett (From the Huntington Digital Library [3])

focus spectrograph. As I explained on page 490, one option to use the telescope was to employ a few extra mirrors to deflect the light outside the telescope tube. There it could be fed into a fixed spectrograph. Since the spectrograph does not have to move along with the telescope, it can be a very large spectrograph, providing very high resolution in wavelength. Since this setup is called the Coudé focus, the instrument itself is called the Coudé spectrograph. The one on the 60-inch telescope provided spectra with 'dispersions' on the photographic plate, on which the spectra are recorded, of a few Ångström (Å) per millimeter. Astronomers often used (and still regularly do) the unit Ångstrom for wavelengths in spectral work, named after the Swedish physicist Anders Jonas Ångström (1814–1874). It is a ten millionth of a millimeter. The modern unit is the nanometer (nm, one millionth of a millimeter), which is 10 Å. The wavelength of visual light is of order 5000 Å or 500 nm (which might also be designated also as 0.5 micron). With the dispersion of the 60-inch Coudé spectrograph, it was possible to measure radial velocities to an accuracy of order one kilometer per second (see Box 15.1). The first results with the 60-inch Coudé were described in an article by Walter Adams, *Some results of a study of the spectra of Sirius, Procyon, and Arcturus with high dispersion* [1]. Since the light of the star had to be spread over a large area, one needed the combination of a large telescope and a bright star. The resulting accuracy was an unbelievable quarter of a km/sec. The sharper the lines in the stellar spectrum, the higher the final accuracy.

Even with the 60-inch telescope, radial velocities to such an accuracy could only be measured for the brightest stars. For studies of the structure of the Milky Way this was also not necessary, as stars move with relative velocities of tens of km/sec and therefore an accuracy of a few km/sec sufficed. Towards the end of 1909 a

> **Box 15.1 Radial velocities**
> Radial velocities are measured on spectra using the *Doppler effect*. We have met Christian Andreas Doppler (1803–1853) already on page 41 in relation to one of the propositions that went along with Kapteyn's PhD thesis. The Doppler effect is the well-known property that sounds from a vehicle moving towards us has a higher pitch than when moving away from us and this also holds for light, where a velocity away from us results in a shift towards redder wavelengths. The positions of the spectral lines in a star's spectrum can be used to determine the radial velocity of that star. The change in wavelength $\Delta\lambda$ compared to the wavelength of the line at rest λ is the same as the radial velocity V_{rad} compared to the speed of light c (at least for velocities much smaller than that of light), or
>
> $$\frac{\Delta\lambda}{\lambda} = \frac{V_{\text{rad}}}{c}.$$
>
> For a medium wavelength in the optical spectrum (say 5000 Å or 500 nanometer) a 1 Å (0.1 nm) shift corresponds to a radial velocity of 60 km/sec, since $c \approx 300,000$ km/sec.

so-called 'intermediate-dispersion spectrograph' was commissioned, located at the Cassegrain focus, where the light is thrown through a central hole in the 60-inch mirror. This spectrograph is then attached to the telescope structure at the back of the primary mirror. Dispersions here ranged from 5 to 18 Å/mm.

The hero of radial velocity determinations was Walter S. Adams. He had studied under Edwin B. Frost (see Fig. 15.1), then at Dartmouth College in Hanover, New Hampshire and upon recommendation by Frost had been hired by Hale at Yerkes in 1902, where Frost also had moved in the mean time. Adams came to Pasadena with Hale in 1904. But before that he and Frost had done important, early work on stellar radial velocities with the Yerkes 40-inch refractor. Again this was for rather bright stars, but an accuracy of order a km/sec was possible. At Mount Wilson, Adams wanted to push this work to fainter magnitudes, partly in the framework of Kapteyn's Plan of Selected Areas (partly, because bright stars generally were not part of the Plan). The first extensive result was published in 1912 by Adams, *The three-prism stellar spectrograph of the Mount Wilson Solar Observatory* [2]. He found many stars to be spectroscopic binaries (from measurements at different times) and a few high proper motion stars to have high radial velocities as well, even up to 170 km/sec towards us and 100 km/sec away from us.

Kapteyn was quick to pick up this work and to undertake studies of the systematics of radial stellar motions. In Kapteyn & Frost (1910), they collected all information he could find in the literature and calculated for each star its 'peculiar' motion, which is the radial velocity corrected for the component of the Sun's motion towards the Antapex of about 20 km/sec. He built on the result by Frost and Adams that the linear motion of helium stars was on average 'exceptionally small', but extended it significantly. Kapteyn found from a sample of 210 stars a 'regular increase in the peculiar radial velocity when we pass from the *Orion* (B) to the Sirian (A) stars, from these to the second-type stars (F to K) and, finally, to those of the third type (M); that is, if we follow what, according to our present knowledge, must be considered as the most probable succession of the different classes.'

This is a valid result even today. Indeed we know now that stars in the disk of the Galaxy near the Sun have an increasing mean random motion (known as velocity dispersion) with age and, since O, B (and A) stars are short-lived, they are on average much younger than G, K and M-stars (the third spectra type), that live as long or even longer than the current age of the Milky Way Galaxy. The cause for this correlation is that stars are born in the interstellar medium from collapsing gas clouds. For the densities and temperatures in this medium, however, gas concentrations of tens to hundreds of thousands of solar masses are required before they can become gravitationally unstable. This follows from a theorem derived by James Jeans and this minimum mass is referred to as the 'Jeans mass'. Stars therefore always form in large groups with masses of this Jeans mass. However, during the collapse the gas heats up and the Jean's mass then decreases. The originally very massive cloud then fragments into smaller pieces, eventually of the size of individual stars. The clouds in which stars form have been identified now as 'Giant Molecular Clouds' (GMC's), since they contain, in addition to hydrogen, helium and other atoms, also molecules (in particular molecular hydrogen H_2, carbon monoxide CO, water H_2O, and many other simple but also very complex ones), that form on the surfaces of dust grains. Stars are then formed with the random velocities that prevail in these GMC's and that is of the order of 10 km/s. But after formation the stars are scattered by the same GMC's or by the spiral structure (both create local disturbances of the gravitational field) and the older the stars in the disk of the Galaxy become, the larger their random motion. For stars of the age of the Sun it amounts to 40–50 km/s, so the Sun actually is relatively slow compared to stars of its generation.

Kapteyn & Frost (1910) went on to discuss various related items. They compared the results extensively with proper motion studies (and found support for his results from radial velocities) and speculated on its cause, concluding that stars must be formed from matter with little or no internal motion. 'Apart from the advantages that we may derive from this result for the classification of the stars in the order of their evolution, it has, I think, a great importance in its bearing upon the question of the generation of the star-streams themselves'. But he refrained from a final conclusion, merely pointing out that a study of stars of different spectral classes separately for each stream needed to be done first.

On November 8, 1909, Kapteyn had asked Frost for as much velocities of A-stars as he could supply (he had only 18 such velocities). In a letter of December 14, 1909, Frost wrote that he had not been able to send Kapteyn radial velocities of A stars, his stars having not been 'adequately observed to serve as a basis for anything. As a partial compensation for this failure to comply with your wishes, I am sending a list of rough values for radial velocity for a number of stars of the Orion type.' These are O or B-type stars and his list had 23 entries. Then he added 16 values

15.1 Radial Velocities

for the velocity of the center of gravity of confirmed binaries. With a list published by Frost and Adams, this should add up to 'between fifty and sixty stars of the Orion type'.

This resulted in the joint paper, Kapteyn & Frost (1910; a correction, retracting a statement quoted from others as incorrect, and 'quite inexcusable', was published as Kapteyn & Frost, 1911), in which an improved value for the velocity of the Sun in space was derived. After all, if 'Orion' or B-stars had low peculiar motions, a derivation of the magnitude of the Sun's speed could be done more accurately and therefore secular parallaxes estimated better. The value was 23 ± 1 km/s. However, they found that separate solutions using stars near the Apex and the Antapex gave significantly different values. This would mean that the stars near the Apex belonged predominantly to one stream and vice versa for the Antapex and pointed at large differences in admixtures of the stars in the two Streams.

Kapteyn demonstrated the use of secular parallax (see Box 15.2) in a paper published shortly thereafter, Kapteyn (1910c). He was interested in distances of stars as a function of their spectral type. Little was known for the case of stars of the 'fourth type'. These were sometimes called N-types, and in fact are a special case of red giants that have large amounts of carbon in their atmospheres. They are now designated 'Carbon stars'. It turned out that at Copenhagen Observatory, where Svante Elis Strömgren was director, an extended observing program had been undertaken (mostly by N.E. Nörlund) of proper motions of fourth-type stars, but of third type, M-stars, as well. Kapteyn used data of 120 of such N-stars and estimated a secular parallax from their proper motions. He then compared this with similar results for other types. The result was that N-type stars had on average extremely small parallaxes and thus large distances. Their formal mean secular parallax was $0''.0007$, which is smaller than the accuracy of the measurement themselves. So they had to be intrinsically very bright stars, as indeed we now know they are. Kapteyn compared this with what he could find in the literature on M-stars (third type). Their parallaxes, based on about 100 stars, were not that small as the Carbon stars.

As we saw, the work at Mount Wilson to measure radial velocities of stars was mainly executed by Walter Adams. This was restricted to relatively bright stars and not necessarily stars in Selected Areas. To give an example, in 1915 Adams published a paper [4] with 500 radial velocities of stars between magnitudes 5 and 8 or so. Using this material and radial velocities measured by Campbell at Lick Observatory [5] on over 900 stars, generally brighter than magnitude 5, Kapteyn and Adams extended the relation between proper motions and radial velocities as a function of spectral type, Kapteyn & Adams (1915). They corrected all radial velocities for the contribution from the Sun's motion and calculated the angle of the star's position with the two vertices of the Star Streams. The total sample of F, G, K and M stars showed that average radial velocities increased for each spectral type with proper motion and the distributions were in good agreement with what was expected from the theory of the two Star Streams. This opened the possibility of a relation between the radial velocity and the absolute luminosity. Adams follows this up shortly thereafter with a paper with himself a sole author in 1915 (reference details below), in which he found also low radial velocities for distant stars of types F to M. This

> **Box 15.2 Secular parallax**
> The secular parallax of a group of stars is the statistical distance indicated by their proper motions on the sky as a reflection of the Sun's motion in space. If the Sun moves away in space in the direction of the Antapex with a velocity $V_\odot$ and if a star's position on the sky is an angle λ away from the Apex the component perpendicular to our line of sight is $T_\odot = V_\odot \sin \lambda$.
> The expected proper motion $v_\odot$ along the line between the star and the A(nta)pex is then is $v_\odot = \pi T_\odot / 4.74$, where π is the parallax in arcsec (and $r = 1/\pi$ the distance in parsecs); the value 4.74 serves to change the units of km/s of T into arcsec/year in v.
> For a group of stars the secular parallax is the π estimated from its proper motion using this formula, but giving it a weight corresponding to what fraction of the Sun's speed is reflected in the proper motion. This weight is $\sin \lambda$. The secular parallax of a group of stars is then
>
> $$\langle \pi \rangle = \frac{4.74 \langle v \sin \lambda \rangle}{V_\odot \langle \sin^2 \lambda \rangle},$$
>
> where the triangular brackets indicate mean values.

progressing work on radial velocities of stars, however, lead to serious disagreements between Kapteyn and Adams, to which I will return below.

15.2 Kapteyn's Developing View of the Sidereal System

In the years between 1910 and 1915, Kapteyn published very few papers in astronomical journals. One reason for this was that he was drawn upon regularly to present lectures for various audience. In 1910 he delivered a major lecture during the Mount Wilson meeting of the International Solar Union, Kapteyn (1910d), discussed already. He presented results of his work on Star Streams and the Milky Way to the Netherlands Academy of Science, Kapteyn (1911a,b,1912b), gave a key-note speech at the 1911 Groningen meeting of the Nederlandsch Natuur- en Geneeskundig Congres (NNGC, see page 284), Kapteyn (1911d), and there was his lecture before the US National Academy on April 22, 1913 that I discussed earlier on (see page 494). The lecture at the NNGC deserves some more attention and I will present that below. Kapteyn coordinated the publication of the first and second report on the progress of the Plan of Selected Areas, Kapteyn (1911c). This, and the constant attention to the execution of the Plan and the measurement of the plates in Groningen, must have taken much of his energy and time. In addition there was the annual trip to Mount Wilson and associated visits to colleagues abroad on his way there and on the way back (also always visiting his son in Colorado), for which the travel alone took a significant part of his time. And a major drain on his time and

15.2 Kapteyn's Developing View of the Sidereal System

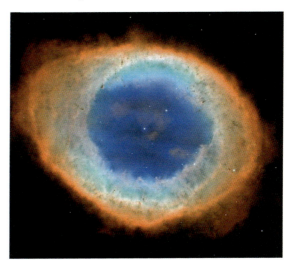

Fig. 15.2 The Ring Nebula in the constellation Lyra is a prime example of a so-called Planetary Nebula. This picture has been taken by Hubble Space Telescope (The Hubble Heritage Project [6])

energy was the teaching load at the University. He was required to make up for his absence at Mount Wilson by an increased teaching load when he was in Groningen.

Kapteyn published one scientific paper in 1912, *On the derivation of the constants for the two star streams*, Kapteyn & Weersma (1912). This was no new work, but the presentation of his formal mathematical treatment of data to arrive at the vertices of the Star Streams. He had never found the time to write this up for publication, which now had been done by his associate H.A. Weersma, who 'generously offered to collect and arrange my notes on the subject, and to prepare the whole for the press'. And in 1915 Kapteyn published a short note, Kapteyn (1915), in which he proposed a new 'device' to reduce magnitude dependent errors in positions and proper motions. He suggested that photographic plates should be taken in two steps. First make an exposure slightly out of focus on which all stars then have images of equal diameter. After developing this plate it should be put back in the telescope and a final plate be exposed through this 'screen-plate', which will attenuate bright stars more than faint stars and therefore make them more comparable in appearance.

Kapteyn's lack of astronomical research papers between 1910 and 1914 also had to do with extensive work he did on 'helium stars', on which he eventually published two papers in the Astrophysical Journal, Kapteyn (1914b) of no less than 85 pages, and one of 91 pages, which was printed in three installments, Kapteyn (1918a,b,c). But before turning to those studies I first discuss some of his more general lectures and work by his PhD students in the intervening time.

The Nederlandsch Natuur- en Geneeskundig Congres (NNGC, see page 284) held its 13-th meeting in Groningen in April, 1911. Kapteyn was a key-note speaker (in Dutch) and his contribution was entitled *Some recent studies on the subject of the evolution of the fixed stars and the Sidereal system*, Kapteyn (1911d). It showed his developing ideas and therefore needs some discussion here.

Fig. 15.3 Illustration by Kapteyn of how he envisages the two Star Streams. The Sun is at rest in this drawing (From Kapteyn, 1911d)

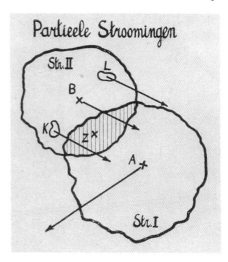

Kapteyn started the presentation by reviewing what was known about types of stars and pointed out that the 'highly probable evolutionary sequence' was from helium stars (B-type) through the first (A) and second type (FG) to the third (M). Although this is not at all an evolutionary sequence according to our current understanding, it remains true that *on average* helium A-stars are indeed the youngest (they have short life-times) and third type M-stars the oldest (their lifetime actually exceeds that of the Milky Way Galaxy and even that of the Universe). So, the general correlation of spectral type with age Kapteyn assumed was correct.

Kapteyn believed that stars were born out of the material of the nebulae. These should then have even smaller random motions than the helium-stars. Proper motions of nebulae are difficult to measure; he had discussed this extensively in Kapteyn (1906a) and again in Kapteyn (1906b), see page 408. Measuring radial velocities of nebulae was in principle an option, but then the brightness was too faint and it only worked when the spectral lines were in emission. For example, he pointed to the planetary nebulae, for which he took the Ring Nebula of Fig. 15.2 as an example. However, he noted that these objects could very well be much later stages in a star's life, such as novae, where a star 'experiences some enormous catastrophe'. Probably Kapteyn's earlier work on Nova Persei (see page 311) and the associated nebula played a role in him coming to this suggestion. Actually, it is quite correct (see page 206); indeed planetary nebulae are late stages of stellar evolution, where the outer shell has been expelled.

Kapteyn briefly mentioned spiral nebulae (he took the trouble to illustrate four very bright ones, among which Messier 51 and Messier 101), (see Fig. 11.11). There must be 'hundreds of thousands, maybe even millions' of them in the sky, but their 'sheer appearance prevents any doubt of a close relationship'. Yet nothing was known about them and he excluded them from further consideration. Kapteyn has not commented on his views on these systems in any communication, paper or letter

15.2 Kapteyn's Developing View of the Sidereal System

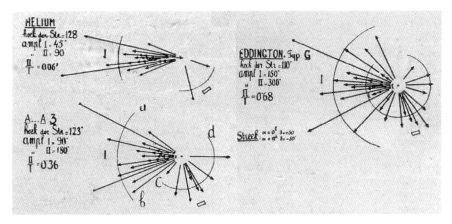

Fig. 15.4 The difference in composition of the two Star Streams in terms of spectral types. Motions of stars in a small area of the sky has been plotted. Stream I is directed to the left (indicated by 'I') and Stream II to the lower-right ('II'). Further see text (Adapted from Kapteyn, 1911d)

that I have seen. How he in his mind related them to the nebulae in which he believed stars formed, is not known.

Having dealt with spectral types and stellar evolution in this way, he then introduced his two Star Streams and depicted them as two penetrating clouds of stars, the Sun lying in the region of overlap (see Fig. 15.3). The most important point was that the two streams had different compositions of spectral type. Figure 15.4 shows at the top-left that A-stars ('helium') seemed restricted almost entirely to Stream I (the ratio of the Streams I/II is 0.06), For type I stars at lower-left this ratio was 0.36, but for all stars together (a mix by Eddington of stars of types I through III) on the right the ratio had become 0.68. Furthermore, the amount of 'parallelism' (the relative domination of peculiar over random motions within each Stream) decreased with the spectral types and therefore in Kapteyn's view according to the stage of their evolution. The same held for the precise direction and magnitude of the stream velocity. Kapteyn drew some remarkable conclusions from this.

But the existence of nebulae as in the Pleiades, (see Fig. 7.11), which as we now know are reflections of starlight, showed an intimate relation with stars. The same held for an object like the Orion Nebula (Fig. 11.8). Here the association was confirmed by measurements that showed that the radial velocity of the nebula and a star in it were very similar. His final conclusion then was: 'The planetary nebulae cannot be regarded as the birth place of stars; infinitely more probable is that one would find this to occur in nebulae with the character of the Pleiades and the Orion Nebula'. A major problem he addressed was that it should be expected that the material out of which the stars have been formed should have even smaller random motions than the helium stars. This lead him to conclude that this primeval matter ('Ur-stof') was not influenced by gravity, 'at least everything happens as if gravity did not exist'.

The Sidereal System by this stage was envisaged by Kapteyn as having formed from two independently moving parts. The directions of streaming were changing with time (as followed from the different average streaming of the stars of different spectral types). This lead him to consider the possibility that gravitational attraction worked differently on a star than it did on the 'Ur-stof'. There are also 'sub-steams', some of which he associated with star groups as the Hyades, etc. The picture became more and more complicated also because he had to assume that for some reason the 'Ur-stof' in one of the Streams was further depleted than in the other.

This speculation shows that at that time Kapteyn already was forming ideas about the relation between distributions of stars and their velocities as regulated by gravity. As we will see this culminated towards the end of his life in his dynamical description of the Sidereal System as consisting of star distributions that are maintained by a balance between their gravity and their kinematics. He already presented a synthesis combining what he knew about stars, their spectra and evolution, about spatial distributions and gravity into a coherent picture about the structure and origin of the sidereal universe.

15.3 In the Mean Time in Groningen

The work of measurement of photographic plates was going on at an unrelenting pace. The routine work on the Harvard and Mount Wilson Durchmusterungs was a major investment of resources. But still other work that Kapteyn felt was necessary for his attack on the problem of the Construction of the Universe, was also progressing. Some of this now was done in the form of PhD thesis research. Remember that Kapteyn had had three students that did their PhD research (mostly) prior to the early years of his Mount Wilson association; de Sitter who finished in 1901, Weersma in 1908 and Yntema in 1909. I have already documented that actually there was a fourth, Adriaan van Maanen who obtained his degree in Utrecht in 1911 under Nijland (see page 490), who had worked in Groningen in the first years that Kapteyn started to visit Mount Wilson. This may be the right place to say a few words about van Maanen's supervisor. Albertus Antonie Nijland (1868–1936; see Fig. 15.5) had obtained two doctorates in Utrecht, one in mathematics under Willem Kapteyn and one in astronomy under Oudemans, respectively in 1896 and 1897. His main interest had been variable stars, but he also performed solar system studies, such as on comets. He organized an extensive Dutch solar eclipse expedition to Sumatra, Netherlands East-Indies, in 1901 [7]. He never published himself on proper motions or parallaxes of stars, nor on counts or distributions of stars on the sky. Yet, he acted as supervisor on van Maanen's thesis research, concerning proper motions of stars in the h and χ Persei star clusters, for which van Maanen measured plates at the Groningen Laboratory (taken for Kapteyn by Sersej Kostinsky at Pulkovo and by Anders Donner at Helsingfors). I have already justified the viewpoint to consider Kapteyn as the actual supervisor.

15.3 In the Mean Time in Groningen

Fig. 15.5 A.A. Nijland as he appears in the *Album Amicorum* presented at the retirement of H.G. van de Sande Bakhuyzen as professor of astronomy and director of Leiden Observatory in 1908 (Archives Leiden Observatory; see caption Fig. 3.7)

Eventually, two other persons obtained PhD's under Nijland in Utrecht on subjects relating more to Kapteyn's primary interests than to Nijland's. The first was Isidore Henri Nort (1872–1943), who defended a thesis in 1917 on *The Harvard Map of the Sky and the Milky Way* [8] According to the Introduction the research was performed following a suggestion by Nijland to look into the major differences in the magnitude scales among brighter stars of Kapteyn and of Pickering and others. Nort used Pickering's *Harvard Map of the Sky*, a set of 55 double contact prints on glass plates covering the whole sky [9], of which Utrecht owned a copy. He used counts from these maps performed by H. Henie at Lund Observatory in Sweden, redetermined for each plate the limiting magnitude and produced star counts in 5000 small areas down to magnitude 11. From this he derived some statistical properties of the distribution of brighter stars. Kapteyn did not figure even in the acknowledgments, so apparently he had little or nothing to do with this work. After his doctorate, Nort left active astronomical research, becoming a teacher and author of text- and popular books in Zutphen and Gouda. He published four further astronomical papers up to 1950 through Utrecht Observatory.

The other person was Jan Cornelis van de Linde (1884–??), who defended a PhD thesis in 1921 on *De verdeeling van de heldere sterren* (the distribution of the bright stars). He used the Revised Harvard Photometry catalogues, again from Pickering. These catalogues had been published in various volumes of the Harvard Annals between 1885 and 1903, and revised in 1908 [10]. He presented maps of counts of stars brighter than 6.25 magnitudes and pointed out irregularities. There is also no evidence of any direct or indirect involvement of Kapteyn.

Kapteyn's student Etine Imke Smid (1883–1960) was the first woman in the Netherlands to obtain a PhD in astronomy. Kapteyn wrote a letter of support for her when she applied to come and study for some time in Leiden. On February 5, 1911 Kapteyn wrote to E.F. van de Sande Bakhuyzen that she 'particularly wanted some experience with practical astronomical work'. He says about her: 'As far as I can see she is not a person with exceptional talent, but she is hard-working and has conscientiously attended all lectures. She seems an extremely modest and unassuming girl. I hope you can do something for her.' There is no mention in her thesis of an extended stay in Leiden, but in the relevant bi-annual report of the Leiden Observatory [11], E.F. van de Sande Bakhuyzen mentioned her in relation to the solar eclipse observations of April 17, 1912. The Royal Academy had in 1901 organized an expedition to the solar eclipse at Sumatra. The committee that coordinated that was still in existence and felt that it would be profitable to have another expedition to the 1912 eclipse in Limburg in the south of the Netherlands in order to study the relative radiation from various parts of the Solar surface and to determine the relative positions and diameters of the images of the Sun and the Moon. A number of students of mathematics and astronomy accompanied the astronomers and one of these was Ms Smid from Groningen.

Her thesis was defended on 13 June, 1914. It concerned the *Bepaling der eigenbeweging in rechte klimming en declinatie van 119 sterren* (Determination of the proper motion in right ascension and declination for 119 stars). It was not a very extensive thesis, combining positions in older and newer catalogues for brighter stars to determine their proper motions. The difficulty that had to be treated was the exact precession (see Box 3.1) correction between the different epochs. In the PhD thesis of Inge de Wilde, *Nieuwe deelgenoten in de Wetenschap (op. cit.)*, there is a picture of Etine Smid on the day of her PhD defense on page 106.

After her doctorate Etine Smid went to Leiden where she worked with among others physicist and later Nobel laureate Heike Kamerlingh Onnes (1853–1926), but she left scientific research in 1916, when she married Cornelis Marinus Hoogenboom and moved to Deventer (see Inge de Wilde: *Nieuwe deelgenoten . . . ,op. cit.*).

Next came Pieter Johannes van Rhijn, whom we met already in relation to his extended stay at Mount Wilson. He defended his thesis on July 9, 1915. The title was *Derivation of the change of colour with distance and apparent magnitude together with a new determination of the mean parallaxes of the stars with given magnitude and proper motion*. This work has been discussed above (see page 414). It contained an extensive chapter in which van Rhijn determined the relation between the average distance of stars as a function of their proper motion and apparent magnitude. This was now done separately for B-stars and FGK-stars. The thesis work was published in the series of publications from the Mount Wilson Observatory, since much of the work was done while van Rhijn was there. This publication, van Rhijn (1916b), in which the section of the distance versus proper motion and magnitude was shortened considerably, is part of the list of Kapteyn publications in Appendix A, because of his strong involvement.

While at Mount Wilson, van Rhijn had been involved in the question of the brightness of the night sky, which Kapteyn had marked as one of the aims of his Plan of Selected Areas. In 1909, L. Yntema (see page 437) had written a PhD thesis on this subject, showing that the actual amount of light measured from the Netherlands was dominated by 'Earth light' rather than by the background from faint stars. In 1913, while he was at Mount Wilson, van Rhijn employed the same instrument

15.3 In the Mean Time in Groningen

that Yntema had used for measurements there. The sky at Mount Wilson should be much darker and the hope was that the background from faint stars could be observed there, serving as an upper limit on the star counts. The result was published, after long delays, first in the Astrophysical Journal, van Rhijn (1919), and more extensively in the Groningen Publications, van Rhijn (1921): *On the brightness of the sky at night and the total amount of starlight*.

He also introduced a new approach by assuming that above Galactic latitude 40° the contribution from the stars to the total sky brightness was small and could be estimated from star counts. The assumption was that the faint background light came from all stars fainter than magnitude 5.5, in other words stars not visible to the naked eye. He proceeded to subtract that from the observed brightness, giving him the distribution of Earth light. He would then extra- and interpolate across the Milky Way to find the star light there. This did not work completely satisfactorily because the background light also showed variations with azimuth and there were variation towards the North Pole. Van Rhijn realized that there was a significant contribution related to the ecliptic. This is the zodiacal light, reflected Sun light from dust in the plane of the Star System, a term that van Rhijn actually used. He traced the zodiacal light by comparing observations at the same zenith distance, and assumed it was symmetric with respect to the ecliptic and decreased along the ecliptic with angular distance from the Sun. The part that depended on zenith distance he called Earth light. To cut a long story short, he corrected the observed surface brightness, by assuming that at larger Galactic latitudes the star counts gave a reliable estimate of the stellar contribution and by assuming symmetry properties as mentioned for zodiacal and Earth light. He also corrected for observed dependencies of the background with azimuth.

In the end, pretty good agreement with Yntema's work resulted. Whereas Yntema found the full sky brightness to be equal to that of 1350 stars of magnitude 1.0, van Rhijn found a value only 6% larger, 1440 stars. But there was disagreement with some of the Groningen star counts, in particular that in Publication 18, Kapteyn (1908d), which pointed at an error in the adopted magnitude scale. This showed the importance of the study by providing a check on star counts, including on the magnitude scales used.

There was a growing insight that the distributions of stellar velocities of different spectral types were very different. There was e.g. the result of Kapteyn (1910b), that the average random motion increased from spectral type B through GKM. This by the way was independently co-discovered by Monck and confirmed by Campbell [12] at Lick Observatory. This and the fact that the Star Streams contained different types of stars, prompted the thesis by Samuël Cornelis Meijering, defended on July 6, 1916.: *On the systematic motions of the K-stars*. For the K-stars there were many data, but contrary to B (Orion) and A (helium) stars, the streams had very unequal numbers. For K-stars this was unknown, but observational data available sufficient quantity.

Meijering used the method developed by Eddington, based on the assumption that there were two streaming motions and that stars had random motion in each stream that were distributed as proposed by Schwarzschild (see page 358). He had

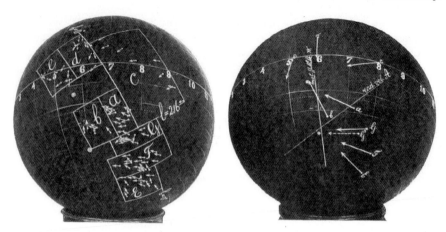

Fig. 15.6 The proper motions of helium stars on the sky. Kapteyn illustrates here the motions on his 'blackboard globes'. Figs. 16.3 and 11.19 show these also (From Kapteyn, 1914b)

1420 stars with proper motion data and 528 with radial velocities. There is no need here to go into the details of his final results, but he did find a solution for the two streams (relative motion 39 km/sec) and their mean (internal) peculiar velocities (about 29 km/sec), plus a solution for the motion of the Sun towards the Antapex (21 km/sec). As far as I am aware the thesis was never published as a journal paper.

15.4 Helium Stars

In 1917 serious friction, which had occurred in milder form before but had been dormant for some time, erupted between Kapteyn and Adams. Before entering into this I need to discuss Kapteyn's extensive work on the parallaxes of 'helium-stars'. It built on his 1910 review at the Mount Wilson Solar Union meeting, Kapteyn (1910d). His first paper on helium stars, Kapteyn (1914b), had the title *On the individual parallaxes of the brighter Galactic helium stars in the southern hemisphere, together with considerations on the parallax of stars in general*. The word *individual* was important here. Kapteyn's starting point was the large degree of 'parallelism' of these stars, which he discussed earlier, and which was confirmed by others. If indeed the stars have small peculiar motions, the Solar Apex motion can in principle be used to estimate individual distances to these stars rather than statistical, secular, ones. The radial velocity program at Mount Wilson was not yet sufficiently advanced so he concentrated on the southern hemisphere. For his study he used stars from the *Preliminary General Catalogue of 6188 stars for the epoch 1900* [13] by Boss, confining himself to stars within 30° from the Milky Way and in a longitude range that in current values would be between about 250° and 30°.

15.4 Helium Stars

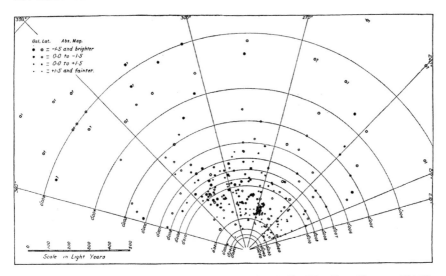

Fig. 15.7 The distribution of helium stars in the plane of the Milky Way (From Kapteyn, 1914b)

Kapteyn nicely illustrated how in his analysis he visualized the streaming patterns on the sky, using two large 'blackboard globes' (see Fig. 15.6), on which he could draw proper motion vectors. These can also be seen in the photograph of his office furniture in Fig. 11.19. The image on the left in Fig. 15.6 showed all individual proper motions of his helium stars and the one on the right the average of these in the boxes indicated on the left. Only the stars that had proper motions exceeding 1″.7 per century, have been included. The bulk of the stars, especially in regions E, F and G moved towards a common vertex with coordinates (estimated as R.A.$\sim 18^h 20^{min}$, $\delta \sim +42°$) rather different from those of the Star Streams, as quoted on page 495. Since radial velocities were also available, this vertex could be used to derive individual parallaxes for each star (following the method outlined in Box 15.2). This then resulted in the map of Fig. 15.7.

This opened the possibility to derive the distance of e.g. a group of early B-stars that had practically the same distance from the Sun, as judged from their mean apparent magnitudes. In modern terms, the data gave the distribution of absolute magnitudes of early B-stars and this could be used to estimate the distance. As an example, Kapteyn derived the distance to the h and χ Persei clusters, discriminating members from non-members on the basis of their radial velocities (measured by Adams and van Maanen). From four member B0-B5 stars, the parallax was found to be 0″.0007. The corresponding distance of 1.4 kpc (1400 pc) was smaller than the current value (2.3 kpc), but it showed the potential power of the method.

In January 1917, Kapteyn sent a manuscript to Mount Wilson Observatory, in which he continued these studies for the interval of (new) Galactic longitude 180° and 250°. This was an even longer paper and eventually it ended up in the Astrophysical Journal in three installments in 1918 as Kapteyn (1918a,b,c). In this area the proper motions were generally small. The radial velocities were very small too, which seemed to indicate that either the velocities were small or the area was near the anti-vertex (defined here as the point of convergence). Moreover they did contain a concentration that Kapteyn called the 'Nebula-group'. The coordinates (R.A.$\sim 5^h30^{min}$, $\delta \sim 0°$) show that on the sky this is the Orion region. Kapteyn mentioned that the Orion Nebula was in this area, and that probably this formed part of it. Strangely, he was incomplete about the sources of his material. He mentioned in detail where the magnitudes, proper motions and spectral types came from, but was silent about the source of the radial velocities.

However, on page 125 he derived a value for the mean peculiar motion of $\bar{u} = 5.78$ km/sec, which was larger than his value in the first paper ($\bar{u} = 3.5$ km/sec). This could be (partly) due to observational error. Kapteyn wrote:

'Firstly, all results were included in [$\bar{u} = 5.78$], whereas in [$\bar{u} = 3.5$] the less reliable values were either excluded or admitted with diminished weight; secondly, [$\bar{u} = 3.5$] was exclusively based on Lick Observatory results, which do not include stars whose spectra are not susceptible of at least fairly good agreement, whereas the Mount Wilson observations, upon which [$\bar{u} = 5.78$] is partly based, include all objects in the original program, irrespective of their difficulty. To secure homogeneity, we limit ourselves to the values obtained by the Lick observers, and if further we exclude all values marked as uncertain, or given without any decimal, or as the estimated velocity of a spectroscopic binary, we find $\bar{u} = 4.24$.

In other words, Kapteyn ignored the Mount Wilson values, which must have been obtained by Adams!

The vertex of the stars in this study was different: (R.A.$\sim 17^h45^{min}$, $\delta \sim +11°$), which he stated 'nearly coincides with the first stream of the non-helium stars'. 'These', he says, '*may*' form a local group. The mean parallax was 0$''$0081, while that for the Nebula-group was 0$''$0054. He derived from his studies that the 'luminosity curve' – the distribution of absolute magnitudes of the helium-stars – was nearly a Gaussian and determined these curves also for subsamples of B-stars (e.g. B0-B5, B8-B9, etc). And then he did the same for A-stars. These two long papers established the distribution of absolute magnitudes for early type stars, which could be used to estimate distances when the spectral type were known.

Early on in the paper, having established the small peculiar motions of the helium stars and the possibility to derive individual parallaxes, Kapteyn noted (page 129, his italics):

'It thus seems not unlikely that further investigation by these methods will give *good determinations of the parallaxes of all the B stars and of a large number of the A stars.*

For the later types the beautiful method proposed by Adams and Kohlschütter and developed in detail by Adams has recently opened the prospect of extensive determination of parallax by spectroscopic means. It awakens the hope that we are at last on the way toward a wholesale determination of the third co-ordinate – distance – the lack of which has been the main obstacle in the way of substantial knowledge of the structure of the stellar system.'

15.5 Friction with Adams

The problem with Adams started with a letter of the latter to Kapteyn on April 24, 1917. This letter is archived in the Hale papers attached to a letter from Kapteyn to Hale of June 26, 1917. The attached Adams' letter is a photographic reproduction provided by Kapteyn. It may be best to start with Adams' letter. It was induced by the manuscript that Kapteyn had sent to Pasadena for publication in January of that year, the second paper on the parallaxes of 'helium-stars', that I just summarized. Adams started by saying the paper was much too long and was not likely to be accepted for the Astrophysical Journal. Next he said he regretted to have to draw Kapteyn's attention to a few matters.

> 'The first of these is the very cavalier fashion in which you refer to the low weight of the Mount Wilson radial velocity results on these stars. Now these observations may not give you the results you might like to have, and it is perfectly true that among the stars included are many with extremely difficult spectra. It is a fact, however, that Professor Campbell on his visit here last August told me he had made a comparison of the B and A stars common to the Lick and Mount Wilson programs, and had found our results to be somewhat more accurate.'

Adams felt unjustly treated by Kapteyn, deserving more credit for his radial velocity work. He had a point here, in that his data were unique and the best available at the time, and indeed it was very difficult work.

> 'Do you not consider it reasonable, whatever you may think of the quality of the results, to show some slight evidence of appreciation of the series of observations which formed nearly the entire spectroscopic program for about two years? In view of your attitude in this matter as well as your letters of rather more than a year ago I can only say that I am thankful that our relations ended when they did. My personal opinion is that you have dealt with subordinates so long that you have forgotten how to act with men engaged in independent work.'

The letter than turned to Kohlschütter. Adams felt accused of not giving the latter sufficient credit in the matter of the determining absolute magnitudes from spectra, or spectroscopic parallaxes, hinted at by Kapteyn.

> 'To the best of my knowledge I have given Kohlschütter full credit at all times for his part in the absolute magnitude method. The original conception, so far as I know, belongs to Hertzsprung. I had many discussion with Kohlschütter about the most suitable lines to investigate based upon my sun-spot spectrum experience. He found certain lines and applied them with success. Had he remained he would, no doubt, have developed the method to its present basis of a working parallax system.'

And Adams finished with

> 'In closing I should like to recall to your memory the various occasions during nearly two years that I called to your attention the fact that the small proper motions stars have small intrinsic velocities and that this might possibly be an absolute magnitude or mass effect. You gave little attention

to it until the end of your stay here when you suddenly realized there might be some basis to it. The Kapteyn and Adams paper of that time I now see referred to as Kapteyn's works. The moral is an obvious one.'

Kapteyn's letter to Hale (of June 16, 1917), to which this was attached in the form of three photographs, was eleven pages long. He immediately comes to the point:

'My dear Hale,

It is with a deep feeling of regret that I write you this letter. A few days ago I got a letter from Adams [...], which together with what passed between us some time ago, has convinced me that any further coming to Mount Wilson would probably do the observatory little good, might do serious damage.

In these circumstances I feel it my duty to resign from my office as a research associate. [...]

You have honoured me by offering me this very exceptional position. I feel strongly that I ought not to reward you by helping to destroy the harmony between the members of your staff.'

Kapteyn next recalled the origin of the problems between him and Adams.

'The trouble began with Adams' publication of his paper *The radial velocity of the more distant stars* in the July Number of 1915 of the Proceedings of the National Academy. In the preceding January number was published a paper by myself and Adams, the origin of which is best explained in connection with a curious passage in Adams letter (10 last lines).

What he says in the 7 first lines of these is in the main true. – True: I had not at first realized the full importance of the matter. When I did, I proposed to Adams to treat this point, together with others, in my opinion certainly not less important points which I had myself in mind, in a joint labour. As he agreed I began to devote all my time to the matter. After having worked for a while, and at your request, I drew up an abstract of the results so far as they had been obtained for the first number of the Proceedings of the National Academy. Meanwhile I continued to work on the matter for more than half a year [...], in fact up to the time that I saw in the proceedings of July 1915 Adams' paper covering the same ground, written without my knowledge under his separate name.'

The joint paper, Kapteyn & Adams (1915), did indeed allude to the fact that 'positive indications have been found of a change of radial velocity with absolute magnitude'. The 1915 paper before the National Academy by Adams alone [14], actually made no firmer claims, except referring to the joint paper and recalling that there it was found that distant stars of types F to M had low average velocity. That paper also pointing out that in contrast to this 'the average radial velocity corrected for the solar motion of such absolutely faint stars as have been observed at Mount Wilson is exceptionally great'. It seemed therefore more a matter of principle than a matter of credit or priority of discovery. Kapteyn related to Hale that he had at the time written to Adams, 'putting it rather mildly':

15.5 Friction with Adams

'Is it not against the rules of the game? We had agreed to write a paper under our joint names and while I am doing my part of the business (rather slowly I admit) you publish a paper by yourself and without my knowledge on the same subject.'

According to Kapteyn 'Adams could not see the point. The result has been that we gave up further joint work on the matter.'

'In fact Adams in his letter now reverses the charge. In the 3 last lines, if I understand them rightly, he accuses me of laying claim to the whole of our joint labour.'

Kapteyn explained that he had in his recent manuscript written that (see page 544): 'For the later types the beautiful method proposed by Kohlschütter or Kohlschütter–Adams recently opened the prospect of extensive determination of parallax by spectroscopic means.' Adams had apparently taken offense to this statement that gave a major share of the credit for this method to Kohlschütter. Before continuing it now would be useful to look into what is involved.

Ever since the spectra of stars were classified in types there was a possibility to judge from the spectrum of a star what its absolute magnitude would be, which would be a quick and easy way of estimating parallaxes of stars. The problem was that there were too few reliable parallaxes to establish a method for this with confidence. Eventually the study of the systematic properties of stars with spectral type led to the discovery of the Hertzsprung–Russell diagram (see Fig. 7.1), which after all is a plot of spectral type versus absolute magnitude. These developments are related to the concept of spectroscopic parallaxes that Kapteyn is referring to. The story of the discovery of the Hertzsprung–Russell diagram (HR-diagram) and the related establishment of the spectroscopic parallax method has been told on other occasions in much detail and I summarize only the highlights. For more details see *DeVorkin's Russell biography, Sandage's Mt. Wilson History* or *The critical importance of Russell's Diagram* by Owen Gingerich (2013) [15].

Hertzsprung was the first to realize that there might be features in stellar spectra related to the absolute magnitude. The Harvard classification scheme of spectral types had resulted in the well-known sequence OBAFGKM (see page 290); but at an earlier stage Antonia Maury had introduced a version that was in the main similar, but contained sub-classes a, b and c. The c-class stars had very narrow spectral lines, unlike the a- and b-classes. In 1905, Hertzsprung, who then still worked as a volunteer at Copenhagen Observatory (and not in Leiden as Gingerich states in the review referred to above) had made the interesting discovery that the narrow-line c-stars seemed to be intrinsically brighter than the rest since they had small proper motions. He published that in an very inaccessible German journal on photography (the 'Zeitschrift für wissenschaftliche Photographie'), followed up by a second paper on the subject in the same journal in 1907. He wrote to Pickering at Harvard, but Pickering would not hear of Maury's sub-classes and seemed not to have responded. Even a summary paper, written by Hertzsprung in 1909 in the Astronomische Nachrichten, *Über die Sterne der Unterabteilungen c und ac nach der Spektralklassifikation von Antonia C. Maury* [16], went mostly unnoticed.

Actually, after having been appointed at Potsdam in 1909 at the Astrophysikalisches Observatorium there by Schwarzschild, Hertzsprung published diagrams for stars in the Pleiades and Hyades clusters, plotting the effective wavelength, a measure for the color, against the apparent magnitude, which since all stars are at the same distance corresponds directly to the absolute magnitude. This was a early version of the HR-diagram. Now, at least in the Hyades there were three stars that were abnormally bright for their spectral class. The matter was shown much more clearly by Russell when he plotted his much more detailed version of the diagram [17] based on trigonometric and on 'hypothetical' parallaxes. The latter are parallaxes of binary stars based on the observed orbits.

This established the existence of separate classes of stars at many spectral types, the 'dwarfs' on the Main Sequence and the 'giants' on what came to be called the Giant Branch (see Fig. 7.1). Giant stars are much larger in diameter and therefore the gasses in the outer atmosphere of the star are more rarefied. The corresponding lower pressure results in narrower spectral lines.

Along these lines, Adams and Kohlschütter at Mount Wilson found spectral characteristics that also depended on absolute magnitudes. Kohlschütter, who had come to Mount Wilson after having been introduced by Kapteyn (see page 490), became involved in the radial velocity program Adams was conducting on the 60-inch telescope. By 1913 they had determined radial velocities of one hundred stars fainter than magnitude 5.5, for which parallaxes had been measured, Adams & Kohlschütter (1914a) [18]. They found, however, that often the spectral classification derived from the ratio of the strengths of lines in their high-resolution spectra was significantly different from the *Draper Catalogue* of Harvard College Observatory. When they compared spectra of these low proper motion stars with those of high proper motion, Adams and Kohlschütter found that there were particular characteristics of both the lines and the intensity of the violet continuum that appeared to correlate with absolute magnitude. Their paper, published in the same year, Adams & Kohlschütter (1914b) [19] showed that for F.G and K stars, the same spectral type obtained for both groups of stars, while the ones with small proper motion showed abnormally great intensity of the hydrogen lines.

The joint work of Adams and Kohlschütter resulted eventually in the method of spectroscopic parallax, but the outbreak of the war in 1914 prevented Kohlschütter from continuing working on this, as he tried to return to Germany (see page 500). He spent the war in relatively comfortable English captivity. Now, Adams then developed by himself the method of estimating absolute magnitudes from spectral characteristics that was started in Adams & Kohlschütter (1914b). Adams wrote a series of four papers on *Investigations in stellar spectroscopy* in the proceedings of the National Academy [20] and a summary paper in an astronomical journal: *A Spectroscopic method of determining stellar parallax* [21] He showed how the spectral characteristics could be used to find whether a star was a Main Sequence dwarf or a giant. Then the Hertzsprung-Russell diagram could be used to estimate its absolute magnitude.

In all these papers Adams did give due credit to Kohlschütter, although only in relation to quantitative relations of finding a star's spectral type from lines in Mount Wilson spectra. Adams did note that early on Kohlschütter had independently from Hertzsprung found a spectral line that appeared to correlate with absolute magnitude. But Adams takes the credit for the development of the spectroscopic parallax method, which may very well be justified. What apparently happened is that Kapteyn from his visits to Mount Wilson felt that Kohlschütter should be credited more prominently with the development of the method than Adams did. He therefore wrote in his manuscript: 'For the later types the beautiful method proposed by Kohlschütter or Kohlschütter–Adams recently opened the prospect of extensive determination of parallax by spectroscopic means.' Adams took offense to this. And probably he was right in claiming that he deserved the credit in the first place. Maybe he would have been willing to share it with Kohlschütter, but being plainly put in second place by Kapteyn infuriated him. Adams certainly seemed to have a point. But he was sometimes (the biography of Russell by DeVorkin has examples of this) described as difficult to deal with and irascible. He was known to have a dislike of foreigners taking up positions in the USA or telling American astronomers what to do (he was sometimes directed by Hale to perform observations for Kapteyn). After World War I, Adams was fiercely opposed to German participation in international organizations. But there is absolutely no indication that Kohlschütter's nationality played a role here, and Kapteyn appears to have given too much of the credit to Kohlschütter. But still, the tone of Adam's letter to Kapteyn was unnecessarily nasty and unfriendly. After Kapteyn's death, Adams did write a few equally nasty letters to van Rhijn (see page 554).

15.6 Hale's Reaction and Seares

Hale replied to Kapteyn on September 21, 1917. He took part of the blame for the developments, referring to his frequent absence.

'Let me say at once that I do not wonder, in view of the tone of Adam's letter, that you consider all further co-operation with him to be impossible. He has expressed to me his regret that he wrote to you in this way. He wrote when irritated over your MS, which he considered to contain direct and unwarranted criticism of himself. And he certainly had no sufficient reason for using such expressions, as I have told him.
Adams says he has seen no impropriety in the publication of the paper which appeared in the Proceedings for July 1915. He gave you on the first page credit for the derivation (in the joint paper of January 1915) of the relationship between proper motion and radial velocity for the K stars, and on the last page he made a second reference to the joint paper, as bearing upon the relationship between radial velocity and absolute luminosity. He did not suppose you would object to the publication of his paper, and was surprised when you did so.

The whole matter seems to turn on the understanding you had with Adams regarding the joint work. For my part, I think he should have sent you the MS of the paper to you before publishing it.'

So Hale started out by admitting that Adams has gone too far in the way he addressed Kapteyn, but then took a tactful position in the case of the 1915 paper, not clearly taking one of the other side. In the Kohlschütter matter he sided definitely with Adams though.

'If you were here I think we could straighten out misunderstandings and get to the bottom of the matter. In the Kohlschütter case, however, I do not see that Adams is to blame. He certainly gave credit to Kohlschütter in his talks with me and also in his published papers. If others have failed to do so, it is not Adam's fault, though it would have been if he had tried to conceal Kohlschtter's part in the work. But Seares has gone into this so fully in his recent letter to you that I need not dwell upon it. I trust that this view of the matter will prove satisfactory to you.'

I will address this Seares letter below. Hale continued by saying he understood Kapteyn's position that he no longer wished to work with Adams, but argued that this should not be a reason to give up his position as Research Associate. It was not necessary to come to Mount Wilson '(keenly as I regret your absence). In fact, I am sure President Woodward will agree that you ought to retain this position even if you do not co-operate with the Observatory or appear as a member of the staff at all.' As a result Kapteyn withdrew his resignation and wrote to Adams on October 18, 1915:

'My dear Adams,
There is no question of any dishonourable actions. I know you too well to think you capable of such a thing. But I think you made a serious mistake. And I think that a man of your keen sense of justice must be able to see when he has made an error and will frankly admit so if he finds he has.
Meanwhile after what has happened I cannot see another way than that proposed by yourself of disconnecting our names in any further treatment of the subject. Then we will be both free to write exactly as we like'

Kapteyn then mentioned he will give up his position as research assistant 'the moment I find further cooperations impossible or even difficult. I most heartily hope it will not come to this'.

The situation was also alleviated by the tactful intervention of Frederick Sears (see Fig. 15.8), who handled the publications emanating from the Mount Wilson Observatory. He wrote on May 2, 1917 to Kapteyn. 'By the time this letter reaches you your manuscript will be in the hands of the editor of the Astrophysical Journal. I hope there will be no difficulty about printing it in the Journal, although I have been a little concerned about the length of the paper, which is considerably in excess of any we have hitherto printed in this manner.' He then stated he has gone through the manuscript attempting to shorten whenever possible, trying 'to preserve your meaning whenever changes have been made'. He identified a few points that 'require special comment'. One of these is Kapteyn's unfortunate wording concerning the quality of Mount Wilson radial velocities. Seares said that

15.6 Hale's Reaction and Seares

Fig. 15.8 Painting of Frederick Hanley Seares (From the Huntington Digital Library [22])

'Campbell himself admits that the Mt. Wilson precision is equal to, if not slightly superior to that measured at Mt. Hamilton' (Lick Observatory). And on the reference to the spectroscopic method for parallaxes, Seared noted that 'Kohlschütter never worked independently on the problem' and told Kapteyn that he adapted the 'text and the footnote accordingly'. He expected that the result would be acceptable to both Kapteyn and Adams.

Seares also addressed the question of the motion of late type stars, which he noted 'might appropriately be modified in accordance with the results of Adams and Strömberg'. Kapteyn wrote back on July 7, 1917, agreeing to some of Seares' changes. But he differed with the view that Kohlschütter never worked independently on the spectroscopic parallax issue. 'Van Rhijn and myself were in California when Kohlschütter found the result and it was, and is, my firm conviction that, if Kohlschütter at that time had published his results, everybody would now credit the parallax method exclusively to him or perhaps to Hertzsprung–Kohlschütter'. But he quoted from a letter to Hale: 'This does not imply any disrespect to what Adams did in the matter. On the contrary, I have little doubt that by Adams' work the value of the method has been very greatly increased and that he further contributed to the elucidation of the underlying phenomena.' Kapteyn asked Seares to check if Adams agreed with this and his original formulation and otherwise leave the text Seares has submitted. Seares wrote on October 7, 1917 that he felt no urge to consult Adams again and proposed to leave the text as he suggested earlier, as Adams raised no objections to that. He reserved judgment on the Adams–Strömberg result as he had not seen that paper. The paper did get published without further change in three parts.

15.7 The Adams: Strömberg Result

But almost immediately a new controversy involving Adams ensued. The paper Seares referred to by Adams and Strömberg on *The relationship of stellar motions to absolute magnitude* [23], and a follow-up paper by Strömberg alone, *A determination of the Solar motion and the Stream-motion from radial velocities and absolute magnitudes of stars of late Spectral types* [24] directly addressed the validity of Kapteyn's formula giving a mean parallax for a given proper motion and apparent magnitude, Kapteyn (1900e). Gustaf Strömberg (1882–1962) was born in Sweden and educated at various places, including Kiel, ending up in 1906 as assistant at Stockholm Observatory. In 1917 he joined the staff of the Mount Wilson Observatory. Strömberg worked on the problem of stellar motions, which eventually led to his discovery of the 'Asymmetric Drift' in stellar motions, that he presented in papers in 1924 and 1925 [25].

This asymmetric drift manifests itself as follows. Groups of stars that have higher random motions also have a different collective, systematic motion with respect to the mean of the stars. It can be understood in the context of a rotating Galaxy. This only came forward in the 1920s after Kapteyn had died, so was unknown at the time the phenomenon was discovered. We have seen that Kapteyn & Adams (1915) already found that stars of later spectral type had on average larger random motions. I also noted that this now is explained as a result of increasing random motion with age, resulting from the deflection of the motions of stars by gravitational irregularities (see page 532). Now think of a collection of stars with high random motion, which moves in the disk of the Galaxy. For any such collections the gravitational pull of the Galaxy as a whole must be compensated by motions; the gravitational energy must be compensated by kinetic energy for the system to be in equilibrium. Most of this kinetic energy is provided by the rotation of this collection of stars as a whole around the center of the Galaxy, but some of it is contained in the random motions. That means that when the random motions are higher, less energy needs to be present in the systematic rotation; the random motions act like a pressure in a gas. Such collections with relatively high random motions then will have a lower systematic rotation and as it were lag behind. This systematic behavior is Strömberg's asymmetric drift. And since older generations of stars have larger random motions, the Strömberg drift also correlates with age.

Adams and Strömberg did confirm the increase in radial velocity on average with absolute magnitude, but Kapteyn's criticism was aimed at their adaption of his formula to find the average parallax of stars from their proper motions and apparent magnitude. Adams and Strömberg argued that this was necessary, since van Maa-

15.7 The Adams: Strömberg Result

nen in an ongoing project to measure parallaxes, had found that stars with small proper motions had generally much larger (up to a factor three) parallaxes than would be derived from Kapteyn's formula. Moreover, Strömberg in his solo paper analyzed the motions independent of assumptions of Star Streams or 'Maxwellian' (Schwarzschild-like) velocity distributions and solved for preferential motions for groups of stars of different luminosities and distances. From his detailed analysis, Strmberg came to the conclusion that it 'perhaps indicates that the stars studied are mainly moving around the center of the galactic system, with a preferential motion in the galactic plane'. This is a prelude to his later detection of the asymmetric drift.

Kapteyn strongly disagreed. He wrote to Hale about the Adams and Strömberg paper on November 3, 1917. 'For what is the use of an associate for certain branches of astronomy, if it does not prevent the appearance of (I cannot help saying plainly my opinion) such a poor paper. Just in these branches? [...] What is the use of having an associate come all the way from Europe, if the consequence is simply quarreling with the members of your staff? For, if I reply to the paper and show that Adams-Strömberg does not understand a bit what Adams-Kapteyn maintains, we have to expect beautiful complications.' As happened before, Kapteyn speaks his mind in a hot-headed manner. Hale, tactfully, asked him to send 'a criticism [...] as I should like to have your view of its weak points'.

Kapteyn did so on May 23, 1918, but added that he was also upset by the solo paper of Strömberg that had appeared in the mean time, and also by a paper by Adams and Joy. Alfred Harrison Joy (1882–1973) was an American spectroscopist, who had joined the staff at Mount Wilson in 1915. In the Adams–Joy paper on *The luminosities and parallaxes of five hundred stars* [26], the authors made use of the method of spectroscopic parallaxes and in particular made a strong case for a clear division between giants and dwarfs, at least for M and K stars (and probably for type G and an indication for it in type F). This was not a very welcome conclusion to Kapteyn, since it would complicate his scheme, as presented at the 1913 address at the National Academy (see page 495), that had the spectral types correspond uniquely to a stellar age sequence. Thus he needed to explain that random velocities increased monotonically with age and the difference between the Star Streams then confirmed his two stream interpretation rather than Schwarzschild's 'ellipsoidal' explanation. Kapteyn's letter, addressing the three papers, was detailed and no less than 14 pages long, and I will not discuss all the details. Kapteyn did note at the outset, that 'it is a great pity that the authors have not thought it worthwhile to correspond with me before publishing their papers. There was every reason to do this even apart from the fact of my position in relation with the Mt. Wilson Observatory, because the results bear strongly on my own work'.

Kapteyn also produced a manuscript for publication *On the parallax of the bright stars having small proper motions*, which ended up in the hands of publication editor Seares at Mount Wilson. Again the tactful manner in which Seares treated all this, saved the day. He wrote a letter to Kapteyn on March 15, 1919. The letter had a few attachments. One was a note by Seares himself evaluating the matter, another Kapteyn's manuscript and a third a note by Strömberg that if Kapteyn agreed could be published to set the record straight. Seares put a part of the blame on Strömberg's command of English and on himself for having not been more alert when the Adams-Strömberg paper passed his desk. To cut a long story short, the note by Strömberg got published and Kapteyn's note did not. The note by Strömberg, however, appeared only in the series of the 'Contributions of the Mount Wilson Observatory', which in almost all other cases except this one consisted of reprints of articles published in other journals. The *Note on the relationship of mean parallax to mean proper motion* [27] is the same as that included by Seares in his letter to Kapteyn, so (although I have found no further correspondence) it must be assumed that Kapteyn gave his consent to publish.

In the note, Strömberg started by mentioning correspondence with Kapteyn and then admitted to having followed an incorrect procedure in taking means, drawing the samples and grouping the stars. He then proceeded to re-derive adaptions for Kapteyn's formula but now restricting it, except for the extrapolation, to small parallaxes. The curves he now derived agreed with the data Kapteyn based his analysis on. Of course the relations between proper motions and radial velocities, as first noted in Kapteyn & Adams (1915), and between radial velocities and absolute magnitude found by Strömberg and Adams, did remain.

So Strömberg's work eventually lead to the discovery of asymmetric drift, to the recognition of a strong asymmetry in the directions of the motions of stars of very high velocities and to the discovery of Galactic rotation by Bertyl Lindblad (1895–1965) in Sweden and particularly Jan Hendrik Oort before the end of the decade of the 1920s. And it eventually caused the realisation that Kapteyn's Star Streams had to be understood in terms of Schwarzschild's ellipsoidal velocity distributions.

The matter did not rest here. In 1922, van Rhijn apparently sent a manuscript to Mount Wilson on a comparison between spectroscopic parallaxes and those from a mean parallax formula. It bore some resemblance to Kapteyn's note, referred to above. Adams reacted to it on November 20, 1922, sending a note by Strömberg and some comments of his own. Adam's tone is not friendly, being irritated that van Rhijn ignored publications by Strömberg and referred to 'true' values when he meant 'his own'. Van Rhijn was surprised: 'I really am sorry if I have hurt you in any way. I owe a great debt of gratitude towards Mount Wilson Observatory and its scientists and it would be very bad if the content of my paper had been disagreeable to you for personal reasons.' On January 22, 1923 there is another letter by Adams and a memorandum by Strömberg. Adams wanted to close the correspondence, which van Rhijn agreed to in a letter of March 5, 1923.

Again in 1926, Adams felt offended by remarks van Rhijn has made in two papers in the Groningen Publications: *On the mean parallax of stars of determined proper motion, apparent magnitude and Galactic latitude for each spectral class* in 1923 and *Comparison between trigonometric, spectroscopic and statistical parallaxes* in 1925 [28]. In a letter of April 14, 1926, van Rhijn wrote to Adams: 'I really am very sorry about the letter you wrote me. I perfectly realize that I owe much of my scientific training to my staying at your observatory and I remember gratefully the many interesting talks I have had with you personally. This is my feeling against you and your observatory and I really have no anomies [enemies or animosities?] against the work of the spectroscopic department of the Mount Wilson Observatory.' He stated he reread the relevant passages (pp. 60–68 in GP34 and 16–18 in GP37) and really didn't 'see where I have used offensive language'. 'Really and truly, Mr Adams I have no personal feeling against you or any of the Mount Wilson people'.

In all honesty, I can see no reason in these texts for Adams to experience them as so offending. These animosities between Adams and Kapteyn (and later van Rhijn) were unfortunate, and both Hale and Seares had tried to resolve these tactfully. There is no need here to establish who is to blame. It is true though that Kapteyn was not always easy to deal with; there is an unconfirmed suggestion that at some stage he had written Eddington as secretary of the Royal Astronomical Society trying to prevent the Gold Medal from being awarded to Adams (allegedly in a letter of January 18, 1921 of H.H. Turner to Hale; Adams received the medal in 1917 anyway and deservedly so). But Adams had a history of falling out with people, in particular prominent American astronomers, such as Harlow Shapley (see Owen Gingerich in the *Legacy*) and Henry Norris Russell. And he had a strong 'sense of propriety' and was a clear example of the xenophobic view and strong aversion among some American astronomers to the hiring of large numbers of European astronomers when young, promising Americans were left unemployed (see *DeVorkin's Russell biography* [29]). Actually, Robert Paul in his *Statistical Astronomy* claims (p. 170) that 'Adams eventually took a great dislike to Kohlschütter (because he was German)', but a few sentences earlier Paul makes the completely incorrect statement that 'Kohlschütter himself remained at the 'mountain' [used in the preceding sentence for Mount Wilson] for many years until his retirement.' The Adams-Kapteyn discordance lasted beyond Kapteyn's lifetime and actually ran so deep that it has been a major cause for Kapteyn not returning to Pasadena after the end of World War I and later after his retirement as professor in Groningen (see page 599).

15.8 Preparations for the First Attempt

Although much work at the Astronomical Laboratory took place in measuring plates in the context of the Plan of Selected Areas, other work went on as well to prepare for a first attempt at the solution of the Sidereal Problem. Kapteyn felt that if he waited for the availability of in particular the deep counts from the Mount Wilson 60-inch survey, he would be well after retirement. Regularly results were published in the series of Groningen Publications, some of which results of special observing campaigns at different observatories: *The proper motions of 3714 stars derived from plates taken at the observatories of Helsingfors and Cape of Good Hope*, Kapteyn & Weersma (1914); *The proper motions of the stars in and near the Praesepe cluster, derived from photographic plates taken at Potsdam*, van Rhijn (1916a), and *The*

> **Box 15.3 The Principal Laws of Statistical Astronomy**
> In the notation as in Schouten (1918):
> h_m, apparent brightness of a star of magnitude m; $h_m = 1$ for $m = 0$.
> i, absolute brightness corresponding to absolute magnitude M.
> $N(h_m)$, observed number of stars of magnitude m.
> $\pi(h)$, mean parallax of stars of magnitude m.
> $D(r)$, density (number per unit volume) of stars at distance r.
> $\varphi(i)di$, percentage of stars between i and $i = di$.
> For any star $i = h_m r^2$; r and $\pi(h)$ need to be determined from proper motions, secular and other parallaxes.
> The $D(r)$ and $\varphi(i)di$ result from:
>
> $$N(h_m) = 4\pi \int_0^\infty \varphi(h_m r^2) r^4 dr,$$
>
> $$N(h_m)\pi(h) = 4\pi \int_0^\infty \varphi(h_m r^2) r^3 dr,$$
>
> which involves simultaneously solving two integral equations.

proper motion of 2380 stars derived from plates taken at the Royal Observatory, Cape of Good Hope, Kapteyn & van Rhijn (1918).

But some of it, mostly done with the assistants, first Weersma and later van Rhijn, prepared the way for making a model for the distribution of stars in space: *On the number of stars of each photographic magnitude in different Galactic latitudes*, van Rhijn (1917), *The secular parallax of the stars of different magnitude, Galactic latitude and spectrum*, Kapteyn, van Rhijn & Weersma (1918) and *Numbers of stars between definite limits of magnitude, proper motion and Galactic latitude for each spectral class.*, Kapteyn & van Rhijn (1920a). This work all was part of the necessary preparatory effort to bring information in catalogues etc. into a form that could be used to study the 'Construction of the Heavens'. And two more PhD theses were prepared and defended.

The PhD thesis by Willem Johannes Adriaan Schouten (1893–1971) *On the determination of the principal laws of statistical astronomy*, Schouten (1918) made a detailed comparison of the various approaches to derive the three 'principal laws' of sidereal astronomy. These are the distributions in space of density, luminosity and three-dimensional velocity. These laws were derived from observations of the numbers of stars as a function of magnitude and proper motion and on the mean parallaxes as a function of the same observables. The problem has been summarized in mathematical form in Box 15.3 (note it is similar to but addresses a different problem than Box 9.1). It involved solving two integral equations, which was a difficult procedure. Schouten first discussed the methods followed by Schwarzschild and by von Seeliger in detail.

I will not discuss the details of the analysis, but restrict myself to summarizing it. A much more involved discussion has been presented in *Paul's Statistical*

15.8 Preparations for the First Attempt

Astronomy. In the case of von Seeliger, a particular mathematical form of the number of stars as a function of magnitude was assumed, which then according to a theorem he developed, resulted in a particular mathematical formula for the number of stars per unit volume as a function of distance from the Sun. Schwarzschild used the mathematical technique of Fourier analysis. Schouten vehemently criticized the approach of von Seeliger. The discussion of Paul in his book just referred to relates how Kapteyn felt embarrassed by the antagonistic tone of the thesis and apparently refused to publish it in the Groningen Publications. In spite of that, it did have an adverse effect on the relation of von Seeliger with 'Kapteyn and his lieutenants' (see *Paul's Statistical Astronomy*, p.177) and was further cause for the two to develop the subject of statistical astronomy separately. The copy of Schouten's thesis at the Kapteyn Institute does not contain his propositions; maybe these were too embarrassing for Kapteyn, so that he had removed them from his copy. There is no sign of them having been cut or torn from the copy, but as often is the case, they might have been loose-leaf.

Schouten became 'privaatdocent' (private lecturer, an in principle unpaid, usually five-year position to teach in a discipline that is not formerly covered) in astronomy at the University of Amsterdam, before he became HBS and Gymnasium teacher in Arnhem; he wrote various popular books on astronomy and related subjects.

Another thesis written under Kapteyn was by Gerrit Hendrik ten Bruggen Cate (1901–1961), *Determination and discussion of the spectral classes of 700 stars mostly near the North Pole*, ten Bruggen Cate (1920). It concerned the reduction and discussion of spectra of 700 stars, obtained by later Nobel laureate Frederik (Frits) Zernike with the 6-inch telescope at Potsdam Astrophysikalisches Observatorium, while he worked as Kapteyn's assistant. Not only did ten Bruggen Cate classify the spectra, he also determined the incidence of spectral types at different magnitudes and the change of color with apparent and absolute magnitude. His most important conclusion was that 'stars with small proper motion are, *ceteris paribus* [all other things being equal] redder than those with large proper motion' and he attributed that to a correlation of absolute magnitude and color. This agreed with the results of van Rhijn's dissertation and with the early results of Adams and Kohlschütter on the brightness of the violet continuum parts of spectra of later type stars.

Ten Bruggen Cate (1883–1963) later changed his name to 'ten Bruggencate'. He should not be confused with Paul ten Bruggencate (1901–1961), who was a German astronomer and a student of von Seeliger. After his degree, the Dutch ten Bruggencate became a physics teacher and director of the 'Hoogere Burgerschool' (HSS) (see page 62) in Sappemeer (near Groningen) and later the Frisian capital Leeuwarden. The thesis was never published in a scientific journal, but ten Bruggencate did in 1925 produce a paper on *The systematic motions of faint stars* [30].

Finally, I mention Egbert Adriaan Kreiken, who studied under Kapteyn, but defended his thesis not long after van Rhijn took over the directorship of the Laboratory. We met Kreiken already on page 14, since he was born in the same house as Kapteyn in Barneveld. Kreiken's PhD thesis had the title *On the colour of the faint stars in the Milky Way and the distance of the Scutum group*; it was defended on February 21, 1923. It was much in the tradition of Kapteyn, using photographic material obtained by Hertzsprung at Mount Wilson. Kreiken felt very much a student of Kapteyn; the thesis has no acknowledgement to van Rhijn, but mentions the generosity of Hertzsprung and assistance at the Laboratory. However, the first page is empty except for a statement 'To the memory of Prof. Dr. J.C. Kapteyn this work is dedicated'. Although also active as a teacher, Kreiken went on the pursue an active and successful career in astronomy, for many years at the Lembang Observatory, later in Liberia and Turkey. He is remembered for his founding of the Ankara Observatory, which in 2013 was named the Ankara University Kreiken Observatory [31]. Kreiken's life has been described in *Prof. Dr. Egbert Adriaan Kreiken* by C. Güner Omay with contributions from nephew Juus Kreiken, who also wrote a very informative article about Kreiken in Dutch [32].

15.9 The Distribution of Stars in Space

In 1920 finally Kapteyn, together with van Rhijn, published a 'provisional' model for the distribution of the stars in space. The paper, *On the distribution of the stars in space especially in the high Galactic latitudes*, Kapteyn & van Rhijn (1920b), started out as follows:

'1. The investigation contained in G.P. [Groningen Publications] 27, 29 and 30 were carried out for the purpose of making possible an elaborate treatment of the arrangement of the stars in space. At least two more publications will be necessary to complete this investigation. Now that, after so many years of preparation, our data seem at last to be sufficient for the purpose, we have been unable to restrain our curiosity and have resolved to carry through completely a small part of the work, even though, by doing so, the rules for strict economy of labor cannot be altogether adhered to.

The present paper is the outcome of this more or less provisional work. In the main it relates to the stars as a whole between Galactic latitudes $\pm 40°$ and $\pm 90°$. It is provisional in that the extremely valuable parallaxes, now placed at our disposal by Mitchell, have not yet been used, and moreover because the thorough discussion which we will have to be made of the measured parallaxes in all Galactic latitudes is not yet complete. We expect little to be changed, however, in the definitive treatment of the subject, which will form a part of the more comprehensive work.

A somewhat less refined investigation of all stars (irrespective of spectrum) in galactic latitudes $0°$, $30°$, $60°$ and $90°$ has also been added.'

Kapteyn was nearing his retirement at the age of 70, which he would reach in 1921, so he was eager to at least perform the work he had been waiting to do for his whole career. It is also here that he finally accepted the parsec and redefined the absolute magnitude:

15.9 The Distribution of Stars in Space

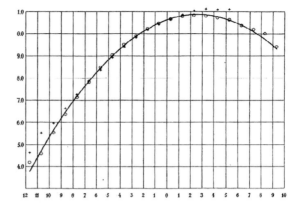

Fig. 15.9 The distribution of absolute magnitudes of stars, the 'luminosity-curve', as derived in Kapteyn and van Rhijn (1920b). The vertical axis is the logarithm of stars per 1000 cubic parsec, and the horizontal one absolute magnitude (at *1(!)* parsec)

'Different units of distance have been used by different astronomers. The most widely used at present is the parsec. For the sake of uniformity we have resolved henceforth not only use this unit but also use the name, which is very convenient (though very ugly). To conform with this new unit, the value of the absolute magnitude, which must still be defined to be the magnitude of a star as it would appear at the unit distance, must also be changed. Its numerical value will now be five less than it was according to former publications, [...].

Most of the work is on star counts at higher latitudes, using the distribution of stars as a function of apparent magnitude and proper motion from Kapteyn & van Rhijn (1920a). First they calibrated the parallax as a function of apparent magnitude and proper motions using measured parallaxes (both trigonometric and secular). Next they had to assume a form for the distribution around these means, which from earlier work in Kapteyn & Donner (1902) was known to be not very critical (but the average values were critical). This resulted, following the methods laid out in Kapteyn (1902d), in the luminosity curve shown in Fig. 15.9. The fitted line is a Gaussian (bell) distribution. This actually is not too bad at all for stars to the left of the peak (at current absolute magnitude $M = +8$); modern determinations do not have the peak but show a rather strong flattening off towards fainter stars on the right (e.g. J.N. Bahcall & R.M. Soneira, 1980 [33]).

The procedure was then to assume this to be independent of distance from the Sun and convert the counts of stars as a function of magnitude to the densities in space (this is solving the last integral equation in Box 15.3). Kapteyn and van Rhijn did this using analytical methods developed by Karl Schwarzschild in 1910 for the case the luminosity curve has a Gaussian shape. This resulted then in a formula relating the density of stars relative to that near the Sun as a function of distance. Within 20 pc it was more or less constant (at 0.0451 stars per cubic parsec) and then dropped, first quickly, reaching half this value at about 200 pc, and then more slowly to one-tenth of that value at some 800 pc from the Sun. The density near the Sun corresponded to a mean separation between stars of 2.8 parsec (8.7 lightyears).

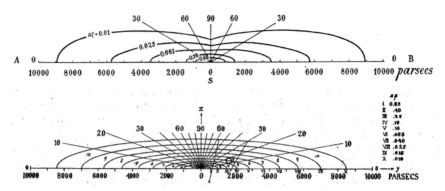

Fig. 15.10 The 'Kapteyn Universe'; the arrangement of stars in space according to the studies of Kapteyn and van Rhijn (1920b) at the top and Kapteyn (1922a) at the bottom. The Sun is shown at about three times the distance from the center as Kapteyn determined it himself

Next they did the same thing on much less complete data for latitudes 0°, 30°, 60° and 90°, using the same luminosity curve, assumed to be the same everywhere in space. To facilitate the calculations they approximated the Sidereal System in this very first attempt by a figure of revolution with the Sun in the center. So, for completeness I summarize the assumptions used for the modeling: (1) the luminosity distribution of stars is the same everywhere in the Sidereal System, (2) the distribution is circularly symmetric about an axis perpendicular to the Milky Way, (3) the Sun is near the center of the distribution, and of course as always (4) there is no absorption of light in space. All these matters have to be revisited, they noted, in future improved attempts. The resulting preliminary view of the distribution of stars in space was that of the upper part of Fig. 15.10. The distribution went out to a density of one-hundredth of that near the Sun (so is significantly extrapolated) and at this level the System measured in round numbers 3500 by 18000 parsecs.

The outer parts obviously were very uncertain, but the analysis made no use yet of the deep star counts to be provided by the Mount Wilson survey in the framework of the Plan of Selected Areas. In Kapteyn and van Rhijn (1922b), *On the upper limit of the distance to which the arrangement of stars in space can at present be determined with some confidence*, it was estimated from simulations, that with counts down to magnitude 17, which the plates taken with the 60-inch would provide, 'the densities should become pretty reliable for the whole of the domain within which the density exceeds 0.1 of that near the Sun'.

How accurate in hindsight were these star counts? In a paper in 1986, already referred to on page 213 (also see my chapter in the *Legacy*) in connection with Herschel's star gauges, I have made a comparison [34] of the star counts produced in Groningen with a modern model for the distribution of stars in the Galaxy, published by J.N. Bahcall & R.M. Soneira (1980). Here I repeat that exercise for the counts published by van Rhijn in 1929 after all the data of the Harvard and Mount Wilson surveys for the Plan of Selected Areas had been incorporated. For the brighter magnitudes the data would be similar to what was used by Kapteyn and van Rhijn

15.9 The Distribution of Stars in Space

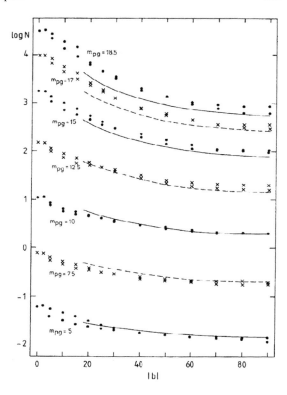

Fig. 15.11 Star counts of van Rhijn (1929), compared to a modern model for the distribution of the stars in the Galaxy by Bahcall & Soniera. The vertical axis shows number of stars brighter than the magnitude indicated, the horizontal scale is Galactic latitude (north and south both plotted) (From P. C. van der Kruit (1986))

in 1920. The paper, *Distribution of stars according to apparent magnitude, Galactic latitude and Galactic longitude*, van Rhijn (1929) [35], is very detailed and I have only show some data from it in Fig. 15.11 at selected magnitudes. For each magnitude we see in the figure the number of stars per square degree in the vertical direction (on a logarithmic scale) as two symbols separately for northern and southern latitudes. The horizontal axis is Galactic latitude. The lines are the counts as they follow from the model for distribution of stars in the Milky Way Galaxy according to Bahcall & Soniera. We see that the lines and the symbols display very closely the same shape, which indicates that the counts are consistent over the sky in a relative sense. But the magnitude scales are not right; for the fainter stars the Groningen magnitudes become systematically too bright by more than half a magnitude. The difference is actually pretty good if one considers the major problems that existed with magnitude scales early (and even much later) in the twentieth century. This figure reveals what a remarkable achievement this part of the Plan of Selected Areas was.

Hugo von Seeliger (Fig. 15.12), working quite independently from Kapteyn, had obtained a similar result. He published his results in a paper in 1920 [36]. Both von Seeliger and Kapteyn found equi-density surfaces that were rather flat. However, the results were only similar in a superficial way, and von Seeliger did not follow it up with a discussion of the dynamics, which Kapteyn did in a subsequent paper. They

Fig. 15.12 Ritter Hugo Hans von Seeliger as he appears in the *Album Amicorum* presented at the retirement of H.G. van de Sande Bakhuyzen as professor of astronomy and director of Leiden Observatory in 1908 (Archives Leiden Observatory; see caption Fig. 3.7)

seem not to have communicated about it, although van Rhijn and von Seeliger did exchange critical remarks in 1921 in two papers in the Astronomische Nachrichten; *Bemerkung zu Dr. H. Seeligers letzten Untersuchungen über das Sternsystem*, van Rhijn (1921), and *Bemerkungen zum Aufsatz des Herrn van Rhijn in AN 5091*, von Seeliger (1921) [37].

15.10 The Crowning Achievement: Galactic Dynamics

In a way the model with the distribution of stars in space with van Rhijn can be seen as representing 'the culmination of his life's work on the sidereal problem' (quote from *Paul's Statistical Astronomy*, p. 157). But Kapteyn would not have seen it this way. What he wanted to achieve was not to merely determine the distribution of the stars in space, but to develop a model that described also the motions in the Sidereal System and linked the space distributions to these motions (the kinematics) and the gravitational force of the stars, in other words he also wanted to understand the dynamics.

For a long time, Kapteyn had been thinking about the dynamics of the stars in space. He wrote about this quite extensively as early as 1915 in a long letter to Hale of September 23, from which I already quoted a part of it that concerned the 'art of discovery' (see page 522 and further). I pick it up again at the ninth page.

'I myself am quite well again and with my health the interest in my work has completely returned. It has led of late to what I think is a very

15.10 The Crowning Achievement: Galactic Dynamics

important result, viz. that the Stream velocity increases with decreasing distance from the Sun. The result seems to me to be well established.

One of the somewhat startling consequences is, that we have to admit that our Solar System seems to be in or near the center of the Universe, or at least to some local center.

Twenty years ago this would have made me very skeptical and in regard to the result of the investigation. Now it is not so. – Seeliger, Schwarzschild, Eddington and myself have found that the number of stars pro unit of volume is greatest near the Sun. I have sometimes felt uneasy in my mind about this result, because in its derivation the consideration of scattering of light in space has been neglected. Still, it appears more and more that the scattering must be too small and also somewhat different in character from what would explain the change in apparent density. This change is therefore pretty surely real.'

At this date Kapteyn may or not have heard of Shapley's result on globular clusters (see page 415) that absorption towards these clusters was very small. Shapley's contribution was submitted in November 1915 and its essential conclusions might be known by others, especially since Shapley at the time worked at Mount Wilson. In any case, at this time Kapteyn seemed to have given up his skepticism about the central position of the Sun and to have accepted that absorption probably was not a very serious problem. Had he heard of Shapley's result by that time, or not long afterwards, it would have strengthened his acceptance of the central solar position and the neglect of absorption, at least he would have felt it could be ignored in a preliminary analysis. The letter to Hale continued:

'Even more important than the central position of the Sun seems to me to be that our result for the first time shows the evidence of <u>force</u> in the great Sidereal System. A rough computation leads to the conclusion that at a distance corresponding with a parallax of $0''.02$ the stars are under the action of a force equal to the attraction of a central mass having 5 million times the Sun's mass.

This number is very considerably higher than the number of stars we assumed up to the present to exist in a sphere of a radius corresponding to $\pi = 0''.02$. But these of course may be completely dark bodies, of which we know nothing.

For some reason, which I cannot well tell at present, I would rather not give my new results for the Proceedings of the National Academy. But I will finish the whole of my paper, which will be pretty extensive, as soon as I can for the Contributions'

It did take quite a bit longer before Kapteyn finished what was to be the culmination of his efforts, a paper in which he combined the distribution of the stars in space with a theory of their motions in a consistent picture of equilibrium in the presence of gravitational forces. The paper, Kapteyn (1922a), entitled *First attempt at a theory of the arrangement and motion of the Sidereal System*, was submitted in September 1921, not long after his retirement.

The title was made more specific in the heading of the abstract: *First attempt at a general theory of the distributions of masses, forces and velocities in the stellar system*. The paper started with a revision of the model for the stellar density that he had determined with van Rhijn to make computation of gravitational forces feasible. He assumed that the surfaces of equal density of stars were ellipsoids (see Fig. 15.10, bottom). In his words: 'by assuming the equi-density surfaces to be *concentric, similar revolution ellipsoids, similarly situated*' [his italics]. The axis ratio of his ellipsoids were all equal. This choice had been made, because then the gravitational field could be calculated rather easily. Why this is can be seen as follows. Take a shell of material between to consecutive equi-density surfaces. The total gravitational force from all these stars together in a point *inside* this shell is zero. All attractions by the stars cancel, when the stars are distributed uniformly. This is a generalization of Isaac Newton's 'shell theorem' for the spherical case. For every point there was no contribution to the gravitational force by all outside shells. So at any position within the Sidereal System only the gravitational forces from the shells inside it contribute and these contributions could furthermore be calculated relatively relatively easily. This was simply done by calculating the forces on any shell from any other one internal to it and adding these. In the best fit to the model of Kapteyn and van Rhijn (1920b) the axis ration was 1 : 5.102. Kapteyn listed the outcome most extensively for the directions towards the pole and in the plane of the Milky Way, and proceeded to find an analytical approximation.

'The results thus far obtained rest, it is true, on provisional data, which even now might be materially improved; they further depend on the supposition, not yet fully demonstrated, that, within the distances here considered, there is no appreciable extinction of light in space, but they are, I think, the legitimate outcome of our data.

For what follows I will now introduce some considerations borrowed from the kinetic theory of gases, the applicability of which might be considered doubtful. At all events I do not pretend to have demonstrated this applicability. The results which will be derived cannot lay claim to be demonstrably correct, but they seem to me to be so remarkable that, after a good deal of hesitation, I have resolved to publish them, in the hope that others, better versed in these matters, may furnish us with a more rigorous solution of the problem involved.'

Kapteyn noted that the flattened system could not be in a steady state unless there was some sort of systematics motion and inferred from that that there must be 'a sort of rotation round the X-axis' (see Fig. 15.10, bottom) to keep it in that shape; along that axis there would then be only peculiar (random) motions. He compared this to molecules in a quiescent atmosphere and derived the relation between the forces and the average motion in that direction. The latter was estimated using published radial velocities of stars and the value for the mean motion was estimated by Kapteyn as about 10 km/sec. The force was related to the number of stars and their average mass. So, Kapteyn came up as the final result that in order to balance the gravitational force and the motions along the vertical axis, the average mass of a star in the neighborhood of the Sun had to be equal to between 1.4 to 2.2 times that of the

15.10 The Crowning Achievement: Galactic Dynamics

> **Box 15.4 Dynamics of stars in the disks of galaxies**
> The distribution of light in disks of external galaxies $L(R,z)$ is currently described by an exponential disk in the radial direction, see Freeman (1970), and an isothermal sheet in the vertical direction, see van der Kruit & Searle (1981) [38]:
> $$L(R,z) = L(0,0)\exp\left(-R/h\right)\operatorname{sech}^2(z/z_\mathrm{o})$$
> At larger distances z from the plane this is an exponential with an e-folding of $z_\mathrm{o}/2$.
> If the matter is distributed similarly as the stars are, the total surface density in a cylinder at a distance R from the center, $\Sigma(R)$, is
> $$\Sigma(R) = \frac{\langle V_z^2 \rangle}{\pi G z_\mathrm{o}}.$$
> The square root of $\langle V_z^2 \rangle$ is the velocity dispersion.
> By measuring the light distribution and this velocity dispersion (first done in van der Kruit & Freeman (1986) [39]) one can perform a calculation analogous to Kapteyn's (1922) and Oort's (1932) early work in our Galaxy.
> The distribution of sech^2 fits the stellar density distribution of Kapteyn & van Rhijn (1920b) very well for $z_\mathrm{o} \sim 650$ parsec; the modern value is about 700 pc. The average velocity of 10 km/sec adopted by Kapteyn corresponds to a velocity dispersion of 12.5 km/sec. The modern value is about 17 km/sec.

Sun, the range in values relating to which shell the analysis was applied to. This, he then noted, is in very good agreement with the determination of the mass of the average binary star of 1.6 times that of the Sun for both components combined. In other words, if the average star is a binary with this combined mass, the total gravity can be explained by the stars present without having to avert to the presence of 'dark matter', which would exert gravity but would not contribute to the luminous stellar mass.

How does this compare the modern determinations? The method followed here by Kapteyn has been perfected not much later by Jan Hendrik Oort in 1932 in a seminal paper, that was entitled *The force exerted by the stellar system in the direction perpendicular to the galactic plane and some related problems* [40], the resulting total mass density in the Solar neighborhood inferred being eventually referred to as the 'Oort-limit'. The current situation is not much different from what Kapteyn concluded, namely that effectively all the mass inferred from such dynamical studies can be attributed to known matter present in the disk of the Galaxy. See Box 15.4 for some more technical remarks. Kapteyn's model in terms of distribution and dynamics along the vertical direction was not at all far off the mark.

For the motions *in* the plane of the Milky Way, Kapteyn resorted to his Star Streams. He first calculated what the (inward) attraction from the stars would be at various positions and concluded that there had to be a significant effect of a centrifugal force. He calculated the required rotation velocity at a number of positions in the Sidereal system, both in the plane and at two latitudes as seen from the center. He found the surprising result that over much of the system the required rotation velocity in the plane of the system was 18–20 (he adopted 19.5) km/sec. So what he proposed then is to assume that the stars are free to move around the center in two opposite directions and what you get is two streams with a relative velocity of about 39 km/sec, very close to the 40 km/sec relative speed of his two Star Streams. So in the radial direction the distribution of stars is kept in equilibrium by the gravity of the stars, together by the centrifugal force of a systematic rotation of the stars, about half in one direction and the other half in the opposite one.

Now, where is the Sun is the Sidereal System? Not in the center and at least 500–600 pc from it for the forces to balance. The directions of the Streams (which are in the plane of the system) should be perpendicular to the direction of the center. Kapteyn could choose between Galactic longitudes 77° and 257° in the coordinate system then in use. In current coordinates, in which have the center of the Galaxy at longitude 0°, this corresponds to 110° and 290°. Kapteyn chose for the latter, which 'nearly all astronomers, who have dealt with the question' would have done. This is because the Milky Way is brighter there – it is in the constellation Carina in the southern hemisphere. Obviously this is off by 70°. But we now interpret the directions of the Star Streams as those in which the local velocity dispersion is largest, and modern theory of the dynamics of the Galaxy predicts that to be in the directions towards and away from the center. So, Kapteyn should have been 90° off. The 70° is quite reasonable and understandable.

From the distribution of Cepheid variables (pulsating stars; see page 372), Hertzsprung had deduced that the Sun is 38 parsecs north of the plane of the Milky Way and Kapteyn adopted that as well. The distance along the plane of the System was more difficult, as he admitted. In the end he settled for 650 parsecs. If it were substantially larger the star counts would have to display a stronger asymmetry than observed. This does not, strangely enough, translate into the Sun's position in Fig. 15.10, bottom. This part of his model was wrong for two reasons: (1) the systematic motions in the stars were not Star Streams, but effects of anisotropy in the distributions in the stellar random motions on top of a much larger rotation around a much more distant Galactic Center, and (2) due to interstellar absorption he was sampling only a small part of the Galaxy. The displacement of the Sun from the center is actually sufficiently large to qualify its position as not very special.

The end-product *is* a consistent picture in which the space distribution of stars and their random and systematic (rotational) velocities correspond to each other as would be required for a system in equilibrium, 'steady state' as Kapteyn put it. Defects that need improvement are according to Kapteyn that the ellipsoids are determined assuming the Sun is in the center and that the average mass of stars in all shells is assumed the same. The latter would be incorrect if the distributions would be different for stars of different spectral class ('hence either the assumptions or the conclusions are wrong').

15.11 The Kapteyn Universe

The 'Kapteyn Universe' (a term coined by James Jeans) was soon to be replaced by a different, much more extended model following the work of Harlow Shapley on globular clusters, the discovery of Galactic rotation by Bertil Lindblad and particularly Jan Hendrik Oort and that of interstellar extinction by Robert Julius Trumpler (1886–1956) in 1930. But the analysis of the vertical dynamics by Kapteyn was in a broad sense correct. His enthusiasm and approach to the problem were the immediate causes for Oort to become interested in astronomy and the structure of the galaxy; in his paper *The development of our insight into the structure of the Galaxy between 1920 and 1940* [41], he describes how 'Eddington's monograph *Stellar movements and the structure of the Universe,* 'together with some parts of Jeans' *Problems of cosmology and stellar dynamics*' was my principal textbook during my student days at Groningen.' Kapteyn's 1922 paper must have been a source of inspiration to Oort also. If so, it even more constitutes the culmination and crowning achievement of Kapteyn's lifetime dedication to the Sidereal Problem.

It is not known how Kapteyn would have related his stellar system to the spiral nebulae. Were these similar systems as part of a grand design in which his universe was only one of many Island Universes? As I mentioned before, I have not been able to locate any statement or thought on this in Kapteyn's publications, lectures, letters or any other source. The 'Great Debate' on *The scale of the Universe* between Shapley and Curtis, where this was one of the two items (see page 592), had taken place in April 1920, but Kapteyn also left no comments on this. In 1984, during IAU Symposium 106 [42], E. Robert Paul in the discussion after his contribution was asked whether Kapteyn did 'have any opinion on the nebulae, the external galaxies?' He replied: 'People who assumed the small galactic system in general asserted the Island Universe theory. Seeliger does argue that the spiral nebulae are island universes. I believe Kapteyn does as well'. Robert Smith concurred that this was indeed Kapteyn's view later in life (see also [43]). This indeed would appear to be a reasonable assumption.

Finally, one might wonder what would have happened had Kapteyn adopted interstellar absorption. That would have been much in violation of Shapley's result and Kapteyn would for some reason or the other have had to assume it depended on latitude. He did find values for extinction in his earlier work that in hindsight are not unreasonable, but he derived that from relatively bright stars all over the sky. No latitude dependence was evident, and it seems unlikely that he would have detected any had he looked for it.

Had Kapteyn applied an absorption correction independent of latitude, he would have found a significantly flatter vertical profile of star densities, which would have lowered the deduced average mass per star, destroying the dynamical consistency that he did find. Similarly, a much flatter radial density profile would have made the picture of the rotation as revealed by the star streams more problematic. It would have required a point of maximum density and center of rotation further away than in the Kapteyn universe and a rotation speed of 20 km/sec as revealed by the Star Streams would be much too small to make sense dynamically.

So there were at least *two* fundamental issues involved. Kapteyn's Universe was not only false because his assumption of transparent space turned out incorrect, but also his conviction that the star streams did not result from an anisotropy in the random stellar motions was essential for his arriving at a consistent picture. The Kapteyn Universe was built on both pillars and both eventually proved wrong. Kapteyn was in good company though. The lack of absorption had been put forward by Shapley on the basis of colors of stars in globular clusters and this would seem very convincing evidence and was widely accepted. The rejection of the Schwarzschild explanation of the star streams seemed likewise well-justified by the very different composition of the two streams and convinced, at least at that time, many astronomers, including Eddington. This observational fact disappeared in later investigations with improved data.

It is more uncertain to speculate what would have happened had Kapteyn lived a decade or so longer and would have been around when Galactic differential rotation and interstellar absorption were established. Much of the progress was the result of the investigations of Jan Hendrik Oort. As early as November 1926, Oort gave an inaugural lecture when he was appointed 'privaat docent' in Leiden. I have translated this lecture, which had the title *Non-light-emitting matter in the Stellar System* and published it as Appendix A in *the Legacy*. In it Oort discussed the disparity of the Kapteyn universe and Shapley's system of globular clusters and concluded: 'The possibility exists that a completely different explanation will be found, but for the moment that of an absorption of light is the least contrived'. This was only four years and a bit after Kapteyn's death and four years before Trumpler established the existence of interstellar absorption observationally beyond any doubt [44] from the discrepancy of the distances of open star clusters determined from the apparent brightness of the stars and the mean angular diameters. And in 1927, only five years after Kapteyn's demise, Oort and Lindblad discovered Galactic rotation and Oort in 1928 [45] described the dynamics of a differentially rotating Galactic disk, in which Kapteyn's Star Streams were replaced by Scharzschild's 'velocity ellipsoid'.

Kapteyn's picture of the Sidereal System was difficult to avoid. Lodewijk Woltjer in the *Legacy* referred to it as *'Kapteyn's unfortunate Universe'*, on account of 'the unlucky moment at which Kapteyn presented his Universe and to the perhaps somewhat unphysical interpretation he gave to the two 'Star Streams', which had contributed so much to his fame'. In a way we might conclude that Kapteyn died at the best moment in history, leaving behind not only a consistent picture, but also the ingredients for extension and improvement. This legacy not only included methods of statistical astronomy and the principles of dynamical analysis, but also Kapteyn's student Jan Hendrik Oort.

Chapter 16
Finale

> *The scientist does not study nature because it is useful;*
> *he studies it because he delights in it,*
> *and he delights in it because it is beautiful.*
> *If nature were not beautiful, it would not be worth knowing,*
> *and if nature were not worth knowing, life would not be worth living.*
> Jules Henri Poincaré (1854–1912).

> *Qui specula caruit, servare unde astra liceret,*
> *sideribus fluctus vidit inesse duos*
> Jacobus van Wageningen (1864–1923)[1]

16.1 Orden pour le Mérite

World War I turned out to have a major impact on Kapteyn, which extended far beyond the impossibility to visit Mount Wilson. Just before the outbreak of the War he was notified that the German Kaiser would bestow upon him the prestigious membership of the Orden Pour le Mérite für Wissenschaften und Künste. I am using the German word 'Orden' (rather than the French 'Ordre') as it is the term used in their Website [1], except where I quote from letters or from the *HHK biography* below. The 'Orden pour le Mérite' (Order for the merit) was indeed prestigious. It had not

---

[1] *'He who lacked a telescope with which to observe the heavens saw that there were two streams of stars.''.* Text on a postcard to Kapteyn on the occasion of his 40th anniversary as a professor (see page 577).

only been awarded to various leading German astronomers such as Argelander in 1874 and von Auwers in 1892, but some very prominent foreign scientists had also been honored by the Kaiser: e.g., Charles Darwin in 1868 and Lord Rayleigh (John William Strutt) in 1903. Dutch scientists, such as the Nobel laureates in chemistry van 't Hoff in 1895 and physicist Lorentz in 1905 received the honor, as well as non-German astronomers like Simon Newcomb in 1905, David Gill in 1910 and Edward Pickering in 1911,. Kapteyn was actually listed by the Orden itself for the year 1915, but he must have been notified in July 1914 (see below). That same year the 'Pour le Mérite' was also bestowed upon Kapteyn's colleague von Seeliger and famous physicist Max Karl Ernst Ludwig Planck (1858–1947), after whom the Max Planck Gesellschaft has been named. No surprise that Kapteyn felt very honored. I first quote the *HHK biography*:

'In July of 1914, not long before the outbreak of the war, the Emperor of Germany bestowed the 'Ordre Pour le Mérite' upon Kapteyn, which he accepted with much joy. 'I don't understand, why this honor is bestowed on me and not someone else.', he said to his assistant van Rhijn. After all, this honor has been awarded to no more than 30 foreigners, and was seen in the scientific community as one of the highest distinctions. Gill wrote in 1892 when he himself received this award: The *Ordre Pour le Mérite* I regard as the highest distinction open to a literary or scientific man.' '

Henriette Hertzsprung-Kapteyn is wrong here in the year; Gill received the 'Pour le Mérite' in 1910. In Kapteyn's case it happened as the War was about to break out.

'When, however, in early August rumours started to go around in America about German violations of the neutrality of the Netherlands, Kapteyn felt obliged to refuse the award after all and he wrote to the German Consulate in Groningen: '6 Aug. 1914. Now that the German army has violated Dutch neutrality, I feel obliged to reconsider my decision of July 18 of this year. In the present circumstances I cannot accept the distinction designated to me by the Emperor of Germany and King of Prussia, the nomination of 'Ausländische Ritter [Foreign Knight] des Ordre Pour le Mérite'. I sincerely hope this letter will reach you in time.' The Consul wrote him back that the rumours were not true and that therefore there was no reason for him to refuse the award.

In England, however, there was a great deal of indignation about the acceptance of Kapteyn of the German award; in particular a few hotheaded individuals among astronomers were reproaching him, [...]

Karl Schwarzschild, the eminent German astronomer, expressed it in a very humane way as: *'When everybody is feral, why would I not be feral?'* [Wenn alle wild sind, warum sollte ich dann nicht wild sein?] One was part of a warfaring, suffering and hating society and nothing else.

Hubrecht, the son of Kapteyn's friend, who was an astronomer in Cambridge at the time, wrote on August 27, 1914: 'Allow me to extend my sincerest congratulations to you on the occasion of this unique distinction, bestowed upon you by the German Emperor. My brother wrote me about this. The news has not yet reached the English newspapers. I think the censoring authorities feel the Emperor is being given too much positive exposure! At least, the few English friends that I have been able to tell the news were all very surprised and had not imagined him [the Kaiser] doing a thing that testifies of taste of culture.'

Jan Bastiaan Hubrecht (1883–1978) was the son of botanist Ambrosius Hubrecht, one of Kapteyn's friends in Utrecht during his student days. Jan Hubrecht was an astronomer.

The situation worsened significantly when the 'Orden Pour le Mérite' (in the military class) was awarded to the captain of the submarine that torpedoed the Lusitania on May 7, 1915. RMS Lusitania was a British passenger ocean liner and more than one thousand of the almost two thousand people aboard lost their lives. This act was a major factor in the eventual entering of the war by the USA (see page 521). The commander of the U-boat, Walther Schwieger (1885–1917), received the Military class 'Pour le Mérite' in July 1917, as the Website of the 'Orden' shows [1].

Kapteyn never returned the medal to the German Kaiser and was never forgiven for it by some of his colleagues in the 'allied countries', particularly in the UK and USA. He could be a stubborn man apparently, who on the one hand thoroughly disapproved of the war and had put the blame for starting it on Germany (see page 521), but on the other firmly believed in neutrality and that scientific matters should not be interfered with by politics or other sentiments.

Kapteyn's honors and distinctions are listed in Appendix A. The most important are: Knight in the Ordre National de la Légion d'Honneur, France (1892), Knight in the Order of the Netherlands Lion in 1903, and – even more prestigious – Commander in the Order of Oranje-Nassau in 1921. I take one comment on this from the *HHK biography*:

'Many Royal distinctions and decorations, medals, honorary doctorates etc. were bestowed upon him in his many years of scientific work. The first order, the Legion d'Honneur of France brought him much happiness; it was the first official recognition of his work, and he wore the red ribbon always in the buttonhole on his lapel. He regarded his further distinctions that he thought highly of, in the first place as means to get what his scientific work required and also to provide a better income for his assistants. This fame and these public honors gave a kind of awareness from which his case would profit and that could make the means available to do his work on an ever expanding scale.'

16.2 Germany and International Organizations

I continue with the *HHK biography*.

'Kapteyn saw with great worry the continuously increasing hatred of the allied states and America towards Germany. His feeling of justice objected to the condemnation of the German people, who were no guiltier than others. Through his mental eye he clearly saw how through the ages history repeated itself, how each country in turn had behaved badly when it was in power. Was the Boer War, that dark page in England's history, not fresh in our minds, and could France be excused after what it did in Morocco or the Dutch in the East Indies? Was the blockade that brought Germany on the brink of starvation, that much more humane? A hard-working nation with many virtues, was being blinded in its too sudden revival and poisoned by a system of military slogans and patriotic phrases, and ruined by an uncritical belief in a government that was not up to its task. That is how he saw the tragedy of this nation: he had no choice but to back the repressed who were not really more barbaric than others. [...]

Kapteyn wrote to Frost at Yerkes: 'November 10, 1918. Except for your admirable President I for one feel but little hope in this issue which to humanity would be worth the awful misery it has gone through. Again, I for one have no fear of Wilson. After Christ he may become the greatest benefactor of the world, I feel sure that he will fight for his ideals or perish in the attempt. Better forgo rightful vengeance, rightful punishment, anything than the hope for some enduring settlement of human affairs, that will make for the real happiness and progress of mankind. There never was such a chance. There would not be now, but for America.'

The English Kapteyn uses in this letter is not phrased too well, but his dislike of the prevailing opinion, and his admiration for the American President Thomas Woodrow Wilson (1856–1924) is obvious.

However, even more serious problems for Kapteyn arose when sentiments to exclude Germany from international organizations after the end of the War became stronger and stronger. The League of Nations was a direct outcome of the decisions of the Peace of Versailles, but Germany was excluded (it did not become a member until 1926, but withdrew again in 1933). These sentiments, however, were not restricted to politics and even the world of scientists was severely affected by this refusal to pave the way towards reconciliation. In October of 1918, the Royal Society of London proposed an inter-allied reunion of scientific academies and organized a meeting in London with representatives exclusively from victorious countries: the UK, USA, Italy, France, Belgium, Serbia and Brazil. The proceedings of this meeting have been published in French in the *Comptes Rendus de l'Academie des Sciences*, Paris, but also in English translation in an astronomical journal [2]. Surely, Kapteyn cannot have missed reading it. The language used by the victors revealed an utterly unconciliatory attitude. It was stated that while 'previous wars had not destroyed the mutual esteem of scientists of belligerent countries for one another', 'today conditions are quite different'. As a result, 'personal relations will for a long time be impossible between the scientists of the allied countries and those of the central empires'. New associations needed to be created 'among the allies, with the eventual cooperation of neutrals'. In no uncertain terms:

'If today the delegates of the scientific academies of the allied nations and of the United States of America find it impossible to take up personal relations again, even in the matter of science, with the scientists of the central nations, inasmuch as these will not be admitted again into the

16.2 Germany and International Organizations

union of civilized nations, they do this with full consciousness of their responsibility, [...]

[...] the central powers have infringed the laws of civilization, disdaining all conventions and unchaining in the human soul the worst passions engendered by the ferocity of the struggle. War is inevitably full of cruelties [...]. These are not the acts we refer to, it is the organized horrors, encouraged and conceived from the beginning, with the sole aim of terrorizing inoffensive populations. The destruction of numberless homes, the violence and massacres on land and on sea, the torpedoing of hospital ships, the insults and tortures inflicted on prisoners of war, will leave in the history of the guilty nations a stain which the mere reparation of material damages will not be able to wash away. In order to restore confidence, without which all fruitful collaboration will be impossible, the central empires will have to repudiate the political methods the practice of which has engendered the atrocities which have roused the indignation of the civilized world'.

The proposal was put forward to constitute a Commission of Study, which delegates from 'the countries at war with the central powers may join' in order to draw up a plan of international organizations. In July 1919 the 'International Research Council' (IRC) was founded, excluding Germany. Kapteyn vehemently opposed this move; it was against all his principles. In the *HHK biography* we read:

'[...] in July 1919 the 'International Research Council' was founded, excluding Germany. Kapteyn discussed with Heymans what they could do about this. He had an unwavering trust in science and its ultimate victory over bitter subjectivity. 'It is my conviction that science must in the long run directly and indirectly become a mighty factor in bringing peace and goodwill among men. If the men of science do give an example of hate and narrow-mindedness, who is going to lead the way?', he wrote in January 1917 to Eddington. In this great English astronomer, a Quaker and strict scientific man, he found a humane and impartial judge, who was not blinded by hate. But that was an exception in those days.

After much deliberation, Kapteyn and Heymans composed an open letter: 'To the Members of the Academies of the Allied Nations and of the United States of America,', in which they adjured that science should be 'the great conciliator and benefactor of mankind.' And they ended: 'We understand how your attention of late has been monopolized by what is temporal and transitory. But now you more than all others are called upon to find again the way to what is eternal. You possess the inclination for objective thought, the wide range of vision, the discretion, the habit of self-criticism. Of you we had expected the first step for the restoration of lacerated Europe. We call on you for cooperation in order to prevent Science from becoming divided, for the first time and for an indefinite period, into hostile political camps.' [3]

To Prof. Korteweg [the Dutchman Diederik Johannes Korteweg (1848–1941) was a professor of mathematics at the University of Amsterdam] Kapteyn wrote: '14 August 1919. I do not expect that this open letter to allied members of academies will end the division between them or the association of academies [Kapteyn refers to the 'Interallied Association of Academies']. I have however not given up all hope that a letter like this, signed by a large number of prominent neutral scientists, will turn many of the allied scientists to a more objective appraisal of the matters concerned, with the result that less haste is being made in founding the interallied societies. In my opinion when time is won, all is won. If this would however not work either, then I will not regret this step that Heymans and I took.'

Before the War, the Association Internationale des Académies (International Association of Academies) had been the main international organization of science. It had been founded in 1899, but abolished in 1913. The developments referred to led to the foundation of the International Research Council in 1919. Via the International Council of Scientific Unions (ICSU) this body eventually evolved into the current International Council for Science, still known by the acronym ICSU. The International Research Council followed the practice of not allowing the former 'central nations' to become members. The organization has national Academies and scientific Unions among its members. The Netherlands, through the Royal Academy of Sciences joined this body in 1922. In the same year, the newly-founded International Astronomical Union IAU also became a member.

The developments in the Netherlands and the Royal Academy have been described in detail by Klaas van Berkel in his history of the Academy, *De stem van de wetenschap* [4], including Kapteyn's role in it. The founding of the IRC in 1919 had taken place with much prevalence of France and Belgium and excluded Germany and former allies. Countries that had been neutral, such as the Netherlands, Scandinavian countries, Spain, etc., were allowed to join, but not until after the Council had been founded and, in consequence, after the by-laws had been written. And these were formulated such that any opportunity to make the accession of Germany possible, was effectively ruled out. The *HHK biography* continued:

'When at the end of 1919 a proposition was sent to all neutral academies to join the International Research Council (or Conseil International de Recherches) with the exclusion of Germany, he opposed this with all his strength, and tried to keep the Amsterdam Academy of Sciences from making this step. In the decisive meeting he and Heymans used all their influence to keep members from voting for it, but they failed. Here it was opportunism against idealism, and also in science idealism lost. He had not expected that and he was so shocked that he and Heymans resigned immediately from the Academy, which according to them had proved to be unable to act fairly and in a scientific manner. It was more than a passing shock; until his death this damaged trust in justice and objectivity remained a painful wound.

At the time when these discussions took place, the department of Natural Science of the Royal Academy was under the chairmanship of Lorentz (see Fig. 16.1), who in the end received much praise for his cautious and prudent leadership. Van Berkel notes that Kapteyn did *not* resign from the Academy, as is usually stated, probably based on the *HHK biography*. He never again attended any of its meetings, although

16.2 Germany and International Organizations

Fig. 16.1 Hendrik Lorentz as he appears in the *Album Amicorum* presented at the retirement of H.G. van de Sande Bakhuyzen as professor of astronomy and director of Leiden Observatory in 1908 (Archives Leiden Observatory; see caption Fig. 3.7)

he did help preparing official opinions by the Academy. But this self-imposed exclusion from the activities from the Royal Academy remained a very painful matter. He had been a member since 1888, and in spite of the distance between Groningen and Amsterdam he had been a very active member, attending almost all meetings. These took place on a Saturday on a monthly basis, and were usually followed by an informal dinner. The meetings were a joy to Kapteyn and it must have been very hard to be forced (in his view) to resign. But he could not reconcile the step the Academy took, i.e., joining an organization from which Germany was excluded, with his conscience. It placed him in an even more vulnerable position among his colleagues, many of whom heartily supported the exclusion of Germany.

I continue with the *HHK biography*.

'There was someone else, however, who joined him in his fight for reconciliation. It was the Swede Strömgren, the director of the Copenhagen Observatory. Like most Swedes he was on the side of the Germans, had an unlimited admiration for German science and his personal sympathies were also with that country. He was upset by the exclusion of Germany, and when in 1919 the French refused to share the data on a comet among Germans, which Harvard (America) did circulate, he became indignant. 'I have two boys of 11 and 9 that would never do such a thing'. Through his arbitration he was able to obtain many results favorable to the Germans in these times of isolation and when it was decided in 1920 to organize a congress of the Astronomische Gesellschaft, he worked with all his powers as its president to make it a success. He proposed to appoint Kapteyn in the Board of the A.G.

'… This decision, as you will understand, will bring you in temporary discord with Baillaud, Lecointe and Turner, for which the entire neutral world will be grateful to you, and a large number of astronomers from allied countries will welcome this decision with sympathy – not to mention the feelings of the German scientific world.... You have given much in your life to the scientific world – I think here you again can give something of lasting importance.'

The French scientists Jules Baillaud, Lecointe and the English astronomer Herbert H. Turner were among the most militant in arguing for the exclusion of German scientists after the war. Jules Baillaud (1876–1960) was the son of Edouard-Benjamin Baillaud, the first president of the International Astronomical Union. George Lecointe (1869–1930) was Belgian. He was director of the Royal Observatory of Belgium and played an important part in the creation of the International Research Council. Herbert Hall Turner, whom we have met before (e.g., as the person who coined the term 'parsec'), was a professor of astronomy and director of the Observatory at Oxford University.

'Kapteyn decided to accept. He wrote [von] Seeliger, the astronomer from München: 'I am fully aware that in general I lack the qualities necessary for a really good member of the Board. That is the reason that I have always declined such offers. Now the situation is in my view somewhat different, since it is the case that he who does not protest the in my opinion irresponsible division, should give all his strength to help to heal this objectionable separation'.

Eddington was the first from the camp of the enemies, who was prepared for reconciliation. He wrote to Strömgren in November 1919: 'I hope to show my interest in the Astronomische Gesellschaft by attending the next meeting – an individual step which no one has any right to object to ... International science is bound to win and recent events – the verification of Einstein's theory – have made a tremendous difference in the last month.'

In 1920 he was the only Englishman that attended the congress of the A.G. And now in 1928, when I write these pages, Germany has become a member of the League of Nations, Wilson's idea is gaining increased acceptance, and time, that great benefactor and conciliator, is busy healing the wounds and the divisions, including those in science.'

16.3 Forty Years as a Professor

In 1918, Kapteyn celebrated his 40th anniversary as a professor at Groningen University. On the occasion his portrait was painted by the Dutch painter Jan Pieter Veth (1864–1925), well-known in the Netherlands and particularly noted for his portraits. In the Kapteyn Room an album is kept presented to Kapteyn on this occasion. The first page (see Fig. 16.2) says: 'On the 20th of February 1918, when it was 40 years after Jacobus Cornelius Kapteyn took up the professorship in astronomy at

16.3 Forty Years as a Professor

Fig. 16.2 First three pages of the album accompanying the presentation of the painting of Kapteyn by Jan Veth in Fig. 16.3. The translation of the text on the first page can be found on p. 577. The names on the second and third pages have been listed in the main text (Kapteyn Astronomical Institute)

the University at Groningen, his friends and students have presented his portrait, painted by Dr. Jan Veth, to Mrs C.E. Kapteyn-Kalshoven'. The album carries the signatures of all contributors, in alphabetic order. The first two pages, however, have been reserved for the signatures of a few persons who were very special to him (see Fig. 16.2): Ursul Philip Boissevain, Gerard Heymans, Adolf Frederik Molengraaff, Jan Willem Moll, Pieter Johannes van Rhijn, Willem de Sitter, (Mrs) Isobel S. Gill, George Ellery Hale, Edward Charles Pickering, Anders Severin Donner, Karl Friedrich Küstner and Robert Thorburn Ayton Innes.

The album also contains a loose, single postcard with the text:

'Ad Jacobum Cornelium Kapteyn
Qui specula caruit, servare unde astra liceret,
sideribus fluctus vidit inesse duos.
a.d. X Kal. Mart. MCMXVIII'

Without abbreviations the bottom line reads: 'ante diem decimum Kalendas Martias 1918'. A loose translation then would be:
'He who lacked a telescope with which to observe the heavens saw that there were two streams of stars. 10 days before March 1, 1918.'
The sender is not identified by name but only by an address (Westersingel 17). I have been informed (by Joan Booth, professor of Latin and Manfred Horstmanshoff, professor of the history of ancient medicine, both in Leiden, that have also provided the translation) that the text shows an excellent command of Latin. It is an 'elegiac couplet', consisting of a 'dactylic hexameter followed by a dactylic pentameter'. On the basis of the address Westersingel 17 in Groningen, I have found – in the records of the city of Groningen – that the author was Jacobus van Wageningen (1864–1932). In 1918, van Wageningen was a professor of Latin linguistics and literature in Groningen.

Ursul Ph. Boissevain delivered a speech on the occasion of the presentation of the painting, which has survived, together with the medals and distinctions of Kapteyn,

Fig. 16.3 The painting of Kapteyn at his desk, presented on the occasion of his 40th anniversary as a professor in Groningen. David Gill is shown at the top-right. The painting hangs in the Kapteyn Room in the Kapteyn Astronomical Institute (Kapteyn Astronomical Institute)

in the Museum Boerhaave in Leiden (see page 629 in Appendix A.). It is in very clear handwriting, each letter penned separately. I quote the final part:

'And we have eagerly seized the opportunity to show our love and affection. We present you with a portrait: we wished that there would be an image of you for future generations appropriate to your dignity. We have found Dr. Jan Veth willing to fulfill our wishes. You can be assured that among those who honour you this way – and among them you will also find the most precious of your foreign friends – there were none who did not fullheartedly join the effort to make this possible. We have only asked your friends to take part in this; I should say: were allowed to take part in this, knowing that you would not have wanted it another way.

And now, dear Kapteyn, accept this as it is offered to you, also a proof of our true affection. May you be blessed with many years of unrelenting energy to investigate and come to greater understanding, by penetrating more and more deeply into the immeasurable spaces of the boundless universe, the sight of which alone fills the simplest of hearts with respect.

16.3 Forty Years as a Professor

Then also the wish will be fulfilled that we cherish for your dear and loyal spouse, for your children and grandchildren, and for ourselves, for whom your friendship belongs to the most highly valued among our possessions.'

The painting is shown in Fig. 16.3. The person in the top-left is David Gill. However, it appears that this was *not* the painting presented at the time of the celebration in 1918. Adriaan Blaauw tells an amusing story in his chapter in the *Legacy*.

'I was told by the late Pieter J. van Rhijn, who was Kapteyn's close collaborator and successor and my predecessor, that Veth was inspired to paint Kapteyn the way we see him here [Fig. 16.3], by a remark made by Mrs Kapteyn. She felt little sympathy for [the original] version, also made by Veth and donated to Kapteyn by friends and colleagues of Kapteyn, which shows Kapteyn posing for the painter. 'This is not how I am used to seeing my husband', she said, as van Rhijn conveyed to me. The way she did see him – at work at his desk – is depicted by the portrait in the Kapteyn Room. In the upper right corner of the painting, Veth sketched David Gill, the close collaborator and a friend of the Kapteyn family. The painting was acquired by the family and donated by Kapteyn's heirs to the University of Groningen around the year 1960, to be placed in the Kapteyn Laboratory. The donation was the result of an approach, initially by van Rhijn in September 1957, to Kapteyn's heirs, in particular to his daughter Mrs Noordenbos–Kapteyn (widow of the Amsterdam professor of surgery W. Noordenbos [this is Kapteyn's elder daughter Jacoba Cornelia]), who at that time lived in England near her daughter Maria Newton–Noordenbos. After consulting Mrs Noordenbos–Kapteyn and the children, it was decided that the painting would be donated to the Laboratory after her decease. A lucky circumstance, which may well have facilitated the transfer, was the fact that the late Maria Newton–Noordenbos, Kapteyn's grand-daughter, was a class-mate of this author [that is Adriaan Blaauw] in grammar school in Amsterdam in the years 1928–1932....'

So there was an earlier version that Mrs Kapteyn did not like. What happened to it? Also, there is a painting of Kapteyn in the Senate Room in the Academy building (see Fig. 16.4), also painted by Jan Veth. Are there three paintings of Kapteyn? I continue with Blaauw's narrative.

'But what became of the 1918 painting donated by friends and colleagues? The walls of the Senate Room of Groningen University are covered with a mosaic of paintings of retired professors. Among them, somewhere near the center of the west wall, we see the one of Kapteyn. According to the rules set by the University for such portraits, it shows Kapteyn dressed in his University gown and cap. It is signed by Jan Veth and carries the year 1921, i.e., that of Kapteyn's retirement. This raises the question: did Veth paint Kapteyn again in 1921, three years after he produced the two paintings mentioned before? The question has puzzled historians – for if indeed Veth did so, where then is the 1918 painting? [...] The most natural solution seems to be that, when the time came for delivering a 're-tirement portrait', the 1918 painting was adapted by Veth himself to the

Fig. 16.4 Formal painting of Kapteyn, also by Jan Veth, which can be found in the Academy Building of Groningen University (University of Groningen)

University's special conditions: he adjusted Kapteyn in the way prescribed. A close inspection of the painting performed in February of the year 1992, in the presence of the curator of the University Museum, Mr F.R.H. Smit, supported this supposition: traces of Kapteyn's head of hair seem to betray Veth's disguising efforts.'

The Kapteyn Room, where the painting of Fig. 16.3 hangs, is in a different location from where it was at the time of the *Legacy* symposium, and is therefore somewhat different from how Adriaan Blaauw described it in the proceedings. But it still contains Kapteyn's desk, the 1918 painting by Jan Veth, the star-map he made as a teenager, his globes and many of his original books (Fig. 11.19).

George Hale sent a congratulatory letter to Kapteyn on the occasion of his 40th anniversary as a professor on January 22, 1918 (also quoted in the *Legacy*):

'You have given a marvelous illustration of the possibilities of work by an observatory without a telescope. You have also stimulated astronomical research throughout the world in an extraordinary way and initiated an undertaking which will be continued far into the future. Under such circumstances you must surely find pleasure in reviewing the years of your world-wide work of cooperation, which has meant so much to astronomers everywhere. Most of all the Mt. Wilson Observatory is deeply indebted to you for the greatly broadened conception of its possibilities, which you have awakened. Every member of the staff would thus join with me in the assurance of their cordial appreciation of the aid and inspiration which we owe to you.'

16.4 Reorganization of Leiden Observatory

On March 3, 1918 (a Sunday) Ernst Frederik van de Sande Bakhuyzen, director of Leiden Observatory, died unexpectedly. On Monday morning, de Sitter was asked to take over the tasks of the director on a temporary basis until a new director could be appointed. By that time de Sitter had a been professor in Leiden (see page 446 and further) for ten years. The resulting reorganization of Leiden Observatory is an important chapter in Kapteyn's life and in the development of astronomy in the Netherlands in the twentieth century. It has been described in some detail in the *Legacy* by Wolter R. de Sitter (1936–2009), grandson of Willem de Sitter. Unfortunately most other publications on the subject are in Dutch, such as two excellent articles by David Baneke: 'Als bij toverslag' [As if by magic] on the reorganization itself and 'Hij kan toch moeilijk de sterren in de war schoppen' [You would not expect him to make a mess of the stars, would you?] on the blocking of the appointment of Pannekoek as deputy director by the Prime Minister because of his socialist views [5]. The issue is addressed only briefly in English by Willem de Sitter in his booklet 'Short history of the Observatory of the University at Leiden 1633–1933' (see below).

There is no question about the state of astronomy at the Observatory (so excluding the work that de Sitter did) when the younger (E.F.) van de Sande Bakhuyzen died (please note that the elder H.G. van de Sande Bakhuyzen was still alive). Antonie Pannekoek (1873–1960) (see Fig. 16.5) had obtained his PhD in Leiden in 1902 (his thesis was on *Lichtwechsel Algols* – brightness variations in Algol –, a well-known variable star also called β Persei, that is an eclipsing binary where the two components occult each other as seen from Earth). He still worked in Leiden, teaching as 'privaatdocent' on the history of astronomy. In his memoirs, *Herinneringen: Herinneringen uit de arbeidersbeweging: Sterrenkundige herinneringen* [6],

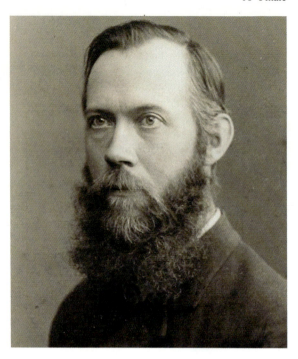

Fig. 16.5 A. Pannekoek as he appears in the *Album Amicorum* presented at the retirement of H.G. van de Sande Bakhuyzen as professor of astronomy and director of Leiden Observatory in 1908 (Archives Leiden Observatory; see caption Fig. 3.7.)

he described the atmosphere at the Observatory under the brothers van de Sande Bakhuyzen as extremely conservative.

'In this environment, where everything happened in the traditions of twenty or thirty years earlier, where there was only endless computation and without anything ever being finished, where the new ways of astronomy were hardly appreciated, all enthusiasm must eventually disappear. Later Kapteyn once remarked to me: I never understood how you kept up with it so long. [...] I dreaded every Monday morning, when I had to attend the weekly conference in the director's office, where there would be some chatter, and every one mentioned what they had done that week – or invented something – and I realized that every week was in large measure the same, just trickled along a bit. I then always felt a smell around me like in catacombs, of deadly rigidity and boredom'.

The focus of astronomy had changed to Groningen, but there the Astronomical Laboratory was minimally staffed. There were computers, but Kapteyn had to obtain external grants for their long-term employment. The yearbooks of the University of Groningen lists the employees. In addition to Kapteyn himself there was one paid assistant, who had a PhD or was preparing for it. This position had been fulfilled by de Sitter (1901–1908), Weersma (1908–1912), Zernike (1912–1914) and van Rhijn (since 1914). Starting in 1918/19, Egbert Kreiken became a second assistant, albeit without a salary from the University. Until 1910 the only further employee was T.W. de Vries with the job description 'clerk' (klerk). In 1910 de Vries' job description

16.4 Reorganization of Leiden Observatory

changed to 'amanuensis', best described as senior laboratory assistant, and the staff was extended with another 'clerk', J.L. Jansen. This was the situation in 1918, when the reorganization in Leiden started. In all fairness, it should be mentioned that from 1919 onwards three computers were added at the Laboratory (J.M de Zoute, D. van der Laan and G.H. Muller), but it is likely that these were already present, but not on permanent or long-term contracts.

Leiden was by far the largest astronomical institution in the Netherlands, not only with two professors, but also a fully equipped observatory. Its personnel had not changed much since 1908, the ten years of the directorship of E.F. van de Sande Bakhuyzen. The 1908 situation is summed up by de Sitter in a *Short history of the Observatory of the University at Leiden 1633–1933, published at the occasion of the celebration of the 300th anniversary of the foundation of the Observatory* [7], as follows: 'In 1908 [...] there were three observators and one 'assistant for the computations' (whose title was changed to that of 'conservator' in 1911), besides a staff of 5 full-time and 3 half-time computers, and three mechanics and instrumentmakers.' This of course, in addition to the director and de Sitter. Most importantly, Leiden University seemed determined to continue to support its observatory and astronomy program. After all, ten years earlier they had hired de Sitter (but left the Observatory in the hands of the younger van de Sande Bakhuyzen). True, the salaries were very low (Kapteyn pointed this out in a letter, noting that some had taken other part-time jobs to support themselves and their families), but there was a clear indication that the Curators were ready to support the continuation of the Observatory.

De Sitter kept a detailed diary of the reorganization, in large part reproduced in the chapter by his grandson in the *Legacy*. He noted he was not feeling well when the director died, but his wife wrote to Kapteyn immediately. As it happened, Kapteyn had been in Leiden the Sunday before (February 24) and had discussed the future of the Observatory with de Sitter. Apparently they both felt it was in poor shape, but decided not to intervene then but wait another year 'until I [de Sitter] had fully regained my health'. De Sitter later became the victim of a serious health problem related to an overdose of ether during surgery for gall-stones and developed tuberculosis. But that surgery did not take place until 1919. Kapteyn wrote back by return of post and urged him to accept the temporary appointment, so that he could remain in charge and have a maximum influence. Of possible candidates to come to Leiden, Kapteyn strongly supported Pannekoek. Next Sunday (March 10), Kapteyn again visited de Sitter in Leiden. Here Hertzsprung (still at Potsdam) was mentioned, but not regarded as a candidate for the directorship. Their conclusion was that the most desirable situation would be for de Sitter to assume the directorship and for Hertzsprung and Pannekoek to become assistant directors.

The solution de Sitter arrived at – surely to a large extent at the advice of Kapteyn – was to have a structure with three departments, each with their own line of research and led by the director and the two deputy directors. He himself would be the director and lead the theoretical department. He had moved into the area of relativity, introducing the work of the physicist Albert Einstein, who was still a German national at the time, to the Anglo-Saxon world. After having met Einstein in Leiden, they later developed the theory of the expanding universe together, eventually in the famous Einstein-de Sitter universe model. He envisaged a situation where Pannekoek would head the department of astrometry, and Hertzsprung a new astrophysics department.

There is no need here to describe the reorganization in detail, so I restrict myself to mentioning that de Sitter, undoubtedly with Kapteyn's help, gained support for this plan. However, financial allocations, in particular with a view to appointments, took a long time to be approved. The Prime Minister of the liberal party, Pieter Willem Adriaan Cort van der Linden (1846–1935), who had been in function since 1913, had doubts about Pannekoek. The latter was a prominent and very active member of the socialist movement. After the general elections of July 1918, however, a new government was formed under Charles Joseph Marie Ruijs de Beerenbrouck (1873–1936). He proved to be a very strong opponent of an appointment of Pannekoek and refused to allow him to be put forward. Eventually, Pannekoek was appointed at the University of Amsterdam, for which the city of Amsterdam was responsible, and the Municipal Council, unlike the national Parliament, had no problem with his socialist background, or at any rate did not let it stand in the way of an appointment. De Sitter and Hertzsprung were only appointed director and deputy director in the course of 1919. No alternative for Pannekoek was identified.

The reorganization of Leiden Observatory was extremely successful, although not all the funds and extra positions for computers etc. were realized (see David Baneke's article 'As if by magic' on the reorganization). Dutch astronomy, in particular that in Leiden, prospered under the leadership of Kapteyn's protégés de Sitter, Hertzsprung and somewhat later Jan Hendrik Oort. A letter by de Sitter to Kapteyn of May 24, 1920, provides entertaining reading. By that time de Sitter was in Arosa on a cure for his tuberculosis. The letter referred to a disagreement about the managerial organization of the Lembang Observatory in the Dutch East Indies, which in Kapteyn's view should be supervised by a national committee, whereas de Sitter was reluctant to give up authority that he felt was part of his directorship.

'The Minister is preparing his plans for specialization and concentration. In Leiden this has been interpreted in such a manner, that each university specializes in a certain area of each discipline. As concerns astronomy, you know my proposal. In short, Leiden specializes in everything except the special items of Groningen and Utrecht. Nijland wrote to tell me that Utrecht for astronomy does not think about specialization, but about concentration, meaning in Leiden. The advice from Utrecht ends: 'The equipment in Leiden must be unstinting, while Utrecht and Groningen, each according to their nature can be furnished on a more modest scale.' Anyhow, to me it seems probable Leiden will be able to obtain equipment and staff on a more liberal scale than Utrecht and Groningen.'

This described the situation well and was an indication of how Dutch astronomy was set to develop during much of the twentieth century.

Kapteyn maintained that he did not really intervene in the case of Hertzsprung, who after all was his son-in-law. But it is not difficult to imagine that he would be pleased to see his daughter Henriette and his granddaughter Rigel leave Potsdam

16.4 Reorganization of Leiden Observatory

and settle in the Netherlands. But I also consider it likely that Kapteyn realized that for the future of *Dutch* astronomy there had to be a strong institution in the country, and that Leiden Observatory, if supported by its university, would be in a much better position to become this leading institute than his less generously supported Laboratory in Groningen, by far the smallest of the three universities (see discussion on page 517). It does not mean he would give up his Laboratory, van Rhijn and Groningen astronomy, but that it would be the best way forward for Dutch astronomy to establish a modern, strong and extended Observatory in Leiden. The correspondence between Hertzsprung and Kapteyn, kept in the files of the University of Aarhus, reveals a consistent picture of Kapteyn advising Hertzsprung about the progress of the reorganization and what his best course of action should be. There is no sign of any interference by Kapteyn with a view to gaining any improper advantage for Hertzsprung. Their tone is interesting. Initially (1906) they correspond in English and address each other as 'Dear sir'. By 1912, Hertzsprung wrote to 'My dear professor Kapteyn', signing with 'Ejnar Hertzsprung', while Kapteyn answered in German to 'Lieber Hertzsprung' (as always signing with 'J.C. Kapteyn'). After the marriage of Ejnar and Henriette, Kapteyn continued for a while in the same fashion, but by 1917 Hertzsprung started to correspond in Dutch (although still based at Potsdam), addressing Kapteyn as 'Lieve Vader' (Beloved father), while Kapteyn addressed him as 'Beste Ejnar' (dear Ejnar), sometimes 'Amice', and signed with 'Vader'. The letters after he had settled in Leiden concerned to a large extent Hertzsprung's efforts to obtain financial support for instrumentation that had been promised to him when he accepted his appointment as (extraordinary) professor at Leiden University.

Henriette Hertzsprung-Kapteyn seemed delighted to settle in Leiden. She wrote to de Sitter, who was in Arosa at the time, on September 20, 1920. The letter shows that they were on very good terms; she addressed him as 'Dear Willem' (Beste Willem) and signed with 'Hetty Hertzsprung-Kapteyn'. She wrote how much she looked forward to the moment when 'after the unpleasant winter the spring will bring you back and 'geht es echt los'.' Actually, Mrs Kapteyn was less formal towards de Sitter than her husband was, who up to the end kept addressing him as 'Amice' and signed letters 'J.C. Kapteyn'. Mrs Kapteyn added a short PS to a letter from Kapteyn to de Sitter (of April 22, 1921), when apparently the latter had just returned from another stay at Arosa. The note is motherly: 'Sweet Willem, [Lieve Willem] A short note to tell you how pleased I am that you are back with your family. Will you please take good care of yourself and avoid draughty corridors? May can be rather treacherous still. With very warm regards, Your E. Kapteyn K.'

What happened after Pannekoek's appointment had been refused? Who was to lead the astrometric department? I return to the *HHK biography*:

'One time when Kapteyn visited the Hertzsprungs in Leiden the matter was discussed over dinner. The daughter had a sudden thought and said: 'Father, would you not take this position temporarily? You know this kind of work intimately. And because of your upcoming retirement you will have sufficient time available.' Hertzsprung immediately supported this unusual idea. Kapteyn laughed and said that the idea had not even occurred to him.

After some thought he said: 'It is not such a strange idea, however, and I will give it some serious thought.' He felt that it was urgently necessary that order was brought to the chaos in the material collected and its old fashioned state. Leiden Observatory had a special place in his heart, he felt at home there, was acquainted with the staff and was highly regarded by all. Someone like that could only succeed. He wrote to Hertzsprung on 18 April 1920:

'Now that you are serious about this, I have given it some thought. I am prepared to accept a position as 'adviser' for one day a week to work on this... This is inspired only because of the extensive plan of yours and only to help the Observatory. Personally there is in this new obligation little that is really attractive to me...' It did take away a major part of his freedom to do what he wanted, which he was looking forward to. 'Es ist gerade diese völlige Freiheit, welche für mich das Emeritat anziehend macht [It is exactly this complete freedom that makes the status of emeritus for me so attractive]', he wrote to Prof. Eberhard in Potsdam, with whom he had developed a good friendship over the last years.'

Gustav Edward Eberhard (1867–1940) received his doctorate under Hugo von Seeliger at Munich, and from 1918 until his retirement he was a senior observer at the Astrophysical Observatory at Potsdam.

'The conclusion was that de Sitter was extraordinarily pleased with the proposal that Kapteyn would temporarily take upon him the deputy directorship, for which he would spend a week in Leiden every two months. He would stay with his children, which was a great joy on both sides. 'Now you will be the assistant of your former assistant', they teased him, but they were also proud for the greatness of his mind which, when it came to matters that he felt were important, did not care about convention. There were some who felt that a special title had to be found for him, as they felt this modest position was not doing justice to his dignity, but such considerations were foreign to him and his unselfish wish to serve was more to him than whatever title. So this way he came back to Leiden Observatory, the place where his astronomical career had started half a century earlier. He made new research plans, which would scrutinize the adopted fundamental systems of star positions, and at the same time oversaw the reduction of older observations at Leiden to final results.'

Indeed, after his retirement in the summer of 1921, Kapteyn took up a part-time, equivalent to one day per week, appointment in Leiden as deputy director and head of the astrometric department.

16.5 Seventieth Birthday

When Kapteyn's 70th birthday was approaching in 1921, de Sitter took the initiative to prepare the publication of Kapteyn's selected works. Another option he had con-

16.5 Seventieth Birthday

sidered was a book about Kapteyn's work with contributions from colleagues (Petra van der Heijden also describes this in the *Legacy*). A committee was formed to this end including George Hale. British Astronomer Royal Frank Dyson (Fig. 16.6), however, quickly ran into opposition in the UK, apparently (as reported in a letter to Hale) because the German Karl Küstner was also involved. So, when Dyson solicited help among British astronomers, he acquired the support of Arthur Eddington, but also met with opposition from others. We have already seen that Herbert Turner was very much opposed to Kapteyn's actions in connection with the 'Orden Pour le Mérite'. In the *Legacy*, Petra van der Heijden quotes the reaction of British astronomer Arthur Hinks. Remember that Hinks was also on the trip to South Africa in 1905, where he, Kapteyn and others 'founded' the Astronomical Society of the Atlantic: (see page 354). The quotes are from letters to William W. Campbell of January 22 and April 3, 1921:

'I hope that you and your colleagues will not think that all of us in England are tumbling over one another to show our admiration for the man who affronted us by accepting the Prussian Ordre Pour le Mérite at the same time with the man who sank the Lusitania. My own feeling is that Kapteyn ought never to be forgiven for that'.

'I hope [...] old Kapteyn may celebrate his 70th birthday unhonoured by those who used to be his friends.'[2]

By the way, both van der Heyden and Hinks incorrectly state that the U-boat captain received the award *at the same time* as Kapteyn did. I quoted above that this captain was awarded the 'Orden' in 1917.

In the end the collected works were never published. I quote another passage from the text of Petra van der Heijden in the *Legacy*.

'Although Hale did not believe that Kapteyn was pro-German, all this trouble around his friend had confused him greatly. He was somewhat reassured by a few letters from Dyson, who believed that Kapteyn's conduct could probably be explained in a 'more favourable light' than the 'unnecessarily violent' accusations of Turner and Hinks (according to a letter from Dyson to Hale of February 28, 1921).

De Sitter, still of the opinion that 'considerations of a political nature should not influence the opinions of scientific men on each other's scientific merits' (from a letter to Hale of March 9, 1921) patiently explained to Hale what he thought Kapteyn's incentives were: he was by no means pro-German, although not as opposed to Germans as others have been, but his circular letter [with Heymans], proposing to resume relations with the Germans he considered 'an unfortunate mistake'. However, de Sitter assured Hale that Kapteyn's actions only originated from an '− in itself most amiable − idealism, refusing to see the real state of things brought about by the war.'

Hale, who now began to understand the real intentions of his friend Kapteyn, did everything he could to explain the situation in letters to

---
[2] Hinks to Campbell, January 22 and April 3, 1921. (Mary Lea Shane Archives of the Lick Observatory)

Fig. 16.6 Frank W. Dyson as he appears in the *Album Amicorum* presented at the retirement of H.G. van de Sande Bakhuyzen as professor of astronomy and director of Leiden Observatory in 1908 (Archives Leiden Observatory; see caption Fig. 3.7)

Turner and Hinks: 'So far as Kapteyn was concerned, I had forgiven him for holding views so contrary to my own, partly because I knew them to result from a distorted scientific idealism of the kind I had once shared, and also because I knew him to be strongly opposed to the German war methods. But I had forgotten how much feeling his mistakes must arouse.' (from Hale to Hinks, March 14, 1921).

So no special celebration took place when Kapteyn turned seventy. Henriette Hertzsprung Kapteyn actually writes why it was not a day of celebrations anyhow:

'January 19, 1921 was Kapteyn's 70th birthday. Is was not a day of joy and festivities, but a day of stress and quiet worry. On that day he and his dearest awaited in Amsterdam the return of his daughter with her husband child, who came back from America. The child had undergone surgery in Boston by Dr. Cushing, the famous neurologist and was expected back that day. [This is Harvey William Cushing (1869–1939)] The future was uncertain; a message had been sent by wire that everything went well during the ocean trip, and everyone breathed in relief. To Kapteyn this first grief in his family had been a severe blow. The 'luck of the Kapteyns had been proverbial in Groningen; everything had been prosperity in this happy family, no great disasters had happened to them, until this dear grandchild had suddenly shown signs of a severe brain illness. He supported

16.6 The Kapteyn Universe and Shapley's Globular Clusters

Fig. 16.7 The Astronomical Laboratory after it had been named after Kapteyn (Illustration reproduced from the *HHK biography*)

the distressed parents with all his power and love, sent them hopeful and comforting letters during their absence, and now waited to welcome them again with love and help carry their burden. It became a day of quiet joy after all as a result of the happy reunion and the togetherness of all who loved each other so much and were so much attached to each other.'

Kapteyn's great-grandson Jan Willem Noordenbos informs me that Greta Noordenbos had a tumor of the brain. At the time of the surgery, she was 14 years old. There was a complication in the form of a rupture of the meninges, so that she lost cerebrospinal fluid through the nose. Eventually this resulted in a meningitis of which she died in 1925 at age 18.

However, there was an important celebration not long after this, when the Government, at the proposal of the University of Groningen, formally named his laboratory after him. The full text *Astronomical Laboratory Kapteyn* was displayed prominently on the building (see Fig. 16.7).

16.6 The Kapteyn Universe and Shapley's Globular Clusters

Even before Kapteyn's model was published, an important development had taken place in the United States. I already mentioned Harlow Shapley's work on globular clusters, that there could not be any appreciable absorption in space (see page 415).

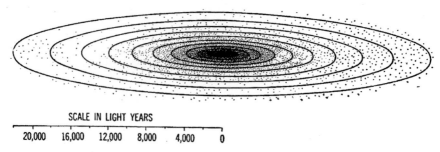

Fig. 16.8 The Kapteyn Universe (Kapteyn Astronomical Institute)

Remember that globular clusters are roughly spherical clusters of stars situated in the spherical halo of the Milky Way and part of the (old) Population II (see page 246). Shapley's studies on globular clusters also involved estimating their distances, using Cepheid stars, which are stars that change their luminosity regularly as a result of radial pulsations. Henrietta Swan Leavitt had discovered that the period of these brightness variations, correlated strongly with the absolute magnitude. From these periods then the absolute magnitude and therefore the distance of any Cepheid could be derived (see page 372). In 1913 Hertzsprung was the first to derive this relation in a paper entitled *Über die räumliche Verteilung der Veränderlichen vom δ Cephei-Typus'* [8] (On the spatial distribution of variables of the δ Cephei type) and used it to obtain distances of Cepheids.

Using this for variable stars in globular clusters, and in addition the apparent magnitude of their brightest stars and apparent diameters on the sky as distance indicators, Shapley mapped the *spatial* distribution of globular clusters (see e.g., his paper *Studies based on the colors and magnitudes in stellar clusters. XII. Remarks on the arrangement of the sidereal universe* [9]). The globular clusters were known to concentrate on the sky towards the constellation Sagittarius in the Milky Way, now known to be the direction towards the center of the Galaxy. The spatial distribution had a strong concentration in space in that direction, but the distance to this center of the globular cluster system was no less than 15 kpc away from us. The dimension of that system was enormous, the total diameter being of the order of 70 kpc or more than 200,000 lightyears. We now know that Shapley's distances were too large by a factor of about two, but still his system was much larger than Kapteyn's Universe. This is illustrated in Fig. 16.9, credited to Jan Hendrik Oort and published in de Sitter's book *Kosmos* in 1934.

The discrepancy is obvious. The problem with Kapteyn's Universe was that he had neglected interstellar absorption. Absorption or extinction results from interstellar dust, but this constituent is constricted very much to a thin layer in the disk of the Galaxy. Consequently, Kapteyn's Universe, which was substantially correct in the vertical direction as we have seen, was much too small in the radial direction. It is ironic that Kapteyn (and van Rhijn) were to a high degree led to ignore absorption on the basis of Shapley's deduction that the reddening towards some globular clusters (at high Galactic latitude away from the Milky Way!) indicated a very small

16.6 The Kapteyn Universe and Shapley's Globular Clusters

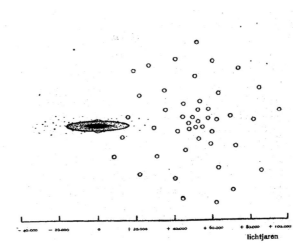

Fig. 16.9 The spatial distribution of stars in the Kapteyn Universe compared to that of globular clusters as determined by Shapley. The figure is from de Sitter's book 'Kosmos', published in 1934 and is credited to Jan H. Oort. For further information see the caption of Fig. 7.21

or negligible amount of extinction. But at the same time, Shapley's system of globular clusters is too large, largely because he did not correct for extinction, which is significant for those clusters towards the center of the Galaxy at low latitude that constitute the central part of the globular cluster system.

The center of this globular cluster system lies towards the constellation Sagittarius, of order 90° away from what Kapteyn inferred in his 'first attempt' that I discussed in the previous chapter. He had concluded that the star streams (roughly directed towards Sagittarius and opposite to that) indicated rotation, which then should be around a center lying in a direction perpendicular to that. Kapteyn felt that this was supported by the fact that in Sagittarius the Milky Way is not particularly bright, in contrast to some other areas like for example the constellation Carina which Kapteyn had identified as the rotation center. What he did not know of course, is that the direction towards the Galactic center happens to be heavily obscured.

What did people make of this? Jan Oort wrote a very enlightening paper about this, *The development of our insight into the structure of the Galaxy between 1920 and 1940* [11]. He wrote: 'When I started astronomical research, in 1922, the relation of the Kapteyn system to Shapley's system of globular clusters was still entirely unclear'. Some people thought that Shapley's globular cluster system was simply too large and assumed that the spiral nebulae were actually systems ('island universes') external to it. Others, including Shapley himself, thought the system was indeed so large, that the spiral nebulae were structures within the larger system and the Kapteyn system might be just one of them. An important, but incorrect, piece of evidence supporting the latter view was the discovery by Adriaan van Maanen, who worked at the Mount Wilson Observatory, of proper motions in some spirals. These indicated rotation, as for example in the pinwheel galaxy M101 (see Fig. 11.11), a study published in 1916 [10]. Then these nebulae could not be far away. Of course it is clear that van Maanen's measurements were wrong, but it has never been fully explained where he went wrong (see e.g. *Adriaan van Maanen's influence on the island universe theory* by R. Berendzen & R. Hart [12]).

As is well documented in many places, these conflicting viewpoints were the basis of a much discussed encounter between Shapley and Heber Doust Curtis (1872–1942), the 'Great Debate', held in April 1920 on 'The scale of the Universe'. Curtis was at Lick Observatory at the time, while Shapley was still at Mount Wilson. Curtis defended the viewpoint that the size of the Galaxy was small and the spirals were external to it. Shapley argued the opposite (see *Paul's Statistical Astronomy* and references therein).

Kapteyn and van Rhijn examined the matter of the Cepheid distances and the consequences for those of globular clusters in one of Kapteyn's last papers, Kapteyn & van Rhijn (1922a). It was published in the 'Bulletin of the Astronomical Institutes of the Netherlands', a joint journal for the Dutch astronomical institutes that was founded in 1921. They claimed that Shapley's distances to globular clusters were wrong by a factor ten or so.

The matter is complicated, but my colleague Jan Willem Pel has provided me with some insight. The calibration of the period-luminosity relation Shapley used, was mostly based on very short-period variables that we now call RR Lyrae stars; these are stars of a few solar masses that have just started helium-burning in their interiors (see page 206). Cepheids are more massive stars that are moving off the Main Sequence. In globular clusters (and the halo in general) these are called W Virginis stars and these differ from the ones near the Sun, that have δ Cephei as their prototype. Shapley had also a few W Virginis stars. But these are considerably fainter than 'classical' ones. This difference was first recognized by Walter Baade in the Andromeda Nebula in the early 1950s and this gave rise to a revision of the distance scale in and age of the Universe by a factor about 2 (in the sense of becoming older). Shapley's calibration of the pulsating variables in globular clusters was wrong by 1 magnitude, so that his distances were about a factor 1.5 too large. In addition some clusters were at low latitude and affected by absorption.

Kapteyn and van Rhijn used proper motions for statistical parallaxes, in particular for RR Lyrae stars that are part of Population II. Since these stars in the halo do not participate in Galactic rotation they have high velocities with respect to the Sun. The parallaxes are then determined by Kapteyn and van Rhijn as if they are due to the peculiar motion of the Sun (about 20 km/s) while it really it is the reflection of the rotation of the Galactic disk (about 220 km/s), which results in a distance error of a factor ten. Of course, they had small samples and the actual error was more like a factor 7.5 or so.

Somewhat earlier, Kapteyn's student Schouten had already criticized Shapley on the basis of the assumption that the distances to globular clusters could be derived by comparing their 'luminosity curve' to that of the solar neighborhood and had proposed that these distances were up to factors of eight too large. In his *Statistical Astronomy*, E. Robert Pail documents how Shapley and Kapteyn corresponded on the issue with great mutual respect, but also how a meeting in Leiden that was organized in May 1922 (weeks before Kapteyn's death), where van Rhijn, Hertzsprung, Schouten and Pannekoek (who actually agreed with Shapley's distances) met Shapley. At this meeting the Dutch astronomers essentially agreed that Shapley was in the right. The differences are perfectly natural in the light of current knowledge. Kapteyn's local luminosity curve is mostly for dwarfs, whereas the stars we see in globular clusters are giants. Many Cepheids used for the calibration of the period-luminosity relation did have large radial velocities, so that in this case large proper motions were not a sign of small distances; Cepheids were giants also.

Eventually, but well after Kapteyn's death, interstellar extinction was discovered and Oort's summary picture of the views around 1920 (Fig. 16.9) was replaced by the latter's drawing of Fig. 7.21.

16.7 Kapteyn's Last Years in Groningen

For a description of the last years of the Kapteyns in Groningen I first return to the *HHK biography*.

'In 1918 the Kapteyns had to leave their beloved house on Ossenmarkt, since the owner wanted to live in it himself. It was in the period of shortage of housing that was prevailing everywhere just after the war, and they decided to live in a hotel the last two years until he was accorded emeritus status, as they were planning to move to Hilversum after his retirement. They moved into the well-known hotel 'de Doelen' on the 'Groote Markt' [the Groote – now spelled Grote – Markt is the central market square in the center of Groningen], where two rooms were allocated to them and with their own furniture was turned into a comfortable home. The large living room at the front side of the hotel, where their grand piano and old-fashioned sofa were placed, looked out over the large market place with the Martini tower to the right and the massive City Hall to the left. Tuesdays and Fridays were market days and these were pleasant and busy happenings. One could spend hours sitting in the deep window sill looking at the hustle and bustle without getting tired. But every year in May, which was usually a month bringing nice weather, all hell broke loose, when the big May fair took place with loud and blaring noises of the merry-go-rounds and 'hippodrômes' [fairground attractions with horses], steam whistles yelled and the music droned and the spring air was filled with the smell of 'poffertjes' [small pancakes served with butter and sugar] and hot engines. Groningen was celebrating and it did not want this privilege to be taken away. The Kapteyns often fled and looked for peace that could not be found at home.
The 'Grote Markt', hotel 'de Doelen' and the view from the room the Kapteyns were occupying, are shown if Figs. 16.10–16.12.

At Martini Kerkhof, not even five minutes from their hotel, lived Prof. Bordewijk, who had not yet become used to living in a strange place like Groningen, which however was very familiar to Mrs Kapteyn, who forty years earlier had had the same difficulties. He and his impulsive wife soon found the way to the hearts of the Kapteyns because of their spontaneous

Fig. 16.10 Hotel 'de Doelen', Groningen (Beeldbank Groningen [13])

openness and their similar minds. At the Kapteyn's they were welcomed so heartily that a close friendship developed between them. Much later people asked Mrs Bordewijk how their close friendship with the Kapteyns had arisen, when they themselves after all were so much younger. 'I don't know, how it came about', she answered, 'but I do know that these were the most beautiful years I had in Groningen.'

Hugo Willem Constantijn Bordewijk (1879–1939) was a professor of constitutional law and economics at Groningen between 1918 and 1938.

'Much more impressive to him than worldly fame was the love of the people around him. It touched his deepest soul and made him happier than any worldly success could do. I have seen him deeply moved by a letter from the German astronomer, Prof. Eberhard from Potsdam, with whom he developed a deep friendship in the last years. He wrote him during his serious illness, how much this friendship meant to him and that it was the best thing that the later years had brought him. 'And I thank destiny most that is has been possible for me to come closer to you in the last two years.' Very appropriately, this sensitive German later wrote about Kapteyn's work: 'Every time when a new publication from you arrived, it was studied thoroughly and with joy.'

16.7 Kapteyn's Last Years in Groningen

Fig. 16.11 The central square in Groningen, 'Grote Markt', and Hotel 'de Doelen" (just left of center in the picture) on a market day (Beeldbank Groningen [14])

This admiration was also very well expressed in a letter from an earlier date from Dr. Innes, the previous director of the observatory at Johannesburg in Transvaal, to whom Kapteyn had proposed they address each other on a less formal basis: April 14, 1906 'It is very kind of you to admit me into the inner circle by addressing me as 'my dear Innes' and more kind to give me permission to address you in the same way. If I do not do so, it is not because I do not value the permission, but because your truly eminent talents place you on a higher sphere, I do hope you will not object to my hero-worship (which for my part I think a very good thing for a man to be capable of)...'

'For his students and for his science he also wanted to provide for the future. He established a fund for which he provided the financial basis himself, and wanted this to be increased by contributions from others. It would open the possibility for talented students with small financial means to study astronomy and to support astronomical research financially. But scientists are not the ones experienced in financial matters and do not have financial means available. The fund increased only very slowly. On one occasion he met a friend of his elder daughter, W. Dekking, a merchant from Rotterdam. He was very impressed with Kapteyn's personality and his ideals and proposed that he turn to the world of traders in Rotterdam and solicit their interest in his fund. Through Dekking's enthusiasm that was very infectious, the fund increased its capital significantly. After Kapteyn's

Fig. 16.12 The Hotel 'de Doelen' overlooked the central square in Groningen, 'Grote Markt', which actually was a market place for at least one day a week (see Fig. 16.11). These two photographs show the view the Kapteyns would have had from the hotel. At the top we see the City Hall of Groningen, the bottom one shows the Martini Church Tower (the church itself is behind the buildings on the right). The top one is dated 'about 1920', the bottom one 'probably 1915' (Beeldbank Groningen [15])

16.7 Kapteyn's Last Years in Groningen

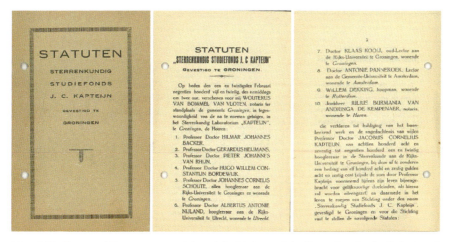

Fig. 16.13 Cover and first two pages of the original bylaws of the Sterrenkundig Studiefonds J.C. Kapteyn; the Kapteyn Fund (Kapteyn Astronomical Institute)

death and the efforts of friends and admirers, but especially through the energetic help of Dekking, it became a fund of standing, that under the name of Kapteyn Fund became an official institution with significant means. It has provided much support and will continue to do so in the future.'

The merchant from Rotterdam was almost certainly Willem Dekking (1873–1923). Indeed Kapteyn's daughter Jacoba Cornelia and her husband Willem Cornelis Noordenbos) lived in Rotterdam for a while (Rigel Hertzsprung was born in their house in 1916). The Fund was formally founded on February 21, 1925, by the signing of a deed at the office of notary van Bommel van Vleuten in Groningen. The first board encompassed van Rhijn, de Sitter, Nijland and Pannekoek, but also Dekking. The deed mentioned a starting capital of 'eleven hundred sixty eight guilders and sixty-eight cents, brought together by Professor Kapteyn during his life', which corresponds to a current purchasing power of 18,650€. The sentence survived in the latest update, of 2010, signed by myself as the chairman of the Board at that time, but the current capital is of order 70,000€. The Fund is called the *Sterrenkundig Studiefonds J.C. Kapteyn* and aims to support studies of students or research by astronomers.

During these last years before retirement, the Kapteyns probably visited Leiden and the Hertzsprung family on a regular basis. Henriette Hertzsprung-Kapteyn relates a story about Kapteyn's interaction with his grandchildren, in this case her own daughter Rigel (1916–1993). Fig. 16.14 shows a rare picture of Rigel with her father (see page 646 for more about Rigel).

'It was moving to see him with his grandchildren. His interest in them and his patience with them, their plans for the future, it all was infinite. For some he started a fund to pay for their studies immediately after their births and he always knew the right advice in cases of difficulties with their upbringing. Even on his final sick bed not long before his death did he make

Fig. 16.14 Hertzsprung and his daughter Rigel, son-in-law and granddaughter of Kapteyn. This picture dates from about 1926, when Hertzsprung and his wife Henriette Kapteyn had already separated (From Privatarchiv Dieter B. Herrmann, Berlin)

plans: every child would plant a tree and see it grow from his house in the countryside, where he planned to live after his retirement. They would enjoy staying with him there away from the cities, he would teach them about the birds, collect flowers with him and he would read them stories. No wonder that his grandchildren loved him very much. The following happened after his death. One of his grandchildren, who was learning history in school, was doing her homework exercises, which involved memorizing dates. The mother explained: 'But Rigel, what does that year 1500 actually stand for?' The child did not know. Apparently the teachers had not explained the meaning of the calendar. And the mother explained: '1500 years ago a man was born and his birth has now been adopted as the start of our calendar. That man was the best, noblest and wisest of all men who had ever lived. Can you guess who that man was?' Without a moment of hesitation the child called out: 'Grandfather!'

The story is a bit confusing, but when Henriette Hertzsprung-Kapteyn says '1500 years ago' she obviously means 1500 years before that year.

16.8 Return to Mount Wilson?

When his retirement was drawing near, Kapteyn started to make plans for his years as a professor emeritus. A thing high on his list was one more visit to Mount Wilson. The issue was brought up first in a letter from Kapteyn to Hale dated August 31,

16.8 Return to Mount Wilson?

1919. After describing how he had enjoyed a visit Seares had made to Groningen, he continued:

'Adams wrote me a letter saying that van Maanen had planned a meeting of the Mount Wilson astronomers with a couple of Dutch scientists and that he too would come if I cared to meet him there. I have much appreciated this generous peace offering and would really have been happy to come to the meeting place (The Hague). Owing to unforeseen circumstances, the meeting had finally to be given up and Adams wrote another tactful letter explaining the circumstance.

After this I have no further objections against coming again to Mount Wilson. Only, next year will be my last year as a professor and director of the Laboratory. [...] It is therefore that I would ask you not to insist on my coming next year (1920). In 1921 I will gladly come and I will stay as long as will seem useful.'

Kapteyn came back to the issue on April 4, 1921.

'My position as a professor comes to an end on the 9th of July. I could then come at once to California, but I would greatly prefer to start end of October or the beginning of November, so that I could be at Pasadena at least about 1 December. There is no other decent reason for this than just my desire to have some rest first and to spend the winter in California rather than frigid, moist Holland. – I suppose you do not care. If you do, please let me know and I will take the first occasion after 9 July.'

However, a letter from Hale of April 1, 1921, must have crossed Kapteyn's. Hale wrote (St. John was Charles Edward St. John (1857–1935), a Mount Wilson staff member and very good friend of Kapteyn):

'After much reflection, and after long discussion of the whole matter with Seares and St. John, I find myself seriously in doubt about the advisability of your returning to Mount Wilson this summer. Upon the advice of my physician, and as my health still remains very uncertain (I have just had to spend another week in bed), I shall be compelled to stay several months away from the observatory, which will leave Adams in charge, so that relations would be very largely with him. For this and other reasons the situation has greatly altered since I wrote you last summer from Montecito, when I was looking forward with such pleasure to your return.

I think that neither you nor Adams have fully understood each other in the past, and I greatly fear that for various reasons (such as differences of opinion about the Entente policy toward the Germans) misunderstandings might arise which would prevent or seriously interfere with your future co-operation with the Observatory.'

Kapteyn was disappointed, of course, but accepted the advice (April 24, 1921).

'I cannot deny that the first reading of your letter of 1 April gave me a shock. I have now thought its content over for a few days and know, not only that you are right, but also that your letter is very generous.

Fig. 16.15 'The professor and his secretary' (Photograph and caption from the *HHK biography*)

Long ago Adams wrote me a letter in which he expressed the hope that my cooperations with Mnt Wilson would not be broken off. The letter contained some other expressions which certainly have softened my feelings towards him a great deal. What you write about his willingness to assist in many of the observations proposed by me proves that these were no mere phrases.

But still I feel, just as Adams seems to feel, that without you to stand between us, misunderstandings might easily arise again. What Heaven forbids. My attitude too in the matter of international cooperation might (as you say) still more complicate matters.

Of course I will not come now. And as you still desire to keep up my cooperation – for which I am cordially grateful – I will do, from here, whatever I can.'

Kapteyn added that indeed the letters crossed, but from his letter it was clear he had not been making arrangements yet for the trip.

Kapteyn wrote one last paper in the Bulletin of the Astronomical Institutes of the Netherlands as a 'communication from the Observatory at Leiden': *On the proper motions of the faint stars and the systematic errors of the Boss fundamental system,* Kapteyn (1922b). It was a discussion of errors in fundamental catalogues and errors in parallax determinations resulting from proper motions errors in the *Preliminary General Catalogue of 6,188 Stars for the Epoch 1900,* which was published in 1910 by Lewis Boss, in particular systematic errors in the system of declinations. It announced that 'the Observatory at Leiden is making preparations for attacking it by two new and independent methods', which will be elaborated on soon by de Sitter and himself. The publication is dated April 25, 1922. By that time Kapteyn had already been confined to what would later prove to be his deathbed for a number of months.

16.9 Retirement and Illness

Upon Kapteyn's retirement in July 1921, his student Pieter Johannes van Rhijn (see Fig. 16.16) was appointed his successor both as a professor of astronomy and as director of the Kapteyn Astronomical Laboratory. From the *HHK biography*:
'In June he delivered his last lectures, and the Kapteyns said goodbye to Groningen. The good friends were invited in turn by them at their table at de Doelen. They felt this was a more intimate way to be together than a big dinner with a large group of people that would bring more glitter than intimacy. And when they said farewell to the hotel it was with much sadness for all they left behind. The manageress was very touched when she said goodbye and said they had been the nicest and easiest guests they had ever had. The appreciation was mutual. They left at seven o'clock in the morning. None of the friends knew about this, so nobody was present to bid them farewell. They had wanted it that way, and quietly they said goodbye to the city where they had lived and worked for 43 years. A long happy life with fruitful labours was behind them, but a new promising life was ahead of this pair, who in the late part of their lives still had the energy and strengths to accept new values. But first they wished to enjoy a vacation in a high, clear climate where Kapteyn could regain his strengths after all his labours of the last few years.'

The Kapteyns had hoped to enjoy a number of pleasant years in Hilversum. He would not have known the statistics we know now. With some extrapolation I find (Centraal Bureau voor de Statistiek [16]) that at birth (for Kapteyn 1851) the life expectancy was no more than 36 years or so, rising at age of ten to 47 years (so age at death of 57). When he reached the age of seventy, 80% of Kapteyn's cohort had died, but he still could statistically expect another 7 years. In addition, he was an astronomer and these statistically have a very long lifespan, even compared to other scientists, at least that was the case in the middle of the 20th century. It is a well-known fact that longevity correlates with the level of education and social status. In *Average age at death of scientists in various specialties* [17], S.M. Luria showed that for a sample of over 2000 scientists who died around 1960, archaeologists and astronomers lived some ten years longer on average (with an error in the mean of 1.5 to 2 years) than physicists, chemists and mathematicians. Some of this is due to the fact that scientists in academia (which astronomers are by definition) lived longer than those in industry by five years or so. In a study aimed at astronomers, *On the life expectancy of astronomers* [18], D.B. Herrmann confirmed the trend for older generations. Kevin Krisciunas in a Web posting with the same title [19] puts

Fig. 16.16 Pieter J. van Rhijn in 1926. This is a crayon drawing donated by his family to the Kapteyn Astronomical Institute

the median age of astronomers born between 1850 and 1860 at 78 years and two decades later even at 80. Kapteyn was a strong man to survive until age seventy, but he did not get the extra ten years or so he might statistically have expected.

In the *HHK biography* we read:

'The Kapteyns decided to spend some time with their daughter and son-in-law, Prof. Noordenbos, in Amsterdam, from where they could easily reach Hilversum to look for a house where they had dreamed to spend their last years. After some searching they found a house that was to their liking in all respects. They inspected the house in the morning and had it checked over and with his usual vigour Kapteyn brought his wife the keys of the house that they owned from that moment onwards.'

This home was the house where Mrs Kapteyn lived for many years after she had become a widow. The address features in the headings of the letters she wrote to George Ellery Hale. It is Oude Amersfoortsche Weg 29. However, at about the same time the first signs of what proved to be his fatal illness appeared.

'And how happy they were with their new home, it was in their eyes a jewel and it promised luck and peace for many years to come. Unfortunately this was not to happen, as the illness got worse. Another half year the family lived between hope and fear for this dear life that slowly wasted away. Half a year of deep togetherness, so rich and happy in mutual love, that Kapteyn was able to say a few days before his death with his radiant tenderness: 'Children, this is still the happiest time of my life.' [...] The large sunny room, where he was lying for months, looked after by his wife and two daughters, was named by him 'the haven of delight', since it was a true haven of peace and happiness for whoever was down or sad. His cheerful face and hopeful plans for the future, his interesting stories and cheerful humour [...]. Many came to visit him, hoping against all hope that their good friend would be saved, since they loved him with a very special love that is rare in life.'

Kapteyn wrote to Hale on January 3, 1922.

'My health has not been of the best of late and I write this from my bed. After leaving Groningen my wife and myself intended to travel during the better part of a year. We could do so without too much expense and it would at the same time give me a good rest after the very busy last year. We had indeed a good time in Switzerland, several parts of Germany, Scotland and England. Since about the first of October, however, I felt unwell and after having been about a month in Leiden, where I have been appointed underdirector (which position takes about 1/7 of my time) we have remained here in Amsterdam at the house of our son in law, Prof. Noordenbos, where I put myself in the hands of the medical faculty. Fortunately nothing serious has been found and the expectation is that by and by I shall again feel quite normal. Only the pains may go on for another month or more.'

But the illness did not get better, and Kapteyn was slowly consumed by a debilitating and fatal disease. What was Kapteyn's illness? Around 1920 infectious and cardiovascular diseases and cancer accounted for about half of the causes of death in men [20], Cancer accounted for 15–20%, about equal to cardiovascular but half of infectious diseases. Kapteyn complained about pain, but does not specify where. In all respects Kapteyn's illness resembles a form of cancer and a good, possible candidate, according to experts I talked to, would be Kahler's disease, a kind of myeloma or cancer of the plasma cells, a type of white blood cell. It is accompanied by pains in the bones. It is consistent with the symptoms described to me, but obviously this cannot be proved with certainty.

16.10 International Astronomical Union

From the *HHK biography*:
'In April the big international astronomical congress was going to be held in Rome [this was the founding congress of the International Astronomical Union (IAU); see Box 16.1]. Since he was unable to be present, Kapteyn – supported by pillows – wrote a program of activities for the coming period, for which he solicited the cooperation of all astronomers. His old fire

blazed again and it ended up being a great piece of work that made a big impression when de Sitter read it out at the congress. Everyone pledged to cooperate and it became a major success for Kapteyn. It was here that the French President of the congress [this was Edouard-Benjamin Baillaud], in his inaugural speech ascribed the progress of astronomy in the last half century, apart from two other factors, to the Groningen Laboratory.'

It is not clear what this proposal of Kapteyn was. There definitely must have been one, as Kapteyn wrote a small postcard to de Sitter with the text:

'Amice, Don't forget that the proposal you will defend is not a proposal by me. I don't want to have anything to do with the Union. In my opinion it is a proposal of the Dutch club. If you do want a name, it is that of van Rhijn, who actually initiated it.'

In previous weeks there were a number of letters of Kapteyn to de Sitter (in Mrs Kapteyn's handwriting), but these concerned the last corrections to the text of

Box 16.1 The International Astronomical Union (IAU)
The IAU was founded in 1919. Its mission is to promote and safeguard the science of astronomy in all its aspects through international cooperation. Its individual members are professional astronomers from all over the world, at the PhD level and beyond, and active in professional research and education in astronomy. In addition, the IAU collaborates with various organizations all over the world. Currently, the IAU has about 10,000 individual members in 90 countries worldwide. Of those countries, 70 are national members.
The scientific and educational activities of the IAU are organized by its Scientific Divisions and, through them, its specialized Commissions and Working Groups. In the past Commissions played a more important role. The IAU Secretariat is hosted by the Institut d'Astrophysique de Paris, France.
The key activity of the IAU is the organization of scientific meetings. Every year the IAU sponsors nine international IAU Symposia. Every three years the IAU holds a General Assembly, which offers six IAU Symposia, some 25 Joint Discussions and Special Sessions, and individual business and scientific meetings of Divisions, Commissions, and Working Groups.
Among the other tasks of the IAU are the definition of fundamental astronomical and physical constants; unambiguous astronomical nomenclature; promotion of educational activities in astronomy; and informal discussions on the possibilities for future international large-scale facilities. Furthermore, the IAU serves as the internationally recognized authority for assigning designations to celestial bodies and surface features on them.
The IAU works to promote astronomical education and research in developing countries through its Program Groups on International Schools for Young Astronomers (ISYA), on Teaching for Astronomy Development (TAD), and on World Wide Development of Astronomy (WWDA), as well as through joint educational activities with COSPAR and UNESCO.
The International Astronomical Union Website is at www.iau.org. A comprehensive history of the first 50 years of the IAU has been written by Adriaan Blaauw, *History of the IAU, Birth and First Half-Century of the International Astronomical Union* [21].

the paper on proper motions of faint stars and systematic errors therein, Kapteyn (1922b). But that probably was not what was referred to. Most likely, what Kapteyn wrote was in one way or another concerning the progress of the Plan of Selected Areas.

The proceedings of the Rome meeting *Transactions of the International Astronomical Union, Vol. I: First General Assembly held at Rome, May 2nd to May 10th, 1922* [22]) have brief summaries of the General Assembly sessions and those of the 32 commissions, but in none of the relevant ones is there any reference to a message from Kapteyn, read by de Sitter, nor is it listed among the presentations at a special meeting on astronomical communications or presentation of slides (presenting recent results). Kapteyn was not listed as a member of the Union or any of the commissions, nor was he (after all he was formally retired) a member of the National Committee for 'Holland', as the Netherlands was called there; the Committee was made up of de Sitter, Hertzsprung, Nijland and Baron Jacob Evert de Vos van Steenwijk (1889–1978) as secretary. The latter had a PhD in astronomy, but coming from a family of politicians he became – after a period as a teacher – mayor of the city of Zwolle and later Haarlem, and subsequently Queen's Commissioner in the province of North-Holland. He remained actively involved in the astronomical community in the Netherlands, however . The most relevant discussion in Rome where Kapteyn's work played a role, was Commission 25 on *Stellar Photometry*, which was chaired by Seares and who in his report mentioned the ongoing work on the Mount Wilson Catalogue.

There is a list of proposals made to the standing committees. Three proposals came from the Netherlands, but all had been received later than the statutory deadline and could only be discussed if half the countries present agreed. Two were from Nijland and were unrelated to Kapteyn's interests. One was by de Sitter to Commission 24 (Stellar Parallax): 'Suggested repetition of parallax plates of P.G.C. stars'. This may very well have involved Kapteyn. The PGC is the *Preliminary General Catalogue of 6188 stars for the epoch 1900, including those visible to the naked eye and other well-determined stars*, prepared by Lewis Boss and published by the Carnegie Institution of Washington in 1910 (*op. cit*). This catalogue brings together fundamental data on well-studied stars, including all those visible to the naked eye, but also fainter ones. It certainly seemed interesting to take parallax plates twenty years after the epoch with best positions available. At the end of the meeting one of the resolutions of the Commission was 'that parallax observers be recommended to obtain two or three plates of each field after a lapse of about ten years [...]'. The aim was to obtain proper motions of comparison stars. Another resolution called on observers 'to include on their working lists, [...] objects for which parallaxes have already been published [...]'. Maybe this is what after discussions was left of the 'Dutch' proposal.

The definition of the absolute magnitude was formulated as being the 'magnitude at the distance of 10 parsecs, at which that star's parallax would be $0''.1$' in the 'Report of the Committee on Notations, Units and Economy of Publications' of the American Section of the International Astronomical Union [23]. This 'American Section' worked under the auspices of the National Research Council, following

Fig. 16.17 Edouard-Benjamin Baillaud as he appears in the *Album Amicorum* presented at the retirement of H.G. van de Sande Bakhuyzen as professor of astronomy and director of Leiden Observatory in 1908 (Archives Leiden Observatory; see caption Fig. 3.7)

the Conference on Interallied Academies in 1918 in London, where the American delegation had proposed that '... Astronomy for example, should be succeeded by a single society ..'. This (and a large number of other proposals on notations etc.) were adopted by the International Astronomical Union, following a recommendation from Commission 3 'des notations, des unités et de l'économie des publications', based on this report, at the first General Assembly in Rome in 1922. However, neither the IAU Transactions of that meeting, nor the PNAS report referred to above mention Kapteyn, although it is generally acknowledged he stood at the basis of the definition.

The address at the opening of the General Assembly of the IAU by its president Edouard-Benjamin Baillaud (see Fig. 16.17), on the other hand, made mention of the Kapteyn Astronomical Laboratory. The description of Henriette Hertzsprung-Kapteyn seems to be based on the text of the (Dutch) obituary Willem de Sitter (1922) wrote: 'And Baillaud, the veteran under the astronomers, mentioned in his opening address as president of the congress held in Rome in May of this year, the three things that had revolutionized the face of science in his more than fifty years of active life as an astronomer, and those three were: photography, the giant telescopes, and the Groningen Laboratory.'

This seems a bit of an exaggeration. To be precise, the relevant paragraph of Baillaud's in the *Transactions* (*op. cit.*) reads in my translation from the French:

'It is always the larger instruments, said Paul Henry and Prosper Henry who have produced the latest discoveries in astronomy. The formula may be a little exaggerated. However, a little bit modified it becomes: larger instruments allow us to do what the other ones did not allow. There is the new revolution happening: after the invention of refracting telescopes, spectroscopy and photography, we witness the arrival of the gigantic instruments, which are the masterpieces of mechanical and optical design, erected at the sites that are most favourable in all respects. At such instruments two or more teams of workers can be successfully employed during the same night. These instruments will often remain sterile if they are not accompanied by all the secondary devices that constitute the astronomical laboratory. Since it is the photographic plates that have to be studied, these laboratories will not need to be next to the telescopes. The laboratory of Groningen has given abundant proof of this. Astronomical life will certainly continue to be tough; but isn't it full of temptation just because of the immensity of problems to be solved and the certainty of reaping a rich harvest. Few sciences are more honourable to humanity.'

I continue with the HHK biography.

'And when a few days later a treatise was sent to him with the title 'The Kapteyn Universe', he smiled happily. His largest wish in life had come true: He had been able to serve science in an important way and he could put his head to rest. The concentrated efforts had made him tired and weak. His body was no longer able to do what his spirit wanted him to do. On the 24th of May he wrote his last letter; it was a beautiful greeting to the Royal Astronomical Society that celebrated its 100th anniversary.'

Indeed, the proceedings of the RAS centenary celebrations [24] do record that a message from Kapteyn had been read.

16.11 The End

On May 8, 1922, Kapteyn wrote a letter to Seares, who was in Rome for the IAU General Assembly by then, asking him to come and visit him after the congress.

'... much to my disappointment I am still lying in bed. Now be good to me, and let this not deter you from coming to us in Hilversum, Oude Amersfoortsche Straatweg 29. If I am not better by the time you come, we cannot talk the whole day, but even in a few hours we can talk over a good part of the Universe. I further arranged that if you stay a couple of days, you can spend [a] good part of the day in seeing the city of Amsterdam under the guidance of my daughter Mrs Hertzsprung, who knows the city and its museums pretty thoroughly. About dinnertime you could then be back in Hilversum and my wife will be delighted to show you the lovely Hilversum things in the evening.'

It is not on record that Seares did in fact come to see Kapteyn. What we do have is a telegram from Charles St. John of June 8, 1922 (10 days before his demise) to Hale: 'WRITE OR WIRE KAPTEYN 10 EMMAPLAIN AMSTERDAM THEME LOVE AND HOPE SERIOUSLY ILL', which was forwarded to Hale in Washington.

Kapteyn died on June 18, 1922.

Obviously, he was not (made) aware of the seriousness of his illness and his rapidly approaching death. Indeed, after his death, his wife wrote to Hale on July 7, 1922, in response to a letter from Hale about Kapteyn's passing away. She also referred to a letter from Hale – maybe triggered by the telegram by St. John – that arrived too late to be read out to Kapteyn. After expressing in much detail her thankfulness for Hale's friendship and his hospitality at Mount Wilson, she described Kapteyn's last months in a very moving paragraph:

'He began to ail in October, when he had had his conference with Einstein, Jeans and the others at Leiden and we were at our eldest daughter in Amsterdam. Mrs Hertzsprung was there first for the weekend, the last month all the time. He was hopeful to the last and full of plans for his beloved work and for his life here in his new home that he bought with such great glee in November. Alas, he saw it, loved it already, but was not to come into it, as I had hoped that he would. I made everything in order in the last part of April but by coming home, my daughter told me that there was no hope of his recovery. This last half year of his illness has been one of such perfect happiness to us all in our great united love that my heart would repeatedly say 'This is my happy illness time' and 'I believe this is the happiest time of my life'. And we could bear up and show our happy faces and the remembrance of this holy time makes me able to bear my loss bravely and be thankful for what has been and what is left to me.'

The last paragraphs of the HHK biography are the following.

'Many trusted friends followed him to the cemetery at Westerveld; no official speeches by dignitaries, according to the wishes of the family who felt that was in his spirit, only true love and friendship accompanied him.

Van Anrooy played on the organ the moving last choir from Bach's Saint Matthew Passion:

> 'Wir setzen uns mit Tränen nieder
> Und rufen dir im Grabe zu:
> Ruhe sanfte, sanfte Ruh.'[3]

And his friend Bordewijk spoke a few tender words with a voice full of emotion. That was all, but the beautiful music played be the friend, the deeply felt words, the quiet pain of all filled that moment with a holy ordination.

In the evening in Groningen Kor Kuiler, the director of the 'Harmonie' orchestra [the clubhouse, 'De Harmonie' was a major building in the center, which housed a magnificent concert hall, the home of this orchestra], performed the Beethoven's Death March as a tribute to this great Groninger.'

Peter Gijsbert van Anrooij was conductor of the Groninger Orkest Vereeniging (1905–1910).

---

[3]In English translation by Z. Philip Ambrose [25]: 'We lay ourselves with weeping prostrate And cry to thee within the tomb: Rest thou gently, gently rest!'

16.11 The End

Fig. 16.18 Tombstones of the burial urns of the Kapteyn family at the cemetery 'Westerveld' at Driehuis, about 6 km north of Haarlem. In the middle respectively Jacobus Cornelius Kapteyn, his granddaughter Greta Noordenbos, his wife Catharina Elisabeth Kapteyn-Kalshoven and his son-in-law Willem Cornelis Noordenbos. On the left Kapteyn daughter Jacoba Cornelia Noordenbos-Kapteyn and her son Willem Noordenbos. On the right Kapteyn's son Gerrit Jacobus and the latter's wife Wilhelmina Henriette van Gorkom (Provided by J.W. Noordenbos, great-grandson of Kapteyn, who currently stewards the graves)

'And all over the world, wherever he was known, people mourned. All realized that a great man had departed, great in spirit and mind, who would be missed sorely; but they also knew that his spirit would remain alive and his works would be continued and extended.'

It is appropriate to conclude this book with a short obituary of Kapteyn. It was written by Eduard Jan Dijksterhuis (1892–1965), known as a prominent Dutch historian of science, particularly for his book *The mechanization of the world picture*, originally published in Dutch in 1950 as *Mechanisering van het wereldbeeld*, and in an English translation in 1961. Klaas van Berkel has written a comprehensive biography of Dijksterhuis [26] (in Dutch). Dijksterhuis had studied mathematics in Groningen, starting in 1911, and had obtained a PhD degree there in 1918. After having been a HBS teacher of mathematics, physics and cosmography, and after having declined an offer of a professorship in Groningen in 1950 (giving his advanced age of 58 as the reason), he was appointed an extraordinary professor of the history of science in Utrecht in 1953 and in Leiden in 1955. Eventually he became an ordinary professor in Utrecht in 1960. During his studies in Groningen, Dijksterhuis had followed lectures by Kapteyn and he regarded Kapteyn as his most important teacher, who inspired him with his lectures and ignited a love for the history of science in him. Dijksterhuis' *in memoriam J.C. Kapteyn* was published in a periodical for secondary education, [27], in the issue of June 21, 1922 – only three

Fig. 16.19 Close-up of the tombstones in Fig. 16.18

days after Kapteyn's death. He must have written it when Kapteyn's death seemed inescapable.

'When I wish to speak a word of grateful remembrance in honour of Prof. Kapteyn, I think in doing so not in the first place of his scientific work, which ensures an honourable place for him among the great men of astronomy, but rather of the personality of the deceased as it has been impressed ineffaceably in the minds of all who studied mathematics and physics in Groningen these last few decades. Kapteyn used to teach a course for first-year students on the historical development of astronomy and then also treat, with nothing except mathematical tools, the principles of spherical astronomy. Who will ever forget the hours in the cluttered lecture room in the old astronomical laboratory, where the professor, after the traditional 'Morning, gentlemen', used to open an old and yellowed notebook, and would then proceed to start his exposition in a never withering liveliness and all-penetrating clearness, and in an instant take his audience to the pure realm of his science. Kapteyn used few drawings or objects to support his presentation, but the imaginative influence of his clear conception was so large that, when he started the conclusions of his exposition with the usual phrase: 'you will, gentlemen, now have no difficulty understanding that . . . ', everybody would indeed understand. As long as he was speaking, difficulties did not exist in astronomy.

The whole scientific world pays tribute to Kapteyn for the enormous labour that he has performed to serve astronomy; his students, however, thank him for the everlasting present that they personally received from him, when he gave the most important thing a teacher can give: the love, the marvel and the enthusiasm for the magnificent science to which he devoted his rich life.'

Appendix A
Publications by and about J.C. Kapteyn, His Honors and Academic Genealogy

> *The compulsion to write should be distrusted, unlike the urge to formulate.*
> Godfried Bomans (1913–1971).[1]

A.1 Publications by J.C. Kapteyn

Below I present a list of all known publications by Jacobus Cornelius Kapteyn. As director he must have been involved in supervising all activities, and therefore I have included all papers in the Publications of the Astronomical Laboratory at Groningen during Kapteyn's lifetime, with the exception of Willem de Sitter's papers on the Galilean satellites of Jupiter. I have also included papers that are directly related to projects Kapteyn was involved in personally, such as those based on PhD theses written under his supervision. PhD theses that have not been published separately in astronomical periodicals or observatory annals, etc., are included in the list as well. Many, but far from all, of these publications have been listed by the NASA Astronomy Data System ADS [1] and usually scanned copies are available there. I have indicated the ADS code at the end of the full reference between square brackets. The publications by the Royal Netherlands Academy of Arts and Sciences (KNAW) are made available on the *Digital Library – Dutch History of science web centre* [2]. The papers Kapteyn (1883) and Kapteyn (1884) are available electronically by downloading a pdf-file of the full volume 3 of *Copernicus* [3]. The table of contents at the end of that file lists the papers with their dates of publication; for these papers the dates are Febr. 28, 1883, and Febr. 12, 1884.

---

[1]*De dwang om te schrijven moet men wantrouwen, de drang om te formuleren niet.* Godfried Bomans was a Dutch writer and television personality.

© Springer International Publishing Switzerland 2015
P.C. van der Kruit, *Jacobus Cornelius Kapteyn*, Astrophysics and Space Science Library 416, DOI 10.1007/978-3-319-10876-6

Fig. A.1 The books of Kapteyn in the Kapteyn Room. From the left the three volumes of the Cape Photographic Durchmusterung plus two of the revisions, four bound volumes with Kapteyn's publications (see Fig. A.2) and one with his Mount Wilson publications up to 1920, Groningen Publications 1 through 37, the three Selected Areas (Systematic Plan) publications with Edward Pickering in the Harvard Annals and the Durchmusterung of the Special Plan (published in 1952). On the extreme right Kapteyn's copy of *Popular Astronomy* by Simon Newcomb (Kapteyn Astronomical Institute)

On my Kapteyn Web-page [4], I provide the list in this Appendix, together with links to electronic copies of every publication. In Figs. A.1 to A.4 I show some pictures of Kapteyn's books. For explanations see the captions.

Kapteyn. J. C. 1875. *Onderzoek der trillende, platte vliezen* [Study of vibrating flat membranes], Ph.D. thesis defended at University of Utrecht on June 24, 1875.

van de Sande Bakhuyzen, E. F. & Kapteyn, J. C. 1877a. *Elemente und Ephemeride des Cometen b. 1877*. Astronomische Nachrichten, 89, 251–252. [1877AN.....89..251V]

van de Sande Bakhuyzen, E. F. & Kapteyn, J. C. 1877b. *Ephemeris of comet b. 1877 (Winnecke's)*. The Observatory 1, 64–64. [1877Obs.....1...64V]

van de Sande Bakhuyzen, H. G. & Kapteyn, J. C. 1879a. *Beobachtungen der Heliometer-Sterne in Perseus*. Astronomische Nachrichten, 95, 5–14. [1879AN.....95....5V]

van de Sande Bakhuyzen, H. G. & Kapteyn, J. C. 1879b. *Beobachtungen von Gill's Mars-Sternen*. Astronomische Nachrichten, 95, 33–42. [1879AN.....95...33V]

Kapteyn, J. C. 1883. *Über das Kepler'sche Problem*. Copernicus 3, 25–34.

Kapteyn, J. C. 1884. *Über eine Methode die Polhöhe möglichst frei von Systematischen Fehlern zu bestimmen*. Copernicus 3, 147–182.

Kapteyn, J. C. & Kapteyn, W. 1884. *Les sinus de quatrième ordre*. Verhandelingen der Koninklijke Akademie van Wetenschappen, vier en twintigste deel, 1–98.

Kapteyn, J. C. & Kapteyn, W. 1886. *Die höheren sinus*. XCIII. Bande der Sitzberichten der Kaiserlichen Akademie der Wissenshaften (Wien). II. Abteilung, April-Heft, Jahrgang 1886, 807–868.

Kapteyn, J. C. 1888a. *Exposé de la méthode parallactique de mesure. Réduction des clichés.* Bulletin de Comité International Permanent de la Carte du Ciel, I, 94–114.

Kapteyn, J. C. 1888b. *Addition au la méthode parallactique de mesure.* Bulletin de Comité International Permanent de la Carte du Ciel, I, 125–127.

Kapteyn, J. C. 1888c. *Bericht über die zur Herstellung einer Durchmusterung des südlichen Himmels ausgeführten Arbeiten.* Vierteljahrsschrift der Astronomische Gesellsschaft, 23, 213–220.

Kapteyn, J. C. 1889a. *Vorläufige Mittheilung betr. Bestimmung von Fixstern-Parallaxen.* Astronomische Nachrichten 123, 105. [1890AN....123..105K]

Kapteyn, J. C. 1889b. *Note relative de mémoire de M. Bakhuyzen sur la mesure de clichés.* Bulletin de Comité International Permanent de la Carte du Ciel, 242–250.

Gill, D. & Kapteyn, J. C. 1889b. *Exposé d'un projet de M. J.-C. Kapteyn relative a la détermination des mouvements propres et des parallaxes d'étoiles.* Bulletin de Comité International Permanent de la Carte du Ciel, 262–264.

Kapteyn, J. C. 1890a. *Über eine photographische Methode der Breitenbestimmung aus Zenithsternen.* Astronomische Nachrichten 125, 81–86. [1890AN....125...81K]

Kapteyn, J. C. 1890b. *Südliche muthmasslich veränderliche Sterne.* Astronomische Nachrichten 125, 165–170. [1890AN....125..165K]

Kapteyn, J. C. 1890c. *De bepaling van de parallaxis van vaste sterren door middel van registreerwaarnemingen.* Verslagen en Mededelingen der Koninklijke Akademie van Wetenschappen, Afdeeling Natuurkunde, derde deel, 114–115.

Kapteyn, J. C. 1890d. *Zweiter Bericht über die zur Herstellung einer Durchmusterung des südlichen Himmels ausgeführten Arbeiten.* Vierteljahrsschrift der Astronomsiche Gesellsschaft, 25, 240–243.

van de Sande Bakhuyzen, H. G., Kapteyn, J. C., et al. 1890a. *Zonen-beobachtungen zwischen $29°50'$ und $35°10'$ Declination.* Annalen van de Sterrewacht te Leiden, 5, I-LII,1–352. [1890AnLei...5....1.]

van de Sande Bakhuyzen, H. G., Kapteyn, J. C., et al. 1890b. *Reductionen der Zenitdistanzen der Fundsmentalsternen etc..* Annalen van de Sterrewacht te Leiden, 6, I-CXXIX,1–412. [1890AnLei...6....1.]

Kapteyn, J. C. 1891a. *Bestimmung von Parallaxen durch Registrir-Beobachtungen am Meridiankreise.* Annalen van de Sterrewacht te Leiden 7, 117–244. [1897AnLei...7..117K]

Boss, L. & Kapteyn, J. C. 1891. *Bestimmung von Parallaxen durch Registrir-Beobachtungen am Meridian Kreise.* Publications of the Astronomical Society of the Pacific 3, 346–353. [1891PASP....3..346K]

Plummer, W. E. & Kapteyn, J. C. 1891. *Stellar parallax, as determined by a transit instrument.* The Observatory, 14, 259–263. [1891Obs....14..259P]

Kapteyn, J. C. 1891b. *De beteekenis der photographie voor de studie van de hoogere delen des hemels*, address at the occasion of the transfer of office of Rector Magnificus at the University of Groningen on 15 September 1891. J.B. Wolters, Groningen.

Fig. A.2 The books of Kapteyn in the Kapteyn Room. Here we see three volumes with reprints of his own papers and one (red) with publications on the Star Streams. After binding all reprints together he had the pages renumbered sequentially. This looks like a streak of vanity in Kapteyn, but may be no more than a appreciation of order and neatness (Kapteyn Astronomical Institute)

Kapteyn, J. C. 1891c *Verslag van de lotgevallen der Rijks-Universiteit te Groningen in het studiejaar 1890–91, gegeven den* 15^{den} *September 1891, door den aftredenden Rector-Magnificus Dr J. C. Kapteyn.* 1891. J.B. Wolters, Groningen.

Kapteyn, J. C. 1892a. *Plan et détails de l'appareil parallactique de mesures.* Bulletin de Comité International Permanent de la Carte du Ciel, I, 377–381.

Kapteyn, J. C. 1892b. *Théorie des erreurs de l'appareil parallactique de mesures et réduction de clichés.* Bulletin de Comité International Permanent de la Carte du Ciel, I, 401–452

Kapteyn, J. C. 1892c. *Difference systématique entre les grandeurs photographiques et visuelles dans le differentes region du ciel.* Bulletin de Comité International Permanent de la Carte du Ciel, II, 131–158.

Kapteyn, J. C. 1892d. *Systematische verschillen tusschen de visueele en photographische helderheid der sterren in verschillende deelen der hemel.* Verslagen en Mededelingen der Koninklijke Akademie van Wetenschappen, Afdeeling Natuurkunde, negende deel, 393–394.

Kapteyn, J. C. 1892e. *De verdeeling an de sterren in de ruimte.* Verslagen en Mededelingen der Koninklijke Akademie van Wetenschappen, Afdeeling Natuurkunde, negende deel, 418–421.

Kapteyn, J. C. 1892f. *To what stellar system does our Sun belong?* (Abstract). Publications of the Astronomical Society of the Pacific 4, 259–260. [1892PASP....4..259K]

Kapteyn, J. C. 1892g. *Dritter Bericht über die zur Herstellung einer Durchmusterung des südlichen Himmels ausgeführten Arbeiten.* Vierteljahrsschrift der Astronomische Gesellsschaft, 27, 218–219.

Kapteyn, J. C. 1893a. *Over de verdeeling van de sterren in de ruimte.* Koninklijke Akademie van Wetenschappen, Zittingsverslag 28 Januari 1893, 125–140.

Kapteyn, J. C. 1893b. *Measuring stellar photographs.* Engineering Jan.27, 1893, 91–92

Kapteyn, J. C. 1893c. *Über den neuen Fleming'schen Stern im Sternbilde Norma.* Astronomische Nachrichten 134, 59. [1893AN....134...59K]

Kapteyn, J. C. 1893d. *Eene nieuwe methode ter bepaling van den afstand van vaste sterren.* Handelingen van het vierde Nederlandsch Natuur- en Geneeskundig Congres, gehouden te Groningen op den 7^{den} en 8sten April 1893, 70–75.

Kapteyn, J. C. 1895. *Over de verdeeling der kosmische snelheden.* Koninklijke Akademie van Wetenschappen, Zittingsverslag 25 Mei 1895, 1–15.

Kapteyn, J. C. 1896a. *Corrections de refraction et d'aberration pour les coordonnées rectangulaires mesurées sur les clichés photographiques.* Bulletin de Comité International Permanent de la Carte du Ciel, III, 17–44.

Gill, D. & Kapteyn, J. C. 1896. *The Cape Photographic Durchmusterung for the equinox 1875. Part I. Zones $-18°$ to $-37°$.* Annals of the Cape Observatory, South Africa, 3, 1–845. [1896AnCap...3....1G]

Kapteyn. J. C. 1896b. *Openbare les gehouden bij de gelegenheid van de opening van het Sterrenkundig Laboratorium te Groningen.* Hoitsema Brothers, Groningen.

Kapteyn, J. C. 1896c. *New southern variable stars.* Astronomische Nachrichten 142, 75–78. [1896AN....142...75K]

Kapteyn, J. C. 1897a. *Verdeeling der kosmische snelheden. Toevoegsel aan de mededeeling van 5 Mei 1895.* Koninklijke Akademie van Wetenschappen, Zittingsverslag 29 Mei 1897, 10 pages.

Kapteyn, J. C. 1897b. *Stern mit grösster bislang bekannter Eigenbewegung.* Astronomische Nachrichten 145, 159. [1898AN....145..159K]

Gill, D. & Kapteyn, J. C. 1897. *The Cape Photographic Durchmusterung for the equinox 1875. Part II. Zones $-38°$ to $-52°$.* Annals of the Cape Observatory, South Africa, 4, 1–702. [1897AnCap...4....1G]

Kapteyn, J. C. 1898a. *Bestimmung von 250 Parallaxen.* Astronomische Nachrichten 145, 289–300. [1898AN....145..289K]

Kapteyn, J. C. 1898b. *Die mittlere Geschwindigkeit der Sterne, die Quantität der Sonnenbewegung und die mittlere Parallaxe der Sterne von verschiedener Grösse.* Astronomische Nachrichten 146, 97–114. [1898AN....146...97K]

Kapteyn, J. C. 1898c. *Bemerkungen zu der Abhandlung des Herrn J. Scheiner "Über die Abhängigkeit der Grössenangaben der Bonner Durchmusterung von der Sternfülle" in Astr. Nachr. Nr. 3505.* Astronomische Nachrichten 147, 305–310. [1898AN....147..305K]

Gill, D. & Kapteyn, J. C. 1898. *The Cape Photographic Durchmusterung.* Viertelsjahrschrift der Astronomiche Gesellsschaft, 33, 192–221.

Kapteyn, J. C. 1899. *Bemerkungen über die Beziehung der photographischen und visuellen Grössen der Sterne.* Astronomische Nachrichten 150, 103–106. [1899AN....150..103K]

Gill, D. & Kapteyn, J. C. 1900. *The Cape Photographic Durchmusterung for the equinox 1875. Part III. Zones* $-53°$ *to* $-89°$. Annals of the Cape Observatory, South Africa, 5, 1–757. [1899AnCap...5....1G]

Kapteyn, J. C. & Donner, A. S. 1900. *The parallax of 248 stars of the region around B.D. 35,4013.* Publications of the Astronomical Laboratory at Groningen 1, 1–87. [1900PGro....1....3K]

de Sitter, W. 1900a. *On the systematic difference, depending on Galactic latitude, between the photographic and visual magnitudes of stars.* Publications of the Astronomical Laboratory at Groningen, 2, 1–22. [1900PGro....2....1D]

de Sitter, W. 1900b. *On isochromatic plates.* Publications of the Astronomical Laboratory at Groningen, 3, 23–36. [1900PGro....3...23D]

Kapteyn, J. C. 1900b. *Over de bepaling van de coördinaten van het apex der zonsbeweging.* Koninklijke Akademie van Wetenschappen, Zittingsverslag 27 Januari 1900, 22 pages.

Kapteyn, J. C. 1900c. *The determination of the apex of the solar motion.* Koninklijke Nederlandse Akademie van Wetenschappen Proceedings Series B Physical Sciences 2, 353–374. [1899KNAB....2..353K, www.dwc.knaw.nl/DL/publications/PU00014484.pdf]

Kapteyn, J. C. & Kapteyn, W. 1900. *On the distribution of cosmic velocities. Part I. Theory.* Publications of the Astronomical Laboratory at Groningen 5, 1–87. [1900PGro....5....1K]

Kapteyn, J. C. 1900d. *Components* τ *and* υ *of the proper motions and other quantities for the stars of Bradley.* Publications of the Astronomical Laboratory at Groningen 7, 5–120. [1900PGro....7D...5K]

Kapteyn, J. C. 1900e. *On the mean parallax of stars of determined proper motion and magnitude.* Publications of the Astronomical Laboratory at Groningen 8, 1–31. [1900PGro....8....1K]

Kapteyn, J. C. 1901a. *De lichtkracht der vaste sterren.* Koninklijke Akademie van Wetenschappen, Zittingsverslag 20 April 1901, 32 pages.

Kapteyn, J. C. 1901b. *Der Apex der Sonnenbewegung, die Constante der Praecession und die Correctionen der Eigenbewegungen in Declination von Auwers-Bradley.* Astronomische Nachrichten 156, 1. [1901AN....156....1K]

Kapteyn, J. C. 1901c. *Beantwoording der kritiek van Dr. J. Stein, S.J..* Koninklijke Akademie van Wetenschappen, Zittingsverslag 26 October 1901, 11 pages.

Kapteyn, J. C. 1901d. *Über die Bewegung der Nebel in der Umgebung von Nova Persei.* Astronomische Nachrichten 157, 201–204. [1901AN....157..201K]

Kapteyn, J. C. 1901e. *Méthode statistique pour la détermination de l'apex du mouvement solaire.* Archives Néerlandaises des Sciences Exactes et Naturelles (Hollandsche Maatschappij der Wetenschappen), Ser. II, Tome VI, 262–284.

Kapteyn, J. C. 1901f. *Over de afstandsbepaling van de hemelichamen.* Popular lecture.

Fig. A.3 In the first volume in Fig. A.2, Kapteyn had pasted a hand-written index to his publications in the three blue volumes. In each he provided a continuous numbering of the pages. The list was systematic by subject. It has been produced around 1916, at least that is the date of the last paper in the collection. Here I reproduce the first page (Kapteyn Astronomical Institute)

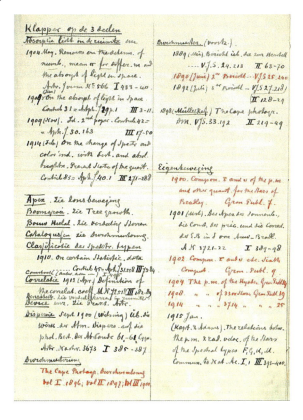

Kapteyn, J. C. 1901g. *On the luminosity of the fixed stars.* Koninklijke Nederlandse Akademie van Wetenschappen Proceedings Series B Physical Sciences 3, 658–689. [1900KNAB....3..658K; www.dwc.knaw.nl/DL/publications/PU00014426.pdf]

Kapteyn. J. C. 1902a. *Reply to the criticism of Dr. J. Stein, S.J.* Koninklijke Nederlandse Akademie van Wetenschappen Proceedings Series B Physical Sciences 4, 232–242. [1901KNAB....4..232K; www.dwc.knaw.nl/DL/publications/PU00014273.pdf]

Kapteyn, J. C. 1902b. *Components τ and υ of the proper motions and other quantities for the stars of Bradley, Sixth Computation.* Publications of the Astronomical Laboratory at Groningen 9, 1–56. [1902PGro....9....1K]

Kapteyn, J. C. & Donner, A. S. 1902. *The parallax of 248 stars of the region around B.D. 35,4013.*, Summary by F. Cohn (1866–1922). Vierteljahrsschrift der Astronomischen Gesellschaft, 36, 6–20.

Kapteyn, J. C. 1902c. *On the mean parallax of stars of determined proper motion and magnitude.*, Summary by K. Kobold. Vierteljahrsschrift der Astronomischen Gesellschaft, 37, 16–24.

Kapteyn, J. C., de Sitter, W. & Donner, A. S. 1902. *Parallaxes of the clusters h and χ Persei, of Groombridge 745, 61 Cygni, and surrounding stars.* Publications of the Astronomical Laboratory at Groningen 10, 1–59. [1902PGro...10....1K]

Kapteyn, J. C. 1902d. *On the luminosity of the fixed stars.* Publications of the Astronomical Laboratory at Groningen 11, 3–32. [1902PGro...11....3K]

Kapteyn, J. C. 1902e. *On the motion of nebulae in the vicinity of Nova Persei.* Popular Astronomy, 10, 124–127. [1902PA.....10..124K]

Kapteyn, J. C. 1902f. *Über die Airy'sche Methode zur Bestimmung des Apex der Sonnenbewegung.* Astronomische Nachrichten 159, 121–126. [1902AN....159..121K]

Gill, D. & Kapteyn, J. C. 1903. *Revision of the Cape Photographic Durchmusterung, Part II. Variable stars, miscellaneous stars, etc..* Annals of the Cape Observatory, South Africa, 9, 2.1–2.188. [1903AnCap...9....2G]

Kapteyn, J. C. 1903a. *Über die Deklination des Apex der Sonnenbewegung.* Astronomische Nachrichten 161, 325–364. [1903AN....161..325K]

Kapteyn, J. C. 1903b. *Skew frequency curves in biology and statistics.* Noordhof, Groningen, 45 pages + figures.

de Sitter, W. 1904. *Investigation of the systematic difference between the photographic and visual magnitudes of stars depending on Galactic latitude, based on photometric observations by W. de Sitter, visual estimates by R.T.A. Innes, and photographs taken at the Cape Observatory, together with catalogues of the photometric and photographic magnitudes of 791 stars..* Publications of the Astronomical Laboratory at Groningen 12, 1–166. [1904PGro...12....1D]

Weersma, H. A. 1904. *The proper motions of 66 stars of the Hyades derived from the observations of 34 catalogues between 1755 and 1900.* Publications of the Astronomical Laboratory at Groningen 13, 1–31. [1904PGro...13....1W]

Kapteyn, J. C., de Sitter, W. & Donner, A. S. 1904. *The proper motions of the Hyades.* Publications of the Astronomical Laboratory at Groningen 14, 1–87. [1904PGro...14D...1K]

Kapteyn, J. C. 1904a. *Remarks on the determination of the number and mean parallax of stars of different magnitude and the absorption of light in space.* Astronomical Journal 24, 115–122. [1904AJ.....24..115K]

Kapteyn, J. C. 1904b. *Statistical methods in stellar astronomy.* Presentation at the Louisiana Purchase Exposition (also known as the fourteenth World's Fair), held at St. Louis, Missouri, International Congress of Arts and Sciences, Vol. VIII: Astronomy and Earth Sciences, 396–425.

Kapteyn, J. C. 1905a. *Star Streaming.* Report of the British Association for the Advancement of Science, South Africa, 257–265.

Kapteyn, J. C. 1905b. *Reply to Prof. Pearsons criticism.* Recueil des Travaux Botaniques Néerlandais. No. 3, 7 pages.

Kapteyn, J. C. 1906a. *Over de parallax van nevelvlekken.* Koninklijke Akademie van Wetenschappen, Zittingsverslag 24 Februari 1906, 9 pages.

Kapteyn, J. C. 1906b. *On the parallax of the nebulae.* Koninklijke Nederlandse Akademie van Wetenschappen Proceedings Series B Physical Sciences 8, 691–699. [1905KNAB....8..691K; www.dwc.knaw.nl/DL/publications/PU00013906.pdf]

Kapteyn, J. C. 1906c. *Sur la parallaxe des nébuleuses*, Archives Néerlandaises des Sciences Exactes et Naturelles (Hollandsche Maatschappij der Wetenschappen), Ser. II, Tome XI, 502–511.

Kapteyn, J. C. 1906d. *Plan of Selected Areas*. Astronomical Laboratory at Groningen, Hoitsema Brothers, Groningen.

Kapteyn, J. C. 1906e. *Courants dans la Systéme Stellaire*, Archives Néerlandaises des Sciences Exactes et Naturelles (Hollandsche Maatschappij der Wetenschappen), Ser. II, Tome XI, 32–54.

de Sitter, W. 1906. *Tables for photographic parallax observations*. Publications of the Astronomical Laboratory at Groningen 15, 1–12. [1906PGro...15....1D]

Kapteyn, J. C. & Kapteyn, W. 1906. *Some useful trigonometrical formulae and a table of goniometrical functions for the four quadrants*. Publications of the Astronomical Laboratory at Groningen 16, 13–19.

Kapteyn, J. C. 1908a. *Over de gemiddelde sterdichtheid op verschillende afstand van het zonnestelsel*. Koninklijke Akademie van Wetenschappen, Zittingsverslag 29 Februari 1909, 10 pages.

Kapteyn, J. C. 1908b. *On the mean star-density at different distances from the solar system*. Koninklijke Nederlandse Akademie van Wetenschappen Proceedings Series B Physical Sciences 10, 626–635. [1907KNAB...10..626K; www.dwc.knaw.nl/DL/publications/PU00013684.pdf]

Kapteyn, J. C. 1908c. *Sur la densité moyenne á des distances différentes du Systéme Solaire*, Archives Néerlandaises des Sciences Exactes et Naturelles (Hollandsche Maatschappij der Wetenschappen), Ser. II, Tome XIII, 458–470.

Kapteyn, J. C. 1908d. *On the number of stars of determined magnitude and determined Galactic latitude*. Publications of the Astronomical Laboratory at Groningen 18, 1–54. [1908PGro...18....1K]

Kapteyn, J. C., de Sitter, W.& Donner, A. S. 1908a. *The proper motions of 3300 stars of different Galactic latitudes*. Publications of the Astronomical Laboratory at Groningen 19, 1–42 and T1-T112. [1908PGro...19D...1K]

Kapteyn, J. C., de Sitter, W. & Donner, A. S. 1908b. *The parallaxes of 3650 stars of different Galactic latitudes, with an appendix containing rules for the treatment of parallax-plates and a graphical table of parallax-coefficients*. Publications of the Astronomical Laboratory at Groningen 20, 1–34 and T1-T136. [1908PGro...20....1D]

Weersma, H. A.. 1908. *A determination of the apex of the Solar motion according to the method of Bravais*. Publications of the Kapteyn Astronomical Laboratory at Groningen, 1–74 and I-XXXI. [1908PGro...21....1W]

Kapteyn, J. C. 1908e. *Recent researches in the structure of the Universe*. Proceedings of the Royal Institution of Great Britain, 16 pages.

Kapteyn, J. C. 1908f. *Recent researches in the structure of the Universe*. The Observatory 31, 346–348. [1908Obs....31..346K]

Yntema, L. 1909. *On the brightness of the sky and the total amount of starlight*. Publications of the Kapteyn Astronomical Laboratory at Groningen, 22, 1–55. [1909PGro...22....1Y]

Kapteyn, J. C., de Sitter, W., Donner, A. S. & Küstner, K. F. 1909. *The parallax of the Hyades*. Publications of the Astronomical Laboratory at Groningen 23, 1–56. [1909PGro...23....1K]

Kapteyn, J. C. 1909a. *Recent researches in the structure of the universe*. Annual Report of the Smithsonian Institution 1909, 301–320. [ia600502.us.archive.org/17/items/annualreportofbo1908smitfo/annualreportofbo1908smitfo.pdf].

Kapteyn, J. C. 1909b. *De theorie van eb en vloed*. Lezing gehouden op 27 Februari 1909, Jaarboekje 1909–1910 de van Mijnbouwkundige Vereeniging te Delft. 5 pages + figures.

Kapteyn, J. C. 1909c. *On the absorption of light in space*. Astrophysical Journal 29, 46–54. Also: Contributions from the Mount Wilson Observatory, Carnegie Institution of Washington No. 31. [1909ApJ....29...46K] or [1909CMWCI..31....1K]

Kapteyn, J. C. 1909d. *On the absorption of light in space*. Second Paper. Astrophysical Journal 30, 284–317. Also: Contributions from the Mount Wilson Observatory, Carnegie Institution of Washington No. 42. [1909ApJ....30..284K] or [1909CMWCI..42....1K]

Kapteyn, J. C. 1909e. *Correction to Professor Kapteyn's article in the November Number*. Astrophysical Journal 30, 398–399. [1909ApJ....30..398K]

Kapteyn, J. C. & Weersma, H. A. 1910. *List of parallax determinations*. Publications of the Astronomical Laboratory at Groningen 24, 1–32. [1910PGro...24....1K]

Kapteyn, J. C. 1910a. *The luminosity curve*. Astronomische Nachrichten 183, 313–332. [1910AN....183..313K]

Kapteyn, J. C. 1910b. *On certain statistical data which may be valuable in the classification of the stars in the order of their evolution*. Astrophysical Journal 31, 258–269. Also: Contributions from the Mount Wilson Observatory, Carnegie Institution of Washington No. 45. [1910ApJ....31..258K] or [1910QB1.C32n45.....]

Kapteyn, J. C. & Frost, E. B. 1910. *On the velocity of the Sun's motion through space as derived from the radial velocity of Orion stars*. Astrophysical Journal 32, 83–90. [1910ApJ....32...83K]

Kapteyn, J. C. 1910c. *On the average parallax of the stars of the fourth type as compared with that of stars of other types*. Astrophysical Journal 32, 91–95. [1910ApJ....32...91K]

Kapteyn, J. C. 1910d. *On the systematic proper motion of the Orion stars*. Evening addresses delivered to the Solar Union, Transactions of the International Union for Cooperation in Solar Research, 3, 215–231. [1910TIUCS...3..201.]

Kapteyn, J. C. 1911a. *De melkweg en de sterstroomen*. Koninklijke Akademie van Wetenschappen, Zittingsverslag 25 November 1911, 8 pages.

Kapteyn, J. C. 1911b. *The Milky Way and the star-streams*. Koninklijke Nederlandse Akademie van Wetenschappen Proceedings Series B Physical Sciences 14, 524–530. [1911KNAB...14..524K; www.dwc.knaw.nl/DL/publications/PU00013176.pdf]

Kapteyn, J. C. & Frost, E. B. 1911. *Correction*. Astrophysical Journal 33, 86. [1911ApJ....33...86K]

Kapteyn, J. C., 1911c. *First and second report on the progress of the Plan of Selected Areas*. Hoisema Brothers, Groningen.

Fig. A.4 Books in the Kapteyn Room. This shelve shows a collection of the Publications of the Kapteyn Astronomical Laboratory Groningen. This set was originally available in the office of the director (respectively Kapteyn, P.J. van Rhijn and A. Blaauw) (Kapteyn Astronomical Institute)

Kapteyn, J. C. 1911d. *Een paar nieuwere onderzoekingen op het gebied der evolutie van de vaste sterren en het sterrenstelsel.* Handelingen van het XIIIe Nederlandsch Natuur- en Geneeskundig Congres, gehouden te Groningen op 20, 21 en 22 April 1911, 44–68.

van Maanen, A., 1911. *The proper motions of 1418 stars in and near the clusters h and χ Persei.* Recherches Astronomiques de l'Observatoire d'Utrecht, 5, v-99. [1911RAOU....5....1V]

Kapteyn, J. C. 1912a. *Definition of the correlation-coefficient.* Monthly Notices of the Royal Astronomical Society 72, 518–525. [1912MNRAS..72..518K]

Kapteyn, J. C. 1912b. *Star systems and the Milky Way.* Koninklijke Nederlandse Akademie van Wetenschappen Proceedings Series B Physical Sciences 14, 909–911. [1911KNAB...14..909K; www.dwc.knaw.nl/DL/publications/PU00013254.pdf]

Kapteyn, J. C. & Weersma, H. A. 1912. *On the derivation of the constants for the two star streams.* Monthly Notices of the Royal Astronomical Society 72, 743–756. [1912MNRAS..72..743K]

Kapteyn, J. C. 1912c. *Tree-growth and meteorological factors.* Private publication, 27 pages.

Kapteyn, J. C. 1913a. *The structure of the Universe.* Science 38, 717–724. [1913Sci....38..717K]

Kapteyn, J. C. 1913b. *On the structure of the Universe.* Scientia 14, 245–357.

Kapteyn, J. C. & Weersma, H. 1914. *The proper motions of 3714 stars derived from plates taken at the observatories of Helsingfors and Cape of Good Hope.* Publications of the Astronomical Laboratory at Groningen 25, 1–28, H1-H13.

Kapteyn, J. C. 1914a. *On the structure of the universe.* Journal of the Royal Astronomical Society of Canada 8, 145–159. [1914JRASC...8..145K]

Smid, E. I. 1914. *Bepaling der eigenbeweging in Rechte Klimming en Declinatie van 119 sterren.*, Ph.D. Thesis, University of Groningen.

Kapteyn, J. C. 1914b. *On the individual parallaxes of the brighter Galactic helium stars in the southern hemisphere, together with considerations on the parallax of stars in general.* Astrophysical Journal 40, 43–126. Also: Contributions from the Mount Wilson Observatory No. 82. [1914ApJ....40...43K] or [1914CMWCI..82 ....3K]

Kapteyn, J. C. 1914c. *Sir David Gill.* Astrophysical Journal 40, 161–172. [1914 ApJ ....40..161K]

Kapteyn, J. C. 1914d. *On the change of spectrum and color index with distance and absolute brightness. Present state of the question.* Astrophysical Journal 40, 187–204. Also: Contributions from the Mount Wilson Observatory, Carnegie Institution of Washington No. 83. [1914ApJ....40..187K] or [1914CMWCI..83....1K]

Kapteyn, J. C. 1914e. *Het sterrekundig laboratorium.* Academia Groningana MDCXIV-MCMXIV: Gedenkboek ter gelegenheid van het derde eeuwfeest der Universiteit te Groningen, 550–552,

Kapteyn, J. C. 1915. *On a device for avoiding systematic errors depending on magnitude in the measurement of stellar photographs.* Astrophysical Journal 41, 77–80. [1915ApJ....41...77K]

Kapteyn, J. C. & Adams, W. S. 1915. *The relations between the proper motions and the radial velocities of the stars of the spectral types F, G, K, and M.* Proceedings of the National Academy of Science 1, 14–21. [1915PNAS....1...14K]

Kapteyn, J. C. & van Uven, M. J. 1916. *Skew frequency curves in biology and statistics, 2nd paper.* Astronomical Laboratory at Groningen, Hoitsema Brothers, Groningen, 69 pages + figures.

Kapteyn, J. C. 1916. *Skew frequency curves in biology and statistics.* Recueil des Travaux Botaniques Néerlandais, XIII, Livr. II, 105–157.

van Rhijn, P. J. 1916a. *The proper motions of the stars in and near the Praesepe cluster, derived from photographic plates taken at Potsdam.* Publications of the Kapteyn Astronomical Laboratory at Groningen, 26, 1–24. [1916PGro...26....1V]

van Rhijn, P. J. 1916b. *The change of color with distance and apparent magnitude together with a new determination of the mean parallaxes of the stars of given magnitude and proper motion.* Astrophysical Journal, 43, 36–42. Also: Contributions from the Mount Wilson Observatory, Carnegie Institution of Washington No. 110. [1916ApJ....43...36V] and [1916CMWCI.110....1V]

Meijering, S. C. 1916. *On the systematic motions of the K-stars.* PhD Thesis, University of Groningen.

van Rhijn, P. J. 1917. *On the number of stars of each photographic magnitude in different Galactic latitudes.* Publications of the Kapteyn Astronomical Laboratory at Groningen, 27, 1–63. [1916PGro...27....1V]

Pickering, E. C. & Kapteyn, J. C. 1918. *Durchmusterung of Selected Areas between $\delta = 0°$ and $\delta = 90°$; Systematic Plan.* Annals of Harvard College Observatory 101, 1–368. [1918AnHar.101....1P]

Kapteyn, J. C. & van Rhijn, P. J. 1918. *The proper motion of 2380 stars derived from plates taken at the Royal Observatory, Cape of Good Hope.* Publications of the Astronomical Laboratory at Groningen 28, 1–51.

Kapteyn, J. C., van Rhijn, P. J. & Weersma, H. 1918. *The secular parallax of the stars of different magnitude, Galactic latitude and spectrum*. Publications of the Astronomical Laboratory at Groningen 29, 1–63. [1918PGro...29Q...1.] and [1918PGro...29...45.]

Kapteyn, J. C. 1918a. *On the parallaxes and motion of the brighter Galactic helium stars between Galactic longitudes* 150° *and* 216°. Astrophysical Journal 47, 104–133. Also: Contributions from the Mount Wilson Observatory, Carnegie Institution of Washington No. 147. [1918ApJ....47..104K] or [1918CMWCI.147....3K]

Kapteyn, J. C. 1918b. *On the parallaxes and motion of the brighter Galactic helium stars between Galactic longitudes* 150° *and* 216° – *Continued*. Astrophysical Journal 47, 146–178. Also: Contributions from the Mount Wilson Observatory, Carnegie Institution of Washington No. 147. [1918ApJ....47..146K] or [1918CMWCI.147....3K]

Kapteyn, J. C. 1918c. *On the parallaxes and motion of the brighter Galactic helium stars between Galactic longitudes* 150° *and* 216° – *Concluded*. Astrophysical Journal 47, 255–282. Also: Contributions from the Mount Wilson Observatory, Carnegie Institution of Washington No. 147. [1918ApJ....47..255K] or [1918CMWCI.147....3K]

Schouten, W. J. A. 1918. *On the determination of the principal laws of statistical astronomy*. PhD Thesis, University of Groningen.

van Rhijn, P. J. 1919. *On the brightness of the sky at night and the total amount of starlight*. Astrophysical Journal, 50, 356–375. Also: Contributions from the Mount Wilson Observatory, Carnegie Institution of Washington No. 173, 1–20. [1919ApJ....50..356V] or [1919CMWCI.173....1V]

Kapteyn, J. C. & van Rhijn, P. J. 1920a. *Numbers of stars between definite limits of magnitude, proper motion and Galactic latitude for each spectral class, together with some other investigations.*, Publications of the Astronomical Laboratory at Groningen 30, 1–110. [1920PGro...30....1.]

Kapteyn, J. C. & van Rhijn, P. J. 1920b. *On the distribution of the stars in space especially in the high Galactic latitudes*. Astrophysical Journal 52, 23–38. Also: Contributions from the Mount Wilson Observatory, Carnegie Institution of Washington No. 188. [1920ApJ....52...23K] or [1920CMWCI.188....1K]

ten Bruggen Cate, G. H. *Determination and discussion of the spectral classes of 700 stars mostly near the North Pole*. PhD Thesis, University of Groningen.

Kapteyn, J. C. 1921. *Proeve eener theorie van de rangschikking en de beweging van het groote sterrenstelsel.*, 'Physica', Nederlandsch Tijdschrift voor Natuurkunde, 1, 352–356.

van Rhijn, P. J. *On the brightness of the sky at night and the total amount of starlight*, Publications of the Kapteyn Astronomical Laboratory at Groningen, 31, 1–83. [1921PGro...31....1V]

Kapteyn, J. C. & van Rhijn, P. J. 1922a. *The proper motions of δ Cephei stars and the distances of the globular clusters*. Bulletin of the Astronomical Institutes of the Netherlands 1, 37–42. [1922BAN.....1...37K]

Kapteyn, J. C. & van Rhijn, P. J. 1922b. *On the upper limit of the distance to which the arrangement of stars in space can at present be determined with some confidence.* Astrophysical Journal 55, 242–271. Also: Contributions from the Mount Wilson Observatory, Carnegie Institution of Washington 229, 1–30. [1922ApJ....55..242K] or [1922CMWCI.229....1K]

Kapteyn, J. C. 1922a. *First attempt at a theory of the arrangement and motion of the Sidereal System.* Astrophysical Journal 55, 302–328. Also: Contributions from the Mount Wilson Observatory, Carnegie Institution of Washington No. 230. [1922ApJ....55..302K] or [1922CMWCI.230....1K]

Kapteyn, J. C. 1922b. *On the proper motions of the faint stars and the systematic errors of the Boss fundamental system.* Bulletin of the Astronomical Institutes of the Netherlands 1, 69–78. [1922BAN.....1...69K]

Pickering, E. C., Kapteyn, J. C. & van Rhijn, P. J. 1923. *Durchmusterung of Selected Areas:* $\delta = -15°$ *and* $-30°$; *Systematic Plan.* Annals of Harvard College Observatory 102, 1–276. [1923AnHar.102....1P]

Pickering, E. C., Kapteyn, J. C. & van Rhijn, P. J. 1924. *Durchmusterung of Selected Areas:* $\delta = -45°$ *and* $-90°$; *Systematic Plan.* Annals of Harvard College Observatory 103, 1–340. [1924AnHar.103....1P]

Seares, F. H., van Rhijn, P. J., Joyner, M. C. & Richmond, M. L. 1925. *Mean distribution of stars according to apparent magnitude and Galactic latitude.* Astrophysical Journal, 62, 320–374. [1925ApJ....62..320S]

Seares, F. H., Kapteyn, J. C., van Rhijn, P. J., Joyner, M. C. & Richmond, M. L. 1930. *Mount Wilson Catalogue of photographic magnitudes in Selected Areas 1–139.* Carnegie institution of Washington.

Mönnichmeyer, C. O. L., Kapteyn, J. C., Hopmann, J. & Schaub, W. 1930. *Katalog von 1172 Sternen in Kapteyn's "Selected Areas" auf Grund der Beobachtungen am Repsoldschen Meridiankreise.* Veröffentlichungen des Astronomisches Institute der Universität Bonn 21, 1–40. [1930VeBon..21....1M]

van Rhijn, P. J. & Kapteyn, J. C. 1952. *Durchmusterung of Selected Areas of the Special Plan.* Kapteyn Astronomical Laboratory at Groningen.

A.2 Publications About Kapteyn

Articles.

Glaisher, J. W. L. 1902. *Address delivered by the President, Dr. J.W.L. Glaisher, on presenting the Gold Medal to Professor J.C. Kapteyn.* Monthly Notices of the Royal Astronomical Society, 62, 334–343 [1902MNRAS..62..329.]

Galloway, J. D. 1902. *Kapteyn's contributions to our knowledge of the stars.* Publications of the Astronomical Society of the Pacific, 14, 97–102 [1902PASP...14 ...97G]

Curtis, H. D. 1913. *Address of the retiring President of the Society in awarding the Bruce Medal to Professor J.C. Kapteyn.* Publications of the Astronomical Society of the Pacific, 25, 15–27 [1913PASP...25...15C]

van Rhijn, R. J. 1921. *De bouw van het sterrenstelsel: Kapteyn's beteekenis voor de moderne astronomie.* De Gids, Juli 1921, 128–144.

van Rhijn, P. J. 1922. *Jacobus Cornelius Kapteyn. In memoriam.* Popular Astronomy, 30, 628–631. [1922PA.....30..628V]

Eddington, S. A. 1922. *Jacobus Cornelius Kapteyn.* The Observatory, 45, 261–265. [1922Obs ....45..261.]

van Maanen, A. 1922. *J.C. Kapteyn, 1851–1922.* Astrophysical Journal, 56, 145–153. [1922ApJ....56..145V]

Seares, F. H. 1922. *J.C. Kapteyn.* Publications of the Astronomical Society of the Pacific, 34, 233–253. [1922PASP...34..233S]

de Sitter, W. 1922. *Jacobus Cornelius Kapteyn †(19 Januari 1851–18 Juni 1922).* Hemel & Dampkring 20, 97–111.

Easton, C. 1922 *Persoonlijke herinneringen aan J.C. Kapteyn.* Hemel & Dampkring, 20, 112–117 and 151–164.

Pannekoek, A. 1922a. *J.C. Kapteyn und sein astronomische Werk.* Die Naturwissenschaften, Heft 45, 1–14.

Pannekoek, A. 1922b. *J.C. Kapteyn en zijn astronomisch werk.* Wetenschappelijke Bladen I., 257–295.

de Sitter, W. 1923. *Jacobus Cornelius Kapteyn. 19. Januar 1851 bis 18. Juni 1922.* Vierteljahrsschrift der Astronomischen Gesellschaft, 58, 162–190 (1923).

Jeans, J. 1923. *Obituary Notices : Associates :- Kapteyn, Jacobus Cornelius.* Monthly Notices of the Royal Astronomical Society, 83, 250–255.

van Rhijn, P. J. 1951. *J.C. Kapteyn Centennial.* Sky & Telescope, 10, 55–57. [1951S&T.... 10...55V]

Dekker, E. 1978. *Kapteyn en de studie van de hoogere deelen des hemels.* Zenit 5, 448–453.

Dekker, E. 1983. *Jacobus Cornelius Kapteyn (1851–1922).* In: 'Sterrenkijken bekeken: Sterrenkunde aan de Groningse universiteit vanaf 1614', by A. Blaauw, J.A. de Boer, E. Dekker and J. Schuller tot Peursum-Meijer, Museum of the University of Groningen

Paul, E. R. 1981. *The death of a research programme – Kapteyn and the Dutch astronomical community.* Journal for the History of Astronomy, 12, 77–94. [1981JHA....12...77P]

Paul, E. R. 1985. *Kapteyn and statistical astronomy.* The Milky Way Galaxy, Proceedings of I.A.U. Symposium No. 106, held May 30-June 3 1983 in Groningen, the Netherlands, p.25–42. [1985IAUS..106...25P]

Paul, E. R. 1986. *Kapteyn and the early twentieth-century universe.* Journal for the History of Astronomy, 17, 155–182 [1986JHA....17..155P]

Tenn, J. S. 1991. *Bruce Medalist Profiles - Kapteyn, Jacobus.* Mercury, 20, 145–147. [1991Mercu..20..145T]

Books.

H. Hertzsprung-Kapteyn: *J.C. Kapteyn; Zijn leven en werken.* P. Noordhoff, Groningen (1928).

E. Robert Paul: *The Milky Way Galaxy and statistical cosmology, 1890–1924.* Cambridge Univ. Press, Cambridge (1993), ISBN 0-5213-5363-7.

P. C. van der Kruit & K. van Berkel: *The Legacy of J.C. Kapteyn: Kapteyn and the development of modern astronomy.* Kluwer Academic Publishers, Dordrecht (2000), ISBN 0-7923-6393-0.

K. van Berkel & A. Noordhof-Hoorn: *Lieve Lize: De minnebrieven van de Groningse astronoom J.C. Kapteyn aan Elise Kalshoven, 1878–1879.* University of Groningen (2008), ISBN 978-90-367-3353-3.

A.3 Kapteyn and His School: PhD Theses up to 1946

Below I present a listing of the PhD theses that can be seen as belonging to the school (students of Kapteyn and students of students) and legacy of Kapteyn. I include theses defended up to 1946 in Groningen, Leiden and Utrecht. Theses under de Sitter on the satellites of Jupiter and of other objects have also been excluded.

In Groningen under the supervision of J. C. Kapteyn:
WILLEM DE SITTER (17 May, 1901): *Discussion of heliometer-observations of Jupiter's satellites made by Sir David Gill, K.C.B. and W.H. Finlay.*
CORNELIS EASTON (1903): *Doctor Honoris Causa.*
HERMAN ALBERTUS WEERSMA (9 July, 1908): *A determination of the solar motion according to the method of Bravais.*
LAMBERTUS YNTEMA (17 December 1909): *On the brightness of the sky and the total amount of starlight.*
ETINE IMKE SMID (13 June, 1914): *Bepaling der eigenbeweging in rechte klimming en declinatie van 119 sterren* (Determination of the proper motion in right ascension and declination for 119 stars).
PIETER JOHANNES VAN RHIJN (9 July, 1915): *Derivation of the change of colour with distance and apparent magnitude together with a new determination of the mean parallaxes of the stars with given magnitude and proper motion.*
SAMUËL CORNELIS MEIJERING (6 July, 1916): *On the systematic motions of the K-stars.*
WILLEM JOHANNES ADRIAAN SCHOUTEN (4 July, 1918): *On the determination of the principal laws of statistical astronomy.*
GERRIT HENDRIK TEN BRUGGEN CATE (24 June, 1920): *Determination and discussion of the spectral classes of 700 stars mostly near the north pole.*

In Utrecht, under the supervision of A. A. Nijland[2]:
ADRIAAN VAN MAANEN (June 2, 1911): *The proper motions of 1418 stars and near the clusters h and χ Persei* (initiated by and with major involvement of Kapteyn).
ISIDORE HENRI NORT (October 15, 1917): *The Harvard map of the sky and the Milky Way.*

---

[2]The thesis by Allard Othumar Holwerda (16 May, 1913): *Frequentiecurven* (Frequency curves) is sometimes attributed to Kapteyn, but in fact it was written under the supervision of his brother Willem. Although Kapteyn's 1903 article on *Skew Frequency Curves* is discussed, no direct involvement of him with this work is apparent.

Fig. A.5 Two of Kapteyn's students: Pieter van Rhijn and Willem de Sitter at the dinner on the occasion of the PhD defense of Jan Oort in Groningen, May 1926. The woman next to de Sitter (sitting) is Mrs de Sitter; the woman standing next to van Rhijn is Oort's future mother-in-law (see also Fig. 12.1) (Photograph Leids Fotoarchief, Sterrewacht Leiden)

JAN CORNELIS VAN DE LINDE (4 November 1921): *De verdeeling van de heldere sterren* (The distribution of the bright stars).

In Groningen under the supervision of van P. J. Rhijn:
EGBERT ADRIAAN KREIKEN (21 February, 1923): *On the colour of the faint stars in the Milky Way and the distance of the Scutum group.*
JAN SCHILT (21 May, 1924): *On a thermo-electric method of measuring photographic magnitudes.*
JAN HENDRIK OORT (1 May, 1926): *The stars of high velocity.*
PETER VAN DE KAMP (30 November, 1926): *De Zonsbeweging met betrekking tot apparent zwakke sterren* (The motion of the Sun with respect to apparently faint stars).
WILLEM JAN KLEIN WASSINK (21 March, 1927): *The proper motion and the distance of the Praesepe cluster.*

BARTHOLOMEUS JAN BOK (6 July, 1932): *A study of the η Carinae region.*
JEAN JACQUES RAIMOND (10 July, 1933): *The coefficient of differential galactic absorption.*
BROER HIEMSTRA (18 May, 1938): *Dark clouds in Kapteyn's special areas 2, 5, 9 and 24 and the proper motions of the stars in these regions.*
ADRIAAN BLAAUW (6 July, 1946): *A study of the Scorpio-Centaurus cluster.*

In Leiden under the supervision of W. de Sitter:
WILLEM HENDRIK VAN DEN BOS (7 July, 1925): *Micrometingen van dubbelsterren* (Micrometer measurements of binary stars).
COERT HENDRIK HINS (15 December, 1925): *Inleiding tot een catalogus van plaatsen en eigenbewegingen van 1533 roode sterren* (Introduction to a catalogue of positions and proper motions of 1533 red stars).

In Leiden under the supervision of E. Hertzsprung:
WILLEM JACOB LUYTEN (1 July, 1921): *Observations of variable stars.*
HENDRIK VAN GENT (8 April, 1932): *Veranderlijke sterren op een veld van 10° bij 10° in of nabij het sterrenbeeld Corona Australis* (Variable stars in a field of 10° by 10° in or near to the constellation Corona Australis).
PIETER THEODORUS OOSTERHOFF (6 April, 1933): *Effectieve golflengten en photographische magnituden van sterren in h en χ Persei* (Effective wavelengths and photographic magnitudes of stars in h and χ Persei).
GERRIT PIETER KUIPER (30 June, 1933): *Statistische onderzoekingen van dubbelsterren* (Statistical studies of double stars).
AERNOUT DE SITTER (29 June, 1936): *Fotovisueele fotometrie van sterren tot 8.0 magnituden benoorden +80° declinatie* (Photovisual photometry of stars to magnitude 8.0 north of declination +80°).
WILLEM CHRISTIAAN MARTIN (11 May, 1937): *Photographische photometrie van veranderlijke sterren in Centaurus* (Photographic photometry of variable stars in Centaurus).
ADRIAAN JAN WESSELINK (13 May, 1938): *Photographische photometrie met toepassing op de veranderlijke ster SZ Camelopardalis* (Photographic photometry with application to the variable star SZ Camelopardalis).
LUCAS PLAUT (2 May, 1939): *Photographische photometrie der veranderlijke sterren CV Carinae en WW Draconis* (Photographic photometry of the variable stars CV Carinae en WW Draconis).
JACOBUS GIJSBERTUS FERWERDA (3 October 1941): *Veranderlijke sterren in de omgeving van Sagittarius* (Variable stars in the neighborhood of Sagittarius).

In Leiden under the supervision of J. H. Oort:
GIJSBERT VAN HERK (1 May, 1936): *Enige uitkomsten van de waarnemingen in de jaren 1931–1933 te Equador* (Some results of observations made in Ecuador from 1931 to 1933).

Fig. A.6 Knighthoods of Kapteyn. From left to right: Knight Ordre National de Légion d'Honneur (1892), Knight Order of the Netherlands Lion (1903), Orden Pour le Mérite (1915) and Commander Order of Oranje-Nassau (1921) (Boerhaave Museum, Leiden [5])

A.4 Kapteyn's Honors

The following list of honors of Kapteyn is translated from the obituary by de Sitter in Hemel & Dampkring 20, 97–111 (1922). Some editing of obvious errors was done, as well as providing more years of award than in the original list.

Kapteyn's great-grandson Jan Willem Noordenbos has shown me a letter from Prof. Adriaan Blaauw to Mrs M.A. Newton in Foxton, Alnmouth, Northumberland, UK, sent from Leiden Observatory on July 14, 1978. Blaauw and Maria Newton–Noordenbos (granddaughter of Kapteyn) had been class mates in high school (see page 579). Apparently replying to a letter from Maria Newton, Blaauw wrote:

'Thank you for your letter with the information on the distinctions of your grandfather Kapteyn. Be assured that these will be very welcome in the Netherlands. As you maybe remember, I have collected various things that I knew of carefully together and put them in the Kapteyn Laboratory, in the so-called Kapteyn Room. That would be a natural place, but there are two other possibilities: the museum of the university in Groningen might certainly be interested, but I think in the first place of the museum for the history of the natural sciences in general in the Netherlands, which is located here in Leiden. Maybe you do have a preference for one of these, but I will look around some more. I myself think of the latter museum in the first place, since in recent years is has developed many activities in the area of the astronomy of the previous century and even employed an astronomer for that.

You write me about the question how to transport the distinctions and the desirability to obtain an export permit. I did not know that it would be so difficult as seen from England; I would have been tempted to just take them in the pocket of my coat, but maybe I am a more experienced smuggler than you are. Do you think there will be an opportunity to bring them to the Netherlands soon?'

Blaauw mentioned that he will travel home late August from Ireland via England. How and when in the end the exchange was accomplished is not clear. In any case, the distinctions ended up in the museum in Leiden mentioned, which now is the Boerhaave Museum. Pictures of the medals are reproduced in Fig.'s A.6 and A.7.

Fig. A.7 Medals of Kapteyn, showing in each case front and back. Top row: Gold medal of the R.A.S. (1902) and the Prix Pontécoulant (1905). Bottom row: Bruce medal (1913) and the James Craig Watson medal (1913) (Boerhaave Museum, Leiden; see Fig. A.6)

The collection was described in a handwritten note by Mrs Newton-Noordenbos, in which she states her mother Jacoba Cornelia Noordenbos-Kapteyn, Kapteyn's oldest daughter, left her all this when she died in September 1961. Mrs Newton-Noordenbos wanted to transfer these items to 'his university'. The inventory also included the text of the speech by Ursul Ph. Boissevain at the occasion of the 40-th anniversary of Kapteyn's professorship (see page 577) and Kapteyn's PhD diploma.

The list in Hemel & Dampkring is preceded by a note from de Sitter: 'Kapteyn's modesty would have objected against a list like this one – that was provided generously by widow Mrs Kapteyn-Kalshoven – would be published. It is however, now a privilege to be able to show how extensively the person and the work of our fellow Dutchman was also respected and treasured.'

Knighthoods and other distinctions:
Knight in the Ordre National de la Légion d'Honneur, France (1892)
Knight in the Order of the Netherlands Lion, the Netherlands (1903)
Member of the Orden Pour le Mérite für Wissenschaften und Künste (1915)
Commander in the Order of Oranje-Nassau, the Netherlands (1921)
Medals:
Gold Medal of the Royal Astronomical Society, London (1902)
Prix de Pontécoulant, French Académie des Sciences de Paris (1905)
James Craig Watson Medal, US National Academy of Sciences, Washington (1913)
Bruce Medal, Astronomical Society of the Pacific, San Francisco (1913)
Honorary Doctorates:
Honorary Doctor of Science, Cape of Good Hope (1905)
Honorary Doctor of Science, Harvard University, USA (1909)
Honorary Doctor of Law, Edinburgh (1921)

Memberships:
The Netherlands:
Royal Netherlands Academy of Arts and Sciences, Amsterdam (1888)
Holland Society of Sciences and Humanities, Haarlem (1893)
Batavian Society for Experimental Philosophy, Rotterdam (member; 1886; honorary member, 1914)
Society of Arts and Sciences of the Province of Utrecht, Utrecht (??)
Etcetera
Foreign:
Foreign Associate of the Royal Astronomical Society, London (1892)
Fellow American Philosophical Society, Philadelphia (1907)
Foreign Associate American National Academy of Sciences, Washington (1907)
Member Imperial Academy of St. Petersburg (1908)
Member Royal Academy Dublin (1908)
Honorary Member Royal Academy Edinburgh (1910)
Corresponding Member British Association, London (1913)
Foreign Member Royal Swedish Academy of Sciences, Stockholm (1914)
Member Royal Swedish Society of Sciences (Uppsala) (1915)
Honorary Member American Society (1915)
Membre-correspondant de l'Académie des Sciences, Paris (19??)
Foreign Member Royal Society, London (1919)
Member Finnish Academy of Sciences, Helsingfors (1921)
Member Royal Physiographical Society (Lund) (1922)

The Prix de Pontécoulant of the French Académie des Sciences de Paris is named after French astronomer Philippe Gustave le Doulcet, Comte de Pontécoulant (1795–1874). The prize is no longer listed on the Website of the Académie.

A.5 Institutions, Objects etc. Named After Kapteyn

Three astronomical objects, one institute, one building, one telescope and various streets have been named after Kapteyn; the astronomical objects are a star, a minor planet and a lunar crater. Currently the International Astronomical Union (IAU; see Box 16.1) is the body that decides on names of astronomical objects and bodies. However, before the founding of the IAU this was not regulated and observers themselves decided how to name an object.

Kapteyn discovered *Kapteyn's Star* while working on the Cape Photographic Durchmusterung and reported it in Kapteyn (1897b) in a communication to the Astronomische Nachrichten with the title *Stern mit grösster bislang bekannter Eigenbewegung*. It had a proper motion (motion on the sky with respect to the fixed stars in general) of about 8.5 arcseconds per year (see Fig. 7.19). For further information on this star see page 237.

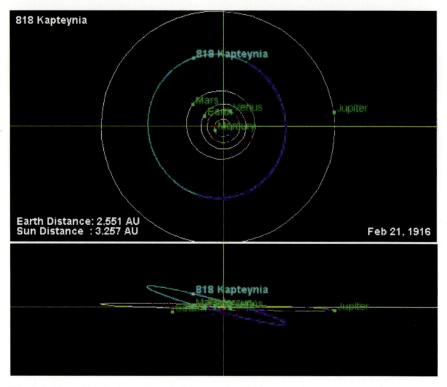

Fig. A.8 The orbit of minor planet 818 Kapteynia (1916 YZ) relative to the inner Solar System. The image at the top is as seen from a location perpendicular to the plane of the ecliptic (the orbit of the Earth), such that the Earth's north pole is directed towards the observer. The horizontal line is that through the equinoxes, when the Sun is directly above the Earth's equator on March 20/21 and September 22/23. The position of the objects is for the date February 21, 1916, the day that the minor planet was discovered by M. Wolf at Heidelberg. The picture at the bottom is the same for an observer in the plane of the ecliptic (Produced with the orbit diagram tool of the JPL Small-Body Database Browser of NASA's Near-Earth Object (NEO) Project [6])

The *minor planet* named after Kapteyn was called Kapteynia by Maximilian Franz Joseph Cornelius Wolf (1863–1932), who discovered it on February 21, 1916 at the Landessternwarte Heidelberg-Königstuhl. The current full designation of the minor planet, is *818 Kapteynia (1916 YZ)*. It moves around the Sun in a somewhat eccentric elliptical orbit with a semi-major axis of 3.2 AU (one Astronomical Unit AU is the mean distance of the Earth from the Sun of about 150 million km) and makes one revolution around the Sun in 5.64 years (see Fig. A.8, the date of the positions is that of the discovery). The orbit makes an angle of about 16° with the plane of the Earth's orbit and the diameter of the minor planet is 49.5 km.

The IAU designated a *crater on the Moon* with the name Kapteyn in 1964. It is near the limb of the Moon and has a diameter of about 50 km (see Fig. A.9).

The institute that is named after Kapteyn is the *'Astronomical Laboratory'* that he founded himself at the University of Groningen (see Fig. A.10). The building it

A.5 Institutions, Objects etc. Named After Kapteyn

Fig. A.9 The crater Kapteyn on the Moon photographed from the Apollo 8 spacecraft (NASA image in the public domain [7])

is housed in together with the Laboratory for Space Research, lecture rooms, etc., is now named Kapteynborg [10].

Quite a few cities have named streets after Kapteyn. Barneveld has a Kapteynstraat at the location of the boarding school 'Benno' of his parents (see Fig. A.12). There is also a Kapteynlaan in Groningen and Utrecht, among other places, and a Kapteynstraat in e.g. Leiden, Amsterdam and Hilversum.

The *Jacobus Kapteyn Telescope* (JKT; see Fig. A.13) is a 1-meter telescope on the Observatorio del Roque de los Muchachos of the Instituto de Astrofísica de Canarias (IAC) at La Palma, Canary Islands, Spain. It is part of the Isaac Newton Group of Telescopes, to which the 2.5 meter Isaac Newton Telescope and the 4.2 meter William Herschel Telescope also belong. This group of telescopes was built by the United Kingdom and the Netherlands, with the co-operation of Spain. The JKT started operations in 1984, but was closed as a common user facility in 2003 (see R. Laing & D. Jones, *The Isaac Newton Group* [11]).

Then, of course, there is the Kapteyn Cottage at Mount Wilson Observatory near Pasadena in California, where Kapteyn and his wife lived during his visits as a Research Associate. The cottage was first used by them in 1910 and their last visit took place in 1914. This is described on page 472 and illustrated in Fig. 13.12. And finally, since 2013 the Royal Natural Sciences Society (see page 276) has organized an prestigious annual lecture on a subject in the natural or medical sciences under the name of Kapteyn Lecture.

Fig. A.10 Part of the Zernike Campus of the University of Groningen, which is located on the northern outskirts of the city. The curved building on the right is Kapteynborg, which houses the Kapteyn Astronomical Institute and the Groningen laboratory of the Netherlands Institute for Space Research. The institute moved to this campus in the late sixties and was originally located in the tall building running from left to right in the middle of the photograph, and moved to its current location in 1983 (Photograph by aerophoto Eelde and University of Groningen [8])

Fig. A.11 The present Kapteynborg. The gray part at the left contains lecture rooms and a restaurant for staff and students of various disciplines located at the Zernike campus; the brown part has the offices and laboratories of the Kapteyn Astronomical Institute of the University of Groningen and the Netherlands Institute for Space Research of the Netherlands Organization for Scientific Research (NWO), which are closely collaborating (Photograph by the author)

Fig. A.12 There are various streets in the Netherlands named after Kapteyn. This sign is from the small town of Barneveld, where Kapteyn was born. This street is the entrance to a parking lot near the center of the town, where the boarding house 'Benno' was located where Kapteyn was born and raised (Photograph by the author)

Fig. A.13 The Jacobus Kapteyn Telescope on the Roque de los Muchachos, La Palma, Canary Islands (From the Website of the Isaac Newton Group of Telescopes [9])

A.6 Academic Genealogy of Kapteyn

An academic genealogy should ideally follow the PhD theses and advisers. This is not always possible. In the text here a broad interpretation has been adopted. So I have accepted Johann Samuel König as a student of the Bernoullis and von Wolff, even though he did not formally submit a PhD thesis; I also ignored the fact that Gerrit Moll received his PhD degree honoris causa. Likewise I have ignored any inconsistencies between the listings in the Mathematics Genealogy Project and biographies available on the Web (including those linked to the MGP). I start with my own PhD supervisor and go back in time. [With thanks to Carlo Beenakker [12], Roelof de Jong [13], Manfred Horstmanshoff [14] and the MGP [15]]

Jan Hendrik Oort: Franeker, April 28, 1900 – Leiden, November 5, 1992 – Student of P.J. van Rhijn; PhD thesis: *The stars of high velocity*, University of Groningen, 1 May 1926. Oort was a professor of astronomy at the University of Leiden 1935–1970 and director of Leiden Observatory 1946–1970. He won the Bruce Medal in 1942.

Pieter Johannes Van Rhijn: Gouda, March 24, 1886 – Groningen, May 9, 1960 – Student of J.C. Kapteyn; PhD thesis: *Derivation of the change of colour with distance and apparent magnitude together with a new determination of the mean parallaxes of the stars with given magnitude and proper motion*, University of Groningen, July 9, 1915. Van Rhijn was a professor of astronomy at the University of Groningen and director of the Astronomical Laboratory 'Kapteyn' 1921–1956.

Jacobus Cornelius Kapteyn: Barneveld, January 19, 1851 – Amsterdam, June 18, 1922 – Student of C.H.C. Grinwis; PhD thesis: *Onderzoek der trillende platte vliezen* (Study of vibrating flat membranes), University of Utrecht, June 24, 1875. Kapteyn was the founder of the Astronomical Laboratory at the University of Groningen, where he was professor of astronomy and theoretical mechanics 1878–1921. He received the Bruce Medal in 1913.

Cornelis Hubertus Carolus Grinwis: Haarlem, March 9, 1831 – Baarn, December 25, 1899 – Student of R. Van Rees; PhD thesis: *De distributione fluidi electrici in superficie conductoris* (On the distribution of electricity over the surface of a conductor), University of Utrecht, 3 July 1858. Grinwis was a professor of mathematics and physics at the University of Utrecht 1867–1896. Two older brothers of Kapteyn also received a PhD from the University of Utrecht with Grinwis as adviser.

Richard van Rees: Nijmegen, May 24, 1797 – Utrecht, August 23, 1875 – Student of G. Moll; PhD thesis: *De celeritate soni per fluida elastica propagati* (On the speed of sound in an elastic fluid), University of Utrecht, December 17, 1819. Van Rees was a professor of mathematics and experimental philosophy at the University of Utrecht 1831–1867.

Gerrit (Gerard) Moll: Amsterdam, January 18, 1785 – Amsterdam, January 17, 1838 – Student of J.T. Rossijn; PhD thesis: the degree was awarded *honoris causa*, University of Utrecht, 28 October 1815. Moll also studied under Jan Hendrik van Swinden (1746–1823) in Amsterdam, where he was a professor of philosophy, physics and astronomy. Moll was a professor of mathematics, astronomy and experimental philosophy 1815–1838 at the University of Utrecht, and director of Utrecht Observatory 1812–1838.

Johannes Theodorus Rossijn: Noordzijpe, December 18, 1744 – Utrecht, December 24, 1817 – Student of A. Brugmans; PhD thesis: *De tonitru et fulmine ex nova electricitatis theoria deducendis* (On thunder and lightning according to the new theory of electricity), University of Franeker, 10 December 1762. Rossijn was a professor of philosophy, mathematics and astronomy at the Universities of Harderwijk 1765–1775 and Utrecht 1775–1815.

Antonius Brugmans: Hantum, October 22, 1732 – Groningen, April 27, 1789 – Student of J.S. König; PhD thesis: *De phaenomeno* (On the phenomenon of), University of Franeker, May 24, 1749 – Brugmans is sometimes credited with 'two philosophical dissertations' (see H.A. Krop, J.A. van Ruler & A.J. Vanderjagt (editors): *Zeer kundige professoren: Beoefening van de filosofie in Groningen van 1614 tot 1996.* [16]) with the titles *Dissertatio metaphysica de essentiarum, idearumque absoluta necessitate earumque origine ex extellectu divino* (Metaphysical dissertation on the absolute necessity of the essentials and of the ideas and their origin in the Divine intellect) (May 25, 1948) and *Dissertatio philosophica inauguralis de phaenomeno* (Philosophical dissertation on the phenomena, 1749). Brugmans was a professor of philosophy at the University of Franeker 1755–1766 and at the University of Groningen 1766–1789.

Johann Samuel König: Büdingen, July 31, 1712 – Zuilenstein, August 21, 1757 – Student of Johann and Daniel Bernoulli and Christian Wolff. König studied in Basel under Johann from 1730 and under Daniel Bernoulli from 1733. He did not formally receive a PhD degree, but is often considered a student of the Bernoullis. In 1735 he went to study under Christian von Wolff in Marburg. König later worked in Bern until he was exiled for his liberal ideas, after which he became a professor of philosophy and mathematics at the University of Franeker 1744–1749. He is the discoverer of König's theorem in kinetics and was the subject of a plagiarism charge related to the principle of least action.

Daniel Bernoulli: Groningen, February 8, 1700 – Basel, March 17, 1782 – Student of Johann Bernoulli(?) ; PhD thesis: *Dissertation physico-medica de respiratione* (Dissertation on the medical physics of respiration), University of Basel, 1721. Daniel Bermoulli was a professor of mathematics at the University of St. Petersburg 1725–1733 and a professor of physics at the University of Basel 1750–1776. He developed Bernoulli's principle in fluid dynamics.

Johann Bernoulli: Basel, July 27, 1667 – Basel, January 1, 1748 – Student of Jacob Bernoulli; PhD thesis: *Dissertatio physico-anatomica de motu musculorum* (Dissertation on the physics and anatomy of muscular motion), University of Basel, 1694. Johann Bernoulli was the younger brother of Jacob. He worked as a professor of mathematics at the University of Groningen 1695–1705. He returned to Basel, where he took over his brother's chair in mathematics. In 1696 he found the solution for the brachistochrone problem in mechanics, laying the foundation for calculus of variations.

Jacob Bernoulli: Basel, December 27, 1654 – Basel, August 16, 1705 – Student of G.W. von Leibniz; PhD thesis: *Solutionem tergemini problematis arithmetici, geometrici et astronomici* (Solutions to a triple problem in Arithmetics, mathematics and astronomy), University of Basel, February 4, 1684. It is widely accepted that he studied the calculus as introduced by Leibniz. The Bernoulli family had a spice business in Amsterdam. Their Calvinist faith forced them to move to Basel (Switzerland) in the 1550s when the Spanish occupied Holland. Jacob Bernoulli became a professor of mathematics at the University of Basel 1687–1705. The Bernoulli numbers in mathematics are named after him.

Christian von Wolff: Breslau, January 24, 1679 – Halle, April 9, 1754 – Student of E.W. von Tschirnhaus and G.W. von Leibniz; PhD thesis: *Dissertatio Algebraica de Algorithmo Infinitesimali Differentiali* (Dissertation on the algebra of solving differential equations using infinitesimals), University of Leipzig, 1704. Christian Wolff was a professor of mathematics and natural philosophy at the University of Halle 1706–1723 and 1740–1754 and at Marburg 1723–1740.

Ehrenfriend Walter von Tschirnhaus: Kieslingswalde, April 10, 1651 – Görlitz, 11 October, 1708 – The MGP lists von Tschirnhaus as having obtained the title of Magister philosophiae and Medicinae Doctor at the University of Leiden in 1669 and 1674. Although he did study in Leiden between these years, there is no further evidence that he actually obtained a PhD there (or anywhere else, for that matter).

Gottfried Wilhelm von Leibniz: Leipzig, July 1, 1646 – Hannover, November 14, 1716 – Student of E. Weigel; PhD thesis: *De casibus perplexis in jure* (On perplexing cases in law), University of Althof, 1666. Leibniz studied at Jena under Weigel from 1663 onward and wrote a habilitation thesis *Dissertatio de arte combinatoria* (Dissertation on the combinatorial art), which was declined by the University of Leipzig. Apparently he received his degree in 1667 at the University of Altdorf, but it is unclear which of the two titles applies and who his adviser was. It seems reasonable to regard Leibniz as a student of Weigel. Leibniz studied mathematics in Paris with Huygens as mentor from 1672 onwards. However, it is unlikely that Huygens had anything to do with a thesis in Altdorf well before 1672. Von Leibniz discovered calculus independently of Newton and invented the mathematical notation we still use. He was a librarian and Court Counselor with the Duke of Hanover from 1676 until his death.

Erhard Weigel: Weiden, December 16, 1625 – Jena, March 20, 1699 – Student of P. Müller; PhD thesis: *De ascensionibus et descensionibus astronomicis dissertatio* (Astronomical dissertation on risings and settings), University of Leipzig, 1650. Weigel was a mathematician, astronomer and philosopher and is credited with the definition of the date of Easter and promoting the introduction of the Gregorian calendar. He was a professor of mathematics at the University of Jena 1653–1699.

Philipp Müller: 1585–1659 – He was the thesis adviser of Erhard Weigel [17] and a professor of mathematics at the University of Leipzig from 1616 onward. Müller may have been a professor of physics as early as 1614 (see H.-J. Girlich and K.-H.Schlote, *Die Entwicklung der Mathematik an der Universität Leipzig* [18]). Müller had a keen interest in astronomy and was one of the first to accept Kepler's laws of planetary motion (see J.L. Russell, *Kepler's Laws of Planetary Motion: 1609–1666* (1964) [19]): 'Kepler had, however, at least one disciple during the 1620s: Philip Müller, professor of mathematics at the Leipzig University'. He actually corresponded extensively with Johannes Kepler (1571–1630). In his biography of Kepler, Max Caspar (see Max Caspar: *Johannes Kepler [20]*) describes Jacob Bartsch (ca 1600–1633) as a pupil of Philip Müller's. Bartsch is known to have made a star chart based on Müller's data and later became Kepler's assistant and son-in-law.

Appendix B
Henriette Hertzsprung-Kapteyn's Biography of J.C. Kapteyn

To my mother,
without whom this life
could never have been so complete.

Henriette Hertzsprung-Kapteyn[1]

In 1928 H. Hertzsprung-Kapteyn, daughter of J.C. Kapteyn, wrote a biography of her father, entitled *J.C. Kapteyn: Zijn leven en werken*, and published it with the publishing company of P. Noordhoff in Groningen. The book is now part of the *Digitale Bibliotheek voor de Nederlandse Letteren* (Digital Library of Dutch Literature), where a complete electronic version of the book is available [1].

B.1 Henriette Hertzsprung–Kapteyn

Henriette Mariette Augustine Albertine Kapteyn (1881–1956) was the second child and second daughter of Jacobus Cornelius Kapteyn (1851–1922) and his wife Catherina Elisabeth Kalshoven (1855–1945). They had married in 1879, after Kapteyn had taken up his professorship in astronomy and theoretical mechanics at the University of Groningen, to which he had been appointed in 1878. Henriette had an older sister Jacoba Cornelia, born in 1880, and a younger brother Gerrit Jacobus, born in 1883.

In historical notes about Kapteyn, whenever this biography is mentioned, the name of the author is almost invariably written as *Henrietta* Hertzsprung-Kapteyn,

---

[1]*Aan mijne moeder, zonder wie dit leven nooit zoo volkomen had kunnen zijn*, Dedication in her biography of her father J.C. Kapteyn.

Fig. B.1 The title page of the biography by Henriette Hertzsprung-Kapteyn (left), and a photograph reproduced from that volume showing the author with her father

so the question arises whether the last letter should be an *'a'* or an *'e'*. The cover of the book only has her first initial *'H'*. On the other hand, a number of genealogies of the Kapteyn and Kalshoven families can be found on the WWW that all spell her name *Henriette* [2]. The City of Groningen provides electronic, public access to its archives, including the birth register for the period from 1811 onward (up to one hundred years prior to the queries). These archives, named *Alle Groningers* [3], contain the birth certificate of Henriette Kapteyn (see Figure B.2). Clearly her name is spelled with an *'e'* at the end. Note also that no dieresis or *umlaut* is used over the second *'e'*. The same is true for the first *'e'* in Mariette.

Henriette studied in Groningen and in Amsterdam and obtained the 'Candidaats' degree (this is the title obtained more or less halfway through academic studies, roughly comparable to what nowadays would be the Bachelors degree) in law and an 'MO' degree in English (see Inge de Wilde: *Nieuwe Deelgenoten in de Wetenschap: Vrouwelijke studenten en docenten aan de Rijksuniversiteit Groningen 1871–1919* [4]). MO stands for Middelbaar Onderwijs (secondary education) and the certificate qualified for teaching at gymnasia and high schools (HBS; see page 62). Henriette Kapteyn married the Danish astronomer Ejnar Hertzsprung (1873–1967) on May 16, 1913. The marriage certificate of Ejnar Hertzsprung and Henriette Kapteyn is also available in the archives of the city of Groningen (see Figures B.3 and B.4). Here her first name is also spelled Henriette.

B.2 Ejnar Hertzsprung and First Marriage

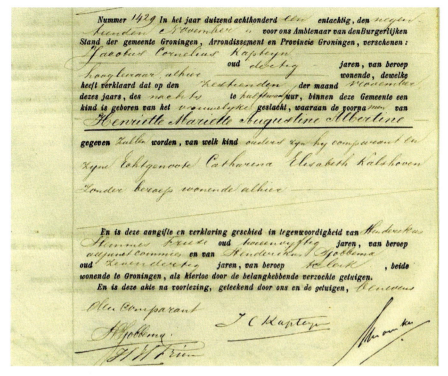

Fig. B.2 Reproduction of Henriette Kapteyn's birth certificate from the archives of the city of Groningen (From Alle Groningers [5])

A similar issue is whether the last name should be spelled Kapteyn or Kapteijn. J.C. Kapteyn always used the *'y'* and his and her name are also spelled in that manner in the biography that Henriette wrote. This seems sufficient reason to adopt it. It is also spelled Kapteyn in the birth certificate of Henriette (see Figure B.2). However, in the marriage certificate in Figures B.3 and B.4 the family name of Henriette herself, of her father and her uncle Frederik Willem Hendrik Kapteyn (1853–1920), who acted as an official witness, was spelled Kapteijn. Note that all signatures do not have dots over the *'y'* which would be required to make it an *'ij'*. The site *Alle Groningers* has more official documents of Kapteyn and his relatives, but most can be found by searching for the name 'Kapteijn'.

B.2 Ejnar Hertzsprung and First Marriage

Hertzsprung, who was senior to Henriette by eight years, was trained as a chemical engineer in Copenhagen and obtained his degree in 1898. A biography of him by Dieter B. Hermann was published in German under the title *Ejnar Hertzsprung,*

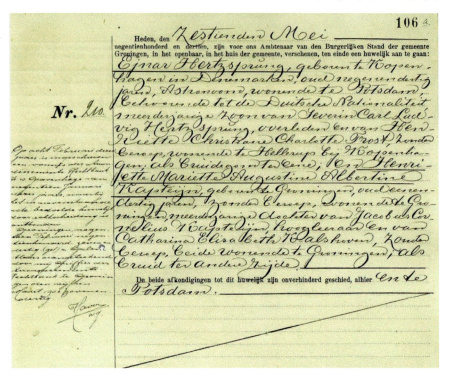

Fig. B.3 Marriage certificate of Ejnar Hertzsprung and Henriette Kapteyn on May 16, 1913. This section shows the top part. The scribbling in the left-hand margin is an entry to the effect that the marriage was dissolved by divorce in 1937 (From Alle Groningers [7])

Pionier der Sternforschung [6], and I have taken some details from this below. Hertzsprung worked in St Petersburg, but in 1901 he went to Leipzig University to study photo-chemistry. He returned to Denmark in 1902 and worked at Copenhagen University Observatory and at the (private) Urania Observatory. In 1909 he was appointed by Karl Schwarzschild at the Astrophysikalisches Observatorium in Potsdam. Before that, Schwarzschild had been director of the Sternwarte Göttingen and moved to Potsdam to manage the local Astrophysikalisches Observatorium in 1909. Schwarzschild is best known for his solution of Albert Einstein's field equations of general relativity of a spherically symmetric case, leading to the Schwarzschild radius, which is the size of the event horizon of a black hole. In 1919 Hertzsprung went to Leiden Observatory and became its director in 1937. He retired in 1946. His greatest contribution is his work on the classification system for stars according to spectral type, stage in their development and luminosity. This resulted in the *Hertzsprung–Russell diagram*, which has been instrumental in understanding stellar evolution.

While Hertzsprung worked in Potsdam, Schwarzschild recommended him to Kapteyn, who decided to introduce him to Hale and bring him along on one of his visits to Pasadena. In 1911 Hertzsprung paid a lengthy visit to Groningen and

B.2 Ejnar Hertzsprung and First Marriage

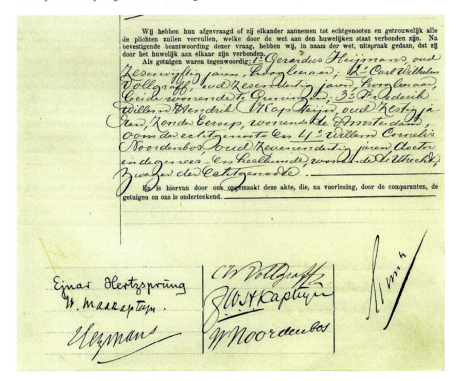

Fig. B.4 Marriage certificate of Ejnar Hertzsprung and Henriette Kapteyn on May 16, 1913. This section shows the bottom part. The signatures below those of Ejnar and Henriette and in the middle are those of the witnesses Gerardus Heymans (professor of psychology and philosophy), Carl Vollgraff (professor of Greek), uncle Frederik Kapteyn and brother-in-law Willem Cornelis Noordenbos (From Alle Groningers)

Kapteyn. Klaas van Berkel notes in the *Legacy* (page 161): 'In Groningen, Hertzsprung however not only got to know Kapteyn much better, but also his daughter Henrietta [should be Henriette], and when he left for the US in June 1912 the two of them were engaged.'

Henriette and Ejnar had one child, a daughter Rigel Hertzsprung (1916–1993), named after the bright star (see Fig. 16.14).

The marriage was not a happy one and they divorced formally on January 19, 1937. They had separated long before that date. In his biography of Hertzsprung, Herrmann quoted from a in a letter of Hertzsprung to his sister of 'May 23, 1923, a few months before his fiftieth birthday', that 'he and Hetty have decided to separate and to dissolve their ten year old marriage. Daughter Rigel was 7 years of age at that time.' Many persons, including Adriaan Blaauw, who had studied with Hertzsprung, have confirmed the widespread opinion that Hertzsprung was a workaholic with little interest in social and other non-professional matters. Klaas van Berkel wrote in the *Legacy*:

Fig. B.5 Tombstone on the graves of Henriette Kapteyn and her second husband Joost Hudig on the Begraafplaats *Leeuwerenk*, Oude Diedenweg 64, 6704 AD Wageningen (From Online begraafplaatsen [9])

'It is unclear however how quickly Kapteyn came to realize just how un-social Hertzsprung was. If we have to believe Luyten, he should have known this before the couple was married. 'For three or four months [after his departure for the US] she [Henriette Kapteyn] never heard a word from Hertzsprung, and finally Kapteyn wrote to the director of Mount Wilson Observatory about it. Eventually Hertzsprung's fiancee received a postcard from him on which he wrote: 'So sorry not to have written before, but the skies here are so beautiful, and, you see, where the stars are, there I'm happy.' ' (Luyten, *My First 72 Years* [8].) Since Kapteyn had traveled to the US. for his yearly stay at Mount Wilson together with Hertzsprung and must have returned in September or early October, this story, if true (and Luyten says he had heard it from Henriette Kapteyn herself), implies that only after returning from the US. Kapteyn was informed about Hertzsprung's lack of attention.'

Willem Jacob Luyten (1899–1995) had obtained his PhD under Hertzsprung in Leiden in 1921.

B.3 Joost Hudig and Second Marriage

A few months after the formal divorce, Henriette married Joost Hudig (1880–1967), in London on April 17, 1937. Hudig was a chemical engineer and specialized in pedology (soil study). He had earlier married Sophia Alida Hudig in Rotterdam in 1905, where both were born (Sophia in 1880; although they shared the family name Hudig, they were not close relatives). Shortly after that they moved to Groningen, where two daughters were born, Johanna Clementina Johanna in 1907, and Johanna Elisabeth in 1909. The elder daughter studied law and became the first female judge in the Netherlands. The site *Alle Groningers* only contains births up to 100 years before the time of consultation, so there may have been more children after 1910. On public record are also the deaths of son Ferrand Whaleij in 1911 (3 days old) and son Joost in 1920 (7 years old). Sophia Alida died in 1924 at the age of 44 and Hudig had been a widower since then.

Joost Hudig worked at the 'Rijkslandbouwproefstation' (State Agricultural Laboratory) in Groningen, of which he also became director. In 1929 he was appointed professor of chemistry and the science of fertilization at what is now the Agricultural University of Wageningen, where he worked until his retirement in 1949. Hudig features in Henriette's book, in Chapter 15. According to her descriptions, Kapteyn and he were neighbors and Hudig acted as secretary to the scientific chapter of the 'Natuurkundig Genootschap' (the current Royal Natural Sciences Society), when Kapteyn was its chair (see page 282). Indeed Hudig moved to the address Oosterhaven ZZ 12 around 1908; Kapteyn lived at number 16a between 1906 and 1910. Henriette was 25 when the Hudigs moved there and she probably no longer lived then with her parents.

Henriette and her second husband were buried in Wageningen. The tombstone is shown in Figure B.5. Note that here there is a dieresis in Henriette's first two names.

B.4 The Biography and Hale

In the Preface I presented the note written by Willem de Sitter requesting help in the preparation of a biography that he and Johan Huizinga we preparing to write. I do not know who of the colleagues responded, but George Hale most certainly did. On March 21, 1927, Hale wrote to Huizinga offering his help and sending a list of letters sent by Kapteyn to himself, Walter Adams and Frederick Seares, with the numbers of pages and the first and the last phrases, prepared by his secretary. In the same letter he mentioned that 'Mrs Hertzsprung' had also contacted him. Huizinga acknowledged the receipt of this letter on April 2, 1927. The ensuing correspondence (see als Petra van der Heijden in the *Legacy* [10]) between Hale and Kapteyn's wife and daughter tells an interesting story.

Well before that date, on October 28, 1925, Henriette Hertzsprung-Kapteyn had written to Hale asking him for 'a few lines about reminiscences of him and about his life at Mt. Wilson'. She wrote this letter from the address Oude Amersfoortsche

Weg 66 in Hilversum. She had moved to Hilversum after her separation from Hertzsprung, together with her daughter Rigel, presumably to be near her mother. Apparently the latter had moved into a house nearby (at no 29 in the same street). She referred to delays 'because of my altered conditions' and also mentions examinations she had been preparing for. Further she noted that she advanced 'but slowly, as I have my child and my household all to myself, the child being a very curious and original specimen of humanity with difficult, but promising properties'. She signed the letter 'Hetty Hertzsprung-Kapteyn'. Rigel (1916–1993) had suffered minor brain damage due to a forceps delivery. She first lived with her mother, but later with her father when he moved to Denmark. She seemed to have had affairs with men, which gave rise to some tensions about her ability to arrange her financial affairs in the family relations, especially with her aunt Jacoba Noordenbos (Kapteyn's oldest daughter), until she married a Danish waiter with the name Thorning. She died in an accident with a bus at a bus-stop. [11]

Hale replied on January 20, 1926, apologizing for his delayed reaction and inquiring into the connection between the biography she was writing and the one by professor de Sitter. Henriette answered on February 11, 1926, again from her address in Hilversum, saying that she had the 'view to write my own little book for his family and friends. Their [she means de Sitter and Huizinga] work will be a standard work of great value no doubt, but mine will be a simple story of his personal life with as much of his science in it, as I can bring in with my ignorance of astronomy'. And again she asked for a contribution with regard to her father's work at Mount Wilson and Hale's personal memories of him.

On August 13, 1926, Hale wrote that he had found himself 'whol[l]y unable to write anything suitable' and apologized for that. He did include a few remarks (that Henriette eventually used in the biography), however, and told her to 'feel entirely free to omit it from the book if it is late or in your eyes unsatisfactory'. Henriette responded on October 19, 1926, explaining that her work on the biography had been delayed 'as my house has been repaired' and also mentioned that her mother was in excellent health. 'We do enjoy living so near to each other'. She also mentioned that Huizinga had been visiting her mother and herself, and conveyed her envy of him for having visited Pasadena, where she had never been and which was to her, 'child of my parents, an almost sacred place'. There was no reference to Huizinga's progress with the biography.

Henriette's biography, obviously, is in Dutch. She wrote to Hale again on August 28, 1929 (after the book had been published), saying that she aimed to publish it in America in English translation and that she had consulted Huizinga and a professor Barnow of Columbia University. This is a misspelling; it must have been Adriaan J. Barnouw (1877–1968), Dutch by birth and a professor of Dutch language and literature on the Queen Wilhelmina Chair, since 1919. Barnouw is particularly known for his translations into English of some Dutch medieval legends, fables and poems. Henriette had hoped to translate her book herself 'as many astronomers asked me to', but found it impossible to make it 'fit for an American public'. Note that this was in spite, or maybe because, of the fact that she had studied English. She asked Hale for help to find a publisher and inquired into the possibility of the book coming out as a 'separate publication of the Mt. Wilson Observatory'. Hale informed her that he had received a copy of the book and said that he had been able to make out enough of it to 'get the chief features of your narrative', but that 'all matters on publication were decided on by the President of the [Carnegie] Institution' and that he had written him about it.

Hale reported back that funds were insufficient, but that he had taken things up with the Publications Committee. On March 19, 1930, Hale wrote that the Carnegie Institution was unable to find the financial means but that it would make an effort to find an American publisher. Henriette acknowledged this on May 7, 1930. However, on July 24, 1930 the editor of the Carnegie Institution (Frank F. Burker) wrote to Hale that he had been unsuccessful in finding a publisher, which was reported to Henriette Hertzsprung-Kapteyn on August 4, 1930. The efforts to publish an English version of the HHK biography had proved to be of no avail.

B.5 Paul's Translation

Much more recently the biography was still translated, by E. Robert Paul, first as a contribution to the journal *Space Science Reviews* [12], later as a book *The Life and Works of J.C. Kapteyn by Henriette Hertzsprung-Kapteyn: An Annotated Translation with Preface and Introduction by E. Robert Paul* [13]. It also contains a list of publications about Kapteyn and notes about astronomy in his time. E.R. Paul was definitely an authority on the subject of Kapteyn and astronomy of his days and published a number of authoritative papers and a scholarly book on the subject (see the Preface, page viii).

The translation of Henriette Hertzsprung-Kapteyn's biography has been seriously criticized. I reproduce the following paragraphs from Appendix B by K. van Berkel and P.C. van der Kruit in the *Legacy*:

'In this volume and in almost any other study of Kapteyn, reference is made to the biography by his daughter *J.C. Kapteyn, Zijn Leven en Werken*, published in 1928. This only biography of Kapteyn has been translated into English (and annotated) by E. Robert Paul. [...] Various Dutch-speaking participants, present at the Symposium, commented on the sometimes misleading translations.

Prof. Adriaan Blaauw informs us as follows: 'In August 1993 I learned by chance from one of the associates of Kluwer Academic Publishers, Dr K. (not an astronomer), about Robert Paul's translation. Publication was in an advanced state, the book was just about to be printed. Dr. K. agreed to slightly delay the printing in order for me to have an opportunity to acquaint myself with the contents. I was impressed by Paul's sympathetic effort to make Mrs Hertzsprung-Kapteyn's work accessible to the English speaking community and his extensive annotations about the personalities – astronomers and university colleagues – occurring in her narratives. However, I was startled by the poor quality of the translation.

I learned that Robert Paul had done this himself, and it was obvious that he had grossly underestimated the pitfalls of the Dutch language. On August 26, I sent Paul by rapid mail 60 proof pages marked with my corrections. My inspection – done under heavy pressure of other work – had been far from exhaustive. In my accompanying letter to Paul I wrote: '[...] The Dutch language, as it is used by some authors, can be of rather complicated structure [...] and thereby sometimes may be misleading to foreigners. Mrs Hertzsprung's writing [...] is no exception. There are cases (as you will note from my corrections), where the imperfect translation then leads to just the opposite from what she meant to say. There are also among my corrections some due to the fact that a Dutch word can sometimes have more than one quite different meanings, and the alternative would have been the correct one. [...]

My suggestions for improvements were acknowledged by Dr. K. on August 31. It was only after publication of the book that, from the complimentary copies I received, I could judge to what extent they had been followed up. It seemed to me that this was the case only for those, that did not require drastic surgery of the page-proofs. So, unfortunately, the book still contains many traces of the imperfect translation.

Alas, Robert Paul died on October 12, 1994, of cancer, at the young age of 51, a little more than a year only after the publication of the book. An obituary, written by Steven J. Dick, appeared in the *Bull. Am. Astron. Soc.* 26, 1606 (1994). Isn't it likely, that Paul, already suffering from his fatal disease during the last stages of the publication of the book, so shortly before he passed away, did not have the strength anymore to thoroughly revise the text?

Wondering about the origin of the present, unsatisfactory situation, it has always been incomprehensible for me that Kluwer accepted the text as it was. One wonders whether Kluwer's editorial advisory committee for the series *Space Science Reviews* had a really serious look at the manuscript. I took the matter up again in February 1996 with Dr K.'s successor at Kluwer, Dr de G., and sent him two sets of copies of the pages of the manuscript – corrected and not corrected – in the hope that Kluwer would make an effort to remedy the situation. This has not been successful.' [...]

The appendix in *the Legacy* contains more details and examples of poor quality translations, but I refrain from repeating this her, except for the final paragraph:

'There is no doubt that E. Robert Paul has done scientific inquiry a great service with his perceptive studies of the history of (statistical) astronomy in Kapteyn's days. However, it is clear from these examples that his translation of Henriette Hertzsprung-Kapteyn's biography is untrustworthy and useless for scholarly purposes. Those who are able to read Dutch do not need his translation, and those who for one reason or another would like to use it after all, will still have to consult the original to check whether Paul provided the correct translation. It would be useless to try and improve this translation, since the work would have to be redone from beginning

B.5 Paul's Translation

to end. Perhaps the publication of the translation made by Peter van de Kamp, which never saw the light of day because Paul's translation was forthcoming, would offer a way out. But the best solution would be to have someone write a new and critical biography of Kapteyn, based not only on Henriette Hertzsprung-Kapteyn's book, but also on the more recent literature concerning the life and work of Kapteyn, including the historical contributions to this volume.'

In a book review of Paul's translation, Derek Jones [14] wrote: 'I am not a scholar of Dutch and cannot comment on the accuracy of the translation, but the resulting English is difficult to read because of the strange usage and vocabulary. [Paul's translation] attempts to convey both the spirit and intent of the original biography and attempts to preserve the many Dutch idiomatic expressions.' Maybe this was not such a good idea after all. I have produced a revised English translation of the biography by Henriette Hertzsprung-Kapteyn and posted it electronically on the Website I maintain on Kapteyn [15]. Unlike Paul, I did not attempt to rewrite the sentences but tried to stay as close to the original text as possible. I have used major parts of this translation in this book, where I felt the texts were important to illustrate certain aspects of Kapteyn's private and non-professional life.

Fig. B.6 Kapteyn at Mount Wilson in 1908 (Photograph Kapteyn Astronomical Institute)

Appendix C
Cornelis Easton: Personal Memories of J.C. Kapteyn

> *His straw hat had lost its original cleanness long ago;*
> *no umbrella in spite of the rain*
> *–'I am not allowed umbrellas anymore', he said jokingly,*
> *'I lose them all the time'–.*
>
> Cornelis Easton.[1]

Not long after Kapteyn's death a tribute was published in two installments in the Dutch journal *Hemel & Dampkring* (Sky & Atmosphere), the magazine of the Dutch association of amateur astronomers and meteorologists, in August and September 1922. In the Kapteyn Room at the Kapteyn Astronomical Institute there is a copy of a reprint of the two parts that were presented by the author to Kapteyn's widow C.E. (Elise) Kapteyn–Kalshoven. Three of the four illustrations in the publication have been reproduced in this appendix; the fourth has been reproduced on page 350. Two of these figures are drawings of Kapteyn by Easton himself.

The author, Cornelis Easton (1864–1929) (see Fig. 11.12), a Dutchman in spite of his name, was a prominent journalist and newspaper editor. He had had no formal training in astronomy, yet conducted original research in that field and in climatology and published a number of papers in professional journals, some of them in conjunction with Kapteyn. In recognition of his contributions, he received an honorary doctorate from the University of Groningen in 1903 with Kapteyn as his 'promotor'. At the time Easton wrote the article, he was editor-in-chief of the newspaper *De Dordrechtsche Courant*.

---

[1] In the article discussed in this appendix.

The full text of the original article in my English translation is available at my Kapteyn-website [1]. In this appendix I reproduce only the parts that concern the person of Kapteyn. It starts with a dedication.

> To be the Eckermann of this Goethe!
> The Forster of this Dickens!...

Johann Peter Eckermann (1792–1854) served for a number of years as Goethe's personal secretary and after Goethe's death published a book *Conversations with Goethe*.. He was also involved in preparing Goethe's works for posthumous publication. John Forster (1812–1876) is well-known for his biography *Life of Charles Dickens*. He is said to have been in the possession of the original manuscripts of Dickens's novels, which he later bequeathed to the South Kensington Museum.

'On a sunny afternoon in the year 1894, now already almost thirty years ago, I found myself in front of prof. Kapteyn, who had come to my hometown as a external examiner for the final high-school exams.

I was met in the bare waiting room in the Gymnasium of Dordrecht by a tall, lean man who was looking remarkably young for his 43 years. My first impression of the man was: How tall he is!

I have addressed the question of Kapteyn's height on page 56.

He seemed to be looking over the heads of almost all his peers and that impression was strengthened by his upright head on a long neck. Already then his eyelids where long and hanging low, which could give the impression that he not only from high up, but also somewhat disparagingly looked down upon the world. His hair and mustache were of a light red color. A handsome face, by the way, with a delicate nose and busy lips, a face that seemed tight from inner emotional tensions.

Compared to professor van de Sande Bakhuyzen with his friendly, charming manners (that this good man has conserved up to the present day) I at first found the young Groningen professor not really forthcoming. There were two reasons for that, as I only later understood. Kapteyn was, except when science was concerned, extremely stringent, and therefore cautious for interference from interested outsiders. He had much appreciation for the work of 'amateurs' in astronomy, such as Franklin Adams, but he then had to be convinced that their work was sufficiently significant to spend time on.'

Easton refers here to John Franklin-Adams (1843–1912), a British amateur astronomer. He was famous for his 'Franklin-Adams plates', which constituted a photographic atlas of the night sky.

To devote his life to the most extensive and complicated of all astronomical problems: the structure of the Universe –that grand resolve must have been developing in his thinking in his junior years when he resolved to dedicate himself to astronomy. 'Le long espoir et les vastes pensées'. are being born at that age. [...]'

In full the quote is *'Quittez le long espoir et les vastes pensées, – Tout cela ne convient qu'à nous'* (Long-growing hope, and lofty plan, Leave thou to us, to whom such things belong [2]) It is a citation from a poem by Jean de la Fontaine (1621–1695) –Fables (1668–1694), Livre onzième, VIII, *Le vieillard et les trois jeunes hommes* (The old man and the three young ones) – about an old man who plants trees for future generations.

'After our first meeting Kapteyn quickly paid me a visit in return. My guest was a heavy smoker and was fond of a glass of wine – later he would give up both of these almost completely, for the benefit of his eyes. Not out of principle. He hated 'Prinzipienreiterei' (harping on about principles), and what little he needed for satisfaction, he did enjoy the less high-minded joys of life. 'One of my sisters', he once said, 'has had her molars removed without anesthesia. I would never do that. One has to accept the pain that must be suffered and I admire that, but why not avoid unnecessary pain?' [...]

The following years our paths did not cross, and only around the turn of the century I was in a position to visit Kapteyn in Groningen.

At that time they lived in the Heerestraat, simple but spacious. It was a pleasure to stay with that family, also later in the remarkable and cozy upper floor on the Ossenmarkt. Immediately after arrival the guest was put at ease, with so much pleasant business and care as if one had just arrived from a train trip from at least St. Petersburg. The professor's soft calmness was happily complemented by the busily moving around of his wife, hopping around it seemed like a bird; both were cordial in their own manner. With them were two daughters and a son. It was one of those households that would have amused Dickens enormously. Dickens, by the way, was present like a good friend in their house; during their first years of marriage husband and wife had read out aloud to each other the complete set of 20 volumes, and a bust of the unforgettable story teller stood in Mrs Kapteyn's bookcase. Moving detail: during his fatal illness, when Kapteyn had to rest most of the time, only a few weeks before his death he was visited by his friend Charles Edward St. John in Amsterdam. Talking with the ill man was not very well possible, but he did read out to him, at his request from Dickens....'

Charles Edward St. John (1857–1935) was an astronomer from Mount Wilson Observatory.

'The man of whom in England recently it was claimed that he had done the most important discovery in the area of stellar astronomy since the time of the Herschels, showed no sign of proud self-consciousness or professorial dignity, when on an evening in June in the year 1906 I picked him up at the Meuse Railway station in Rotterdam. He looked extra unassuming. His straw hat had seen its original cleanness long ago; no umbrella in spite of the rain – 'I do not get any umbrella's anymore, he said jokingly, I loose them all the time' – and we moved slowly, but not unnoticed, through the crowd to our house.'

Fig. C.1 Illustration accompanying the article of C. Easton in Hemel & Dampkring. This is a drawing of Kapteyn made by Easton himself. The year is 'about 1890' and Kapteyn must have been about 40 years of age

The Meuse railway station in Rotterdam is no longer in use.

'There was only one thing that was more pleasant than staying with the Kapteyns; to have Kapteyn staying with you. This extraordinary man was able to neutralize, through his not particularly exuberant, but fully natural good spirits, the oppressing influence of a modern street in Rotterdam under a dark, wet sky in the summer. At home, during a cold dinner, the subject of conversation turned to life in Groningen.'

Easton here probably means a dinner that consisted of cold dishes, and not a cooked meal that had turned cold.

'Rob' [Kapteyn's son Gerrit Jacobus] had left home. 'A boy with many talents and a good brain (said his father), but he has the characteristic habit, inherited from his ancestors, to do exactly what he wants to. Well, I do not give up hope; we were with ten brothers and five sisters and we all ended up well.'

'But maybe it was ornithology to which – second to astronomy – he felt attracted most. Had astronomy left him more spare time he would have, it seems to me, been inclined to spend each free afternoon lying on his stomach somewhere in the fields or hiding in the reeds watching birds with a copy of Jac. Thijsse's book in his pocket.'

Jacobus Pieter Thijsse (1864–1945) was a botanist, who wrote a little book *Het Vogeljaar*, published in 1903, which became the most popular book on birds in the Dutch language and was reprinted many times.

'He knew a lot about that, and whoever has tried to tell a few dozen or even one dozen Dutch songbirds apart, only by listening to their songs, will realize that it must have taken much dedication and diligence for one who had so few hours, even minutes, to spare, to have gone that far. I remember a walk with him and with our Jan and Titia in the woods near Bloemendaal. What a delightful companion he was for children! 'Children', he had comforted them that Sunday morning, when he had noticed two pairs of eyes looking disappointingly at a wet, gray sky filled with clouds, 'children, the gardener of my uncle always said: if there is only one little patch of sky big enough to cut a pair of trousers out of, then it will become nice weather!' – And indeed, the weather turned nice. In his manner we wandered through alleys passing high beeches and he was the first among us who noticed an estrildid finch in the distance sitting on a fence. And we *heard* the estrildid finches, and chiffchaffs (who ricochet pebblestones against each other) and the garden warbler, who goes on and on, and the willow warbler, about which Thijsse writes so emotionally, and then we laughed about this, as wise city dwellers, but deep in our hearts we found that lingering dying out of the soft song of this bird very moving. And we came home merry, like large and small children (the great Kapteyn included) that had relished a rare free day. [...]

The next days of his stay in Rotterdam he served again as external examiner for the final exams of the grammar school. When he came back home we were treated of course with the corresponding jokes. One pupil had translated 'ici-bas' as 'here stocking', somebody else translated the sentence 'Platon et Socrate sont deux philosophes' as: 'One great philosopher and one great philosopher makes two great philosophers'. (I don't doubt the authenticity of this, but also cannot resist to suppose it was made up at home using a dictionary.) 'How much money (another question at the exams) does one need to invest in order to pay the tuition of 1000 Guilders per year for four years of studies of his son, who now is four years old? One candidate found an answer of 87 Guilders and 42 cents; but there also was another candidate who found such an enormous number that he had to write diagonally on his piece of paper in order to fit it on one line.'

When the laughing had died away he started to tell the most hilarious stories, such as how he once was staying somewhere as a guest, had opened the double doors of his bedroom to get some fresh air on the balcony, only to find that by holding on to the door handle he could barely avoid falling down three stories, as there was no floor to the balcony!...

He was the most pleasant and entertaining companion when he was among good friends. He and my wife would roar with laughter about little things. Once she came to bring him a cup of coffee in the morning: the tall man – it was summer and hot – was sitting in the open window, staring over the landscape behind him, softly humming or softly talking to himself, deeply absorbed in thought, probably about his luminosity function or selective absorption in space. On the sound of the door opening he turned to look inside but heavily bumped his illustrious head. While rubbing that part of his body he made such a strange face that his hostess burst into laughter so much that she had to put the cup down so as not to spill the coffee, and while he shouted 'So, what is there to laugh about' he laughed himself along with her until he had tears in his eyes.

Sometimes he told about his youth. Of the prowess of one of his brothers, who was posing as 'little Mercury' on the ridge of the house's roof, standing on one leg, while his father looked up and was not able to move being filled with fear. – Or: that he on an occasion had sat down on the hanging frame that house painters were using, upon which the frame started to come down in small steps – how he escaped from it he did not remember. Or how he and one of his brothers in the times before the 'Rover safety' exultantly cycled into the hometown Barneveld; there nobody had seen bicycles and the populace all gathered in great numbers to see them so that the two cyclists fell from their two-wheelers 'like ripe pears'.'

The 'Rover safety' is the model for our modern bicycle with a diamond shaped frame and two wheels in the same line, the rear one driven by a chain.

'Kapteyn was not a born speaker. He never had done any training in the art of elocution – he would not hear of it – his performance was simple and far from impressive, his voice somewhat high-pitched and nasal. Also he was not able to, as Lorentz could in an unsurpassed way, drive his presentation so in your brains, so to speak, that for a moment you understand the most difficult things, that is to say seem to understand – as long as the impression lasted. Kapteyn's mind was in the other hand so completely clear and logical, his choice of wording so excellent, his gestures so simple and still expressive that he needs to be counted among the best in explaining abstract matter. It was most difficult for him, who lived with his subjects from early in the morning until late at night, to judge the extent the background knowledge and ability to grasp things among his listeners; but even those that did not fully comprehend it all, still left for home with a 'profit'.

His mind was preoccupied more with important things than caring for his appearance, and whoever would see him *en négligé* [in casual dress as if at home with no guests around] could take a photograph of him as seen in the previous issue of this journal, but that does not mean he did not care about his appearance. Whenever he felt it was worth the trouble to dress with care, he would tend to prefer the modern over the old-fashioned in his suits. His tall figure made it easy to dress well. When he wore his professors' gown he even could look like a dignified professor. Presenting himself officially he did through his manners, just as he would be a worldly person on other occasions. Through his broad education, his knowledge of languages, his fine sense and tact, being worldly was natural and easy for him – although it was not part of his character to develop relationships out of vanity or as a "bridge', a means to go where he wanted to be. In those case he traveled the 'royal' road; he accomplished what he wanted simply through his own extraordinary and undeniable merits.

Was he indifferent to recognition? Absolutely not. I think that in recent years, and shortly after his death, there has been too much reference to his modesty. One should not think that he drifted over the vanities of the world with downy wings of angels in childish naivety. He was extremely well aware of the social orders and used these in practical manners whenever he could make use of them –in the first place to promote his work. In scientific matters he was never short of funds for his research in this ungenerous rather than free-handed society of ours and what he lacked in guilders he knew how to supplement in dollars. That he did not aspire personal gain, nor honors, goes without saying: people do not regard such altruism with scientists and artists as very normal, but take their own self-interest for granted. Of course the honors did come his way, but only in later years. They pleased him enormously, and not just because it reflected upon those around him; he was too sensitive a person to be insensitive to wide appreciation. Vanity in the petty sense of the word, was foreign to him; others, not himself, were annoyed that the Netherlands awarded him 'the Lion' officially for his having been ...... a teacher for 25 years; the fact that the state was not represented at his 40-th anniversary as a professor; that this great Dutchman, had he been born in England undoubtedly would have been 'Lord Kapteyn' long before his death, at his retirement was honored(!) with just a little more than an average Royal distinction in a second rate order. [...]'

The 'lion' here refers to the Royal distinction of Knight in the Order of the Dutch Lion. In the last sentence, Easton refers to Kapteyn's Royal distinction as Commander in the Order of Oranje-Nassau, which is nowadays seen as a significantly higher distinction than average. Even the slightly less prestigious Knight in the Order of the Dutch Lion is nowadays well above average.

'Therefore he often made the impression of an almost funny calmness, with a bit of indifference for the minor things in life. It sounds a bit comical when we see him described in an American newspaper after a lecture he had given somewhere in California as 'a quiet, modest, almost bashful man'. Maybe this journalist on the other side of the Big Pond thought that an ordinary scientist had to be a sort of stubborn idiot and an extraordinary scientist, a noisy boaster. He did make a compliment on the stranger's English: 'Like nearly all educated foreigners who essay to speak the English language, he does so with an accent (well! listen to the American English! – E.), but with a pureness of idiom and diction, which makes of it really classical English. His literary productions in English are marvels in their strong, vigorous simplicity'. – Indeed, he did know how to write in other languages, English in particular, as easily as he spoke them; also in this area he displayed his versatility. [...]

Fig. C.2 Illustration accompanying the article of C. Easton in Hemel & Dampkring. It is a drawing of Kapteyn by Easton in 1920 (see text)

A few years before the war, when I had sent him a set of photographs that was published as a Photographic Chart of the Milky Way, he wrote to me: 'I have put this map on the wall next to my work table, but for the moment I feel that the profound implication works *de*pressing. While working with papers I sometimes feel a certain satisfaction and especially hope – the appearance of the details that all need to be explained, subsequently reduces this to a proper feeling of insignificance.'

'For that 'certain satisfaction' there were very good reasons. Only few researchers are blessed with the opportunity to see the result of their labors grow in their hands into a magnificent edifice, *aere perennius*. He saw it and it must have given him the greatest and purest joy and 'intellectual' experience.'

Easton refers to a paper he published in the Astrophysical Journal in 1913, *A photographic chart of the Milky way and the spiral theory of the Galactic System* [3], see Fig. 11.13. The quotation is *'Exegi monumentum aere perennius'*; I have erected a monument more lasting than bronze. It is from Horace (Quintus Horatius Flaccus; 65BC–8BC).

'Since he never wasted time or energy and in later years more and more became aware of his health, he made everything subservient to his

Fig. C.3 Illustration accompanying the article of C. Easton in Hemel & Dampkring. It is a photograph of Kapteyn at the age of 70

scientific mission. He did have an interest in the world around him, but he intentionally did not actively participate in social activities. He never was a member of a city council; he was, if I am right, not a member of the board of a society or club and he did not participate in conferences on a general subject.'

This is not correct, since Kapteyn was a very active member of the board of the 'Natuurkundig Genootschap', see page 276.

'He had become immune for 'distraction', which was enhanced by the fact that his wife took care of all other daily matters and he was not of a weak character. It even applied these to his social contacts; how cordial he may have been for his friends, the remark by 'Larouchefoucauld' was definitely applicable to him: 'J'aime mes amis.... je n'ai pas des grandes inquiétudes en leur absence'.'

François de la Rochefoucauld (1613–1680) was a French writer, well known for his collection of aphorisms. The full text is in English: 'I love my friends, and I love them in such a way that I do not hesitate to sacrifice my interests to theirs, I support them, I patiently accept their bad moods and I do apologize for everything; I just do not show my affection all the time, and I have no great problems with their absence.'

'For one who was not very fond of attending meetings, he came relatively often to the sessions of the Astronomers Club that was founded in October 1918 at the initiative of Nijland.

The drawing that I show here was made during one of those meetings in Utrecht. He was sitting more or less opposite from me, leaning somewhat backwards on his chair, but with his head straight up he had crossed his legs and as usual dangled one leg intensely on the knee of the other. He looked better then and took part in the discussions. His retirement was not very far away. The nice, (in an old-fashioned way) intimate house on the Ossenmarkt would have to be vacated.'

The following Easton presents is a footnote.

That house deserves an article by itself. Strangers had to search for the entrance for a long time. There was a narrow passage next to the house between it and the next house and there one would find a small door that was almost never locked; next to it on a nameplate *i.*C. Kapteyn, rather than J.C. A dark winding staircase led to a beautiful spacious hall; the front living room was large also and the room in the back, that was furnished as a study – but in the evening everybody sat there, since K. did not work then to spare his eyes –, looked out over the gardens and the Nieuwe Kerk. Concerning his further plans: maybe still another trip to America next year, or a visit to Switzerland for a while; the dearest wish of Kapteyn and his wife though was their own house somewhere in 't Gooi, with five rooms or so, no maid, but with a dog... [...]

It has not been given to him to spend the last part of his life quietly in the countryside. Still, the words of Wordsworth came true:

> And I should like my days to be
> Bound each to each by natural piety.

He truly has been one of the high priests of natural science. Only death was able to extinguish in him the drive to uncover the truth, which is another form of religion. At the top of one of his rare popular articles he wrote this citation as epigraph, which describes his drive perfectly:

≪ If God held in His right hand all truth, and in his left nothing but the ever ardent desire for truth, even with the condition that I should err forever, and told me choose, I would bow down to his left, saying: 'Oh, Father, give; pure truth can be but for Thee alone. ≫

William Wordsworth (1770–1850) was an English poet. I have not been able to find an article with this quote at the top; however, Kapteyn (1913a) has it at the end and I quote it from there. It is from Gotthold Ephraim Lessing (1729–1781).

Appendix D
Notes and References

> *If you don't know where you're going,*
> *you'll wind up somewhere else.*
> Yogi (Lawrence Peter) Berra (1925–present).

Below are further details on sources that were used or quoted in the text. These do not contain any remarks, further background or explanations, but are restricted to the bibliographic details of books, articles and other material, and provides, whenever possible, directions to websites where electronic versions can be found. The notes have been referred to in the text with the numbers between square brackets, starting at '1' in each chapter or appendix. The first set applies to the preface.

Many astronomical journal papers that I refer to are available from the **NASA Astronomy Data System ADS**, which can be accessed at adsabs.harvard.edu in the USA or a number of mirror nodes on other continents. Whenever available, references to entries in ADS are included in these notes *by their ADS designation*. So if for example the note refers to the paper *A photographic chart of the Milky way and the spiral theory of the Galactic System* by Cornelius Easton, published in 1913 in the Astrophysical Journal, 37, 105–118, the annotation [1913ApJ....37..105E] has been added. This means that this paper is listed in ADS and the full URL to access this paper is adsabs.harvard.edu/abs/1913ApJ....37..105E in the USA node, esoads.eso.org/abs/1913ApJ....37..105E in Europe, etc. For many, especially older publications, scanned versions of the papers are provided by ADS in .pdf or .gif formats, but in other (mostly recent) cases electronic subscriptions are required to download the full text from the journal or publisher's site.

Papers by Kapteyn himself or his close collaborators are in the text referred to by author and years of publication. These references are not given in this appendix, but appear in the listing of Kapteyn's publications in Appendix A. Thus for example Kapteyn (1922a) is Kapteyn's famous paper *'First attempt at a theory of the*

Fig. D.1 Kapteyn in 1908 in the 'Monastery' (the residence of astronomers during their observing sessions) at Mount Wilson Observatory [4]

arrangement and motion of the Sidereal System', published in the Astrophysical Journal, vol. 55, pages 302–328 (1922). For papers by Kapteyn for which no electronic versions are available on the Web, I provide scanned version on my Kapteyn homepage www.astro.rug.nl/JCKapteyn.

References

[1] From *The Progress of Science*, the Scientific Monthly, June 1921.
[2] The Observatory, vol. 48, 293–294 (1925). [1925Obs....48..293D]
[3] She published it with P. Noordhoff in Groningen. The book is now part of the *Digitale Bibliotheek voor de Nederlandse Letteren* (Digital Library of Dutch Literature). See www.dbnl.org/tekst/hert042jcka01_01/, where versions in .pdf and .txt formats are provided.
[4] Space Science Reviews, vol. 64, 1–92 (1993) (with a preface and introduction on pp. x-xix). [1993SSRv...64D...5P]
[5] Kluwer (1993), ISBN 978-07-923-2603-8.
[6] Journal for the History of Astronomy, 12, 77–94 (1981). [1981JHA....12...77P]
[7] IAU Symp. 106, 25–42 (1985). [1985IAUS..106...25P]
[8] Journal for the History of Astronomy, 17, 155–182 (1986). [1986JHA....17..155P]
[9] Cambridge Univ. Press (1993), ISBN 0-5213-5363-7.
[10] Kluwer (2000), ISBN 0-7923-6393-0.
[11] See www.astro.rug.nl/JCKapteyn/HHKbiog.pdf.
[12] University of Groningen (2008), ISBN 978-90-367-3353-3.
[13] Princeton University Press (2000), ISBN-10 06-910-4918-1.
[14] *Centennial History of the Carnegie Institution of Washington: Volume 1*, Cambridge University Press (2004), ISBN-13: 978-05-218-3078-2.
[15] See the press release www.astro.rug.nl/vdkruit/jea3/homepage/031209fwn nie uws.html.
[16] See www.astro.rug.nl/~vdkruit/jea3/homepage/address65.pdf.
[17] See magazine.dutchancestrycoach.com/converting-dutch-historic-currencies.
[18] Website at www.iisg.nl/hpw/calculate.php.
[19] See www.astro.rug.nl/~vdkruit/# Challenges symposium and www.elmerspaa rgaren-fotografie.nl/.
[20] New York, G.P. Putnam's sons (1901), www.archive.org/details/starsstudyof univ00newciala; footnote on p.49.

Chapter 1. Growing up in Barneveld

[1] The Observatory, 45, 261–265 (1922). [1922Obs....45..261.]
[2] De Gids, Jaargang 86. 130–133, 1922. See www.dbnl.org/tekst/_gid00119 2201_01/_gid001192201_01_0081.php.
[3] See www.clker.com/clipart-kaart-nederland-jan.html; online royalty free public domain clip art.
[4] See www.barneveld.incijfers.nl/.
[5] Nicolette M. Hijweege: *Bekering in bevindelijk gereformeerde kring*, PhD Thesis, University of Amsterdam (2004), ISBN 90-435-0995-7, dare.uva.nl/record/121886.; Saskia Reuzenkamp (ed.): *Steeds gewoner, nooit gewoon*, Sociaal en Cultureel Planbureau (2010), ISBN 978 90 377 0501 0, www.scp.nl/Publicaties/Alle_publicaties/Publicaties_2010/Steeds_gewoner_no oit_gewoon.; Fred van Lieburg (ed.): *Refogeschiedenis in perspectief.*

Opstellen over de bevindelijke traditie. Groen (2007) ISBN 978 90 5829 780 8, www.dutchbiblebelt.org/fileadmin/uploads/2007Refogeschiedenis_bundel.pdf.
[6] See *Waarom daar?* by Jan Dirk Snel in 'Refogeschiedenis' (ed. F. van Lieburg, *op. cit.*).
[7] See www.barneveld.nl/document.php?m=6&fileid=11445&f=56889951195fd cd171c14fc2108e5ffe&attachment=1&c=2724.
[8] Website at www.biografischwoordenboekgelderland.nl/bio/6_Gerrit_Jacobus_ Kapteyn.
[9] Nr. 04.0394; beeldbank.ede.nl/atlantispubliek/default.aspx?modules=beeldban ken#media/410011/1
[10] Nr. 03.0553; beeldbank.ede.nl/atlantispubliek/default.aspx?modules=beeldban ken#media/498575/1.
[11] Nr. 03.2280; beeldbank.ede.nl/atlantispubliek/default.aspx?modules=beeldba nken#media/464635/1.
[12] See www.barneveld.nl/gemeentearchief/personen_3619/item/gemeentelijke-jo ngens-en-meisjeskostscholen_2711.html.
[13] Website of 'het Nut' is www.nutalgemeen.nl/.
[14] E.g. www.middel.org/canon-van-groningen/kapteyn/88-genealogie-kapteyn.
[15] See genver.nl/index.htm.
[16] Nr. 03.0934; beeldbank.ede.nl/atlantispubliek/default.aspx?modules=beeldban ken#media/434609/1.
[17] FamilySearch, www.genver.nl.; Gelderland, Barneveld, file 5808804 (Geboorten 1849–1852), image 108/210.
[18] Barneveldse Krant, 17 Febr. 1983. Gemeentearchief Barneveld.
[19] Nr. ANS2.0095; beeldbank.ede.nl/atlantispubliek/default.aspx?modules= bee ldbanken#media/436904/1.
[20] From *Sterrenkijken bekeken*, by A. Blaauw, J.A. de Boer, E. Dekker & J. Schuller tot Peursum-Meijer, Universiteits Museum Groningen (1983).

Chapter 2. Studies in Utrecht
[1] Verloren (1997), ISBN 90-655-0557-1.
[2] Tijdschrift voor de Geschiedenis van de Geneeskunde, Natuurwetenschappen, Wiskunde en Techniek, vol. 7, p. 32–48 (1984).
[3] See web.science.uu.nl/NatuurkundigGezelschap/.
[4] Boom, Meppel/Amsterdam (1985) www.dbnl.org/titels/titel.php?id=berk003 voet01.
[5] Jensma & de Vries, *op. cit.*
[6] Jensma & de Vries, *op. cit.*
[7] See dap.library.uu.nl/.
[8] See www.genealogy.ams.org/.
[9] See www.knmi.nl/klimatologie/metadata/043_utrecht.html.
[10] Bekking, Amersfoort, ISBN: 978-90-6109-323-7 (1993).
[11] See www.sonnenborgh.nl/page=site.treenode/tree=english, from which part of this text has been taken.

[12] See www.universiteitsmuseum.nl/Collectie/Detail/UG-5223?q=Oudemans en www.universiteitsmuseum.nl/Collectie/Detail/UG-5234?q=Hoek.
[13] Astronomical Society of the Pacific Conference Series, No. 470 (2012). [aww.aspbooks.org/a/volumes/table_of_contents/?book_id=520]
[14] See www.universiteitsmuseum.nl/Collectie/Detail/UG-5212?q=Buys+Ballot and www.universiteitsmuseum.nl/Collectie/Detail/UG-5226?q=Grinwis.
[15] Orbits of minor planets and comets can be studied using the NASA orbits tool of the Near-Earth Object Program at neo.jpl.nasa.gov/orbits/.
[16] See www.conspiracyoflight.com/Hoek/Hoek_Experiment.html.
[17] Monthly Notices of the Royal Astronomical Society, 28, 131–150 (1868). [1868MNRAS..28..131H]
[18] Copies of the Utrechtse Studenten Almanak are available in the library of Utrecht university.
[19] Marcel Decker Inc. (2004), ISBN 0-8247-5629-0; e-book by Taylor & Francis e-Library, ISBN 0-203-02630-6.
[20] See X5548mcise.uri.edu/sadd/mce565/notes.htm.
[21] Annales de Chimie et de Physique, Troisième Série, Tome LX, p. 449–478 (1860); eBook: books.google.nl/books?id=0eI3AAAAMAAJ.
[22] University of Southampton, Southampton SO17 1BJ, United Kingdom. resource.isvr.soton.ac.uk/spcg/tutorial/tutorial/Tutorial_files/Web-standing-membrane.htm. See also www.southampton.ac.uk/soundwaves.
[23] Soedel *(op. cit.)*, Sadd *(op. sit.)*, Jimin He & Zhi-Fang Fu, *Modal Analysis*, Butterworth-Heinemann, www.scribd.com/doc/50637604/44/Vibrations-of-membranes.
[24] See ru.wikipedia.org/wiki/%D0%A4%D0%B0%D0%B9%D0%BB:Zoellner_Photometer.png (Wikimedia Commons) or www.vehi.net/brokgauz/all/006/6172.shtml.111.
[25] Journal for the History of Astronomy, 31, 323–338, 2000, www.astronomyca.com/z/zollner-photometer.htm, [2000JHA....31..323S] and British Journal for the History of Science, 34, 439–451 (2001) respectively: apps.we bofknowledge.com/full_record.do?product=WOS&search_mode=full_mode=GeneralSearch&qid=1&SID=3Fnc1fKGHPbB7CA4hhh&WOS&searchpage=1&doc=1.
[26] Bulletin of the Astronomical Institutes of the Netherlands, 1, 167–169 (1922). [1922BAN.....1..167V]
[27] In six installments in 'Popular Astronomy' vol. 52 (1946), starting with [1946PA.....54..211W].
[28] MNRAS 19, 175–180 (1859) [1859MNRAS..19..175A], MNRAS 23, 166–169 (1859), [1863MNRAS..23..166D], and MNRAS 30, 9–18 (1869). [1869MNRAS..30....9P]

Chapter 3. Astronomer in Leiden

[1] Observatory, , 71, 128 (1951). [1951Obs....71..128]
[2] From FamilySearch, www.genver.nl., Utrecht, Utrecht, File (6580428 Huw.-bijlagen 1879–1880 v/a mei t/m apr), Images 247–253/2518.

[3] *De lotingsinstrumenten voor de Nederlandse dienstplicht* by F.Staarman and *De dienstplicht op de markt gebracht; het fenomeen dienstvervanging in de negentiende eeuw* by by E.W.R. van Roon. See alfredstaarman.nl/wp-content/uploads/Alfred-Staarman-De-lotingsinstrumenten-voor-de-dienstplicht-in-de-collectie-van-het-Legermuseum-in-jaarboek-Legermuseum-Armamentaria-nr.-31-1996-41-47.pdf and www.knhg.nl/bmgn2/R/Roon_E._W._R._van_-_De_dienstplicht_op_de_markt_gebracht._H.pdf

[4] See also www.regionaalarchiefzutphen.nl/informatiebladen/47-militieregisters.

[5] Gemeentearchief Barneveld.

[6] Astrophysical Journal 56, 145–153 (1922). [1922ApJ....56..145V]

[7] Wetenschappelijke Bladen I., 257–295 (1923), see also *J.C. Kapteyn und sein astronomisches Werk* in Die Naturwissenschaften, Heft 45, 1–14 (1922).

[8] Published by Leiden Observatory (1965), Waanders/De Kler, Zwolle (1983) and Leiden University (2011).

[9] Studium, 4, 195–126 (2011); www.gewina-studium.nl/index.php/studium/article/view/1545/7241.

[10] Respectively www.dbnl.org/tekst/will078twee01_01/andsss.sagepub.com/content/21/3/503.abstract.

[11] See also www.strw.leidenuniv.nl/album1908/book_info.html.

[12] Astronomische Nachrichten 60, 273–286 (1863). [1863AN.....60..273K]

[13] Respectively pages i-xlii [1868AnLei...1D...1.], xliii-lii [1868An Lei...1D..43.], liii-lxv, [1868AnLei...1D..53.] and lxvi-lxxxvi.[1868An Lei...1D..66.]

[14] nl.m.wikipedia.org/wiki/Bestand:Prof._H.G._van_de_Sande_Bakhuyzen.jpg.

[15] *Verslag van de Staat der Sterrenwacht te Leiden en de aldaar volbrachte werkzaamheden, 1875–1876.* [1876VeLdn..12....1V]

[16] www.strw.leidenuniv.nl/outreach/strwarchief/observatoryarchives.php.

[17] Verslag van den Staat der Sterrewacht te Leiden en van de aldaar volbrachte werkzaamheden (1875–1876). [1876VeLdn..12....1V]

[18] See www.nationaalarchief.nl/.

[19] Annalen van de Sterrewacht te Leiden, 1, liii-lxv (1868). [1868AnLei ...1D..53]

[20] See www3.uni-bonn.de/.

[21] Celestial Mechanics 36, 207–239 (1985). [1985CeMec..36..207F]

[22] Journal of the Royal Astronomical Society of Canada, 85, 43–50 (1991). [1991JRASC..85...43B]

[23] These Annual Reports (in Dutch) are available on ADS as [1875VeLdn ..11....1V, 1876VeLdn..12....1V, 1877VeLdn..13....1V and 1879VeLdn..15 ....1V]

[24] www.museumboerhaave.nl/object/universeelinstrument-v10680/.

[25] Verzeichniss der Instrumente der Sternwarte in Leiden, beim Anfange des Jahres 1868. [1868AnLei...1D..53.]

[26] Published in two installments in Astronomische Nachrichten, 107, 49–60 and 97–112 (1883). [1883AN....107...49V] and [1883AN....107...97V]

[27] From *The Dutch Transit of Venus Expeditions of 1874 and 1882* by Robert van Gent, Anne Zandstra, Hans Hooijmaijers & Klaus Staubermann. [www.phys.uu.nl/$\sim$vgent/venus/cmm/dutch_expeditions] Collectie Universiteitsmuseum Utrecht, inv.nr. 0285–3994.
[28] Astronomische Nachrichten, 95, 81–96 (1879). [1879AN.....95...81V]
[29] Published by the U.S. Naval Observatory (1881). [1881USNOM..17C...1.]
[30] Astronomische Nachrichten, 96, 119–128 (1879). [1879AN.....96..119D]
[31] Summarized in Monthly Notices of the Royal Astronomical Society, 41, 317–324 (1881). [1881MNRAS..41..317G]

Chapter 4. Professor in Groningen
[1] From *Sterrenkijken bekeken, op. cit.*
[2] See Klaas van Berkel: *In het voetspoor van Stevin*; Bastiaan Willink: *De tweede Gouden Eeuw* and *Origins of the Second Golden Age* ...; Jensma & H. de Vries, *Veranderingen in het hoger onderwijs*...; P.A.J. Caljé: *Student, universiteit en samenleving*; all *op. cit.*
[3] See www.rug.nl/museum/galerij/portretten/hoogleraar/mulerius.
[4] See hoogleraren.ub.rug.nl/?page=showPerson&type=hoogleraar&hoogleraar_id=26&lang=nl.
[5] See hoogleraren.ub.rug.nl/?page=showPerson&type=hoogleraar&hoogleraar_id=91&lang=nl.
[6] See irs.ub.rug.nl/ppn/108936023 and Digitale collecties van de Bibliotheek Rijksuniversiteit Groningen at facsimile.ub.rug.nl.
[7] *Jaarboek Rijks-Universiteit Groningen, 1877–1878*, Wolters (1879), p. 39. [archive.org/details/jaarboekderrijk00grongoog]
[8] Original in Universiteits Bibliotheek Groningen.
[9] FamilySearch, www.genver.nl., Utrecht, Utrecht, File (6580300 Huwelijken 1879–1880), Image 166/576.
[10] FamilySearch, www.genver.nl.
[11] *Lijst van (voornamelijk hervormde en lutherse) predikanten, hulppredikers, kandidaten, proponenten en theologanten*. This belongs to J. Vree, *Overschot op de Nederlandse kandidatenmarkt: een bron van overzeese predikanten, hulppredikers, enz. (1829–1872)*, Documentatieblad voor de Nederlandse Kerkgeschiedenis na 1800 (DNK), 66, 17–53 (2007); www.hdc.vu.nl/nl/onderzoek-en-publicaties/publicaties/documentatieblad/index.asp.

Chapter 5. From Kepler to Parallax
[1] The full book is available at books.google.com.hk/books?id=KUE1AAAAcAAJ&printsec=frontcover&hl=zh-CN&source=gbs_ge_summary_r&cad=0#v=onepage&q&f=false.
[2] Second Edition, The Macmillan Company, London (1914).
[3] Engelmann, Leipzig, 2 volumes (1870/1880).
[4] The volumes can be accessed electronically on the *Internet Archive* through archive.org/details/handwrterbuchdnnvale with $nn = 01, 02, 31, 32, 04$.
[5] Willmann-Bell (1993), ISBN 978-0943396408.

[6] See www.dunsink.dias.ie/index.php?option=com_content&view=category& id=97%3Astro-dunsink-observatory&layout=blog&Itemid=139&lang=en.
[7] See also www.fingaldublin.ie/interior-pages/activities-attractions-amp-confer ence/visitor-attractions/dunsink-observatory/.
[8] Cambridge University Press (2010), ISBN 978-0-521-19267-5, page 236.
[9] Available as preprint as arXiv-math/0510050v1; arxiv.org/pdf/math/0510050 .pdf
[10] Respectively in Annales Scientifiques de l'É.N.S., troisième série, tome 10, 91–122 (1893) [archive.numdam.org/ARCHIVE/ASENS/ASENS_1893_ 3_10_/ASENS_1893_3_10_91_0/ASENS_1893_3_10_91_0.pdf] and Nieuw Archief voor de Wiskunde, xx, 116–127 (1893).
[11] Newspaper archived at Research Library and Archives at the Pasadena Museum of History.
[12] Reproduced with permission from the Department of Manuscripts & University Archives, University Library, Cambridge, UK.
[13] Published in Memoirs of the Royal Astronomical Society, 48, 1–198 (1885). [1885MmRAS..48....1G]
[14] These reports were published in the Monthly Notices of the Royal Astronomical Society. The reports of 1885 to 1990 are in vol. 46, p.221 (1886); 47, 164 (1887); 48, 180 (1888); 49, 183 (1889); 50, 192 (1990) and 51, 212 (1991). [1886MNRAS..46..221.] [1887MNRAS..47..164.] [1888MNRAS..48..180.] [1889MNRAS..49..183.] [1890MNRAS..50..192.] [1891MNRAS..51..212.]
[15] Monthly Notices of the Royal Astronomical Society, 55, 34–36 (1894). [1894MNRAS..55...34G]
[16] Inaugural-Dissertation Kaiser-Wilhelms-Universität Strassburg (1901).
[17] Internet Archive at archive.org/details/cu31924004071688.
[18] Journal for the History of Astronomy, xxii, 267–296 (1991). [1991JHA....22..267K]
[19] Published in the Observatory, 40, 271–273 (1917). [1917Obs....40..271S]
[20] Published in Bulletin of the Astronomical Institutes of the Netherlands, 3, 1–6 (1925). [1925BAN.....3....1D]
[21] Abhandlungen der Königlichen Akademie der Wissenschafter zu Berlin, 1867, p.19 (1867)
[22] *Beschreibung der Registrir-Apparate der Sternwarte in Leiden*, Annalen van de Sterrewacht te Leiden, 2, 6–18 (1870). [1870AnLei..2....6.]
[23] The full reference is Sitzungsberichte der Mathematisch-Naturwissenschaftlichen Classe der Kaiserlichen Akademie der Wissenschaften, Abtheilung II (Mathematik, Physik, Chemie, Mechanik, Meteorologie und Astronomie, LII, 546–546 (1866). [home.us.archive.org/ detailssitzungsbericht193klasgoog]
[24] See www.museumboerhaave.nl/object/meridiaankijker-v03471h/.
[25] Astronomy & Astrophysics, 65, 77–81 (1978). [1978A&A....65...77L]

[26] Maintained at the Centre de Données Astronomiques de, Strasbourg (simbad.u-strasbg.fr/simbad/). Ackowledgement: This research has made use of the SIMBAD database, operated at CDS, Strasbourg, France.
[27] Memoirs of the Royal Astronomical Society, 48, 191 (1885)
[28] Science, 3, 617–620 (1896). [1896Sci.....3..617F]
[29] Publications of the Washburn Observatory, XI (1902). [archive.org/details/meridiobservations11flinrich]
[30] Astronomische Nachrichten, 101, 69–72 (1881). [1881AN....101...69G]
[31] See www.canonsociaalwerk.eu/1950_BenjaminSpockNL/img/cover%20Gerard%20allebe.jpg.
[32] See fleximap.groningen.nl/gnmaps/monumenten/.

Chapter 6. Cape Photographic Durchmusterung

[1] New York, Macmillan (1961)
[2] Fourth edition, Macmillan (1977), ISBN 00-236-0190-6.
[3] See American Institute of Physics Website: www.aip.org/history/cosmology/tools/pic-spectroscopy-orion.htm.
[4] In *Mapping the Sky: Past Heritage and Future Directions*, IAU Symposium 133, Paris, 143–148 (1988). [1988IAUS..133..143M]
[5] The European Physical Journal: Historical Perspectives on Contemporary Physics. [2012arXiv1209.3563P]
[6] Published in two parts in The Observatory, 10, 267–272 and 283–294 (1888). [1887Obs....10..267G] and [1887Obs....10..283G]
[7] See gallica.bnf.fr/ark:/12148/cb343481087/date1882.
[8] See www.saao.ac.za/public-info/pictures/comet/
[9] See www.nationalarchives.gov.uk/currency/results.asp#mid.
[10] The Milky Way Galaxy, IAU Symposium 106, 25–42 (1985). [1985IAUS..106...25P]
[11] Thomas Maclear & David Gill, London: Eyre & Spottiswoode for H.M.S.O., 1884. [1884cseo.book.....M]
[12] See assa.saao.ac.za/html/his-obs-cape-gall_dev.html.
[13] See www.rug.nl/museum/geschiedenis/hoogleraren/kapteyn?lang=en.
[14] *Notes on visits to some continental observatories*, The Observatory, 12, 344–349 (1889). [1889Obs....12..344S]
[15] It appeared in French in the *Bulletin Astronomique* Series I, 4, 361–380 (1887). [1887BuAsI...4..361G]
[16] *Naissance et Developpement de la Carte du Ciel en France* in IAU Symposium 133: *Mapping the Sky: Past Heritage and Future Directions*, 29–32 (1988). [1988IAUS..133...29W]
[17] London: Murray; available at archive.org/details/greatstarmapbein00turnuoft.
[18] *Bulletin Astronomique* Serie I, 3, 161–164 (1886). [1886BuAsI...3..161G]
[19] *Bulletin Astronomique* Serie I, 3, 321–324 (1886). [1886BuAsI...3..321G]
[20] Paris : Gauthier-Villars (1887). See visualiseur.bnf.fr/Visualiseur?Destination=Gallica&O=NUMM-94857.

[21] The full title is *La Carte du Ciel – Histoire et actualité d'un projet scientifique international*, EDP Sciences SBN (2008), ISBN 978-2-7598-0057-5.
[22] *'Petite Histoire' du Congres Astrophotographique de 1887* by A.M. Motais de Narbonne, in IAU Symposium 133: *Mapping the Sky: Past Heritage and Future Directions*, 129–133 (1988). [1988IAUS..133...129M]
[23] See the paper by Motais de Narbonne, *op. cit.*
[24] See answers.yahoo.com/question/index?qid=20070225055803AAc6VDO.
[25] Journal of Astronomical History and Heritage, 12m 119–124 (2009). [2009JAHH...12..119L]
[26] *Réunion du Comité International Permanent pour l'Exécution de la Carte Photographique du Ciel à l'Observatoire de Paris* by G. Bigourdan, Bulletin Astronomique, Serie I, 6, 531–538 (1889) and 8, 461–468 (1891). [1889BuAsI...6..531B and 1891BuAsI...8..461B]
[27] *Mesure des Clichés d'après la Méthode des Coordonnées Rectangulaires*, Bulletin de Comité International Permanent de la Carte du Ciel, 164–204 (1889)
[28] *A revisit to the region of Collinder 132 using Carte du Ciel and Astrographic Catalogue plates* by R.B. Orellana, M.S. de Biasi, I.H. Bustos Fierro & J.H. Calderón, Astronomy & Astrophysics, 521, A39 (2010). [2010A&A...521A..39O]
[29] *Carte du Ciel: Call for the Compilation of a Complete Inventory of Publications*, ASP Conference Series, 377, 369 (2007) [2007ASPC..377..369R] and her website at www.astropa.unipa.it/Library/CarteduCiel/index.htm.
[30] *David Gill and celestial photography*, *op. cit.*
[31] Annals of the Cape Observatory, South Africa, 9, 0.1–1.63 (1903). [1903AnCap...9....1I]
[32] Annals of the Cape Observatory, South Africa, 9, 3.1–3.86 (1903). [1903AnCap...9....3G]
[33] Volume 7, pp.235–240.
[34] *Cf.* Groningen, Town and Provincial Archives, inventory no 83: prison archives, documents concerning the employment of prisoners.
[35] A.S. Eddington, *Jacobus Cornelius Kapteyn*, The Observatory, 45, 261–265 (1922), [1922Obs....45..261.], quotation p. 262.
[36] For Domela's term in prison, see for instance Jan Meyers (1993), *Domela, een Leven op Aarde. Leven en Streven van Ferdinand Domela Nieuwenhuis* (Amsterdam), 163–176.
[37] A.S. Eddington, *"Jacobus Cornelis Kapteyn, 1851–1922"*, Proceedings Royal Society London, Section A, 102, xxix-xxxv (1922), quotation p.xxxi.

Chapter 7. An Astronomical Laboratory

[1] See en.wikisource.org/wiki/Herschel,_William_(DNB00).
[2] www.atlasoftheuniverse.com/hr.html and en.wikipedia.org/wiki/File:HRDiagram.png. This file is licensed under the Creative Commons Attribution-Share Alike 2.5 Generic license.

References

[3] See www.eso.org/public/images/eso1118a/ and www.noao.edu/outreach/aop/observers/n891.html (Minimum credit line: Dale Cupp/Flynn Haase/NOAO/AURA/NSF).

[4] Respectively New York, Norton [1964, c1963] (DLC) 64010566; Princeton Univ. Press, ISBN 978-0-691-1483-5; Cambridge Univ. Press, ISBN 798-1-01838-9.

[5] See commons.wikimedia.org/wiki/File:Bolton-herschel.jpg, commons.wikimedia.org/wiki/File:Herschel_Caroline_1829.jpg and commons.wikimedia.org/wiki/File:John_Herschel_South_African_expedition.png. These images are in the public domain (copyright has expired).

[6] See commons.wikimedia.org/wiki/File:PSM_V09_D079_Herschel_40_foot_telescope_at_slough.jpg (see previous note).

[7] Philosophical Transactions of the Royal Society of London, 75, 213–266 (1785). [1785RSPT...75..213H]

[8] Astronomy & Astrophysics, 157, 230–244 (1986), [1986A&A...157..230V]

[9] Philosophical Transactions of the Royal Society of London, 107, 302–331 (1817). [1817RSPT..107..302H]

[10] Imperial Academy of Sciences of Saint Petersburg, re-issued recently under ISBN-10 11-4404-207-0.

[11] The book is available on archive.org/details/popularastronomy031620mbp.

[12] The original image is out of copyright. I gratefully acknowledge the preservation and scanning by 'University of Cambridge, Institute of Astronomy Library'. See www.repository.cam.ac.uk/handle/1810/225982.

[13] Verslag van den Staat der Sterrewacht te Leiden en van de aldaar volbrachte werkzaamheden (1886–1888). [1888VeLdn..24....1V]

[14] Verslag van den Staat der Sterrewacht te Leiden en van de aldaar volbrachte werkzaamheden (1890–1892). [1892VeLdn..28....1V]

[15] See www.grunn.nl/fotoalbum/img/oude_boteringestraat_gerechtsgebouw.gif

[16] Monthly Notices of the Royal Astronomical Society, 46, 237–238 (1886). [1886/MNRAS..46..237V]

[17] Bulletin Astronomique, Ser. I, 12, 97–106 (1895). [1895BuAsI..12...97V]

[18] Astronomische Nachrichten, 146, 209–21 (1898). [1898AN....146..209V]

[19] www.beeldbankgroningen.nl Identificatienummer: NL-GnGRA_1785_610.

[20] Astronomische Nachrichten, 124, 177 (1890). [1890AN....124..177H]

[21] Published in the Observatory, Vol. 25, pp. 158–161 (1902). [1902Obs....25..158C]

[22] The Observatory 21, 106 (1898). [1898Obs....21..103.]

[23] The Observatory, 21, pp. 409–410 (1898). [1898Obs....21..409.]

[24] DSS2 is available at the Space Telescope Science Institute Website at stdatu.stsci.edu/dss/. For a Web animation, see groups.google.com/a/googleproductforums.com/forum/?fromgroups#!msg/gec-sky/kulxqfbRZvQ/JaolC-O5EDkJ.

[25] The Observatory, 22, 99–101 (1899). [1898Obs....22...99G]

[26] E. Kotoneva, E., K. Innanen, K., P.C. Dawson, P.R. Wood & M. de Robertis: *A study of Kapteyn's star*, Astronomy & Astrophysics, 438, 957–962 (2005). [2005A&A...438..957K]
[27] Astronomical Journal 112, 1595–1613 (1996). [1996AJ....112.1595E]
[28] Press release heic0809, 2 April 2008; see www.spacetelescope.org/news/heic0809/.
[29] Boston, Harvard Univ. Press (1932). The 1934 Dutch version, published by van Stockum & Zoon, Den Haag, is available electronically from the 'Digital Library of Dutch Literature' at www.dbnl.org/tekst/sitt003kosm01_01/.
[30] E. Wylie-de Boer, K.C. Freeman and M. Williams, Astronomical Journal, 139, 636–645 (2010). [2010AJ....139..636W]
[31] Guillem Anglada-Escudé, *et al.*, *Two planets around Kapteyn's star: a cold and a temperate super-Earth orbiting the nearest halo red-dwarf*, eprint arXiv. [2014arXiv1406.0818A]
[32] *Letters of J.C. Kapteyn*. [www.helsinki.fi/astro/history/donner/jck.html]
[33] Monthly Notices of the Royal Astronomical Society, 95, 343–347 (1935). [1935MNRAS..95..343.]
[34] Homepage of the Share Initiative is www.tsieuropean.co.uk.
[35] Monthly Notices of the Royal Astronomical Society, 59, 341–345 (1899). [1899MNRAS..59..341D]
[36] Annals of the Cape Observatory, 8, 2.i-2.175 (1900). [1900AnCap...8....2G]
[37] The thesis is not available in electronic form. However, it was published in a different form as *Determination of the mass of Jupiter and elements of the orbits of its satellites'* by W. de Sitter, D. Gill & W.H. Finlay, Annals of the Cape Observatory, 12, 1.1–1.173 (1915). [1915AnCap..12....1D]
[38] Monthly Notices of the Royal Astronomical Society, 91, 706–738 (1931). [1931MNRAS..91..706D]
[39] From www.noao.edu/image_gallery/html/im0552.html. Credit line: National Optical Astronomy Observatory/Association of Universities for Research in Astronomy/National Science Foundation
[40] W. de Sitter, Publications of the Astronomical Laboratory Groningen, 15, 3–12 (1906). [1906PGro....15...1D]

Chapter 8. Colors and Motions
[1] See kaarten.abc.ub.rug.nl/ and atlas1868.nl/dr/vries.html, also at facsimile.ub.rug.nl.
[2] Published in Science, 5, 777–785 (1897). [1897Sci.....5..777N]
[3] *The Autocrat of the Breakfast–Table* is available at archive.org/details/autocratbreakfa11holmgoog.
[4] See archive.org/details/writingsoliverw14holmgoog and archive.org/details/writingsoliverw11holmgoog.
[5] The website of the Koninklijk Natuurkundig Genootschap is www.kng-groningen.nl.
[6] See nl.wikipedia.org/wiki/Hendrik_Jacob_Herman_Modderman_sr.

References

[7] Groningen: Gebroeders Hotsema. [www.dwc.knaw.nl/pub/tjadenmodderman 1901.pdf]

[8] Profiel, Bedum, ISBN 90-529-4220-X, see also www.kng-groningen.nl/boekjub.htm.

[9] Verloren (2012), ISBN 97-8908-704-194-6.

[10] See www.rug.nl/museum/galerij/portretten/hoogleraar/huizinga.

[11] The website of the Nederlandsch Natuur- en Geneeskundig Congres is www.nngc.nl.

[12] Astronomische Nachrichten, 147, 1–12 (1898). [1898AN....147....1S]

[13] Publications of the Astronomical Laboratory Groningen, 3, 23–26 (1900). [1900PGro...23D]

[14] en.wikipedia.org/wiki/File:Obafgkm_noao_big.jpg. 'This file is in the public domain because it was solely created by NASA'.

[15] Annals of Harvard College Observatory, 27, 1–388 (1890). [1890AnHar..27....1P]

[16] Philosophical Transactions of the Royal Society of London, 73, 247–283 (1783). [1783RSPT...73..247H]

[17] *Local kinematics and the local standard of rest* by R. Schönrich, J. Binney & W. Dehnen, Monthly Notices of the Royal Astronomical Society, 403, 1829–1833 (2010). [2010MNRAS.403.1829S].

[18] See www.iau.org/public/themes/constellations/. See also www.iau.org/ copyright/.

[19] Knowledge, April 1, 66–68, and May 1, 84–85 (1898).

[20] Credit: ESA/Hubble & Digitized Sky Survey 2. Davide De Martin (ESA/Hubble). [spacetelescope.org/images/heic1112f/]

[21] Publikationen des Astrophysikalischen Observatoriums zu Potsdam, vol. 7 (1892). [articles.adsabs.harvard.edu/cgi-bin/iarticle_query?journal=POPot&volume=0007&type=SCREEN_THMB]

[22] *Versuch einer Ableitung der Bewegung des Sonnensystems aus den Potsdamer spectrographischen Beobachtungen*, Astronomische Nachrichten, 132, 81 (1893). [1893AN....132...81V]

[23] Astronomical Journal, 17, 41–44 (1896). [1896AJ.....17...41N]

[24] Astronomical Journal, 20, 1–6 (1899). [1899AJ.....20....1N]

[25] New York, G.P. Putnam's sons (1901), *op. cit.*

[26] There are two installments in the Observatory, Vol. XXX, 299–306 and 335–339 (1907). [1907Obs....30..299G] and [1907Obs....30..335G]

[27] Celestial Mechanics, 36, 207–239 (1985). [1985CeMec..36..207F]

[28] Sdu Uitgevers, Den Haag (2000), ISBN 90 12 08622 1; www.dbnl.org/tekst/bank003190001_01/bank003190001_01_0010.php.

[29] Heidelberg: Carl Winter (1910).

[30] *Persoonlijke Herinneringen aan de Stad Groningen rond de Eeuwwisseling*, Groningsche Volksalmanak voor 1957, pp. 33–52.

[31] See www.genealogieonline.nl/genealogie-baert-cornelis-kalshoven/I284.php.

[32] In: *Universitas Groningana MCMXIV – MCMLXIV*, Groningen. 176–183.

[33] In: *Verzamelde werken*, VI, 336–338 (Haarlem, 1950), 336–338.

Chapter 9. Star Streams

[1] Respectively *The old-nova GK Per (1901). IV – The light curve since 1901*, Astronomy & Astrophysics Supplement Series, 54, 393–403 (1983), [1983A&AS...54..393S], and *The old-nova GK Per (1901). I – Determination of the orbital period*, Astronomy & Astrophysics, 99, 392–393 (1981). [1981A&A....99..392B]

[2] *Photographs and measures of the nebula surrounding Nova Persei*, Astrophysical Journal, 16, 249–256 (1902). [1902ApJ....16..249.] Reproduced from plates relocated in the archives of Lick Observatory and scanned.

[3] Astrophysical Journal, 16, 198–202 (1902). [1902ApJ....16..198H]

[4] www.noao.edu/outreach/aop/observers/GKper.html.

[5] Annales d'Astrophysique, 2, 271–302 (1939). [1939AnAp....2..271C]

[6] The volumes up to 1950 can be examined through www.dwc.knaw.knaw.nl/toegangen/digital-library-knaw/?pagetype=bundel.

[7] www.astro.rug.nl/$\sim$vdkruit/jea3/homepage/voetbal.pdf, 2006.

[8] KNAW Proceedings, 4, 221–232 (1902). [www.dwc.knaw.nl/DL/publications/PU00014272.pdf]

[9] Astronomische Nachrichten, 158, 167–174 (1902). [1902AN....158..167S]

[10] Astronomical Journal, 21, 161–168 (1901). [1901AJ.....21..161B]

[11] *Precession and solar motion, First* and *Second Paper*, Astronomical Journal, 26, 95–99 & 111–122 (1910). [1910AJ.....26...95B] and [1910AJ.....26..111B]

[12] The 1888 version is available on archive.org/details/textbookofgenera00youn, the 1898 version as atextbookgenera05youngoog on the same site.

[13] Astrophysical Journal, 151, 393–409 (1968). [1968ApJ...151..393S]

[14] Inflation calculator at www.davemanuel.com/inflation-calculator.php.

[15] *The effects of the 1904 North Atlantic fare war upon migration between Europe and the United States* by Drew Keeling, www.iga.ucdavis.edu/Research/All-UC/conferences/2006-fall/Keeling.pdf.

[16] See archive.org/details/cu31924015340114.

[17] Volume VIII, Astronomy and Earth Sciences, edited by Howard J. Rogers (1908). [archive.org/details/internationalcon08inteiala]

[18] See exhibits.slpl.org/lpe/data/LPE240025391.asp?thread=240029400.

[19] The URL is archive.org/details/internationalcon08inteiala.

[20] The URL is archive.org/details/reportofbritisha06scie.

[21] Popular Astronomy, 31, 429–440 (1923). [1923PA.....31..429M]

[22] Monthly Notices of the Royal Astronomical Society, 82, 432–438 (1922). [1922MNRAS..82..432E]

[23] The Observatory, 45, 261–265 (1922). [1922Obs....45..261.]

[24] See en.wikisource.org/wiki/Popular_Science_Monthly/Volume_66/November_1904/The_Progress_of_Science.

[25] www.beeldbankgroningen.nl/beeldbank/; identificatienummer: NL-GnGRA _1986_2563), Foto collectie RHC Groninger Archieven (1986–2563).

[26] www.groningeninbeeld.nl/Markten/images/MRK_010.jpg.

Chapter 10. Selected Areas

[1] Monthly Notices of the Royal Astronomical Society 62, 334–343 (1902). [1902MNRAS..62..334.]

[2] Publications of the Astronomical Society of the Pacific, 14, 97–102 (1902). [1902PASP...14...97G]

[3] *A history and description of the Royal Observatory, Cape of Good Hope*, by Sir David Gill, K.C.B.; London: His Majesty's Stationary Office (1913).

[4] Bulletin Astronomique, 23, 480 (1906). [1906BuAsI-23..480]

[5] See www.eso.org/public/images/yb_southerncross_cc/.

[6] The Observatory, 29, 129–134 (1906). [1906Obs....29..129.]

[7] Monthly Notices of the Royal Astronomical Society, 67, 34–63 (1906). [1906MNRAS..67...34E]

[8] Monthly Notices of the Royal Astronomical Society, 65, 428–457 (1905). [1905MNRAS..65..428D]

[9] Nachtrichten der Königliche Gesellschaft der Wissenschaften zu Göttingen, Mathematisch-physikalische Klasse (1907).

[10] Proceedings of the Royal Society of Edinburgh, 28, 231–238 (1908), summary in The Observatory, 31, 200–204 (1908). [1908Obs....31..200D]

[11] Astrophysical Journal, 14, 297–312 (1901). [1901ApJ....14..297N]

[12] Cambridge University Press, ISBN 0 521 35363 7 (1993).

[13] Transactions of the International Union for Cooperation in Solar Research, 1, 5–10 (1906). [1906TIUCS...1.....5.]

[14] bookhistory.harvard.edu/takenote/sites/default/files/attachments/ Edward\Pickering.jpg,reproducedwithpermission.

[15] Hoitsema Brothers, Groningen (1923).

[16] Bulletin of the Astronomical Institutes of the Netherlands, 6, 75–81 (1930). [1930BAN.....6...75V]

[17] B.T. Lynds in: Astronomical Society of the Pacific Leaflets, Vol. 9, No 412, 89–96 (1963), [1963ASPL....9...89L], A. Blaauw & T. Elvius, T. in the compendium *Stars and Stellar Systems*, Volume V: *Galactic Structure*. ed. A. Blaauw and M. Schmidt, University of Chicago Press, 589–597 (1965).

[18] See dasch.rc.fas.harvard.edu/telescopes.php, reproduced with permission.

[19] From *Sterrenkijken bekeken, op. cit.*

[20] From A. Schwassmann & P.J. van Rhijn (1935), [1935bsdn.book.....S.]

[21] Bd.1: Eichfeld 1 bis 19, Deklination +90 deg., +75 deg., +60 deg. (1935); Bd.2: Eichfeld 20 bis 43, Deklination +45 deg. (1938); Bd.3: Eichfeld 44 bis 67, Deklination +30 deg. (1947); Bd.4: Eichfeld 68 bis 91, Deklination +15 deg. (1951); Bd.5: Eichfeld 92 bis 115, Deklination 0 deg. (1953). Published by the Hamburger Sternwarte in Bergedorf.

[22] Bd.I. Pol und Zone -75 deg. (1929); Bd. II. Zone -60deg. (1930); Bd. III. Zone -45deg. (1931); Bd. IV. Zone -30deg. (1935); Bd. V. Zone -15deg. (1938); published by the Potsdam Astrophysikalische Observatorium.

[23] *Faint standards of photographic magnitude for the Selected Areas*, Publications of the Astronomical Society of the Pacific, 26, 51–52 (1914). [1914PASP...26...51S]

[24] *Studies based on the colors and magnitudes in stellar clusters*, Astrophysical Journal, 45, 118–141 (1917). [1917ApJ....45..118S]
[25] *Stellar populations and the distance scale: the Baade-Thackeray correspondence*, Journal for the History of Astronomy, 31, 29–36 (2000). [2000JHA....31...29F]
[26] Publications of the Kapteyn Astronomical Laboratory at Groningen, 38, 1–77 (1925). [1925PGro...38D...1V]
[27] In *Sterrenkijken bekeken*, A. Blaauw, J.A. de Boer, E. Dekker & J. Schuller tot Peursum-Meijer, Universiteits Museum Groningen (1983).
[28] Photograph at 194.171.109.12/cat_toon_foto.php?registratiecode=VFOTNL02 9009&exact=JA&cat=VFOT&zoekterm=Kapteyn, reproduced with permission.

Chapter 11. Extinction

[1] Annals of the Astronomical Observatory of Harvard College, 48, 149–185 (1903). [1903AnHar..48..149P].
[2] Popular Astronomy, 1, 224–226 (1893). [1893PA......1..224V]
[3] M.A.C. Perryman, A.G.A. Brown, Y. Lebreton, A. Gomez, C. Turon, G. Cayrel de Strobel, J.C. Mermilliod, N. Robichon, J. Kovalevsky & F. Crifo, Astronomy & Astrophysics, 331, 81–120 (1998). [1998A&A...331...81P]
[4] See www.spacetelescope.org/images/heic0515c/.
[5] See hoogleraren.ub.rug.nl/?page=showPerson&type=hoogleraar&hoogleraar_id=1876&lang=en\?iframe=true.
[6] Algemeen Nederlands Tijdschrift voor Wijsbegeerte en Psychologie, 53, 113–114 (1961).
[7] Astronomical Journal, 26, 31–36 (1908). [1908AJ.....26...31B]
[8] Astronomical Journal, 24, 43–49 (1904). [1904AJ.....24...43C]
[9] *Darkness at Night: A riddle of the Universe*, Harvard University Press (1987), ISBN 978-0-674-19270-6.
[10] Page 185, bottom in [1903AnHar..48..149P].
[11] T. Credner & S. Kohle, www.allthesky.com. With permission.
[12] Press Release ESO1103, Jan. 2011; www.eso.org/public/news/eso1103/.
[13] I quote from the 1893 edition, §§788 and 792. Available at archive.org/details/outlinesofastron00hersuoft.
[14] See www.spacetelescope.org/projects/fits_liberator/fitsimages/john_corban_4/.
[15] ESO Press Release 9934, Febr. 1999; www.eso.org/public/news/eso9934/.
[16] Both *op. cit.*, respectively www.archive.org/details/starsstudyofuniv00newciala and archive.org/details/handwrterbuchd31vale.
[17] Popular Astronomy, 14, 475–488 (1906). [1906PA......14..475P]
[18] See hubblesite.org/gallery/album/galaxy/pr2005012a/ and /pr2006010a/.
[19] Astrophysical Journal, 12, 136–158 (1900). [1900ApJ....12..136E]
[20] English translations at www.dwc.knaw.nl/toegangen/digital-library-knaw/?pagetype=publDetail&pId=PU00013940 and pId=PU00013941.
[21] Popular Astronomy, 14, 579–583 (1906). [1906PA......14..579B]
[22] Astrophysical Journal, 37, 105–118 (1913). [1913ApJ....37..105E]

[23] Veröffentlichungen der Königlichen Sternwarte zu Bonn, no. 1, (1895), 1–97. Listed in ADS [adsabs.harvard.edu/abs/1895VeBon...1....1M] but no electronic copy available to my knowledge.
[24] Annals of Harvard College Observatory, 28, 1–128 (1897). [1897AnHar..28....1M]
[25] *A Plot of UBV Diagram* by B. Nicolet, Astronomy & Astrophysics Supplement, 42, 283–283 (1980). [1980A&AS...42..283N]
[26] Annals of the Astronomical Observatory of Harvard College, 50, 1–252 (1908). [1908AnHar..50....1]
[27] Monthly Notices of the Royal Astronomical Society, 69, 61–72 (1908), [1908MNRAS..69...61T] and Annals of Harvard College Observatory, 59, 157–186. [1912AnHar..59..157K]
[28] Astrophysical Journal, 36, 169–227 (1912). [1912ApJ....36..169P.]
[29] The British Journal for the History of Science, 30, 337–355 (1997). [dash.harvard.edu/bitstream/handle/1/3716614/Voskuhl_Recreating.pdf?sequence=2]
[30] Proceedings of the National Academy of Sciences of the United States of America, 2, 12–15 (1916). [1916PNAS....2...12S]
[31] Proceedings of the Royal Society of London. Series A, Containing Papers of a Mathematical and Physical Character, 111, 424–456 (1926). [1926RSPSA.111..424E, see also rspa.royalsocietypublishing.org/content/111/759/424.full.pdf]
[32] Bulletin of the Astronomical Institutes of the Netherlands, 4, 123–128 (1928). [1928BAN.....4..123V]
[33] See www.rigb.org/registrationControl?action=home.
[34] Publications of the Astronomical Society of the Pacific, 42, 214–227 (1930). [1930PASP...42..214T]
[35] www.beeldbankgroningen.nl/beeldbank/;identificatienummer:NL-GnGRA_1986_372. With permission from Aviodrome Lelystad Airport.
[36] www.beeldbankgroningen.nl/beeldbank/; identificatienummer: NL-GnGRA_1785_6866. Foto P. Kramer, RHC Groninger Archieven (1785–6866).
[37] www.groningerarchieven.nl, reproduced on groningertram.com/de-groninger-tram-vroeger/).
[38] www.beeldbankgroningen.nl/beeldbank/; identificatienummer: NL-GnGRA_1986_2921), Foto collectie RHC Groninger Archieven (1986–2921).

Chapter 12. Students

[1] Also on Jet Katgert-Merkelijn's homepage: home.strw.leidenuniv./nl~merke/lyn/1926_Groningen.html.
[2] *De Bouw der Sterrenstelsels* (1931), Wolters, Groningen.
[3] Annual Reviews of Astronomy & Astrophysics, 19, 1–5 (1981). [1981ARA&A..19....1O]
[4] *Universitas Groningana MCMXIV-MCMLXIV*, Groningen (1964), 176–183.

[5] The home page of the Mathematics Genealogy Project is at genealogy.math.ndsu.nodak.edu/index.php. Kapteyn is #112114, Haga #114664 and Schoute #49650. The completeness of the MGP is far from guaranteed.
[6] Astronomische Nachrichten, 16, 43–50 (1838). [1838AN.....16...43A]
[7] Kindly made available by Dr. Matsuoka, matsuoka@astro.princeton.edu.
[8] Lund panorama reproduced with permission, www.astro.lu.se/Resources/Vintergatan/; van der Kruit (1986) *op. cit.*.
[9] Astronomy & Astrophysics, 157, 230–244 (1986). [1986A&A...157..230V]
[10] Astrophysical Journal, 14, 297–312 (1901). [1901ApJ....14..297N]
[11] *Cosmic optical background: The view from Pioneer 10/11* by Y. Matsuoka, N. Ienaka, K. Kawara & S. Oyabu, Astrophysical Journal, 736, article id. 119, 14 pp. (2011). Picture kindly made available by Dr. Matsuoka.
[12] Digital Special Collections. socrates. leidenuniv.nl/R?func=search-simple&local_base=gen01-disc.
[13] University of Chicago Press; 1st edition (1997), ISBN-13: 978-0226468860.
[14] *Information handling in astronomy – Historical vistas*, Ed. A. Heck, Astrophysics and Space Science Library, 285, 267–273, Dordrecht: Kluwer (2003). [2003ASSL..285..267J]

Chapter 13. Mount Wilson

[1] See en.wikipedia.org/wiki/Mount_Wilson_Toll_Road.
[2] See hdl.huntington.org/cdm/singleitem/collection/p15150coll2/id/1868/rec/26.
[3] Respectively: The MIT Press (1972), ISBN-10 0262230496 and American Institute of Physics (1994), ISBN-10 1563962497.
[4] Respectively W.S. Adams, Astrophysical Journal, 87, 369–388 (1938) and H.D. Babcock, Publications of the Astronomical Society of the Pacific, 50, 156–165 (1938). [1938ApJ....87..369A and 1938PASP...50..156B]
[5] Proceedings of the American Academy of Arts and Sciences, 57, 478–482 (1922).
[6] See www.dudleyobservatory.org/archives/archives_astrojrnl.htm.
[7] Astronomical Journal, 25, 169–175 (1907). [1907AJ.....25..169C]
[8] See hdl.huntington.org/cdm/singleitem/collection/p15150coll2/id/500/rec/87.
[9] See hdl.huntington.org/cdm/singleitem/collection/p15150coll2/id/1143/rec/12.
[10] From www.astro.caltech.edu/palomar/images/w100.jpg, reproduced with permission.
[11] Carnegie Institution of Washington *Year Book*, vol. 6 (1907), page 136–137.
[12] See The Observatory, 30, 243–245 (1907). [1907Obs....30..243]
[13] See www.measuringworth.com/datasets/exchangeglobal/result.php.
[14] See hdl.huntington.org/cdm/singleitem/collection/p15150coll2/id/399/rec/69.
[15] See hdl.huntington.org/cdm/singleitem/collection/p15150coll2/id/431/rec/91.
[16] www.flora-and-sam.com/pages/ImmigrationShips.htm#rotterdam.
[17] See en.wikipedia.org/wiki/Transcontinental_Express
[18] Taken from wikimapia.org/9752676/Kapteyn-Cottage.

[19] See hdl.huntington.org/cdm/singleitem/collection/p15150coll2/id/1124/rec/20.
[20] *The Fourth Conference of the International Union for Co-operation in Solar Research*, H.C. Wilson, Publications of the Astronomical Society of the Pacific, 22, 169–179 (1910); *The Mount Wilson Conference of the Solar Union*, C.A. Chant, C. A. Journal of the Royal Astronomical Society of Canada, 4, 356–372 (1910). [1910PASP...22..169W and 1910JRASC...4..356C]
[21] See [1911TIUCS...3....1.] for opening pages; use ADS to find further parts.
[22] According to the Photographic Archive of the University of Chicago Library, see photoarchive.lib.uchicago.edu/db.xqy?one=apf6--04419-017.xml.
[23] See hdl.huntington.org/cdm/singleitem/collection/p15150coll2/id/1140/rec/6, identification of McBrids photoarchive.lib.uchicago.edu/db.xqy?one=apf6-04419-017.xml.
[24] See hdl.huntington.org/cdm/singleitem/collection/p15150coll2/id/855/rec/3.
[25] Picture and permission provided by Susanne Elisabeth Nørskov, AU Library, Fysik & Steno, Institut for Fysik og Astronomi, Aarhus Universitet, Denmark.
[26] See www.phys-astro.sonoma.edu/brucemedalists.
[27] Astronomische Nachrichten, 292, 142 (1970). [1970AN....292..142S]

Chapter 14. Tides, Statistics and the Art of Discovery

[1] From The Chemical News and Journal of Physical Science (22 Sep 1916).
[2] See www.rijkswaterstaat.nl/geotool/astronomisch_getij.aspx.
[3] International Statistical Review 77, 96–117 (2009).
[4] See hoogleraren.ub.rug.nl/?page=showPerson&type=hoogleraar&hoogleraar_id=108&lang=enand&hoogleraar_id=125&lang=en.
[5] E. Limpert, W.A. Stahel & M. Abbt, BioScience 51, 341–352 (2001).
[6] J. Aitchison & J.A.C. Brown, Cambridge University Press (1966).
[7] Annals of the Astronomical Observatory of Harvard College, 48, 149–185 (1903). [1903AnHar..48..149P]
[8] Philosophical Transactions of the Royal Society, London, A186, 343–414 (1895).
[9] Biometrika, 4, 169–212 (1905).
[10] Journal of the Royal Statistical Society, 61, 670–700 (1898).
[11] Biometrika, 5, 168–171 (1906).
[12] Statistica Neerlandica, 255–25813 (2008); onlinelibrary.wiley.com/doiu/10.1111/j.1467--9574.1959.tb00870.x/pdf.
[13] See www.vanuven.nl, homepage of the Van Uven Stichting.
[14] Gerwina, 15, 195–207 (1992).
[15] Extrait du T. IX des Mémoires présentés par divers Savants à l'Académie Royale des Sciences de l'Institut de France, 9, 255–332. (1846). A version from 1884 can be downloaded using download.digitale-sammlungen.de/BOOKS/pdf_download.pl?id=bsb10053322.
[16] Biometrika, 13, 25–45 (1920).
[17] In 3 volumes, published in 4 parts, Cambridge University Press.

[18] www.beeldbankgroningen.nl/beeldbank/; identificatienummer: NL-GnGRA_1986_446. Foto collectie RHC Groninger Archieven (1986–446).
[19] www.beeldbankgroningen.nl/beeldbank/; identificatienummer: NL-GnGRA_1785_5645. Foto collectie RHC Groninger Archieven (1785–5645).

Chapter 15. First Attempt

[1] Astrophysical Journal, 33, 64 (1911). [1911ApJ....33...64A]
[2] Astrophysical Journal, 35, 163 (1912). [1912ApJ....35..163A]
[3] See cdm16003.contentdm.oclc.org/cdm/singleitem/collection/p15150coll2/id/1136/rec/3
[4] Astrophysical Journal, 42, 172–194 (1915). [1915ApJ....42..172A]
[5] Lick Observatory Bulletin 229 (1913). [1913LicOB...7..113C]
[6] NASA, ESA, and the Hubble Heritage Team (STScI/AURA), heritage.stsci.edu/2013/13/index.html.
[7] See www.exploratorium.edu/eclipse/1901.html for a brief report.
[8] Recherches Astronomiques de l'Observatoire d'Utrecht, 7, iii-P4.1 (1917). [1917RAOU....7....1N]
[9] See Harvard College Observatory Circular, 185, 1 (1914). [1914HarCi.185....1P]
[10] E.C. Pickering: *Revised Harvard photometry : a catalogue of the positions, photometric magnitudes and spectra of 9110 stars, mainly of the magnitude 6.50, and brighter observed with the 2 and 4 inch meridian photometers*, Annals of the Astronomical Observatory of Harvard College, 50, i-194 (1908). [1908AnHar..50....1P]
[11] Verslag van den staat der Sterrenwacht te Leiden en van de aldaar volbrachte werkzaamheden (1910–1912), vol. 49. [1913VeLdn..49....1V]
[12] *Some peculiarities in the motions of the stars*, Lick Observatory bulletin 196 (1911). [1911LicOB...6..125C]
[13] Washington, D.C.: Carnegie Institution (1910). [1910pgcs.book.....B], available at archive.org/details/preliminarygener00carnrich.
[14] Proceedings of the National Academy of Sciences, 1, 417 (1915). [1915PNAS....1..417A]
[15] ASP Conference Proceedings, 471, 205–215 (2013). [2013ASPC..471..205G]
[16] Astronomische Nachrichten, 179, 373–380 (1909). [1909AN....179..373H]
[17] *Relations between the spectra and other characteristics of the stars*, Popular Astronomy, 22, 275–294 and 331–351 (1914). [1914PA.....22..275R and 22..331R]
[18] *The radial velocities of one hundred stars with measured parallaxes*, Astrophysical Journal, 39, 341–349 (1914). [1914ApJ....39..341A]
[19] *Some spectral criteria for the determination of absolute stellar magnitudes* Astrophysical Journal, 40, 385–398 (1914). [1914ApJ....40..385A]
[20] *I. A quantitative method of classifying stellar spectra*, Proceedings of the National Academy of Sciences, 2, 143–147; *II. A spectroscopic method of determining stellar parallaxes*, ibid., 147–152; *III. application of a spectroscopic method of determining stellar distances to stars of measured parallax, ibid.*,

References

152–157; *IV. Spectroscopic evidence for the existence of two classes of M type stars*, ibid., 157–163 (1916). [1916PNAS....2..143A], [2..147A], [2..152A] and [2..157A]

[21] Publications of the Astronomical Society of the Pacific, 28, 61–69 (1916). [1916PASP...28...61A]

[22] See hdl.huntington.org/cdm/singleitem/collection/p15150coll2/id/1874/rec/3.

[23] Astrophysical Journal, 45, 293–305 (1917). [1917ApJ....45..293A]

[24] Astrophysical Journal, 47, 7–37 (1918). [1918ApJ....47....7S]

[25] Astrophysical Journal, 59, 228–251 (1924) and *ibid.*, 61, 363–388 (1925). [1924ApJ....59..228S and 1925ApJ....61..363S]

[26] Astrophysical Journal, 46, 313–339 (1917). [1917ApJ....46..313A]

[27] Contributions from the Mount Wilson Observatory / Carnegie Institution of Washington, No 170, 1–12 (1919). [1919CMWCI.170....1S]

[28] Publications of the Kapteyn Astronomical Laboratory Groningen, 34, 1–80 (1923) and 37, 1–31 (1925). [1923PGro...34....1V] and [1925PGro...37....1V]

[29] Citations resp. from pp. 226, 227 of the *Legacy* and pp. 223, 326 and 331 in *DeVorkin's Russell biography*.

[30] Bulletin of the Astronomical Institutes of the Netherlands, 3, 35–41 (1925). [1925BAN.....3...35T]

[31] See rasathane.en.ankara.edu.tr/.

[32] Ankara, ISBN 978-605-87419-0-4; www.bitav.org.tr/TR/Genel/dg.ashx?DIL=1&BELGEANAH=176&DOSYAISIM=EAK_Life_Story_Book_CGO3.pdf and *Egbert Adriaan Kreiken: pionier in de Turkse sterrenkunde*, Zenit, januari 2013, 14–18.

[33] *The universe at faint magnitudes. I - Models for the galaxy and the predicted star counts*, Astrophysical Journal Supplement Series, 44, 73–110 (1980). [1980ApJS...44...73B]

[34] Astronomy and Astrophysics, 157, 230–244 (1986), section 5.5. [1986A&A...157..230V]

[35] Publications of the Kapteyn Astronomical Laboratory Groningen, 43, 1–104 (1929). [1929PGro...43....1V]

[36] Sitzungsberichte der Mathematisch-Physikalischen Classe der Königlich-Baierische Akademie der Wissenschaften zu Muünchen, Jahrgang 1920, 87–114.

[37] Respectively Astronomische Nachrichten, 213, 45–48 (1921) and 214, 145–150 (1921). [1921AN....213...45V] and [1921AN....213...45]

[38] Astrophysical Journal, 160, 811–830 (1970). [1970ApJ...160..811F]; Astronomy and Astrophysics, 95, 105–115. (1981). [1981A&A....95..105V]

[39] Astrophysical Journal, 303, 556–572. (1986). [1986ApJ...303..556V]

[40] Bulletin of the Astronomical Institutes of the Netherlands, 6, 249–287 (1932). [1932BAN.....6..249O]

[41] Annals of the New York Academy of Sciences, 198, 255–266 (1972). [1972NYASA.198..255O]

[42] *The Milky Way Galaxy*, IAU Symposium 106, H. van Woerden et al., eds., Dordrecht: Kluwer, 1985)

[43] *Beyond the Galaxy: the development of extragalactic astronomy 1885–1965, Part 1* Journal for the History of Astronomy 39, 91–119 (2008). [2008JHA....39...91]

[44] *Absorption of light in the Galactic System* by Robert J. Trumpler, Robert J., Publications of the Astronomical Society of the Pacific, 42, 214–227 (1930). [1930PASP...42..214T]

[45] See *Observational evidence confirming Lindblad's hypothesis of a rotation of the Galactic System* by J.H. Oort, Bulletin of the Astronomical Institutes of the Netherlands, 3, 275–282 (1927) and *Dynamics of the Galactic system in the vicinity of the Sun, ibid.*, 4, 269–284 (1928). [1927BAN.....3..275O] and [1928BAN.....4..269O]

Chapter 16. Finale

[1] See www.orden-pourlemerite.de.

[2] Publications of the Astronomical Society of the Pacific, 30, 331–335 (1918). [1918PASP...30..331M]

[3] A full scan of the printed version can be found on my Kapteyn homepage at www.astro.rug.nl/JCKapteyn/Statement_USAcad.pdf.

[4] *De stem van de wetenschap; Geschiedenis van de Koninklijke Nederlandse Academie van Wetenschappen*, deel 2, 1914–1918, Amsterdam: Bert Bakker, ISBN 978-90-351-360104 (2011).

[5] *Als bij toverslag, De reorganisatie van de Leidse Sterrewacht, 1918–1924* and *Hij kan toch moeilijk de sterren in de war schoppen. De afwijzing van Pannekoek als adjunct-directeur van de Leidse Sterrewacht*. Respectively BMGN/Low Countries Historical Review 120, 207–225 (2005); and Gewina 27, 1–13 (2004).

[6] By Anton Pannekoek, B.A. Sijes, E.P.J. van den Heuvel, J.M. Welcker & J.R. van der Leeuw, Amsterdam, van Gennep, (1982). Published posthumously.

[7] Publisher was Enschedé en Zonen, Haarlem (1933).

[8] Astronomische Nachrichten, 196, 201–210 (1913). [1913AN....196..201H]

[9] Astrophysical Journal, 49, 311–336 (1919). [1919ApJ....49..311S]

[10] *Preliminary evidence of internal motion in the spiral nebula Messier 101*, Astrophysical Journal, 44, 210–228 (1916). [1916ApJ....44..210V]

[11] Annals of the New York Academy of Sciences, 198, 255–266 (1972). [1972NYASA.198..255O]

[12] Journal for the History of Astronomy, 4, 46–56 and 73–98 (1973). [1973JHA.....4...46B] and [1973JHA.....4...73B]

[13] www.beeldbankgroningen.nl/beeldbank/; identificatienummer: NL-GnGRA_1986_1694, Foto M. Th. Koop, collectie RHC Groninger Archieven (1986–1694).

[14] www.beeldbankgroningen.nl/beeldbank/; identificatienummer: NL-GnGRA_1785_7624, Foto P. Kramer, collectie RHC Groninger Archieven (1785–7624).

[15] www.beeldbankgroningen.nl/beeldbank/; identificatienummer: NL-GnGRA_1986_1825 and NL-GnGRA_1986_1346, Foto collectie RHC Groninger Archieven (1986–1825 and 1986–1346).1

[16] Levensverwachting; geslacht en leeftijd, vanaf 1861; statline.cbs.nl/StatWeb/selection/?DM=SLNL&PA=37450&VW=T.
[17] Public Health Reports 84, 661–664 (1969); www.ncbi.nlm.nih.gov/pmc/articles/PMC2031509/.
[18] The Messenger, 67, 62–63 (1992), [1992Msngr..67...62H]
[19] See people.physics.tamu.edu/krisciunas/astrs.html.
[20] *Tijd en toekomst* by A.H.P. Luijben & G.J. Kommer, Rijks Instituut voor Volksgezoindheid en Milieu Rapport 270061008 (2010); see www.rivm.nl/bibliotheek/rapporten/270061008.html.
[21] Kluwer (1994), ISBN 0-7923-2979-1.
[22] Edited by A. Fowler, London, Imperial College Bookstall.
[23] See *Report on the Organization of the International Astronomical Union* by W.W. Campbell and Joel Stebbins, in: 'Proceedings of the National Academy of Sciences (USA)', 6, 349–396 (1920). [www.pnas.org/content/6/6/349.full.pdf+html]
[24] *Royal Astronomical Society centenary celebrations, 1922 May 29 - June 3*, The Observatory, 45, 201–212 (1922). [1922Obs....45..201.]
[25] See www.uvm.edu/~classics/faculty/bach/.
[26] *Dijksterhuis: een biografie*, Bakker, Amsterdam (1996), ISBN 90-351-1694-1.
[27] Weekblad voor Gymnasiaal en Middelbaar Onderwijs, 18-e jaargang, N0. 42, 1909–1910.

Appendix A. Publications by and about J.C. Kapteyn, his honors and academic genealogy

[1] See adsabs.harvard.edu/abstract_service.html, or any other mirror site.
[2] See www.dwc.knaw.nl/toegangen/digital-library-knaw/?pagetype=publist&search_authorPE00001201.
[3] At archive.org/details/copernicus03dubluoft.
[4] My Kapteyn Webpage can be found at www.astro.rug.nl/JCKapteyn/publications.
[5] See www.museumboerhaave.nl/Adlib/search/simple, search for 'Kapteyn'.
[6] The NASA Near-Earth Object Project is at neo.jpl.nasa.gov/orbits/.
[7] From en.wikipedia.org/wiki/Kapteyn_(crater). 'This file is made available under the Creative Commons CC0 1.0 Universal Public Domain Dedication.'
[8] See www.rug.nl/corporate/universiteit/luchtfoto.
[9] See www.ing.iac.es/PR/. Reproduced with permission.
[10] The Websites of the Kapteyn Astronomical Institute are www.rug.nl/sterrenkunde and www.astro.rug.nl.
[11] Vistas in Astronomy, 28, 483–503 (1985). [1985VA.....28..483L]; see also www.ing.iac.es/PR/jkt_info/.
[12] See www.lorentz.leidenuniv.nl/history/explosion/HKO_tree.html.
[13] See www-int.stsci.edu/ dejong/stamboom.html.
[14] See www.hum.leiden.edu/icd/organisation/members/horstmanshoffhfj.html.
[15] See genealogy.math.ndsu.nodak.edu.; Kapteyn is listed as number 112114.
[16] Verloren (1997), ISBN 90-6550-543-1.

[17] See Rienhard Klette; www.tcs.auckland.ac.nz/~rklette/acad_ancestors.html.
[18] See www.math.uni-leipzig.de/preprint/2007/p1-2007.pdf (p.4).
[19] The British Journal for the History of Science, 2, 1–24 (1964). Quote from p.7.
[20] Abelard-Schumann (1959), ISBN 0-484-67605-6.

Appendix B. Henriette Hertzsprung-Kapteyn's biography of J.C. Kapteyn

[1] See www.dbnl.org/tekst/hert042jcka01_01/, where versions in .jpg and .txt formats are provided.
[2] See for example as www.genealogieonline.nl/genealogie-baert-cornelis-kalshoven/I225.php, www.middel.org/canon-van-groningen/kapteyn and www.lamartin.com/genealogy/kalshoven.htm.
[3] The Website is www.allegroningers.nl.
[4] Groninger Historische Reeks (1998).
[5] See www.allegroningers.nl, Geboorteregister Groningen 1881, Aktenummer 1429.
[6] Berlin, Springer (1994) ISBN 3-540-56788-6.
[7] See www.allegroningers.nl, Huwelijksregister Groningen 1913, Aktenummer 210.
[8] *My first 72 years of astronomical research. Reminiscences of an astronomical curmudgeon, revealing the presence of human nature in science.* Privately published by Willem J. Luyten, 1940 East River Road, Minneapolis, MN 55414, USA. 20+203 pp. (1987). [1987fsya.book.....L]
[9] See www.online-begraafplaatsen.nl/zerken.asp?command=showgraf&grafid=200874.
[10] The *Legacy*, pages 38–40.
[11] Private communication in this paragraph from Jan Willem Noordenbos, great-grandson of Kapteyn, who remembered 'Aunt Rigel' well.
[12] Space Science Reviews, 64, 1–92 (1993) (with a preface and introduction on pp. x-xix)
[13] Kluwer (1993), ISBN 978-07-923-2603-8.
[14] The Observatory, 115, 284–5 (1995). [1995Obs...115..284J]
[15] See www.astro.rug.nl/JCKapteyn; the URL of the English translation is www.astro.rug.nl/JCKapteyn/HHKbiog.pdf.

Appendix C. Cornelius Easton: Personal memories of J.C. Kapteyn

[1] The URL is www.astro.rug.nl/JCKapteyn/Easton.pdf.
[2] See www.readbookonline.net/readOnLine/20049/.
[3] Astrophysical Journal, 37, 105–118 (1913). [1913ApJ....37..105E].
[4] See hdl.huntington.org/cdm/singleitem/collection/p15150coll2/id/569/rec/3.

Index

Symbols
α Centauri, 135, 139, 354
δ Cephei, 590, 592
ω Centauri, 245, 246
100-inch telescope (Mount Wilson), 458, 459
1882, Great comet of, 154, 157, 158
60-inch telescope (Mount Wilson), 375, 376, 378, 455, 457, 458, 460, 463, 465, 468, 482, 488, 548
61 Cygni, 139, 252–254, 260

A
Abbot, Charles Greeley, 468, 485
aberration of light, 66, 67, 74, 128, 236
Absolute Magnitude, 46, 327
   definition, 326–328, 558, 605
absorption of light, *see* extinction, interstellar
Academic Statute, 88
Académie des Sciences, 185, 630, 631
Academy Building, 62, 86, 92, 98, 518–520
actinometer, 414
Adams, Walter Sydney, 368, 376, 454, 468, 484, 485, 487, 495, 496, 533, 542–555, 557, 599, 622, 645
Aethra (minor planet), 117
airglow, 438
Airy, George Biddell, 48, 53, 81, 196, 234, 236, 321, 323, 324, 436
Albany Observatory, 471
Aldebaran, 384, 385
Alexander von Humbold Stiftug, 200
Algol (β Persei), 581
Allebé, Gerardus Arnoldus Nicolaus, 149, 150
Allegheny Observatory, 471, 487, 491
Almagest, 74

altitude, 64, 77, 126
American Philosophical Society, 631
American Society, 631
Anderson, Thomas David, 311
Andrée Wiltens, Albert John, 32
Andrée Wiltens, Elisabeth Henriette, 108
Andrée Wiltens, Henry Maximiliaan, 32, 109
Andrée Wiltens, Henry William, 32, 108
Andrée Wiltens, Jacob Willem Gerard Hendrik, 32
Andrée Wiltens, Maximiliaan Leonard, 32
Andromeda Nebula, 208, 294, 404
Ångström, Anders Jonas, 530
Ankara Observatory, 558
annual parallax, 138
anomalistic month, 124
Antapex, 291
Apex, *see* Solar Apex
apparent magnitude, 46
Arago, François Jean Dominique, 46
Argelander, Friedrich Wilhelm August, 74, 75, 155, 163, 231, 321, 351, 436, 523, 570
Arminius, Jacobus, 90
Ascension (island), 81
Astrographic Catalogue, 194, 196, 249, 397, 489
Astronomical Journal, 118, 478
Astronomical Laboratory Kapteyn, 589
Astronomical Society of the Atlantic, 353, 362, 587
Astronomical Society of the Pacific, 203, 493, 630
Astronomische Gesellschaft, 68, 75, 575, 576
Astronomische Gesellschaft Katalog (AGK), 75, 78, 156, 231, 523

Astronomische Nachrichten, 118
Astronomisches Rechen-Institut (Berlin), 62
Asymmetric Drift, 552, 554
Ausfeld, Hermann, 73
Auwers, *see* von Auwers
Auwers-Bradley stars, 74, 295, 321, 322, 326, 327, 342, 343
azimuth, 64, 77, 126

B

Baade, Wilhelm Heinrich Walter, 378, 380, 592
Baart de la Faille, Jacob, 90, 91, 277
Babcock, Harold Delos, 454, 468, 488
Bach, Johann Sebastian, 608
Backlund, Jöns Oskar, 336, 345, 353
Bacon, Francis, 425
Bacon, Roger, 27
Baillaud, Edouard-Benjamin, 604, 606
Baillaud, Jules, 576
Bakhuyzen, *see* van de Sande Bakhuyzen
Barnard 68 (dark cloud), 402, 403
Barnard's Loop, 399, 401
Barnard's Star, 238
Barnard, Edward Emerson, 238, 407, 475
Barneveld, 1–18
Barnouw, Adriaan Jacob, 646
Bartsch, Jacob, 638
Batavia (Jakarta), 59
Batavian Society for Experimental Philosophy (Rotterdam), 168, 631
Battista, Orlando Aloysius, 1
Bauschinger, Julius, 62
Becker, Ernst Emil Hugo, 96
Becker, Wilhelm, 377
Benno, 5, 9–12, 14, 21, 22, 633, 635
  observatory, 22
Bergedorfer Spektral-Durchmusterung, 377, 379, 381
Bergedorfer Sternwarte, *see* Hamburg Observatory
Berlage, Hendrik Petrus, 304
Berlin (Babelsberg) Observatory, 74, 95, 96, 295, 376
Berlin Academy, 187, 190
Bernoulli, Daniel, 28, 40, 637
Bernoulli, Jacob, 637
Bernoulli, Jacob (II), 39, 40
Bernoulli, Johann, 28, 91, 637
Berra, Yogi (Lawrence Peter), 27, 661
Bessel, Friedrich Wilhelm, 74, 139, 216, 295
Bethe, Hans Albrecht, 270
Bible Belt, 3
Big Dipper, 46

Bijlsma, Dirk Klazes, 100
binary star, 67, 76
Blaauw, Adriaan, v, ix, x, xii, xiii, xvii, 204, 369, 371, 372, 377, 379, 380, 420, 579, 580, 604, 621, 628, 629, 643, 647–648, 664
Black Hole, 209
blackboard globes, 420, 542, 543
Boer wars, 274
Boerhaave, Herman, 28
Boissevain, Ursul Philip, 56, 266, 423, 475, 577, 630
Bok globule, 402
Bok, Bartholomeus Jan, 377, 628
Bolland, Gerardus Johannes Petrus Josephus, 425
Boltzmann, Ludwig Eduard, 366
Bonn Observatory, 75, 96, 155, 231, 362, 367, 376, 387, 408, 487
Bonner Durchmusterung, 74, 75, 155, 163, 286, 523
Bordewijk, Hugo Willem Constantijn, 593, 608
Bos, Pieter Roelf, 281
Boss, Benjamin, 478
Boss, Lewis, 146, 324, 336, 343, 388, 389, 478, 542, 600, 605, 613, 624
Boswell, James, 425
Bouquet de La Grye, Jean Jacques Anatole, 189
Boveri, Theodore Heinrich, 477
Boyden Station, 369
Boyden, Uriah Atherton, 369
Bradley, James, 67, 74, 295, 299, 300, 318, 616, 617
Brandt, Catharina Elisabeth, 6
Bravais, Auguste, 292, 300, 322, 324, 388, 408, 435, 514, 626
Bravais, Louis François, 292
British Association (for the Advancement of Science), 184, 297, 342, 353, 515, 631
British Society, 298
Broerstraat, 419
Brouwer, Hendrik Albertus, 268
Brouwer, Seerp, 91
Brouwer, Simon, 49, 108
Bruce Medal, 203, 493, 630
Brück, Hermann Alexander, 377
Brugmans, Antonius, 29, 637
Bruno, Giordano, 221
Bulletin Astronomique, 118
Burck, William, 81, 108
Bussy-Rabutin, Roger, xiv

Index

Buys Ballot, Christophorus Henricus Didericus, 29–31, 36, 37, 39, 41, 58, 71, 93, 122, 284

C
Calvinism, 90
Campbell, William Wallace, 301, 337, 338, 487, 530, 533, 541, 545, 551, 587
Camper, Petrus, 86, 518, 520
Cannon, Annie Jump, 289
Cape Observatory, *see* Royal Observatory, Cape of Good Hope
Cape Photographic Durchmusterung, 156–204, 223, 246, 612, 615, 616, 618, 631
Capteyn, Paulus, 6
Carbon star, 533
Carnegie Institution, 365, 454, 458
Carnegie, Andrew, 365, 454
Carte du Ciel, 119, 184–196, 223, 225, 226, 230, 233, 247, 249, 250, 255, 269, 270, 375, 397, 446, 489, 523
Cassegrain focus, 490, 531
Cassini, Giovanni Domenico, 221
Cassiopeia, 46
Catalogue of Nebulae and Clusters, 403
Cauchy, Augustin Louis, 39, 40, 476
celestial sphere, 64
Celsius, Anders, 190
Center for High Angular Resolution Astronomy, 473
Centre de Données Astronomiques, 669
Cepheid, 372, 566, 590
Ceres, 38
Chandler, Seth Carlo, 478
charge-coupled device (CCD), 153
Chladni, Ernst Florens Friedrich, 39, 40
Christ, Jesus, 3
Christie, William Henry Mahoney, 187, 190, 192, 193, 196–198, 202, 203
Cincinnati Observatory, 322
Clerke, Agnes Mary, 205, 238, 293
Coalsack, 354, 362, 402
coelostat, 458
Cohn, Fritz, 617
comet, 38
Comstock, George Cary, 146, 336, 339, 362, 391, 394, 395, 397, 479
Conférence Générale des Poids et Mesures, 190
Congress of Vienna, 28
Convention du Métre, 190
Cook, Frederick Albert, 511
Cookson, Bryan, 257, 353
Copeland, Ralph, 118

Copenhagen Observatory, 491, 533, 547, 575, 642
Copernicus (journal), 116, 118, 126, 131
Copernicus, Nicolaus, 74, 91, 263
Córdoba Durchmusterung, 74, 156, 238, 240, 285
Córdoba Observatory, 195, 231, 376
Corneille, Pierre, 12
Cort van der Linden, Pieter Willem Adriaan, 584
Coudé focus, 490
Coudé spectrograph, 530
Couderc, Paul, 314
Courvoisier, Leopold, 137
Cruls, Luís Ferdinand, 189
Crux (constellation), 354
Curtis, Heber Doust, 567, 592
Cushing, Harvey William, 588

D
Daguerre, Louis Jacques Mandé, 154
daguerreotype, 154
Dartmouth College, 531
Darwin, Charles Robert, 30, 110, 231, 510, 570
Daudet, Alphonse, 25
de Boer, Floris, 88, 433
de Boer, Petrus, 88
de Cock, Hendrik, 15
de Genestet, Petrus Augustus, 10
de la Fontaine, Jean, 652
de la Rochefoucauld, François, 659
de Rabertin, Roger, 496
de Rabutin, Roger, xiv
de Sitter, Aernout, ix, 628
de Sitter, Willem, vii–ix, 3, 138, 201, 233, 244, 246, 256–261, 282, 287–288, 305, 309, 353, 372, 379, 385, 387, 434, 445–448, 500, 538, 577, 581–586, 597, 604–606, 616, 618–620, 625–629, 645, 646
de Sitter, Wolter Reinold, 581
de Vos van Steenwijk, Jacob Evert (Baron), 605
de Vries, Hugo, 344, 507
de Vries, Teunis Willem, 170, 173, 216, 260, 371, 582
de Zoute, J.M., 371, 372
declination, 63, 64, 126
Dekking, Willem, 597
Delambre, Jean-Baptiste Joseph, 29
Descartes, René, 524
DeVorkin, David Hyam, x, 490, 547, 555
Dickens, Charles, 110, 652, 653
Dijksterhuis Eduard Jan, 609
disk, 210

Domela Nieuwenhuis, Ferdinand, 204
Donders, Franciscus Cornelis, 30
Donner, Anders Severin, 187, 188, 217, 233, 248–253, 260, 269, 302, 353, 359, 364, 384, 385, 387, 490, 514, 538, 577, 616–620
Doppler, Christian Andreas, 41, 284, 531
Dorpat Observatory, 139
double refraction, 47
Downing, Arthur Matthew Weld, 81
draconic month, 124
Draper Catalogue, 289, 296, 324, 411, 420, 548
Draper, Henry, 154, 289
Draper, John William, 110
Dreyer, John Louis Emil, 118
Dubois, Eugène, 283
Dudley Observatory, 146, 324, 336, 388, 478
Dun Echt Observatory, 118, 131
Dunsink Observatory, 118
Durchmusterung, 74
dynamic range, 154
Dyson, Frank Watson, 342, 357, 359, 368, 587, 588

E
Eastman, George, 154
Eastman, John Robie, 80
Eastman-Kodak Company, 154
Easton, Cornelis, x, 56, 122, 262, 303, 404–407, 625, 626, 651–660
Eberhard, Gustav Edward, 586, 594
eccentric anomaly, 112
Eckermann, Johann Peter, 652
ecliptic, 64
ecliptic latitude, 64
ecliptic longitude, 64
École Polytechnique (Paris), 25, 40
Eddington, Arthur Stanley, 1, 204, 342, 356, 357, 416, 418, 459, 471, 500, 537, 541, 555, 563, 567, 568, 573, 576, 587
Eddington, Arthus Stanley, 343
Edgeworth, Francis Ysidro, 510
Eecen, Adrianus, 69, 73
Ehrenfest, Paul, 425, 526
Eighty Years War, 86
Eijkman, Johan Fredrik, 281
Einstein, Albert, vii, xi, 312, 424, 576, 583, 608, 642
Elkin, William Lewis, 135
elliptical galaxy, 211
Emma, Queen-regent of the Netherlands, 218, 226
Emmius, Ubbo, 86, 90
Enschedé, Willem Adriaan, 71, 92, 93, 96, 106

equatorial mount, 66
equinox, 64, 74
Eratosthenes of Cyrene, 60
Ermerins, Jan Willem, 92
Escher, Maurits Cornelis, 205
European Southern Observatory, 208, 354, 400, 402, 671
European Space Agency, 196, 384, 390
extinction, interstellar, 216, 300, 339, 391–392, 403, 563, 566–568, 590
Eyjafjallajökull volcano, xvii

F
Fahrenheit, Daniel Gabriel, 191
Fath, Edward Arthur, 468, 488
Faye (comet), 117
Faye, Hervé, 189, 192
Feast, Michael, 163, 166, 196, 257, 259, 379
Fechner, Gustav Theodor, 510
Ferwerda, Jacobus Gijsbertus, 628
Feynman, Richard Phillips, 501
Finlay, William Henry, 135, 258, 626
Finnish Academy of Sciences, 631
Firework nebula, 315
Fleming, Williamina Paton Stevens, 237, 289, 484, 615
Flint, Albert Stowell, 146, 300, 325
Flourens, Émile, 192
Flower Observatory, 270
Flower, Reese Wall, 270
Forster, John, 652
Foucault, Jean Bernard Lón, 309
France, Anatole, 351
Franklin-Adams, John, 652
Freeman, Kenneth Charles, 565
Fresnel, Augustin-Jean, 46
Frieseman, H.M.A.A., *see* Kalshoven–Frieseman, Henriëtte Mariëtte Augustine Albertine
Frost, Edwin Brant, 362, 365, 366, 471, 482, 486, 487, 491, 530, 531, 572, 620
Fundamental Catalog, 63, 74, 76

G
GAIA satellite, 390
Galactic disk, 209, 244
Galactic equator, 64
Galactic halo, 209, 210, 244, 246
Galactic latitude, 64
Galactic longitude, 64
galaxy, 210
    classification, 211
    formation, 210
Galilean satellites (Jupiter), 258

Galilei, Galileo, 74, 211, 258
Galle, Johann Gottfried, 38
Galton, Francis, 506, 509, 510, 513, 514
Gauss, Johann Carl Friedrich, 322
Gautama Siddhartha, 111
Geertsema, Carel Coenraad, 226
Georgia State University, 473
Germain, Marie-Sophie, 39, 40
Giant Molecular Clouds, 532
Gill, David, ix, 81, 126, 131–137, 145, 147, 154–204, 223, 224, 231, 236–244, 246–248, 258, 269, 271, 272, 275, 286, 297, 302, 305, 337, 352, 353, 355, 360, 362, 364, 367, 368, 447, 448, 463, 471, 480, 491, 497, 515, 516, 570, 578, 579, 612, 613, 615, 616, 618, 622, 626
Gill, Isobel S., 577
Gingerich, Owen, 327, 547, 555
Glaisher, James Whitbread Lee, 351
Gleuns Jr., Willem, 280
Gleuns, Willem, 70, 280
globular cluster, 209, 244–246, 415, 590, 591
Goethe, see von Goethe
Gold Medal Royal Astronomical Society, 203, 351, 630
Golius, Jacobus, 59
Gomarus, Franciscus, 90
Gotha Observatory, 73
Grinwis, Cornelis Hubertus Carolus, 25, 29–31, 34, 39, 58, 71, 93, 97, 636
Groneman, Florentius Goswin, 146, 277, 278
Groneman, Hendrik Jan Herman, 146
Groningen, city of, 85
Groombridge, Stephen, 259, 357
Grubb, Howard, 194
Gunning, Johannus Hermanus, 10

H

h and χ Persei, 259
Haga, Hermanus, 226, 252, 434
Hagen, Johann Georg, 397
Hale, George Ellery, 5, 118, 302, 336, 344, 364–368, 375, 453–500, 517, 519–527, 531, 562, 577, 581, 587, 599, 602, 608, 642, 645–647
Hale, William Ellery, 458
Hall, Asaph, 146
Halley's comet, 38, 61
Halley, Edmond, 38, 139, 221
Hamburg Observatory, 376, 490
Hamburger, Hartog Jacob, 171, 279, 418
Hamerling, Robert, 26
Hamilton, James, 202
Harmonie, de (concert hall), 102, 426, 427, 608

Harting, Pieter, 30
Harvard College Observatory, 48, 144, 194, 229, 289, 302, 336, 362, 368–370, 376, 377, 454, 465, 471, 478, 483, 495, 548
Harvard Map of the Sky, 539
Harvard Northern Durchmusterung, 495
Harvard University, 270, 630
Harvard-Groningen Durchmusterung, 369–375, 377, 489, 560
Hassler, Ferdinand Rudolph, 345
Hassler, Mary Caroline, 274, 345, 469, 471, 480
Havank (Hendrikus Frederikus van der Kallen), 263
Hearn, Patrick Lafcadio, 311
Hegel, Georg Wilhelm Friedrich, 387
heliometer, 67, 135, 139, 160
helium burning, 207
helium stars, 486
Helmert, Friedrich Robert, 235
Helsingfors Observatory, 187, 217, 249, 251, 252, 300, 386, 388, 538
Hemel & Dampkring, 122, 302, 350, 629, 630, 651
Henderson, Thomas James, 139
Henry Draper Catalogue, 144
Henry, Mathieu-Prosper, 158, 185, 187, 193, 194, 222, 237, 497
Henry, Paul-Pierre, 158, 185, 187, 194, 222, 237, 497
Herrmann, Dieter Bernhard, 598, 601, 643
Herschel, Caroline Lucretia, 211, 231
Herschel, Frederick William, 38, 205, 211, 212, 214, 215, 231, 263, 291, 315, 352, 363, 391, 397, 403, 428, 653
  20-foot telescope, 211, 215
  40-foot telescope, 212, 213, 215
  Construction of the Heavens, 213
  equalisation of starlight, 214
  limiting magnitude, 214
  star counts, 212
Herschel, John Frederick William, 46, 212, 231, 352, 391, 397, 402, 403, 414, 653
Hertzsprung, Ejnar, 5, 149, 206, 233, 283, 379, 490, 491, 493, 494, 545, 547, 549, 551, 558, 566, 583–585, 590, 592, 598, 605, 628, 640–644
Hertzsprung, Rigel, 5, 527, 597, 598, 643, 646
Hertzsprung–Russell diagram, 206, 207, 288, 410, 547, 548, 642
Hertzsprung-Kapteyn, see Kapteyn, Henriette Mariette Augustine Albertine
Heymans, Gerardus, 56, 279, 303, 387, 423, 424, 510, 524, 526, 573, 574, 577, 643

Heynsius, Adriaan, 10
Hiemstra, Broer, 628
Hill, Octavia, 480, 481
Hilversum, 5
Hinks, Arthur Robert, 314, 342, 353, 354, 587
Hins, Coert Hendrik, 628
Hipparchus of Nicaea, 64, 74
Hipparcos, 144, 196, 292, 384, 388, 390
Hoek, Martinus, 37, 39, 62
Hoek, Paulus Peronius Cato, 81
Hohwü, Andreas, 68
Holland Society of Sciences and Humanities, 631
Holleman, Arnold Frederik, 282
Holmes, Oliver Wendell, 273
Holwerda, Allard Othumar, 626
Hondsrug, 85
Hoogere Burgerschool, 62, 71, 108, 146, 264, 277, 557
Hook and Cod Wars, 4
Hooker, John Daggett, 458
Hopmann, Josef, 624
Horrebow, Peder, 128
Horrebow-Talcott method, 128, 130
Horsehead nebula, 401
Hotel De Doelen, 594–596
hour angle, 126
Hubble classification, 211
Hubble, Edwin Powell, 211, 378
Hubrecht, Ambrosius Arnold Willem, 81, 82, 97, 571
Hubrecht, Jan Bastiaan, 570
Hudig, Johanna Clementina, 645
Hudig, Joost, 149, 283, 426, 644, 645
Hudig, Sophia Alida, 645
Huggins, William, 198, 367, 404
Huizinga, Dirk, viii, 171, 223, 226, 233, 281, 282, 418
Huizinga, Johan, vii–ix, 1, 645, 646
Humason, Milton Lasell, 376
Huygens, Christiaan, 38, 47
Hyades, 252, 269, 318, 359, 384, 385, 387, 388, 460, 495, 548, 618
hydrogen burning, 206

I
Ilcken, Ada Christina Mathilde, 14
Imperial Academy of St. Petersburg, 631
Innes, Robert Thorburn Ayton, 203, 238–244, 258, 287, 577
Innes, Robert Thorburn Ayton, 595, 618
Instituto de Astrofísica de Canarias, 633
Interallied Association of Academies, 574
intermediate dispersion spectrograph, 531

International Association of Academies, 574
International Astronomical Union, 574
International Astronomical Union (IAU), 327, 369, 381, 603, 604
International Council for Science, 574
International Council of Scientific Unions, 574
International Research Council, 573, 574, 576
interstellar absorption, *see* extinction, interstellar
interstellar reddening, *see* extinction, interstellar
Isaac Newton Telescope, 633

J
Jackson, Jesse Louis, 1
Jacobs, Aletta Henriëtte, 303
Jacobus Kapteyn Telescope, 633, 635
James Clerk Maxwell, 131
James Ludovic Lindsay, 118, 131
Janssen, Pierre Jules César, 192
Jeans mass, 532
Jeans, James, 98, 532, 567, 608, 625
Johns Hopkins University, 270
Jones, Derek, 649
Jost, Ernst Heinrich Rudolph, 300
Joy, Alfred Harrison, 553
Joyner, Mary Cross, 488, 624
Juliana, Queen of the Netherlands, 276
Julius, Willem Henri, 366, 458, 482, 483
Juno (minor planet), 117
Jupiter, 154

K
König, Johann Samuel, 28
Kaiser, Frederik, 37, 60–68, 225
Kalshoven, Catharina Elisabeth, x, 5, 6, 25, 49–660
Kalshoven, Catherina Elisabeth, 272
Kalshoven, Jacobus Wilhelmus, 49
Kalshoven, Jacobus Wilhelmus (Jacques), 50, 52, 108
Kalshoven, Jacqueline Wilhelmine, 49, 107
Kalshoven, Johan Christaan, 108
Kalshoven, Maria Gabrielle, 50
Kalshoven, Marie Gabrielle, 50, 148
Kalshoven–Frieseman, Henriëtte Mariëtte Augustine Albertine, 6, 49–52
Kam, Nicolaas Mattheus, 62
Kamerlingh Onnes, Heike, 34, 106, 540
Kapteyn
    spelling of name, 641
Kapteyn (crater), 632, 633
Kapteyn Astronomical Institute, ix, 21, 279, 632, 634

Index

Kapteyn Astronomical Laboratory, 589
Kapteyn Cottage, 473–475
Kapteyn Fund, 597
Kapteyn Group, 245, 246
Kapteyn homes
  Eemskanaal, 98, 266, 348, 480
  Grote Markt, 98, 594–596
  Heerestraat, 98, 152, 266, 346, 347, 424, 653
  Hilversum, 593, 602
  Hotel de Doelen, 593, 601
  Oosterhaven, 98, 266, 480
  Oosterstraat, 98, 151, 421
  Ossenmarkt, xiv, 98, 348, 349, 428, 593, 653, 660
  Peperstraat, 97, 98
  Vries, 265, 472, 480
  Winschoterkade, 98, 105, 106, 148, 151
Kapteyn Lounge, 279, 283
Kapteyn Room, ix, x, 21, 122, 141, 230, 279, 420, 422, 612, 614, 621, 629
Kapteyn's Star, 238–246, 615, 631
Kapteyn, Adriaan Pieter Marinus, 16, 17, 107
Kapteyn, Agatha, 16
Kapteyn, Albertina Maria, 20, 104, 107, 144, 200
Kapteyn, Albertus Philippus, 17
Kapteyn, Cornelia Louise Alexandra, 16, 104, 106, 107
Kapteyn, Frederik Willem Hendrik, 641, 643
Kapteyn, Gerrit Jacobus, 5–14, 21, 22, 107, 110, 151, 307, 639, 654, 656
Kapteyn, Gerrit Jacobus (son), 148, 298, 303, 306, 307, 347, 466, 467, 480, 502, 609, 610
Kapteyn, Henriette Mariette Augustine Albertine, viii, 5, 148, 151, 202, 206, 298, 303, 347, 481, 492, 526, 585, 607, 625, 639–641
  first name, 639
  marriage to Hertzsprung, 494, 642–644
  marriage to Hudig, 644, 645
Kapteyn, Huibert Paulus, 16, 22
Kapteyn, Jacoba Cornelia, 148–150, 202, 264, 298, 303, 347, 480, 579, 588, 597, 609, 610, 630, 639
Kapteyn, Jacoba Cornelia (Aunt Ko), 18
Kapteyn, Jacobus Cornelius, 1–660
  70th birthday, 586, 588
  birth certificate, 19
  brothers and sisters, 16, 17, 19
  Bruce Medal, 630
  children, 149
  church wedding, 106
  definition of absolute magnitude, 326, 327, 338, 605
  fatal illness, 603
  Gold Medal Royal Astronomical Society, 351, 630
  height, 56, 652
  inaugural lecture, 33, 99, 100, 102
  marriage, 107
  marriage certificate, 107–109
  military service, 53
  Natuurkundig Genootschap, 276–284
  Orden Pour le Mérite, 569–571, 587
  PhD thesis, 34, 41–49, 58
  portrets, 420, 578–580
  Rector Magnificus, 218, 220
  resigns from Royal Academy, 574
  retirement, 601
  Royal distinctions, 569, 571, 630, 657
  spelling of name, 18, 641
  star map, 21
  statistics, 505–516
  teacher, 429–434
  tides, 501–504
  tree rings, 122–126
  women rights, 303
Kapteyn, Johannes Catharinus, 16
Kapteyn, Machtelina Elisabeth, 16
Kapteyn, Maria Adriana, 16
Kapteyn, Marius, 16
Kapteyn, Nicolaas Pieter, 16, 30, 31, 34, 35
Kapteyn, Paulus Huibert, 16
Kapteyn, Willem, 30, 31, 34, 35, 39–41, 82, 83, 118, 121, 147, 317, 538, 612, 616, 619, 626
Kapteyn-Kalshoven, see Kalshoven, Catharina Elisabeth
Kapteyn-Koomans, see Koomans, Elisabeth Cornelia
Kapteynborg, 632–634
Kapteynia, 632
Kapteynstraat, 633, 635
Karslruhe, Sternwarte, 318
Keats, John, 428
Keeler, James Edward, 118
Kelvin, Lord (William Thomson), 270
Kempf, Paul Friedrich Ferdinand, 296, 301
Kepler's equation, 112–119, 121, 147
Kepler's laws, 113
Kepler, Johannes, 28, 74, 85, 112, 392, 612, 638
Keunen, Johannes Petrus, 445
Keyser, Jan Frederik, 61
Kinman, Thomas D., 376, 378
Kirchhoff, Gustav Robert, 39, 40

Kladboeken, ix, 372
Klein Wassink, Willem Jan, 627
Kluyver, Jan Cornelis, 445
Knipperdolling, Bernhard, 26
Kobold, Hermann Albert, 322
Kohlschütter, Ernst Arnold, 490, 500, 519, 544–552, 557
König, Johann Samuel, 637
Königsberg Observatory, 139
Königstuhl Observatory, 96, 632
Koninklijk Natuurkundig Genootschap (KNG), see Natuurkundig Genootschap (Groningen)
Koninlijke Academie van Kunsten en Wetenschappen (KNAW), see Royal Netherlands Academy of Arts and Sciences (KNAW)
Koomans, Elisabeth Cornelia, 5, 8–14, 107
Korteweg, Diederik Johannes, 573
Kostinsky, Sergej Konstantinovich, 300, 490, 538
Kreiken Observatory, 558
Kreiken, Egbert Adriaan, 14, 557, 582, 627
Kreiken, Willem Rudolph, 14
Krul, Wessel, 101, 104, 112, 204, 220, 302, 304, 307
Kuiler, Kor, 608
Kuiper, Gerrit Pieter, 628
Küstner, Karl Friedrich, 362, 367, 368, 387, 389, 487, 577, 587, 620

L

La Plata Observatory, 376
La Silla Observatory, 400
Lagrange, Joseph-Louis, 39, 114
Lamé, Gabriel Léon Jean Baptiste, 41, 45
Laplace, Pierre-Simon, Marquis de, 503
Law on Higher Education of 1876, 87–89
le Verrier, Urbain, 38
League of Nations, 572, 576
Leavitt, Henrietta Swan, 372, 590
Lecointe, George, 576
Legendre, Adrien-Marie, 322
Légion d'Honneur, 571
Leibniz, see von Leibniz
Leiden Observatory, 58–81, 96, 100, 121, 197, 225, 231, 248, 252, 258, 338, 355, 357, 371, 376, 386, 406, 444–449, 469, 491, 539, 562, 575, 582, 588, 600, 606
    reorganisation, 581–584
    specializes in everything, 584
Leiden University, 27, 59, 61, 203
Lembang Observatory, ix, 14, 558
Lessing, Gotthold Ephraim, 12, 25, 426, 660

Liccione, Anthony, 383
Lick Observatory, 312, 313, 337, 376, 487, 533, 551, 680
light-year, 139
Lindblad, Bertil, 554, 567, 568
Lindsay, James Ludovic, 202
lines of nodes, 42
Linnæus, Carl Nilsson, 28
Liouville, Joseph, 300, 435
Littrow, see von Littrow
Lixk Observatory, 592
Local Standard of Rest, 139, 290, 291, 295, 315
Lockyer, Joseph Norman, 367
Lœwy, Maurice, 193
log-normal probability distribution function, 505
Lord Kelvin, see Thomson, William
Lord Rayleigh, see Strutt, John William
Lord Rosse, see Parson, William
Lorentz, Hendrik Antoon, 34, 492, 525, 526, 570, 574, 575, 656
Louisiana Purchase Exposition, 333
luminosity curve, 559
Lunar eclipse, 75, 124
Lunar occultation, 75, 78
Lund Observatory, 439, 539
Lusitania, 521, 571, 587
Luyten, Willem Jacob, 628, 644

M

M101, 210, 404–406, 525, 591
M51, 404–406
Macpherson, Hector Carswell, 342
Magellanic Clouds, 208, 438, 439
magnitude, 46
    absolute, 46, 327, 559
    apparent, 46
Main Sequence, 206, 312
Mannheim Observatory, 69, 95, 96
Mars, 76, 79, 131, 612
Martin, Willem Christiaan, 628
Massachusetts Institute of Technology, 527
Mathieu, Émile Léonard, 41
Maury, Antonia Caetana de Paiva Pereira, 289, 410, 547
Maxwell, James Clerk, 358
Mayer, Christian, 221
McAldie, Alexander George, 125
McGee-Newcomb, Anita, 345
Mees, Rudolf Adriaan, 71, 91, 92, 106
Meijer, Johan Hendrik, 10
Meijering, Samuël Cornelis, 541, 622, 626
meridian circle, 65, 66, 68, 77, 142

Mesdag, Ferdinand Taco, 502
Messier Catalogue, 294, 400, 403, 404
Messier, Charles, 403
Meudon Observatory, 192
Michelson, Albert Abraham, 38
Michelson-Morley experiment, 38
Milky Way, 209
Milky Way Galaxy, 209
Milton Lasell Humason, 376, 488
minor planet, 37, 38, 62
Modderman, Hendrik Jacob Herman, 277
Modderman, Rudolph Sicco Tjaden, 277
Molengraaff, Adolf Frederik, 268, 577
Molière (Jean-Baptiste Poquelin), 12
Moll, Gerard (Gerrit), 29, 36, 61, 62, 509, 636
Moll, Jan Willem, 507, 508, 577
Monastery (Mount Wilson), 465, 662
Monck, William Henry Stanley, 294, 541
Mönnichmeyer, Carl Otto Louis, 408, 624
Moon, 75
Morley, Edward Williams,, 38
Mouchez, Ernest Amédée Barthélemy, 157, 158, 184–193, 497
Moulton, Forest Ray, 481
Mount Wilson Catalogue, 375, 378, 487–489, 560
Mount Wilson Observatory, x, 302, 365, 367, 376, 378, 414, 415, 453, 455–500, 540, 545–555, 592, 644, 653, 662
Mount Wilson Toll Road, 453, 464
moving cluster method, 388
Muir, John, 311
Mulder, Gerardus Johannes, 30
Mulerius, Nicolaus, 89
Müller, Karl Hermann Gustav, 241
Müller, Philipp, 638
Multatuli (Eduard Douwes Dekker), 45, 110
München Observatory, 96, 576

N
Napoleon (Bonaparte), 28
National Academy of Arts and Sciences (US), 365, 367, 517, 521, 606, 631
National Academy of Sciences (US), 493, 494
Natuurkundig Genootschap (Groningen), 104, 146, 276–284, 512, 633, 659
  scientific chapter, 281
  scientific department, 645
Natuurkundig Gezelschap (Utrecht), 29
Nautical Almanac Office, 270
Nederlandsch Natuur- en Geneeskundig Congres, 284
Nederlandse Astronomen Club, 660
Neptune, 38

neutron star, 209
New–England Magazine, 273
Newcomb (McGee), Anita, 345
Newcomb, Simon, xiii, 53, 74, 137, 221, 269–274, 286, 296, 297, 300, 322, 333, 335, 344, 361, 404, 439, 463, 467, 469, 471, 570, 612
Newcomb-Hassler, Mary Caroline, *see* Hassler, Mary Caroline
Newton, Isaac, 42, 113, 425, 453, 503, 564, 638
Newtonian focus, 490
NGC6744, 210
NGC891, 210, 403
Nijland, Albertus Antonie, 170, 434, 490, 538, 539, 584, 597, 605, 626, 660
Nolst Trenité, Jean Gédéon Lambertus, 107
Noordenbos, Greta, 298, 589, 609, 610
Noordenbos, Jan Willem, xvi, xvii, 589, 609, 610
Noordenbos, Willem, 579, 588, 602, 603
Noordenbos, Willem Cornelis, 149, 298, 480, 597, 609, 610, 643
normal probability distribution function, 505, 506
Nort, Isidore Henri, 539, 626
North Polar Sequence, 370, 375, 488
nova, 237, 312
Nova Persei, 284, 311–315, 616, 618
Noyes, Arthur Amos, 527
nucleosynthesis, 209, 210
nutation, 64, 74

O
O'Grady Marcella, 477
objective prism, 376, 414
obliquity of the ecliptic, 64
observator, 61
Observatorio del Roque de los Muchachos, 633
Observattoire de Paris, *see* Paris Observatory
Occam's Razor, 416
Olbers paradox, 392
Olbers, Heinrich Wilhelm, 392
Ommelanden, 86
Oort, Jan Hendrik, 138, 233, 244, 359, 374, 379, 380, 431, 554, 565, 567, 568, 584, 591, 627, 628, 636
Oosterhoff, Pieter Theodorus, 628
Oppolzer, *see* von Oppolzer
opposition, 76
Orange Free State, 274
Orden Pour le Mérite, 569–571
Order of Oranje-Nassau, 571, 657

Order of the Dutch Lion, 571, 657
Organiek Besluit of 1815, 28, 87
Orion, 46, 399, 401, 486
Orion Nebula, 154, 208, 215, 362, 399, 400, 404, 544
Orion stars, 486
ornithology, 654
Osborn, Henry Fairfield, 468
Osiander, Andreas, 91, 263
Oude Boteringestraat, 98, 226, 227, 234, 235
Oudemans, Jean Abraham Chrétien, 36, 37, 62, 78, 79, 97, 100, 102, 187, 189, 233, 286, 538
Oxford Observatory, 576

P

Palomar Observatory, 401
Palomar Observatory Sky Survey, 195
Pannekoek, Antonie, 137, 204, 446, 581–584, 592, 597, 625
parallactic method, 174, 193, 194, 613, 614
parallactic motion, 295, 316
parallax, 138, 139, 300
    secular, 49, 139, 292, 388, 391, 408, 534
    Solar, 75, 76
    spectroscopic, 547–549
    trigonometric, 376, 387, 390, 548
Paranal Observatory, 402
Paris Observatory, 119, 158, 184, 185, 187, 188, 190, 253, 376
Parkhurst, John Adelbert, 414
parsec, 138, 139
Parson, William, 404
Pasadena Star (newspaper), 122, 123, 125
Passage, de (Café-restaurant), 423, 425
Paul, Erich Robert, viii, 363, 557, 567, 592, 625, 647–649
Pearson, Karl, 506, 508, 510–515, 618
Peary, Robert Edwin, 511
Pease, Francis Gladhelm, 468
Perdok, Wiepko Gerhardus, 433
perihelion, 61, 79, 113
period-luminosity relation, 372, 590
Perrault, Charles, 188
Perrault, Claude, 188
Perrier, François, 189
Perrine, Charles Dillon, 312
Perry, Stephen Joseph, 189
Perryman, Michael, 155, 384
photographic process, 153–154
Physiological Laboratory, 98, 171, 172, 226, 233, 418
Piazzi, Giuseppe, 38

Pickering, Edward Charles, 194, 237, 289, 302, 336, 352, 362, 368, 369, 372, 375, 390, 392, 394, 395, 397, 410, 411, 454, 482, 484, 487, 495, 539, 547, 570, 577, 612, 622, 624
Pinwheel galaxy, 404–406, 525, 536, 591
Pioneer 10/11, 438, 440
Plan of Selected Areas, 359–381, 446
Planck, Max Karl Ernst Ludwig, 570
planet, 38
planetary nebula, 207, 536
Plaskett, John Stanley, 530
plate constants, 174
Plaut, Lukas, 377, 379, 380, 628
Pleiades, 78, 208, 222, 223, 318, 384, 385, 402, 495, 537, 548
Plummer, William Edward, 146, 613
Pogson, Norman Robert, 46
Poincaré, Jules Henri, 352, 366, 404, 569
Poisson, Siméon Denis, 39–41, 43
Polaris, 46, 65
polarization of light, 46
Polytechnical School Delft, 94, 96, 512
Pontécoulant, Philippe Gustave le Doulcet, 631
Porter, Russell Williams, 322
Potsdam Astrophysikalisches Observatorium, 193, 206, 241, 250, 286, 296, 301, 376, 377, 490, 519, 548, 557, 586, 642
Praesepe, 622
precession, 64, 74, 291, 540
precessional constant, 64, 299, 300
prime focus, 490
prime vertical, 130
Princeton Observatory, 471
Proctor, Richard Anthony, 48, 231
proper motion, 48, 74, 139, 300
    parallactic, 295
    reduced, 436
Protestant church, 3, 15, 25
Proxima Centauri, 135
Ptolemaeus, Claudius, 74
Ptolemy, 74
Pulkovo Observatory, 62, 185, 187, 190, 216, 231, 300, 336, 339, 353, 362, 376, 538

Q

quadrant, 59
quantum efficiency, 153
quasar, 333
Quetelet, Lambert Adolphe Jacques, 506
Quintus Horatius Flaccus, 658

R

Racine, Jean, 12

Index 695

Radcliffe Observatory, 362, 376, 378
Raimond, Jean Jacques, 628
Ravensperger, Hermann, 90
Rayet, Georges Antoine Pons, 253
Rayleigh scattering, 412
Rayleigh, Lord (John William Strutt), 412
reciprocity failure, 154
red giant, 207, 312, 410
Reduction of Groningen, 86
reflection nebula, 402, 537
reflex zenith-tube, 236
refraction (atmospheric), 66, 67, 127, 128
Repsold factory, 77, 129, 135, 199, 226
Republic Observatory, 204
Republic of the Seven United Netherlands, 86
Réunion, 79
Reynolds, John Henry, 211
Rhodes. Cecil, 354
Richmond, Myrtle L., 488, 624
Riemann, Georg Friedrich Bernhard, 41
right ascension, 63–65, 126
Ristenpart, Frederich Wilhelm, 318
Ritchey, George Willis, 312, 458
Roberts, Isaac, 189, 222, 223
Rondeel, E.J., 371, 372
Roosevelt, Franklin Delano, 529
Roosevelt, Theodore, 344
Rossijn, Johannes Theodorus, 29, 636
Rotterdam, 653
Rousseau, Jean-Jacques, 12, 25, 45, 149
Rover safety (bicycle), 307, 656
Royal Academy Dublin, 631
Royal Academy Edinburgh, 631
Royal Astronomical Society, 131, 342, 555, 607, 630, 631
  Gold Medal, 351
Royal Geographical Society, 256
Royal Greenwich Observatory, 74, 81, 187, 190, 196, 203, 231, 236, 295, 356, 376
Royal Military Academy, 94, 96
Royal Natural Sciences Society, see Natuurkundig Genootschap (Groningen)
Royal Netherlands Academy of Arts and Sciences (KNAW), 94, 119, 270, 284, 286, 290, 293, 295, 297, 319, 321, 323, 326, 365, 398, 405, 408, 483, 534, 540, 574, 611, 631, 674
Royal Observatory of Belgium, 506, 576
Royal Observatory, Cape of Good Hope, 131, 139, 167, 238, 243, 256, 287, 352, 353, 362, 376, 491, 671
Royal Physiographical Society (Lund), 631
Royal Society, 156, 159, 163, 166–168, 185, 196, 198, 200, 202–204, 631

Royal Society of London, 298, 515, 572
Royal Swedish Academy of Sciences, 631
Royal Swedish Society of Sciences (Uppsala), 631
RR Lyrae stars, 379, 592
Rueb, Adolf Stephanus, 36
Ruijs de Beerenbrouck, Charles Joseph Marie, 584
Russell, Henry Norris, x, 206, 415, 487, 548, 555

S

Sandage, Allan Rex, x, 454, 458, 460, 465, 468, 491, 547
Saros period, 124
satellites (moons), 38
Saturn, 154
Savornin Lohman, Witius Henrik de, 92
Sawerthal, Henry, 178
Schönfeld, Eduard, 163, 189, 231, 351, 523
Schaub, Werner, 624
Scheiner, Julius, 250, 286, 615
Schiller, see von Schiller
Schilt, Jan, 627
Schlesinger, Frank, 487, 530
Schmidt, Maarten, 369
Schols, Charles Mathieu, 96
Schoute, Pieter Hendrik, 433, 434
Schouten, Willem Johannes Adriaan, 556, 592, 623, 626
Schwarzschild, Karl, 357, 358, 368, 418, 485, 487, 490, 530, 541, 548, 554, 556, 563, 568, 570, 642
Schwassmann, Friedrich Karl Arnold, 376
Schwearzschild, Karl, 568
Schwieger, Walther, 571
Seares, Frederick Hanley, 378, 468, 469, 487, 488, 499, 521, 550–555, 599, 605, 624, 625, 645
Searle, Leonard, 565
Secchi, Pietro Angelo, 288
secession of 1834, 15
secular parallax, 48, 139, 292, 388, 391, 408
Seeberg-Sternwarte, 73
Seeliger, see von Seeliger
Selected Areas
  photometric sequences, 378–379
Shapley, Harlow, 377, 378, 415, 488, 555, 567, 568, 589–592
sidereal day, 65
sidereal time, 65
Sijmons, Barend, 104
Simbad data base, 669
Simons, Gerrit, 62

sine functions of higher orders, 119–121
Sirius, 46, 135, 213, 215, 291
Sitter, de, *see* de Sitter
Skew Curve Machine, 508, 509
skew frequency curves, 505–516
sky brightness
    total starlight, 361, 437, 540
Slipher, Vesto Melvin, 485
Smid, Etine Imke, 539, 621, 626
Smith, Robert W., 567
Smithsonian Astrophysical Observatory, 468, 485
Snellius, Willibrord, 59
Snow Telescope (Mount Wilson), 456, 458
Snow, George Washington, 458
Society of Arts and Sciences of the Province of Utrecht, 168, 631
Solar Apex, 139, 290, 299, 301, 316, 321, 323, 324, 389, 434, 534, 542
Solar constant, 485
Solar eclipse, 124, 540
Solar parallax, 75, 76
Solar System, 37, 38
Solar Towers (Mount Wilson), 456, 458
Sonnenborgh, 36
Southern Cross, 153, 354
special relativity, 312
Spectral types, 288
spectroscopic binary, 387
spectroscopic parallax, 547
speed of light, 67, 312
Spencer Jones, Harald, 256
spiral galaxy, 211, 404
St. John, Charles Edward, 599, 607, 653
Staatkundig Gereformeerde Partij, 3
star
    classification, 288
    evolution, 76, 536
    formation, 206
    magnitude, 46
    mass, 76, 207
    spectroscopy, 289
Star Streams, 315–349, 352–359, 434, 486, 494, 523, 533, 535, 537, 542, 568, 614
Statistical astronomy, Fundamental Equation of, 332
Stebbins, Joel, 378
Stein, Johannes Wilhelmus Jacobus Antonius, 323, 436, 617
Stellar Populations, 209, 244
Stellar Statistics, Fundamental Equation of, 332
Sternwarte Göttingen, 357, 490, 642
Sternwarte Karlsruhe, 300

Stieltjes, Thomas Joannes, 69
Stockholm Observatory, 552
Stokes, George Gabriel, 163, 168, 198
Stone, Ormond, 336
Strasbourg Observatory, 96
Stratingh, Sibrandus, 277
Stratonoff, W., 404
Strömberg, Gustaf, 552–555
Strömgren, Bengt Georg Daniel, 432
Strömgren, Svante Elis, 432, 491, 533, 575, 576
Strutt, John William, 403, 570
Strutt, John William (Lord Rayleigh), 412
Struve, *see* von Struve
Sturm, Charles-François, 25
Südliche Durchmusterung, 74, 156, 163, 285, 523
Suermondt, Eleonora, 258
Sullivan, Woodriff Turner III, 380
supergiant, 209
superluminal velocity, 314
supernova, 209, 401
synodic month, 124

T

Tammes, Jantina, 507, 513
Tashkent Observatory, 404
Taurus (constellation), 384
Technical University of Delft, 502
ten Bruggen Cate, Gerrit Hendrik, 557, 623, 626
ten Bruggencate, Paul, 557
Tennant, James Francis, 189
Tennyson, Alfred, 497
Teyler Foundation, 168
Teyler van der Hulst, Pieter, 168
Thackeray, Andrew David, 378
Thackeray, William Grasett, 357
Thijsse, Jacobus Pieter, 654
Thomson, Joseph John, 403, 525
Thomson, William (Lord Kelvin), 270
tides, 501–504
Tienemann, Johannes, 283
Tietjes, Friedrich, 62
Tisserand, François Félix, 192
Towsend, Euphemia Clementina, 32
Transvaal Republic, 274
Treaty of the Metre, 190
tree rings, 122–126, 147
Trépied, Charles, 193
triangulation, 60
true anomaly, 112, 113
Trumpler, Robert Julius, 417, 567, 568

Index 697

Turner, Herbert Hall, 139, 184, 191, 327, 336, 338, 555, 576, 587
Twain, Mark, 153, 354
Two-body problem, 113

U

Undina (minor planet), 37
Union of Utrecht, 86
universal instrument, 77, 129
University of Amsterdam, 27
University of Capetown, 630
University of Edinburgh, 630
University of Franeker, 27
University of Groningen, 27, 86
  tercentennial, 517
University of Harderwijk, 27
University of Lausanne, 219
Urania Observatory, 642
Uranus, 38, 211
Ursa Major, 46
US Naval Observatory, 74, 80, 270
Utrecht Observatory, 36, 100, 170, 490
Utrecht University, 27
Utrechts Studenten Corps, 31
Utrechtse Studenten Almanak, 39
Uylenbroek, Pieter, 61

V

Valentiner, Karl Wilhelm, 68, 79, 96, 117, 137, 404
van Ankum, Hendrik Jan, 93
van Anrooij, Peter Gijsbert, 305, 427, 608
van Bell, Frederik Willem Bernard, 98
van Berkel, Klaas, ix, x, xv, xvii, 30, 31, 102, 107, 109, 493, 609, 626, 647, 667
van Calker, Friedrich Julius Peter, 88
van de Kamp, Peter, 627
van de Linde, Jan Cornelis, 539, 627
van de Sande Bakhuyzen, Ernst Frederik, 69, 70, 73, 77, 79, 138, 162, 305, 447, 448, 491, 539, 540, 581, 583, 612
van de Sande Bakhuyzen, Hendricus Gerardus, 65, 68–81, 93, 97, 100, 102, 104, 118, 121, 129, 140, 143, 144, 162, 187, 189, 192, 193, 197, 225, 232, 248, 252, 258, 286, 305, 309, 320, 323, 338, 355, 357, 371, 386, 405, 406, 444, 445, 447, 448, 469, 539, 562, 575, 581, 582, 588, 606, 612, 613, 652
van de Snepscheut, Johannes Lambertus Adriana, 27
van den Bos, Willem Hendrik, 628
van den Broeke, Henrik, 108

van der Heijden, Petra, ix, 249, 270, 365, 366, 587
van der Kallen, Hendrikus Frederikus (Havank), 263
van der Kruit, Pieter Corijnus, ix, xi, xii, 213, 215, 317, 437, 560, 561, 565, 597, 626, 647
van der Kruit-Arends, Cornelia, v, xii, xiii
van der Waals, Johannes Diderik, 34
van der Wijck, Bernhard Hendrik Cornelis Karel, 85, 98, 101
van Dijk, Isaac, 466
van Gent, Hendrik, 628
van Gorkom, Wilhelmina Henriette, 148, 306, 609, 610
van Heiden Reinestein, Louis (Graaf), 92
van Hennekeler, Andreas, 68
van Herk, Gijsbert, 628
van Hulsteijn, Johanna Josephine, 32, 109
van Lokhorst, Jacobus, 220, 225
van Maanen, Adriaan, 98, 434, 490, 492, 493, 525, 538, 543, 553, 591, 599, 625, 626
van Rappard, Anthony Gerhard Alexander (Ridder), 62
van Rees, Richard, 29, 39, 636
van Rhijn Luminisity Function, 379
van Rhijn, Pieter Johannes, 14, 170, 204, 302, 332, 361, 369, 372, 376, 380, 413–416, 420, 430, 488, 490, 493, 540, 551, 554, 556–561, 570, 577, 579, 582, 592, 597, 602, 621–627, 636
van Schaffelaar, Jan, 4
van Swinden, Jean Henri, 29
van Swinderen, Theodorus, 276, 279
van 't Hoff, Jacobus Henricus, 34, 570
van Uven, Marie Johan, 512, 622
van Wageningen, Jacobus, 577
van Württemberg, Sophia Frederika Mathilde, 218
variable star, 76
Vatican Observatory, 288, 323, 397
Vega, 139, 291, 327
velocity dispersion–age relation, 532, 552
Venus, 76, 79, 131, 158
vernal equinox, 65
Versailles, Peace of, 572
Vertex of Star Streams, 341
Verweij, Albert, 111
Veth, Jan Pieter, 420, 576–580
Vindicat Atque Polit, 432
Voûte, Joan George Erardus Gijsbertus, 491
Vogel, Hermann Carl, 193, 296
Vollgraf, Carl, 643
Voltaire (François-Marie Arouet), 233

von Auwers, Arthur Julius Georg Friedrich, 570
von Auwers, Georg Friedrich Julius Arthur, 74, 139, 187, 190, 192, 197, 200–202, 295, 299, 300, 616
von Beethoven, Ludwig, 608
von Goethe, Johann Wolfgang, 12, 25, 652
von Helmholtz, Hermann Ludwig Ferdinand, 219, 270
von Humboldt, Friedrich Wilhelm Heinrich Alexander, 200
von Leibniz, Gottfried Wilhelm, 638
von Littrow, Karl Ludwig, 140
von Oppolzer, Theodor, 115, 117
von Schiller, Johann Christoph Friedrich, 12, 33
von Seeliger, Hugo Hans (Ritter), 96, 231, 415, 418, 556, 557, 561–563, 567, 570, 576, 586
von Seeliger, Hugo Hans, Ritter, 363, 479
von Steinheil, Carl August, 46
von Struve, Friedrich Georg Wilhelm, 139, 216, 231, 403, 439, 498
von Struve, Otto Wilhelm, 185, 187, 190, 192, 197, 202, 216, 339, 432
von Tschirnhaus, Ehrenfriend Walter, 638
von Wolff, Christian, 638
Vulcanus, 78

W
W Virginis stars, 592
Wagner, Wilhelm Richard, 387
Walloon Reformed Church, 107
Washburn Observatory, 146, 300, 325, 336, 362
Washington University, St. Louis, 333, 337, 344
Watt, James, 131
weather, 122–126
Weersma, Herman Albertus, 372, 386, 388, 434, 450, 535, 538, 556, 582, 618–621, 623, 626

Weigel, Erhard, 638
Weldon, Walter Frank Raphael, 506, 510
Wesselink, Adriaan Jan, 628
Wheatstone, Charles, 40
Whirlpool galaxy, 404–406, 536
white dwarf, 207, 312
White House, 344
Whitehead, Alfred North, 501
Wilhelmina, Queen of the Netherlands, 218, 226
Wille, Sara, 109
Willem III, King of the Netherlands, 97, 218, 226
William Herschel Telescope, 633
Wilson, Thomas Woodrow, 572, 576
Wisse, Anna, 304
Wolf, Charles Joseph Étienne, 253
Wolf, Maximilian Franz Joseph Cornelius, 632
Wolf-Rayet star, 253
Woltjer, Lodewijk, 568
Wood, De Volson, 383
Woodward, Robert Simpson, 463, 467

Y
Yale Observatory, 135, 376, 471
Yerkes Actinometry, 414
Yerkes Observatory, 48, 312, 336, 344, 362, 365, 376, 454, 458, 471, 486, 491, 531, 572
Yntema, Lambertus, 361, 434, 437–443, 538, 540, 541, 619, 626
Young, Charles Augustus, 327

Z
Zeeman, Pieter, 483
zenith distance, 127
Zernike campus, 634
Zernike, Frederik (Frits), 308, 372, 374, 522, 557, 582
zodiacal light, 438, 541
Zöllner photometer, 46–48, 73, 258, 287, 441
Zöllner, Johann Karl Friedrich, 46